This book presents the first comprehensive exposition of the interpretation of quantum mechanics pioneered by Louis de Broglie and David Bohm. The purpose is to explain how quantum processes may be visualized without ambiguity or confusion in terms of a simple physical model.

Developing the theme that a material system such as an electron is a particle guided by a surrounding quantum wave, a detailed examination of the classic phenomena of quantum theory is presented to show how the spacetime orbits of an ensemble of particles can reproduce the statistical quantum predictions. The mathematical and conceptual aspects of the theory are developed carefully from first principles and topics covered include self-interference, tunnelling, the stability of matter, spin $\frac{1}{2}$, and nonlocality in many-body systems. The theory provides a novel and satisfactory framework for analysing the classical limit of quantum mechanics and Heisenberg's relations, and implies a theory of measurement without wavefunction collapse. It also suggests a strikingly novel view of relativistic quantum theory, including the Dirac equation, quantum field theory and the wavefunction of the universe.

This book provides the first comprehensive technical overview of an approach which brings clarity to a subject notorious for its conceptual difficulties. The book will therefore appeal to all physicists with an interest in the foundations of their subject, and will stimulate all students and research workers in physics who seek an intuitive understanding of the quantum world.

THE QUANTUM THEORY OF MOTION

An account of the de Broglie–Bohm causal interpretation of quantum mechanics

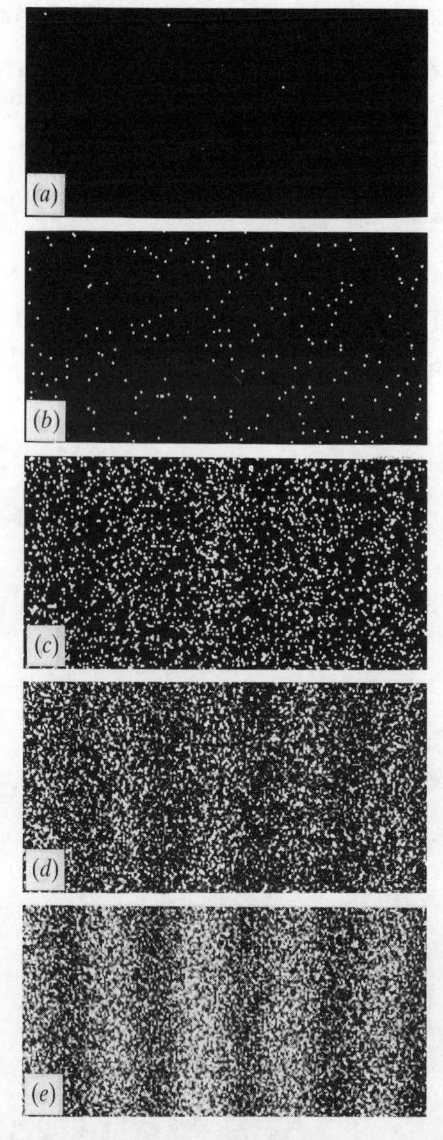

Frontispiece. The cumulative pattern generated by the 'self-interference' of electrons sent one by one through a two-slit interferometer. Number of electrons: (*a*) 10, (*b*) 100, (*c*) 3000, (*d*) 20 000, (*e*) 70 000 (from Tonomura *et al.* (1989)).

THE QUANTUM THEORY OF MOTION

An account of the de Broglie–Bohm causal interpretation of quantum mechanics

PETER R. HOLLAND
Professor in the Foundations of Physical Sciences
University of the West of England, Bristol

Published by the Press Syndicate of the University of Cambridge
The Pitt Building, Trumpington Street, Cambridge CB2 1RP
40 West 20th Street, New York, NY 10011-4211, USA
10 Stamford Road, Oakleigh, Melbourne 3166, Australia

First published 1993
First paperback edition 1995

Printed in Great Britain at the University Press, Cambridge

A catalogue record for this book is available from the British Library

Library of Congress cataloguing in publication data

Holland, P. R.
The quantum theory of motion : an account of the de Broglie–Bohm
causal interpretation of quantum mechanics / P.R. Holland.
p. cm.
ISBN 0 521 35404 8 (hbk) ISBN 0 521 48543 6 (pbk)
1. Motion. 2. Quantum theory. 3. Relativistic quantum theory.
I. Title.
QC133.H65 1993
530.1′2–dc20 92-20205 CIP

ISBN 0 521 35404 8 hardback
ISBN 0 521 48543 6 paperback

KW

To Jackie, Lucette & Jack
who gave far more
than they know

Contents

Preface

The *frontispiece* portrays a sequence of pictures in which a pattern reminiscent of the interference of waves is progressively built up by a series of individual events in which one electron at a time is sent through a two-slit interferometer and arrives at a detecting screen. Each event is unpredictable, yet over time a definite and reproducible pattern is formed. It is not just arbitrary. What causes the electrons to aggregate in this way? Is there some force acting on each individual electron as it passes through the device which impels it, on the average, to land in certain regions of the screen rather than others?

The historically dominant view in quantum mechanics regards this seemingly natural question as meaningless. For the query rests on the supposition that matter can be at least conceptually analysed, and that when we speak of an 'electron' we really mean an autonomous entity that can move in spacetime and be subject to forces. In contrast, it is generally asserted that our theoretical account of physics in the microdomain must stop short at predicting the likely outcomes when observations at the classical macroscopic scale are performed (the density of dots on the screen). That two electrons in the same quantum state appear at different points does not permit one to infer that they are, in fact, physically distinguished in some way. The uncertainty is postulated to be intrinsic to the system.

This implies the following difficulty: our most basic physical theory contains no account of the constitution and structure of matter, corresponding to the interacting particles and fields of classical physics. It is a means to compute the statistical results of macroobservations carried out on systems that are unspecified and, indeed, unspecifiable. The word 'electron' does not actually *mean* anything at all – it is simply shorthand for a mathematical function. Quantum mechanics is the subject where we never know what we are talking about.

Yet there is a way to know what we are talking about. It was suggested by Louis de Broglie in 1927 and developed into a physical theory by David Bohm in 1952, and has been almost universally ignored since. According to de Broglie and Bohm, the novelty of quantum mechanics is not that we have to revise our customary notions regarding the reality of physical systems, but that their conception must be extended. In describing the electron the classical corpuscle may be retained but must be supplemented by a new type of physical field mathematically described by Schrödinger's wavefunction. The wave guides the particle so that it performs quantum motions rather than classical ones, which gives us the title of this book. The probability distribution of quantum mechanics, such as that displayed in the *frontispiece*, is reproduced because the particle orbits tend to congregate where the wave is most intense. Complete specification of the state of an individual system then requires both aspects: not wave *or* particle but wave *and* particle. The wave-particle composite continuously evolves according to a set of deterministic laws.

This theory of motion is not at all foreign to physicists' practice, for it is tacitly invoked whenever one desires to discuss things to which quantum mechanics potentially applies that no one doubts have *positions* (e.g. meters). Perhaps part of the reason why this approach has been spurned is that its motivation has been misunderstood. The aim of the de Broglie–Bohm theory is *not* to attempt a return to classical physics, or even particularly to invent a deterministic theory. *Its goal is a complete description of an individual real situation as it exists independently of acts of observation.* According to Einstein, that is the programmatic aim of physics. *Determinism is a means to this end.* What emerges is a highly nonclassical theory in which the parts of a larger system are subject to organization by the whole, something not anticipated in the classical paradigm.

There are unsatisfactory aspects of the de Broglie–Bohm model but it has been discarded by the wider scientific community in the absence of a sustained technical examination. This book is intended as a contribution to filling that vacuum. We do not wish to appear too evangelical in our prosecution of the theory, but close study does reveal that the gains in conceptual clarity over the current rivals far outweigh the drawbacks. In the first instance what is at issue is the task of *imagining* the quantum world and that is our principal concern. There now exists a core of established results in this field, some of which are reported here for the first time, and as a research area there are rich prospects. A recurring theme in the book is how the classical and quantum paradigms are connected, a problem that seems to be insoluble in the conventional approach since the usual quantum notion of state does not contain the classical conception as a special case. Another problem where there is great scope for a rational input from the causal model is the interpretation of relativistic quantum theory. The usual presentation of this is extraordinarily vague and provides a classic example of how the rules regarding the meaning of the quantum formalism as pertaining to measurement outcomes are quietly put aside and the physical disposition of matter discussed as if it were objectively real. Most importantly, there is the possibility of bringing the de Broglie–Bohm theory into the experimental arena, a subject that is currently under investigation.

It is a pleasure to thank the following friends and colleagues for their support and encouragement during the writing of this book, and for their courtesy in providing valuable information and/or constructive criticism of the final manuscript: Michael Berry, Piotr Garbaczewski, Dipankar Home, Phil Jacobs, Tassos Kyprianidis, Mioara Mugur-Schächter, Chris Philippidis, Helmut Rauch, Euan Squires and Jean-Pierre Vigier. I have tried to take into account their suggestions but the responsibility for the outcome is mine alone. I am indebted also to D. & J. Canavan for help with the typing at a critical moment, to L. Holland and M. Anderson for preparing many of the figures, and to A. Tonomura for providing the *frontispiece*.

The volume was written while I was a visiting fellow at the Laboratoire de Physique Théorique, Institut Henri Poincaré, Paris, to whom thanks are due for hospitality and generous provision of facilities. Grants from the Royal Society, the SERC and the French government provided partial support in the early stages and the final chapter was written during the tenure of a Leverhulme Trust fellowship. All these agencies are cordially acknowledged. During the bulk of the writing I worked as an editor for Elsevier Science Publishers BV. Thanks are due to Elsevier for its generosity, and to Joost Kircz and Jean-Pierre Vigier without whose assistance and understanding the book would not have been finished.

Paris *PRH*
May, 1992

Note added to Preface for paperback edition

According to the theory of evolution the path of natural phenomena may be regarded as explicable yet unpredictable. Darwin's theory comprises an observer-independent, causal reconstruction of events whose random character precludes evolutionary forecasting. Of course, the historically important issue at the heart of the theory of evolution is the quality of its *explanation* rather than its immediate predictive power. The achievement of the de Broglie–Bohm theory in the domain of quantum physical phenomena is the production of a science similar in its scope and conceptual structure to the elements of Darwinism just cited. Yet what is perceived as an intellectually satisfying and fertile mode of explanation in evolutionary biology has been fiercely resisted in quantum physics. Darwin would not fare well if judged by the criteria of prediction and control currently dominant in physics. But how can we conceive of a testable prediction of an explanatory theory unless we contemplate the explanation it offers?

It is instructive how impractical, inhibiting ideas came to dominate and distort the entire development of a fundamental field of physics. The early quantum physicists attributed to nature a limitation we can now see was simply a deficiency of contemporary thought. The biological comparison shows the arbitrariness in the historical development of physical ideas, but it is not merely a source of analogy. Quantum mechanicists often claim their theory is universal and applicable, in principle, to the entire universe. But biological processes, which are surely part of the same universe, lie beyond the reach of physical explanation as we currently understand it. As far as we can tell, they are not instances of quantum mechanics. Actually, the failure of quantum mechanics to reproduce the valid results of other sciences already occurs at a more basic level of macro-experience. The difficulty arises in attempts to derive classical particle and field theory from the quantum theory in cases where the former are known to be valid. The de Broglie–Bohm theory suggests that this programme is generally unrealisable; generic classical processes are inaccessible starting from quantum ones. Even in those cases where the Correspondence Principle of the de Broglie–Bohm model is obeyed (as gauged by the relative effectiveness of the quantum potential) only a subclass of admissible classical behaviour may be recovered in general. Taking into account the entire spectrum of physical processes to which they have been applied, we cannot therefore assert that the quantum theory of matter and motion is *better* than the corresponding classical theory. Rather, they emerge as *different* theories whose domain of overlap is smaller than is customarily believed. Some examples illustrating this point, hitherto unnoticed even by partisans of the de Broglie–Bohm model but which may prove to be its most significant new insight, are given in Chap. 6.

It is pleasing to report that since this book was completed many new workers have been drawn to the de Broglie–Bohm theory. Work is still in progress so I have restricted myself in this reprint to the correction of minor errors. The comments of many correspondents are much appreciated.

Paris, *PRH*
July 1994

1

Quantum mechanics and its interpretation

1.1 The nature of the problem

The quantum world is inexplicable in classical terms. The predictions pertaining to the interaction of matter and light embodied in Newton's laws for the motion of particles and Maxwell's equations governing the propagation of electromagnetic fields are in flat contradiction with the experimental facts at the microscopic scale. A key feature of quantum effects is their apparent indeterminism, that individual atomic events are unpredictable, uncontrollable and literally seem to have no cause. Regularities emerge only when one considers a large ensemble of such events. This indeed is generally considered to constitute the heart of the conceptual problem posed by quantum phenomena, necessitating a fundamental revision of the deterministic classical world view.

Some of the principal phenomena, among those discussed in this book, are as follows:

(*a*) *Self-interference:* a beam of electrons sent one at a time through a barrier containing two apertures builds up through a series of localized detection events on a screen an interference pattern characteristic of waves (cf. the *frontispiece* and §5.1).

(*b*) *Tunnelling:* an α-particle trapped in a nucleus can pass through a potential barrier in a manner forbidden to a classical particle (§5.3).

(*c*) *The stability of matter:* atoms and molecules are found to exist only in certain discrete, or 'stationary', energy states. For this reason they do not 'collapse', the result predicted by classical electrodynamics. During transitions between stationary states (quantum jumps) an atom exchanges a discrete quantity of energy with the electromagnetic field (§§4.5, 7.6).

(*d*) *Spin* $\frac{1}{2}$*:* a beam of atoms sent through an inhomogeneous magnetic field is split into a discrete set of subbeams (the Stern–Gerlach experiment). This reveals a novel type of nonclassical internal angular momentum (§9.5).

(*e*) *Nonlocal correlations:* the properties of one system can depend on those of another arbitrarily distant system with which it has interacted in the past (Chap. 11).

The discrete, statistical and nonlocal character of these phenomena is clearly in conflict with the continuous, determinist and generally local structure of the world according to classical particle and field physics.

It is remarkable that a single coherent mathematical theory, quantum mechanics, could be devised to correlate the heterogeneous empirical data just cited (and many more). Through the Schrödinger equation, quantum mechanics describes the laws of evolution of statistical ensembles of similarly prepared systems. To date, the results obtained are in accord with all experimental evidence. But, insofar as it only predicts the outcomes of measurements performed on statistical aggregates of physical systems, quantum mechanics does not in itself provide an *explanation* of the experimental facts. What is missing is *a description of the actual individual events of experience*, of which the statistical phenomena would be functions.

The challenge is to develop a theory of individual material systems, each obeying its own law of motion, whose mean behaviour over an ensemble reproduces the statistical predictions of quantum mechanics. The empirical record would then be explained as the outcome of a sequence of well-defined processes undergone by systems possessing properties that exist independently of acts of observation.

A way to do this was found by de Broglie and Bohm. It turns out that a causal representation of quantum phenomena may be constructed which leaves intact the basic mathematical formalism of the theory, if this is reinterpreted and extended in a certain way. The detailed working out of the de Broglie–Bohm idea is the subject of this book.

1.2 The wavefunction and the Schrödinger equation

In classical physics the state of an individual material or field (e.g. electromagnetic) system is uniquely defined by the position $\mathbf{x}(t)$ of the object in the first case and the real or complex amplitude $\phi(\mathbf{x}, t)$ in the second, as functions of the time t. Here $\mathbf{x} = (x, y, z)$ represents the Cartesian coordinates of a point in space. The equations of motion of classical physics, Newton's laws and Maxwell's equations, specify these state variables for all time if they and the corresponding momenta are precisely determined at one instant.

The quantum theory developed in the 1920s is connected with its classical predecessor by the mathematical procedure of 'quantization', in which classical dynamical variables are replaced by operators. In the process, a new entity appears which the operators act on, the *wavefunction.* For a single-body matter system this is a complex function, $\psi(\mathbf{x}, t)$, and for a field it is a complex functional, $\psi[\phi, t]$. The wavefunction constitutes a new notion of the state of a physical system.

The historical problem of interpreting quantum theory may be formulated as follows: In prosecuting their quantization procedure, the Founding Fathers introduced the new notion of state not in *addition* to the classical state variables but *instead* of them (Fig. 1.1). They could not see, and finally did not want to see even when presented with a consistent example, how to retain in some form the assumption at the heart of the classical paradigm that matter has substance and form independently of whether or not it is observed. The ψ-function alone was adopted as characterizing the state of a system. Since there is no way to describe individual processes using just the wavefunction, it seemed natural to claim that these are indeterminate and unanalysable in principle.

How the state function is to be interpreted, at least partially, and how it evolves in time are questions addressed by the axioms of quantum mechanics, which we now summarize in a fairly informal way. These are generally agreed upon and will be accepted here. It is not possible to state the rules without some interpretative elements intruding, particularly Born's probabilistic interpretation of ψ. But the further significance of ψ, if any, is open to debate and will be discussed later. Although the formalism is interwoven with the interpretation, the latter is not unique and one should not equate either of

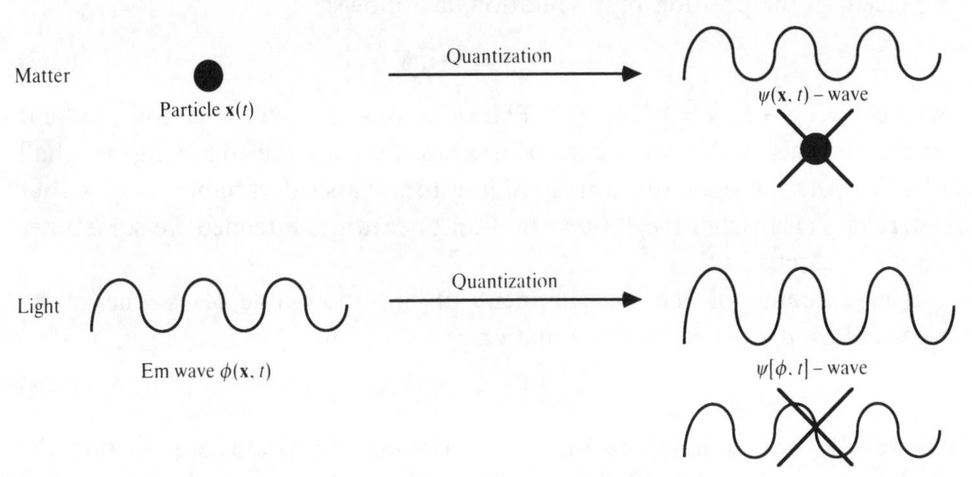

Fig. 1.1 When the wavefunction was introduced, the classical particle and field variables characterizing the states of individual physical systems were discarded.

them with the Copenhagen interpretation (§1.3). More detailed and rigorous presentations will be found in the standard sources such as von Neumann (1955), Messiah (1961), Schiff (1968) and Dirac (1974). Summaries are given by Bjorken & Drell (1964), Jammer (1974, Chap. 1) and d'Espagnat (1976).

We shall consider just the one-body problem. The wavefunction introduced above is referred to as 'the state in the position representation'. This means the following. We denote the state of the system by the ket $|\psi\rangle$, an abstract vector in a linear vector space. In this space we introduce a set of axes $|\mathbf{x}\rangle$, one axis for each value of $\mathbf{x}$. Then the wavefunction is the set of components (one for each $\mathbf{x}$) of the state vector with respect to this basis: $\psi(\mathbf{x}, t) = \langle \mathbf{x}|\psi(t)\rangle$.

The set of complex numbers, i.e., the wavefunction, has the following interpretation. Suppose that the wavefunction is square-integrable, so that it lies in a Hilbert space $\mathscr{H}$, and is normalized:

$$\langle \psi|\psi\rangle = \int_{-\infty}^{\infty} |\psi|^2 \, \mathrm{d}^3x = 1. \tag{1.2.1}$$

Then if at time t a measurement is performed to determine the position of the system described by the function $\psi(\mathbf{x}, t)$, the probability that the result lies in the element of space d^3x around the point $\mathbf{x}$ is given by

$$P(\mathbf{x}, t) \, \mathrm{d}^3x = |\psi(\mathbf{x}, t)|^2 \mathrm{d}^3x. \tag{1.2.2}$$

This probability interpretation may be generalized and applied to the measurements of other physical quantities. We first note that all observables in quantum mechanics are represented by linear Hermitian operators acting in the Hilbert space. For example, the classical canonical momentum $\mathbf{p}$ is replaced, in the position representation, as follows:

$$\mathbf{p} \rightarrow \hat{\mathbf{p}} = -\mathrm{i}\hbar\nabla \tag{1.2.3}$$

where $\mathrm{i} = \sqrt{-1}$, $\hbar = h/2\pi$, h is Planck's constant and ∇ is the gradient operator. This is the procedure of quantization. In the following we shall often denote the operator corresponding to a classical variable A by $\hat{A}$, but where it is clear from the context that an operator is intended we sometimes omit the caret.

The outcome of the measurement of an observable $\hat{A}$ is one of its eigenvalues, a, defined by the equation

$$\hat{A}|a\rangle = a|a\rangle \tag{1.2.4}$$

where a is real and $|a\rangle$ is the corresponding eigenstate (we assume the spectrum is discrete and ignore degeneracy where several independent eigenstates may correspond to the same eigenvalue). In the position representation we shall write $\langle \mathbf{x}|a\rangle = \psi_a(\mathbf{x})$. The eigenfunctions corresponding to distinct eigenvalues a, a' are orthonormal,

$$\int_{-\infty}^{\infty} \psi_{a'}^*(\mathbf{x})\psi_a(\mathbf{x})\,\mathrm{d}^3x = \delta_{aa'}, \tag{1.2.5}$$

and form a complete set so that an arbitrary wavefunction can be expanded in terms of them:

$$\psi(\mathbf{x}) = \sum_a c_a\psi_a(\mathbf{x}), \tag{1.2.6}$$

where c_a are complex numbers. We have $c_a = \langle a|\psi\rangle$ so that these numbers are the components of the vector $|\psi\rangle$ with respect to the basis $|a\rangle$ in $\mathscr{H}$.

If now a measurement of $\hat{A}$ is performed on a system in the general normalized state (1.2.6), the probability of the outcome a is given by $|c_a|^2$. As a result of the measurement the system has 'collapsed' into the state $|a\rangle$. Notice that $|c_a|^2$ does not refer to the probability that the system is *in* the state $|a\rangle$, independently of the performance of a measurement.

The expectation value of the operator $\hat{A}$ in the state $|\psi\rangle$ is given by

$$\langle\hat{A}\rangle = \langle\psi|\hat{A}|\psi\rangle = \sum_a a|c_a|^2. \tag{1.2.7}$$

By a suitable choice of operators, all the testable predictions of quantum mechanics can be expressed in terms of expectation values in the state $|\psi\rangle$. To do this, we introduce the 'projection operator' $\hat{P}_a = |a\rangle\langle a|$. Then the probabilities are given by

$$|c_a|^2 = \langle\psi|\hat{P}_a|\psi\rangle. \tag{1.2.8}$$

Similarly, we find for the eigenvalues the expressions

$$\left.\begin{aligned} a &= \langle a|\hat{A}|a\rangle \\ &= \langle\psi|\hat{P}_a\hat{A}\hat{P}_a|\psi\rangle/\langle\psi|\hat{P}_a|\psi\rangle. \end{aligned}\right\} \tag{1.2.9}$$

In quantum mechanics the classical Hamiltonian function, $H = \mathbf{p}^2/2m + V(\mathbf{x})$ where V is the external potential energy and m is the mass of the system, is replaced by the Hermitian operator $\hat{H} = \hat{\mathbf{p}}^2/2m + \hat{V}(\hat{\mathbf{x}})$. This determines the evolution of the quantum state via the Schrödinger equation $\mathrm{i}\hbar\,\partial|\psi\rangle/\partial t = \hat{H}|\psi\rangle$. The eigenvalues of $\hat{H}$ are the possible energies of the system corresponding to stationary states. In the position representation the Schrödinger equation becomes the partial differential equation

$$\mathrm{i}\hbar\,\frac{\partial\psi(\mathbf{x},t)}{\partial t} = -\frac{\hbar^2}{2m}\nabla^2\psi(\mathbf{x},t) + V(\mathbf{x})\psi(\mathbf{x},t). \tag{1.2.10}$$

This law of motion has two notable features: (1) It admits a linear superposition principle. If ψ_1 and ψ_2 are two solutions, then the function

$$\psi(\mathbf{x}, t) = c_1\psi_1(\mathbf{x}, t) + c_2\psi_2(\mathbf{x}, t), \tag{1.2.11}$$

where c_1, c_2 are complex constants, is also a solution. (2) The hermiticity of $\hat{H}$ implies a conservation law for the flow of probability:

$$(\mathrm{d}/\mathrm{d}t) \int_{-\infty}^{\infty} |\psi|^2 \, \mathrm{d}^3x = 0. \tag{1.2.12}$$

To solve (1.2.10) for all time we must specify the initial state function $\psi_0(\mathbf{x}) = \psi(\mathbf{x}, 0)$. In addition, ψ and its derivatives must obey certain boundary and subsidiary conditions. Thus, it is required in regions where V is finite that ψ and $\nabla\psi$ be bounded, continuous and single-valued functions of $\mathbf{x}$. If the potential V has a discontinuity along a surface with normal unit vector $\mathbf{n}$, ψ and $\nabla\psi \cdot \mathbf{n}$ must be continuous across it. If, however, the potential step is infinite, $\psi = 0$ along the surface and $\mathbf{n} \cdot \nabla\psi$ is indeterminate.

Further details of the formalism, such as the extension to many-body systems, systems with spin, density matrices and the relativistic formulation, will be introduced as we need them.

As presented so far, quantum mechanics appears essentially as a set of working rules for computing the likely outcomes of certain as yet undefined processes called 'measurements'. One might well ask what happened to the original programme embodied in the old quantum theory of explaining the stability of atoms as objective structures in spacetime. In fact, quantum mechanics leaves the primitive notion of 'system' undefined; it contains no statement regarding the objective constitution of matter corresponding to the conception of particles and fields employed in classical physics. There are no 'electrons' or 'atoms' in the sense of distinct localized entities beyond the act of observation. These are simply names attributed to the mathematical symbol ψ to distinguish one functional form from another. The original quest to comprehend atomic structure culminated in a set of rules governing laboratory practice.

It is not clear that one can apply the formalism to concrete problems without at least some mental image of the systems studied. For example, in applying quantum mechanics to the formation of molecules (§7.6), we have in mind some informal picture of the physical distribution of matter in space so that certain interaction energies may be deemed 'small' and treated as perturbations. And many orthodox accounts often slip into describing the motion of a wave packet through space as if it were something 'real'. Thus, in applying the theory physicists tacitly consider, and perhaps have to consider, quantum mechanics to be something more than just a means of correlating experimental results, and attribute to it the ability to describe some kind of reality beyond the phenomena.

The quantum formalism is not an entirely well-defined, closed structure. For example, there is no general rule for defining the order of the operators when products of classical variables are quantized, and the treatment of time is an open question (cf. §5.5). But even for that part which is uncontroversial, including the rules just stated, the 'interpretations' that may be put on the formalism are open-ended constructions about which there is no global consensus. Their limits and consistency are not established. Thus the common assumption that experiment cannot decide between the various interpretations if these reproduce all the usual quantal predictions should be viewed critically.

1.3 The completeness assumption

It was observed in §1.2 that the rules of quantum calculus cannot be stated without some reference to the physical interpretation of the mathematical symbols. In this connection it may be noted that the notion of the 'quantum state' used in the sense of some attribute *of* a physical system, although widely employed in presentations of the subject, is not a concept that appears in Bohr's analysis of the interpretative problems of quantum mechanics (e.g. Bohr, 1934, 1949, 1958). Bohr emphasizes the necessity of a complete specification of the experimental conditions for an unambiguous application of the formalism. Moreover, he requires that these conditions must be expressed in terms drawn from classical physics, for it is only in this way that he considers unambiguous communication in physics is possible. The predictions embodied in the ψ-function then pertain to the total classically-described experimental phenomenon and not to the independent properties of an 'object'. Bohr (1948) summarized his view thus:

> The entire formalism is to be considered as a tool for deriving predictions, of definite or statistical character, as regards information obtainable under experimental conditions described in classical terms and specified by means of parameters entering into the algebraic or differential equations of which the matrices or the wave-functions, respectively, are solutions. These symbols themselves, as is indicated already by the use of imaginary numbers, are not susceptible to pictorial interpretation; and even derived real functions like densities and currents are only to be regarded as expressing the probabilities for the occurrence of individual events observable under well-defined experimental conditions.

Words like 'position' and 'spin' are not meant to be understood in the way we might think, or indeed how we really do continue to think, but as convenient shorthands for macroscopic and technically complex instruments designed to 'measure' these 'properties'. And the following words are attributed to Bohr (quoted by Jammer (1974, p. 204)): 'There is no quantum world, there is only an abstract quantum physical description. It is wrong to think the task of physics is to find out how nature *is*. Physics concerns what we can say about nature.'

Much of Bohr's argument is concerned with demonstrating that any attempt to construct a physical explanation of the formalism inevitably leads to ambiguity and confusion. He saw in quantum mechanics ground for an application of his philosophy of complementarity, but we emphasize that this view cannot be *derived* from quantum mechanics. In this approach the wave and particle aspects of matter are supposed to be revealed by complementary but mutually exclusive experimental arrangements. The purpose of complementarity is not to go beyond classical concepts but to preserve them by limiting their range of application. In this sense quantum mechanics is claimed to introduce no new physical concepts. Nevertheless, in the process of arguing for this programme Bohr brought out, perhaps in a rather indirect manner, what may be the really novel feature of the quantum description, namely its holistic character, that the observer cannot be excluded from an account of the physics she observes.

Bohr assumes that everyday language, and its refinement in classical physics (particles, fields, position, momentum, energy), is a natural and necessary mode of discourse for human beings to communicate their experiences unambiguously, and that it is unproblematic. This seems to ignore an essential component of language that it is a function of, and contributor to, a changing social context. Classical physics took millennia to develop and takes years of schooling to learn. The range and content of 'everyday experience' is constantly being enriched and altered and it is hard to see what is natural about its state in our particular epoch. Why should everyday language not eventually come to include concepts that do enable quantum phenomena to be conceptually analysed? Actually, this insistence of Bohr on the necessity of a classical mode of discourse is only of historical interest, for we now know that we can indeed develop concepts to analyse the individuality of quantum experience (for example, through the theory described in this book), *if we countenance the introduction of new concepts that go beyond the classical paradigm.* In this connection we might note that the appearance of complex numbers in quantum mechanics, adduced by Bohr in the passage cited above as indicative of the impossibility of a pictorial representation of the phenomena, is actually irrelevant to this question (cf. the presentation of classical mechanics in terms of complex fields in Chap. 2). Further difficulties with Bohr's approach are discussed in Chap. 8.

Although the claim is widely disseminated that Bohr's 'Copenhagen interpretation' is the most consistent and satisfactory interpretation of quantum mechanics available, it is remarkable that there is no source one can turn to for an unambiguous rendering of Bohr's position about which there is general agreement. Scholars give varying accounts of his work (cf., e.g., Folse (1985) and Murdoch (1989)). Textbooks do not apply Bohr literally. When one investigates the properties of the hydrogen atom, for example, one solves the Schrödinger equation for two charged particles interacting via a Coulomb potential, i.e., for a system in its own right. One hardly bothers to specify the experimental arrangements necessary to verify the predictions contained within the wavefunction that are calculated in this way. Indeed, the natural inclination of scientists is to attempt to visualize the inner workings of an atom, for example by treating $|\psi|^2$ as a kind of objective 'charge density' (cf. §4.5). In doing this they move beyond the interpretation of ψ that pertains simply to the relative frequencies of measurement results to attempt a description of the essence behind the phenomena. This is the unspoken contradiction at the heart of quantum physics: physicists do want to find out 'how nature is' and feel they are doing this with quantum mechanics, yet the official view which most workers claim to follow rules out the attempt as meaningless!

The views of the principal exponents of the conventional interpretation of quantum theory (Bohr, Heisenberg and von Neumann) will be examined in the course of this book, particularly in Chap. 8. Although their analyses differ in certain important respects, they share a common theme which is at the core of the conventional interpretation presented in most textbooks. This may be stated as the following assumption of 'completeness':

> The wavefunction is associated with an individual physical system. It provides the most complete description of the system that is, in principle, possible. The nature of the description is statistical, and concerns the probabilities of the outcomes of all conceivable measurements that may be performed on the system.

In practice one checks these statistical predictions through the relative frequencies of the results of measurements carried out on an ensemble of identically prepared (i.e., same ψ) systems. What this assumption boils down to is the proposal that if ψ is held as descriptive of a single system, the rules set out in §1.2 constitute *in themselves* the entire physical theory. Notice that it is asserted that not only is this postulate sufficient for interpreting quantum theory, but that it is *necessary*.

We emphasize the speculative character of the completeness assumption. It is not forced upon us by the experimental facts and the more detailed theory of individual processes that it forbids is not excluded by the generally agreed formalism. A difficulty with it is that it takes for granted that the formalism is essentially closed and unambiguous. But the latter is not completely well defined and in an area such as the problem of time (cf. §5.5) the completeness assumption becomes vulnerable and open to question.

Because it is postulated that predicting the results of experiments exhausts the possibilities of description of an individual system that are even conceivable, the indeterminism implied in the statistical interpretation of ψ is not of the kind encountered in classical statistical mechanics. There the evolution of a system is unpredictable due to our ignorance of an, in fact, well-defined physical state of an individual system. Here no meaning is to be attributed to the notion of a state beyond that encoded in ψ and hence the indeterminism is in some sense supposed to be intrinsic to the very nature of the system. Of course, if one relinquishes the assumption that material systems of the same kind (e.g., electrons) are distinguished from one another by virtue of having a well-defined location in spacetime, physical processes involving them will appear to be inherently indeterminist since systems that are prepared identically in the quantum mechanical sense (same initial wavefunction) and subjected to the same forces (same Hamiltonian) behave differently (points appear apparently at random on a detecting screen).

Following Einstein (§1.4) we shall locate the origin of the difficulties and paradoxes encountered in quantum theory in the attempt to squeeze the entire physical content of the theory of an individual system into the straitjacket of the statistically interpreted ψ-function. It has allowed quantum mechanics to acquire an aura of 'magic' in which 'smoky dragons' are invoked in discussions of two-slit interference effects and mind is held actively to influence generally occurring physical processes. The various difficulties of interpretation largely evaporate if the completeness assumption is relinquished in favour of less restrictive assumptions where one admits the possibility that other physical elements may enter the theory and that ψ has a significance beyond the mere specification of probabilities. Then one can indeed use quantum mechanics to find out 'how nature is'.

The historical triumph of the Copenhagen interpretation seems to be a rather fortuitous affair and we do not consider that the exclusion of other valid points of view has had a beneficial influence on the development of theoretical science. The legacy of the orthodox view has been a lopsided presentation of the subject where great emphasis is laid on what we *cannot* know about nature. The only research it appears to have stimulated is the attempt to understand and/or refute it, and the latter has hardly been encouraged. Because there is nothing in the theory to discuss but the results of experiments, it has also contributed to the notion that science is essentially concerned with the prediction and control of physical phenomena, and that progress in physics is most likely to come about through the manipulation of formalisms rather than the sharpening of our conception of reality. This is somewhat odd in a field that prides itself on the clarity and exactness of its thought and expression. The elegance of modern theoretical physics is largely to be found in its formal languages, not in the images with which it seeks to comprehend the world. Too often a concept is judged on what it 'predicts' and on whether a test can be proposed at its moment of inception to demonstrate its 'truth'. But analysed in these terms the transformation between Ptolemaic and Copernican cosmology is incomprehensible, for what was at issue there was a novel perception of how nature is, and this could not be immediately 'proved' by a new 'prediction'. To find a test, ideas must first be nurtured; in the meantime, they should be assessed according to different criteria, such as their explanatory power.

In the following the important issue is not so much the denial of causality in the processes governed by quantum mechanics, but the claim that no model at all can be constructed of an individual system. This latter point is a fundamental component in Einstein's critical analysis, as we now see.

1.4 Einstein's point of view

There is a popular view of Einstein in relation to the quantum theory which holds that he was unable to assimilate the revolutionary changes in world view apparently required by the new theory, that what bothered him most was the elimination of determinism from fundamental physics ('God does not play dice'), and that he 'wasted' the last 30 years of his scientific life in a fruitless quest to reestablish old-fashioned classical determinism as the ground of physical theory, having been 'beaten' by Bohr in their famous dialogue. Contrary to this view, consideration of what Einstein actually said in his public and private writings in the quantum period from the late 1920s to his death reveals a rather different picture. He, in fact, offered a carefully crafted critique that has never been satisfactorily answered, or in some ways even addressed, by the quantum establishment, and which, moreover, endures today. Pertinent references are the articles of Einstein (1936, 1940, 1948, 1949), the letters in Prizbram (1967) and Born (1971), and the analyses of Ballentine (1972) and Fine (1986).

A consistent theme in Einstein's commentary is the incompleteness of the theoretical description provided by the ψ-function, and his advocacy of a definite counter-interpretation.

To illustrate what he means by 'incomplete', Einstein (1949, p. 667) considers the case of radioactive decay in which an α-particle is emitted from an atom localized practically at a point. This system may be modelled by a closed potential barrier which at $t = 0$ encloses the α-particle. As time passes, the wavefunction, initially finite only inside the barrier, leaks into the surrounding space. According to the usual prescription this function yields the probability that at some instant the particle is found in a certain portion of the external space. Yet the wave may take many centuries to expand into the outer space whereas the particle may be found there after only a relatively short time. The wavefunction does not therefore imply any assertion concerning the time instant of the disintegration of the radioactive atom. That is, it does not describe the actual individual event revealed by the detector, including its cause.

If it is reasonable to suppose that the individual atom really does have a definite moment of decay, one may conclude, according to Einstein, that the ψ-function does not provide a complete description of the individual; it must be considered incomplete.

An objection may be raised that one is concerned here with a microscopic system of which we can have no direct knowledge, and that one can only claim there is a definite moment of disintegration if this can be determined empirically. To counter this, Einstein amplifies his example of incompleteness to the macroscopic scale (his version of the cat problem of Schrödinger (1935b)) by including the detector (Geiger counter) and a registration-strip, upon which a mark is made when the detector fires, in the entire system to which quantum mechanics is to be applied. After a suitable time has elapsed we expect to find a single mark on the strip. Yet the theory only offers the relative probabilities for the location of the mark *if this is observed*. It does not describe the objective definiteness of the mark as a property of the total system *per se*.

In this example the lack of description of the actual moment of decay is translated into a failure to describe the location of the mark on the strip. Einstein admits as a logical possibility that the mark becomes definite only when the strip is observed but states that, because we are now entirely within the sphere of macroscopic concepts (Einstein, 1949, p. 671),

> ... there is hardly likely to be anyone who would be inclined to consider it seriously ... One arrives at very implausible theoretical conceptions, if one attempts to maintain the thesis that the statistical quantum theory is in principle capable of producing a complete description of an individual physical system.

Einstein gave further arguments that quantum mechanics works with an inadequate description of individual systems in connection with the classical limit of the theory (a similar example to the macroscopic example above – see §6.5) and correlations in many-body systems (Chap. 11). His method was to show that those who adhere to the completeness assumption are compelled to adopt 'unnatural theoretical interpretations', in which according to him no one seriously believes.

It will be noted that in the example just described Einstein's concern is the *indeterminateness* of the world according to the usual quantum description of physical events, i.e., its failure to describe a reality comprising independent substantial objects beyond the phenomena. The complex function ψ simply does not exhibit any feature that could be put into correspondence with the (presumed) real state of affairs *viz.* that matter has substance and form independently of whether or not it is observed. Einstein's argument does not signify a general distaste for statistical theories in physics (he did after all develop the theory of Brownian motion). In his view, the indeterministic aspect of quantum mechanics follows from the failure to provide a complete description and not because it is an intrinsic characteristic of matter. In a letter written to Schrödinger in 1950 he says (Prizbram, 1967, p. 40) '... it seems certain to me that the fundamentally statistical character of the theory is simply a consequence of the incompleteness of the description.' In Einstein's programme, resolving the difficulty of describing a determinate reality entails constructing a causal (determinist) description, because he felt that this is a basic requirement of a complete physical theory (cf. Fine (1986, p. 103)). That is, in the process of making microphysics determinate, it would cease to be intrinsically statistical.

Of course, although linked in Einstein's critique, the requirements of determinism and determinateness are logically distinct. It is useful to separate the problem of completing quantum mechanics as a broad research programme from the specific brand of completion that invokes a deterministic description. As an analogy we may think of the classical theory of Brownian motion (Ballentine, 1972). This provides an objective description of a particle where individual events have definite antecedents, but the process is innately indeterminist and furnishes only statistical predictions.

As a way of resolving the interpretative dilemma within the terms of reference employed in quantum mechanics, Einstein proposed his own interpretation in which he *advocates* Born's statistical postulate, but interpreted in the sense that ψ pertains not to a single physical system but rather to an *ensemble* of systems. In this view ψ is admitted to be an incomplete representation of actual physical states and plays a role roughly analogous to the distribution function in classical statistical mechanics. Unfortunately, Einstein did not develop this idea sufficiently far for it to be clear exactly what it entails. He did not present it as an independent point of view but rather invoked it in the context of his examples designed to illustrate the untenability of the completeness assumption. He consistently claimed that the 'ensemble interpretation' dissolves the various difficulties and paradoxes flowing from the postulate of completeness, but he never explained precisely how.

Einstein's interpretation of the ψ-function has subsequently been developed into the 'statistical interpretation of quantum mechanics' (Ballentine, 1970; for a review see Home & Whitaker, 1992). Here it is asserted that ψ refers only to ensembles and that these are composed of particles pursuing definite (but unknown) spacetime trajectories. The only meaning attributed to $|\psi|^2$ is a statistical one, that it determines the relative frequency with which positions are realized in an ensemble of similarly prepared systems. Although it adopts a more modest and temperate line than the Copenhagen interpretation, a difficulty with this proposal is that one is not told what the laws are that govern the motions of the particles which are now admitted to exist objectively. Clearly, they cannot be the classical Newtonian laws because the ensemble of individual motions must reproduce the predictions of quantum mechanics, which in general contradict those of classical mechanics. In remaining agnostic on the issue of the laws obeyed by the particles, the statistical interpretation does not seem to offer any greater insight into the nature of quantum phenomena than the usual view. It is akin to treating the evolution of systems in classical statistical mechanics as defined just by the Liouville equation and not enquiring into the latter's origin in Hamilton's equations governing the underlying ensemble elements. This gap in the statistical interpretation is, of course, filled by the causal interpretation. But in the process it is found that the wavefunction cannot have the significance simply of encoding statistical information, but actually acquires the highly nonclassical property of being a physical component of each ensemble element, something not envisaged in the statistical interpretation. In this sense, the causal interpretation may be viewed as conceptually lying somewhere between the Copenhagen and statistical interpretations; it posits that ψ is in itself an incomplete description of an individual system, but in completing the theory, ψ is associated with the individual.

Einstein's critique, particularly his insistence that a deeper explanation of the phenomena correctly correlated by quantum mechanics is possible and necessary, provides the context for the theory presented in this book. Many statements made subsequently evoke his sentiments, without always directly referring to one of his utterances (for his reaction to the causal interpretation see §1.5.2).

1.5 The causal interpretation

1.5.1 De Broglie and Bohm

There was a general movement in theoretical physics in the 1920s against the idea that individual atomic events could be visualized as parts of causally connected sequences of spacetime processes. In the paper where he proposed the probability interpretation of the wavefunction, Born (1926) wrote: 'I myself am inclined to give up determinism in the world of atoms. But that is a philosophical question for which physical arguments alone are not decisive.'

It seems to have been regarded as almost axiomatic that the trajectory concept of classical mechanics is incompatible with wave mechanics. An argument in this direction was offered by Schrödinger (1926a,c, 1928), based on Hamilton's analogy between the rays of geometrical optics and mechanical paths. Schrödinger first asserted that in optics the conception of rays is well defined only in the geometrical limit and that in domains where the finiteness of the wavelength becomes relevant it loses nearly all significance because, even in homogeneous media, the 'rays' would be curved and appear mutually to influence one another. Then by analogy he claimed that the notion of the path of a mechanical system in ordinary mechanics likewise becomes inapplicable in cases where the de Broglie wavelength is comparable to characteristic lengths associated with the orbit, and indeed that it entails a contradiction. Although he was not a supporter of the emerging Copenhagen interpretation, Schrödinger proposed that the path should be replaced by the wave and have only an approximate significance in what would be the analogue of the geometrical optics limit. An obvious reply to this is as follows. Whatever the merits of the argument for excluding the ray concept from undulatory optics may be (and this is not a closed subject, see §12.6), the analogy drawn by Schrödinger between wave mechanics and wave optics is not an exact one (the mathematical theories are not in one-to-one correspondence) and the case against the meaningfulness of material paths in wave mechanics cannot therefore be regarded as proved, at least not on these grounds.

The idea that one should introduce the new matter wave not instead of the material point but as coexisting with it was advanced in this period by Louis de Broglie (1926a,b, 1927a,b,c). In the nonrelativistic Schrödinger case de Broglie suggested that the wavefunction is associated with an ensemble of identical particles differing in their positions and distributed in space according to the usual quantum formula, $|\psi|^2$. But he recognized a dual role for the ψ-function; not only does it determine the likely location of a particle, it also *influences* the location by exerting a force on the orbit. It thus acts as a 'pilot-wave' that guides the particles (only one of which actually accompanies each wave) into regions where ψ is most intense. For a scalar external potential the law of motion of the system point is that of the classical Hamilton–Jacobi theory according to which the possible paths are orthogonal to the surfaces of constant phase (de Broglie generally presented his theory in terms of the relativistic Klein–Gordon equation which was subsequently found to present severe interpretative problems – see §12.1). In fact, the pilot-wave idea is a truncated version of de Broglie's complete proposal in which he envisaged the particle as being represented by a singularity in a second field introduced in addition to the ψ-wave, rather than just treating it as a classical-style point. We shall not discuss this 'double-solution' interpretation further (see e.g. de Broglie (1956); Jammer (1974, p. 44)).

One may say that the contradiction perceived by Schrödinger, that the ensemble of paths seem mutually to influence one another, had been removed in de Broglie's approach by treating the wave itself as an agent that causes the paths to curve (in addition to classical forces).

De Broglie presented his proposal at the 1927 Solvay conference (*Electrons et Photons*, 1928, pp. 105–32). In particular, he applied his guidance formula to compute the orbits for the hydrogen atom stationary states (see §4.5). The approach met with a general lack of enthusiasm. Although it was discussed, only Einstein said de Broglie was right to search in the direction of a particle interpretation, although he did not endorse the specific model described (*Electrons et Photons*, 1928, p. 256). Kramers, while not questioning the possibility of tracing precise orbits, remarked that he could see no *advantage* in doing so (*Electrons et Photons*, 1928, p. 266). Pauli presented a detailed objection (*Electrons et Photons*, 1928, pp. 280–2; see §7.5.2) which de Broglie attempted to answer. However, the unfavourable climate, presumably compounded by Heisenberg's discovery of the 'uncertainty' relations, eventually led him to abandon his programme and indeed he soon began to propagate arguments against it (de Broglie, 1930). De Broglie returned to research in this area only 25 years later when in 1952 David Bohm rediscovered the approach and developed it to the level of a fully fledged physical theory (the story is told in de Broglie (1953a, 1956)).

In the intervening period Rosen (1945) had noted the possibility of retaining the particle picture in the quantum domain but was led to give up his proposed interpretation because he felt it was inconsistent with the existence of interference phenomena.

Bohm's classic pair of papers (Bohm, 1952a,b) remain the starting point for anyone wishing to find out about the de Broglie–Bohm theory. Bohm showed conclusively by developing a consistent counterexample that the assumption of completeness described in §1.3, a notion that pervaded practically all contemporary quantal discourse, was not logically necessary. One *could* analyse the causes of individual atomic events in terms of an intuitively clear and precisely definable conceptual model which ascribed reality to processes independently of acts of observation, *and* reproduce all the empirical predictions of quantum mechanics. Bohm's model is essentially de Broglie's pilot-wave theory carried to its logical conclusion.

But its significance goes beyond a mere existence proof, a kind of theoretical game of no direct value to the practising quantum physicist. A phrase such as 'an electron moves along the x-axis' is no longer simply an aid to calculating the wavefunction but refers to an objective process engaged in by a material system possessing its own properties through which the appearances (the results of successive measurements) are continuously and causally connected. It is thus very much a 'physicist's theory' and indeed puts on a consistent footing the way in which many scientists instinctively think about the world anyway. This comes about not merely through an *extension* of classical notions but requires the development of a new physical intuition. Bohm locates the novelty of quantum mechanics not in its statistical or discrete aspects but in a new physical conception of the state of a system (mathematically described by Schrödinger's wavefunction) that manifests itself in the motion of particles through a new type of potential, the quantum potential. The resulting theory stands in a clear and obvious relation to its classical counterpart (Chap. 6). The principal feature it shares with the classical paradigm is that the individuality of experience is comprehensible, but it diverges in its new notion of state.

Bohm applied his interpretation to a range of examples drawn from nonrelativistic quantum mechanics and speculated on possible alterations in the particle and field laws of motion such that the predictions of the (modified) theory continue to agree with those of quantum mechanics where this is tested but disagree in as yet unexplored domains. In Bohm's theory Born's statistical postulate is not dropped but is incorporated as a special case of the more general conception that $|\psi|^2$ represents the likely *current* location of a particle. Born's assumption follows when the theory is applied to measurement processes, the analysis of which is a significant feature of Bohm's contribution. He lays great emphasis on the creation of the outcome of the measurement through the interaction between the system and apparatus; in general, this does not passively reveal the premeasurement value of a physical quantity. Bohm extended his approach to include the second quantized electromagnetic field and also answered the earlier objections of Pauli, de Broglie and Rosen cited above. It should be noted that Bohm took issue with de Broglie's conception of light in which 'photons' are conceived as massive corpuscles moving within the electromagnetic guiding field, and proposed instead that the only 'real' parameters are the field coordinates and their conjugate momenta (see Chap. 12).

Because the *causes* of microevents could be analysed in their individuality and were no longer treated as irreducibly indeterminist and inexplicable, the de Broglie–Bohm proposal came to be known as the 'causal interpretation' of quantum mechanics. But while it indeed sets up a correspondence between all the mathematical symbols appearing in the formalism and physical properties of the wave-particle composite, the theory actually goes beyond the mere interpretative debate concerning the assignation of meanings to the symbols because it adds not just to the concepts but to the formalism itself, through the particle law of motion (cf. §5.5). Its description as a causal 'interpretation' therefore appears to be inadequate. Since it is essentially a novel theory of material motion (the first in quantum mechanics, in fact), a more appropriate title is 'the quantum theory of motion'. We shall use the various terms 'pilot-wave', 'causal interpretation' and 'quantum theory of motion' interchangeably in this book. But however we think of it, this theory of motion is not presented as a conceptually closed edifice offering the final word on quantum mechanics, and its originators never intended it to be this. Rather, it is a view worth developing for the insight it provides, and as a clue for possible future avenues of enquiry.

A few papers appeared in the wake of Bohm's articles raising technical questions connected with his approach, notably by Takabayasi (1952, 1953; replied to by Bohm, 1953b). The issues raised in these various articles will be discussed in later chapters. In this period de Broglie developed a small school of collaborators, including Jean-Pierre Vigier (see de Broglie (1956), Vigier (1956)). A good review of developments in the theory up to the mid-1950s is given by Freistadt (1957) and a qualitative account by Bohm (1957). A point to note is the attempt of Bohm (1953a) to demonstrate within the framework of the causal axioms why the probability distribution should be $|\psi|^2$, and also the attempts to include in the theory nonrelativistic systems with spin (Chap. 9) and systems governed by the Dirac equation (Bohm, 1953b; de Broglie, 1956; Takabayasi, 1957; see §12.2).

In the main the commentary of other physicists, particularly the Founding

Fathers (see below), was, where it existed, negative, and the theory did not enter the mainstream of physics either as a research topic or in textbooks. In fact, in the following 25 years only occasional and sporadic references were made to the de Broglie–Bohm theory. Books were written, courses taught and research conducted as if what Bohm had demonstrated to be possible was still, in fact, impossible. Although de Broglie continued to advertise the idea in his books, and Bohm worked on hidden-variable theories, no development or application of the pilot-wave theory was made.

A solitary and notable exception is the work of Bell (1966) who made reference to the nonlocality inherent in the de Broglie–Bohm picture (Chaps. 7, 11). Raising the question of whether this is a generic feature of all 'hidden-variable' completions of quantum mechanics, Bell (1964) was led to his inequality distinguishing a class of local theories from quantum mechanics, a step which brought the issue within the realm of experimental physics.

Belinfante (1973) and Jammer (1974) respectively published technical and nontechnical abridged accounts of the theory but it was not until the late 1970s that serious interest was rekindled when the trajectories corresponding to the two-slit experiment were explicitly displayed in computer graphics (Philippidis, Dewdney & Hiley, 1979; §5.1). During the 1980s developments took place in several branches of the theory, as will be discussed in detail later. This was motivated in part by continued dissatisfaction with the conventional solution to the problems of interpretation, and a certain thawing in the attitude that these matters were all settled long ago. Also, what had for a generation been *gedanken* experiments, such as single-photon interferometry and Einstein–Podolsky–Rosen correlations, could now actually be performed. In this recent period the de Broglie–Bohm idea has featured in some popular accounts of quantum mechanics (e.g., Squires (1986)) and Bohm has returned to actively develop the theory, the principal new element in his work being the proposal that the quantum potential may be interpreted as a kind of 'information potential' (Bohm, 1987). The approach has been eloquently defended in several articles by Bell (1987). Some criticisms of the approach have appeared (e.g., Tipler (1984, 1987) (for a reply see Dewdney, Holland, Kyprianidis & Vigier (1986)), de Muynck (1987)) and comparisons have been drawn between the pilot-wave theory and interpretations other than the 'conventional' one, such as the 'many-worlds' picture (e.g., Bell (1987), Bohm & Hiley (1987), Zeh (1988)), a subject we shall not discuss.

Nevertheless, at the time of writing it is fair to say that the de Broglie–Bohm theory of motion is still marginalized and, when it is referred to, often misrepresented. The sustained technical examination of its novel features, a prerequisite before any decision can be taken regarding its value, has not been undertaken. There has been some discussion of why this should be so (e.g., Bohm & Peat (1989) and the following two subsections) but there clearly remains considerable scope for analysing the social relations of science in the context of this physical theory.

The causal interpretation may be viewed in the context of Einstein's critique as a concrete proposal for how a complete description of individual events may be obtained, but the deterministic model employed is by no means the only possible completion that is conceivable. As observed in the Preface, reintroducing determinism into microphysics was a means to the end defined as accounting for the individuality of physical systems, but other avenues are open through which one may achieve the same goal. An attempt at providing a complete description along indeterministic lines was proposed by Bohm & Vigier (1954). The idea is to suppose that the particle is constantly subjected to random perturbations coming from some background source, such as random fluctuations in the ψ-field, so that its motion is akin to a kind of Brownian movement and hence deviates from the deterministic law of the basic pilot-wave theory. The latter now only describes the mean motion and the particle may jump between the mean flow lines. By means of this further postulate, it can be proved that an arbitrary initial distribution of positions will decay in the course of time to the $|\psi|^2$-distribution of quantum mechanics (for the proof see also Belinfante (1973, p. 186); see also Valentini (1991)). This approach is connected with the subsequently developed stochastic interpretation of quantum mechanics (Nelson, 1966, 1985; Jammer, 1974, Chap. 9; Vigier, 1982) in which it is demonstrated how the Schrödinger equation is implied if particles in a stochastic process are subjected to a particular kind of force law. It is to be emphasized that the Bohm–Vigier and Nelson programme of deriving the laws of quantum mechanics from the laws obeyed by particles at some deeper or subquantum-mechanical level, whatever its merits, is logically unconnected with the basic de Broglie–Bohm pilot-wave idea which constitutes in itself a self-contained and consistent theory of motion that does not *require* the assumption of a further as yet unrevealed layer of physical reality. In this book we shall be concerned solely with working out the original deterministic de Broglie–Bohm proposal and do not discuss possible stochastic extensions.

1.5.2 What the great men said

The reaction to Bohm's work by the Copenhagen establishment was generally unfavourable, unrestrained and at times vitriolic (e.g. Rosenfeld (1958)).

We first consider the response of Heisenberg (1955; 1962, Chap. 8) to

Bohm's contribution, and see in outline how the points he raises may be answered (see also Bohm (1962) for a reply). Heisenberg first questions what it means to say that a wave propagating in configuration space is 'real'. His objection to this notion is based on his assertion that only 'things' in three-dimensional space are 'real'. He offers no logical or scientific argument to show that examining the possiblity of the physical reality of multi-dimensional spaces is a fruitless enterprise. It is useful to recall here the Kaluza–Klein programme in general relativity where physicists contemplate spacetimes of dimension greater than four as a valuable aid to comprehending and unifying the basic physical interactions.

Heisenberg goes on to bemoan what he considers an asymmetrical treatment of position and momentum in the causal interpretation, and the apparent breaking of a fundamental symmetry of the quantum theory. This criticism seems to confound the quite reasonable asymmetry in the physical interpretation, which assumes that the preferred arena for physical processes is position space, with the symmetry exhibited by the mathematical theory (cf. §3.12). The asymmetry is also connected with the nature of measuring processes which generally entail active transformations and do not passively reveal preexisting states (Chap. 8). Classical physics exhibits in the canonical formalism an analogous feature of mathematical symmetry but physical asymmetry.

The 'hidden parameters', i.e., the particle orbits, are denounced by Heisenberg as a "superfluous 'ideological superstructure'" having little to do with immediate physical reality because the causal formulation generates the same empirical results as the Copenhagen view. But in Bohm's theory it is precisely the positions of particles that are recorded in experiments; they are the immediately sensed 'reality'.

Finally, Heisenberg mentions Bohm's tentative proposals for modifying the quantum laws of motion so as to permit an experimental test of the trajectory interpretation in a domain where the quantum theory may conceivably break down. This possibility he dismisses as akin to the 'strange hope' that someday it will turn out that sometimes $2 \times 2 = 5$. That is, Heisenberg believes that any alternative or even modification to the current quantum theory is logically impossible. He considers that Bohm's purpose was to return to classical physics and thus misses the key point about the de Broglie–Bohm proposal: it refutes the view that the actual individual facts of experience are in principle unimaginable.

The objection pertaining to the asymmetrical role of the position and momentum variables in the causal interpretation had been raised in an earlier paper by Pauli (1953). He considers that this feature renders the theory 'artificial metaphysics' and that any proposal to modify the formalism so as to demonstrate empirically the existence of particle tracks will inevitably conflict with established experimental facts. A point of interest raised by Pauli is why, in a theory which treats the laws of the ensemble as functions of individual laws, the probability distribution should be given by the quantum formula $|\psi|^2$ and not arbitrarily specified (a similar observation is made by Keller (1953)). Although a physical justification of this assumption is not a logical requirement of a causal theory (one may just take the probability to be $|\psi|^2$ with no further discussion, as indeed the orthodox view does), Bohm & Vigier (1954) did, in fact, supply an argument justifying the $|\psi|^2$-distribution through a development of the original model, as noted above. Commenting on Pauli's article, Born (1971, p. 207) wrote to Einstein that '... Pauli has come up with an idea ... which slays Bohm not only philosophically but physically as well'!

At a 1957 conference in Bristol, Rosenfeld repeated the charge that Bohm was engaged in 'metaphysics'. Bopp summarized the discussion thus: '... we say that Bohm's theory cannot be refuted, adding ... that we don't believe in it' (Körner, 1957, p. 51; Jammer, 1974, p. 296).

Viewed in retrospect, what is striking about the reaction to Bohm is not that the proponents of the orthodoxy should attempt to defend themselves vigorously, but the inadequate character of the arguments they adduced in their defence. Personal distaste regarding specific features of the theory, such as its asymmetry, or the *ad hoc* manner in which the quantum potential is introduced, or its intrinsic nonlocality, seems to us to be beside the point (they are among the criticisms that have been levelled by advocates of the approach). The point is that it demonstrates that quantum phenomena need not be sealed in black boxes and forever hidden from our conceptual gaze, as claimed by the Copenhagen lobby. Such a demonstration is not invalidated by dubbing it 'metaphysical' (this term carries with it the implied rebuke that Bohm was doing 'mere philosophy', the physicist's ultimate censure). In a climate more disposed to a spirit of free enquiry, Bohm's work would have been acclaimed rather than treated as an inconvenience and then ignored. As an example of how new physics may be generated by a consideration of alternative theories, we have already cited the case of Bell's theorem which followed directly from contemplation of the de Broglie–Bohm model and is widely regarded as one of the seminal discoveries of twentieth century physics.

The fact is, of course, that the assumptions made by the Founding Fathers in regard to the possibilities of visualizing the origin of quantum phenomena were equally 'metaphysical' and devoid of empirical support. One can indeed argue that Bohm had provided at least theoretical evidence against the orthodox view, whereas there was no counterevidence forthcoming, either theoretical or experimental, to exclude the pilot-wave. In fact, the establishment did have in its possession theoretical evidence which indeed purported to demonstrate the impossibility of theories, of the type propounded by Bohm, that add supplementary variables to the quantum formalism while reproducing all its empirical predictions: the 1932 theorem of von Neumann (1955). Given the existence of Bohm's counterexample, one might have expected that the causal interpretation and von Neumann's theorem would both have been subjected to a sustained theoretical analysis to discover in which the error lay. While the quantum nobility expressed (private) doubts that Bohm had been able to circumvent von Neumann's theorem (Hanson, 1969, p. 174), this analysis simply was not carried out (the story is told by Pinch (1977, 1979)). More than 10 years elapsed before von Neumann's assumptions were critically analysed and found to be wanting in that he supposes that the probability distribution of the supplementary variables has the same properties as the quantum mechanical distribution (Mugur-Schächter, 1964). This supposition entails, for example, that the mean values of the new variables are linearly superposable in the same way that the means of quantum mechanical observables are, and hence that the outcome of the measurement of a linear sum of operators is the linear sum of the outcomes of measurements of the operators individually (Bell, 1966). The causal interpretation does not fall within the scope of this assumption because for it measurements entail transformations of the system under investigation. These transformations are very specific to the observable 'measured' and, as we have said, such processes do not passively reveal premeasurement values. Bell (1966, 1982) identified similar problems with other 'impossibility proofs'. Eventually, it proved possible to show that quantum mechanics can always be supplemented by 'hidden variables' (Gudder, 1970).

In letters written in the early 1950s, Einstein (1989, p. 60) expressed solidarity with de Broglie in his search for a complete representation of physical reality. But he does not seem to have been overly impressed by the specific solution advanced in the pilot-wave theory. To Born (1971, p. 192) he wrote in 1952 'Have you noticed that Bohm believes (as de Broglie did, by the way, 25 years ago) that he is able to interpret the quantum theory in deterministic terms? That way seems too cheap to me.' While the de Broglie–Bohm proposal might be considered as a deterministic completion carried out in accordance with Einstein's critique, Einstein did not think that fundamental progress towards the discovery of a deterministic substructure could be made by a completion from within, i.e., simply by appending supplementary physical variables to an essentially unmodified formalism.

He expressed this in 1954 as follows (quoted by Fine (1986, p. 57)): 'I think it is not possible to get rid of the statistical character of the present quantum theory by merely adding something to the latter, without changing the fundamental concepts about the whole structure.'

Einstein was after a more radical completion which countenanced going beyond the classical concepts that Bohr had retained in his interpretation and which de Broglie and Bohm were using to describe the individual process (material points and forces). Generally, he felt that the quantum theory did not serve as a useful point of departure. De Broglie's notion of the 'double-solution' was closer in spirit to Einstein's field-theoretic approach than the basic pilot-wave model but was at that stage (and is still today) largely an unfulfilled programme. Still, despite his reservations, Einstein (1953) took Bohm's work seriously enough to offer an objection which raised an interesting issue, as we discuss in §6.5.

It might be argued that Einstein's negative attitude was a tactical mistake and that, whatever he perceived as its drawbacks, some model such as that of de Broglie and Bohm was better than none at all in countering the prevailing vagueness in interpretation, at least as a makeshift before a more satisfactory foundation could be found (the view taken by de Broglie and Bohm towards their theory). But it seems unlikely that Einstein's endorsement would have made much historical difference given that for 25 years he had been branded as a quantum dissident and that the precise nature of his own critique was not widely understood. After all, his solution to the problem of interpretation within the quantum scheme, the view that ψ refers only to ensembles, was widely advertised in his public writings and was itself ignored (although admittedly he never developed the idea very far).

1.5.3 Some objections

There is a sense in which physicists do not understand how it is even *possible* to have a causal theory of the de Broglie–Bohm type, and feel uneasy that there must be some inconsistency in it if it really does what it claims and covers all the empirical ground accounted for by quantum mechanics. But beyond aesthetic displeasure, which has always loomed large in discussions of it, to our knowledge no serious technical objections have ever been raised against the de Broglie–Bohm world view. Anticipating our subsequent detailed analysis, we here summarize and offer preliminary answers to some of the typical objections that have been advanced against this theory.

(1) You cannot prove the trajectories are there

Insofar as the quantum theory of motion reproduces the assertion of quantum mechanics that one cannot perform a precise measurement of position simultaneously with a precise measurement of momentum, this statement is true (§8.4). But this cannot be adduced as evidence against the tenability of the trajectory concept. Science would not exist if ideas were only admitted when evidence for them already exists. One cannot after all empirically prove the completeness postulate. The argument in favour of the trajectory lies elsewhere, in its capacity to make intelligible a large swathe of empirical facts.

(2) It predicts nothing new

First we note that the postulate of completeness predicts nothing about the details of a process beyond the distribution of measurement results. In contrast, the quantum theory of motion permits more detailed predictions to be made pertaining to the individual process (cf. $5.1.3). Whether these may be subjected to an experimental test is an open question (cf. §§5.5, 8.8).

(3) It attempts to return to classical physics

The deterministic model of wave and particle causally evolving from the past into the future is a particular solution to the problem of describing a *determinate* reality. Although de Broglie and Bohm have often been chastised for reintroducing the classical paradigm, this misses the key point that they are invoking a concept not anticipated in classical physics, that of a 'state' of a mechanical system that lies beyond the material points. The role of the trajectory is to bring out this new concept so sharply it cannot be ignored. This essentially nonclassical programme should be contrasted with that of Bohr who strove to leave intact as far as possible the classical concepts by restricting their applicability.

(4) The price to be paid is nonlocality

Nonlocality is an intrinsic feature of the de Broglie–Bohm theory (Chaps. 7, 11). This property does not contradict the statistical predictions of relativistic quantum mechanics (Chap. 12) but it is considered to be in some way a defect, the implication being that entirely local theories are preferable. Yet nonlocality seems to be a small price to pay if the alternative is to forego any account of objective processes at all (including local ones). Also, it is inconsistent to deny the logical possibility of a pictorial representation of the phenomena, and then lay down conditions for what such a picture should consist of when one is produced.

(5) *It is more complicated than quantum mechanics*

Mathematically, the quantum theory of motion requires a reformulation of the quantum formalism (not an alteration). The reason is that the usual presentation of the theory is not the one most appropriate to the physical interpretation. But, mathematically, the theory remains quantum mechanics. In particular, the quantum potential is implicit in the Schrödinger equation.

(6) *It is counterintuitive*

It certainly runs counter to classical intuition. The concept of 'intuition' is like that of 'human nature': it is a function of history and not eternally frozen. The notion that a body persists in a state of uniform motion unless acted upon by a resultant force would be counterintuitive to Aristotle but natural for Galileo. Quantum phenomena require the creation of quantum intuition.

(7) *There is no reciprocal action of the particle on the wave*

In classical physics there is a dialectical interplay between particle and field, each generating the dynamics of the other. In the pilot-wave model the dynamical connection is one way. Among the many nonclassical properties exhibited by this theory (cf. §3.3), one is that the particle does not react dynamically on the wave it is guided by. But while it may be reasonable to require reciprocity of actions in classical theory, this cannot be regarded as a logical requirement of all theories that employ the particle and field concepts, especially one involving a nonclassical field.

It will be noted how, having been chided for its classical pretensions (point (3)), the causal interpretation is admonished in points (4), (6) and (7) because it is not classical enough!

2

Hamilton–Jacobi theory

2.1 The need for a common language

If we wish to compare the methods, content, claims and experimental predictions of two physical theories we have to find some common ground between them. What is needed ideally is a *language* which embraces the essential elements of each theory as parts of a broader structure which transcends them both. There are two components to such a language, which might be called the formal and the informal. Briefly, by 'formal' we mean a precisely defined set of concepts and their relationships from which one can deduce unambiguous conclusions by a series of logical steps (mathematics); by 'informal' we mean the intuitive concepts and pictures that a theory employs in order to render intelligible the 'reality' for which it seeks to account. These two aspects are naturally closely connected. The possibility of constructing a language of this type is not given *a priori.* It may turn out that it is possible to develop only one of the two components, and the theories we want to compare may or may not be commensurable, or only partly so.

We shall examine this question in connection with the relation between two physical theories of matter: classical mechanics and quantum mechanics. This relation is subtle and operates on several levels and we shall return to it throughout the book. For the present we confine ourselves to some general remarks. Roughly speaking, we may say that classical mechanics as a distinct discipline constitutes a language possessing both formal and informal aspects. Newton's laws (or their refinements in the Lagrangian, Hamiltonian and Hamilton–Jacobi formalisms) allow us not only to predict the results of experiments on fields and particles, but also provide an *explanation* of these results in terms of a definite world view – that of mass points pursuing well-defined trajectories in space and time and interacting via preassigned potentials. Formally this is a theory of continuous functions in configuration or phase space and this aspect derives its physical meaning from a definite theory of matter and motion with which it is inextricably linked.

Quantum mechanics on the other hand possesses a sophisticated and highly developed formal language of linear operators in Hilbert space but, in the orthodox interpretation, only provides hints of what an informal language which would explain the results the formalism predicts might be like. Indeed, the absence of a clear physical picture has tended to lead to an identification of 'quantum reality' with Hilbert space, i.e., with the formal language.

At the present stage in the development of physics the possibility of an intuitive account is closely connected with the possibility of a theory of matter and motion as a process in space and time. In this book we shall therefore be mainly concerned with developing a language which allows us to comprehend microphysical phenomena in this way. Naturally, on its own the spacetime theory of matter and motion of classical mechanics is not broad or deep enough to include quantum mechanics. We have to develop a new language.

Several avenues of enquiry are open to us. We might try to express classical mechanics in a Hilbert space language with the hope that eventually this may lead to new insights into the quantum theory. Formally this can indeed be done; for example, canonical transformations may be shown to be equivalent to unitary transformations in a Hilbert space of square integrable functions on phase space (Koopman, 1931). While this implies the possibility of a formal comparison between classical and quantum mechanics, it does not seem to lead to a clear physical conception of the latter. In particular it does not lead to a spacetime picture of quantum processes.

Alternatively, one may try to introduce a kind of 'phase space' into quantum mechanics. Again, this can indeed be done and leads to a formulation of quantum mechanics in terms of, for example, Wigner functions (§8.4.3). This is a more promising approach but again it does not result in a theory of material processes in space and time.

We pursue here a different approach, one that associates a well-defined phase space (i.e., simultaneously real position and momentum variables) with a quantum mechanical system, but which is immediately tied to a picture of physical events as processes in spacetime. To locate what is new in quantum mechanics in relation to classical mechanics we formulate both as particular instances of (a suitably generalized) Hamilton–Jacobi theory. It turns out that the Hamilton–Jacobi theory admits a natural generalization which provides a language broad enough to embrace both theories, formally and informally. Formally there is a clear mathematical procedure as to how one passes from the quantum to the classical domain (see Chap. 6), and informally the language provides an unambiguous physical picture of waves, rays (trajectories) and their interrelationships. Although historically connected with just classical mechanics and field theory and the semiclassical approximation to quantum mechanics, the Hamilton–Jacobi method transcends its origins.

The aim of this chapter is to provide the relevant classical background so as to put the generalization required by quantum mechanics into context. It is not an intrinsic requirement of a causal spacetime theory of quantum processes that we employ the Hamilton–Jacobi language of waves and rays. Indeed a 'minimalist' version of the causal interpretation of quantum mechanics can be formulated without reference to it (Chap. 3). However, the Hamilton–Jacobi language affords considerable insight into the formal and conceptual structure of quantum mechanics, in particular its relation with classical mechanics, and we shall make extensive use of it in this book.

2.2 The Hamilton–Jacobi method in classical mechanics

2.2.1 Hamilton's principal function

We begin by reviewing some basic results in classical mechanics, in particular how a certain function, a solution of the Hamilton–Jacobi partial differential equation, facilitates the solution of the equations of motion. This method is due to Jacobi. For definiteness we shall talk in terms of the motions of particles, i.e., material points (in three-dimensional space or configuration space), but all our remarks and results apply to any physical system as usually treated by the methods of classical mechanics (e.g., rigid bodies and fields). Our treatment is nonrelativistic and we use absolute time as an evolution parameter. Further details may be found in the texts of Synge (1954), Landau & Lifschitz (1960), Lanczos (1970), Arnold (1978) and Goldstein (1980).

The following notation is used. The generalized coordinates are denoted by q_i, $i = 1, \ldots, n$, where n is the number of degrees of freedom of the system, $\mathbf{x} = (x, y, z)$ are the coordinates of a body in a Cartesian system, p_i represents the momenta in both general and Cartesian coordinates, and t is the time. Often we shall neglect the indices and write just q and p. These then stand for the points whose coordinates are q_i and p_i. It is assumed that a metric is given on configuration space so that we can form invariants. Under coordinate transformations q_i transforms as the components of a contravariant vector and p_i as the components of a covariant vector, but we shall not distinguish this in the notation (all indices are subscripts). No attempt is made at a rigorous presentation.

Starting from the Lagrangian $L = L(q, \dot{q}, t)$, where along a trajectory

$q = q(t)$ and $\dot{q} = \mathrm{d}q/\mathrm{d}t$, we define the momentum canonical to the coordinate q_i to be

$$p_i = \partial L/\partial \dot{q}_i. \tag{2.2.1}$$

The Hamiltonian is defined as a function on the phase space (q, p) by a Legendre transformation:

$$H(q, p, t) = \sum_i p_i \dot{q}_i - L(q, \dot{q}, t). \tag{2.2.2}$$

The equations of motion may be derived as follows. Consider two points in state space, (q_0, t_0) and (q, t), and the paths that may join them. Then the actual path traversed by the physical system is that which extremizes the action integral

$$I(q, t; q_0, t_0) = \int_{q_0, t_0}^{q, t} L(q, \dot{q}, t)\,\mathrm{d}t. \tag{2.2.3}$$

This is Hamilton's Principle which we write as

$$\delta I = 0 \tag{2.2.4}$$

and it implies the n second-order Euler–Lagrange equations

$$\frac{\mathrm{d}}{\mathrm{d}t}\frac{\partial L}{\partial \dot{q}_i} - \frac{\partial L}{\partial q_i} = 0, \qquad i = 1, \ldots, n. \tag{2.2.5}$$

Alternatively, we can substitute for L in (2.2.3) from (2.2.2) and write the action function as

$$I = \int \sum_i p_i\,\mathrm{d}q_i - H\,\mathrm{d}t. \tag{2.2.6}$$

Varying this with respect to the $2n$ independent variables q_i and p_i yields Hamilton's set of $2n$ first-order differential equations:

$$\dot{q}_i = \partial H/\partial p_i|_{q_j = q_j(t), p_j = p_j(t)}, \quad \dot{p}_i = -\partial H/\partial q_i|_{q_j = q_j(t), p_j = p_j(t)}. \tag{2.2.7}$$

for all i, j. Solution of eqs. (2.2.5) or (2.2.7) requires that we specify the initial coordinates and the initial velocities or canonical momenta.

In the Hamiltonian formulation of the theory we may replace the independent variables q_i, p_i by a new set of $2n$ independent variables Q_i, P_i:

$$Q = Q(q, p, t), \qquad P = P(q, p, t). \tag{2.2.8}$$

This is a coordinate transformation in phase space. The new set of coordinates is canonical if Hamilton's equations (2.2.7) retain their form under the transformation. If $K = K(Q, P, t)$ is the new Hamiltonian, we require

$$\dot{Q}_i = \partial K/\partial P_i|_{Q=Q(t),P=P(t)}, \qquad \dot{P}_i = -\partial K/\partial Q_i|_{Q=Q(t),P=P(t)}. \tag{2.2.9}$$

These equations may be derived from a variational principle applied to an action function of the form (2.2.6):

$$I = \int \sum_i P_i \, \mathrm{d}Q_i - K \, \mathrm{d}t. \tag{2.2.10}$$

The integrands in (2.2.6) and (2.2.10) then differ by a total time derivative:

$$\sum_i P_i \dot{Q}_i - K = \sum_i p_i \dot{q}_i - H - \mathrm{d}F/\mathrm{d}t, \tag{2.2.11}$$

where F is an arbitrary differentiable function of q, p, Q, P and t. By (2.2.8) only $2n$ of the total of $4n$ coordinates q, p, Q and P are independent and so F depends on just $2n$ of these coordinates, together with t. For example, it may have the form $F(q, Q, t)$, $F(q, P, t)$, $F(p, Q, t)$ or $F(p, P, t)$ or some other mixture of q, p, Q and P. Note that the new coordinates may not have the dimensions of position and momentum – they may represent some conjugate pair such as energy and time or action-angle variables.

Consider the case where $F = F(q, Q, t)$. Then from (2.2.11) we deduce that

$$p_i = \partial F/\partial q_i, \tag{2.2.12a}$$

$$P_i = -\partial F/\partial Q_i, \tag{2.2.12b}$$

$$K = H + \partial F/\partial t. \tag{2.2.12c}$$

Given F we can reconstruct from these relations the canonical transformation (2.2.8). To do this we solve (2.2.12a) for Q in terms of p, q and t and then (2.2.12b) for P. The new and old Hamiltonians are related by (2.2.12c). In view of this, F is called the *generating function* of the canonical transformation. Notice that F is not a function on phase space but relates two sets of coordinate systems on that space.

In order that we may use the relations (2.2.12) to solve the dynamical problem we need a further result: the motion of the system in time is equivalent to *the continuous unfolding of a canonical transformation.* To see this, consider the infinitesimal canonical transformation $Q_i = q_i + \delta q_i$, $P_i = p_i + \delta p_i$, where δq_i, δp_i are infinitesimal changes in the coordinates and momenta. To the first order in small quantities, the generating function can in this case be taken to be a function on the phase space labelled by q, p and it may be shown that

$$\delta q_i = \varepsilon \, \partial G/\partial p_i, \qquad \delta p_i = -\varepsilon \, \partial G/\partial q_i$$

where ε is an infinitesimal parameter and $G = G(q, p, t)$. Consider the case where $\varepsilon = \mathrm{d}t$, a small time interval, and $G = H$, the Hamiltonian. Then

$$\delta q_i = \mathrm{d}t\, \partial H/\partial p_i = \mathrm{d}t\, \dot{q}_i = \mathrm{d}q_i,$$
$$\delta p_i = -\mathrm{d}t\, \partial H/\partial q_i = \mathrm{d}t\, \dot{p}_i = \mathrm{d}p_i.$$

In other words, the infinitesimal canonical transformation generated by the Hamiltonian is precisely the physical change undergone by the generalized canonical coordinates of the system during the time interval $\mathrm{d}t$. Since an arbitrary canonical transformation can be built from a succession of infinitesimal transformations we conclude that the actual motion during a finite time interval of any system governed by Hamilton's equations is a continuous canonical transformation:

$$q = q(q_0, p_0, t), \qquad p = p(q_0, p_0, t) \tag{2.2.13}$$

where q_0, p_0 are the initial canonical coordinates. We shall find the generating function of this *finite* transformation below.

Inverting (2.2.13) to give q_0, p_0 in terms of q, p and t we see that the problem of motion is solved if we can find a canonical transformation from coordinates q, p at time t to a set of constant (in time) coordinates q_0, p_0 at some initial time t_0. Returning to (2.2.9), we see that this may be achieved if $K = 0$, for this ensures that the new coordinates are constant in time:

$$\dot{Q} = \dot{P} = 0. \tag{2.2.14}$$

From (2.2.12c) K will be zero if F satisfies the equation $\partial F/\partial t + H(q, p, t) = 0$. It is usual to denote F by S in this case. If we suppose for definiteness that F is a function of q, Q and t, and substitute for p in H from (2.2.12a), we obtain the *Hamilton–Jacobi equation*:

$$\partial S(q, Q, t)/\partial t + H(q, \partial S(q, Q, t)/\partial q, t) = 0. \tag{2.2.15}$$

The function S is called *Hamilton's principal function.*

Eq. (2.2.15) is a first-order partial differential equation in the $(n + 1)$ variables $q_1, \ldots, q_n, t$. Since S itself does not appear in the equation, a complete solution involves n nonadditive constants $\alpha_1, \ldots, \alpha_n$: $S = S(q, \alpha, t)$. We can take the αs to be the new coordinates as in (2.2.15): $\alpha_i = Q_i$. But more generally the αs may be any function of the Qs or, if some other form is assumed for the generating function, of some combination of the Qs and Ps.

In order to solve the equations of motion by the Hamilton–Jacobi method we proceed as follows. Given a Hamiltonian $H = H(q, p, t)$ we substitute $p = \partial S/\partial q$ and write down the Hamilton–Jacobi equation

$$\partial S(q, \alpha, t)/\partial t + H(q, \partial S(q, \alpha, t)/\partial q, t) = 0. \tag{2.2.16}$$

We seek a complete solution for S in terms of n nontrivial integration constants $\alpha_1, \dots, \alpha_n$. The new constant coordinates Q_i may be chosen as any n independent functions of α_i: $Q_i = \gamma_i(\alpha_1, \dots, \alpha_n)$, so that we may write $S = S(q, \gamma, t)$. The transformation equation (2.2.12b) introduces the new constant momenta

$$P_i = -\partial S/\partial \gamma_i = \beta_i. \tag{2.2.17}$$

This relation is the Jacobi law of motion: it can, in principle, be solved algebraically for q_i in terms of t and the $2n$ constants β_i, γ_i. This is possible if $\det(\partial^2 S/\partial q_i\, \partial \gamma_j) \neq 0$; p_i is given by (2.2.12a). The solution is completed by expressing β_i, γ_i in terms of the actual initial conditions q_{0i}, p_{0i} of the problem. Evaluating the transformation equation (2.2.12a) at time t_0 gives

$$p_0 = \partial S/\partial q|_{q=q_0, t=t_0} \tag{2.2.18}$$

which implies a relation between q_0, p_0 and γ, and evaluating (2.2.17) at t_0 gives a relation between q_0, γ and β. Finally then we obtain

$$q = q(q_0, p_0, t), \qquad p = p(q_0, p_0, t). \tag{2.2.19}$$

By an identical procedure we can derive (2.2.19) if we use a generating function in which the nonadditive constants are expressed in terms of the new momenta: $P_i = \gamma_i(\alpha_1, \dots, \alpha_n)$. Then the transformation equation (2.2.12a) is unchanged and (2.2.12b) is replaced by an expression for the new coordinates, $Q_i = \partial S/\partial \gamma_i = \beta_i$, which is now the Jacobi law of motion.

We have thus constructed a function, Hamilton's principal function, which generates a canonical transformation to coordinates and momenta that are constant along a trajectory. This establishes an equivalence between the $2n$ first-order Hamilton equations and the single first-order Hamilton–Jacobi partial differential equation. This type of relation is well known in the theory of differential equations where the mechanical paths are the characteristics of the Cauchy problem associated with the Hamilton–Jacobi equation.

2.2.2 The action function

An alternative perspective on the meaning of the Hamilton–Jacobi function S may be gained as follows. In formulating Hamilton's Principle (2.2.4) we considered the motion of a system between two given instants t_0 and t at which its coordinates are q_0 and q respectively. The values of I for neighbouring paths linking these fixed limits are compared, and the true path is the one for which I is an extremum. We shall now regard the action function I as a quantity associated with just the actual path traversed by the system and consider the change in I induced by variations in the final coordinate q and time t and the initial coordinate q_0, keeping the initial time t_0 fixed. That is, define

$$I(q, t; q_0, t_0) = \int_\gamma L(q, \dot{q}, t)\,\mathrm{d}t, \tag{2.2.20}$$

where the integral is evaluated along the extremal γ joining the point (q_0, t_0) to (q, t).

Taking the differential of I as a function of the variable coordinates it is possible to show that

$$\mathrm{d}I = p\,\mathrm{d}q - H\,\mathrm{d}t - p_0\,\mathrm{d}q_0, \tag{2.2.21}$$

where $p = \partial L/\partial \dot{q}$, $H = p\dot{q} - L$, $\dot{q}$ is the terminal velocity of the trajectory γ, and similar relations hold for q_0 and p_0. Writing

$$\mathrm{d}I = (\partial I/\partial q)\,\mathrm{d}q + (\partial I/\partial t)\,\mathrm{d}t + (\partial I/\partial q_0)\,\mathrm{d}q_0$$

and comparing with the right hand side of (2.2.21) we deduce that

$$p = \partial I/\partial q, \qquad H = -\partial I/\partial t, \qquad p_0 = -\partial I/\partial q_0$$

and hence that I satisfies the Hamilton–Jacobi equation:

$$\partial I/\partial t + H(q, \partial I/\partial q, t) = 0. \tag{2.2.22}$$

Indeed, (2.2.21) has the form of a canonical transformation (2.2.11) which trivializes the motion ($K = 0$), where I is the generating function and $q_0 \equiv Q$, $p_0 \equiv P$. The action function I is therefore a particular complete integral of the Hamilton–Jacobi equation, the n nonadditive constants being the initial positions q_{0i}, $i = 1, \ldots, n$ and the new momenta being the actual initial momentum coordinates. In view of the identity of the action function with a form of Hamilton's principal function we shall henceforth denote both by the symbol S.

We started by showing how Hamilton's equations could be solved by integrating the Hamilton–Jacobi equation. The result just proved enables us to do the converse – to solve the Cauchy problem for the Hamilton–Jacobi equation (i.e., solve (2.2.16) subject to the initial condition $S(q, t_0) = S_0(q)$, ignoring the constants) by writing down the equivalent set of Hamilton equations and solving these by some other method. The resulting trajectory is a characteristic of the partial differential equation and $S(q, t)$ may be constructed by integrating the Lagrangian (obtained from the Hamiltonian by an inverse Legendre transformation) along the characteristic. The result is

$$S(q, t) = S_0(q_0(q, t), t_0) + \int_\gamma L(q, \dot{q}, t)\,\mathrm{d}t. \tag{2.2.23}$$

Notice that the function S obtained in this way depends on S_0 and may be quite different in form from a complete integral for the same problem found by a separation of variables, even though the paths associated with the two functions are the same. This point is considered in more detail in §2.3.

2.2.3 A single particle

In the following we shall often refer to the case of a single body of mass m in a scalar external potential V and described by a Cartesian coordinate system. The Lagrangian is given by

$$L(\mathbf{x}, \dot{\mathbf{x}}, t) = \tfrac{1}{2}m\dot{\mathbf{x}}^2 - V(\mathbf{x}, t) \tag{2.2.24}$$

from which the canonical momentum (2.2.1) is found to be

$$\mathbf{p} = m\dot{\mathbf{x}} \tag{2.2.25}$$

and the Hamiltonian (2.2.2) is

$$H(\mathbf{x}, \mathbf{p}, t) = \mathbf{p}^2/2m + V(\mathbf{x}, t). \tag{2.2.26}$$

Eq. (2.2.26) is evidently the total energy. The Euler–Lagrange equations (2.2.5) imply the equation of motion in Newton's form

$$m\ddot{\mathbf{x}} = -\nabla V|_{\mathbf{x}=\mathbf{x}(t)}. \tag{2.2.27}$$

From (2.2.26) we obtain the Hamilton–Jacobi equation

$$\partial S/\partial t + (\nabla S)^2/2m + V = 0. \tag{2.2.28}$$

When the potential is time-dependent the energy of the particle, $(-\partial S/\partial t)$ evaluated along the trajectory $\mathbf{x} = \mathbf{x}(t)$, is not conserved in general.

We shall also study the case where the particle carries a charge e and is acted upon by an external electromagnetic field with scalar potential $A_0(\mathbf{x}, t)$ and vector potential $\mathbf{A}(\mathbf{x}, t)$. If c is the speed of light, the Lagrangian, canonical momentum and Hamiltonian are respectively

$$\left.\begin{aligned} L &= \tfrac{1}{2}m\dot{\mathbf{x}}^2 + (e/c)\mathbf{A}\cdot\dot{\mathbf{x}} - eA_0 - V \\ \mathbf{p} &= m\dot{\mathbf{x}} + (e/c)\mathbf{A} \\ H &= (1/2m)[\mathbf{p} - (e/c)\mathbf{A}]^2 + eA_0 + V. \end{aligned}\right\} \tag{2.2.29}$$

The equation of motion includes the Lorentz force:

$$m\ddot{\mathbf{x}} = -\nabla V + e(\mathbf{E} + c^{-1}\dot{\mathbf{x}} \times \mathbf{B}) \tag{2.2.30}$$

where $\mathbf{E} = -\nabla A_0 - (1/c)\,\partial\mathbf{A}/\partial t$, $\mathbf{B} = \nabla \times \mathbf{A}$ and the Hamilton–Jacobi equation becomes

$$\partial S/\partial t + (1/2m)[\nabla S - (e/c)\mathbf{A}]^2 + eA_0 + V = 0. \qquad (2.2.31)$$

The case of a rotator, for which the generalized coordinates are the Euler angles, is treated in Chap. 10.

2.3 Properties of the Hamilton–Jacobi function

2.3.1 The nonuniqueness of S for a given mechanical problem

The Hamilton–Jacobi theory is often taught as a technique for solving the equations of motion of mechanics, and this is indeed how we have introduced it in the last section. Yet it presents certain features that point the way to a development of the conceptual framework of Newtonian mechanics, and in this and the next sections we call attention to some of these.

To solve a dynamical problem in classical mechanics using Jacobi's transformation theory as set out in §2.2, we require a complete integral of the Hamilton–Jacobi equation depending on as many nonadditive constants as there are degrees of freedom in the system. Such a solution may be found for example by a separation of variables. In order to understand clearly the difference with the analogous case of solutions to the quantum Hamilton–Jacobi equation studied later, it is important to emphasize that *any* complete solution will do in determining the particle motion. A given Hamilton–Jacobi equation may have many different complete integrals, for fixed particle initial conditions, in the sense that the functional dependence of S on q, α and t may vary. But, for the given initial coordinates and momenta, the particle motion associated with all these functions will be the same.

While the S-function corresponding to a given mechanical problem (i.e., prescribed Hamiltonian, q_0 and p_0) is not unique, the various solutions are nevertheless distinguished by virtue of the fact that S is evidently a field function in configuration space. Thus, while the aim of Jacobi's method is the computation of a single orbit, S is actually connected with an infinite set of potential trajectories pursued by an ensemble of identical particles. This set is obtained by varying the constants β for fixed α. The relation (2.2.12a), $p = \partial S/\partial q$, gives at each point q at each instant the canonical momentum of a system that may potentially pass through that point. The physical momentum follows from (2.2.1). The ensemble of motions is characterized by the form of S as a function of q and t and the values of the constants α.

The different functional forms of S connected with the same Hamiltonian imply different types of ensemble. However, at one point at one moment the various S-functions will have the same gradient and they will all imply the same subsequent (and preceding) motion for a system passing through that point. For other space points where the gradients of the S-function do not coincide, the motions of the other ensemble elements are different.

These points are easily illustrated by a simple example. It suffices to treat a free particle in Cartesian coordinates, for which (2.2.28) becomes

$$\partial S/\partial t + (\nabla S)^2/2m = 0. \tag{2.3.1}$$

First of all we solve this equation by separating the variables, i.e., we treat S as a sum of four functions depending on x, y, z and t respectively. The result is

$$S(x, y, z, P_1, P_2, P_3, t) = -(1/2m)(P_1^2 + P_2^2 + P_3^2)t + P_1 x + P_2 y + P_3 z, \tag{2.3.2}$$

where the nonadditive constants P_1, P_2, P_3 are the components of a momentum vector. Eq. (2.2.12a), $p_i = \partial S/\partial x_i$, $i = 1, 2, 3$, yields $p_i = P_i$. The trajectory is found by writing $\partial S/\partial P_i = Q_i$, where Q_i are constant coordinates. We obtain $-(\mathbf{P}/m)t + \mathbf{x} = \mathbf{Q}$, showing that $\mathbf{Q}$ is the initial (we choose $t_0 = 0$) coordinate vector. Writing $\mathbf{x}_0 = \mathbf{Q}$ and $\mathbf{v} = \mathbf{P}/m$ we therefore find

$$\mathbf{x}(t) = \mathbf{x}_0 + \mathbf{v}t. \tag{2.3.3}$$

The motion is uniform and rectilinear, starting from the point $\mathbf{x}_0$ with velocity $\mathbf{v}$. The ensemble described by (2.3.2) pursues a set of parallel straight line motions of momentum $\mathbf{P}$ generated by varying $\mathbf{x}_0$ (the βs of this example).

To obtain a different functional form for S we construct the latter from (2.2.20) by integrating along the trajectory (2.3.3). The Lagrangian is $L = \frac{1}{2}m\dot{\mathbf{x}}^2$ and so the free particle action function is given by the expression

$$S(\mathbf{x}, t; \mathbf{x}_0, 0) = (m/2t)(\mathbf{x} - \mathbf{x}_0)^2. \tag{2.3.4}$$

Here the nonadditive constants are the initial coordinates $\mathbf{x}_0$. Applying the Jacobi theory we can reconstruct the path from (2.3.4): $\partial S/\partial x_{0i} = -P_i$ implies $-m(\mathbf{x} - \mathbf{x}_0)/t = -\mathbf{P}$. Writing again $\mathbf{P}/m = \mathbf{v}$ we recover (2.3.3). In this case, the function (2.3.4) describes an ensemble of particles which all emanate from the point $\mathbf{x}_0$ with a range of momenta $\mathbf{P}$ (i.e. β). The motions generated by (2.3.2) and (2.3.4) coincide when $\mathbf{x}_0$ and $\mathbf{P}$ are chosen to be the same in both cases.

We shall see that in quantum theory two S-functions are distinguished not only globally through the ensembles they generate but in a stronger sense: two particles that start with the same $\mathbf{x}_0$, $\mathbf{p}_0$ do *not* in general pursue the same trajectory in the given potential V.

2.3.2 The basic law of motion

It is by no means guaranteed that we will have available a complete integral of the Hamilton–Jacobi equation. There is no rule for obtaining such a solution; the possibility of solving by a separation of variables depends on finding a suitable coordinate system and for many problems of physical interest (e.g. the three-body problem) this is impossible. It may be possible to find a solution depending on less than the required number of nonadditive constants, but even this may not be feasible. It follows that if, for whatever reason, we do not possess a complete solution of the Hamilton–Jacobi equation, the Jacobi law of motion $\partial S/\partial \gamma = \beta$ cannot be used to solve the dynamical problem completely, and perhaps cannot be used at all.

Suppose though that we have available a *general* solution to the Hamilton–Jacobi equation, that is a function of $S(q, t)$ not depending explicitly on any constants. Then, although we cannot use $\partial S/\partial \gamma = \beta$ to solve for the motion of the system point algebraically, we may employ the other set of transformation equations (2.2.12a). The covariant momentum is

$$p_i = \partial S/\partial q_i \tag{2.3.5}$$

and expressing p in terms of q, $\dot{q}$ and t using (2.2.1) we may solve for $q(t)$ by directly integrating (2.3.5) and specifying $q_0 = q(0)$.

Of course, there should be consistency between the two laws of motion ($\beta = \partial S/\partial \gamma$ and $p = \partial S/\partial q$) in the case where we have a complete integral and both are applicable. This is ensured by (2.2.18), which is simply (2.3.5) evaluated at a definite point q_0, and indeed we must always use this in conjunction with (2.2.17) to solve for the motion completely. Thus, however the trajectory is calculated, we always make use of the relation (2.3.5), either to fix the initial conditions in the case of a complete integral or to integrate directly when we have no complete integral. It follows that, since $p = \partial S/\partial q$ applies in all cases, it is natural to regard this as the basic law of particle motion in the Hamilton–Jacobi theory.

We now give illustrations of this technique (albeit applied to complete integrals) for the case of the two S-functions given in §2.3.1. There $\mathbf{p} = \nabla S$ and from (2.2.25), $\mathbf{p} = m\dot{\mathbf{x}}$.

For S of the form (2.3.2) we have

$$m\dot{\mathbf{x}} = \nabla S = \mathbf{P}, \tag{2.3.6}$$

which is a constant. If the initial condition is $\mathbf{x}(0) = \mathbf{x}_0$ this integrates immediately to give (2.3.3). On the other hand, the form (2.3.4) gives

$$m\dot{\mathbf{x}} = \nabla S = m(\mathbf{x} - \mathbf{x}_0)/t. \tag{2.3.7}$$

This is again readily integrated to yield (2.3.3) if $\mathbf{x}(0) = \mathbf{x}_0$. Notice the quite different functional dependence of ∇S in (2.3.6) and (2.3.7). In one case it is constant in space and time and in the other it is variable. Nevertheless, both forms imply the same physical trajectory if the initial canonical coordinates are the same, and coincide along that trajectory.

Indeed, all functions of S representing particle properties will take the same value when evaluated along the same trajectory. Apart from the momentum ∇S, these include the energy $(-\partial S/\partial t)$ and the angular momentum $\mathbf{x} \times \nabla S$. The two functions differ in their global properties.

What lies behind the more general method of solution is the formulation of dynamical evolution in terms of a Cauchy problem for the Hamilton–Jacobi equation:

$$\partial S/\partial t + H(q, \partial S/\partial q, t) = 0. \tag{2.3.8}$$

Specification of the initial S-function S_0 for all q implies a unique solution for $S(q, t)$. It also has the effect of fixing the initial canonical momentum everywhere in configuration space:

$$p_0 = \partial S_0/\partial q. \tag{2.3.9}$$

Obviously, the specification of an ensemble of p_0s considerably overdetermines the mechanical problem since all that is required is p_0 at one point q_0. But, of course, one is tacitly making a choice of $S_0(q)$ for all q when one seeks a complete integral of the Hamilton–Jacobi equation. For example, for a conservative one-body system we may seek a solution by separating the time variable:

$$S(\mathbf{x}, E, t) = W(\mathbf{x}, E) - Et. \tag{2.3.10}$$

The function $W(\mathbf{x}, E)$ is called *Hamilton's characteristic function* and satisfies the equation

$$E = (\nabla W)^2/2m + V. \tag{2.3.11}$$

Clearly, W is nothing more than the initial S-function, $W(\mathbf{x}) = S_0(\mathbf{x})$. The function (2.3.10) describes an ensemble of particles with the same energy E and variable momentum $\mathbf{p} = \nabla W$. An example is the function (2.3.2) which corresponds to choosing $S_0 = \mathbf{P}\cdot\mathbf{x}$. Any solution to the Hamilton–Jacobi equation for the given time-independent potential V will have the property that $(-\partial S/\partial t) = E$ when evaluated along the trajectory.

To summarize so far, we have shown the following. The problem of dynamics as defined by Hamilton's canonical equations may be formulated in terms of a partial differential equation (2.3.8) determining the evolution of a field $S(q, t)$. This function determines at each point and at each instant the momentum of a system that may be potentially placed there through the relation (2.3.5). For one body the basic law of motion is $\dot{\mathbf{x}} = \nabla S/m$. The function S is thus connected with an ensemble of identical systems rather than a single orbit. It is in this way that the S-functions may be physically distinguished. For fixed q_0, p_0 all S-functions imply the same time development $q(t)$. This reflects the fact that the state of a material system is completely exhausted by specifying its position and momentum – the S-function plays no role in either defining the state or in determining the dynamics.

2.3.3 *Multivalued trajectory fields*

The set of classical orbits moving in a given potential forms a single-valued congruence when represented in phase space, i.e., only one trajectory may pass through each phase space point. It is a common property of classical force fields that when mapped into configuration space the trajectory field is multivalued: at an instant t more than one orbit may pass through a point q. The degree to which this happens depends on the nature of the force and the particular ensemble chosen (i.e., on $S_0(q)$) and is reflected in the value of $S(q, t)$ (which may, for example, include square roots and hence possess different branches). Most interesting ensembles (bound or scattering) in most interesting force fields are of this sort.

An example is an ensemble of one-dimensional harmonic oscillators of frequency ω having potential energy $V = \frac{1}{2}m\omega^2 x^2$ and energy E. Each orbit is distinguished by the initial position x_0:

$$x(t) = x_0 \cos \omega t + (a^2 - x_0^2)^{1/2} \sin \omega t \tag{2.3.12}$$

where $a = (2E/m\omega^2)^{1/2}$ is the amplitude of the motion. Two trajectories cross each point x with equal and opposite velocities (see Fig. 2.1 which shows

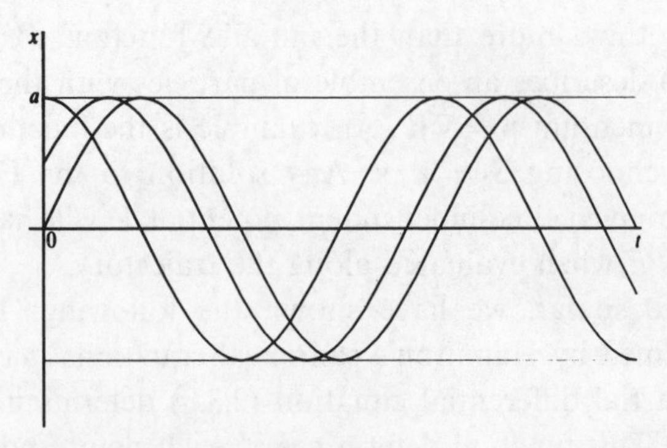

Fig. 2.1 Three orbits of a harmonic oscillator (corresponding to $x_0 = 0, a/2, a$). Each orbit has two segments, each belonging to a single-valued trajectory field.

three orbits). The total set of trajectories is composed of two single-valued subsets corresponding to each of the solutions for $\partial S/\partial x$ implied by the Hamilton–Jacobi equation:

$$\partial S/\partial x = \pm m\omega(a^2 - x^2)^{1/2}. \tag{2.3.13}$$

This should be compared with the quantum oscillator (§4.9).

Trajectories may cross even if the initial ensemble is single-valued, as discussed later (§§2.5, 2.6). An example is attractive Coulomb scattering with the potential source at the origin and all trajectories initially coming in from infinity parallel to the z-axis. Each orbit eventually crosses the z-axis and hence each space point will support two trajectories. Then the ensemble of S-functions will consist of two elements: one generating the motions prior to crossing the z-axis and one after (Korsch & Möhlencamp, 1978b; Rowe, 1987).

Note that in these examples different segments of a single orbit are assigned to different single-valued subensembles. We stress the generality of multi-valued momentum fields in classical mechanics as they cannot occur in quantum mechanics (for pure states) and hence they are of some significance when we come to consider the classical limit of quantum mechanics (Chap. 6).

2.4 The propagation of the S-function

We shall deduce here some general characteristics of solutions to the Hamilton–Jacobi equation, in relation to the system motion, which do not depend on the precise functional form of S.

It is convenient to fix attention on a single particle in Cartesian coordinates. Consider a solution $S(\mathbf{x}, t)$ and for the moment let us ignore any possible dependence on constants. At a given instant of time t the equation $S(\mathbf{x}, t) = c$, where c is a constant, defines a surface in Euclidean space. As t varies, the surface traces out a volume (Fig. 2.2). At each point of the moving surface we may construct the gradient of the function, ∇S, which is, of course, orthogonal to any vector lying in the surface. Now in the case of an external scalar potential we know from (2.2.25) that the particle trajectories associated with S are given by the solutions to $m\dot{\mathbf{x}} = \nabla S$. We therefore have the following theorem: The mechanical paths of a moving point are perpendicular to the surfaces S = constant for all $\mathbf{x}$ and t. It follows that a family of trajectories may be obtained by constructing the normals to a set of surfaces, each orbit being generically distinguished by its starting point $\mathbf{x}_0$. This result remains valid in the quantum theory of motion (§3.2).

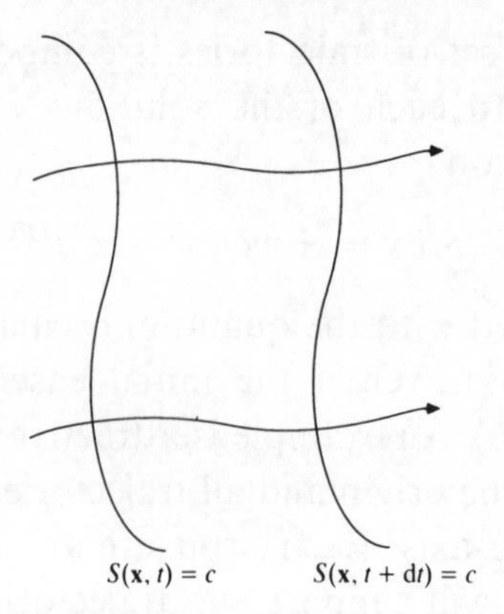

Fig. 2.2 Propagation of a general solution to the Hamilton–Jacobi equation showing orthogonal trajectories.

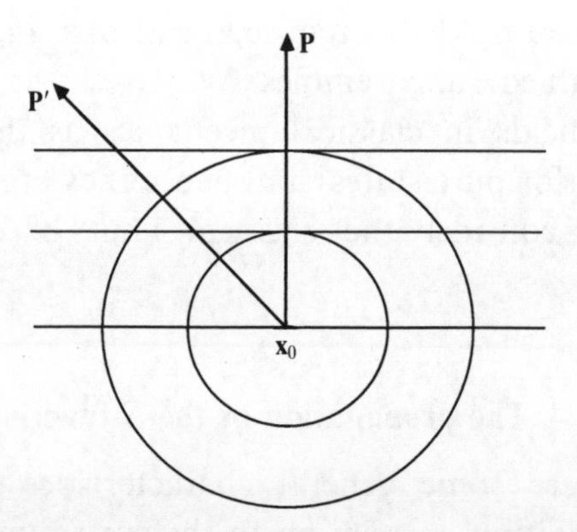

Fig. 2.3 The propagation of plane and spherical Hamilton–Jacobi wave surfaces which have a common tangent at one point $\mathbf{x}_0$. Both functions imply the same motion for a particle starting from $\mathbf{x}_0$ with momentum $\mathbf{P}$. The trajectory with momentum $\mathbf{P}'$ is generated by the spherical wave but not the plane wave.

As an example consider the action function (2.2.20). The surface is initially concentrated at the point $\mathbf{x}_0$ and expands in a series of concentric closed surfaces. The motion is similar to that of a shock wave emanating from a sudden disturbance at the source point $\mathbf{x}_0$ and we may think of the travelling surface as a wavefront. Typical trajectories originating at $\mathbf{x}_0$ are sketched in Fig. 2.3. In the particular case of a free particle, (2.3.4) gives

$$(\mathbf{x} - \mathbf{x}_0)^2 = ct, \qquad c = \text{constant}, \tag{2.4.1}$$

and the wavefronts are spherical, of radius $(ct)^{1/2}$.

If S depends on some arbitrary parameters α other than $\mathbf{x}_0$, the surface propagates as in Fig. 2.2, i.e., it starts from a certain initial surface which does not degenerate to a point. An example is provided by the function (2.3.2) whose wavefronts are planes:

$$\mathbf{P}\cdot\mathbf{x} - Et = c. \tag{2.4.2}$$

We can give a simple pictorial interpretation of the ensemble property of solutions to the Hamilton–Jacobi equation described in §2.3. This is embodied in the following theorem: Suppose the Hamiltonian is fixed and consider two solutions $S_1(\mathbf{x}, t)$ and $S_2(\mathbf{x}, t)$ of the Hamilton–Jacobi equation which have at an instant a common tangent at the point $\mathbf{x}_0$ (i.e. $\nabla S_1(\mathbf{x}_0) = \nabla S_2(\mathbf{x}_0)$). Then the propagating surfaces have a common tangent for all time. The theorem is depicted in Fig. 2.3 for the case of the plane and spherical waves (2.4.2) and (2.4.1). Both solutions describe an ensemble of trajectories of which the one passing through $\mathbf{x}_0$ with momentum $\mathbf{P}$ is common. As noted previously, this result does not remain valid in the quantum theory of motion; there, different S-functions whose gradients coincide at a point at one time do not generally imply identical subsequent particle motions.

Further insight into the propagation of the wave surfaces and their relation with the paths may be gained by deriving a formula for the wave velocity. We have said that for each t the relation $S(\mathbf{x}, t) = c$ defines a surface in three-dimensional space. After an interval $\mathrm{d}t$ the surface satisfies the condition $S(\mathbf{x}, t + \mathrm{d}t) = c$ which defines a new surface in space. Now, as a result of arbitrary displacements $\mathrm{d}\mathbf{x}$ and $\mathrm{d}t$ in space and time, the function S, in general, changes by an amount

$$\mathrm{d}S = \nabla S\cdot\mathrm{d}\mathbf{x} + (\partial S/\partial t)\,\mathrm{d}t. \tag{2.4.3}$$

But, in the evolution envisaged, S has not changed its value $(=c)$ so that here $\mathrm{d}S = 0$. Eq. (2.4.3) becomes

$$0 = |\nabla S|\mathbf{n}\cdot\mathrm{d}\mathbf{x} + (\partial S/\partial t)\,\mathrm{d}t \tag{2.4.4}$$

where $\mathbf{n} = \nabla S/|\nabla S|$ is a unit vector perpendicular to the surface $S = c$ at the point $\mathbf{x}$. Now, $\mathbf{n}\cdot\mathrm{d}\mathbf{x} = \mathrm{d}s$, the component of $\mathrm{d}\mathbf{x}$ lying along the normal to the surface. Thus (2.4.4) implies that at time t the speed of the wavefront at the point $\mathbf{x}$ is given by

$$u(\mathbf{x}, t) = \mathrm{d}s/\mathrm{d}t = -(\partial S/\partial t)/|\nabla S|. \tag{2.4.5}$$

The law of propagation (2.4.5) implies a relation between the wave and particle velocities (since the vector ∇S characterizes both). Thus, for the action function (2.3.4) the formula (2.4.5) gives

$$u = |\mathbf{x} - \mathbf{x}_0|/2t. \tag{2.4.6}$$

The wavefront therefore expands at a rate equal to half the particle velocity.

If instead we restrict attention to a conservative system and separate the time variable as in (2.3.10), then (2.4.5) becomes

$$u = E/|\nabla W| = E/p(\mathbf{x}), \tag{2.4.7}$$

where W is Hamilton's characteristic function and p is the magnitude of the particle momentum. The Hamilton–Jacobi equation (2.3.11) implies that

$$u = E/\{2m[E - V(\mathbf{x})]\}^{1/2}. \tag{2.4.8}$$

Writing $v = p/m$ for the particle velocity, we obtain from (2.4.7) the following relation between the velocity of a point of the moving surface and the velocity of the associated particle placed at that point:

$$uv = E/m. \tag{2.4.9}$$

We see again that the particle does not keep step with the motion of the wavefront (in an interval $\mathrm{d}t$ it does not move a distance $\mathrm{d}s$), although both have the same ordering in time.

Of course, these properties of the function S only capture some of the features of a genuine wave motion. The propagation of S proceeds according to the laws of geometrical optics; the trajectory corresponds to a light ray and the Hamilton–Jacobi equation corresponds to the eikonal equation. What is missing is any counterpart of the Principle of Linear Superposition.

It is sometimes stated that there is no notion of periodicity, i.e., wavelength and frequency, in the classical Hamilton–Jacobi theory. This is not strictly true; the theory already tacitly contains the germ of these concepts. Consider for example the one-dimensional plane wave $S = Px - Et$ measured in units of some constant h having the dimension of action. Then trigonometric functions such as $\cos(S/h)$ are spatially and temporally periodic functions of their arguments corresponding to a wavelength $2\pi h/P$ and frequency $E/2\pi h$. These definitions are consistent with the definition (2.4.7) of wave speed which results from their product. The frequency and wavelength of such Hamilton–Jacobi waves would only become relevant to physical phenomena in a theory which describes the linear superposition of waves and hence interference between them. No mechanism for this exists in classical mechanics and so the value of the constant h cannot be determined. This is achieved in quantum mechanics where, following de Broglie's proposal, h is identified with Planck's constant.

It is not always true that the paths are orthogonal to the surfaces S = constant. The orthogonality is a consequence of the particular external potential assumed above, but there is still a well-defined relationship between the moving surfaces and the paths for other types of external potential. For example, in an electromagnetic field the physical particle momentum is given by (2.2.29):

$$m\dot{\mathbf{x}} = \nabla S - (e/c)\mathbf{A}. \qquad (2.4.10)$$

Clearly $\dot{\mathbf{x}}$ is not collinear with ∇S. Actually, we can restore the orthogonality between wave and path in cases such as this if we suitably generalize the geometry of the space in which motions occur, by introducing non-Euclidean metrics (Holland, 1982; Asanov, 1985). We will not go into this here, but in the electromagnetic case the type of metric required is anisotropic, i.e., it depends on the particle characteristics; such a geometry is a function of the motion and not merely a passive backdrop.

More generally still, the general features of the S-function and its relation to paths outlined above extend to arbitrary many-body systems in time-dependent potentials. S propagates in configuration space in accordance with the laws of geometrical optics, and the rays (paths) are either orthogonal to the wavefronts or can be treated as such in a generalized geometry. It is in this way that classical mechanics anticipates a wave theory of matter, and the necessity of employing functions irreducibly defined in configuration space to account for many-body phenomena.

2.5 Classical statistical mechanics†

2.5.1 Conservation of probability

A basic property of the classical mechanical laws is that they map a precisely specified initial state (x_0, p_0) into a sharp final state $(x(t), p(t))$. While, in principle, there is no restriction on the precision with which we may control the initial phase space coordinates, practical considerations inevitably intrude and limit the extent of our knowledge of x_0 and p_0 (even though they are always, in fact, precisely defined). We have then to resort to a consideration of an ensemble of systems possessing a range of initial conditions consistent with our imperfect knowledge of the actual system of interest. We admit the possibility that however arbitrarily close the initial coordinates of the ensemble elements may be, the subsequent motion may exhibit instabilities so that in many cases it is effectively unpredictable. Yet the evolution of the statistical ensemble, i.e., our knowledge, is entirely a function of the underlying causal dynamics of individual material objects pursuing well-defined orbits

† Parts of this and the following two sections assume some familiarity with the elements of quantum mechanics, in particular its de Broglie–Bohm formulation. Reference should be made to Chap. 3 for background.

according to Newton's laws. Classical statistical mechanics is a theory in which motion is determinist but unpredictable.

We have seen that the Hamilton–Jacobi theory is naturally connected with an ensemble description. A solution of the Hamilton–Jacobi equation determines a velocity field via the relation (for one body)

$$\mathbf{v}(\mathbf{x}, t) = \nabla S(\mathbf{x}, t)/m. \tag{2.5.1}$$

Integrating the equation $\mathbf{v} = \dot{\mathbf{x}}$ we obtain an infinite ensemble of possible motions, corresponding to the same S-function, labelled by the initial position: $\mathbf{x} = \mathbf{x}(t, \mathbf{x}_0)$.

Classical statistical mechanics assigns a probability distribution function $f(\mathbf{x}, \mathbf{p}, t)$ $(\geqslant 0)$ to the phase space of a mechanical system, the evolution of which is governed by Liouville's equation.† Here we shall be concerned with a slightly less general problem in which the initial momenta of an ensemble of particles are fixed once and for all by a definite choice of Hamilton–Jacobi function and we are only free to choose the initial position coordinates as we please. We assume that the variable $\mathbf{x}_0$ is distributed according to some probability law and shall investigate the evolution in time of this distribution. This evolution will evidently be determined by the type of ensemble described by S_0 (i.e., the initial momentum field). The momentum of the particle, which is unknown only because the position is, will be distributed according to a law induced by the position probability function. The case where position and momentum are independently distributed will be commented upon in §2.5.3.

The set of particles connected with S forms a fictitious ensemble in the sense of Gibbs, i.e., a cloud of identical noninteracting particles which differ from one another only in virtue of their initial locations and velocities. The motion of each particle is completely determined once we have prescribed S and $\mathbf{x}_0$. Suppose that the density of particles per unit volume in an element d^3x surrounding the point $\mathbf{x}$ at time t is given by a function $\rho = \rho(\mathbf{x}, t)$ $(\geqslant 0)$. We will derive an equation to be satisfied by this function which expresses the local conservation of particles in the ensemble. It is instructive to prove this in two ways.

In the first method, we consider a volume Ω fixed in space, enclosed by a surface A. The total number of particles in this volume at time t is $\int_\Omega \rho \mathrm{d}^3x$. The number of particles moving through a surface element $\mathrm{d}\mathbf{A}$ in unit time is $\rho\mathbf{v}\cdot\mathrm{d}\mathbf{A}$, so the total number flowing out of the volume Ω in unit time is

† This is usually applied to systems of great complexity (Tolman, 1938) but its methods are equally applicable to a system having a few degrees of freedom such as a single particle.

$\int_A \rho\mathbf{v}\cdot\mathrm{d}\mathbf{A}$. In order that there be no creation or destruction of ensemble elements, this expression should be equal to the decrease per unit time of the total number of particles contained in Ω:

$$-\frac{\mathrm{d}}{\mathrm{d}t}\int_\Omega \rho \mathrm{d}^3x = \int_A \rho\mathbf{v}\cdot\mathrm{d}\mathbf{A}.$$

Applying Green's theorem this becomes

$$\int_\Omega [\partial\rho/\partial t + \nabla\cdot(\rho\mathbf{v})]\,\mathrm{d}^3x = 0.$$

The requirement that this latter relation be valid for an arbitrary volume Ω then entails

$$\frac{\partial\rho}{\partial t} + \nabla\cdot(\rho\mathbf{v}) = 0. \quad (2.5.2)$$

This is the required local equation of conservation. It determines the evolution of ρ in time corresponding to a prescribed velocity field once we have specified the initial function $\rho_0(\mathbf{x})$.†

In the first method of proving (2.5.2) we considered a fixed volume and a variable number of ensemble points in it. In the second method we consider a volume Ω whose boundary A is composed of representative points of the ensemble. Then the boundary moves in accordance with the law of motion obeyed by each ensemble element, but no element can leave or enter the volume if $\mathbf{v}$ is a single-valued field (for in order to do so two elements would have to occupy the same position in space with different velocities). Thus the total number of ensemble elements remains fixed while the volume varies. This condition may be expressed as the requirement that $(\mathrm{d}/\mathrm{d}t)\int_\Omega \rho\,\mathrm{d}\Omega = 0$. In reformulating this equation we have to take into account that the shape of the volume varies with time. Thus

$$\int_\Omega [(\mathrm{d}\rho/\mathrm{d}t)\,\mathrm{d}\Omega + \rho\,\mathrm{d}(\mathrm{d}\Omega)/\mathrm{d}t] = 0.$$

Employing the results $\mathrm{d}\rho/\mathrm{d}t = \partial\rho/\partial t + \mathbf{v}\cdot\nabla\rho$ and

$$\mathrm{d}(\mathrm{d}\Omega)/\mathrm{d}t = (\nabla\cdot\mathbf{v})\,\mathrm{d}\Omega$$

we again recover (2.5.2).

Now, the idea of the Gibbs approach is that the ensemble of bodies of density $\rho(\mathbf{x}, t)$ provides a picture of all the motions potentially accessible to

† One may deduce the conservation law as satisfied by a certain function of S directly from the Hamilton–Jacobi equation (van Vleck, 1928).

the single body of interest. The function ρ may be treated as representing a probability density: the probability that the particle lies between the points $\mathbf{x}$ and $\mathbf{x} + \mathrm{d}\mathbf{x}$ at time t is given by $\rho\, \mathrm{d}^3x$. Since the particle must be somewhere in space it is natural to require that

$$\int_{\text{all space}} \rho\, \mathrm{d}^3x = 1 \tag{2.5.3}$$

(although there are cases where we relax this condition). Eq. (2.5.3) is preserved by the conservation law (2.5.2).

We emphasize the obvious point that, regardless of our degree of knowledge of the initial conditions $\mathbf{x}_0$, these are always, in fact, well defined and the particle motion is entirely determined once we know the Hamilton–Jacobi function. The density ρ merely reflects the extent of our knowledge of the system and does not in any way enter into the determination of the motion. The behaviour of $\rho(\mathbf{x}, t)$ on the other hand is determined by the field $\mathbf{v}(\mathbf{x}, t)$ and hence by S. Obviously, the different Hamilton–Jacobi functions corresponding to a given Hamiltonian imply a variety of velocity fields and therefore different functions $\rho(\mathbf{x}, t)$ (for a given $\rho_0(\mathbf{x})$). In particular, if S depends on a set of nonadditive constants, so does ρ.

To see the types of statistical evolution that are possible when the velocity field is prescribed, we look at two examples in one dimension.

To begin with, suppose that the velocity field is time-independent: $v = v(x)$. Then the conservation law (2.5.2) becomes

$$\frac{\partial \rho}{\partial t} + \frac{\partial}{\partial x}(\rho v) = 0 \tag{2.5.4}$$

and this has the general solution

$$\rho(x, t) = [1/v(x)]\alpha\left(t - \int \mathrm{d}x/v\right), \tag{2.5.5}$$

where α is an arbitrary function of its argument such that $\rho \geqslant 0$. We may express (2.5.5) in terms of the initial distribution $\rho_0(x)$. We have

$$\alpha(\phi(x)) = v(x)\rho_0(x), \qquad \phi(x) = -\int \mathrm{d}x/v(x).$$

As a relation between functions of ϕ this is

$$\alpha(\phi) = v[x(\phi)]\rho_0[x(\phi)]$$

where $x(\phi) = \phi^{-1}(x)$. Then

$$\alpha(t + \phi) = v[x(t + \phi)]\rho_0[x(t + \phi)]$$

and substituting in (2.5.5) we obtain

$$\rho(x, t) = [1/v(x)]v\left[x\left(t - \int dx/v\right)\right]\rho_0\left[x\left(t - \int dx/v\right)\right]. \qquad (2.5.6)$$

As a particular case, if we require that ρ should be independent of time, or that $\rho_0(x) \propto 1/v(x)$, it follows from (2.5.5) that α = constant and hence that

$$\rho(x, t) = A/|v(x)| \qquad (2.5.7)$$

where A is a normalization constant. Then the probability of the particle being in an element dx around the point x is inversely proportional to its velocity at that point. In general, though, there is no such simple relation between probability and velocity.

Suppose instead that the velocity depends only on the time: $v = v(t)$. Then (2.5.2) reduces to

$$\partial\rho/\partial t + v\,\partial\rho/\partial x = 0, \qquad (2.5.8)$$

which has the general solution

$$\rho(x, t) = \rho_0\left(x - \int v\,dt\right), \qquad (2.5.9)$$

where the initial distribution ρ_0 is an arbitrary function of its argument. Eq. (2.5.9) prescribes a wave-type development for ρ; while varying in space and time, it maintains the same value for coordinates and times related by $x - \int v\,dt$ = constant, i.e., ρ moves without change of shape along particle trajectories.

2.5.2 Connection with Liouville's equation

The particular statistical mechanics described above assigns a unique momentum to each space point. The corresponding phase space distribution function takes the initial form

$$f_0(\mathbf{x}, \mathbf{p}) = \rho_0(\mathbf{x})\delta[\mathbf{p} - \nabla S_0(\mathbf{x})] \qquad (2.5.10)$$

and at time t

$$f(\mathbf{x}, \mathbf{p}, t) = \rho(\mathbf{x}, t)\delta[\mathbf{p} - \nabla S(\mathbf{x}, t)]. \qquad (2.5.11)$$

Single-valued ensembles of this kind will be called 'pure' because, as we shall see later, they correspond to the pure states of quantum mechanics. Here we examine their relation with Liouville's equation describing the evolution of arbitrary classical statistical ensembles. Ensembles not expressible in the form (2.5.11) will be termed 'mixed'.

For a single particle moving in an external scalar potential the Liouville equation is

$$\frac{\partial f}{\partial t}+\sum_{i=1}^{3}\left(\frac{p_i}{m}\right)\frac{\partial f}{\partial x_i}-\sum_{i=1}^{3}\frac{\partial V}{\partial x_i}\frac{\partial f}{\partial p_i}=0 \tag{2.5.12}$$

with $f = f(\mathbf{x}, \mathbf{p}, t)$. Given the initial distribution $f_0(\mathbf{x}, \mathbf{p})$, $f(\mathbf{x}, \mathbf{p}, t)$ is uniquely induced by the mechanical laws obeyed by each ensemble element. Because (2.5.12) is a linear equation its solutions may be linearly superposed to yield new solutions, and $f(t)$ is a linear function of f_0.

We project (2.5.12) into a set of equations in physical space by defining the spatial functions

$$\left.\begin{aligned} P(\mathbf{x}) &= \int f\,\mathrm{d}^3p, \qquad \overline{p_i(\mathbf{x})} = \int p_i f\,\mathrm{d}^3p/P(\mathbf{x}) \\ \overline{p_ip_j(\mathbf{x})} &= \int p_ip_j f\,\mathrm{d}^3p/P(\mathbf{x}). \end{aligned}\right\} \tag{2.5.13}$$

These are respectively the mean density, mean momentum and mean momentum stress at each space point. Integrating (2.5.12) with respect to the variables p_i we obtain a conservation law for the spatial probability:

$$\frac{\partial P}{\partial t}+\frac{1}{m}\sum_i\frac{\partial}{\partial x_i}(P\bar{p}_i)=0. \tag{2.5.14}$$

On the other hand, multiplying (2.5.12) by p_i and integrating over the momentum variables, we obtain

$$\frac{\partial}{\partial t}(P\bar{p}_i)+\frac{1}{m}\sum_j\frac{\partial}{\partial x_j}(P\overline{p_ip_j})+P\frac{\partial V}{\partial x_i}=0, \tag{2.5.15}$$

which is a momentum transport equation (we integrate by parts and assume $f \to 0$ as $p_i \to \infty$).

An equivalent statement of Liouville's theorem is that for a cloud of representative points in phase space whose boundary is defined by the individual phase space orbits, the volume and the total number of ensemble elements in the volume are constants. Hence $\mathrm{d}f/\mathrm{d}t = 0$ or f is a constant along a phase space path. This should be compared with the case of the space-projected density P for which $\mathrm{d}P/\mathrm{d}t \neq 0$ unless $\nabla\cdot\bar{\mathbf{p}} = 0$ (we write $\bar{\mathbf{p}} = m\dot{\mathbf{x}}$). Thus P is not conserved along a mean spacetime orbit. This reflects the fact that in the proof of (2.5.2) either the volume or the number of elements are constant, but not both together.

If the ensemble is described by the expression (2.5.11) for all t, insertion into (2.5.13) yields

$$P = \rho, \qquad \overline{p_i} = \partial_i S, \qquad \overline{p_i p_j} = \partial_i S\,\partial_j S \tag{2.5.16}$$

so that $\overline{p_i}$ and $\overline{p_i p_j}$ become actual values rather than means. From (2.5.14) and (2.5.15) we then recover the conservation law (2.5.2) and the field-theoretic version of Newton's second law:

$$\partial\rho/\partial t + \nabla\cdot(\rho\nabla S/m) = 0 \tag{2.5.17}$$

$$\left[\partial/\partial t + (1/m)\sum_j \partial_j S\,\partial_j\right]\partial_i S = -\partial_i V. \tag{2.5.18}$$

These are, of course, the correct equations relating ρ and S and we conclude that (2.5.11) is a consistent subsidiary condition in statistical mechanics.

The average value over the ensemble of an arbitrary function $F(\mathbf{x}, \mathbf{p})$ on the phase space is given by

$$\begin{aligned}\langle F\rangle &= \int F(\mathbf{x}, \mathbf{p}) f(\mathbf{x}, \mathbf{p})\,\mathrm{d}^3x\,\mathrm{d}^3p \\ &= \int F(\mathbf{x}, \nabla S)\rho(\mathbf{x})\,\mathrm{d}^3x \end{aligned}\tag{2.5.19}$$

if (2.5.11) holds. The momentum distribution is given by

$$\begin{aligned} g(\mathbf{p}) &= \int f(\mathbf{x}, \mathbf{p})\,\mathrm{d}^3x \\ &= \int \rho(\mathbf{x})\delta(\mathbf{p} - \nabla S)\,\mathrm{d}^3x. \end{aligned}\tag{2.5.20}$$

The distribution functions of other physical quantities may be likewise deduced from the position distribution for ensembles described by (2.5.11).

However, it is the exception rather than the rule that a distribution function retains the form (2.5.11) for all t if it started out as (2.5.10) (for examples where it does see §2.5.3). It was pointed out in §2.3.3 that generic classical ensembles are spatially multivalued and the Liouville evolution reflects this by mapping pure states into mixed states (this has nothing to do with irreversibility but represents a reversible change in the global character of an ensemble). Liouville's equation implies the spatial conservation law (2.5.14) but the momentum $\bar{\mathbf{p}}$ is not generally a gradient field and does not represent the actual velocity of an ensemble element.

In taking account of ensembles which display an arbitrary distribution of momentum at each space point, we may enquire whether a general phase space distribution function can be expressed as a linear sum of pure states (2.5.11), in analogy with the decomposition of the density matrix in quantum mechanics. That is, suppose that S_i, $i = 1, \ldots, N$, denotes a set of N Hamilton–Jacobi functions potentially associated with a mechanical system (all obeying the same Hamilton–Jacobi equation) which each define a single-valued momentum field ∇S_i (i may be replaced by a continuous index). The position probability distribution implied by the ith function on solving (2.5.2) will be denoted $\rho_i(\mathbf{x}, t)$, where $\rho_{i0}(\mathbf{x})$ is arbitrary. Then, attributing a constant relative weight P_i with $\sum_i P_i = 1$ to each pure subensemble, define a phase space distribution function

$$f(\mathbf{x}, \mathbf{p}, t) = \sum_{i=1}^{N} P_i \rho_i(\mathbf{x}, t) \delta[\mathbf{p} - \nabla S_i(\mathbf{x}, t)]. \tag{2.5.21}$$

At each point $\mathbf{x}$ we now evidently have a distribution of momenta.

Now, although there are certainly cases where (2.5.21) is a meaningful expression (cf. §2.5.3), it is not possible to decompose a general phase space function in this way for the reason already stated, and as is shown in an explicit example in §2.6.2: the pure subensembles are not individually conserved by the Liouville equation. The nonconservation of pure states has no parallel in quantum statistical mechanics where, although the decomposition of the density matrix is not unique, each pure state is individually conserved by the underlying wave equation and the latter is of primary significance. In classical statistical mechanics the Liouville equation is primary and as a consequence the notion of pure state is not generally a useful one. It has indeed been pointed out (Berry, private communication) that the distinction drawn here between pure and mixed classical states is not canonically invariant. Nevertheless, from a formal point of view the single-valued classical pure case provides a key to understanding the connection with quantum mechanics, as we discuss further in §2.6 (there is a related invariant concept, the Lagrangian submanifold (Arnold, 1978, p. 440)).

Note that the terminology used above has been chosen to coincide with that used in quantum mechanics. 'Pure' and 'mixed' refer to global properties of the ensemble under consideration and not to the state of a single particle which is, of course, always 'mixed', unless the distribution happens to be a δ-function.

2.5.3 Pure and mixed states

Liouville's equation has the property that precise knowledge of initial conditions is not lost in the course of time. That is, if the initial distribution function takes the form of a δ-function,

$$f_0(\mathbf{x}, \mathbf{p}) = \delta(\mathbf{x} - \mathbf{x}_0)\delta(\mathbf{p} - \mathbf{p}_0), \tag{2.5.22}$$

then it does so at all times:

$$f(\mathbf{x}, \mathbf{p}, t) = \delta[\mathbf{x} - \mathbf{x}(t, \mathbf{x}_0, \mathbf{p}_0)]\delta[\mathbf{p} - \mathbf{p}(t, \mathbf{x}_0, \mathbf{p}_0)]. \tag{2.5.23}$$

This is readily confirmed by insertion in (2.5.12). A δ-distribution is preserved for any pure ensemble.

It goes without saying that the existence of a δ-distribution has nothing at all to do with the issue of whether the particle actually has a well-defined position and momentum at each instant – this is always so. It merely tells us that our knowledge of the system is precise and remains so.

We pass now to a consideration of free particles in one dimension, and let the initial normalized distribution and S-function take the form

$$\rho_0(x) = (2\pi\sigma^2)^{-1/2}\,\mathrm{e}^{-x^2/2\sigma^2}, \tag{2.5.24}$$

$$S_0(x) = px, \tag{2.5.25}$$

where σ and p are constants. This describes a case where the particles in the ensemble all have the same initial velocity $v = p/m$ and a Gaussian distribution of positions of rms width σ about the origin. Since the particles are free their speeds remain v and the distribution function (2.5.9) at time t is given by

$$\rho(x, t) = (2\pi\sigma^2)^{-1/2}\,\mathrm{e}^{-(x-vt)^2/2\sigma^2}. \tag{2.5.26}$$

The probability function thus retains its form; there is no diffraction or 'spreading' of this classical 'wave packet' which would show up as a change in the relative separation of trajectories with time. For a comparison with the corresponding solutions to the quantum equations solved for identical initial conditions (2.5.24) and (2.5.25) see §4.7.

We can obtain a spreading Gaussian packet by inserting a time-dependent S-function into the conservation law corresponding to an ensemble of variable momentum. To this end we shall use the free action function (2.3.4) (with $x_0 = 0$), $t > 0$. Then (2.5.2) becomes

$$t\,(\partial\rho/\partial t) + x\,(\partial\rho/\partial x) + \rho = 0. \tag{2.5.27}$$

This is readily integrated to yield the general solution

$$\rho(x, t) = t^{-1}\beta(x/t), \qquad t > 0, \tag{2.5.28}$$

where β is an arbitrary nonnegative function of its argument. At an instant $t = T$ suppose that the distribution function is given by the Gaussian (2.5.24).

Then equating (2.5.28) to (2.5.24) at this moment gives an expression for β and hence the density function at time $t > T$ is

$$\rho(x, t) = (2\pi\sigma^2)^{-1/2}(T/t)\,\mathrm{e}^{-x^2T^2/2\sigma^2t^2}. \tag{2.5.29}$$

Clearly, this function spreads with time and this reflects the underlying ensemble in which the larger momentum particles lie at larger distances. The corresponding phase space distribution is given by

$$f(x, p, t) = \rho(x, t)\delta(p - mx/t). \tag{2.5.30}$$

We consider now an example of a free mixed ensemble that exhibits the emergence of correlations between x and p from an initially uncorrelated product function of Gaussian distributions in each variable:

$$f_0(x, p) = (2\pi\sigma_0\pi_0)^{-1}\,\mathrm{e}^{-x^2/2\sigma_0^2}\,\mathrm{e}^{-p^2/2\pi_0^2} \tag{2.5.31}$$

where σ_0, π_0 are half-widths. The general solution of the free Liouville equation is

$$f(x, p, t) = f_0(x - pt/m, p) \tag{2.5.32}$$

and hence (2.5.31) evolves into

$$f(x, p, t) = (2\pi\sigma_0\pi_0)^{-1}\,\mathrm{e}^{-(x - pt/m)^2/2\sigma_0^2}\,\mathrm{e}^{-p^2/2\pi_0^2}. \tag{2.5.33}$$

The correlations show up in the spatial density:

$$P(x, t) = \int f\,\mathrm{d}p = (2\pi\sigma^2)^{-1/2}\,\mathrm{e}^{-x^2/2\sigma^2}, \tag{2.5.34}$$

where $\sigma^2 = \sigma_0^2 + \pi_0^2t^2/m^2$ is the half-width at time t. The dispersion of (2.5.34) obviously reflects the fact that faster particles move further in a given interval.

Eq. (2.5.33) is an example of a mixed state that may be decomposed into conserved pure states, as in (2.5.21). We have

$$f(x, p, t) = \int \mathrm{d}\alpha\, P(\alpha)\rho(x, \alpha)\delta[p - \partial S(x, \alpha)/\partial x], \tag{2.5.35}$$

where

$$\left.\begin{aligned} P(\alpha) &= (2\pi\pi_0^2)^{-1/2}\,\mathrm{e}^{-\alpha^2/2\pi_0^2}, \qquad \partial S/\partial x = \alpha, \\ \rho(x, \alpha) &= (2\pi\sigma_0^2)^{-1/2}\,\mathrm{e}^{-(x - \alpha t/m)^2/2\sigma_0^2}. \end{aligned}\right\} \tag{2.5.36}$$

Thus, f may be built from a continuous set of nonspreading Gaussian position distributions of the type (2.5.26) whose momenta α have a Gaussian distribution.

This example and that of (2.5.29) form classical analogues of the dispersion of a free quantal Gaussian packet (§4.7).

2.6 Classical mechanics as a field theory

2.6.1 The wave equation of classical mechanics

The equations of the statistical mechanics of a single particle in a scalar external potential, as expressed for a single-valued ensemble in the Hamilton–Jacobi formalism, are as follows:

$$\partial S/\partial t + (\nabla S)^2/2m + V(\mathbf{x}, t) = 0 \tag{2.6.1}$$

$$\partial \rho/\partial t + \nabla\cdot(\rho\nabla S/m) = 0. \tag{2.6.2}$$

To obtain a solution we must specify $S(\mathbf{x}, 0) = S_0(\mathbf{x})$ and $\rho(\mathbf{x}, 0) = \rho_0(\mathbf{x})$. The choice of $S_0(\mathbf{x})$ fixes the initial momentum field and a trajectory is found by solving the law of motion $\dot{\mathbf{x}} = (1/m)\nabla S|_{\mathbf{x}=\mathbf{x}(t)}$ given the initial position $\mathbf{x}_0$. The dynamics is independent of the choice of initial distribution ρ_0, but the development of the latter is conditioned by S_0.

In this formulation classical statistical mechanics becomes a field theory in spacetime, (2.6.1) and (2.6.2) being field equations determining the evolution of the partially coupled fields $S(\mathbf{x}, t)$ and $\rho(\mathbf{x}, t)$. Applying the operator ∇ to (2.6.1) and rearranging we deduce the field-theoretic equation of motion (2.5.18) already deduced from Liouville's equation. Putting $\mathbf{v} = \dot{\mathbf{x}}$ we obtain Newton's second law

$$m\ddot{\mathbf{x}} = -\nabla V|_{\mathbf{x}=\mathbf{x}(t)}. \tag{2.6.3}$$

In this connection it is helpful to think in terms of a hydrodynamic analogy. In the above theory S/m plays the role of the velocity potential, and ρ the density, of a compressible and irrotational fluid. The basic laws of hydrodynamics are the equation of continuity (2.6.2) (evolution of ρ) and the Euler equation (evolution of $\mathbf{v}$). These are coupled partial differential equations in which the force term entering in the Euler equation generally depends on ρ. The fluid described by our 'Euler equation' (2.5.18) is special in that $\mathbf{v}$ is independent of ρ. A complete coupling of the fields is achieved in quantum mechanics.

We can derive the field equations (2.6.1) and (2.6.2) by standard techniques from a variational principle in which the fields $S(\mathbf{x}, t)$ and $-\rho(\mathbf{x}, t)$ are treated as canonically conjugate 'position' and 'momentum' variables in a phase space of $2 \times (3 \times \infty)$ dimensions ($\mathbf{x}$ is a continuously variable coordinate label). The Lagrangian density is

$$\mathscr{L} = -\rho\left[\frac{\partial S}{\partial t} + \frac{(\nabla S)^2}{2m} + V\right]. \tag{2.6.4}$$

An interesting light is thrown on this approach if we write classical mechanics in the 'language' of quantum mechanics by introducing the classical 'wavefunction'

$$\psi = R\,\mathrm{e}^{\mathrm{i}S/\hbar}, \tag{2.6.5}$$

where $R = +\rho^{1/2}$ and $\hbar$ is a constant having the dimension of action (later to be identified with Planck's constant divided by 2π). The function ψ and its complex conjugate ψ^* form a new set of canonical variables; eq. (2.6.5) and its complex conjugate define a canonical transformation from the old canonical coordinates $(S, -\rho)$ to the new set proportional to (ψ, ψ^*).

In terms of these field variables the Lagrangian density (2.6.4) becomes

$$\mathscr{L} = (\mathrm{i}\hbar/2)(\psi^*\,\partial\psi/\partial t - \psi\,\partial\psi^*/\partial t) + (\hbar^2|\psi|^2/8m)(\psi^{-1}\,\nabla\psi - \psi^{*-1}\,\nabla\psi^*)^2 - V|\psi|^2. \tag{2.6.6}$$

Varying with respect to ψ and ψ^* we deduce the 'wave equation of classical mechanics' (Schiller, 1962a,b; Rosen, 1964b, 1986):

$$\mathrm{i}\hbar\frac{\partial\psi}{\partial t} = -\frac{\hbar^2}{2m}\nabla^2\psi + V\psi + \frac{\hbar^2}{2m}\frac{\nabla^2|\psi|}{|\psi|}\psi \tag{2.6.7}$$

and its complex conjugate, with $|\psi| = R$. Substituting (2.6.5) into (2.6.7) and separating into real and imaginary parts we, of course, recover eqs. (2.6.1) and (2.6.2).

This equation will be recognized to be similar in form to the Schrödinger equation, and, in fact, differs from it only in virtue of the last term on the right hand side. The most obvious and important feature of (2.6.7) is its nonlinearity – given two solutions ψ_1 and ψ_2 at the same spacetime point we cannot generally construct a third solution by a linear combination of them. $\psi_1 + \psi_2$ will be a solution if $\psi_1 = a\psi_2$ where a is a constant or if ψ_1 and ψ_2 have no common support (i.e., R_1 and R_2 do not overlap). In fact, the term responsible for the nonlinearity, the last one on the right hand side, is proportional to the quantum potential. It is necessary that it be present in order that we recover the classical Hamilton–Jacobi equation, i.e., so that $\hbar$ does not appear in the latter.

On the other hand the absence of the quantum potential term in the wave equation corresponds to its presence in the Hamilton–Jacobi equation. If we regard classical mechanics as a limit of quantum mechanics (as we shall do in Chap. 6), this means that a nonlinear differential equation may be a limiting case of a linear equation.

The fact that there is no *linear* superposition principle in classical mechanics does not rule out the possibility of constructing new solutions by some nonlinear process of combination. The existence of such a principle is likely to be of limited physical interest however since, as we have seen, the actual motion of a system is independent of which S-function is used to generate it. Nevertheless it may be of mathematical interest. The quantum analogue of the complete integral of the classical Hamilton–Jacobi equation is a solution of the Schrödinger equation labelled by a set of quantum numbers (eigenvalues of Hermitian operators). A general solution to the Schrödinger equation, which does not generally depend on constants, may be found by linearly superposing the eigensolutions. It would be interesting to know if a general solution to the classical wave equation could be formed by an analogous procedure in which complete integrals are nonlinearly combined.

We have thus reformulated the classical statistical mechanics of a single particle in a form which closely resembles the formal structure of quantum mechanics. A field function, $\psi(\mathbf{x}, t)$, has been introduced which satisfies the 'wave equation' (2.6.7). The phase of the wavefunction, $S(\mathbf{x}, t)$, is the classical Hamilton–Jacobi function and the normals to the surfaces $S =$ constant are the physical paths. The square of the amplitude of the wavefunction, $R^2(\mathbf{x}, t)$, gives the probability density of an ensemble of trajectories associated with the same S-function. In particular, the classical probability current, defined as $\mathbf{j} = \rho\, \nabla S/m$, is identical in form to its quantum counterpart (see §3.2):

$$\mathbf{j} = (\hbar/2mi)\psi^* \overleftrightarrow{\nabla} \psi. \tag{2.6.8}$$

The canonical formalism based on the Lagrangian density (2.6.6) may be developed in order to define a stress tensor of the field, energy and momentum densities etc., but we shall not do this here since the resulting expressions are special cases of those derived later in the quantum case (see §3.9). All of this can be readily generalized to the many-body case where we introduce a function $\psi(\mathbf{x}_1, \ldots, \mathbf{x}_n, t)$.

In this scheme ψ has a purely descriptive or mathematical significance. There is a wave-particle duality in classical mechanics, but the wave is purely passive. We may think of ψ as actually distributed in space, but it is not a causal agent of change and it does not contribute to the physical state of a system, which is completely defined by its position and momentum at each instant.

Being of a purely descriptive nature, the classical wavefunction is not required to satisfy the boundary and subsidiary conditions imposed on the quantum wavefunction (§1.2). For example, it is not single-valued and the conditions that S and $\rho\mathbf{n}\cdot\nabla S$ are continuous across a discontinuity in V are weaker than the analogous requirements imposed in the quantum case (where ψ and $\mathbf{n}\cdot\nabla\psi$ are continuous). Another important difference with quantum mechanics is that the classical ψ-function is derived from the functions R and S whereas the quantum ψ is primary and R and S are deduced from it. This means in particular that the classical S-function may be well defined in nodal regions, i.e., where $R = 0$, and the definition of ψ breaks down there. In the quantum case $\psi = 0$ at nodes and S is undefined.

We saw in §2.5 that some statistical problems involving multivalued velocity fields may be treated by combining the pure states to form a mixed state, (2.5.21). Translating this technique into the classical wavefunction language we may, closely following the analogy with quantum mechanics, introduce a density matrix (Rosen, 1965):

$$\rho(\mathbf{x}, \mathbf{x}') = \sum_i P_i \psi_i(\mathbf{x}) \psi_i^*(\mathbf{x}'). \tag{2.6.9}$$

Each pure state is associated with a wavefunction $\psi_i = R_i \, \mathrm{e}^{\mathrm{i}S_i/\hbar}$. However, for the reasons already stated in §2.5 and as we shall see below, a classical statistical state cannot generally be expressed in the form (2.6.9).

2.6.2 The potential step

Here we shall illustrate how a classical ensemble problem may be formulated in a language similar to that employed in quantum mechanics, using the classical wavefunction introduced above. We will see why the wavefunction approach is of limited practical use in the classical domain, following our remarks on pure ensembles in §2.5.

We shall study the transmission and reflection of a one-dimensional beam (i.e., an ensemble of identical noninteracting particles) incident on a potential step (see Rosen (1965)):

$$V(x) = \begin{cases} V, & x > 0 \\ 0, & x < 0. \end{cases} \tag{2.6.10}$$

A choice of incident wavefunction in the region $x < 0$ which describes a set of particles having the same momentum p_0 but a variable spatial distribution is the plane wave modulated by a variable amplitude: $\psi_0(x) = R_0(x) \, \mathrm{e}^{\mathrm{i}p_0 x/\hbar}$. The evolution of each ensemble element (uniform motion with speed p_0/m)

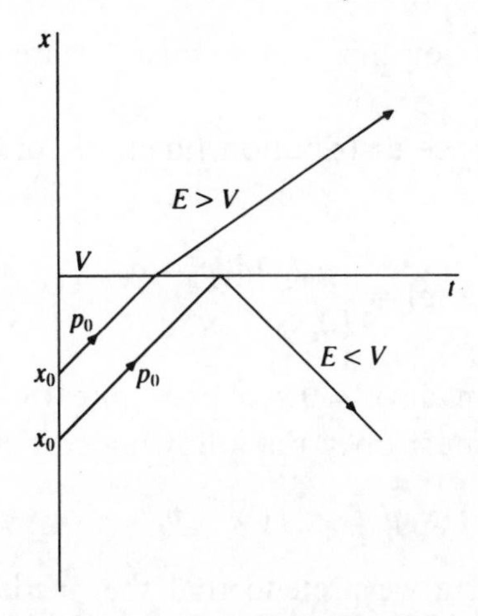

Fig. 2.4 A particle incident on a potential step V is reflected or refracted depending on the magnitude of the energy E. The momentum and density are discontinuous at $x = 0$.

implies that at time t the incident wavefunction is given by

$$\psi(x, t) = R_0(x - p_0 t/m)\,\mathrm{e}^{\mathrm{i}(p_0 x - Et)/\hbar}, \qquad x < 0 \tag{2.6.11}$$

where $E = p_0^2/2m$.

Suppose first that $E > V$. Then there will be a transmitted beam in the constant potential region $x > 0$ and no reflected beam (Fig. 2.4). The transmission part of the wavefunction will be

$$\psi_T(x, t) = R_0'(x - p_0' t/m)\,\mathrm{e}^{\mathrm{i}(p_0' x - Et)/\hbar}, \qquad x > 0 \tag{2.6.12}$$

where $p_0' = [2m(E - V)]^{1/2}$. Continuity at $x = 0$ implies that the current is conserved: $p_0 R_0^2(-p_0 t/m) = p_0' R_0'^2(-p_0' t/m)$. Hence the transmitted component will be

$$\psi_T(x, t) = (p_0/p_0')^{1/2} R_0[(p_0/p_0')(x - p_0' t/m)]\,\mathrm{e}^{\mathrm{i}(p_0' x - Et)/\hbar}, \tag{2.6.13}$$

which is normalized if ψ is. Notice that both $\partial S/\partial x$ and R are discontinuous at $x = 0$, something that cannot occur in quantum mechanics. The functions (2.6.11) and (2.6.13) generate the trajectory

$$x = \begin{cases} x_0 + p_0 t/m, & 0 < t < -m x_0/p_0 \\ x_0 + p_0' t/m, & t > -m x_0/p_0. \end{cases} \tag{2.6.14}$$

The case $E > V$ is thus adequately accounted for using pure states (at each space point the momentum is unique). If instead $E < V$ there will be a reflected beam but no transmitted one and hence at each point an ensemble element may have one of two possible momentum values. The appropriate theoretical description of this is the mixed state. To find this we solve Liouville's equation.

The initial phase space distribution function corresponding to the wavefunction (2.6.11) is

$$f_0(x, p) = \begin{cases} \rho_0(x)\delta(p - p_0), & x < 0 \\ 0, & x > 0. \end{cases} \tag{2.6.15}$$

To treat the reflection at $x = 0$ we first note that because $\mathrm{d}f/\mathrm{d}t = 0$ the distribution function must obey the following condition:

$$f(x, p, t) = f(x, -p, t - 2mx/p). \tag{2.6.16}$$

To find its explicit form we pretend that the barrier is absent and take its effect into account by imposing the symmetry condition (2.6.16) on a solution to the free Liouville equation. Denoting the general solution of the latter by $F(x, p, t)$ we have

$$F(x, p, t) = F_0(x - pt/m, p) \tag{2.6.17}$$

and hence (2.6.16) implies

$$F_0(x, p) = F_0(-x, -p). \tag{2.6.18}$$

In order to make this condition consistent with our initial function (2.6.15) we introduce a fictitious or 'virtual' distribution function in the region $x > 0$ and write

$$F_0(x, p) = \rho_0(x)\delta(p - p_0) + \rho_0(-x)\delta(p + p_0). \tag{2.6.19}$$

This function evidently satisfies (2.6.18) and coincides with (2.6.15) when $x < 0$. In the domain $x < 0$, the free evolution of (2.6.19) yields the sought-for solution to Liouville's equation in the given potential (for another example of the technique of 'virtual distributions' see §6.5). From (2.6.17) we have then

$$\begin{aligned} f(x, p, t) &= F(x < 0, p, t) \\ &= \rho_0(x - p_0t/m)\delta(p - p_0) + \rho_0(-x - p_0t/m)\delta(p + p_0), \qquad x < 0. \end{aligned} \tag{2.6.20}$$

Given that ρ_0 is normalized it is easy to see that the distribution function in the domain accessible to the particle is also:

$$\int_{-\infty}^{0} \mathrm{d}x \int_{-\infty}^{\infty} \mathrm{d}p\, f(x, p, t) = 1. \tag{2.6.21}$$

If $\rho_0(x)$ is a packet function peaked around the point $x = -y$, (2.6.20) describes two packets moving towards one another, overlapping perfectly at $t = my/p_0$, and passing through one another. It is at this juncture that we see the inadequacy of the classical wavefunction formalism in accounting for this process. We might for example expect that the mixed state (2.6.20) may be composed from two wavefunctions in the domain $x < 0$: the incident function (2.6.11) and a reflected one,

$$\psi_{\mathrm{R}}(x, t) = R_0(-x - p_0 t/m)\, \mathrm{e}^{\mathrm{i}(-p_0 x - Et)/\hbar}, \tag{2.6.22}$$

with corresponding density matrix (cf. (2.6.9))

$$\rho(x, x') = \tfrac{1}{2}\psi(x)\psi^*(x') + \tfrac{1}{2}\psi_{\mathrm{R}}(x)\psi_{\mathrm{R}}^*(x'). \tag{2.6.23}$$

But clearly, while the mean density

$$P(x, t) = \rho_0(x - p_0 t/m) + \rho_0(-x - p_0 t/m) \tag{2.6.24}$$

is a conserved quantity, neither of the individual densities connected respectively with the incident and reflected wavefunctions is conserved. Thus ψ and ψ_{R} are not solutions to the classical wave equation.

This example illustrates the fundamental difference between the quantal and classical treatments of statistical problems. One cannot account for generic classical processes using just a pure state, i.e., a wavefunction; we must introduce a mixed state because forces bring about the spatial mixing, or crossing, of trajectories. In contrast, quantum mechanics exhibits in the linear superposition principle a new method of combining solutions which allows a process to be entirely accounted for using just the pure state. One can certainly treat the quantum problem using a mixed state, and hence obtain multivalued velocity fields, but this is now a derived concept and not a basic feature of the theory. A consequence of the quantum approach is that the independence of the particle motion from the choice of wavefunction (or distribution function) that is a characteristic of the classical theory no longer holds. The quantum mechanical description of the square potential using pure states is given in §4.5 and §5.3.

2.7 Generalization – internal potentials

We wish to propose that the Hamilton–Jacobi theory of waves and rays offers a deeper approach to the theory of motion than the Lagrangian and Hamiltonian formulations from which it may be derived. It is suggested that we invert the usual procedure by which particle dynamics is defined: instead of introducing trajectories through the Lagrangian or Hamiltonian and then showing that the formula

$$p(q, \dot{q}, t) = \partial S/\partial q|_{q=q(t)} \tag{2.7.1}$$

is consistent with this, we start with a Hamilton–Jacobi equation and *define* the trajectories to be the solutions $q(t)$ of (2.7.1). Such an approach then provides the starting point for a natural generalization of classical mechanics where one considers Hamilton–Jacobi systems which cannot, in general, be transformed into an equivalent Lagrangian or Hamiltonian problem. The latter will be defined in a formal sense, and indeed are needed so as to ensure consistency (e.g., in order to give p in terms of q, $\dot{q}$ and t so that (2.7.1) may be solved), but they will not be completely defined until the Hamilton–Jacobi equation has been solved. Such a theory provides a context within which it has meaning to think of the S-function as a physically real field, for it implies the possibility of a mechanism by which S may manifest itself in the motions of particles. We shall denote this more general conception 'generalized Hamilton–Jacobi theory' in order to distinguish it from the classical case where no such mechanism exists.

What we have in mind is the following, which we describe for one body. Suppose that, in addition to the usual external potential $V(\mathbf{x}, t)$, a new type of energy $V'(\mathbf{x}, t)$ is introduced into the Hamilton–Jacobi equation which depends explicitly on S and on various orders of the derivatives of S:

$$\partial S/\partial t + (\nabla S)^2/2m + V + V'(S, \partial_i S, \partial_{ij} S, \partial_{ijk} S, \ldots) = 0 \tag{2.7.2}$$

where $\partial_{ij} S = \partial^2 S/\partial x_i\, \partial x_j$, $i, j = 1, 2, 3$, etc. Conditions are imposed on V' so that it satisfies the usual symmetry properties, e.g., that it transforms as a scalar under Galilean coordinate transformations, and we must specify the initial function $S_0(\mathbf{x})$. If we continue to define the orbits as solutions to the equation

$$\dot{\mathbf{x}} = (1/m)\nabla S|_{\mathbf{x}=\mathbf{x}(t)} \tag{2.7.3}$$

then it may be expected that the presence of the term V' will bring about deviations in the motion found in the analogous classical case (where $V' = 0$), since S now satisfies a different equation and moreover enters explicitly into the equation of motion as a kind of force. To see this, we take the gradient of (2.7.2) and, identifying $\dot{\mathbf{x}}$ with $(1/m)\nabla S$, find

$$m\ddot{\mathbf{x}} = -\nabla(V + V')|_{\mathbf{x}=\mathbf{x}(t)}. \tag{2.7.4}$$

As regards the particle motion it is therefore consistent to regard V' as a potential energy on the same footing as V, i.e., as similar in its action to an external potential which gives rise to a force $-\nabla V'$ at the location of the particle. Notice that a particle which is classically free ($V = 0$) will not be so here since it is subject to V'. The classical limit is obtained when $V' \to 0$.

To think in these terms is misleading however. For V' is not an external potential of the usual type, which is a preassigned function of the coordinates and independent of the dynamics it determines – it depends on the function (S) conventionally associated with the motion of the particle through the momentum (∇S) and the kinetic energy ($(1/2m)(\nabla S)^2$). It is therefore more appropriate to call V' an 'internal' potential in order to distinguish it from the more common preassigned potentials of classical dynamics. It is consistent to retain the word 'potential' though since, as we have seen, V' formally enters the equation of motion (2.7.4) on the same footing as V.

In order to give a consistent physical interpretation of V' we have to generalize our concept of what constitutes 'matter'. We must suppose that there is associated with a particle pursuing a well-defined trajectory in some region of space a new type of physical field mathematically described by functions of $S(\mathbf{x}, t)$. But this field is not external to the particle in the sense that it may or may not be present. It is objectively real and always accompanies the particle in its motion – indeed it guides the latter in accordance with the laws of motion (2.7.3) or (2.7.4). When we speak of a 'material system' we therefore mean a composite entity comprising simultaneously real wave and corpuscular aspects. The value of the field at a spacetime point contributes as much to the definition of the 'state' of a system as does the position of the particle. In fact, given that the field S is now to be considered as a component of a physical system, it is somewhat arbitrary to associate $(1/2m)(\nabla S)^2$ with just the kinetic energy of the corpuscle and V' with the potential energy. From the point of view of the system as a whole (wave plus particle) this division is arbitrary and it is better to treat V' as a new form of energy which is not reducible to kinetic or potential energies of the conventional type. Nevertheless, it is useful to distinguish formally between the kinetic and potential energy-like terms in (2.7.2) in connection with the particle motion.

Notice that in this most simple of possible generalizations of classical Hamilton–Jacobi theory the particle plays a passive role. The S-function has been elevated from a passive to an active role and now physically acts to affect particle motion, but the particle does not in turn influence the wave. One can certainly envisage a theory in which the particle and wave aspects of matter interact in a dialectical fashion but we will not discuss this possibility here since it is not necessary for an understanding of quantum mechanics.

Notice also that the appearance of derivatives of S of an order higher than the first is not something that can be readily embraced by some kind of generalization of classical mechanics which retains the basic structure of that theory *viz.* that matter has simply a corpuscular aspect with no active wave component. For example one may be tempted to interpret the higher derivatives as implying the relevance of higher-order derivatives of q with respect to t in some generalized Lagrangian formalism ($L = L(q, \dot{q}, \ddot{q}, \ldots, t)$). This, however, is not so. The generalization of classical mechanics which employs such Lagrangians leads to a Hamilton–Jacobi equation which is still of the first order in the derivatives of S, but where S depends on an extended range of variables (León & Rodrigues, 1985). The introduction of higher-order derivatives of S is thus indicative of something qualitatively new in the theory of matter and motion.

In order that our purpose is not misconstrued it should be emphasized that presenting classical and quantum mechanics in terms of the same theoretical structure is not an attempt at reducing the latter to a version of the former. The notion of a well-defined particle trajectory is not a concept peculiar to the classical paradigm. From our new perspective we see that it has a wider applicability and indeed that classical mechanics is a particular case of quantum mechanics not in the sense of the emergence of the trajectory concept in some limit, but in the types of motions that particles can perform. Quantum mechanics can be encompassed in a generalized Hamilton–Jacobi theory but it certainly cannot be squeezed into a classical framework in which the S-function is physically emasculated.

We can also envisage a less direct influence of S on the motion if S couples to other field quantities (through further field equations) which themselves appear in the Hamilton–Jacobi equation. These other fields will also enter into the definition of 'matter', as S does. As an example, let us return to the statistical mechanics of a single particle, but now in the context of a generalized Hamilton–Jacobi theory in which the internal potential V' depends on the density $\rho(\mathbf{x}, t)$ and its derivatives rather than S directly:

$$\frac{\partial S}{\partial t} + \frac{(\nabla S)^2}{2m} + V + V'(\rho, \partial_i \rho, \partial_{ij} \rho, \ldots) = 0. \tag{2.7.5}$$

We assume that the continuity equation (2.5.2) remains valid so that together with (2.7.5) we have a pair of coupled partial differential equations through which ρ and S codetermine one another (this is the case of quantum theory). In the event that ρ simply has the meaning of a probability density its appearance in the dynamical equations would appear to make little physical sense in that it would entail an interaction between members of a fictitious ensemble. But suppose that we give ρ another, more fundamental, property

– that it is an objectively real field which is, along with S, a component of a *single* material system. This will be its primary physical significance. Then, as we shall show in detail in this book, it is possible to assume that the probabilistic aspect of ρ is a *secondary* property which is consistent with, and does not contradict, its role as a component of an individual system. Endowing matter with a field aspect, mathematically described by ρ and S which no longer have a purely descriptive significance, enables us to avoid the paradox of an individual's properties apparently depending on an ensemble.

To summarize the approach proposed here, we may say that the programme of classical mechanics is:

Particle dynamics $\Rightarrow$ evolution of field S in configuration space.

The new programme is:

Field dynamics of S (and other coupled fields) in configuration space $\Rightarrow$ particle dynamics.

Corresponding to this *new theory of motion* is a *new conception of matter*. The classical theory of matter, of mass points interacting via preassigned potentials whose natures are independent of and external to the points acted upon, has been replaced by a more general conception in which matter has an intrinsic field aspect, the mass points moving and interacting under the influence of the new internal energy as well as the more familiar potentials of classical dynamics. This leads to a synthesis of the wave and particle characteristics of matter. These properties are not mutually exclusive, but simultaneously existent. The generalization envisaged here is not of a statistical character and it is not tied to any particular scale (microscopic, macroscopic or something between). It is a causal determinist theory of individuals. The internal potential is an organizational or self-referential form of energy which brings about an 'inner tension' in the material system to which the mass points respond.

3

Elements of the quantum theory of motion

3.1 The basic postulates

The statistical interpretation of the wavefunction is in accord with experimental facts. An interference pattern on a screen is built up by a series of apparently random events, and the wavefunction correctly predicts where the particle is most likely to land over an ensemble of trials (cf. the *frontispiece*). Yet the interpretation of the wavefunction which ascribes to it a purely statistical significance is not forced on us by the experimental results – the latter show only that it is consistent to regard the wavefunction as containing information on probability; they do not exclude the possibility that it may possess other properties. On the contrary, one may take the view that the characteristic distribution of spots on a screen which build up an interference pattern is evidence that the wavefunction indeed has a more potent physical role than a mere repository of information on probabilities, for how are the particles guided so that statistically they fall into such a pattern? Such a question is naturally ruled out by the purely probabilistic interpretation. But the latter is appropriate only if we wish to reduce physics to a kind of algorithm which is efficient at correlating the statistical results of experiments. If we wish to do more, and attempt to understand the experimental results as the outcome of a causally connected series of individual processes, then we are free to enquire as to the possible further significance of the wavefunction (beyond its probabilistic aspect), and to introduce other concepts in addition to the wavefunction.

This is our approach here. But not only do we entertain the notion that the wavefunction has a direct physical significance in each individual process, we go further to propose that such action is its *primary* feature. In our formulation of quantum mechanics probability only enters as a subsidiary condition on a causal theory of the motion of individuals; the statistical meaning of the wavefunction is a *secondary* property. Failure to recognize this has been the source of much confusion in understanding the causal interpretation.

The additional element that we introduce apart from the wavefunction is a particle, conceived in the classical sense as pursuing a definite continuous track in space and time. In the last chapter we saw how the Hamilton–Jacobi theory provides a general theory of waves and rays and their interrelationships which transcends its particular historical association with classical mechanics. Our aim in this chapter is to develop this view and to show how underlying quantum mechanics is a causal theory of the motion of waves and particles which is consistent with probability, but does not require it. It is convenient to summarize here the basic postulates of our approach (Bohm, 1952a,b) which form the core of the theory presented in this book (for the one-body problem):

(1) An individual physical system comprises a wave propagating in space and time together with a point particle which moves continuously under the guidance of the wave.
(2) The wave is mathematically described by $\psi(\mathbf{x}, t)$, a solution to Schrödinger's wave equation.
(3) The particle motion is obtained as the solution $\mathbf{x}(t)$ to the equation

$$\dot{\mathbf{x}} = (1/m)\nabla S(\mathbf{x}, t)|_{\mathbf{x}=\mathbf{x}(t)} \tag{3.1.1}$$

where S is the phase of ψ. To solve this equation we have to specify the initial condition $\mathbf{x}(0) = \mathbf{x}_0$. This specification constitutes the only extra information introduced by the theory that is not contained in $\psi(\mathbf{x}, t)$ (the initial velocity is fixed once we know S). An ensemble of possible motions associated with the same wave is generated by varying $\mathbf{x}_0$.

These three postulates on their own constitute a consistent theory of motion. In order to ensure the compatibility of the motions of the ensemble of particles with the results of quantum mechanics, we make a further postulate:

(4) The probability that a particle in the ensemble lies between the points $\mathbf{x}$ and $\mathbf{x} + \mathrm{d}\mathbf{x}$ at time t is given by

$$R^2(\mathbf{x}, t)\, \mathrm{d}^3x \tag{3.1.2}$$

where $R^2 = |\psi|^2$.

This postulate has the effect of selecting from all the possible motions implied by the law (3.1.1) those that are compatible with an initial distribution $R^2(\mathbf{x}, 0) = R_0^2(\mathbf{x})$. It is thus a consistent subsidiary condition imposed on a causal theory of motion and has no more fundamental status than that. Notice that we have formulated postulate (4) in terms of the probability that a particle actually *is* at a precise location at time t. The usual interpretation of $|\psi(\mathbf{x}, t)|^2$ is that it determines the probability density of *finding* a particle in the volume d^3x at time t if a suitable 'measurement' is carried out. In fact, our notion is applicable to all states of the particle and in particular to those interactions which are characterized by the term 'measurement' (if the theory is consistently extended to include such interactions); i.e., the 'probability of finding' is a special case of the 'probability of being' (see Chap. 8).

Note that our definition of an 'individual physical system' is not restricted to atomic or subatomic particles. It applies to all matter, regardless of scale, although the wave aspect is generally apparent only in phenomena involving microscopic particles.

One of the merits of the theory is that it applies to the world as a whole – there is no arbitrary division into subject and object, or observer and observed. It would be more logical therefore to present first of all the many-body theory and treat the one-body theory with external potentials as a special instance of this when we can neglect the 'rest of the world'. Instead, in order to explain the many unfamiliar features of this theory of motion, we begin with the simplest system. This procedure is particularly useful because the many-body problem possesses properties whose novelty can best be understood by comparison with the properties of the simpler, one-body system.

3.2 Wave and particle equations of motion

3.2.1 Reformulation of the Schrödinger equation

In order to motivate the introduction of the particle concept into quantum mechanics we shall decompose the Schrödinger equation into two real equations, in a particular way, by expressing the wavefunction in polar form:

$$\begin{aligned}\psi &= R\cos(S/\hbar) + iR\sin(S/\hbar)\\ &= R\,e^{iS/\hbar}.\end{aligned} \tag{3.2.1}$$

Here $R = R(\mathbf{x}, t)$ is a real amplitude function and $S = S(\mathbf{x}, t)$ is a real phase function, of space and time, and $\hbar = h/2\pi$ where h is Planck's constant. S thus has the dimension of action and is measured in units of $\hbar$, and R, in keeping with its definition below (§3.6) as the square root of a density function, has the dimension $L^{-3/2}$ (in three dimensions). Writing ψ in terms of its real and imaginary parts

$$\psi = \psi_1 + i\psi_2 \tag{3.2.2}$$

we have

$$\left.\begin{aligned} R &= (\psi_1^2 + \psi_2^2)^{1/2} = (\psi^*\psi)^{1/2} \\ S/\hbar &= \tan^{-1}(\psi_2/\psi_1) = (1/2\mathrm{i})\log(\psi/\psi^*). \end{aligned}\right\} \tag{3.2.3}$$

The amplitude function is by definition nonnegative:

$$R \geqslant 0 \qquad \forall \mathbf{x}, t. \tag{3.2.4}$$

Henceforth by the phrase 'the wave ψ' we shall mean the pair of real fields ψ_1 and ψ_2 (which are coupled by the wave equation). The word 'wave' is justified because ψ is a continuous function that obeys a linear superposition principle (and not because it is periodic in space and time which generally it is not).

The wavefunction is a solution of the Schrödinger equation

$$\mathrm{i}\hbar\frac{\partial\psi}{\partial t} = \left(-\frac{\hbar^2}{2m}\nabla^2 + V\right)\psi, \tag{3.2.5}$$

where m is the inertial mass and $V = V(\mathbf{x}, t)$ is the potential energy due to an external classical potential field (which we assume to be a real and purely multiplicative function). The wave equation (3.2.5) is stated as a postulate and justified by its consequences, as we would state Newton's laws. There is surprisingly little attention paid in the books to the possibility of deriving Schrödinger's equation from physical first principles (some works even try to dispense with it altogether, e.g., Lévy-Leblond & Balibar (1990)). A simple but rather cheap method is to turn the nonlinear-field formulation of classical mechanics (cf. §2.6) into a linear-field theory by deleting the quantum potential-like term and imposing on the ψ-function physically reasonable boundary conditions. This is hardly a 'derivation', though it fares no worse than the original plausibility argument of Schrödinger (1926a) in which de Broglie's expression for wavelength is inserted into a classical wave equation. Another more physical route is offered by stochastic mechanics (Nelson, 1966, 1985).

Inserting (3.2.1) into (3.2.5) and separating into real and imaginary parts we obtain the following field equations for the fields R and S (the calculation is explicitly performed in Appendix A). The real part gives

$$\frac{\partial S}{\partial t} + \frac{(\nabla S)^2}{2m} - \frac{\hbar^2}{2m}\frac{\nabla^2 R}{R} + V = 0 \tag{3.2.6}$$

and the imaginary part may be brought to the form

$$\frac{\partial R^2}{\partial t} + \nabla\cdot\left(\frac{R^2\nabla S}{m}\right) = 0. \tag{3.2.7}$$

Eqs. (3.2.6) and (3.2.7) are a pair of coupled partial differential equations in which the fields R and S codetermine one another. The wave equation determines ψ only up to a constant factor. If ψ is normalized this uniquely fixes R but S is defined only up to an additive constant. In order that the theory based on (3.2.6) and (3.2.7) be mathematically equivalent to that based on (3.2.5) we have to translate the conditions imposed on ψ that make (3.2.5) physically meaningful into conditions on R and S.

To begin with, to obtain a unique solution to (3.2.5) for all t we have to specify the initial wavefunction $\psi_0(\mathbf{x}) = \psi(\mathbf{x}, 0)$ for all $\mathbf{x}$ (this is arbitrary apart from the conditions below). Equivalently, we must specify the independent real functions

$$R_0(\mathbf{x}) = R(\mathbf{x}, 0), \qquad S_0(\mathbf{x}) = S(\mathbf{x}, 0). \tag{3.2.8}$$

These are unique apart from respective multiplicative and additive constants since all ψ_0s connected by constant factors are physically equivalent. At points where $\psi_0 = 0$, S_0 is undefined. Next, the continuity and finiteness requirements on ψ and $\nabla\psi$ stated in §1.2 are readily translated into conditions on R, S and their derivatives. We shall not state them, however, since in practice it is more convenient to solve problems directly for ψ (due to the linearity of the wave equation), except to note that unless otherwise stated we assume that $R \to 0$ at infinity – most proofs of results involving integrals rely on this.

The linearity of (3.2.5) is a very important property, for it implies that there is an infinite number of possible wave motions associated with the same potential V, which may be generated at will by superposing known solutions with arbitrary constant complex weights.

3.2.2 Single-valuedness of the wavefunction

There is a final condition on ψ that is worth stating explicitly in terms of the new variables – its single-valuedness.

This means the following: with each point in space, at each instant, there is associated a unique value of the complex-valued function ψ. To see the significance of this requirement, we observe from (3.2.1) that R must be a single-valued function of position but that the value of the phase function at each point is not uniquely fixed: any two functions S, S' which differ by an integral multiple of $2\pi\hbar$,

$$S' = S + nh, \qquad n \text{ integer}, \tag{3.2.9}$$

give rise to the same ψ.

We can express this multivaluedness of S in another way. Consider a closed loop C in space. This must be continuous and not cross itself or pass through nodal regions (see below) but is otherwise arbitrary. Let us (mentally) traverse this circuit and return to our starting point. The net change in phase that accumulates during this trip, which is the sum of all the infinitesimal changes $\mathrm{d}S$ associated with each element of arc $\mathrm{d}\mathbf{x}$, will by (3.2.9) be given by

$$\oint_C \mathrm{d}S = \oint_C \nabla S \cdot \mathrm{d}\mathbf{x} = nh, \qquad n \text{ integer}. \tag{3.2.10}$$

If $n \neq 0$, this implies that somewhere within C the function S must become discontinuous. In view of the smoothness of ψ such discontinuities can only occur at points where $\psi = 0$ (the 'nodes' of the function ψ) for here the phase is indeterminate and may undergo a discrete jump. The nodal regions may be found by solving the simultaneous equations (cf. (3.2.2))

$$\psi_1 = \psi_2 = 0 \tag{3.2.11}$$

which, amounting as they do to the intersection of two surfaces, implies that the nodes are generically lines in space which sweep out surfaces as time progresses (for an example where the nodal regions are spatial surfaces see §4.3). The nodal lines extend from $-\infty$ to ∞ or form closed loops. In view of (3.2.3) we can write (3.2.11) succinctly as the single condition

$$R = 0 \qquad \text{(two real conditions)}. \tag{3.2.12}$$

It should be noted that (3.2.12) is a necessary condition for the appearance of phase singularities (if $R \neq 0$ within C they cannot occur) but it is not sufficient – it should be checked in each case that indeed $n \neq 0$ in (3.2.10). Clearly, in formulating (3.2.10) we must require that the loop C does not pass through points where $R = 0$.

Insight into the meaning of (3.2.10) is gained if we study the structure of the wavefronts. The latter are the surfaces in space defined at each instant by:

$$S(\mathbf{x}, t) = \text{constant} \tag{3.2.13}$$

and, in particular, the wavecrests are defined by

$$S = nh, \qquad n \text{ integer}. \tag{3.2.14}$$

We may then interpret (3.2.10) as stating that the net number of wavecrests that come to an end within C is $|n|$ (Berry, 1981). The wavefronts end, of course, on the nodal lines.

In order to be consistent with (3.2.10), the vector field $\mathbf{p} = \nabla S$, which is a single-valued function, will be irrotational except along the nodal lines

(Takabayasi, 1983; Holland, 1987):

$$\nabla \times \mathbf{p} = \sum_a \Gamma_a \int_{L^{(a)}} \delta(\mathbf{x} - \mathbf{x}^{(a)})\, \mathrm{d}\mathbf{x}^{(a)} \tag{3.2.15}$$

where $\mathbf{x}^{(a)} = \mathbf{x}^{(a)}(\lambda^a)$ are the coordinates of a point on the ath line $L^{(a)}$ and $\Gamma_a = n_a h$.

Relation (3.2.10) holds for one instant of time and a circuit C fixed in space. As the wave evolves it may happen that a nodal line crosses C. Then the value of $\oint \mathrm{d}S$ will undergo a discrete jump and subsequently (3.2.10) will be valid again, with n equal to a different integer.

Finally, we point out that (3.2.10) is valid independently of the Schrödinger equation. Interpreting ∇S as a momentum field, it is reminiscent of the Bohr–Sommerfeld quantization condition, but, of course, in the old quantum theory S was treated as a solution of the classical Hamilton–Jacobi equation, rather than of (3.2.6). In pure classical mechanics there is no requirement that the wavefunction be single-valued (§2.6).

3.2.3 Introduction of the particle

We have rewritten the field equation (3.2.5) for the complex amplitude ψ as a pair of real equations (3.2.6) and (3.2.7). The latter are mathematically equivalent to the former if we impose conditions such as (3.2.10) on the new variables. The purpose of this exercise is to facilitate the physical interpretation of the Schrödinger equation, and it is to this that we now turn. We shall be concerned here mainly with (3.2.6) and shall leave until §3.6 the interpretation of (3.2.7).

We propose to associate with the physical wave ψ propagating in space a point particle of mass m which pursues a trajectory $\mathbf{x} = \mathbf{x}(t)$. Since they both depend on the same parameter m (and more generally on the charge e, magnetic moment μ etc.) and, as we shall see, the former exerts a force on the latter, it is appropriate to treat the wave and particle as aspects of a *single* physical system. To see the connection between the wave and particle we treat (3.2.6) as a particular instance of the generalized Hamilton–Jacobi theory set out in §2.7. Eq. (3.2.6) has the form of the classical Hamilton–Jacobi equation apart from the extra term

$$Q(\mathbf{x}, t) = -\frac{\hbar^2}{2m}\frac{\nabla^2 R}{R}. \tag{3.2.16}$$

The function Q will be called the 'quantum potential energy' or simply the 'quantum potential'. Using this notation (3.2.6) becomes

$$\partial S/\partial t + (\nabla S)^2/2m + Q + V = 0. \tag{3.2.17}$$

Following the discussion in §2.7 let us construct from the vector field $\mathbf{p} = \nabla S$ a vector field $\mathbf{v} = (1/m)\mathbf{p}$ and assume that the latter defines at each point of space at each instant the tangent to a possible particle trajectory passing through that point. This naturally provides a description of an ensemble of particles, which is fictitious in the sense that only one track is realized in any given field, and is quite independent of statistical considerations. The trajectories are orthogonal to the surfaces $S =$ constant and may be found by integrating the differential equations†

$$m\dot{\mathbf{x}} = \mathbf{p}(\mathbf{x}, t)|_{\mathbf{x}=\mathbf{x}(t)} \tag{3.2.18}$$

or, in terms of the velocity field $\mathbf{v}$,‡*

$$\begin{aligned}\dot{\mathbf{x}} &= \mathbf{v}(\mathbf{x}, t)|_{\mathbf{x}=\mathbf{x}(t)} \\ &= (1/m)\nabla S(\mathbf{x}, t)|_{\mathbf{x}=\mathbf{x}(t)}\end{aligned} \tag{3.2.19}$$

to yield $\mathbf{x} = \mathbf{x}(t)$, where we must specify the initial position $\mathbf{x}_0$. The single-valuedness of the function $\mathbf{p}$ ensures that $\mathbf{x}$ is a single-valued function of t. From the point of view of the particle we may then interpret (3.2.17) as an equation for its total energy. The latter is defined by

$$E(\mathbf{x}(t), t) = -\partial S(\mathbf{x}, t)/\partial t|_{\mathbf{x}=\mathbf{x}(t)} \tag{3.2.20}$$

and is given by the sum of the kinetic energy

$$(\nabla S)^2/2m|_{\mathbf{x}=\mathbf{x}(t)} = \tfrac{1}{2}m\dot{\mathbf{x}}^2 \tag{3.2.21}$$

and the total (quantum plus classical) potential energy

$$[V(\mathbf{x}, t) + Q(\mathbf{x}, t)]|_{\mathbf{x}=\mathbf{x}(t)} = V(\mathbf{x}(t), t) + Q(\mathbf{x}(t), t). \tag{3.2.22}$$

In this way we introduce a *phase space*, with coordinates $(\mathbf{x}, \mathbf{p})$, into quantum mechanics. The system trajectory is defined by the equation $\mathbf{p} = \nabla S(\mathbf{x}, t)$.

Now clearly, as anticipated in the previous chapter, Q is not a preassigned

† To avoid any confusion we emphasize that the arguments of ψ, $\mathbf{x}$ and t, are four *independent* coordinates labelling a point in Newtonian spacetime. The time coordinate also plays the role of parametrizing the particle trajectory, along which $\mathbf{x}$ becomes a function of t.

‡ In an arbitrary system of coordinates q^i we have

$$\dot{q}^i = (1/m)g^{ij}\partial_j S(q^i, t)|_{q^i=q^i(t)}$$

where g^{ij} is a metric characterizing the coordinate system.

* For a demonstration of how this law follows from general requirements including Galilean invariance (§3.11.1) see Dürr, Goldstein & Zanghi (1992).

function of the coordinates in the way that V is. It depends on the total quantum state and, although it is expressed in terms of just R, Q is inextricably coupled to $S(\mathbf{x}, t)$. Indeed if we specify $R_0(\mathbf{x})$ and treat ∇S as a given function of $\mathbf{x}$ and t, we can, in principle, solve (3.2.7) for $R(\mathbf{x}, t)$ in terms of ∇S. Substituting the result into (3.2.16) yields Q in terms of R_0 and the derivatives of S. We obtain then for (3.2.17) an equation of the type (2.7.2). For example, we can give an explicit expression for Q in the one-dimensional case when the velocity field is time-independent using the solution to (3.2.7) given in (2.5.6) (for another explicit example see §4.1).

To show that it is nevertheless consistent to regard Q as a potential on the same footing as V *in respect of the particle motion*, we apply the operator ∇ to (3.2.6) and find, after some rearrangement, the field equation

$$[\partial/\partial t + (1/m)\nabla S \cdot \nabla]\nabla S = -\nabla(V + Q). \tag{3.2.23}$$

Making the identification of $\nabla S/m$ with particle velocity as in (3.2.19) we find by evaluating (3.2.23) along the trajectory

$$\frac{\mathrm{d}}{\mathrm{d}t}(m\dot{\mathbf{x}}) = -\nabla(V + Q)|_{\mathbf{x}=\mathbf{x}(t)} \tag{3.2.24}$$

where

$$\mathrm{d}/\mathrm{d}t = \partial/\partial t + \dot{\mathbf{x}} \cdot \nabla \tag{3.2.25}$$

represents the time rate of change with respect to a point moving with the particle. Eq. (3.2.24) has the form of Newton's second law, in which the particle is subject to a quantum force $-\nabla Q$ in addition to the classical force $-\nabla V$. The effective potential acting on the particle is $(Q + V)$, not V. Newton's first law, that a body persists in a state of uniform motion unless acted on by a resultant force, follows as a special case of (3.2.24).

Either (3.2.19) or (3.2.24) may be considered as the particle 'law of motion'. In practice, it is easier to solve the former.

In this theory of motion the initial velocity of the particle is determined by the initial field $\psi_0(\mathbf{x})$. The latter implies an initial single-valued momentum field

$$\mathbf{p}_0(\mathbf{x}) = \nabla S_0(\mathbf{x}) \tag{3.2.26}$$

so that, if the initial position is specified, the initial velocity is uniquely given by

$$\dot{x}(0) = (1/m)\nabla S_0(\mathbf{x})|_{\mathbf{x}=\mathbf{x}_0}. \tag{3.2.27}$$

It follows that the only information that must be added to that contained in the fields in order to obtain a unique orbit for all time is $\mathbf{x}_0$. Specification of

$\mathbf{x}$ and ψ at a certain time then defines the 'state' of an individual system (although we shall occasionally use the expression 'quantum state' to denote just the wave). Given $\mathbf{p}_0$, we may, in principle, invert (3.2.26) to find the corresponding $\mathbf{x}_0$s. Note that the choice of $\mathbf{x}_0$ is arbitrary except that it cannot lie in a nodal region, for at these points ∇S_0 is undefined (§3.3.8).

The initial phase function thus has, through its gradient, a meaning for the particle. At this stage the initial field $R_0(\mathbf{x})$ has no significance for the particle; as a special case of the general theory it may be ascribed the meaning of a density of particles in the ensemble, as we shall see in §3.6 (in the case that $R(\mathbf{x}, t) = R_0(\mathbf{x})$ for all t, this field defines the quantum potential).

It is emphasized that the law of motion (3.2.19) is *postulated* and then shown to be consistent – it cannot be derived from the Schrödinger field theory. Why then should it be relevant to quantum mechanics? The reason is simple: the velocity field $\mathbf{v}(\mathbf{x}, t)$ is proportional to what is termed in quantum mechanics the 'probability density current', defined by

$$\mathbf{j} = (\hbar/2mi)(\psi^* \overleftrightarrow{\nabla} \psi). \quad (3.2.28)$$

We have

$$\mathbf{j}(\mathbf{x}, t) = R^2 \mathbf{v} \quad (3.2.29)$$

so that the current lies along the tangent to each point of a trajectory. Hence the trajectories $\mathbf{x}(t)$ are the lines along which, conventionally, probability 'flows'. The detailed demonstration that the motion determined by the law (3.2.19) leads to the empirical results of quantum mechanics is given in §3.7 and in later chapters.

Having defined the momentum and total energy of the particle, it is natural to define its angular momentum about a point at the origin of coordinates as the value of the field quantity

$$\mathbf{L}(\mathbf{x}, t) = \mathbf{x} \times \nabla S(\mathbf{x}, t) \quad (3.2.30)$$

evaluated along the trajectory:

$$\mathbf{L}(\mathbf{x}(t), t) = \mathbf{x}(t) \times m\dot{\mathbf{x}}(t). \quad (3.2.31)$$

We shall define the *total* angular momentum later (§3.5). The equation of motion of $\mathbf{L}$ is given in §3.9.3.

To summarize, an individual material system comprises the following two components:

(*a*) A wave having an amplitude and phase at each spacetime point. Its law of motion is (3.2.5).

(*b*) A mass point pursuing a spacetime trajectory. Its law of motion is (3.2.19) or (3.2.24).

An ensemble of mass points is associated with each wave. In the usual interpretation each system in the ensemble described by ψ is considered to be in the same state. Here they are distinguished by their initial locations.

From this definition we may associate various properties with the respective system components (e.g., field energy (cf. §3.9) and particle momentum). To establish the connection between the two aspects of matter we have rewritten the complex Schrödinger equation in a manner that is suggestive of a physical interpretation, as a coupled system of equations for the real fields R and S. These fields play several roles simultaneously:

(i) They are associated with two physical fields propagating in spacetime and define, along with the particle, an 'individiual physical system'.

(ii) They enter into the definition of properties associated with a particle (momentum, energy and angular momentum). These are not arbitrarily specified but are a specific combination of field variables and are closely related to the associated quantum mechanical operators (see §3.5). To obtain a particle property one evaluates the relevant function along a trajectory.

(iii) They enter as causal agents in the particle equation of motion, via the quantum potential.

(iv) They have other meanings which ensure the consistency of the theory with its usual interpretation and its connection with classical mechanics. S is a generalized Hamilton–Jacobi function.

The wave equation does two things: it determines the spacetime dependence of the ψ-field, and it tells how the physical properties of a particle embedded in that field evolve (by eqs. (3.2.20) and (3.2.21) for example). In practice, it is easier to solve the wave equation in its linear form (3.2.5) and treat the variables R and S as derived quantities.

Since the set of equations (3.2.5) and (3.2.19) uniquely specify the future (and past) continuous evolution of the field and particle system, the theory forms the basis of a 'causal interpretation' of quantum mechanics. But because it involves physical assumptions that are not usually made in quantum mechanics, we prefer to treat this system of equations as defining a new theory of motion, which is appropriately called the 'quantum theory of motion'. We do not expect that the simple-minded notion of causality employed here (that the future deterministically flows out of the past) encompasses all the phenomena encountered in experience, but it is sufficient for the quantum mechanical processes described in this book (it would not account for the development of an organism, for example, where matter apparently aggregates in certain patterns to achieve a definite end).

3.2.4 What about the commutation relations?

One of the basic features of quantum mechanics is the association of Hermitian operators with physical 'observables', and the consequent appearance of noncommutation relations between the operators. Thus $\hat{x}$ and $\hat{p}$ satisfy

$$[\hat{x}, \hat{p}] \equiv \hat{x}\hat{p} - \hat{p}\hat{x} = i\hbar. \tag{3.2.32}$$

As a result, a wavefunction cannot be simultaneously an eigenfunction of $\hat{x}$ and $\hat{p}$. Since according to the quantum theory of measurement (Chap. 8) the outcome of a 'measurement' of an 'observable' involves the transformation of the wavefunction into an eigenfunction of the associated operator, it apparently follows that a system cannot simultaneously be in a state in which its position and momentum are precisely known. How do we reconcile this argument with our assumption that a particle may indeed be ascribed simultaneously well-defined position and momentum variables as properties that exist during all interactions, including those that we designate 'measurements'?

To answer this we note that our *knowledge* of the state of a system should not be confused with what the state *actually is*. Heisenberg's programme for matrix mechanics was to formulate physical theory solely in terms of observable quantities but this cannot be carried out. There are all sorts of assumptions and postulates in quantum mechanics which according to the theory itself cannot be made subject to experimental verification (examples are: the continuity of space and time, the detailed structure of the ψ-wave such as its nodes, and so on). It is the theory that tells us what is 'observable' but we should be careful not to conclude from this that the theory is therefore incompatible with an ontology more detailed than that revealed just by the results of experiments. Quantum mechanics is constructed so that we cannot 'observe' position and momentum simultaneously but this fact *per se* does not have any bearing on the issue of whether a particle has a well-defined track *in reality*. If we treat measurement as a particular application of the theory in which the wave undergoes a certain type of interaction with other waves, then we can interpret noncommutativity as an expression of the different types of motion accessible to a particle when the wave undergoes the peculiar types of interaction appropriate to a position or momentum 'measurement'. According to the quantum theory of motion the particle has a position and momentum prior to, during, and after the measurement, whether this be of position, momentum or any other 'observable'.

Notice that the above discussion does not rely on the notion of probability, which we have not introduced yet.

The properties we have associated with the particle are actually closely related to the corresponding quantum operators and indeed serve to bring out the meaning of the latter, i.e., the sense in which they can be 'associated' with a physical system (see §3.5).

3.2.5 Minimalist causal interpretation

Although we do not take this point of view, it might be argued that the generalized Hamilton–Jacobi theory and the quantum potential are not necessary components of a causal interpretation of quantum mechanics. A 'minimalist' version may be developed based on the following considerations.

One may define, as we have done, an individual system as comprising a Schrödinger wave ψ and a corpuscle traversing a track $\mathbf{x}(t)$. One then defines the path as the solution to (3.2.19), given $\mathbf{x}(0)$. This may be justified since, as we pointed out above, ∇S is proportional to the current (3.2.28). One regards (3.2.19) as the law of motion and for the purpose of calculating trajectories it is not necessary to mention Hamilton–Jacobi theory or the quantum potential (see e.g., Bell (1987, 1990a)).

This approach is unsatisfactory for several reasons. Since the purpose of the causal interpretation is to offer an explanation of quantum phenomena it seems strange to ignore theoretical structures which may aid that objective. The introduction of the quantum potential as a causal agent has explanatory power which one unnecessarily foregoes by concentrating on just (3.2.19). It represents the difference between classical and quantum mechanics. It is hard to see how the classical limit of quantum mechanics can be achieved (Chap. 6) unless quantum particles have energy, angular momentum etc., in addition to a location. Interpreting (3.2.6) as a generalized Hamilton–Jacobi equation does not imply an attempt at squeezing quantum mechanics into a classical language. Rather, the Hamilton–Jacobi theory provides an appropriate context in which to discuss classical and quantum mechanics in the *same* language, as instances of a general theory of waves and rays, and not a means of reducing one to the other. To cite just one example of the usefulness of the quantum potential, it is not possible to give an account of the conservation laws without its intervention (see §3.9).

3.3 Properties of the model

3.3.1 Context dependence

When we solve the wave equation in any given situation, the solution $\psi(\mathbf{x}, t)$ will in general depend on a set of parameters which characterize the environment that the propagating wave encounters – such as slit width, radius of scattering centre etc. A different choice of these parameters will, in general, imply different values of ψ at *each* point in space and time (i.e., at all points and not just those in the vicinity of the physical objects with which the parameters are associated)†. We shall refer to this feature of encoding at each point information on the whole context as 'context dependence' or 'wholeness'.

Now the field is not only a function of environmental parameters (which is after all a property of any wave motion), it depends also on parameters such as the mass (and more generally charge, magnetic moment and so on) which are classically associated with a localized material body. That is, even though the particle is located along a curve in space, the value of ψ at all points depends on m (and e, μ, etc.). From the point of view of the Hamilton–Jacobi approach to classical mechanics the circumstance that the function $S(\mathbf{x}, t)$ depends on the mass is perfectly reasonable. What is novel here is that this dependence occurs in the context of a physical wave. The phrase 'an electron of mass m' is therefore to be interpreted that both ψ and the corpuscle are associated with the parameter m; ψ may be said to be 'massive' (and 'charged' etc.). This association is different in kind to the manner in which, for example, a gravitational field depends on the mass of its source, since there is no 'source' of the ψ-field in the conventional sense of a distinct localized entity.

3.3.2 Relative status of field and particle

Ontologically the wave and particle are on an equal footing (i.e., they both objectively exist). Yet in the model described in §3.2 the wave is the senior partner in the composite physical system in the following formal and physical senses.

First of all, the particle equation of motion (3.2.24) is a deduction from the Schrödinger equation once we postulate (3.2.19). We cannot reverse our steps and deduce the wave equation from the particle equation.

Secondly and physically, the particle simply responds to the local value of the field in its vicinity (via Q) – there is no reciprocal action of the particle on the wave. (This suggests that Newton's third law (to every action there is an equal and opposite reaction) is violated.) It is for this reason that de Broglie called ψ a 'pilot-wave' and (3.2.19) the 'guidance formula'. The initial position $\mathbf{x}_0$ is freely specifiable and has no effect on the temporal development of

† We do not imply here that the parameters are altered as part of the physical process – each different set of values is considered for the time being as a distinct physical situation. The problem of changes in boundary conditions treated as part of the process is discussed later (§11.6).

R and S. The novelty of the theory therefore lies in the characteristics of the ψ-wave and the implications this has for the behaviour of the particle. Because we do not regard the formal and conceptual structure of the theory as closed we can envisage a more active role for the particle, something which is not even admitted as conceivable in the conventional view. This may, for instance, enter as a 'source' of the ψ-field through an inhomogeneous term in the wave equation but we shall not pursue this here.

In this connection it is to be noted that ψ is not intended to be in any sense a 'model' of the particle it guides. This theory is to be sharply distinguished from the attempts of Schrödinger (1926b) and de Broglie (1956; Jammer, 1974) to fashion the corpuscle out of the pure field. The particle is always treated as a point within the wave (regardless of the form of the latter); they are logically distinct ultimates. Of course, the wave sometimes displays particle-like properties (e.g., certain kinds of packets) but the particle does not generally move with the group velocity of the wave.

Finally, these remarks on the primacy of the wave should be qualified by the observation that it is the *particle* that one detects in experiments. This point is discussed in detail in Chap. 8.

3.3.3 Classically-free motion

When $V = 0$ the particle equation of motion is

$$m\ddot{\mathbf{x}} = -\nabla Q|_{\mathbf{x}=\mathbf{x}(t)}. \qquad (3.3.1)$$

We immediately conclude from this that *classically-free motion is not free quantum mechanically*, due to the fact that for generic solutions of the free Schrödinger equation the quantum potential is nonvanishing. The accessible motions may take on a variety of forms depending on our choice of free wavefunction. This contrasts with the uniform and rectilinear motion uniquely implied by classical mechanics in this case.

3.3.4 Effect of external potential on field and particle

When we insert a classical potential into Schrödinger's equation, it is evidently the *field* that responds to it and hence the quantum potential carries information about it. As a result the particle is influenced in two ways by the external potential: referring to (3.2.24), it is acted upon by the classical force $-\nabla V$ and by the quantum force $-\nabla Q$ which, as we have just remarked, generally depends on V. This, indeed, explains how quantum particles in classical potentials display significantly nonclassical behaviour, since Q

provides a nonclassical vehicle for V. Thus:

$$m\ddot{\mathbf{x}} = -\nabla V - \nabla Q \tag{3.3.2}$$

$-\nabla Q$: effect of V via Q – purely quantum mechanical

$-\nabla V$: classical effect of V

Notice that if V is localized in space (e.g., if it is defined by a screen with two slits), Q may propagate information on it to regions where $V = 0$ and the equation of motion is given by (3.3.1). It is very important to bear in mind the V-dependence of Q for it provides the explanation according to this model of all typical quantum phenomena involving classical potentials (such as tunnelling in classically forbidden regions, stationary states and, when electromagnetic potentials are involved, the Aharonov–Bohm effect). Naturally, there are special cases where Q contains no information on V in the sense that $-\nabla Q$ maintains its value as $V \to 0$ (cf. §4.8).

This feature is connected with the fact that the basic dynamical law of quantum mechanics (the Schrödinger equation) is formulated in terms of an interaction of the wave directly with the potential V and not the force $-\nabla V$. This implies that the value of the potential at a point has a more direct physical meaning in quantum mechanics than it has in classical mechanics. For example, in the latter we can add to V any function of the time to form a new potential

$$V' = V + f(t). \tag{3.3.3}$$

Then $\nabla V' = \nabla V$ and the physical motion implied by both potentials is the same. In quantum mechanics the choice of potential is more closely connected with the physical situation under consideration. For example, the possibility of defining stationary states is related to the time-independence of V. If we chose some other time-dependent potential V' then it would not be possible, in general, to separate variables and define stationary states. If the physical situation involves essentially static fields (e.g., the hydrogen atom) then we should employ a static potential in the Schrödinger equation (Furry, 1963, p. 97).

The extent to which the information in V is translated into effects on the particle motion via Q requires a more detailed discussion since it relates to the gauge invariance of the theory. For the time being, we may say that *classical potentials have nonclassical effects in quantum mechanics because their influence is made manifest in the motion of particles via the mediating role of the quantum potential.*

Since the motion depends on the total quantum state we can generate an infinite number of possible motions by linearly superposing solutions to the wave equation (for the given V). This is in stark contrast to classical physics where the specification of the initial position and velocity uniquely fixes the particle motion in a given external potential.

3.3.5 Factorizable and nonfactorizable wavefunctions

An interesting example of the nonclassical effect of classical potentials is that *a classical force may act in directions orthogonal to its line of action.* To see this we first distinguish between factorizable and nonfactorizable wavefunctions.

A factorizable wavefunction is one which may be written as a product:

$$\psi(\mathbf{x}, t) = \psi_1(x, t)\psi_2(y, t)\psi_3(z, t). \tag{3.3.4}$$

Then the phase and quantum potential decompose into a sum of functions:

$$\left.\begin{aligned} S(\mathbf{x}, t) &= S_1(x, t) + S_2(y, t) + S_2(z, t), \\ Q(\mathbf{x}, t) &= Q_1(x, t) + Q_2(y, t) + Q_3(z, t). \end{aligned}\right\} \tag{3.3.5}$$

The motions in the three coordinate directions are thus completely independent; e.g., $m\dot{x} = \partial S_1/\partial x$ whose solution depends only on the initial x-coordinate.

The wavefunction (3.3.4) is a possible solution to the wave equation when the potential is of the form

$$V(\mathbf{x}) = V_1(x) + V_2(y) + V_3(z) \tag{3.3.6}$$

(including when $V = 0$) where, in general, the factorizability will be preserved for all t (if it is true at one instant) only if V is time-independent. Potential functions that are time-dependent, or more complicated functions of the coordinates, will tend to produce *nonfactorizable* wavefunctions from an initial state (3.3.4) as time progresses. In that case $S(\mathbf{x}, t)$ and $Q(\mathbf{x}, t)$ cannot be decomposed as in (3.3.5) and the associated motions in orthogonal directions are not independent.

In classical physics the independence of orthogonal motions when the potential takes the form (3.3.6) is guaranteed. In quantum mechanics we can superpose solutions of the type (3.3.4) to generate nonfactorizable wavefunctions even though V may be time-independent and be a sum as in (3.3.6). If, for example, $V = V_1(x)$, then, in general, $S(\mathbf{x}, t)$ and $Q(\mathbf{x}, t)$ are nonadditive functions of x, y and z and the y- and z-components of the motion are influenced by V (via Q) and depend on x_0. The classical force $(-\partial V_1/\partial x)$

therefore influences the forces $(-\partial Q/\partial y)$ and $(-\partial Q/\partial z)$ in directions orthogonal to its line of action.

The factorizability condition (3.3.4) is connected with the use of rectangular Cartesian coordinates. A wavefunction may factorize in other orthogonal coordinate systems (e.g., spherical polar) but we will not, in general, obtain the decomposition (3.3.5) into independent orthogonal motions in these systems.

3.3.6 Comparison with other field theories

In the simple model we have presented the manner in which the ψ-field couples to the particle via a quantum force does not on the face of it involve any concepts of cause and effect beyond those of classical physics. It is not essentially different from the action of an electromagnetic field on a charged particle, or a gravitational field on a massive particle, insofar as the 'particles' may be treated as structureless points (although there is no obvious 'quantum charge' associated with the particle which responds to the quantum potential. The reason is that Q is not an external potential of the conventional type, as discussed in §2.7). The description given in Chap. 5 of two-slit interference has many features in common with the case of a surface water wave negotiating a pair of slits in a barrier, the particle being analogous to a speck of dust following a path on the water's surface. It is tempting then to suppose that a classical explanation of quantum mechanics might be possible. However, although the analogies with more familiar theories are valuable in certain cases, they become seriously misleading if pushed too far. In the domain to which the quantum potential model applies we have to develop a new intuition. The radical departure from classical notions inherent in the causal interpretation lies in the details of the model, in the types of motions that are accessible to particles. There is great scope for speculation on the internal structure of the particle we have identified as a point and on how Q exerts a force at its location, but we shall not pursue this.

Some novel features of the theory in addition to those outlined in §§3.3.1–3.3.5 are as follows:

(1) There is no 'source' of the ψ-field in the conventional sense of a localized entity whose motion 'generates' it. As a result ψ is not 'radiated'.

(2) At this level of the theory no 'ether' is introduced which would support the propagation of ψ. As with electromagnetism, ψ may be thought of as a state of vibration of empty space.

(3) The influence of the wave on the particle, via Q, is independent of its *intensity* (cf. §3.4.1).

(4) The initial velocity of the particle is fixed by the initial wavefunction and not arbitrarily specified as in electromagnetic and gravitational theories.
(5) The Schrödinger equation is at one and the same time an equation for the evolution of the wave and an equation that determines the particle equation of motion. This should be contrasted with classical electromagnetism, for example, where Maxwell's equations and the Lorentz force law are logically distinct. The quantum case is closer to that of general relativity where in certain circumstances the geodesic law of motion can be deduced from the gravitational field equations and need not be independently postulated. This analogy is imperfect, however, because the classical notion of a 'test particle' whose characteristics are independent of and external to the guiding field does not exist in quantum mechanics – ψ is a function of 'particle-like' parameters such as m. Moreover, we assume the relation (3.2.19) and then deduce (3.2.24) as a consequence. It is thus not strictly true to say that the equation of motion of a particle follows from the wave equation in quite the same way as in general relativity.
(6) The wave equation describes the propagation of a complex amplitude ψ, or equivalently two coupled real fields. Complex waves are often introduced in other field theories for mathematical convenience, but one always takes the real part at the end of a calculation. In quantum mechanics two real fields are required.
(7) The ψ-field is finite and carries energy, momentum and angular momentum throughout space, far from where the particle is located. This is the case with classical field theories. However, the conservation laws obeyed by the field (cf. §3.9) are independent of the particle since the latter does not physically influence the former.
(8) The 'four known interactions' (electromagnetic, weak, strong and gravitational) are apparently mediated by the exchange of 'particles' (e.g., photons). The ψ-field does not act in this way, although at the level of first quantization that we are discussing we would not expect it to. The question as to whether one obtains 'quantons' at the second quantized level is discussed in Chap. 12.

These differences with conventional theories should certainly not be construed as evidence against the theory of motion we are investigating. After all, it may be the more familiar theories that are deficient. Can Maxwellian electrodynamics claim to be so free of conceptual problems that a mutual influence of the field and particle should be automatically demanded in all theories that employ these concepts (cf. the objection of d'Espagnat (1983, p. 91) and Shimony (1989))? And more importantly, is not some insight into individual processes better than none, regardless of our aesthetic predilections?

3.3.7 Are there quantum jumps?

The idea that material systems can only possess certain values of physical quantities corresponding to the spectra of Hermitian operators is, according to the quantum potential model, incorrect. The reason is that these quantities are well defined and continuously variable for all quantum states, and the values they take on when we consider the subset of eigenstates have no *fundamental* physical significance. One of the characteristic features of quantum mechanics, the existence of discrete energy levels, is thus a particular instance of the restriction of a basically continuous theory to motion associated with a subclass of eigenfunctions. Such states may possess particular physical importance in relation to the stability of matter, for example, but the particle momentum and energy are just as unambiguously defined when the wave is a superposition of eigenstates. There are then no 'quantum jumps' in the sense of a process that is instantaneous or beyond analysis (see also §3.7).

3.3.8 Trajectories do not cross, or pass through nodes

Although the phase function S is multivalued, ∇S is a single-valued function of position. This means that at each point in space at each instant there is a unique tangent vector associated with ∇S and consequently only one trajectory passes through that point, for each t. Trajectories cannot therefore cross or even touch. A given space point may have more than one trajectory pass through it at different times, of course. A single-valued trajectory field is called a *congruence*. In the sense of forming congruences, quantal paths have a simpler global structure than their generic classical counterparts. Multivalued fields are described by density matrices (§3.6).

There is an exception to this rule in that particles cannot pass at all through nodal regions (where $\psi = 0$), for there ∇S is undefined and does not define the tangent to a curve. This property is consistent with the conservation-type law (3.2.7) which maps nonnodal regions into themselves along trajectories. To see this we write (3.2.7) in the form

$$\partial R^2/\partial t + \mathbf{v}\cdot\nabla R^2 = -R^2\nabla\cdot\mathbf{v}. \qquad (3.3.7)$$

Hence, using (3.2.25), we have along a path

$$(\mathrm{d}/\mathrm{d}t)\log R^2(\mathbf{x}(t), t) = -\nabla\cdot\mathbf{v}(\mathbf{x}, t)|_{\mathbf{x}=\mathbf{x}(t)}, \qquad (3.3.8)$$

whence

$$R^2(\mathbf{x}(t), t) = R_0^2(\mathbf{x}_0)\,\mathrm{e}^{-\int_0^t \nabla\cdot\mathbf{v}(\mathbf{x}(t),t)\,\mathrm{d}t}. \qquad (3.3.9)$$

It follows that $R_0^2(\mathbf{x}_0) \neq 0 \Rightarrow R^2(\mathbf{x}(t), t) \neq 0$ for all t. Note that the property that a particle cannot be initially located at a node or ever pass through such points holds independently of the probability hypothesis introduced in §3.6.

The effective potential $(Q + V)$ always acts so as to preserve the properties of single-valuedness and nonnodalness of quantal congruences, which follow most directly from properties of the phase. Because of the requirements of boundedness and continuity satisfied by ψ, nodes are the only points where Q becomes singular.

3.3.9 *Empty waves*

It is implicit in our discussion so far that a particle is always accompanied by a wave. This is so even when the wave is physically ineffective (in the sense that $Q + V = 0$), for the phase function S propagates in spacetime and the particle moves within it. The converse, however, is not always true: there are cases, and they arise in practically all situations of physical interest, where the wave splits up into a set of parts which have no appreciable spatial overlap (i.e., they are separated by regions where $\psi = 0$). By the definition of a 'particle', the latter can only be located within one member of the set (and not in the nodal regions) and so it is reasonable to name the remaining waves 'empty'. This term does not carry with it any implication other than that there is no associated particle – the empty waves possess energy-stress-momentum (see §3.9), may be affected by external potentials, and if recombined (superposed) with the wave containing the particle will, of course, influence the subsequent particle motion.

3.3.10 *Conditions for interference*

If we have two waves $\psi_1(\mathbf{x})$ and $\psi_2(\mathbf{x})$ associated with the same physical system propagating in some region, their superposition $\psi = \psi_1 + \psi_2$ is also a solution of the Schrödinger equation (we assume that $\psi_1 \neq a\psi_2$, $a =$ constant). Its amplitude is given by

$$R^2(\mathbf{x}) = R_1^2 + R_2^2 + 2R_1R_2\cos[(S_1 - S_2)/\hbar]. \tag{3.3.10}$$

The third term on the right hand side characterizes *interference* between the component waves – its sign is variable depending on the phase difference and this implies that R may be greater or less than the amplitudes R_1 and R_2. This term is finite only where the component waves appreciably overlap in space. In regions where they do not we have $R_1R_2 \approx 0$ and hence

$$R^2 \approx R_1^2 + R_2^2. \tag{3.3.11}$$

Similarly, we find interference terms appearing in the momentum field determined by ψ. We have for the phase of ψ

$$\tan(S/\hbar) = \frac{R_1 \sin(S_1/\hbar) + R_2 \sin(S_2//\hbar)}{R_1 \cos(S_1/\hbar) + R_2 \cos(S_2/\hbar)}, \tag{3.3.12}$$

from which it is easy to show that

$$\begin{aligned} \nabla S = R^{-2}\{R_1^2\nabla S_1 + R_2^2\nabla S_2 + R_1R_2 \cos[(S_1 - S_2)/\hbar]\nabla(S_1 + S_2) \\ - \hbar[R_1\nabla R_2 - R_2\nabla R_1] \sin[(S_1 - S_2)/\hbar]\}. \end{aligned} \tag{3.3.13}$$

Eq. (3.3.13) implies that the motion of the particle in the region of overlap is qualitatively distinct from that generated by either of the component waves. It is not simply a kind of 'linear superposition' of the motions generated by the partial waves, a feature which appears in the simplest cases (e.g., §4.3).

If R_1 and R_2 do not overlap appreciably, $R_1R_2 \approx 0$ and (3.3.13) reduces to

$$\nabla S \approx (R_1^2 + R_2^2)^{-1}(R_1^2\nabla S_1 + R_2^2\nabla S_2). \tag{3.3.14}$$

Then $\nabla S \approx \nabla S_1$ or ∇S_2 depending on the region of space under consideration.

More generally we can superpose any number of waves with arbitrary constant complex coefficients

$$\psi = \sum_i c_i\psi_i \tag{3.3.15}$$

to form a new admissible solution. As a mathematical converse to this physical superposition, it holds that any (suitably well-behaved) ψ can be expanded in terms of the eigenfunctions of any Hermitian operator. It should be borne in mind that the physical wave is ψ itself and that the particular decomposition we may use to express it has only a mathematical significance.

We emphasize the difference between forming a *product* of wavefunctions (which implies the physical independence of the associated motions (§3.3.5)) and taking their *sum* (which implies interference and new effects if the summands overlap).

3.3.11 Generalization of the de Broglie relations

The de Broglie relations

$$\mathbf{p} = \hbar\mathbf{k} \qquad \text{and} \qquad E = \hbar\omega \tag{3.3.16}$$

played a pivotal role in the development of quantum mechanics. Here $\mathbf{p}$ and E are the momentum and energy of a corpuscle associated with a vibration of wave vector $\mathbf{k}$ and frequency $\omega = \hbar\mathbf{k}^2/2m$. These relations are essentially connected with a plane wave whose phase satisfies the classical Hamilton–Jacobi equation (§4.2). In the case of an external potential the de Broglie wavelength is conventionally defined as

$$\lambda = 2\pi/|\mathbf{k}| = h/|\mathbf{p}| = h/[2m(E - V)]^{1/2}. \tag{3.3.17}$$

Here $\mathbf{p}$ is the momentum of a classical particle in the external force and hence λ is associated with a *classical* Hamilton–Jacobi wave. Curiously, the true generalization of de Broglie's basic formulae (3.3.16) has never been written down, presumably because it requires a notion of momentum that is valid for all quantum states. Actually, the expression (3.3.17) which employs a variable momentum already goes beyond what is meaningful in the usual interpretation. But in the case of a general Schrödinger wavefunction this definition has no particular significance and one should invoke the causal definition of momentum.

Now, the classical definition of wavelength, which in the general case we shall denote λ_Q, is the distance over which the phase advances by $2\pi\hbar$:

$$S(\mathbf{x} + \lambda_Q \mathbf{n}, t) - S(\mathbf{x}, t) = 2\pi\hbar \tag{3.3.18}$$

where $\mathbf{n} = \nabla S/|\nabla S|$ is a unit vector. Taylor expanding and assuming that λ_Q is small, or $\nabla^n S = 0$ for all $n > 1$, (3.3.18) implies

$$\lambda_Q(\mathbf{x}, t) = h/|\nabla S(\mathbf{x}, t)|. \tag{3.3.19}$$

We shall adopt (3.3.19) as the definition of the local or 'actual' wavelength, even when the wave is not periodic. We shall see later in concrete examples that it is a consistent definition. The subscript Q emphasizes that S is the quantal phase ((3.3.19) holds outside nodes) so that

$$\lambda_Q(\mathbf{x}, t) = h/[2m(-\partial S/\partial t - Q - V)]^{1/2}. \tag{3.3.20}$$

It will be noted that the formula (3.3.19) is not only variable in spacetime but is *state-dependent*. It is not something associated with a particle independently of the context, unlike (3.3.17). Its principal new feature is to take into account the existence of the quantum potential energy.

Similarly, we define the frequency of the wave at a point to be

$$\nu_Q(\mathbf{x}, t) = -(1/h)\,\partial S(\mathbf{x}, t)/\partial t. \tag{3.3.21}$$

These quantities may be evaluated along a particle trajectory. For example, from (3.3.19) the wavelength will be

$$\lambda_Q(\mathbf{x}(t), t) = h/m|\dot{\mathbf{x}}|. \tag{3.3.22}$$

3.4 Some properties of the quantum potential

3.4.1 Dependence on form rather than amplitude

The quantum potential, being constructed from the wavefunction, obviously possesses many of the properties of ψ set out in §3.3. We shall now look at some further properties of the model which relate specifically to qualitative features of the quantum potential.

In keeping with its usual definition, we define the *intensity* I of the wave ψ to be proportional to the square of the amplitude: $I = R^2$. If we multiply the amplitude by a real constant a (to yield a new solution to the Schrödinger equation) we see that I is scaled but that the quantum potential is left unchanged:

$$Q = \frac{-\hbar^2}{2m} \frac{\nabla^2(aR)}{aR} = \frac{-\hbar^2}{2m} \frac{\nabla^2 R}{R}. \tag{3.4.1}$$

Thus, contrary to what one might expect in a classical wave, a particle does not respond to the intensity of the wave in its vicinity, but rather to its *form* (Bohm & Hiley, 1987).

An important consequence of this is that the boundary condition $\psi \to 0$ as $r \to \infty$, which we impose on any physically reasonable wave, does not necessarily imply that the quantum potential becomes physically ineffective in this region. Indeed the latter may have a complicated fluctuating behaviour at large distances, something which is particularly evident in the interference phenomena described in Chap. 5. Another consequence is that we can multiply ψ by a normalization factor without altering its physical effects.

Of course, the notion of a force which is independent of the magnitude of an amplitude is not entirely foreign to classical physics, in certain special cases. In §3.10 we discuss a hydrodynamical analogy to the quantum theory of motion and write down Euler's equation of motion for a fluid of density ρ and velocity $\mathbf{v}$, (3.10.3). There the force on a particle of mass m is proportional to $-(1/\rho)\nabla p$ where p is the pressure. If we assume that $p \propto \rho$ then the force is proportional to $\nabla(\log \rho)$ and so depends on the form of ρ rather than on its magnitude. However, in the case of the quantum potential this feature is generic rather than a special case. Q is invariant under other transformations of the wave, as shown in §3.11.

3.4.2 Comparison with classical potentials

The essential feature of a classical external potential is that it remains aloof from the process it influences. It is therefore a preassigned function of the coordinates. This limited conception of a potential has to be generalized when one comes to quantum mechanics. The quantum potential is derived from the total quantum state and thus an infinite number of different forms (as a function of $\mathbf{x}$ and t) associated with the same physical problem (i.e., a given external potential V) may be generated at will by linearly superposing ψ-solutions.

While the quantum potential is not a preassigned function of the coordinates, it nevertheless does have certain universal characteristics, such as that outlined in §3.4.1, but these features are essentially nonclassical.

The total effective potential, $(V + Q)$, is generally a considerably more complicated function than the simple potentials of classical physics. This should not be considered a defect, however; general relativity introduced new types of potentials satisfying nonlinear equations of much greater complexity than the Schrödinger equation.

Finally, Q depends on the mass etc. of the particle, a property it has in common with classical potential energies. The difference lies in the functional complexity of the dependence.

3.4.3 Superposition

It was pointed out in §3.3.10 that the particle motion associated with a superposition of two waves is very different from that implied by the component waves individually. This feature is reflected in the form of the quantum potential, derived from the total wavefunction, which is a very complicated function of the partial amplitudes and the phase difference and their derivatives. In the case where the waves do not appreciably overlap, however, the effective quantum potential may be simply expressed as

$$-\frac{\hbar^2}{2m}\frac{\nabla^2 R}{R} = -\frac{\hbar^2}{2m}\frac{\nabla^2 R_1}{R_1} - \frac{\hbar^2}{2m}\frac{\nabla^2 R_2}{R_2} \tag{3.4.2}$$

i.e., a simple superposition of the two spatially nonoverlapping partial quantum potentials. This is, of course, to be expected from our considerations in §3.3.10.

An important property is that the addition of a small amplitude wave to ψ, perhaps excited by a weak external perturbation, can trigger a big change in Q and imply the appearance of instabilities in the particle dynamics.

3.4.4 Radar wave analogy

It has been proposed that one should interpret Q as an 'information potential' (Bohm & Hiley, 1984). This means that particles move under their 'own' energy, but are guided by Q, much as a ship on automatic pilot may be steered under the influence of radar waves of considerably less energy than the ship's power source. This analogy is valuable in bringing out certain novel features of the quantum potential model. There are indeed examples where Q is numerically very small in relation to the particle kinetic energy (cf. §5.1). It is also true that a classically-free particle may have zero quantum potential energy, and that in time-dependent problems changes in Q are not generally balanced by corresponding changes in the kinetic energy (cf. §3.9), so it is appropriate to speak of the particle as moving under its 'own' energy.

With regard to this analogy it should be emphasized that while the quantum field does not push on the particle as we might expect a classical wave to, it does nevertheless guide the particle by exerting a direct force on it via Q. The particle responds to more subtle features of ψ than the intensity. The notion of force is as important as the energy concept in the causal model; it explains, for instance, the channelling of trajectories to yield interference patterns (§5.1) and the stability of atoms where the quantum force can bring an electron to rest (this is an example where $Q \to 0$ as $r \to \infty$ – see §4.5). Naturally the word 'potential' already encompasses to some extent the notion of an entity spread out in space that encodes at each point information on the whole which can become active if a body able to respond to the force is placed there. Moreover, the quantum potential energy is not generally small in comparison with the kinetic and classical potential energies, it depends on particle parameters, and it does not emanate from the corpuscle.

3.5 The relation between particle properties and quantum mechanical operators

We have interpreted eq. (3.2.6) in two ways: as one half of a pair of field equations, and as an expression for the total energy $-\partial S/\partial t$ of a particle with kinetic energy $(\nabla S)^2/2m$ and potential energy $(V + Q)$. In doing this we have associated with the particle properties, such as momentum and energy, whose mathematical form is given by certain combinations of the fields R, S and their derivatives. These expressions are, of course, closely connected with the corresponding classical expressions in the Hamilton–Jacobi theory, but here we wish to show the close connection with the quantum mechanical operators customarily associated with these physically relevant quantities. It is this close association which, in fact, guarantees that the causal interpretation implies the experimental results predicted by quantum mechanics (see §3.7) and gives us a physical interpretation of expectation values (see §3.8).

Consider an Hermitian operator $\hat{A}$ which is a function of the operators $\hat{\mathbf{x}}$ and $\hat{\mathbf{p}}$: $\hat{A} = \hat{A}(\hat{\mathbf{x}}, \hat{\mathbf{p}})$. In the position representation the quantum mechanical expectation value of this operator in the normalized state $\psi(\mathbf{x}, t) = \langle \mathbf{x} | \psi(t) \rangle$ is given by

$$\langle \hat{A} \rangle = \langle \psi | \hat{A} | \psi \rangle = \frac{\int \psi^*(\mathbf{x})[\hat{A}(\hat{\mathbf{x}}, -i\hbar\nabla)\psi](\mathbf{x})\, \mathrm{d}^3x}{\int \psi^*(\mathbf{x})\psi(\mathbf{x})\, \mathrm{d}^3x} \tag{3.5.1}$$

where

$$(\hat{A}\psi)(\mathbf{x}) = \int \hat{A}(\mathbf{x}, \mathbf{x}')\psi(\mathbf{x}')\, \mathrm{d}^3x' \tag{3.5.2}$$

and, although the state is normalized, we have explicitly included the denominator. The hermiticity of $\hat{A}$ implies that only the real part of the integrand contributes to (3.5.1) and we may write without loss of generality

$$\langle \hat{A} \rangle = \frac{\int \mathrm{Re}\, \psi^*(\mathbf{x})(\hat{A}\psi)(\mathbf{x})\, \mathrm{d}^3x}{\int \psi^*(\mathbf{x})\psi(\mathbf{x})\, \mathrm{d}^3x}. \tag{3.5.3}$$

It is then reasonable to call the following expression the 'local expectation value' of the operator $\hat{A}$ in the state $|\psi\rangle$ in the position representation:

$$A(\mathbf{x}, t) = \mathrm{Re}\, \psi^*(\mathbf{x}, t)(\hat{A}\psi)(\mathbf{x}, t)/\psi^*(\mathbf{x}, t)\psi(\mathbf{x}, t), \tag{3.5.4}$$

where we should insert (3.5.2).† As is evident, (3.5.4) is a field function of $\mathbf{x}$ and t and combines information about the operator and the wavefunction. We have introduced $|\psi|^2$ in the denominator in order to turn a density into a function having the dimension of the operator. Then (3.5.4) may be interpreted as a property of the *particle*, or rather an ensemble of particles, as we shall now see by example.

(i) Position

To begin with, for the position operator in the position representation $\hat{\mathbf{x}}(\mathbf{x}, \mathbf{x}') = \mathbf{x}\, \delta(\mathbf{x} - \mathbf{x}')$ and (3.5.4) reduces to just the spatial argument of the wavefunction, for all states:

$$\mathbf{x} = \psi^*\mathbf{x}\psi/\psi^*\psi. \tag{3.5.5}$$

Evaluating this along a trajectory ($\mathbf{x} = \mathbf{x}(t)$) gives the position of the particle;

† The imaginary part of the integrand is not without interest, in particular with regard to stochastic processes (Nelson, 1966).

By insertion of suitable δ-functions, (3.5.4) may be expressed as a global expectation value of the type (3.5.1).

the local expectation value of the position operator for any state evaluated along a trajectory is just that trajectory. An ensemble of trajectories is obtained by varying the initial point $\mathbf{x}_0$. This result readily extends to any multiplicative function of $\mathbf{x}$.

(ii) Linear momentum

Inserting $\hat{\mathbf{p}} = -i\hbar\nabla_{\mathbf{x}}\,\delta(\mathbf{x}-\mathbf{x}')$ for $\hat{A}(\mathbf{x},\mathbf{x}')$ in (3.5.4) yields

$$\begin{aligned}\mathbf{p}(\mathbf{x},t) &= \operatorname{Re}\psi^*(-i\hbar\nabla)\psi/|\psi|^2\\ &= (\hbar/2mi|\psi|^2)[\psi^*\nabla\psi - (\nabla\psi^*)\psi] \qquad (3.5.6)\\ &= \nabla S \qquad (3.5.7)\end{aligned}$$

(writing $\psi = R\,e^{iS/\hbar}$), i.e., the momentum field of an ensemble of particles. Choosing $\mathbf{x}_0$, *the local expectation value of the momentum operator evaluated along a trajectory is the particle momentum,* $m\dot{\mathbf{x}}(t)$. It has the same form as in classical mechanics.

(iii) Total energy

Consider first a classically-free particle ($V = 0$). Then the Hamiltonian operator $\hat{H} = \hat{\mathbf{p}}^2/2m$ and we have by a simple calculation

$$\begin{aligned}E(\mathbf{x},t) &= \operatorname{Re}\psi^*[-(\hbar^2/2m)\nabla^2\psi]/|\psi|^2\\ &= (\nabla S)^2/2m - \hbar^2\nabla^2 R/2mR \qquad (3.5.8)\\ &= \mathbf{p}^2/2m + (1/R)(\hat{\mathbf{p}}^2/2m)R. \qquad (3.5.9)\end{aligned}$$

Evaluated along a trajectory, the first term on the right hand side of (3.5.8) is the kinetic energy and the second term is the quantum potential energy (the form (3.5.9) will be used below). The local expectation value of $\hat{H}$ is therefore the total energy of the particle as we defined it in (3.2.20). Notice that we do not obtain just the kinetic energy as we would for a free classical particle – the physical significance of the definition (3.5.1) of average for an energy *operator* is that it introduces an additional (quantum potential) energy.

If now we include the external potential in $\hat{H}$ we easily obtain

$$\begin{aligned}E(\mathbf{x},t) &= \operatorname{Re}\psi^*[-(\hbar^2/2m)\nabla^2 + V]\psi/|\psi|^2\\ &= (\nabla S)^2/2m + Q + V. \qquad (3.5.10)\end{aligned}$$

The local expectation value of the Hamiltonian operator evaluated along a trajectory is the total particle energy, $\tfrac{1}{2}m\dot{\mathbf{x}}^2 + V + Q$. It has a similar form to the classical expression, apart from the addition of Q.

(iv) *Orbital angular momentum*

The orbital angular momentum operator about the origin is defined by $\hat{\mathbf{L}} = \hat{\mathbf{x}} \times \hat{\mathbf{p}}$. In the position representation we obtain from (3.5.4)

$$\begin{aligned} \mathbf{L}(\mathbf{x}, t) &= \operatorname{Re} \psi^*[\mathbf{x} \times (-i\hbar\nabla)]\psi/|\psi|^2 \\ &= \mathbf{x} \times \nabla S, \end{aligned} \tag{3.5.11}$$

which coincides with the expression (3.2.30). *The local expectation value of the orbital angular momentum operator evaluated along a trajectory is the particle orbital angular momentum,* $\mathbf{x} \times m\dot{\mathbf{x}}$. This is the classical expression.

(v) *Total orbital angular momentum*

By now it should be apparent that the expression (3.5.4) has the general significance that it generates from the wavefunction expressions for physically meaningful properties of an ensemble of particles, in a nonarbitrary way. These expressions agree with what we would expect from our general kinematical and dynamical theory and are defined for arbitrary quantum states (and not just for eigenstates of the relevant operator). What this suggests is that we should adopt (3.5.4) as a generally valid *definition* of the physical properties of a particle associated with a given operator and state. In this way we bring out in a visualizable way the physical significance of operators via their effects on states. We shall call the local expectation values the 'actual properties' of a particle.

Let us apply this method to find an expression for the total orbital angular momentum of a classically free system ($V = 0$). We have not obtained this before, but we expect that it will contain a term additional to the square of (3.5.11) just as the energy (3.5.8) comprises the quantum potential. Denoting the local expectation value of $\hat{\mathbf{L}}^2$ by $l(\mathbf{x}, t)$, we have

$$\begin{aligned} l(\mathbf{x}, t) &= \operatorname{Re} \psi^*[\mathbf{x} \times (-i\hbar\nabla)]^2\psi/|\psi|^2 \\ &= (\mathbf{x} \times \nabla S)^2 - \hbar^2(\mathbf{x} \times \nabla)^2 R/R && (3.5.12) \\ &= (\mathbf{x} \times \mathbf{p})^2 + (1/R)\hat{\mathbf{L}}^2 R && (3.5.13) \end{aligned}$$

(the calculation is given in Appendix B). The first term on the right hand side of (3.5.12) represents the total orbital angular momentum expected on classical grounds in the absence of an external torque (rotational analogue of the kinetic energy of translation) and the second term is a quantum mechanical addition (rotational analogue of the quantum potential energy for translational motion). The form of this latter term as given in (3.5.13) should be compared with (3.5.9): in each case the quantum mechanical addition comes about from the application of the operator under consideration to the amplitude function. Notice also the characteristic appearance of R in the denominator.

In principle, one could extend the results given above, which deal with the 'natural' physical properties of particles, to any operator combination of $\hat{\mathbf{x}}$ and $\hat{\mathbf{p}}$. That is, if some such operator combinations were found to have physical significance, we could define an associated property of the particle via definition (3.5.4). More generally we could do this for arbitrary operators, including those that cannot be expressed as $\hat{A}(\mathbf{x})\delta(\mathbf{x} - \mathbf{x}')$ in the position representation. In this book we shall not discuss any operator functions of $\hat{\mathbf{x}}$ and $\hat{\mathbf{p}}$ other than those above, but in §3.12.2 we show how (3.5.4) forms the basis of a general method by which one can generate 'causal interpretations' of other quantum theories.

3.6 Introduction of probability

3.6.1 Ensemble of identical systems

An individual physical system comprises a field and a particle. The laws governing the evolution of the variables intrinsically connected with these entities are entirely determinist in character – if we give their values at a particular instant ($\psi_0(\mathbf{x})$ and $\mathbf{x}_0$), their subsequent (and previous) values are uniquely determined for all time. These initial values are freely specifiable so long as they satisfy the conditions set out in §3.2 (ψ_0 is bounded, continuous etc. and $\mathbf{x}_0$ may not be chosen to lie in nodal regions). Now for practical purposes the notion of 'giving' a precisely defined coordinate or field distribution is an idealization which can never be realized operationally. The grossness of our instruments, the presence of random background noise, and the intervention of many other outside influences, imply that we will never know *precisely* the initial conditions, even though we may conceive of these as being well defined in fact. Thus, in the world of experience, we have to deal with a range of possible initial values of the relevant physical properties of our system.

To treat this problem we introduce the notion of an ensemble of identical systems which differ from one another only insofar as each is associated with a different choice of the initial conditions potentially available to the actual system of interest, in accordance with our knowledge of these conditions. Bearing in mind our definition of an individual physical system, we have to consider an ensemble of both particles *and waves* (i.e., each element of the ensemble comprises a particle and a wave). By 'identical systems' we mean then that the waves and particles are, respectively, physically indistinguishable (apart from the freedom in the initial conditions already referred to) in the sense that each wave obeys the same Schrödinger equation (same potential, boundary conditions etc.), and each particle satisfies the same equation of motion

$$\dot{\mathbf{x}}_i = (1/m)\nabla S_i|_{\mathbf{x}_i = \mathbf{x}_i(t)}, \tag{3.6.1}$$

where S_i is the phase of the ith wave in the ensemble.

We can characterize the ensemble by introducing a function which determines the number of systems which lie in each of the available states. This will reflect the extent of our knowledge concerning the actual system of interest. To this end we suppose that $\psi_0(\mathbf{x})$ is one of a discrete set of states $\psi_i(\mathbf{x}), i = 1, 2, \ldots$, and we assume the existence of a function $P_0(\mathbf{x}, i)$ satisfying the condition

$$P_0(\mathbf{x}, i) \geqslant 0, \qquad \forall \mathbf{x}, i. \tag{3.6.2}$$

The number of systems in the ensemble whose initial wavefunction is given by $\psi_i(\mathbf{x})$ and whose initial particle position lies in a volume d^3x around the point $\mathbf{x}$ is given by $P_0(\mathbf{x}, i)\,\mathrm{d}^3x$. The function $P_0(\mathbf{x}, i)$ may thus be thought of as a density function in space for each i. We further require that the number function be well defined and satisfy (3.6.2) for all time: $P(\mathbf{x}, i, t) \geqslant 0$. Its evolution will be assumed to be determined by the laws of motion of each member of the ensemble, which separately evolve into $\psi_i(\mathbf{x}, t)$ and $\mathbf{x}_i = \mathbf{x}_i(t)$ for each i (as in classical statistical mechanics).

Each member of the ensemble is representative of the possible behaviour of the system of interest, but the ensemble is 'fictitious' in the sense that we conceive of all the elements being present at the same time, each independently evolving according to the deterministic laws of motion, even though, in fact, we only ever have one system present. In practice we wish to apply the notion of ensemble to a sequence (in time) of processes in each of which one of the set of identical systems of interest takes on an arbitrary but definite choice of the available initial conditions (including the possibility that the same initial conditions may be realized more than once in a succession of trials). Then the expression $P_0(\mathbf{x}, i)\,\mathrm{d}^3x$ will be proportional to the probability that at time $t = 0$ an *individual* system will have its wavefunction given by $\psi_i(\mathbf{x})$ and its particle lies in a volume d^3x around the point $\mathbf{x}$. It is natural then to require that the function P_0 be normalized to unity in order that it represents a genuine probability function:

$$\sum_i \int P_0(\mathbf{x}, i)\,\mathrm{d}^3x = 1. \tag{3.6.3}$$

It is required that (3.6.3) be invariant with respect to the equation of evolution of $P(\mathbf{x}, i, t)$. The expression $P(\mathbf{x}, i, t)\,\mathrm{d}^3x$ will then give the probability of the state of an individual system at time t.

Probability is interpreted here as the relative frequency of occurrence of states in a large number of trials (e.g., a beam of noninteracting electrons). The processes under consideration are any that the system of interest may normally partake in and are not restricted to that class termed 'measurements'. Thus the probability relates to the state that the system is actually *in* and not just to what we would find if we performed a measurement to determine that state (although it applies to the latter as well – see Chap. 8).

The ensemble average of some physical property evaluated at a definite time may be identified with the average property obtained over a succession of trials (see §3.8). Notice that the average does not always give an indication of the likely behaviour of an individual system.

We shall now arrive at the particular instance of this statistical theory that corresponds to quantum mechanics by a series of steps involving ever more restrictive assumptions which culminate in the $|\psi|^2$-distribution of particles and the density matrix distribution of waves.

To begin with, we assume that the particle and wave variables are uncorrelated, i.e., independently distributed. Then the joint probability function $P(\mathbf{x}, i)$ factorizes:†

$$P(\mathbf{x}, i) = p_i P(\mathbf{x}), \tag{3.6.4}$$

where $P(\mathbf{x}) \geqslant 0$, $0 \leqslant p_i \leqslant 1$, and $\sum_i p_i = \int P(x)\,\mathrm{d}^3x = 1$. Whichever wave $\psi_i(\mathbf{x})$ is actually realized (with probability p_i), the particle coordinates are distributed with a probability density $P(\mathbf{x})$ determined by all the waves in the ensemble.

3.6.2 *Ensemble of particles*

Let us restrict attention to the case where we have maximal knowledge of the state of the wave part of the system, but only partial knowledge of the particle position. Thus the initial form of the wave $\psi_0(\mathbf{x})$ is fixed and all that varies in a sequence of trials is the initial particle position, distributed according to $P_0(\mathbf{x})$. The latter is assumed to satisfy the same boundary conditions as the function R (§3.2.1). Since all the particles in the single-valued ensemble are associated with the same wave for all time, it is natural to require that $P(\mathbf{x}, t)$ satisfies an equation of continuity, as in the

† The factorizable case reflects the physical independence of the wave from the particle, i.e., that although the wave guides the particle there is no reciprocal action. If we incorporated the latter in the dynamical theory then we should have to consider more general, nonfactorizable, distribution functions.

Hamilton–Jacobi theory of classical statistical mechanics (§2.5):

$$\partial P/\partial t + \nabla\cdot(P\nabla S/m) = 0 \tag{3.6.5}$$

where S is the phase of the wave ψ. Given $P_0(\mathbf{x})$ and $S(\mathbf{x}, t)$, the distribution $P(\mathbf{x}, t)$ is uniquely fixed by (3.6.5) for all time. The choice of P_0 must be consistent with the dynamical condition that particles cannot lie at nodal points of ψ_0 because ∇S_0 is undefined, and hence it must assign zero probability to these points.

By Green's theorem we have

$$\begin{aligned}(\mathrm{d}/\mathrm{d}t)\int_{\text{all space}} P(\mathbf{x}, t)\,\mathrm{d}^3x &= -\int \nabla\cdot(P\nabla S/m)\,\mathrm{d}^3x\\ &= -\int (P\nabla S/m)\cdot\mathrm{d}\mathbf{A}\\ &= 0,\end{aligned} \tag{3.6.6}$$

since the surface integral at infinity vanishes due to the boundary conditions on P. It follows that the normalization condition

$$\int P(\mathbf{x}, t)\,\mathrm{d}^3x = 1 \tag{3.6.7}$$

is preserved for all t if it is true at one instant. The result (3.6.6) may be interpreted as stating that the total number of particles in the ensemble remains constant in the course of time; there is no 'creation' or 'annihilation' of trajectories (they never come to an end).

Notice that in this theory we have a *joint probability density for position and momentum variables.* Since the initial momentum is uniquely fixed by $S_0(\mathbf{x})$ at each point, we have the initial phase space distribution

$$f_0(\mathbf{x}, \mathbf{p}) = P_0(\mathbf{x})\,\delta[\mathbf{p} - \nabla S_0(\mathbf{x})] \tag{3.6.8}$$

and at time t

$$f(\mathbf{x}, \mathbf{p}, t) = P(\mathbf{x}, t)\,\delta[\mathbf{p} - \nabla S(\mathbf{x}, t)]. \tag{3.6.9}$$

Condition (3.6.7) becomes

$$\int f(\mathbf{x}, \mathbf{p}, t)\,\mathrm{d}^3x\,\mathrm{d}^3p = 1 \tag{3.6.10}$$

and the marginal distributions are given by

$$\int f(\mathbf{x}, \mathbf{p}, t)\,\mathrm{d}^3p = P(\mathbf{x}, t) \tag{3.6.11}$$

which is the position probability density, and

$$\int f(\mathbf{x}, \mathbf{p}, t)\, \mathrm{d}^3x = g(\mathbf{p}, t) \tag{3.6.12}$$

which is the probability density of momentum. The latter clearly satisfies $g \geqslant 0$ and from (3.6.10)

$$\int g(\mathbf{p}, t)\, \mathrm{d}^3p = 1. \tag{3.6.13}$$

Since $\mathbf{p}$ is functionally related to $\mathbf{x}$ (via ∇S), the probability distribution in momentum is a consequence of that in $\mathbf{x}$.

3.6.3 A special assumption

Apart from natural assumptions of smoothness, normalizability and that it be zero in nodal regions, the function $P_0(\mathbf{x})$ is freely specifiable. Its introduction represents our ignorance of the precise initial state of the particle and in no way impinges on the underlying dynamical process in which the particle is guided by the ψ-field. We now make a particular choice of $P_0(\mathbf{x})$, which is the one that characterizes quantum mechanics.

Postulate. Assume that the square of the initial field function $R_0(\mathbf{x})$ is normalizable:

$$\int R_0^2(\mathbf{x})\, \mathrm{d}^3x = 1. \tag{3.6.14}$$

Then the probability that a particle in an ensemble associated with the same wave ψ_0 is in a volume d^3x around the point $\mathbf{x}$ is given by $\mathrm{R}_0^2(\mathbf{x})\, \mathrm{d}^3x$. That is,

$$P_0(\mathbf{x}) = R_0^2(\mathbf{x}). \tag{3.6.15}$$

Since the function $R^2(\mathbf{x}, t)$ satisfies the law of motion (3.2.7),

$$\frac{\partial R^2}{\partial t} + \nabla \cdot \left(\frac{R^2 \nabla S}{m} \right) = 0, \tag{3.6.16}$$

it follows from our remarks in §3.6.2 that $R^2(\mathbf{x}, t)$ determines the probability distribution at time t if (3.6.15) holds at $t = 0$ (i.e., $P(\mathbf{x}, t) = R^2(\mathbf{x}, t)$ and R^2 is normalized for all t if (3.6.14) holds).

Now there is nothing in the theory of motion of individual systems described in §3.2 that would completely justify this postulate, although features of it certainly provide support for such an assumption. The fact that $R^2 \geqslant 0$, satisfies a conservation equation and may be normalized, and that particles cannot pass through nodes (where $R = 0$), are all conditions we would expect in a theory of probability. But on their own they do not uniquely imply such an interpretation. If outside the nodes of ψ_0 the initial distribution of particles $P_0(\mathbf{x}) \neq R_0^2(\mathbf{x})$ then it is easy to see that $P(\mathbf{x}, t) \neq R(\mathbf{x}, t)$ outside nodes for all t and hence R^2 will never represent the particle distribution. Eq. (3.6.15) is an additional postulate which is consistent since, of all possible nonnegative, normalizable and conserved functions $P_0(\mathbf{x})$ that we could choose, we can certainly choose $R_0^2(\mathbf{x})$. Such a choice of particle distribution imposes on the function $R(\mathbf{x}, t)$ an additional role, although this in no way introduces an inconsistency into the logical structure of the theory (i.e., in the relation between the underlying dynamical theory and the introduction of probabilistic concepts into both of which R enters); since R is a component of an individual system, it follows that the ensemble depends on the individual, and not the other way round (this answers an objection raised by Wigner (1983a)).

The circumstance that a single function should be assigned more than one meaning is hardly new in physics. An electromagnetic wave influences the motion of a charged particle placed in it, and its amplitude determines the intensity of the wave as a function of position. These are logically distinct properties since *a priori* the particle need not respond to the intensity of the wave, but may be affected by some other combination of field components. What is novel here is the type of properties R has; it describes both the *actual* situation and our *knowledge* of that situation. The dual role of R implies that expressions into which it enters, e.g., averages, will have two distinct meanings – one to do with an individual wave and one to do with an ensemble of particles (see §3.8). We emphasize that the probability interpretation of R^2 is merely consistent with and not derivable from the theory of motion of §3.2.

The relations (3.6.8)–(3.6.13) for the phase space probability density of position and momentum remain valid with P replaced by R^2. Thus the joint distribution function is given by

$$f(\mathbf{x}, \mathbf{p}, t) = R^2(\mathbf{x}, t)\, \delta[\mathbf{p} - \nabla S(\mathbf{x}, t)], \tag{3.6.17}$$

which is normalized

$$\int f(\mathbf{x}, \mathbf{p}, t)\, \mathrm{d}^3x\, \mathrm{d}^3p = 1, \tag{3.6.18}$$

and the marginal distributions are given by

$$\int f(\mathbf{x}, \mathbf{p}, t)\, \mathrm{d}^3 p = R^2(\mathbf{x}, t) \tag{3.6.19}$$

$$\int f(\mathbf{x}, \mathbf{p}, t)\, \mathrm{d}^3 x = g(\mathbf{p}, t) \tag{3.6.20}$$

with

$$\int R^2(\mathbf{x}, t)\, \mathrm{d}^3 x = \int g(\mathbf{p}, t)\, \mathrm{d}^3 p = 1. \tag{3.6.21}$$

The functional dependence of momentum on position means that our lack of knowledge of the actual position of a particle implies a corresponding lack of knowledge of the actual momentum. The conditional probability of $\mathbf{p}$ given $\mathbf{x}$ is $\delta(\mathbf{p} - \nabla S)$.

It follows from the field equations (3.2.7) and (3.2.23) that the distribution function (3.6.17) obeys the law of evolution

$$\frac{\partial f}{\partial t} + \frac{\mathbf{p}}{m} \cdot \nabla f + \nabla(V + Q) \cdot \nabla_{\mathbf{p}} f = 0. \tag{3.6.22}$$

This should be compared with the classical Liouville equation (§2.5) to which it reduces when $Q = 0$. Unlike the latter, (3.6.22) is a nonlinear equation in that Q and f both involve R. It should also be compared with the Wigner equation (§8.4.3).

Because all physical quantities are functions of $\mathbf{x}$, their probability distributions are induced by $|\psi|^2$ via formulae such as (3.6.20). For example, the partial distributions of angular momentum and energy are given by

$$g(\mathbf{L}) = \int R^2(\mathbf{x})\, \delta(\mathbf{L} - \mathbf{x} \times \nabla S)\, \mathrm{d}^3 x, \tag{3.6.23}$$

$$g(E) = \int R^2(\mathbf{x})\, \delta(\mathrm{E} + \partial S/\partial t)\, \mathrm{d}^3 x. \tag{3.6.24}$$

The distribution function (3.6.4) for the wave-particle system becomes now

$$P(\mathbf{x}, i) = p_i R_i^2(\mathbf{x}) \tag{3.6.25}$$

which expresses the dependence of the particle distribution on the wave which is actually realized.

To summarize, we have developed a determinist theory of particle motion based on the Schrödinger equation. Our complete lack of reference to probability hitherto indicates that the wave equation is by no means necessarily tied to a statistical interpretation. We can nevertheless impose a subsidiary condition on the theory of motion in which certain initial positions are favoured over others. This introduction of probability is no more intrinsic to the basic theory of motion than it is in classical mechanics, but is postulated for practical reasons. As to why the probability density should be given by R^2 and not some other function is at this stage left unexplained. In this regard we are in the same position as the conventional approach where $|\psi|^2$ is postulated to represent a probability density for no theoretical reason at all – it is justified *a posteriori* by comparison with experiment. Since probability is not intrinsic to the basic theory, an advantage of our approach is that we are able to raise the question of whether an explanation for $|\psi|^2$ may emerge at a deeper level, rather than accepting this as an inexplicable property of matter.

3.6.4 Ensemble of waves. The density matrix

We now relax the requirement that the quantum state (wave) be precisely known and consider the case where we only have knowledge of the probability distribution p_i. Such a situation is usually described by a density matrix which may be introduced as follows. Written out explicitly using matrix indices, the expectation value (3.5.1) is

$$\begin{aligned}\langle\hat{A}\rangle = \langle\psi|\hat{A}|\psi\rangle &= \int \psi^*(\mathbf{x})\hat{A}(\mathbf{x}, \mathbf{x}')\psi(\mathbf{x}')\,\mathrm{d}^3x\,\mathrm{d}^3x' \\ &= \int \hat{A}(\mathbf{x}, \mathbf{x}')\rho(\mathbf{x}', \mathbf{x})\,\mathrm{d}^3x\,\mathrm{d}^3x' \\ &= \mathrm{Tr}(\hat{\rho}\hat{A}),\end{aligned} \tag{3.6.26}$$

where

$$\rho(\mathbf{x}, \mathbf{x}') = \psi(\mathbf{x})\psi^*(\mathbf{x}') \tag{3.6.27}$$

is the pure state density matrix, the density operator $\hat{\rho} = |\psi\rangle\langle\psi|$ in the position representation. The diagonal elements give the particle distribution: $\rho(\mathbf{x}, \mathbf{x}) = |\psi(\mathbf{x})|^2$. Now if the system may be potentially in one of a number of states $|\psi_i\rangle$ with probability p_i, $i = 1, 2, \ldots$ (assumed discrete), we can retain the formula (3.6.26) for $\langle\hat{A}\rangle$ if we define the density matrix to be

$$\rho(\mathbf{x}, \mathbf{x}') = \sum_i p_i\psi_i(\mathbf{x})\psi_i^*(\mathbf{x}'). \tag{3.6.28}$$

Then, in this mixed state,

$$\begin{aligned}\langle\hat{A}\rangle &= \mathrm{Tr}(\hat{\rho}\hat{A}) \\ &= \sum_i p_i \int \psi_i^*(\mathbf{x})\hat{A}(\mathbf{x}, \mathbf{x}')\psi_i(\mathbf{x}')\,\mathrm{d}^3x\,\mathrm{d}^3x'.\end{aligned} \tag{3.6.29}$$

Now it is usually claimed, and at first sight it seems reasonable, that the average (3.6.26) over an ensemble of particles is of a fundamentally different character to the average (3.6.28) over an ensemble of waves. The latter is performed in an obviously classical way by summing the various possible states with weights which represent the frequency with which they appear in a sequence of trials. The particle average on the other hand (associated with a given wave) is apparently performed in a highly nonclassical way using operators acting on amplitudes, rather than expressions of the form

$$\langle \hat{A} \rangle = \int R^2(\mathbf{x}) A(\mathbf{x})\, \mathrm{d}^3 x, \tag{3.6.30}$$

where A is a physical variable associated with the operator $\hat{A}$. It turns out, however, that we can express (3.6.26) generally in the classical form (3.6.30), as is shown in §3.8. This means that we can treat the ensemble of waves and particles in an entirely symmetrical manner. Accepting for the present that (3.6.30) is valid, we have, returning to (3.6.29), the following expression for an expectation value:

$$\langle \hat{A} \rangle = \sum_i \int P(\mathbf{x}, i) A(\mathbf{x}, i)\, \mathrm{d}^3 x \tag{3.6.31}$$

where $P(\mathbf{x}, i) = p_i |\psi_i(\mathbf{x})|^2$.

The statistical distribution in ψ is reflected in the particle distribution which is now:

$$\rho(\mathbf{x}, \mathbf{x}) = \sum_i p_i |\psi_i|^2 = \sum_i P(\mathbf{x}, i). \tag{3.6.32}$$

Since the phases are statistically distributed there is an uncertainty in the actual momentum of the particle additional to that induced by the position distribution. The phase space probability distribution (3.6.17) for the particle becomes in the case of a distribution of waves:

$$f(\mathbf{x}, \mathbf{p}) = \sum_i p_i R_i^2(\mathbf{x})\, \delta(\mathbf{p} - \nabla S_i). \tag{3.6.33}$$

Integrating (3.6.33) over $\mathbf{p}$ we recover (3.6.32). Integrating over $\mathbf{x}$ yields the momentum distribution

$$g(\mathbf{p}) = \sum_i p_i g_i(\mathbf{p}) \tag{3.6.34}$$

where $g_i(\mathbf{p})$ is given by (3.6.20).

Each of $\rho(\mathbf{x}, \mathbf{x}')$ and $f(\mathbf{x}, \mathbf{p})$ contains the same information and given one we can construct the other. The density matrix is a more useful entity though since it obeys a linear equation of motion. Notice that a given mixed state $\rho(\mathbf{x}, \mathbf{x}')$ may be decomposed into pure states in an infinite number of ways so we cannot uniquely deduce from it the set of waves ψ_i in the ensemble and their respective weights p_i. A similar remark applies to the decomposition of $f(\mathbf{x}, \mathbf{p})$.

The density matrix formalism is therefore a particular type of *statistical mechanics of waves and rays.* Each element in the ensemble of individual systems comprises a wave ψ_i and an associated particle whose momentum is given by $\mathbf{p}_i = \nabla S_i$. Because ρ represents a fictitious ensemble all the waves may be considered to occupy, simultaneously, overlapping regions of space without interfering. Only one wave and one particle is present in any one trial. The density matrix describes both an 'ensemble of ensembles' of particles, and an ensemble of waves and particles.

3.6.5 Remarks on our notion of probability

The essential difference between our notion of probability and the one usually employed in quantum mechanics is that we are describing the likely state of matter as it actually is, whatever processes it may be part of. Both R^2 and p_i refer to our partial knowledge of the true state of a system which is in itself well defined. In contrast, in the Born interpretation $|\psi|^2$ does not represent our ignorance of an actual state but concerns rather the distribution of values found if one performs a 'measurement'. As we shall see, we recover Born's interpretation as a special case for the particular processes characterized as measurements (§3.7 and Chap. 8).

The notion of probability used here (and in conventional treatments) requires that there exists in the real world an unlimited number of the systems under consideration (electrons, atoms etc.). This is so in order that the concept of a fictitious representative ensemble is equivalent to a sequence of actual individual processes, and hence that the results obtained by averaging over the ensemble are equal to those given by an average over the sequence. At the level of theory and experiment normally treated by quantum mechanics this seems to be a valid assumption – there do indeed appear to exist physical systems such as electrons for which an effectively unlimited number of copies exist.

3.7 Agreement with the results of quantum mechanics

3.7.1 Eigenvalues and probabilities

Henceforth we shall make the assumption introduced in the last section that an ensemble of particles associated with the same ψ-wave is distributed according to $R^2(\mathbf{x}, t)$ in order to ensure that the aggregate of individual particle motions will reproduce the experimental results predicted by quantum mechanics.

The detailed treatment of the process of measurement is given in Chap. 8. Here we point out in general terms some pertinent features, in particular how the actual properties of a particle (position, momentum, ...) are connected with the results of measurements designed to 'measure' the corresponding 'observables'.

According to quantum mechanics the determination of the 'state' of a system, e.g., its momentum, is achieved by a process part of whose outcome is that the system is left in an eigenstate of the associated (momentum) operator (the mutual transformations of other systems with which the system of interest interacts are ignored here). The system may then be said to 'have' a definite momentum (an eigenvalue of the momentum operator) with its conjugate variable, the position, being completely unknown, or randomly fluctuating, or undefined, depending on one's interpretation. Yet this is physically unclear since it is not usually stated what it *is* that has this well-defined attribute. How do the actual values of position, momentum, angular momentum and energy of the quantum theory of motion relate to the eigenvalues of the corresponding operators? The answer is that if the system is left in an eigenstate of the operator under consideration, the actual value coincides with an eigenvalue of that operator. We have, using (3.5.5), (3.5.7), (3.5.10) and (3.5.11),

Position $\mathbf{x} \to \mathbf{x}_0$ when $\psi(\mathbf{x}) \to \delta(\mathbf{x} - \mathbf{x}_0)$ with $\hat{\mathbf{x}}\psi = \mathbf{x}_0\psi$

Momentum $\mathbf{p} = \nabla S \to \hbar\mathbf{k}$ when $\psi(\mathbf{x}) \to \mathrm{e}^{\mathrm{i}[\mathbf{k}\cdot\mathbf{x} - (\hbar \mathbf{k}^2 t/2m)]}$ with $\hat{\mathbf{p}}\psi = \hbar\mathbf{k}\psi$

Energy $-\partial S/\partial t \to E$ when $\psi(\mathbf{x}) \to f(\mathbf{x})\,\mathrm{e}^{-\mathrm{i}Et/\hbar}$ with $\hat{H}\psi = E\psi$

Angular momentum $L_z = x\,\partial_y S - y\,\partial_x S \to m_z\hbar$ when $\psi(\mathbf{x}) \to \mathrm{e}^{\mathrm{i}m_z\phi}$ with $\hat{L}_z\psi = m_z\hbar\psi$

and similarly for the other components of $\mathbf{L}$, where $\phi = \tan^{-1}(y/x)$ is the azimuthal angle. A similar result is obtained for $\hat{\mathbf{L}}^2$.

This establishes the connection between the particle properties of the quantum theory of motion, which are well defined for all quantum states and continuously variable, and the eigenvalues of operators which are usually treated as the only definite properties that a system may possess and constants in spacetime. That ψ may be a superposition of eigenstates represents a new state of matter and not a lack of definition of particle properties.

The key point is that the particle does not acquire these properties only when it undergoes special types of interactions ('measurements') but possesses them throughout its entire history. In this view, the hallowed 'measurements' are really rather ordinary interactions which are typically occurring all the time in real processes, and during which the actual values evolve continuously into an appropriate eigenvalue. The whole process is governed by the Schrödinger equation.

Suppose that prior to the measurement interaction the wavefunction may be written

$$\psi(\mathbf{x}) = \sum_a c_a \psi_a(\mathbf{x}), \tag{3.7.1}$$

where $\psi_a(\mathbf{x})$ are eigenfunctions of the operator $\hat{A}$ to be measured, in the position representation, and c_a are complex constants with a an eigenvalue. The inteference between the summands in (3.7.1) implies that this sum does not express our ignorance of a current state described by one of the partial waves realized with relative probability $|c_a|^2$ in an ensemble – that would be a mixture. The aim of the measurement process is to separate the ψ_as so that they no longer overlap by coupling to a further system. Then the superposition behaves as *if* it were a mixture (it is physically distinguished from a genuine mixture in that the partial waves coexist). For the actual value (3.5.4) we then obtain

$$A(\mathbf{x}) = \mathrm{Re}\, \psi^* \hat{A} \psi / |\psi|^2 = a \tag{3.7.2}$$

in the domain where ψ_a is finite and hence for the probability distribution of the actual values we find

$$g(a) = \int |\psi(\mathbf{x})|^2 \, \delta[a - A(\mathbf{x})] \, \mathrm{d}^3 x = |c_a|^2. \tag{3.7.3}$$

Thus, Born's postulate that $|c_a|^2$ represents the probability that a system in the state ψ will be *found* in the state ψ_a is a special case of the probability that it *is* in a certain state (for the complete theory see Chap. 8).

The particle trajectory is initially determined by ψ and evolves continuously so that it is finally guided by one of the ψ_as.

3.7.2 What are the hidden variables?

The quantum theory of motion has been historically treated as an example of a 'hidden-variables' theory. The reason for this is as follows. If we believe that the most complete specification of the state of an individual system is contained in the wavefunction, that the latter has a purely probabilistic significance, and that these assumptions are sufficient to account for all conceivable experimental results, then any further specification of the state of an individual (such as the position of a particle in an ensemble associated with the same wavefunction) can at best only have theoretical significance and the extra variables must be forever 'hidden'. The problem with this point of view is that what one observes as the results of experiments are localized events; an interference pattern is built up over a period of time by, say, the blackening of definite regions of an emulsion, one after another. The confrontation between this fact and the conventional view on the meaning of ψ leads to the problems and paradoxes of the 'wave packet collapse' hypothesis. But from the point of view of the quantum theory of motion, the localized events reveal the current position $\mathbf{x}(t)$ of the particle (when suitably amplified through other processes). *The observed position is what the actual position evolves into* in the circumstances of a measurement interaction. It is therefore misleading to term $\mathbf{x}(t)$ a 'hidden variable' – on the contrary, it is ψ that is 'hidden' in that we only derive information about it by observing the particle (just as we only know about electromagnetic fields by the effect they have on charged particles). If anything is a 'hidden-variable' interpretation of quantum mechanics, it is the conventional one that only works with ψ!

Since the specification of the initial position implies a unique position at all other times, the quantum theory of motion makes more detailed predictions about the behaviour of a particle than is given by $|\psi|^2$ alone. For example, in two-slit interference (Chap. 5) it predicts that a particle passes through one slit and subsequently moves towards the screen without ever crossing the axis of symmetry of the apparatus. These more detailed predictions concerning an individual process naturally do not contradict the statistical predictions of quantum mechanics (since it is precisely the accumulation of all the individual processes which yields the pattern described by $|\psi|^2$, and indeed explains it). The question of whether the predictions of this theory of motion may be subjected to an experimental test is discussed in §5.5 and Chap. 8.

On the face of it one can distinguish three different senses in which a physical variable enters into quantum mechanics, and a further sense defined by the quantum theory of motion. These are as

(*a*) an Hermitian operator
(*b*) an eigenvalue of (*a*)
(*c*) the expectation value of (*a*)
(*d*) the local expectation value of (*a*)

The property (*d*) might, following the common parlance, be termed the 'hidden variable' associated with the corresponding physical variable. In our view it is the most useful and important function of the four – it gives insight into the physical significance of (*a*), (*b*) is a special case of it, and (*c*) is derived from it.

3.8 The formation of averages

3.8.1 Averages as individual properties

We remarked in §3.6 that the dual role of the function R (as a field component of an individual system and as a statistical distribution of particles associated with the same wave) implies that many expressions into which it enters will naturally have at least two interpretations. Particularly important examples of such expressions are averages. We shall consider first the interpretation of these as yielding information on the overall structure of a wave as a single physical entity propagating in space.

Suppose we have a wave packet, for which the amplitude is appreciable only in a limited region. Then, at time t, the quantity

$$\langle \hat{\mathbf{x}} \rangle = \int R^2 \mathbf{x} \, \mathrm{d}^3 x \tag{3.8.1}$$

gives the position of the centre of the packet, and

$$(\Delta \hat{x})^2 = \int R^2 (x - \langle x \rangle)^2 \, \mathrm{d}^3 x \tag{3.8.2}$$

gives a measure of its mean square width in the x-direction. Introducing the Fourier transform of the wavefunction,

$$\varphi(\mathbf{k}) = (2\pi)^{-3/2} \int \psi(\mathbf{x}) \, \mathrm{e}^{-\mathrm{i}\mathbf{k}\cdot\mathbf{x}} \, \mathrm{d}^3 x, \tag{3.8.3}$$

we have that

$$\langle \hat{\mathbf{p}} \rangle = \int R^2 \nabla S \, \mathrm{d}^3 x \tag{3.8.4}$$

$$= \hbar \int \mathbf{k} |\varphi|^2 \, \mathrm{d}^3 k, \tag{3.8.5}$$

which therefore gives the centre of the packet at time t in momentum space, and

$$(\Delta \hat{p}_x)^2 = \int \psi^* (\hat{p}_x - \langle \hat{p}_x \rangle)^2 \psi \, \mathrm{d}^3 x$$

$$= \int R^2 \left[\frac{1}{2m} \left(\frac{\partial S}{\partial x} \right)^2 - \frac{\hbar^2}{2mR} \frac{\partial^2 R}{\partial x^2} \right] \mathrm{d}^3 x - \langle \hat{p}_x \rangle^2 \tag{3.8.6}$$

$$= \int |\varphi|^2 (\hbar k_x - \langle \hat{p}_x \rangle)^2 \, \mathrm{d}^3 k, \tag{3.8.7}$$

which gives the mean square width in momentum space.

This is not the only interpretation that we may give to the overall structure of a wave. Just like any classical field, the ψ-wave associates a density of energy and momentum with each point in space (see §3.9). The expression (3.8.4) then represents the total field momentum at time t

3.8.2 Ensemble averages and quantum mechanical expectation values

We consider an ensemble of particles associated with the same wavefunction ψ and some function $A(\mathbf{x}, t)$ which represents, when evaluated along a trajectory, some physically meaningful property of a particle (momentum, energy etc.). Since the ensemble density is given by $R^2(\mathbf{x}, t)$ it is natural to define the ensemble average of the quantity A at time t to be

$$\langle A \rangle = \int R^2(\mathbf{x}, t) A(\mathbf{x}, t)\, \mathrm{d}^3x. \tag{3.8.8}$$

This is the same expression we discussed in §3.8.1, but we now give it a quite different interpretation. How does this definition relate to the quantum mechanical definition of average, or expectation value, defined by (3.5.3) (with (3.5.2)):

$$\langle \hat{A} \rangle = \int \mathrm{Re}\, \psi^*(\mathbf{x})(\hat{A}\psi)(\mathbf{x})\, \mathrm{d}^3x, \tag{3.8.9}$$

where $\hat{A}$ is the Hermitian operator corresponding to the physical property described by $A(\mathbf{x}, t)$? It turns out that (3.8.8) and (3.8.9) may be identified if the actual value $A(\mathbf{x}, t)$ is correctly identified. This follows easily from the demonstration in §3.5 that properties such as $A(\mathbf{x}, t)$ may be identified with the local expectation value of $\hat{A}$ in the state $\psi(\mathbf{x}, t)$, i.e., with the integrand of (3.8.9), if we admit that it depends on ψ as well as $\hat{A}$.

Thus, for the examples studied in §3.5 we have the following results (cf. (3.5.5), (3.5.7), (3.5.10), (3.5.11) and (3.5.12)):

$$\langle \mathbf{x} \rangle = \int R^2 \mathbf{x}\, \mathrm{d}^3x = \int \psi^* \mathbf{x} \psi\, \mathrm{d}^3x = \langle \hat{\mathbf{x}} \rangle, \tag{3.8.10}$$

$$\langle \mathbf{p} \rangle = \int R^2 \nabla S\, \mathrm{d}^3x = \int \psi^*(-i\hbar\nabla)\psi\, \mathrm{d}^3x = \langle \hat{\mathbf{p}} \rangle, \tag{3.8.11}$$

$$\begin{aligned}\langle E \rangle &= \int R^2[(\nabla S)^2/2m + Q + V]\, \mathrm{d}^3x \\ &= \int \psi^*[-(\hbar^2/2m)\nabla^2 + V]\psi\, \mathrm{d}^3x = \langle \hat{H} \rangle,\end{aligned} \tag{3.8.12}$$

$$\langle \mathbf{L} \rangle = \int R^2 \mathbf{x} \times \nabla S \, d^3x$$

$$= \int \psi^*(\mathbf{x} \times -i\hbar\nabla)\psi \, d^3x = \langle \hat{\mathbf{L}} \rangle, \tag{3.8.13}$$

$$\langle l \rangle = \int R^2[(\mathbf{x} \times \nabla S)^2 - \hbar^2(\mathbf{x} \times \nabla)^2 R/R] \, d^3x$$

$$= \int \psi^*(\mathbf{x} \times -i\hbar\nabla)^2\psi \, d^3x = \langle \hat{\mathbf{L}}^2 \rangle. \tag{3.8.14}$$

Some remarks on these relations are in order. (3.8.10) gives a measure of the mean displacement in the ensemble at time t. Note though that there may be no individual element of the ensemble which actually follows the track pursued by $\langle \mathbf{x} \rangle$. Notice also that the expectation value of the kinetic energy and total angular momentum operators are not equal to the ensemble averages of just the kinetic energy and total angular momentum, respectively, as in classical mechanics, but include additional quantum terms. As we have pointed out several times before, the reason for this is that a classically free particle is not free quantum mechanically. We have to keep in mind when forming ensemble averages the physical meaning of the terms summed over, for example that a particle possesses quantum potential energy as well as kinetic energy.

Will the quantum mechanical expectation value of an Hermitian operator coincide with the ensemble average of a suitably defined local quantity for all operators? It is easy to see that formally this will be so. We simply define the function $A(\mathbf{x}, t)$ in (3.8.8) to be the local expectation value (3.5.4). Thus, although only a few operators are of interest physically, we could if we wished extend our particle interpretation of operators acting on states to arbitrary combinations of the operators $\hat{\mathbf{x}}$ and $\hat{\mathbf{p}}$, and still other types of operators. We would expect in the typical case to find contributions to the local expectation value analogous to Q, in addition to the classical terms (for example, for $\hat{\mathbf{p}}^n$ the latter is $(\nabla S)^n$ etc.), and these would have physical significance.

Having established this it is now a simple matter to write expectation values as ensemble averages over a phase space, where the joint distribution function is given by (3.6.17):

$$\langle \hat{A} \rangle = \int R^2(\mathbf{x}, t) \, \delta[\mathbf{p} - \nabla S(\mathbf{x}, t)] A(\mathbf{x}, t) \, d^3x \, d^3p. \tag{3.8.15}$$

The local expectation values of some operators used in quantum mechanics do not have an obvious physical interpretation however. The operators we have discussed above have a matrix representation of the form $A(\mathbf{x}, \mathbf{x}') = f(\mathbf{x})\,\delta(\mathbf{x} - \mathbf{x}')$ where $f(\mathbf{x})$ is a differential and/or multiplicative operator. An operator that is not of this type is a projection operator

$$\hat{A} = |\psi_i\rangle\langle\psi_i| \tag{3.8.16}$$

where $|\psi_i\rangle$ is the eigenstate of some operator. We have

$$A(\mathbf{x}, \mathbf{x}') = \psi_i(\mathbf{x})\psi_i^*(\mathbf{x}') \tag{3.8.17}$$

and hence from (3.5.4) and (3.5.2)

$$A(\mathbf{x}, t) = \mathrm{Re}\, \psi^*(\mathbf{x}, t)\psi_i(\mathbf{x}, t)c_i(t)/|\psi|^2, \tag{3.8.18}$$

where $c_i = \int \psi_i^*(\mathbf{x})\psi(\mathbf{x})\, \mathrm{d}^3x$. This expression is not readily interpretable, although it may certainly be employed formally (see also Wan & Sumner (1988)).

To summarize, we note that conventionally the method of forming expectation values in quantum mechanics by bracketing operators with the wavefunction is considered not to be assimilable to a classical-style ensemble average. On the contrary we have shown here that one may retain the classical definition of ensemble average, but the entities averaged over may contain contributions (constructed from ψ) characteristic of quantum mechanics, in addition to the classically expected terms. This leads to a consistent interpretation of averages since what in classical physics is just a distribution function (R^2) is for us a physical field which enters into the definition of the particle properties $A(\mathbf{x}, t)$.

3.8.3 *Ehrenfest's theorem*

The time rate of change of the expectation value of an arbitrary Hermitian operator $\hat{A}$ is given by (Schiff, 1968, p. 170):

$$\mathrm{d}\langle\hat{A}\rangle/\mathrm{d}t = \langle\partial\hat{A}/\partial t\rangle + (\mathrm{i}/\hbar)\langle[\hat{H}, \hat{A}]\rangle. \tag{3.8.19}$$

For the Hamiltonian $\hat{H} = \hat{\mathbf{p}}^2/2m + V(\hat{\mathbf{x}})$ we find by inserting $\hat{A} = \hat{\mathbf{x}}$ and $\hat{\mathbf{p}}$ in (3.8.19), in turn,

$$\mathrm{d}\langle\hat{\mathbf{x}}\rangle/\mathrm{d}t = \langle\hat{\mathbf{p}}\rangle/m \tag{3.8.20}$$

and

$$\mathrm{d}\langle\hat{\mathbf{p}}\rangle/\mathrm{d}t = -\langle\nabla V\rangle. \tag{3.8.21}$$

These two relations are known as Ehrenfest's theorem. According to §§3.8.1 and 3.8.2 they have two interpretations: as describing the overall kinematics and dynamics of a single system, e.g., a wave packet, and as describing the average motion of an ensemble of particles.

Eqs. (3.8.20) and (3.8.21) have the form of the classical equations describing the position and momentum of a single particle:

$$\mathrm{d}\mathbf{x}/\mathrm{d}t = \mathbf{p}/m \tag{3.8.22}$$

$$\mathrm{d}\mathbf{p}/\mathrm{d}t = -\nabla V. \tag{3.8.23}$$

In the classical case we also have the relations

$$\left.\begin{aligned} \mathrm{d}\langle\mathbf{x}\rangle/\mathrm{d}t &= \langle \mathrm{d}\mathbf{x}/\mathrm{d}t\rangle \\ \mathrm{d}\langle\mathbf{p}\rangle/\mathrm{d}t &= \langle \mathrm{d}\mathbf{p}/\mathrm{d}t\rangle \end{aligned}\right\} \tag{3.8.24}$$

where the mean value is defined as above, i.e., for an ensemble of density R^2 and momentum $\mathbf{p} = \nabla S$. These are proved by partial integration and involve the requirement that $R \to 0$ at infinity. In the quantum case the right hand sides of (3.8.24) have no meaning in the usual approach.

In the causal interpretation they do have a meaning. Eqs. (3.8.24) remain valid except that (3.8.23) is replaced by (3.2.24):

$$\mathrm{d}\mathbf{p}/\mathrm{d}t = -\nabla(V + Q). \tag{3.8.25}$$

Taking the expectation value of (3.8.25) and using (3.8.24) we obtain

$$\mathrm{d}\langle\mathbf{p}\rangle = -\langle\nabla(V + Q)\rangle. \tag{3.8.26}$$

It is easy to see by partial integration that $\langle\nabla Q\rangle = 0$:

$$\begin{aligned} -(2m/\hbar^2)\langle\nabla Q\rangle &= \int R^2\nabla(\nabla^2 R/R)\,\mathrm{d}^3x \\ &= -2\int(\nabla R)\nabla^2 R\,\mathrm{d}^3x \\ &= +2\int\nabla^2 R(\nabla R)\,\mathrm{d}^3x \\ &= 0 \end{aligned}$$

assuming that $R \to 0$ at infinity. We therefore recover (3.8.21).

An important question is the following: Will a mass point initially placed at the centre of a packet, $\mathbf{x}_0 = \langle\mathbf{x}\rangle(0)$, move with the centre point subsequently, $\mathbf{x}(t) = \langle\mathbf{x}\rangle(t)$? This will be so if $\langle\mathbf{x}\rangle(t)$ is a solution of the guidance law $m\dot{\mathbf{x}} = \nabla S|_{\mathbf{x}=\mathbf{x}(t)}$. Equivalently we require that

$$\langle\mathbf{p}(\mathbf{x})\rangle = \mathbf{p}(\langle\mathbf{x}\rangle). \tag{3.8.27}$$

This relation holds if $\mathbf{p}$ is a linear function of $\mathbf{x}$, or

$$S(\mathbf{x}, t) = \tfrac{1}{2}a\mathbf{x}^2 + \mathbf{b}\cdot\mathbf{x} + c \tag{3.8.28}$$

where a, b and c may be time-dependent (this includes a free Gaussian packet, cf. §4.7). But, in general, (3.8.27) is not valid and the actual and mean motions will part company.

3.8.4 *Ensemble averages and the density matrix*

In the case that the actual wave realized in each trial is unknown (mixed state) the quantum mechanical expectation value may be written as (cf. (3.6.29)):

$$\langle \hat{A} \rangle = \sum_i p_i \langle \psi_i | \hat{A} | \psi_i \rangle. \tag{3.8.29}$$

Since according to §3.8.2 each $\langle \psi_i | \hat{A} | \psi_i \rangle$ can be written as an ensemble average, it follows that (3.8.29) can also. Notice though that the p_is have a purely probabilistic significance and hence (3.8.29) cannot be interpreted as a property of an individual system, as we did for the pure state in §3.8.1.

3.9 Conservation laws for the field and particle

3.9.1 *Canonical formalism for the Schrödinger field*

The connection between the field and particle aspects of matter will be investigated further here with regard to the conservation laws satisfied by each. To begin with we set up a canonical formalism for the wavefunction, treated as a complex classical field (Takabayasi, 1952).

The Schrödinger equation may be derived by requiring that the action $\int L \, \mathrm{d}t$, where

$$L = \int \mathscr{L} \, \mathrm{d}^3x \tag{3.9.1}$$

is the Lagrangian, be stationary with respect to variations in the independent field coordinates $\psi(\mathbf{x})$ and $\psi^*(\mathbf{x})$, where $\mathscr{L} = \mathscr{L}(\psi, \dot{\psi}, \partial\psi, \text{cc})$ is the Lagrangian density given by

$$\mathscr{L} = (\mathrm{i}\hbar/2)(\psi^*\dot{\psi} - \dot{\psi}^*\psi) - (\hbar^2/2m)\, \partial_i\psi\, \partial_i\psi^* - V\psi\psi^*. \tag{3.9.2}$$

Here $\dot{\psi} = \partial\psi/\partial t$ and repeated indices are summed over ($i = 1, 2, 3$). Denoting the functional derivative applied to a functional $F = F(\psi, \partial\psi)$ by

$$\frac{\delta F}{\delta \psi} = \frac{\partial F}{\partial \psi} - \partial_i \frac{\partial F}{\partial(\partial_i \psi)}$$

the Euler–Lagrange equations are

$$\frac{\partial}{\partial t}\frac{\partial \mathscr{L}}{\partial \dot{\psi}} = \frac{\delta \mathscr{L}}{\delta \psi}, \qquad \frac{\partial}{\partial t}\frac{\partial \mathscr{L}}{\partial \dot{\psi}^*} = \frac{\delta \mathscr{L}}{\delta \psi^*}. \tag{3.9.3}$$

Inserting (3.9.2) in (3.9.3) yields the Schrödinger equation and its complex conjugate (cc).

We may achieve the same result from Hamilton's equations if we introduce a canonical formalism for the fields together with Poisson brackets etc. The canonical momenta conjugate to the field coordinates are defined by

$$\left.\begin{aligned} \pi_\psi &= \partial \mathscr{L}/\partial \dot{\psi} = (\mathrm{i}\hbar/2)\psi^* \\ \pi_{\psi^*} &= \partial \mathscr{L}/\partial \dot{\psi}^* = -(\mathrm{i}\hbar/2)\psi. \end{aligned}\right\} \tag{3.9.4}$$

The variables ψ, ψ^* thus define a $2 \times (3 \times \infty)$-dimensional phase space. We shall not pursue the Hamiltonian formalism for the fields here (see Chap. 12).

From the Lagrangian we can define a stress tensor and energy and momentum densities of the field and derive conservation laws. In order to do this it is convenient to transform to a new set of real variables $\rho = R^2$ and S which leads to more readily interpretable expressions. Define then the transformation

$$\psi = \rho^{1/2}\,\mathrm{e}^{\mathrm{i}S/\hbar}, \qquad \psi^* = \rho^{1/2}\,\mathrm{e}^{-\mathrm{i}S/\hbar}. \tag{3.9.5}$$

The Lagrangian density (3.9.2) becomes

$$\begin{aligned} \mathscr{L} &= \mathscr{L}(\rho, \partial\rho, \dot{\rho}, S, \partial S, \dot{S}) \\ &= -\rho[\dot{S} + (\nabla S)^2/2m + V] - \hbar^2(\nabla\rho)^2/8m\rho. \end{aligned} \tag{3.9.6}$$

The Euler–Lagrange equations are (3.9.3) with ψ replaced by ρ and ψ^* by S respectively. They yield, respectively, the generalized Hamilton–Jacobi equation (3.2.6) and the conservation law (3.2.7). Once again we may define conjugate momenta:

$$\left.\begin{aligned} \pi_\rho &= \partial \mathscr{L}/\partial \dot{\rho} = 0 \\ \pi_S &= \partial \mathscr{L}/\partial \dot{S} = -\rho. \end{aligned}\right\} \tag{3.9.7}$$

3.9.2 Field conservation laws

In terms of the fields ρ and S, the stress–energy–momentum tensor is defined by

$$T^\mu{}_\nu = -\left[\frac{\partial \mathscr{L}}{\partial(\partial_\mu \rho)}\,\partial_\nu \rho + \frac{\partial \mathscr{L}}{\partial(\partial_\mu S)}\,\partial_\nu S\right] + \delta_{\mu\nu}\mathscr{L} \tag{3.9.8}$$

and it satisfies

$$\partial_\mu T^\mu{}_\nu = \frac{\partial \mathscr{L}}{\partial x^\nu}, \tag{3.9.9}$$

where the right hand side is evaluated for constant ρ, S, $\partial_\mu \rho$ and $\partial_\mu S$. Here we have used a relativistic notation with $\mu = 0, 1, 2, 3$ and $\delta_{\mu\nu} = 1$ if $\mu = \nu$ and 0 if $\mu \neq \nu$. When the right hand side vanishes, (3.9.9) is a local conservation law.

Substituting for $\mathscr{L}$ from (3.9.6), the tensor (3.9.8) has the following components (for a discussion of their physical meaning see Misner, Thorne & Wheeler (1973, Chap. 5)):

Energy density

$$-T^0{}_0 \equiv \mathscr{H} = \rho[(\nabla S)^2/2m + V] + \hbar^2(\nabla\rho)^2/8m\rho \tag{3.9.10}$$

Momentum density

$$T^0{}_i \equiv \mathscr{G}_i = \rho\,\partial_i S \tag{3.9.11}$$

Energy current density

$$-T^i{}_0 \equiv \mathscr{S}_i = -(\rho\dot{S}\,\partial_i S/m + \hbar^2\dot{\rho}\,\partial_i\rho/4m\rho) \tag{3.9.12}$$

Momentum current density or stress tensor

$$T^i{}_j = \rho\,\partial_i S\,\partial_j S/m + \hbar^2\,\partial_i\rho\,\partial_j\rho/4m\rho + \delta_{ij}\mathscr{L}. \tag{3.9.13}$$

The appellations attributed to these quantities are justified by the form of (3.9.9) which splits up into a law of energy transport,

$$\partial\mathscr{H}/\partial t + \partial_k \mathscr{S}_k = \rho\,\partial V/\partial t, \tag{3.9.14}$$

and a law of momentum transport,

$$\partial\mathscr{G}_i/\partial t + \partial_k T^k{}_i = -\rho\,\partial_i V. \tag{3.9.15}$$

These equations contain no more information than the generalized Hamilton–Jacobi equation (3.2.6) and the conservation equation (3.2.7). Indeed they can be easily derived from the latter: for example, (3.9.14) follows by applying $\partial/\partial t$ to the identity $(\partial\rho/\partial t)(\partial S/\partial t) - (\partial S/\partial t)(\partial\rho/\partial t) = 0$ where we substitute for $\partial S/\partial t$ from (3.2.6) in the first term and for $\partial\rho/\partial t$ from (3.2.7) in the second. And, of course, we may use (3.2.6) and (3.2.7) to derive alternative expressions for (3.9.10)–(3.9.13).

When the external potential is independent of time (position) we obtain from (3.9.14) ((3.9.15)) a local law of conservation of field energy (momentum). We can turn these into global laws by integrating over all space. Then (3.9.14) implies

$$(\mathrm{d}/\mathrm{d}t)\int \mathscr{H}\,\mathrm{d}^3x = \int (\partial \mathscr{H}/\partial t)\,\mathrm{d}^3x$$

$$= \int (-\partial_k \mathscr{S}_k + \rho\,\partial V/\partial t)\,\mathrm{d}^3x$$

$$= \int \rho(\partial V/\partial t)\,\mathrm{d}^3x \tag{3.9.16}$$

by Green's theorem and assuming that the field vanishes at infinity. Similarly (3.9.15) gives

$$(\mathrm{d}/\mathrm{d}t)\int \mathscr{G}_i\,\mathrm{d}^3x = \int (\partial \mathscr{G}_i/\partial t)\,\mathrm{d}^3x$$

$$= \int (-\partial_k T^k{}_i - \rho\,\partial_i V)\,\mathrm{d}^3x$$

$$= -\int \rho\,\partial_i V\,\mathrm{d}^3x \tag{3.9.17}$$

by partial integration. It follows that if the mean external power vanishes the total field energy is conserved:

$$\int \mathscr{H}\,\mathrm{d}^3x = \text{constant}, \tag{3.9.18}$$

and if the mean external force is zero the total field momentum is conserved:

$$\int \mathscr{G}_i\,\mathrm{d}^3x = \int \rho\,\partial_i S\,\mathrm{d}^3x = \text{constant}. \tag{3.9.19}$$

To complete the set of field conservation laws we define the angular momentum density about the origin to be the tensor

$$\mathscr{A}^\mu{}_{ij} = x_i T^\mu{}_j - x_j T^\mu{}_i \tag{3.9.20}$$

which is antisymmetric in i and j. This has the components

$$\mathscr{A}^0{}_{ij} = x_i \mathscr{G}_j - x_j \mathscr{G}_i, \tag{3.9.21}$$

$$\mathscr{A}^k{}_{ij} = x_i T^k{}_j - x_j T^k{}_i, \tag{3.9.22}$$

where we should substitute (3.9.11) and (3.9.13). From (3.9.15) we may derive the following equation of angular momentum flow

$$\frac{\partial \mathscr{A}^0{}_{ij}}{\partial t} + \partial_k \mathscr{A}^k{}_{ij} = -\rho(x_i\,\partial_j V - x_j\,\partial_i V). \tag{3.9.23}$$

In the case that the external torque $-\mathbf{x} \times \nabla V$ vanishes, (3.9.23) reduces to a local angular momentum conservation law. The global law follows as usual:

$$(\mathrm{d}/\mathrm{d}t) \int \mathscr{A}^0{}_{ij}\, \mathrm{d}^3x = -\int \rho(x_i\, \partial_j V - x_j\, \partial_i V)\, \mathrm{d}^3x \qquad (3.9.24)$$

by partial integration. Thus if the mean external torque is zero, the total field angular momentum is conserved:

$$\int \mathscr{A}^0{}_{ij}\, \mathrm{d}^3x = \text{constant}. \qquad (3.9.25)$$

The field energy, momentum and angular momentum are distributed throughout space and locally influence a particle, via the quantum potential.

3.9.3 Particle conservation laws

It will be noted that the field momentum and angular momentum densities (3.9.11) and (3.9.21) are respectively proportional to the momentum and angular momentum fields associated with the particle. As a result, the total values of the former quantities coincide with the mean particle momentum and angular momentum over an ensemble. In addition, although the local energy density of the field (3.9.10) is not proportional to the particle energy (3.2.20), a simple integration by parts shows that the global values of both are the same. We therefore have the following results:

Total field energy

$$\int \mathscr{H}\, \mathrm{d}^3x = \text{mean particle energy } \langle E(\mathbf{x}, t)\rangle$$

$$= \langle \hat{H} \rangle \qquad (3.9.26)$$

Total field momentum

$$\int \mathscr{G}\, \mathrm{d}^3x = \text{mean particle momentum } \langle \mathbf{p}(\mathbf{x}, t)\rangle$$

$$= \langle \hat{\mathbf{p}} \rangle \qquad (3.9.27)$$

Total field angular momentum

$$\int \mathbf{x} \times \mathscr{G}\, \mathrm{d}^3x = \text{mean particle angular momentum } \langle \mathbf{x} \times \mathbf{p} \rangle$$

$$= \langle \hat{\mathbf{L}} \rangle \qquad (3.9.28)$$

where we have used (3.8.12), (3.8.11) and (3.8.13).

Once again we notice the interplay between real physical quantities and ensemble averages due to the dual interpretation of the function R^2 as a field and as an ensemble density.

In view of these results, eq. (3.9.17) for the rate of change of total field momentum will be seen to be nothing but Ehrenfest's theorem (3.8.21). Relations (3.9.16) and (3.9.24) may also be interpreted as Ehrenfest-type theorems. The global field conservation laws (3.9.18), (3.9.19) and (3.9.25) can therefore be immediately translated into ensemble conservation laws. Thus

$$\langle \partial V/\partial t \rangle = 0 \Rightarrow \langle E \rangle = \langle \hat{H} \rangle = \text{constant} \tag{3.9.29}$$

$$\langle \nabla V \rangle = 0 \Rightarrow \langle \mathbf{p} \rangle = \langle \hat{\mathbf{p}} \rangle = \text{constant} \tag{3.9.30}$$

$$\langle \mathbf{x} \times \nabla V \rangle = 0 \Rightarrow \langle \mathbf{x} \times \mathbf{p} \rangle = \langle \hat{\mathbf{L}} \rangle = \text{constant}. \tag{3.9.31}$$

These relations are the usual conservation laws of quantum mechanics. They may be equivalently formulated in terms of commutation relations with the Hamiltonian operator. In the Heisenberg picture the equation of motion of an operator $\hat{A}$ is given by

$$\mathrm{d}\hat{A}/\mathrm{d}t = \partial \hat{A}/\partial t - (1/i\hbar)[\hat{H}, \hat{A}]. \tag{3.9.32}$$

If the operator is not an explicit function of time, then the condition

$$[\hat{H}, \hat{A}] = 0 \tag{3.9.33}$$

states a 'conservation law'. For $\hat{H} = \hat{\mathbf{p}}^2/2m + V$, and writing in turn $\hat{A} = \hat{H}$, $\hat{\mathbf{p}}$ and $\hat{\mathbf{L}}$ in (3.9.32), we have

$$\left.\begin{aligned} \mathrm{d}\hat{H}/\mathrm{d}t &= \partial V/\partial t \\ \mathrm{d}\hat{\mathbf{p}}/\mathrm{d}t &= -\nabla V \\ \mathrm{d}\hat{\mathbf{L}}/\mathrm{d}t &= -\mathbf{x} \times \nabla V \end{aligned}\right\} \tag{3.9.34}$$

(showing that the Heisenberg operators satisfy classical-type relations). Since in this picture $\langle \mathrm{d}\hat{A}/\mathrm{d}t \rangle = \mathrm{d}\langle \hat{A} \rangle/\mathrm{d}t$, we recover from (3.9.34) the conservation laws (3.9.29), (3.9.30) and (3.9.31).

Thus (3.9.33) allows us to draw conclusions about average values, but does not yield detailed information on the variation in time of the properties of an individual ensemble element. For example, even though the operator $\hat{\mathbf{p}}$ commutes with the free Hamiltonian ($V = 0$) and from (3.9.34), $\mathrm{d}\hat{\mathbf{p}}/\mathrm{d}t = 0$, the actual momentum of the particle is not necessarily conserved, as we shall see shortly. Conversely, the actual momentum may be conserved when $\hat{\mathbf{p}}$ does *not* commute with $\hat{H}$ (for an example, see §7.4).

The above considerations refer to an arbitrary state ψ. If ψ happens to be an eigenstate of the operator $\hat{A}$ in question, then, of course, $\langle[\hat{H}, \hat{A}]\rangle = 0$ and $\mathrm{d}\langle\hat{A}\rangle/\mathrm{d}t = 0$ automatically.

What, then, about the *local* conservation laws for each particle in the ensemble? These are readily obtained from the equations of motion derived from the generalized Hamilton–Jacobi equation. Applying the operator $\partial/\partial t$ to (3.2.17) we obtain

$$[\partial/\partial t + (1/m)\nabla S \cdot \nabla]\, \partial S/\partial t = -\partial(V + Q)/\partial t. \quad (3.9.35)$$

Identifying $\nabla S/m$ with $\dot{\mathbf{x}}$ and $-\partial S/\partial t$ with the energy $E(\mathbf{x}, t)$ we have

$$\frac{\mathrm{d}E}{\mathrm{d}t} = \frac{\partial}{\partial t}(V + Q)|_{\mathbf{x}=\mathbf{x}(t)}, \qquad E = \tfrac{1}{2}m\dot{\mathbf{x}}^2 + Q + V. \quad (3.9.36)$$

Similarly, applying ∇ to (3.2.17) we obtain (3.2.24):

$$\mathrm{d}\mathbf{p}/\mathrm{d}t = -\nabla(V + Q)|_{\mathbf{x}=\mathbf{x}(t)}, \qquad \mathbf{p} = m\dot{\mathbf{x}}, \quad (3.9.37)$$

and the rate of change of particle angular momentum, $\mathbf{L} = \mathbf{x} \times \mathbf{p}$, is easily found to be

$$\mathrm{d}\mathbf{L}/\mathrm{d}t = -\mathbf{x} \times \nabla(V + Q)|_{\mathbf{x}=\mathbf{x}(t)}, \qquad \mathbf{L} = \mathbf{x} \times m\dot{\mathbf{x}}. \quad (3.9.38)$$

We see that in each of these equations there enters on the right hand side a term characteristic of quantum mechanics. In (3.9.37) we have the familiar quantum force, $-\nabla Q$, in (3.9.36) we have a 'quantum power' term, $\partial Q/\partial t$, and in (3.9.38) a quantum torque, $-\mathbf{x} \times \nabla Q$. Clearly, then, the local conservation laws will involve conditions not only on the external potential but also on the quantum potential:

$$\partial(V + Q)/\partial t = 0 \Rightarrow E(\mathbf{x}(t), t) = \text{constant (energy conservation)} \quad (3.9.39)$$

$$\nabla(V + Q) = 0 \Rightarrow \mathbf{p} = m\dot{\mathbf{x}} = \text{constant (momentum conservation)} \quad (3.9.40)$$

$$\mathbf{x} \times \nabla(V + Q) = 0 \Rightarrow \mathbf{L} = \mathbf{x} \times m\dot{\mathbf{x}} = \text{constant (angular momentum conservation)} \quad (3.9.41)$$

for all elements of the ensemble. These conservation laws are therefore much more stringent than the global ones which merely require that V satisfies certain conditions. Even if the particle is classically free ($V = 0$), the particle energy and momentum will, in general, be variable due to the time and space dependence of the quantum potential. *Classically conserved motion is not conserved quantum mechanically for generic solutions of the Schrödinger equation.* As examples, the momentum is conserved when ψ is a plane wave (eigenfunction of $\hat{\mathbf{p}}$), the energy when ψ is a stationary state (eigenfunction of $\hat{H}$) and an angular momentum component is conserved when ψ is an eigenfunction of the relevant component of $\hat{\mathbf{L}}$. Each of the conservation laws may hold for more general states. Moreover, there are cases where some of the particles in the ensemble connected with ψ obey conservation laws but not all. In that case the local conditions on the left hand sides of (3.9.39)–(3.9.41) hold only in restricted regions.

One might expect the conservation laws would apply to the total field plus particle system in interaction, as in classical electrodynamics. The reason they do not is that the particle does not react back on the wave; the field satisfies its own conservation laws as set out in §3.9.2. Correspondingly the particle satisfies its own conservation laws, but here the field enters as an 'external source', via Q. Thus, we may say that if we can neglect the 'environment' of the field (i.e., $V = 0$) then this forms a closed system with corresponding energy, momentum and angular momentum conservation laws. The particle, however, even in this case, is an open system since it receives energy and momentum from the field.

Of course, the single-body theory with an external potential is an idealization since it neglects the physical system constituting the 'source' of V. Even if we include the latter to form a classically closed many-body system, however, the particle properties will still in general not be conserved due to the dynamical quantum field (see Chap. 7).

From the standpoint of general theoretical principles this feature of the causal interpretation may appear as unsatisfactory, calling for a development of the theory to include a more symmetrical relation between wave and particle. At present we have no idea how a source term for the ψ-field could be consistently introduced into the dynamical equations in such a way that it does not disturb the empirically well-verified predictions of quantum theory while going beyond it to yield new testable results.

3.10 Hydrodynamic analogy

One of the earliest physical interpretations of the Schrödinger equation is due to Madelung (1926). This is derived by representing the wavefunction in the polar form (3.2.1) and so is closely connected with the quantum theory of motion. The basic elements of the Madelung approach are the fields $\rho = R^2$ and $\mathbf{v} = \nabla S/m$ satisfying the field equations (3.2.23) and (3.2.7):

$$\partial \mathbf{v}/\partial t + (\mathbf{v}\cdot\nabla)\mathbf{v} = -(1/m)\nabla(V + Q) \tag{3.10.1}$$

$$\partial \rho/\partial t + \nabla\cdot(\rho\mathbf{v}) = 0. \tag{3.10.2}$$

These equations are similar to those of classical hydrodynamics if we identify $m\rho$ with the density of a fluid extending throughout space, $\mathbf{v}$ with the velocity field of the fluid and m with the mass of a speck of dust travelling in the fluid. Then (3.10.2) expresses the conservation of fluid. For the purposes of comparison, the classical Euler equation is

$$\partial v_i/\partial t + (\mathbf{v}\cdot\nabla)v_i = -(1/m)\,\partial_i V - (1/m\rho)\,\partial_j(p\delta_{ij}) \tag{3.10.3}$$

where p is the pressure (Landau & Lifschitz, 1959). Now, we can rearrange the quantum force term in (3.10.1) so that this equation reads

$$\partial v_i/\partial t + (\mathbf{v}\cdot\nabla)v_i = -(1/m)\,\partial_i V - (1/m\rho)\,\partial_j\sigma_{ij}, \tag{3.10.4}$$

where

$$\sigma_{ij} = -(\hbar^2\rho/4m)\,\partial_{ij}\log\rho \tag{3.10.5}$$

is a stress tensor. Eq. (3.10.4) may be interpreted as describing a particular hydrodynamics in which the pressure tensor $p\delta_{ij}$ in the classical theory (3.10.3) is replaced by the tensor σ_{ij} (Takabayasi, 1952; Halbwachs, 1960).

The interpretation is completed by imposing on the fields ρ and $\mathbf{v}$ the appropriate boundary and subsidiary conditions. The velocity field is irrotational except at nodes where it is undefined, and the single-valuedness condition (3.2.10) becomes

$$\oint \mathbf{v}\cdot d\mathbf{x} = nh/m. \tag{3.10.6}$$

A fluid for which $n \neq 0$ is interpreted as possessing 'quantized vortices'.

The path of a fluid particle is defined as a solution to the differential equation $\dot{\mathbf{x}} = \mathbf{v}$; it therefore coincides with the trajectory of the quantum theory of motion. It is important to distinguish the paths from the fluid streamlines. The latter are defined by the solution to the equations

$$\frac{dx}{v_1(\mathbf{x}, t)} = \frac{dy}{v_2(\mathbf{x}, t)} = \frac{dz}{v_3(\mathbf{x}, t)} \tag{3.10.7}$$

at each instant of time. The tangents to the streamlines give the directions of the velocities of fluid particles at various points in space at a given instant. In contrast, the tangents to the paths give the directions of the velocities of given fluid particles at various times. The two concepts only coincide when $\mathbf{v}$ is time-independent (steady flow).

The tensor σ_{ij} is not unique since we can add to it any tensor whose divergence is zero. Another form follows if we derive (3.10.1) from the momentum relation (3.9.15). The Lagrangian density (3.9.6) may be simplified by substituting from the Hamilton–Jacobi equation to obtain

$$\mathscr{L} = -(\hbar^2/4m)\nabla^2\rho. \tag{3.10.8}$$

Inserting this into (3.9.13) gives for the tensor $T^i{}_j$

$$T^i{}_j = m\rho v_i v_j + (\hbar^2/4m)(\partial_i\rho\, \partial_j\rho/\rho - \delta_{ij}\nabla^2\rho). \tag{3.10.9}$$

Eq. (3.9.15) then becomes

$$\partial(\rho v_i)/\partial t + \partial_k(\rho v_k v_i) = -(\rho/m)\, \partial_i V - (\hbar^2/4m^2)\, \partial_k(\partial_i\rho\, \partial_k\rho/\rho - \delta_{ik}\nabla^2\rho), \tag{3.10.10}$$

which may be brought to the form

$$\partial v_i/\partial t + (\mathbf{v}\cdot\nabla)v_i = -(1/m)\, \partial_i V - (1/m\rho)\, \partial_j\sigma'_{ij}, \tag{3.10.11}$$

where

$$\sigma'_{ij} = (\hbar^2/4m)(\partial_i\rho\, \partial_j\rho/\rho - \delta_{ij}\nabla^2\rho). \tag{3.10.12}$$

3.11 Invariance properties

3.11.1 Galilean invariance

Here we shall confirm that the kinematical properties of a particle indeed obey the expected transformation rules with respect to transformations of the coordinate system by deriving them from the Galilean invariance of the Schrödinger equation.

It is sufficient to consider the free wave equation. Consider an inertial reference system Σ with respect to which the coordinates of a spacetime point P are x_i and t. The wave equation in this system is given by

$$i\hbar\frac{\partial\psi(\mathbf{x}, t)}{\partial t} = -\frac{\hbar^2}{2m}\nabla^2\psi(\mathbf{x}, t). \tag{3.11.1}$$

Galilean invariance requires that (3.11.1) should be independent of the origin, orientation and uniform motion of Σ with respect to some other inertial system Σ', and independent of the choice of the zero in time. Thus we require that with respect to the system Σ' which assigns to the point P the coordinates and time

$$\left.\begin{aligned} x'_i &= a_{ij}x_j - V_i t + a_i \\ t' &= t + t_0 \end{aligned}\right\} \tag{3.11.2}$$

the wave equation takes the form

$$i\hbar\frac{\partial\psi'(\mathbf{x}', t')}{\partial t'} = -\frac{\hbar^2}{2m}\nabla'^2\psi'(\mathbf{x}', t'), \tag{3.11.3}$$

where a_{ij} is a constant orthogonal rotation matrix ($a_{ij}a_{ik} = \delta_{jk}$), $\mathbf{V}$ is a constant vector representing a uniform motion, $\mathbf{a}$ is a constant vector representing a displacement in space and t_0 represents a displacement in time. The transformation (3.11.2) defines the 'Galilean group'. The wavefunctions in the two systems Σ and Σ' are related by a unitary representation of the Galilean group:

$$\psi'(\mathbf{x}', t') = U(\mathbf{x}, t)\psi(\mathbf{x}, t), \qquad U(\mathbf{x}, t) = e^{if(\mathbf{x},t)/\hbar},$$
$$f(\mathbf{x}, t) = \tfrac{1}{2}m\mathbf{V}^2 t - mV_i a_{ij} x_j + c, \tag{3.11.4}$$

where c is a constant. (3.11.4) is easy to prove using the results

$$\left.\begin{aligned} \partial_{i'} &= a_{ij}\,\partial_j, \qquad \nabla'^2 = \nabla^2 \\ \partial/\partial t' &= \partial/\partial t + V_i a_{ij}\,\partial_j \end{aligned}\right\} \tag{3.11.5}$$

and the linear independence of ψ and $\nabla\psi$ (Kaempffer, 1965, p. 341).

It is apparent that the wavefunction is not a scalar quantity with respect to uniform translations ($\mathbf{V} \neq 0$). The appearance of the phase factor (3.11.4) is connected with the asymmetrical manner in which the t and $\mathbf{x}$ variables enter in the Schrödinger equation (it does not arise in the ordinary wave equation where this symmetry is restored). The amplitude and phase functions therefore transform as follows under a Galilean transformation:

$$\left.\begin{aligned} R'(\mathbf{x}', t') &= R(\mathbf{x}, t) \\ S'(\mathbf{x}', t') &= S(\mathbf{x}, t) + \tfrac{1}{2}mV^2 t - mV_i a_{ij} x_j + c. \end{aligned}\right\} \tag{3.11.6}$$

Using the relations (3.11.5) we find the following connections between the momentum and energy fields in the two frames:

$$\partial_{i'} S'(\mathbf{x}', t') = a_{ij}\partial_j S(\mathbf{x}, t) - mV_i, \tag{3.11.7}$$

the well-known transformation law of a momentum vector, and

$$-\frac{\partial S'(\mathbf{x}', t')}{\partial t'} = -\frac{\partial S(\mathbf{x}, t)}{\partial t} - V_i a_{ij}\,\partial_j S(\mathbf{x}, t) + \tfrac{1}{2}mV^2, \tag{3.11.8}$$

which, although less familiar, is the correct law of transformation of energy in classical mechanics. The transformation properties of other quantities, such as angular momentum, follow straightforwardly. As a particular case of (3.11.7), when ψ is a plane wave (§4.2), we see that the eigenvalues of the momentum operator $\hat{\mathbf{p}} = -i\hbar\nabla$ transform as a vector with respect to uniform translations but that from (3.11.5) the operator itself does not.

It follows from (3.11.7) that if momentum is conserved in one frame, $\nabla S|_{\mathbf{x}=\mathbf{x}(t)} =$ constant, it is conserved in all frames, $\nabla' S'|_{\mathbf{x}'=\mathbf{x}'(t')} =$ (different) constant. In contrast, the conservation of energy in one frame does not necessarily translate into all frames due to the in-general variable momentum term on the right hand side of (3.11.8).

The fact that ψ is not a scalar has sometimes been adduced as evidence against the physical 'reality' of the de Broglie waves, but this is not a convincing argument. First, we see from (3.11.4) that ψ possesses the fundamental property of a tensor that if it vanishes in one frame it vanishes in all Galilean-related frames. Indeed, treating $\psi = \psi_1 + i\psi_2$ as a two-component real vector, (3.11.4) is simply a rotation of the vector. Moreover, it is evident from (3.11.6) that the quantum potential is a Galilean scalar, $Q'(\mathbf{x}', t') = Q(\mathbf{x}, t)$, which explicitly demonstrates the Galilean invariance of the individual processes treated by the causal interpretation.

It is interesting to consider the transformation properties of the generalized de Broglie relation (3.3.19), $\lambda_Q = h/|\nabla S|$. If we look at waves on the surface of the sea from a moving ship, we shall observe the same wavelength as a gull hovering above the water's surface; the wavelength of a classical wave motion is a Galilean invariant. The quantal wavelength on the other hand obviously does not have this property. From (3.11.7), $\lambda'_Q(\mathbf{x}', t') \neq \lambda_Q(\mathbf{x}, t)$; there is, in general, a complicated relation between the wavelengths in the two frames. One should be careful not to conclude from this result, however, that there is some inconsistency in the theory, since one cannot 'observe' the de Broglie wavelength in the way one would a classical wavelength. It is another example of the nonclassical nature of quantum waves.

An interesting feature of the unitary transformation (3.11.4) is its mass-dependence. It may be shown from this that one cannot superpose states corresponding to different masses without destroying Galilean invariance. It follows from this 'superselection rule' that two waves corresponding to different mass-values propagating in the same region will not interfere or affect one another in any way.

For completeness we add that the causal model obeys the discrete symmetries displayed by the Schrödinger equation, including time reversal invariance.

3.11.2 Gauge invariance. Introduction of the electromagnetic field

A different type of phase transformation arises when we consider the invariance of the Schrödinger equation in the presence of an external electromagnetic field with respect to a gauge transformation. For a system of mass m and charge e, the wave equation is (Bohm, 1951, p. 358):

$$i\hbar\frac{\partial\psi}{\partial t}=\left\{-\frac{\hbar^2}{2m}\left[\nabla-\left(\frac{ie}{\hbar c}\right)\mathbf{A}\right]^2+eA_0+V\right\}\psi, \tag{3.11.9}$$

where A_0 and $\mathbf{A}$ are the electromagnetic potentials and c is the speed of light. Because quantization proceeds from knowledge of the canonical momenta in classical electrodynamics it is the potentials that enter into the Schrödinger equation rather than the electric and magnetic field strengths:

$$\mathbf{E}=-\nabla A_0-(1/c)\,\partial\mathbf{A}/\partial t,\qquad \mathbf{B}=\nabla\times\mathbf{A}. \tag{3.11.10}$$

In classical physics the field strengths are considered to represent a complete description of the electromagnetic field, the potentials being mere mathematical auxiliaries. The reason is that the physical effect of the field on charged particles is mediated by the Lorentz force

$$\mathbf{F}=e[\mathbf{E}+(1/c)\mathbf{v}\times\mathbf{B}], \tag{3.11.11}$$

where $\mathbf{v}$ is the particle velocity. Now the fields (3.11.10) are left unchanged by the following transformation of the potentials

$$\mathbf{A}\to\mathbf{A}'=\mathbf{A}-\nabla\varphi,\qquad A_0\to A_0'=A_0+(1/c)\,\partial\varphi/\partial t \tag{3.11.12}$$

where φ is an arbitrary single-valued scalar function of $\mathbf{x}$ and t. Since an unlimited number of functionally distinct potentials are compatible with the same force (3.11.11), it is concluded that the potentials have in themselves no physical significance.

In quantum theory the situation is more subtle. For, while the theory as a whole is gauge invariant (as we shall see shortly), this does not imply that the potentials are devoid of any physical significance beyond that already captured by the fields, as is assumed in classical physics. It was pointed out in §3.3.4 that classical potentials have a more direct physical significance in quantum mechanics. The celebrated example of this in the electromagnetic case is the Aharonov–Bohm effect, discussed in §5.2.

The Schrödinger equation (3.11.9) is form invariant with respect to the gauge transformation (3.11.12) if we suppose that the wavefunction undergoes a suitable unitary transformation. That is, we have

$$i\hbar\frac{\partial\psi'}{\partial t}=\left\{-\frac{\hbar^2}{2m}\left[\nabla-\left(\frac{ie}{\hbar c}\right)\mathbf{A}'\right]^2+eA_0'+V\right\}\psi', \tag{3.11.13}$$

where $\psi'=U\psi$. It is easy to show that

$$U=e^{ie\varphi/\hbar}. \tag{3.11.14}$$

Conversely, we may view the introduction of a set of potentials which undergo the inhomogeneous transformation (3.11.12) as the necessary requirement for quantum mechanics to be invariant with respect to local phase transformations (3.11.14) (i.e. the group $U(1)$).

To see the consequences of this, let us first substitute $\psi = R\,\mathrm{e}^{\mathrm{i}S/\hbar}$ into (3.11.9). The result is a generalized Hamilton–Jacobi equation and an equation of conservation:

$$\partial S/\partial t + (1/2m)[\nabla S - (e/c)\mathbf{A}]^2 + eA_0 + V + Q = 0, \qquad (3.11.15)$$

$$\partial R^2/\partial t + \nabla\cdot\{R^2[\nabla S - (e/c)\mathbf{A}]/m\} = 0, \qquad (3.11.16)$$

where Q is the usual quantum potential (3.2.16). Following the approach of §3.2 we *postulate* that the particle momentum field is given by

$$\mathbf{p}(\mathbf{x}, t) = \nabla S - (e/c)\mathbf{A} \qquad (3.11.17)$$

just as in classical Hamilton–Jacobi theory (§2.2). The trajectories are then found by solving the law of motion

$$m\dot{\mathbf{x}} = [\nabla S - (e/c)\mathbf{A}]|_{\mathbf{x}=\mathbf{x}(t)}. \qquad (3.11.18)$$

The paths are thus no longer orthogonal to the surfaces S = constant. The single-valuedness requirement (3.2.10) becomes

$$\oint_C \mathrm{d}S = \oint_C \mathbf{p}\cdot\mathrm{d}\mathbf{x} + (e/c)\Phi = nh \qquad (3.11.19)$$

where $\Phi = \oint_C \mathbf{A}\cdot\mathrm{d}\mathbf{x}$ is the electromagnetic flux linking the circuit C. (3.11.17) is proportional to the quantum mechanical 'probability current density':

$$\mathbf{j} = (\hbar/2m\mathrm{i})[\psi^*\nabla\psi - (\nabla\psi^*)\psi] - (eR^2/mc)\mathbf{A}. \qquad (3.11.20)$$

The particle energy is equal to

$$E(\mathbf{x}(t), t) = -\partial S/\partial t|_{\mathbf{x}=\mathbf{x}(t)} \qquad (3.11.21)$$

and comprises as usual kinetic energy and external and internal (Q) potential energy. We can put the law of motion in Newtonian form by applying the operator ∇ to (3.11.15). The result is:

$$m[\partial\mathbf{v}/\partial\mathbf{t} + (\mathbf{v}\cdot\nabla)\mathbf{v}] = -\nabla(V + Q) + \mathbf{F}. \qquad (3.11.22)$$

Identifying $\mathbf{v}$ with $\dot{\mathbf{x}}$ yields

$$m\ddot{\mathbf{x}} = [-\nabla(V + Q) + \mathbf{F}]|_{\mathbf{x}=\mathbf{x}(t)}. \qquad (3.11.23)$$

The particle is subject to the classical forces $-\nabla V$ and $\mathbf{F}$ and the quantum force $-\nabla Q$.

Returning to the effect of a gauge transformation, we have from (3.11.14)

$$\left.\begin{aligned} R'(\mathbf{x}) &= R(\mathbf{x}) \\ S'(\mathbf{x}) &= S(\mathbf{x}) + e\varphi(\mathbf{x}). \end{aligned}\right\} \tag{3.11.24}$$

Hence,

$$\mathbf{v}' = \nabla S' - (e/c)\mathbf{A}' = \nabla S - (e/c)\mathbf{A} = \mathbf{v}, \tag{3.11.25}$$

so that the probability density and velocity field are gauge invariant quantities, as is the quantum potential. Although the kinetic energy is gauge invariant, the total energy (3.11.21) is not due to the potential energy term eA_0 in (3.11.15).

It follows that the particle motion is in no way influenced by the particular choice of gauge that it may be found convenient to employ for the potentials. In this regard the situation is the same as in classical electrodynamics. Yet Q is determined by an equation, (3.11.15), which directly involves the potentials rather than the fields. Hence, although gauge invariant, the quantum potential may contain information on the potentials that is more detailed than that picked up by the field strengths. This result is of some importance, for it implies that in the theory of charged particles the Lorentz force does not exhaust the possibilities for an electromagnetic field to influence the motion. The quantum force may be finite in a region of space where $\mathbf{F} = 0$ but $\mathbf{A} \neq 0$ (cf. §5.2).

In conclusion, we may say that quantum mechanics is not only *statistically* Galilean and gauge invariant, but it obeys these symmetries at the level of the individual processes that make up the ensemble.

3.12 Relation with quantum mechanical transformation theory

3.12.1 The position representation and the Schrödinger picture

Quantum mechanics is mathematically formulated so that its relationships are covariant with respect to (unitary) coordinate transformations in Hilbert space, i.e., transformations of *representation*, and to transformations of *picture* (Schrödinger, Heisenberg etc.). These invariances are considered to be a major feature of the theory both as a means of exhibiting its formal elegance, and as providing further support for the argument against attempts at constructing a spacetime model of material processes. For, if all representations are equivalent, why should we single out for special attention the one in which the position operator is diagonal? And, moreover, the nonexistence of a coordinate system in Hilbert space which corresponds to both the position and momentum operators being diagonal (a phase space) seems to mitigate against a particle picture.

To what extent is the causal interpretation compatible with these symmetries? To examine this question let us first recall the analogous situation in classical mechanics.

The classical arena for physics is three-dimensional Euclidean space. Nevertheless, it is often convenient to introduce 'fictitious' spaces such as phase space as an alternative means of visualizing spatial processes, and indeed this leads to a deeper formulation of dynamics. Now clearly the existence of the canonical approach to mechanics in no way implies that Euclidean space plays a less privileged role physically. We demand only that the causal description of phenomena in physical space should have a well-defined *mathematical* equivalence with its representation in other spaces.

The situation is very similar in respect of the causal formulation of quantum mechanics. Unitary invariance of the quantum formalism has in itself no bearing on the problem of whether or not a causal representation is possible, and what is the appropriate space in which to develop it if it is possible. In this book we postulate that the Schrödinger picture and the position representation are of special significance since we aim to treat quantum mechanics as a physical theory of the dynamics of particles and waves in space and time (in the case of many-body systems we generalize this conception (Chap. 7)). The mathematical symmetry is therefore disrespected for physical reasons. Of course, the possibility of such a treatment is not given *a priori* and it must be demonstrated that this postulate does not conflict with any compelling principles or experimental results. The transformation theory meanwhile retains its calculational usefulness, but it has no fundamental physical significance beyond that.

In support of this point of view, it should be noted that in actual applications of quantum mechanics the Hamiltonian operator is not a symmetrical function of the operators $\hat{\mathbf{x}}$ and $\hat{\mathbf{p}}$ (apart from the harmonic oscillator). Its typical form is

$$\hat{H}(\hat{\mathbf{x}}, \hat{\mathbf{p}}) = \tfrac{1}{2}(m_{ij}\hat{p}_i\hat{p}_j + \hat{p}_i\hat{p}_j m_{ij}) + \tfrac{1}{2}(a_i\hat{p}_i + \hat{p}_i a_i) + V, \qquad (3.12.1)$$

where m_{ij}, a_i and V are arbitrary functions of $\hat{\mathbf{x}}$. In the position representation this will be a multiplicative and differential function of $\mathbf{x}$. The resulting wave equation has a natural interpretation as describing the continuous local evolution of a wave in space (inserting $\psi = R\,\mathrm{e}^{\mathrm{i}S/\hbar}$ yields simple generalizations of (3.2.6) and (3.2.7)). But suppose we tried to develop a causal interpretation in the momentum representation ($\hat{\mathbf{p}} \to \mathbf{p}$, $\hat{\mathbf{x}} \to -\mathrm{i}\hbar\nabla_{\mathbf{p}}$) (for a discussion see Epstein (1952, 1953), Bohm (1952d), and Freistadt (1957)). Then (3.12.1) would involve integro-differential operators and the substitution of $\varphi(\mathbf{p}) = R\,\mathrm{e}^{\mathrm{i}S/\hbar}$ (where R and S are functions of $\mathbf{p}$) into the wave equation would not lead to a clear physical interpretation. Moreover, we shall see on many occasions that the momentum representation cannot be consistently associated with the current momentum of a system. Of course, excluding the momentum representation does not imply that other representations may not be valuable for a causal interpretation in some cases (cf. §5.5).

Another example where transformation symmetry is broken is the quantization of the electromagnetic field which is carried out in the momentum representation and apparently cannot be formulated in configuration space.

Summarizing, we may say that the quantum mechanical transformation theory is valuable in facilitating the mathematical solution of the wave equation, but this does not imply that all coordinate systems in Hilbert space are physically equivalent. In seeking a causal representation of phenomena in spacetime, we postulate that the Schrödinger picture and the position representation are preferred. It would be wrong to turn this into a dogma, but it seems most appropriate at the level of elementary wave mechanics.

3.12.2 The method of the causal interpretation

In §3.8 we pointed out that the quantum mechanical expectation values of the operators $\hat{\mathbf{x}}$, $\hat{\mathbf{p}}$, $\hat{H}$, $\hat{\mathbf{L}}$ and $\hat{\mathbf{L}}^2$ could be expressed as ensemble averages (with respect to R^2) of certain expressions (the 'local expectation values' of §3.5) formed from the fields R and S which may be associated with physically meaningful properties of a particle. This formulation of averages can be extended to arbitrary operators $\hat{A}(\hat{\mathbf{x}}, \hat{\mathbf{p}})$ if we introduce further properties defined in terms of R and S.

Consideration of the expectation value of an operator provides the key to formalizing a 'method of the causal interpretation' which is intended as a guide to how we might proceed in formulating causal interpretations of other quantum theories, besides elementary wave mechanics.

Consider a complete set of commuting operators and work in the representation $|\xi\rangle$ in which they are simultaneously diagonal (ξ labels the eigenvalues of the set of operators). This is a 'position representation' understood in the broad sense that the commutation relations of the theory are realized by differential operators acting on continuous functions. An individual system has an associated wavefunction which in this representation has the components $\psi(\xi) = \langle\xi|\psi\rangle$ and satisfies the Schrödinger equation

$$\mathrm{i}\hbar\,\frac{\partial\psi(\xi)}{\partial t} = (\hat{H}\psi)(\xi) = \int \hat{H}(\xi, \xi')\psi(\xi')\,\mathrm{d}\xi', \tag{3.12.2}$$

where $\hat{H}(\xi, \xi') = \langle \xi | \hat{H} | \xi' \rangle$ is the Hamiltonian. The hermiticity of the latter implies that

$$(\mathrm{d}/\mathrm{d}t) \int |\psi(\xi)|^2 \, \mathrm{d}\xi = 0. \tag{3.12.3}$$

Thus $|\psi(\xi)|^2$ may be interpreted as the probability density in ξ-space.

Consider an Hermitian operator $\hat{A}$ whose matrix representation is $\hat{A}(\eta, \eta')\,\delta(\varepsilon, \varepsilon')$ where η is a subset of the indices ξ and ε is the complement of η in $\{\xi\}$: $\{\xi\} = \{\eta\} \cup \{\varepsilon\}$. The expectation value of $\hat{A}$ in the state $|\psi\rangle$ is given by

$$\begin{aligned}\langle \hat{A} \rangle &= \int \mathrm{Re}\, \psi^*(\xi)\hat{A}(\eta, \eta')\,\delta(\varepsilon, \varepsilon')\psi(\eta', \varepsilon')\, \mathrm{d}\eta\, \mathrm{d}\varepsilon\, \mathrm{d}\eta'\, \mathrm{d}\varepsilon' \\ &= \int \mathrm{Re}\, \psi^*(\xi)\hat{A}(\eta, \eta')\psi(\eta', \varepsilon)\, \mathrm{d}\eta\, \mathrm{d}\varepsilon\, \mathrm{d}\eta' \\ &= \int |\psi(\xi)|^2[\mathrm{Re}\, \psi^*(\xi)\hat{A}(\eta, \eta')\psi(\eta', \varepsilon)/|\psi(\xi)|^2]\, \mathrm{d}\eta'\, \mathrm{d}\xi.\end{aligned}$$

We define

$$A(\xi) = \mathrm{Re} \int \psi^*(\xi)\hat{A}(\eta, \eta')\psi(\eta', \varepsilon)\, \mathrm{d}\eta'/|\psi(\xi)|^2 \tag{3.12.4}$$

as the 'local expectation value' of $\hat{A}$ in the state $|\psi\rangle$ in the representation $|\xi\rangle$, since

$$\langle \hat{A} \rangle = \int |\psi(\xi)|^2 A(\xi)\, \mathrm{d}\xi. \tag{3.12.5}$$

It is 'local' in virtue of being a function in ξ-space. It would be misleading to term this expression a 'local observable' since it will be ascribed an objective significance which makes no reference to observation. Another name might be 'beable' (Bell, 1987). It is a function that may be associated with a single system in the ensemble since the numerator, which is a density in ξ-space, has been divided by the denominator, which determines the density of systems.

The idea is now to derive from the Schrödinger equation an equation of motion for $A(\xi, t)$.

To see how we may obtain this, consider the case where (3.12.4) reduces to

$$A(\xi) = \mathrm{Re}\, \psi^*(\xi)\hat{A}(\xi)\psi(\xi)/|\psi(\xi)|^2 \tag{3.12.6}$$

with $\hat{A}(\xi, \xi') = \hat{A}(\xi)\,\delta(\xi, \xi')$. Then from (3.12.2)

$$\partial A/\partial t = (1/|\psi|^2)\,\mathrm{Re}\,\psi^*\{\partial\hat{A}/\partial t + (1/i\hbar)[\hat{A}, \hat{H}]\}\psi - (A/|\psi|^2)\,\partial|\psi|^2/\partial t$$
$$+ (1/|\psi|^2)\,\mathrm{Re}(1/i\hbar)[\psi^*\hat{H}\hat{A}\psi - (\hat{H}\psi)^*\hat{A}\psi]. \quad (3.12.7)$$

The task is now to express the right hand side of (3.12.7) in terms of A and its derivatives together with other known functions. This cannot be done explicitly unless we specify $\hat{A}$ and $\hat{H}$, but it is worth noting that the first term on the right hand side is proportional to the integrand in the Ehrenfest theorem and that the remaining local terms disappear when we take averages. As an example, in the case studied in this chapter where $\xi = \mathbf{x}$, putting $\hat{A} = \hat{\mathbf{p}}$ yields $A(\mathbf{x}, t) = \nabla S$ and (3.12.7) becomes (3.2.23). If the variables ξ represent some physically meaningful property of the system, we define the 'trajectory' of the system to be the solutions $\xi(t)$ to a set of equations

$$\mathrm{d}\xi/\mathrm{d}t = f(\psi) \quad (3.12.8)$$

(given $\xi(0)$) and hence associate $A(\xi(t))$ with a further property of the system.

The systems to which this method may be applied are of a very general character. The ξs need not necessarily include the coordinates of space and hence may not describe the track of a particle. They may be a set of configuration space variables of a many-body system or something else entirely, such as field coordinates. In all cases, $\psi(\xi)$ is envisaged as a component of an individual physical system, the other component being the entity with the property $\xi(t)$. Although we argue that physical space has a primary significance in elementary wave mechanics, extension of the causal interpretation to other types of systems requires that we study processes in other types of spaces.

Finally, we note that application of the method does not always lead to meaningful results. The 'method' is, in any case, only a guide and we shall deviate from it in certain details in extending the quantum theory of motion to include the Pauli theory of spin (Chap. 9) and field theory (Chap. 12).

3.13 Comparison with classical Hamilton–Jacobi theory

The essential difference between the classical and quantum treatments of the motion of a particle may be expressed as follows.

In classical mechanics the mechanical problem has a unique solution once the external potential is specified and the initial position and velocity of a particle are given. If one formulates the problem in phase space and solves for the motion using the Hamilton–Jacobi equation, all of the (infinite number of) S-functions corresponding to the given initial conditions generate the same motion. The system evolution is independent of the 'state' defined by the classical wavefunction. The S-functions are distinguished by the global ensembles they generate (cf. §2.3).

In the quantum theory of motion, specification of the external potential and the initial position and velocity of a particle is not sufficient to determine the motion uniquely; one must specify in addition the quantum state. A set of particles in classically identical initial states (same $\mathbf{x}_0$, $\mathbf{p}_0 = \nabla S_0(\mathbf{x}_0)$, $V(\mathbf{x})$) are no longer in identical states as their subsequent motions generically differ in infinitely variable ways. This implies that the quantal S-functions are distinguished in a much stronger way than their classical counterparts, at the level of individual ensemble elements. (Examples are free particles with the same initial momentum in plane wave and Gaussian states (§§4.2 and 4.7) and harmonic oscillators in stationary and packet states which start from rest (§4.9).)

An objection has been raised against this treatment of motion in quantum mechanics (Halpern, 1952; Bohm, 1952c) which, although we have essentially already answered it in Chap. 2, is worth repeating. The argument centres around the legitimacy of applying techniques of Hamilton–Jacobi theory in a context where the S-function is basically a field. For it might be argued that a solution of the problem of motion requires a complete integral depending on a set of nonadditive constants, which, in general, the quantum S-function is not. Moreover, if S does depend on a set of constants (i.e., quantum numbers) these will also appear in the function R and hence in the (quantum) potential acting on the particle, something that cannot happen in classical mechanics.

This objection does not represent so much a criticism of quantum Hamilton–Jacobi theory as a too narrow conception of Hamilton–Jacobi theory in classical mechanics. For even in the latter we are not constrained to formulate the problem in terms of an S-function depending on arbitrary constants – if we do not possess such a solution, but have rather a general solution, we can still solve for the motion by integrating $\dot{\mathbf{x}} = \nabla S/m$ and specifying $\mathbf{x}(0)$. Therefore, the fact that the quantum phase function may not depend on a set of constants cannot be counted as an argument against the quantum theory of motion. S may, of course, depend on a set of quantum numbers. But these constants are generally discrete and are not of the type employed in the classical Jacobi method, which are continuously variable. In this case again we solve for the motion by integrating $\dot{\mathbf{x}} = \nabla S/m$ and the fact that the total potential acting on the particle is a function of the quantum numbers is of no consequence for the formal problem of solving the equation of motion. Having a potential depend on the same constants as the S-function is, of course, a novel feature, but this is relevant only to the varieties of motions that are implied by ψ (since changing the constants alters the character of the motion) rather than to the method of solving for them.

To make this point clearer let us fix attention on a solution to the Schrödinger equation and treat the quantum potential derived from it as a given function of $\mathbf{x}$ and t (on the same footing as the external potential). Then we may try to solve the equation of motion (3.2.24),

$$m\ddot{\mathbf{x}} = -\nabla(V + Q)|_{\mathbf{x}=\mathbf{x}(t)}, \tag{3.13.1}$$

directly by the Jacobi method, as in classical mechanics (Kyprianidis, 1988a). That is, we define the Hamiltonian to be

$$H(\mathbf{x}, \mathbf{p}) = \mathbf{p}^2/2m + V(\mathbf{x}) + Q(\mathbf{x}), \tag{3.13.2}$$

corresponding to a Lagrangian $L(\mathbf{x}, \dot{\mathbf{x}}, t) = \frac{1}{2}m\dot{\mathbf{x}}^2 - V - Q$, and seek solutions $\varphi = \varphi(\mathbf{x}, \alpha_i)$ to the Hamilton–Jacobi equation

$$\partial\varphi/\partial t + H(\mathbf{x}, \nabla\varphi) = 0 \tag{3.13.3}$$

which depend on a set of nonadditive constants α_i. Let us suppose that we find such a solution and that the α_is are the initial momentum coordinates. In general, $\nabla\varphi \neq \nabla S$, where S is the phase of the wavefunction which determines Q. In order to ensure consistency with the Schrödinger equation, which yields Q uniquely given S_0 (and R_0), we require that at the initial position $\mathbf{x}_0$ of the particle

$$\boldsymbol{\alpha} = \nabla\varphi_0|_{\mathbf{x}=\mathbf{x}_0} = \nabla S_0|_{\mathbf{x}=\mathbf{x}_0}.$$

But still, in general, φ and S will not coincide globally (i.e., φ does not satisfy the conservation law), because S is not generally a special case of a complete integral. Nevertheless, applying the Jacobi method of solution to φ, or directly integrating $\dot{\mathbf{x}} = \nabla S/m$, we will deduce the same trajectory from both approaches for the given initial conditions $(\mathbf{x}_0, \boldsymbol{\alpha})$. The complete integral method is, of course, of no practical use since in order to find Q we need to know $S_0(\mathbf{x})$, and hence $S(\mathbf{x}, t)$, globally.

The classical S-function propagates in the manner of a wavefront in geometrical optics (§2.4) – each point on a wavefront maps into unique points on previous and successive wavefronts. In the quantum case the propagation of the wavefronts follows from the law of propagation of the wave which, because it is linear and first order in time, takes the form

$$\psi(\mathbf{x}, t) = \int G(\mathbf{x}, t; \mathbf{x}_0, t_0)\psi_0(\mathbf{x}_0)\, \mathrm{d}^3x_0, \tag{3.13.4}$$

where G is the Green function (Huygens' principle). What (3.13.4) implies is that at time t each point $\mathbf{x}$ receives contributions from the whole of space at time t_0, and that each point $\mathbf{x}_0$ maps into many points $\mathbf{x}$. We can find $S(\mathbf{x}, t)$ from (3.13.4) via $S = (1/2\mathrm{i}) \log(\psi/\psi^*)$ and it depends on $R_0(\mathbf{x}_0)$ as well as $S_0(\mathbf{x}_0)$, as we expect.

In §2.4 it was shown that the wavefronts propagate with speed $u = -(\partial S/\partial t)/|\nabla S|$. That demonstration remains valid here and one may arrive at the same result by forming the product λv using (3.3.19) and (3.3.21).

Appendix A

Inserting (3.2.1) into (3.2.5) and multiplying through by $e^{-\mathrm{i}S/\hbar}$ we obtain

$$\mathrm{i}\hbar\left[\frac{\partial R}{\partial t} + \left(\frac{\mathrm{i}R}{\hbar}\right)\frac{\partial S}{\partial t}\right] = -\frac{\hbar^2}{2m}\left\{\nabla^2 R - \left(\frac{R}{\hbar^2}\right)(\nabla S)^2 + \mathrm{i}\left[\left(\frac{2}{\hbar}\right)\nabla R\cdot\nabla S + \left(\frac{R}{\hbar}\right)\nabla^2 S\right]\right\} + VR.$$

Separating into real and imaginary parts yields respectively (3.2.6) and (3.2.7) of the text if $R \neq 0$ and we divide the real part by R and multiply the imaginary part by $2R/\hbar$.

Appendix B

In Cartesian coordinates

$$\hat{L}_i = (\hat{\mathbf{x}} \times \hat{\mathbf{p}})_i = \varepsilon_{ijk}\hat{x}_j\hat{p}_k, \qquad i, j, k = 1, 2, 3$$

(sum over repeated indices) where ε_{ijk} is the antisymmetric symbol, with $\varepsilon_{123} = 1$. Then

$$\hat{\mathbf{L}}^2 = (\hat{\mathbf{x}} \times \hat{\mathbf{p}})^2 = \hat{x}_j\hat{p}_k\hat{x}_j\hat{p}_k - \hat{x}_j\hat{p}_k\hat{x}_k\hat{p}_j,$$

where we have used the identity

$$\varepsilon_{ijk}\varepsilon_{inm} = \delta_{jn}\delta_{km} - \delta_{jm}\delta_{kn}.$$

Inserting $\hat{p}_i = -\mathrm{i}\hbar\,\partial_i$ this becomes

$$\hat{\mathbf{L}}^2 = -\hbar^2(\mathbf{x}^2\nabla^2 - 2\mathbf{x}\cdot\nabla - x_i x_j\,\partial_{ij}). \tag{B1}$$

We then easily find

$$\mathrm{Re}\,\psi^*\hat{\mathbf{L}}^2\psi/|\psi|^2 = \mathbf{x}^2(\nabla S)^2 - (\mathbf{x}\cdot\nabla S)^2 - \hbar^2\mathbf{x}^2\nabla^2\mathrm{R}/\mathrm{R} + \hbar^2 x_i x_j\,\partial_{ij}R/R + 2\hbar^2\mathbf{x}\cdot\nabla R/R. \tag{B2}$$

To derive (3.5.12) of the text we combine the first two terms on the right hand side of (B2) using the identity

$$(\mathbf{x} \times \mathbf{p})^2 = \mathbf{x}^2\mathbf{p}^2 - (\mathbf{x}\cdot\mathbf{p})^2$$

and combine the last three terms using the expression (B1) for $-\hbar^2(\mathbf{x} \times \nabla)^2$.

4

Simple applications

4.1 Stationary states

4.1.1 General properties

We shall consider first some general characteristics of the 'stationary states'. These are eigenfunctions of the Hamiltonian operator:

$$\hat{H}\psi(\mathbf{x}, t) = E\psi(\mathbf{x}, t), \qquad \hat{H} = \frac{-\hbar^2}{2m}\nabla^2 + V. \tag{4.1.1}$$

The requirement that ψ satisfies the Schrödinger equation fixes its time dependence:

$$\psi(\mathbf{x}, t) = \psi_0(\mathbf{x})\,\mathrm{e}^{-\mathrm{i}Et/\hbar}. \tag{4.1.2}$$

Notice that (4.1.2) can only be a solution if the external potential V is independent of the time: $V = V(\mathbf{x})$. The initial function ψ_0 is not arbitrary but must satisfy the time-independent Schrödinger equation (4.1.1).

Writing $\psi = R\,\mathrm{e}^{\mathrm{i}S/\hbar}$ one thus has

$$\left.\begin{aligned} R(\mathbf{x}, t) &= R_0(\mathbf{x}) \\ S(\mathbf{x}, t) &= S_0(\mathbf{x}) - Et. \end{aligned}\right\} \tag{4.1.3}$$

The following deductions may be made from (4.1.3):

(*a*) The probability density is independent of the time: $|\psi|^2 = R_0^2(\mathbf{x})$.

(*b*) The quantum potential $Q = (-\hbar^2/2mR_0)\nabla^2 R_0$ is time-independent. Therefore so is the total effective potential:

$$\partial(V + Q)/\partial t = 0. \tag{4.1.4}$$

(*c*) The velocity field $\mathbf{v} = (1/m)\nabla S_0$ is independent of time and hence the trajectories, found by solving $m\dot{\mathbf{x}} = \mathbf{v}(\mathbf{x})|_{\mathbf{x}=\mathbf{x}(t)}$, are solutions of

$$\mathrm{d}x/v_1(\mathbf{x}) = \mathrm{d}y/v_2(\mathbf{x}) = \mathrm{d}z/v_3(\mathbf{x}). \tag{4.1.5}$$

Moreover, $\psi_0(\mathbf{x})$ is often a real function for many bound state problems of interest (although not for scattering states) which implies that the velocity is zero. The particle is at rest where one would classically expect it to move since the quantum force $(-\nabla Q)$ cancels the classical force $(-\nabla V)$.

(*d*) The energy of all particles in the ensemble, $-\partial S/\partial t$, is a constant of the motion and equal to the energy eigenvalue:

$$-\partial S/\partial t = E. \tag{4.1.6}$$

Comparing with (4.1.4), we have an example of the conservation of energy (cf. §3.9.3).

(*e*) The total field energy is conserved: $\langle \hat{H} \rangle = E$.

(*f*) A stationary state is not preserved under Galilean transformations. From (3.11.8) the energy in the transformed frame is given by

$$-\frac{\partial S'(\mathbf{x}', t')}{\partial t'} = -\frac{\partial S(\mathbf{x}, t)}{\partial t} - mV_i a_{ij}\, \partial_j S + \tfrac{1}{2}m\mathbf{V}^2$$

$$= E - mV_i a_{ij}\, \partial_j S_0 + \tfrac{1}{2}m\mathbf{V}^2$$

using (4.1.3), and this is not a constant unless $\partial_j S_0(\mathbf{x}) = \text{constant}$ (a requirement that is valid for plane waves).

(*g*) The generalized de Broglie wavelength and frequency, (3.3.20) and (3.3.21), are time-independent:

$$\left.\begin{aligned} \lambda_Q(\mathbf{x}) &= h/[2m(E - V - Q)]^{1/2} = h/|\nabla S_0| \\ \nu_Q(\mathbf{x}) &= E/h. \end{aligned}\right\} \tag{4.1.7}$$

The speed of a wavefront is therefore

$$u(\mathbf{x}) = \lambda_Q \nu_Q = E/[2m(E - V - Q)]^{1/2} = E/|\nabla S_0|. \tag{4.1.8}$$

If the wavefunction is a superposition of stationary states none of the properties (*a*)–(*g*) hold in general. However, the stationary states are not the only states in which energy is conserved. Particle energy is conserved whenever $-\partial S/\partial t|_{\mathbf{x}=\mathbf{x}(t)} = \text{constant}$. The requirement $-\partial S/\partial t = E = \text{constant}$ for the entire ensemble is satisfied by any wavefunction of the type

$$\psi(\mathbf{x}, t) = R(\mathbf{x}, t)\, e^{i(S_0(\mathbf{x}) - Et)/\hbar}, \tag{4.1.9}$$

so long as $\partial(V + Q)/\partial t = 0$. But such a solution is not an energy eigenstate (4.1.1) and hence E does not have the significance of an energy *level* (for an example see §4.3). Still more general solutions displaying energy conservation for individual ensemble elements are possible if we allow the energy to vary over the ensemble (see §6.3).

Substituting (4.1.3), the Hamilton–Jacobi and conservation equations reduce to

$$E = (\nabla S_0)^2/2m + Q + V \tag{4.1.10}$$

$$\nabla\cdot(R_0^2\nabla S_0) = 0. \tag{4.1.11}$$

In the one-dimensional case we can explicitly find R_0 in terms of S_0 and so write (4.1.10) purely in terms of S_0. From (4.1.11), $(\mathrm{d}/\mathrm{d}x)(R_0^2 S_0') = 0$ so that

$$R_0(x) = c(\mathrm{d}S_0/\mathrm{d}x)^{-1/2} \tag{4.1.12}$$

where c = constant (assumed to be nonzero). Substituting (4.1.12) into the expression for the quantum potential, (4.1.10) becomes

$$E = (1/2m)S_0'^2 + (\hbar^2/4m)(-\tfrac{3}{2}S_0'^{-2}S_0''^2 + S_0'^{-1}S_0''') + V, \tag{4.1.13}$$

which is independent of c. This equation is of the type discussed in §2.7. The surfaces $S(x, t)$ = constant propagate as wavefronts, but notice how the law of propagation is quite different from the classical case (§2.4). We cannot express the wave velocity (4.1.8) in terms of known functions of x (i.e., V) alone because ∇S_0 is a function of the derivatives of S_0 itself. Correspondingly, the particle motions have, in general, little in common with classical conservative motion in the potential V.

Alternatively, we can eliminate S_0 from the one-dimensional versions of (4.1.10) and (4.1.11) to leave a differential equation for R_0. From (4.1.10)

$$S_0' = [2m(E - V - Q)]^{1/2} \tag{4.1.14}$$

and substituting this into (4.1.11) yields

$$R_0'' + (2m/\hbar^2)(E - V)R_0 = \text{constant} \times R_0^{-3}. \tag{4.1.15}$$

This nonlinear differential equation is of known form, being an Ermakov system (Reid & Ray, 1984; Nassar, 1990).

4.1.2 *Is the ground energy the lowest possible?*

Suppose the system is in a bound state and the energy spectrum, which is discrete, has a minimum value $E_1 < 0$ (the ground state). Then the total energy of the particle according to the quantum theory of motion is, in this state, E_1. Is this the lowest energy that the particle may ever possess in the course of its motion? Certainly if the system is confined to stationary states this will be so, but it is easy to construct examples where the total energy may be *less* than E_1, in nonstationary states.

Consider the wavefunction

$$\psi(\mathbf{x}, t) = A(\mathrm{e}^{-\mathrm{i}E_1 t/\hbar} + aB\,\mathrm{e}^{-\mathrm{i}E_2 t/\hbar}), \tag{4.1.16}$$

where A and B are prescribed real functions of $\mathbf{x}$, a is an arbitrary real constant, and E_2 ($>E_1$) is the energy of an excited state. We shall show that we may choose a so that $-\partial S/\partial t < E_1$ at a certain spacetime point, where S is the phase of (4.1.16) which satisfies

$$\tan[(S + \tfrac{1}{2}(E_1 + E_2)t)/\hbar] = [(1 - aB)/(1 + aB)] \tan[(E_1 - E_2)t/2\hbar]. \quad (4.1.17)$$

Differentiating (4.1.17) we obtain

$$-\frac{\partial S}{\partial t}(\mathbf{x}, t) = \tfrac{1}{2}(E_1 - E_2)(aB - 1)(aB + 1)^{-1} \times \{\cos^2[(E_1 - E_2)t/2\hbar] + [(1 - aB)^2/(1 + aB)^2] \times \sin^2[(E_1 - E_2)t/2\hbar]\}^{1/2}.$$

Consider the instants of time, $t = T$, at which the square root factor is unity. Then we require that, instantaneously,

$$-\frac{\partial S}{\partial t}(\mathbf{x}, T) = \tfrac{1}{2}(E_1 - E_2)(aB - 1)/(aB + 1) < E_1. \quad (4.1.18)$$

If we fix attention on a definite point $\mathbf{x}$ it is obvious that we may choose a so that (4.1.18) is satisfied whatever the values of B, E_1 and E_2. We take

$$a < (E_2 - 3E_1)/B(E_1 + E_2). \quad (4.1.19)$$

Assuming that the point $\mathbf{x}$ at which (4.1.18) is evaluated is not a node, a particle may pass through it and, if (4.1.19) is satisfied, its energy will be instantaneously lower than the ground state energy.

The fact that in nonstationary states the actual energy may be less than the minimum energy eigenvalue does not contradict spectroscopic data since in a measurement of energy the system is left in an energy eigenstate and E_1 is the lowest value that will be found.

4.2 Plane and spherical waves

An eigenfunction of the momentum operator corresponding to the eigenvalue $\hbar\mathbf{k}$ has the general form

$$\psi(\mathbf{x}, t) = A(t)\, e^{i\mathbf{k}\cdot\mathbf{x}}, \qquad \hat{\mathbf{p}}\psi = \hbar\mathbf{k}\psi. \quad (4.2.1)$$

This will be a solution of the wave equation if the system is free and ψ is a stationary state:

$$\psi(\mathbf{x}, t) = L^{-3/2}\, e^{i(\mathbf{k}\cdot\mathbf{x} - \omega_{\mathbf{k}}t)}, \qquad \omega_{\mathbf{k}} = \hbar\mathbf{k}^2/2m. \quad (4.2.2)$$

The function (4.2.2) has been normalized in a box of side L. Since the amplitude is constant the quantum potential vanishes and the wave is of 'classical' type (i.e., it satisfies the classical Hamilton–Jacobi and conservation equations). The phase function is

$$S(\mathbf{x}, t) = \hbar\mathbf{k}\cdot\mathbf{x} - \hbar\omega_{\mathbf{k}}t \tag{4.2.3}$$

apart from a constant. The wavefronts $S =$ constant are planes propagating in the $\mathbf{k}$-direction with speed $\omega_{\mathbf{k}}/|\mathbf{k}| = \hbar|\mathbf{k}|/2m$. The particle orbits, which are orthogonal to the S-surfaces, are evidently straight lines in space. This is confirmed by applying the guidance formula; we have

$$\mathbf{p} = \nabla S = \hbar\mathbf{k}, \tag{4.2.4}$$

$$E = -\partial S/\partial t = \hbar\omega_{\mathbf{k}}, \tag{4.2.5}$$

which are the original de Broglie relations and, writing $\mathbf{p} = m\dot{\mathbf{x}}$, we have from (4.2.4)

$$\mathbf{x}(t) = \mathbf{x}_0 + \hbar\mathbf{k}t/m \tag{4.2.6}$$

where $\mathbf{x}_0$ may lie anywhere in space (Fig. 4.1). The path is therefore uniform and rectilinear and the energy is constant.

In the usual interpretation the state (4.2.2), being an eigenstate of momentum, is considered to be one in which the momentum of the particle is definite $(=\hbar\mathbf{k})$ but the position is completely 'unknown' (or even 'does not exist'). But as we see, the particle may be conceived as having a well-defined position when the wave is a momentum eigenfunction, and the 'uncertainty' relation does not have the implication normally ascribed to it (cf. §§6.3, 8.5).

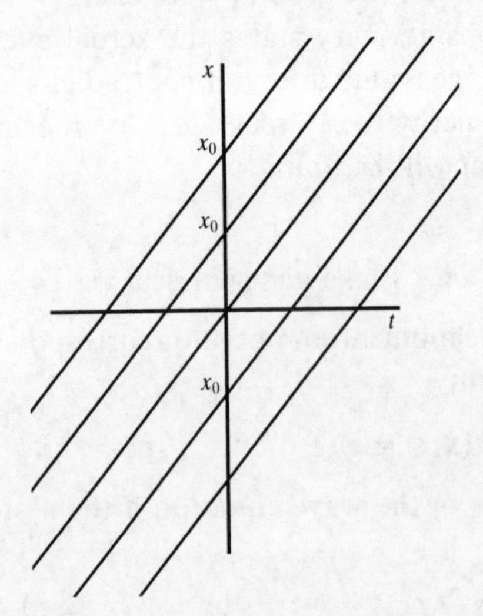

Fig. 4.1 A selection of trajectories (4.2.6) in one dimension (x) for different choices of x_0. The gradient is fixed by the particle velocity, $\hbar k/m$.

The expression (4.2.2) is not the only function whose wavefronts are planes. We may construct other solutions whose phases are given by (4.2.3) but whose amplitudes are variable (cf. §§4.3, 4.11). The quantum potentials associated with these waves are constant and the motion is still uniform and rectilinear, but the waves are no longer momentum eigenfunctions. Just as energy may be conserved in nonstationary states, so momentum may be conserved when the wavefunction is not a momentum eigenstate. A similar result is obtained for a spherical wave

$$\psi = (A/r)\, \mathrm{e}^{\mathrm{i}(kr - Et/\hbar)}, \tag{4.2.7}$$

where $r = (x^2 + y^2 + z^2)^{1/2}$ and $E = \hbar^2 k^2/2m$. The particle moves uniformly along a radius and the direction of the momentum varies over the ensemble.

4.3 Superposition of plane waves

The principle of superposition allows one to construct solutions to the wave equation whose properties are qualitatively distinct from the interfering partial waves. Here we shall examine the trajectories associated with the superposition of two plane waves, each of whose associated motions are of course straight lines. Consider

$$\psi(\mathbf{x}, t) = A[\psi_1(\mathbf{x}, t) + c\psi_2(\mathbf{x}, t)] \tag{4.3.1}$$

where

$$\psi_j(\mathbf{x}, t) = \mathrm{e}^{\mathrm{i}(\mathbf{k}_j \cdot \mathbf{x} - \omega_{\mathbf{k}_j} t)}, \qquad \omega_{\mathbf{k}_j} = \hbar \mathbf{k}_j^2/2m, \tag{4.3.2}$$

$j = 1, 2$, and $A = |A|\, \mathrm{e}^{\mathrm{i}\varphi}$ and $c = |c|\, \mathrm{e}^{\mathrm{i}\delta}$ are complex constants. $|c|$ determines the relative amplitude of the waves. We assume throughout that $\mathbf{k}_1 \neq \mathbf{k}_2$; if $\mathbf{k}_1 = \mathbf{k}_2$ we recover a simple plane wave. In §3.3.10 we gave the formulae for the amplitude and phase of ψ:

$$R = |A|(1 + |c|^2 + 2|c| \cos \xi)^{1/2} \tag{4.3.3}$$

$$\begin{aligned}\tan(S/\hbar - \varphi) = {} & [\sin(\mathbf{k}_1 \cdot \mathbf{x} - \omega_{\mathbf{k}_1} t) + |c| \sin(\mathbf{k}_2 \cdot \mathbf{x} - \omega_{\mathbf{k}_2} t + \delta)] \\ & \times [\cos(\mathbf{k}_1 \cdot \mathbf{x} - \omega_{\mathbf{k}_1} t) + |c| \cos(\mathbf{k}_2 \cdot \mathbf{x} - \omega_{\mathbf{k}_2} t + \delta)]^{-1},\end{aligned} \tag{4.3.4}$$

where

$$\xi = (\mathbf{k}_1 - \mathbf{k}_2) \cdot \mathbf{x} - (\omega_{\mathbf{k}_1} - \omega_{\mathbf{k}_2})t - \delta \tag{4.3.5}$$

is the relative phase. Clearly, ψ is not a stationary state (unless $\mathbf{k}_1 = -\mathbf{k}_2$).

The function (4.3.3) displays a set of interference fringes whose maxima

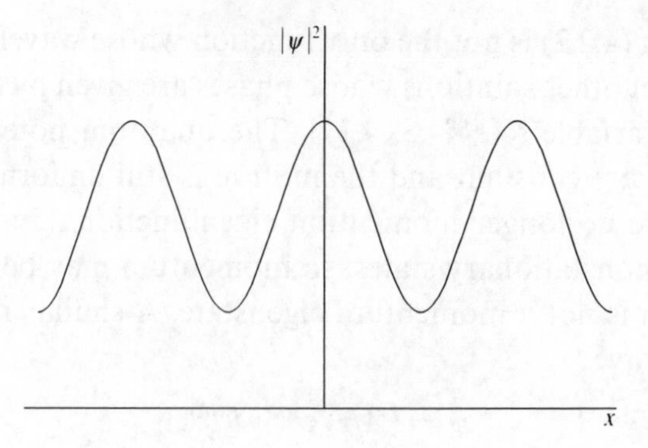

Fig. 4.2 A snapshot of the function (4.3.3) in one dimension. As time passes the pattern moves to the right.

(minima) occur where $\cos \xi = 1(-1)$ (see Fig. 4.2). The difference between the maxima and minima is proportional to $|c|$. The wavefunction has strict nodes only in the case $|c| = 1$ (the partial waves have the same amplitude) and $\cos \xi = -1$. In this case the nodes are a set of propagating planes:

$$(\mathbf{k}_1 - \mathbf{k}_2)\cdot\mathbf{x} - (\omega_{\mathbf{k}_1} - \omega_{\mathbf{k}_2})t - \delta = (2n + 1)\pi, \tag{4.3.6}$$

where $n \in \mathbb{Z}$ labels which surface.

From (4.3.4) we find for the momentum field

$$\nabla S = \hbar[\mathbf{k}_1 + |c|^2\mathbf{k}_2 + |c|(\mathbf{k}_1 + \mathbf{k}_2)\cos\xi](1 + |c|^2 + 2|c|\cos\xi)^{-1} \tag{4.3.7}$$

which is a well-defined function of $\mathbf{x}$ and t everywhere except at the nodes (where the denominator vanishes). In the one-dimensional case we can give an explicit expression for the trajectories as a function of time, for $m\dot{x} = \partial S/\partial x$ is satisfied by the relation (Takabayasi, 1953)

$$\begin{aligned}\sin[(k_1 - k_2)x - (\omega_{k_1} - \omega_{k_2})t - \delta]\\ + [(k_1 - k_2)/2|c|][(1 + |c|^2)x - (\hbar t/m)(k_1 + |c|^2k_2)] = \text{constant}.\end{aligned} \tag{4.3.8}$$

To see the character of the trajectories implied by this relation we rewrite it in the form

$$\xi + \varepsilon \sin\xi = \omega t + a \tag{4.3.9}$$

where

$$\xi = (k_1 - k_2)x - (\omega_{k_1} - \omega_{k_2})t - \delta \tag{4.3.10}$$

as defined in (4.3.5),

$$\varepsilon = 2|c|/(1 + |c|^2), \qquad 0 \leqslant \varepsilon \leqslant 1, \tag{4.3.11}$$

$$\omega = [\hbar(k_1 - k_2)^2/2m](1 - |c|^2)/(1 + |c|^2), \tag{4.3.12}$$

which may be positive, negative or zero, and a is a constant defined in terms of the initial position x_0 by

$$a = (k_1 - k_2)x_0 - \delta + \varepsilon \sin[(k_1 - k_2)x_0 - \delta]. \tag{4.3.13}$$

x_0 is arbitrary except that in the case $|c| = 1$ it cannot lie on the nodes (the particle remains between a given pair of nodes for all time). If we think for a moment of the motion referred to the axes $(\xi, \omega t)$ we see from (4.3.9) that the particle oscillates symmetrically about the line $\xi = \omega t + a$ with period $1/\omega$ (i.e., when $\omega t \to \omega t + 2\pi$, $\xi \to \xi + 2\pi$) and amplitude ε. Translating this into a spacetime diagram (for an example see Fig. 4.3), the particle oscillates about the line

$$x = vt + (\delta + a)/(k_1 - k_2), \tag{4.3.14}$$

where

$$v = \hbar(k_1 + k_2|c|^2)/m(1 + |c|^2), \tag{4.3.15}$$

with an amplitude $\varepsilon/(k_1 - k_2)$. When $t \to t + 2\pi/\omega$, $x \to x + 2\pi v/\omega$ as is clear from (4.3.9) and (4.3.10) and confirmed by (4.3.19) below. Different choices of a correspond to different x_0s via (4.3.13). If we consider two lines (4.3.14) characterized by the constants a and a', their separation along the x-axis is given by $(a - a')/(k_1 - k_2)$. The distance between the corresponding initial

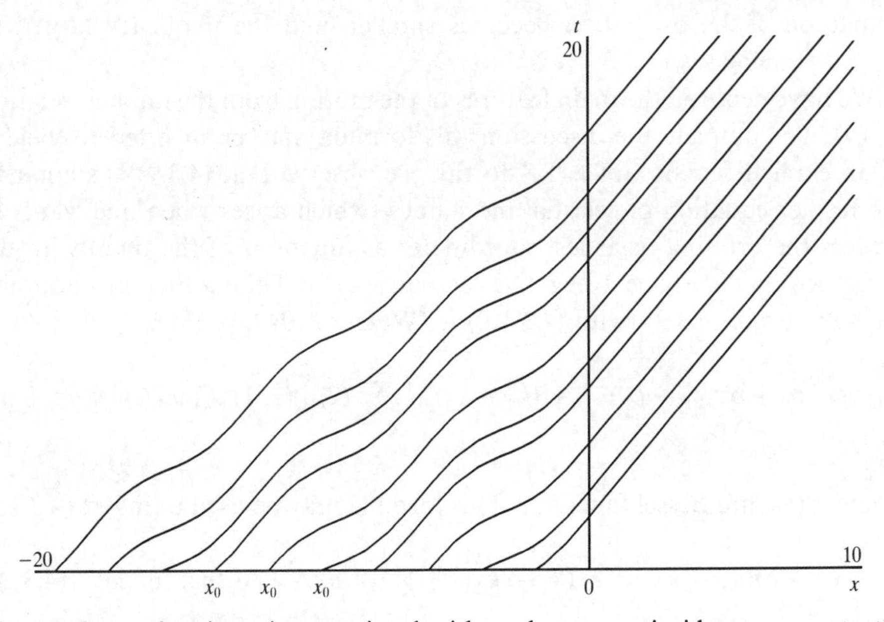

Fig. 4.3 Quantal trajectories associated with a plane wave incident on a potential step V at the origin. Energy $E > V$ (see §4.4).

positions x_0 and x_0' on the orbits associated with these two lines is not given by this expression, however, as may be seen from (4.3.13). Put differently, the particle need not start at the same point of the cycle for each orbit.

For a large disparity in the relative amplitudes of the component waves ($|c|$ very large or very small) the amplitude of oscillation of the particle is small and the path approaches the uniform motion implied by a pure plane wave:

$$x = \hbar k_1 t/m + x_0 \quad \text{or} \quad x = \hbar k_2 t/m + x_0 \tag{4.3.16}$$

where from (4.3.13), $x_0 = (\delta + a)/(k_1 - k_2)$ when $\varepsilon = 0$.

One might expect that the maximum oscillation will occur at the other extreme, when the subwaves are of equal amplitude ($|c| = 1 \Rightarrow \varepsilon = 1$). In fact, we obtain in this case again uniform motion since from (4.3.12) $|c| = 1$ implies $\omega = 0$ and so the period is infinite. The trajectory lies parallel to the line (4.3.14) but does not, in general, coincide with it:

$$x = \hbar(k_1 + k_2)t/2m + x_0. \tag{4.3.17}$$

From (4.3.6) it is clear that the nodes travel at the same speed:

$$x = \hbar(k_1 + k_2)t/2m + [\delta + (2n + 1)\pi](k_1 - k_2)^{-1} \qquad \text{(nodal eq.).} \tag{4.3.18}$$

The particle therefore remains at a constant distance from the nodes.

We can change the relative phase (4.3.10) by varying $(k_1 - k_2)$ or δ. The amplitude of the oscillation becomes smaller, and the frequency larger, as $(k_1 - k_2)$ increases.

We have deduced the main features of the motion from the implicit relation (4.3.9). To complete the discussion this formula may be inverted to yield x as an explicit function of t. To do this we observe that (4.3.9) is similar to the Kepler equation of celestial mechanics (which arises when one wants to express the orbiting particles' coordinates as functions of the time) with the difference that we have a negative 'eccentricity' ε. Taking this into account, the solution of (4.3.9) (with (4.3.10)) is (Watson, 1944, p. 551):

$$x = vt + (a + \delta)(k_1 - k_2)^{-1} + (k_1 - k_2)^{-1} \sum_{n=1}^{\infty} (2/n)(-1)^n J_n(n\varepsilon) \sin[n(\omega t + a)] \tag{4.3.19}$$

where $J_n(n\varepsilon)$ are Bessel functions. This formula may be used to invert (4.3.13):

$$x_0 = (a + \delta)(k_1 - k_2)^{-1} + (k_1 - k_2)^{-1} \sum_{n=1}^{\infty} (2/n)(-1)^n J_n(n\varepsilon) \sin na. \tag{4.3.20}$$

It is apparent that when $|c| = 1$, so that $\omega = 0$, we have from (4.3.19) and

(4.3.20) the equation (4.3.17) where x has a simple dependence on x_0. We readily confirm also from (4.3.19) the uniform motion (4.3.16) when $|c|$ is very large or very small. In all other cases $x(t)$ is a complicated function of x_0.

It will be observed that v in (4.3.15) is a weighted average of the velocities associated with the component plane waves and represents the velocity we might expect to obtain if we formally 'superposed' two classical motions of speeds $\hbar k_1/m$ and $\hbar k_2/m$ with suitable weights. The infinite sum of Bessel functions corresponds to the effects of interference between the plane waves and accounts for the oscillations about the classical superposition.

The case of equal amplitudes ($|c| = 1$) is interesting for another reason, for it provides an example where the energy and momentum are conserved even though ψ is not a stationary state or a momentum eigenfunction (cf. §§4.1, 4.2). To see this, we evaluate the quantum potential. From (4.3.3) its general form (in three dimensions now) is given by

$$\begin{aligned} Q &= -\hbar^2\nabla^2 R/2mR \\ &= \hbar^2|A|^2|c|(\mathbf{k}_1 - \mathbf{k}_2)^2(R^2\cos\xi + |A|^2|c|\sin^2\xi)/2mR^4 \end{aligned} \tag{4.3.21}$$

which is a function of $\mathbf{x}$ and t. When $|c| = 1$, however, it reduces to a constant:

$$Q = \hbar^2(\mathbf{k}_1 - \mathbf{k}_2)^2/8m. \tag{4.3.22}$$

Since from (4.3.7) the kinetic energy is also a constant,

$$(\nabla S)^2/2m = \hbar^2(\mathbf{k}_1 + \mathbf{k}_2)^2/8m, \tag{4.3.23}$$

it follows that the total energy (the sum of (4.3.22) and (4.3.23)) is a constant:

$$E = [(\hbar\mathbf{k}_1)^2 + (\hbar\mathbf{k}_2)^2]/4m. \tag{4.3.24}$$

This is simply the average of the kinetic energies associated with the component plane waves. This expression for the energy is confirmed by (4.3.4) which yields when $|c| = 1$

$$S = \tfrac{1}{2}\hbar(\mathbf{k}_1 + \mathbf{k}_2)\cdot\mathbf{x} - \hbar^2(\mathbf{k}_1^2 + \mathbf{k}_2^2)t/4m \tag{4.3.25}$$

apart from a constant. The wavefunction is thus a plane wave of the type (4.1.9) with a time- and space-dependent amplitude, and Q satisfies our energy conservation criterion ($\partial Q/\partial t = 0$) and our momentum conservation criterion ($\nabla Q = 0$).

Finally, we note that the above results may be used to give a simple model of an interferometer if we take as wavefunction the stationary state

$$\psi = A(\mathrm{e}^{\mathrm{i}kx} + |c|\,\mathrm{e}^{-\mathrm{i}kx + \mathrm{i}\delta})\,\mathrm{e}^{\mathrm{i}k'y}\,\mathrm{e}^{-\mathrm{i}(\omega_k + \omega_{k'})t}. \tag{4.3.26}$$

This describes a state in which the x- and y-motions are independent (the factorizability of ψ is preserved for all t by the free wave equation). The particle trajectory in the x-direction is given by (4.3.19) with $k_1 = -k_2 = k$ and in the y-direction by $y = \hbar k't/m$. Further details are given in Chap. 5.

4.4 Interference and the potential step

Introductory accounts of the phenomenon of 'tunnelling', i.e., the appearance of particles in what are classically forbidden regions of space, employ plane waves incident on potentials idealized as having simple shapes such as square barriers or wells. Significance is given to reflection and transmission currents which apparently give some kind of insight into the tunnel effect. We shall use the results of the last section to bring out the assumptions tacit in this treatment and show why the terminology used is somewhat misleading in that it does not accurately reflect what is actually happening to the particle.

We consider the one-dimensional potential step whose classical analogue was studied in §2.6.2 (for the relativistic version see §12.3): $V > 0$ $(=0)$ for $x > 0$ (<0). Suppose first that the energy of the corresponding classical particle is greater than the potential energy: $E > V$. The nonclassical feature in this case is usually claimed to be the apparent possibility that particles may be reflected by the potential. To see whether this is so, we write down the wavefunction which, in the domain $x < 0$, will be composed of incident and reflected components and in $x > 0$ contains just a transmitted part:

$$\psi(x, t) = \begin{cases} A(\mathrm{e}^{ikx} + c\,\mathrm{e}^{-ikx})\,\mathrm{e}^{-iEt/\hbar}, & x < 0 \qquad (4.4.1) \\ Ab\,\mathrm{e}^{\mathrm{i}(k'x - Et/\hbar)}, & x > 0 \qquad (4.4.2) \end{cases}$$

where $\hbar k = (2mE)^{1/2}$ and $\hbar k' = [2m(E - V)]^{1/2}$ are the classical momenta in the two regions. Continuity of ψ and $\partial\psi/\partial x$ at $x = 0$ yields for the constant coefficients

$$c = (k - k')(k + k')^{-1}, \qquad b = 2k(k + k')^{-1}, \tag{4.4.3}$$

so that c and b are real and $0 < c < 1$. The interference of the incident and reflected waves implies a standing wave pattern $R^2 = |A|^2(1 + c^2 + 2c \cos 2kx)$ which is a matter-wave analogue of Wiener's optical fringes (Born & Wolf, 1970, p. 279). The reflectivity and transmissivity are defined as the quantities c^2 and b^2k'/k respectively. These epithets suggest that a fraction c^2 of incident particles will be reflected and a fraction b^2k'/k transmitted. What is the actual motion of the ensemble elements?

In front of the step the trajectory is oscillatory and follows from (4.3.19) on putting $k_1 = -k_2 = k$. The orbit oscillates about the straight line (4.3.14) defining the 'classical mean' with a frequency and amplitude dependent on V. The total energy of the particle is conserved and coincides with the classical value (E) but is composed of variable kinetic and quantum potential energies. Inside the step the trajectory is a straight line of velocity $\hbar k'/m$ as in the classical case. Making an arbitrary choice of constants $\hbar = 1$, $k = \frac{1}{2}$, $m = \sqrt{3}/4$ and $E = 3V/2$, the complete trajectory is

$$t = \begin{cases} x + \frac{1}{2}\sin x - a, & x \leqslant 0 \\ \frac{3}{2}x - a, & x \geqslant 0 \end{cases} \tag{4.4.4}$$

where $a = x_0 + \frac{1}{2}\sin x_0$, $x_0 \leqslant 0$. A selection of orbits corresponding to uniformly spaced x_0 is shown in Fig. 4.3. Clearly, there are no quantum turning points (points where the particle is brought to rest). Whatever its initial position, the particle always moves in the direction of increasing x and no reflection occurs. The finite 'reflectivity' does not therefore define an attribute of the actual motion. In this regard the motion is similar to the classical case and is, of course, in accord with the constancy of the current throughout the x-axis. Note that unlike the classical case the trajectory is not refracted at $x = 0$.

Suppose now that $E < V$. Solution of the Schrödinger equation together with the boundary conditions yields:

$$\psi(x, t) = \begin{cases} 2A\,e^{i\beta/2}\cos(kx - \beta/2)\,e^{-iEt/\hbar}, & x < 0 \quad (4.4.5) \\ A(1 + e^{i\beta})\,e^{-(\alpha x + iEt/\hbar)}, & x > 0 \quad (4.4.6) \end{cases}$$

where $\beta = 2\tan^{-1}(-\alpha/k)$ and $\alpha = [2m(V - E)]^{1/2}$ is real. Eq. (4.4.6) exhibits the tunnel effect; there is a finite exponentially decaying probability that the particle lies in the classically forbidden domain $x > 0$. The particle behaviour corresponding to this is rather surprising. Because the spatial part of (4.4.6) is real the causal interpretation implies that, if the particle is in this region, it is at *rest*, with negative quantum potential energy $Q = E - V$. The motion in the classically accessible region ($x < 0$) is also nonclassical. The wavefunction (4.4.5) exhibits a static distribution of nodes and, because the phase outside the nodes is constant, the particle is again at rest. The total energy here has the classical value but is of completely nonclassical origin: $E = Q(=\hbar^2k^2/2m)$. The reflectivity and transmissivity are 1 and 0 respectively and, as above, do not accurately reflect the actual details of the motion. The notions that the particle is 'incoming' or 'reflected' simply do not apply.

These results indicate the danger of importing classical intuition into highly nonclassical situations. For there is nothing in the plane wave description of the potential step corresponding to an initially free incident particle that only after a finite time encounters the step and is either reflected or transmitted. A reflected wave does not gradually form as a consequence of scattering by the step but rather is already defined for all x at $t = 0$, on an equal footing with the incident wave. Indeed, referring to e^{ikx} as the 'incident' wave and e^{-ikx} as the 'reflected' one as if these and their associated currents have some independent significance is liable to mislead – the physical wave in the domain $x < 0$ is the superposition (4.4.1) or (4.4.5) for all t and this has quite different properties from either summand. At $t = 0$ the wave is already accommodated to the boundary conditions at $x = 0$ and so a particle wherever it is initially placed moves under the influence of the step; it is never 'free'. Moreover, $|\psi|^2$ is finite for all x at $t = 0$, so x_0 could lie in the region $x > 0$, the 'transmission' region.

These remarks do not point to an inconsistency in the use of plane waves in this context or in the causal treatment of the problem. Rather, the latter brings out how far the plane wave theory fails to conform to the usual description of tunnelling processes and the mental images we may harbour about them. It explains why if we wish to start with an initially free system incident on a barrier we should employ wave packets (see §5.3). Packets are not, of course, required by the causal interpretation as a condition of its applicability. In this connection we point out that in the case $E < V$ the state of rest described above is not inconsistent with the fact that if we measured the momentum of the particle we would find a finite value with a probability given by the usual expression (modulus squared of the Fourier coefficient of the wave). We shall have occasion to emphasize many times that the results of a measurement, although causally and continuously connected with the premeasurement value, in general differ from the latter due to the disturbance caused by interaction with the measuring device.

The plane wave treatment of the potential *barrier* when $V > E$ has been discussed from the point of view of the causal interpretation by Hirschfelder, Christoph & Palke (1974a) and the trajectories plotted by Spiller *et al.* (1990). Remarks similar to those made above apply to that case. Although now the particle has finite motion, no actual reflection occurs. Our results for the step correspond to the limit of an infinitely wide barrier in the domain $x > 0$.

4.5 The hydrogen-like atom

4.5.1 Rotating plane waves and electron trajectories

We shall now examine the rotational analogue of the linear translation of the wavefronts corresponding to a momentum eigenfunction. To do this we consider the eigenfunctions of a component of orbital angular momentum, which for convenience we take to be $\hat{L}_z$. Since $[\hat{L}_z, \hat{\mathbf{L}}^2] = 0$ we seek simultaneous eigenfunctions of $\hat{L}_z$ and $\hat{\mathbf{L}}^2$. In spherical polar coordinates (r, θ, ϕ) where

$$x = r \sin\theta \cos\phi, \qquad y = r \sin\theta \sin\phi, \qquad z = r\cos\theta, \tag{4.5.1}$$

we have (Schiff, 1968, p. 76)

$$\left.\begin{aligned}
\hat{L}_x &= \mathrm{i}\hbar\left(\sin\phi \frac{\partial}{\partial\theta} + \cot\theta\cos\phi\frac{\partial}{\partial\phi}\right)\\
\hat{L}_y &= \mathrm{i}\hbar\left(-\cos\phi \frac{\partial}{\partial\theta} + \cot\theta\sin\phi\frac{\partial}{\partial\phi}\right)\\
\hat{L}_z &= -\mathrm{i}\hbar\,\partial/\partial\phi\\
\hat{\mathbf{L}}^2 &= -\hbar^2\left[\frac{1}{\sin\theta}\frac{\partial}{\partial\theta}\left(\sin\theta\frac{\partial}{\partial\theta}\right) + \frac{1}{\sin^2\theta}\frac{\partial^2}{\partial\phi^2}\right]
\end{aligned}\right\} \tag{4.5.2}$$

and

$$\left.\begin{aligned}
\hat{L}_z Y_{lm}(\theta,\phi) &= m\hbar Y_{lm}(\theta,\phi),\\
\hat{\mathbf{L}}^2 Y_{lm}(\theta,\phi) &= l(l+1)\hbar^2 Y_{lm}(\theta,\phi),
\end{aligned}\right\} \tag{4.5.3}$$

where $Y_{lm}(\theta, \phi)$ are the spherical harmonics, l is the orbital angular momentum number, $l = 0, 1, 2, \ldots$, and m is the azimuthal quantum number, $-l \leqslant m \leqslant l$ (in this section we denote mass by m_0). We have

$$Y_{lm}(\theta,\phi) = f_{lm}(\theta)\,\mathrm{e}^{\mathrm{i}m\phi}, \tag{4.5.4}$$

where $f_{lm}(\theta)$ are a set of real functions (proportional to the Legendre polynomials).

To obtain a solution of the Schrödinger equation we assume that the system is either free or that the external potential is spherically symmetric ($V = V(r)$) so that the wave equation separates in these coordinates. A stationary state corresponding to energy E is given by

$$\psi_{Elm}(r,\theta,\phi) = g_{Elm}(r) f_{lm}(\theta)\,\mathrm{e}^{\mathrm{i}(m\phi - Et/\hbar)}, \tag{4.5.5}$$

where the function $g_{Elm}(r)$ is real. The phase function is therefore

$$S(r,\theta,\phi,t) = m\hbar\phi - Et \tag{4.5.6}$$

apart from a constant.

For each t and $m \neq 0$, the wavefronts S = constant are planes parallel to, and ending on, the z-axis (which is a nodal line of ψ when $m \neq 0$ since the spherical harmonics are proportional to $\sin\theta$). As t increases the planes rotate

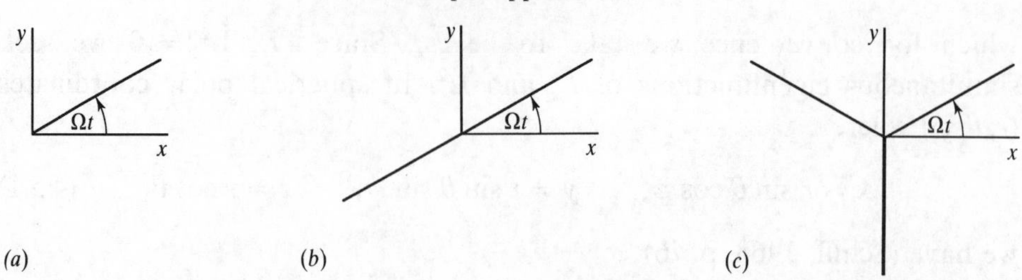

Fig. 4.4 The wavecrests $S = nh$, $n \in \mathbb{Z}$, for the state (4.5.5) when (*a*) $m = 1$, (*b*) $m = 2$, (*c*) $m = 3$. The crests rotate anticlockwise with frequency $\Omega = E/m\hbar$ about the z-axis (a nodal line). The wavefronts of states for which $m < 0$ rotate clockwise. The wavelength is $\lambda_Q = |2\pi r \sin\theta/m|$ where r and θ specify a point on a wavefront.

about the z-axis with angular velocity $\Omega = E/m\hbar$ (see Fig. 4.4). The number of wavecrests, defined by $S = nh$, $n \in \mathbb{Z}$, that come to an end on the z-axis is equal to $|m|$. This illustrates the interpretation of the single-valuedness requirement,

$$\oint \mathrm{d}S = mh, \tag{4.5.7}$$

described in §3.2.2.

It is evident from the orthogonality of the paths to the surfaces of constant S that the trajectories will be circles lying in planes parallel to the xy-plane. This is easily confirmed by solving the guidance equation (de Broglie, 1956, p. 119; Belinfante, 1973, p. 190). In spherical polar coordinates

$$v_r = m_0^{-1}\,\partial S/\partial r, \qquad v_\theta = (m_0 r)^{-1}\,\partial S/\partial\theta, \qquad v_\phi = (m_0 r\sin\theta)^{-1}\,\partial S/\partial\phi \tag{4.5.8}$$

with

$$v_r = \dot{r}, \qquad v_\theta = r\dot{\theta}, \qquad v_\phi = r\sin\theta\,\dot{\phi}. \tag{4.5.9}$$

Substituting (4.5.6) into (4.5.8) yields

$$v_r = v_\theta = 0, \qquad v_\phi = m\hbar/m_0 r\sin\theta \tag{4.5.10}$$

and hence from (4.5.9)

$$r = r_0, \qquad \theta = \theta_0, \qquad \phi = \phi_0 + m\hbar t/m_0 r_0^2 \sin^2\theta_0, \tag{4.5.11}$$

where (r_0, θ_0, ϕ_0) are the initial coordinates. The result (4.5.11) is valid for all m. The particle orbits the z-axis along a circle of constant radius $(r_0 \sin\theta_0)$ and with constant angular speed, which is a multiple of $\hbar/m_0 r_0^2 \sin^2\theta_0$ (Fig. 4.5). The initial coordinates are as usual arbitrary except that they

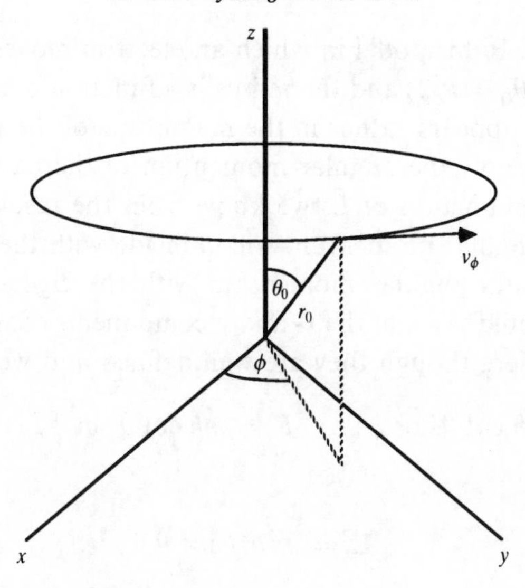

Fig. 4.5 The particle trajectory for a spherically symmetric external potential when the wave is a stationary state and an eigenfunction of $\hat{L}_z$ corresponding to the eigenvalue $m\hbar$. When $m = 0$ the particle is at rest. $|m|$ wavelengths fit into one orbit.

cannot be chosen to lie in nodal regions.† In these coordinates the probability of a particle being between the points $\mathbf{x}$ and $\mathbf{x} + \mathrm{d}\mathbf{x}$ is given by

$$P = |\psi|^2 r^2 \sin\theta \, \mathrm{d}r \, \mathrm{d}\theta \, \mathrm{d}\phi. \tag{4.5.12}$$

From the generalized de Broglie relation (4.1.7), $\lambda_Q = h/m_0|v_\phi|$ and we see that along the orbit the wavelength is given by $\lambda_Q = |2\pi r_0 \sin\theta_0/m|$. Since the circumference is $2\pi r_0 \sin\theta_0$ it follows that we can fit exactly $|m|$ wavelengths into one quantum orbit. This should be compared with the usual de Broglie wavelength connected with Bohr–Sommerfeld quantization (Bohm, 1951, p. 70) which is defined in terms of the momentum of a classical particle in the potential V. There, an integral number of wavelengths fit into a *classical* orbit.

The angular speeds of the wavefronts and the particle are related by

$$\Omega\dot{\phi} = E/m_0 r_0^2 \sin^2\theta_0. \tag{4.5.13}$$

The higher the quantum number m the faster the particle moves, and the slower the wavefronts rotate.

The fact that the radius of the circle is freely specifiable means in particular that it is independent of E or m. The picture is therefore somewhat different

† In (4.5.6) we have assumed that the real coefficient in (4.5.5) is positive. In domains where the coefficient is negative we should add π to (4.5.6).

from the primitive Bohr model in which an electron moves in a circle in the equatorial plane ($\theta_0 = \pi/2$) and the radius is a function of m (see below). The quantization here appears rather in the magnitude of the particle velocity.

Let us consider now the angular momentum of the particle, $\mathbf{L} = \mathbf{x} \times \nabla S$. Since ψ is an eigenfunction of $\hat{L}_z$ we know from the results of §3.5 that the z-component of angular momentum will coincide with the eigenvalue of $\hat{L}_z$, and the total orbital angular momentum with the eigenvalue of $\hat{\mathbf{L}}^2$. Conventionally one would say that the x- and y-components of angular momentum are 'undefined'. Here though they are well defined and we have

$$L_x = -m\hbar \cot\theta \cos\phi, \qquad L_y = m\hbar \cot\theta \sin\phi, \qquad L_z = m\hbar \quad (4.5.14)$$

and hence

$$\mathbf{L}^2 = m^2\hbar^2/\sin^2\theta. \quad (4.5.15)$$

The total orbital angular momentum is given by:

$$\mathbf{L}^2 - \hbar^2(\mathbf{x} \times \nabla R)^2/R = l(l+1)\hbar^2. \quad (4.5.16)$$

Evaluated along a trajectory (4.5.11) we see that L_z and the total angular momentum are conserved, but that L_x and L_y are not unless $\theta_0 = \pi/2$. Thus, although the classical force is central the motion is asymmetrical, which reflects the fact that a particular direction in space has been singled out as the quantization axis.

This suggests that the effective potential acting on the particle is not symmetric. To find its form we write down the quantum Hamilton–Jacobi equation, which reduces to

$$E = m^2\hbar^2/2m_0 r^2 \sin^2\theta + Q + V(r). \quad (4.5.17)$$

This enables us to find the effective force without knowing the explicit expression for R. We have

$$-\nabla(Q + V) = -(m^2\hbar^2/2m_0)\nabla(-1/r^2 \sin^2\theta), \quad (4.5.18)$$

which shows that the effective potential is indeed not central. Notice that this potential does not depend on any feature of V other than that it is symmetric. The component of the total force (in direction $\hat{\boldsymbol{\theta}}$) orthogonal to the line of action of the classical force (in direction $\hat{\mathbf{r}}$) is therefore independent of the latter.

So far we have worked with an arbitrary V and unrestricted E. In the case of the hydrogen-like atom, with the nucleus placed at the origin of coordinates, $V = -Ze^2/r$, and m_0 the reduced mass which may be approximately identified with the electron mass, we have

$$E_n = -m_0 Z^2 e^4 / 2\hbar^2 n^2, \tag{4.5.19}$$

where $n \geqslant l + 1$ and the stationary states (4.5.5) will be denoted ψ_{nlm}. The most general stationary state corresponding to the quantum number n may be written

$$\begin{aligned}\psi_n(r, \theta, \phi, t) &= \sum_{l=0}^{n-1} \sum_{m=-l}^{l} c_{lm} \psi_{nlm}(r, \theta, \phi, t) \\ &= \left(\sum_{l=0}^{n-1} \sum_{m=-l}^{l} c_{lm} F_{nl}(r, \theta)\, \mathrm{e}^{\mathrm{i}m\phi} \right) \mathrm{e}^{-\mathrm{i}E_n t/\hbar},\end{aligned} \tag{4.5.20}$$

where c_{lm} are arbitrary complex constants and $F_{nl}(r, \theta)$ are real functions. The phase of ψ_n will be a complicated function of r, θ and ϕ and the trajectory will have a correspondingly complex structure (although of constant energy E_n). In fact, we may generate an infinite set of possible motions by choosing different values of the constants c_{lm}. If the atom is in a stationary state it will remain there unless it interacts with another system. If this happens and the atom is left in another stationary state, i.e., a transition occurs, then in general the electron motion initially and finally will be determined by waves of the general form (4.5.20) rather than the simple eigenstates of angular momentum we have discussed above (during the transition the motion is guided by a superposition of stationary states).

4.5.2 The stability of matter. Relation with Bohr orbits

Returning to the angular momentum eigenstates, it will be noted from (4.5.10) that in states for which the azimuthal quantum number m is zero the particle is at rest:

$$\mathbf{v} = 0 \Rightarrow r = r_0, \qquad \theta = \theta_0, \qquad \phi = \phi_0. \tag{4.5.21}$$

We have

$$V + Q = E_n \tag{4.5.22}$$

so that the quantum force exactly balances the classical force (Fig. 4.6). For the hydrogen-like atom this will be so in particular for the ground state ψ_{100}. This result provides the explanation according to the quantum theory of motion for the stability of matter. For if the particle is at rest relative to the nucleus it is evidently not accelerating, and hence does not radiate. Therefore it does not lose energy and it will not spiral into the nucleus, the famous outcome predicted by classical electrodynamics.

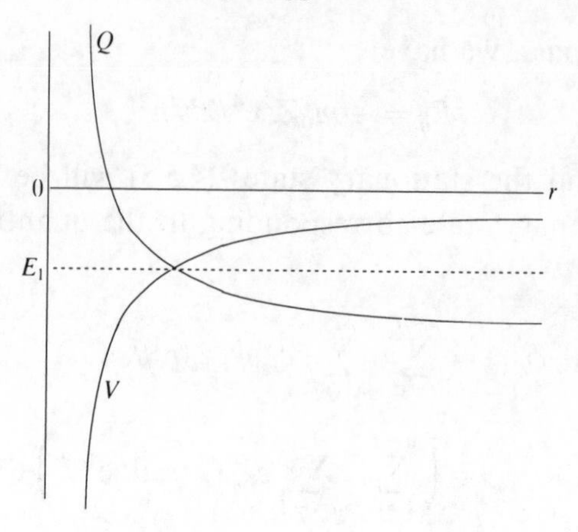

Fig. 4.6 The Coulomb potential V and the quantum potential Q for the ground state.

This account of why matter does not collapse is different from that advanced by Bohr in the old quantum theory. It was assumed by Bohr that the electron moves along circles in the equatorial plane under the influence of the central Coulomb force, the allowed orbits being determined by a quantization condition imposed on solutions to the classical Hamilton–Jacobi equation. This was essentially a postulate and no explanation of its physical significance was offered. The ground state corresponds to a minimum radius. For us on the other hand the electron is not confined to the equatorial plane and there is no minimum 'radius' – the initial coordinates (4.5.21) are arbitrary.

Nevertheless, is there a relation between the trajectories of Fig. 4.5 and the Bohr orbits when $m \neq 0$? Since the latter are associated with the classical Hamilton–Jacobi equation, we might expect to recover them when the effective force acting in the equatorial plane is just that due to $-\nabla V$. From (4.5.17) we have, when $\theta = \pi/2$,

$$Q = E_n - m^2\hbar^2/2m_0r^2 + Ze^2/r, \tag{4.5.23}$$

so that

$$-\partial Q/\partial r = -m^2\hbar^2/m_0r^3 + Ze^2/r^2. \tag{4.5.24}$$

The requirement that the quantum force (4.5.24) vanishes implies that

$$r = m^2\hbar^2/m_0Ze^2 \tag{4.5.25}$$

for $r \neq 0$. This indeed yields the set of Bohr orbits for $|m| = 1, 2, \ldots$ but they do not have quite the same meaning as in Bohr's theory. As we have seen, for us the ground state is characterized by $m = 0$ and a state of rest, whereas for Bohr it is characterized by a circular orbit at a radius given by putting

$|m| = 1$ in (4.5.25). For us, an orbit at a radius (4.5.25) with $|m| = 1$ corresponds to an excited state. Moreover, inserting (4.5.25) in (4.5.23) shows that the particle has finite quantum potential energy. And, of course, in the causal theory (4.5.25) is just a subclass of the permissible orbits and the particle may pass through regions where the quantum potential is not stationary, independently of the quantum numbers characterizing the state.

Readers who have read this far and exclaim 'I don't believe it' when confronted with a stationary electron in $m = 0$-states should bear in mind two points.

The first is that quantum mechanics was developed precisely to deal with the problem of the stability of matter. We have remarked previously that one has to be prepared to put aside expectations based on acquaintance with classical physics (i.e., habit and tradition) in order to develop a new intuition appropriate to the theory of motion put forth here. Actually, if we have two forces that balance in classical mechanics then uniform motion (including rest) is what we expect. What might be surprising at first sight here is that such an equilibrium condition arises for a particle in a Coulomb potential where we might anticipate that some kind of motion is taking place. But clearly a radical change is necessary in the types of motion accessible to particles in order to account for stationary states, which are among the most 'nonclassical' states there are. And this explanation should be compared with the usual interpretation which offers no physical explanation at all for the stability of matter. In that view there are no 'atoms' that are 'there' as objective localized structures. It is our informal image that puts in a nucleus and electron as quasi-local objects.

Of course, one finds in books informal models of the hydrogen atom in which $|\psi|^2$ is interpreted as a 'charge density' in a 'cloud of probability', with the implication that the electron is somehow 'everywhere' that the probability is finite. Indeed, since for us the wave is a component of the electron, it is true that the latter is extended, but we also have a corpuscle that occupies a definite position in space. The 'charge density' is then both a characteristic of the physical field and an indication of the likely position of the corpuscle. But the latter refers to an ensemble of identical atoms – only one position is actually occupied by the particle in each atom in the ensemble. A consistent application of the conventional view only says that if we look for an electron a device will click in a small region. During transitions, it is the ψ-wave that interacts with the wavefunction of the electromagnetic field.

The second point is that the velocity (4.5.21) is not what one would necessarily find if one performed a measurement. If the atom is in the ground state

$$\psi_{100} = (\pi a^3)^{-1/2} \mathrm{e}^{-r/a}, \qquad a = \hbar^2/m_0 Z e^2, \tag{4.5.26}$$

then the probability of finding the particle with a momentum in the volume element $\mathrm{d}^3 p$ around the point $\mathbf{p}$ is given by $|\varphi(\mathbf{p})|^2 \, \mathrm{d}^3 p$, where (Belinfante, 1973, p. 194):

$$\begin{aligned} |\varphi(\mathbf{p})|^2 &= \left| h^{-3/2} \int \psi_{100}(\mathbf{x}) \, \mathrm{e}^{-\mathrm{i}\mathbf{p}\cdot\mathbf{x}/\hbar} \, \mathrm{d}^3 x \right|^2 \\ &= (8a^3/\pi^2 \hbar^3)(1 + \mathbf{p}^2 a^2/\hbar^2)^{-4}. \end{aligned} \tag{4.5.27}$$

The actual momentum found is causally but not directly related to (4.5.21) and depends on the initial position of the particle in the atom prior to the measurement. The probability of finding a momentum in the neighbourhood of $\mathbf{p} = 0$ is $8a^3/\pi^2\hbar^3$. The function $|\varphi(\mathbf{p})|^2$ thus has nothing directly to do with the probability density $g(\mathbf{p})$ (cf. §3.6) which refers to the actual momentum of the particle in the ground state:

$$\begin{aligned} g(\mathbf{p}) &= \int R^2(\mathbf{x}) \delta(\mathbf{p} - \nabla S) \, \mathrm{d}^3 x \\ &= \delta(\mathbf{p}) \int |\psi_{100}|^2 \, \mathrm{d}^3 x \\ &= \delta(\mathbf{p}) \end{aligned} \tag{4.5.28}$$

since $\nabla S = 0$ where $R \neq 0$ and ψ_{100} is normalized. The connection between the two functions is that $g(\mathbf{p}) \to |\varphi(\mathbf{p})|^2$ during a measurement (§§8.3–8.6).

Finally, an objection has been raised against the causal treatment of the $m = 0$-states in that the state of rest (4.5.21) apparently conflicts with the possibility that the orbital angular momentum quantum number l is nonzero (Rosen, 1945, 1974). This difficulty is overcome if we observe that the actual total orbital angular momentum, given by (4.5.16), contains a term (the second) other than that directly connected with the motion, just as the particle at rest possesses quantum potential energy. The electron may therefore have finite angular momentum although it is not moving. This example illustrates that the quantum numbers do not directly represent dynamical properties.

4.6 Wave packets

The superposition of plane waves studied in §4.3 did not imply any restriction on the regions of space through which a particle could potentially travel. We can construct a state which confines the particle to a particular region if we superpose many plane waves and arrange their amplitudes and phases so that they constructively interfere in a restricted region of space, and destructively interfere outside this region.

The function

$$\psi(\mathbf{x}, t) = (2\pi)^{-3/2} \int \mathrm{d}^3k \, \varphi(\mathbf{k}) \, \mathrm{e}^{\mathrm{i}(\mathbf{k}\cdot\mathbf{x} - \omega_{\mathbf{k}} t)}, \tag{4.6.1}$$

where $\omega_{\mathbf{k}} = \hbar \mathbf{k}^2/2m$, satisfies the free Schrödinger equation. Writing $\varphi(\mathbf{k}) = |\varphi(\mathbf{k})| \, \mathrm{e}^{\mathrm{i}\chi(\mathbf{k})}$ we suppose that $|\varphi|$ is only appreciable for $\mathbf{k}$ near to some wave vector $\mathbf{k}_0$. The phase of the integrand in (4.6.1) is

$$\Phi(\mathbf{k}) = \chi(\mathbf{k}) + \mathbf{k}\cdot\mathbf{x} - \omega_{\mathbf{k}} t. \tag{4.6.2}$$

The requirement that all waves with values of $\mathbf{k}$ near $\mathbf{k}_0$ have nearly the same phase ('principle of stationary phase') is expressed by

$$\nabla_{\mathbf{k}} \Phi|_{\mathbf{k}=\mathbf{k}_0} = 0 \tag{4.6.3}$$

whence, substituting (4.6.2), we have

$$\nabla_{\mathbf{k}} \chi|_{\mathbf{k}=\mathbf{k}_0} + \mathbf{x} - t\nabla_{\mathbf{k}} \omega_{\mathbf{k}}|_{\mathbf{k}=\mathbf{k}_0} = 0 \tag{4.6.4}$$

or $\mathbf{x} = \mathbf{x}_0 + \mathbf{u}t$ where $\mathbf{x}_0$ is minus the first term in (4.6.4) and

$$\mathbf{u} = \nabla_{\mathbf{k}} \omega_{\mathbf{k}}|_{\mathbf{k}=\mathbf{k}_0} \tag{4.6.5}$$

is the group velocity. We have $\mathbf{u} = \hbar \mathbf{k}_0/m$ and so the centre of a free packet pursues a classical path, as will any particle placed there (but see §3.8.3).

The trajectory of a particle lying at any other point in the packet will not be rectilinear and uniform in general, however, due to dispersion: each plane wave in the spectral decomposition of the packet has a different phase velocity ($\omega_{\mathbf{k}}/|\mathbf{k}| = \hbar|\mathbf{k}|/2m$). This implies that the relation between the initial and subsequent amplitudes is not of the form

$$|\psi(\mathbf{x}, t)|^2 = |\psi_0(\mathbf{x} - \mathbf{u}t)|^2. \tag{4.6.6}$$

This property is reflected in nonuniform trajectories.

Further information on the structure of the packet is contained in the rigorous relation (we work in one dimension now)

$$\Delta x \, \Delta k \geqslant \tfrac{1}{2} \tag{4.6.7}$$

which expresses the reciprocal nature of the root mean square (rms) spatial extension of the packet, Δx, and the rms spread of the wave vectors entering the superposition (4.6.1), Δk. We may say that the approximate range of k-values for which $\varphi(k)$ is appreciable is $k_0 - \tfrac{1}{2}\Delta k < k < k_0 + \tfrac{1}{2}\Delta k$. Let

$\lambda_0 = 2\pi/|k_0|$ be the 'average' wavelength of the packet. Then the spread in wavelength, $\Delta\lambda$, satisfies $\lambda_0 \mp \frac{1}{2}\Delta\lambda = 2\pi/|k_0 \pm \frac{1}{2}\Delta k|$ which implies that $\Delta k = 2\pi\Delta\lambda/\lambda_0^2$. Then (4.6.7) implies the following formula:

$$\Delta x \geqslant \lambda_0^2/4\pi\,\Delta\lambda. \tag{4.6.8}$$

The quantity $\lambda_0^2/\Delta\lambda$ is called the 'coherence length' of the packet.

We can intuitively understand (4.6.7) as the condition that the plane wave of length λ_0 in the superposition (4.6.1) is out of phase with the wave of length $\lambda_0 + \frac{1}{2}\Delta\lambda$ at the edge of the packet. The coordinate of a point approximately on the border of the packet is $x_0 + ut + \frac{1}{2}\Delta x$. The phases of the two waves under consideration evaluated at this point are, from (4.6.2), given by

$$\Phi(k_0) = \chi(k_0) + k_0(x_0 + ut + \tfrac{1}{2}\Delta x) - \hbar k_0^2 t/2m, \tag{4.6.9}$$

$$\begin{aligned}\Phi(k_0 + \tfrac{1}{2}\Delta k) = \chi(k_0) - \tfrac{1}{2}x_0\,\Delta k + (k_0 + \tfrac{1}{2}\Delta k)(x_0 + ut + \tfrac{1}{2}\Delta x)\\ - \hbar(k_0^2 + k_0\,\Delta k)t/2m,\end{aligned} \tag{4.6.10}$$

where in (4.6.10) we have Taylor expanded $\chi(k_0 + \frac{1}{2}\Delta k)$ for small Δk and used $\chi'(k_0) = -x_0$. Taking the difference between (4.6.9) and (4.6.10) and using (4.6.5), the condition for destructive interference reduces to

$$\Phi(k_0 + \tfrac{1}{2}\Delta k) - \Phi(k_0) = \tfrac{1}{4}\Delta x\,\Delta k = |n + \tfrac{1}{2}|\pi, \qquad n \in \mathbb{Z}, \tag{4.6.11}$$

or

$$\Delta x\,\Delta k \geqslant 2\pi. \tag{4.6.12}$$

The right hand side of (4.6.12) is larger than that of (4.6.7) since the waves do not exactly cancel at the point considered.

4.7 Diffraction at a Gaussian slit

4.7.1 Properties of the Gaussian packet

The initial form of the normalized Gaussian wave packet is as follows (Belinfante, 1973, p. 194):

$$\psi_0(\mathbf{x}) = (2\pi\sigma_0^2)^{-3/4}\,\mathrm{e}^{\mathrm{i}\mathbf{k}\cdot\mathbf{x} - \mathbf{x}^2/4\sigma_0^2}. \tag{4.7.1}$$

This is a product of three identical functions in each of the coordinate directions and the square of the amplitude of the factor in the x-direction is plotted in Fig. 4.7. The factorizability is preserved for all t and so the components of the particle motion in each direction are independent. σ_0 is the rms width of the packet in each direction: $\sigma_0^2 = \langle x^2\rangle = \langle y^2\rangle = \langle z^2\rangle$ with $\langle x\rangle = \langle y\rangle = \langle z\rangle = 0$. We have $\langle\hat{\mathbf{p}}\rangle = \hbar\mathbf{k}$ so the centre of the packet has initial (group) velocity $\mathbf{u} = \hbar\mathbf{k}/m$.

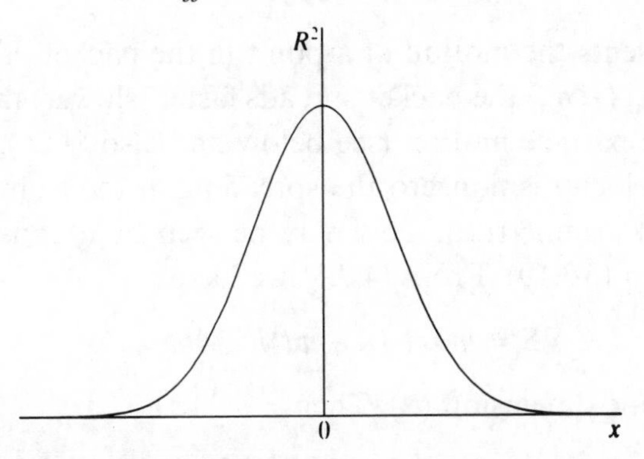

Fig. 4.7 The amplitude squared of a Gaussian packet. The magnitude of the force acting on a particle is proportional to $|x|$.

The free Schrödinger evolution of (4.7.1) leads to the following function at time t:

$$\psi(\mathbf{x}, t) = (2\pi s_t^2)^{-3/4}\, e^{i\mathbf{k}\cdot(\mathbf{x} - \frac{1}{2}\mathbf{u}t) - (\mathbf{x} - \mathbf{u}t)^2/4s_t\sigma_0} \tag{4.7.2}$$

where

$$s_t = \sigma_0(1 + i\hbar t/2m\sigma_0^2). \tag{4.7.3}$$

The amplitude and phase functions are as follows:

$$R(\mathbf{x}, t) = (2\pi\sigma^2)^{-3/4}\, e^{-(\mathbf{x} - \mathbf{u}t)^2/4\sigma^2} \tag{4.7.4}$$

$$S(\mathbf{x}, t) = -(3\hbar/2)\tan^{-1}(\hbar t/2m\sigma_0^2) + m\mathbf{u}\cdot(\mathbf{x} - \tfrac{1}{2}\mathbf{u}t) + (\mathbf{x} - \mathbf{u}t)^2\hbar^2 t/8m\sigma_0^2\sigma^2, \tag{4.7.5}$$

where

$$\sigma = |s_t| = \sigma_0[1 + (\hbar t/2m\sigma_0^2)^2]^{1/2} \tag{4.7.6}$$

is the rms width of the packet at time t in each coordinate direction ($\sigma^2 = \langle x^2 - \langle x\rangle^2\rangle$ etc.). The packet thus has an overall uniform motion with group velocity $\mathbf{u}$; the mean position $\langle \mathbf{x}\rangle = \mathbf{u}t$ moves like a classical free particle. It spreads in each direction at a rate

$$d\sigma/dt = \hbar^2 t/4m^2\sigma\sigma_0^2 \xrightarrow{t\to\infty} \hbar/2m\sigma_0. \tag{4.7.7}$$

It follows that prior to $t = 0$ the packet contracts (as t decreases in magnitude through negative values) and subsequently it expands. The more sharply peaked the packet is at $t = 0$, the faster it contracts (for $t < 0$) and spreads (for $t > 0$). Putting in some numbers: for an electron with $\sigma_0 = 0.1$ cm, $d\sigma/dt \sim 5\ \text{cm s}^{-1}$ whereas with $\sigma_0 = 5$ Å, $d\sigma/dt \sim 100\ \text{km s}^{-1}$.

$x = \sigma(t)$ represents the motion of a point in the packet initially at $x = \sigma_0$. For points $x > \sigma_0$ ($<\sigma_0$) the packet spreads faster (slower) than $\dot{\sigma}$, behaviour mimicked by the particle motion (see below and also §11.6).

If the group velocity is nonzero the spreading in the front and the rear of the packet is not symmetrical. This may be seen by a consideration of the actual wavelength (3.3.19). From (4.7.5) we have

$$\nabla S = m\mathbf{u} + (\mathbf{x} - \mathbf{u}t)\hbar^2 t/4m\sigma_0^2\sigma^2. \tag{4.7.8}$$

Let us work in one dimension (x). Then

$$\lambda_Q(x, t) = h/|\partial S/\partial x|. \tag{4.7.9}$$

Defining the average wavelength of the packet to be $\lambda_0 = 2\pi/|k|$, $k = mu/\hbar$, we see from (4.7.9) that at the centre of the packet ($x = ut$) the local and average wavelengths coincide. Assuming $u > 0$, in the front of the packet ($x > ut$) the local wavelength satisfies $\lambda_Q < \lambda_0$ and gets smaller the further one moves from the centre. On the other hand, in the rear of the packet ($x < ut$) λ_Q at first increases as we move away from the centre, but beyond a certain point (where $\partial S/\partial x = 0$) it starts to decrease. To see this asymmetry pictorially one has to plot Re ψ or Im ψ – it does not show up in $|\psi|^2$ (Brandt & Dahmen, 1985, Chap. 3). For a wave packet at rest ($u = 0$), λ_Q decreases in the same way on both sides of $x = 0$.

The momentum space representation of the wavefunction (4.7.2) is:

$$\begin{aligned}\varphi(\mathbf{p}) &= h^{-3/2} \int \mathrm{d}^3x\, \psi(\mathbf{x})\, \mathrm{e}^{-\mathrm{i}\mathbf{p}\cdot\mathbf{x}/\hbar} \\ &= [(2/\pi)^{1/2}\sigma_0/\hbar]^{3/2}\, \mathrm{e}^{-\sigma_0^2(\mathbf{p}/\hbar - \mathbf{k})^2 - \mathrm{i}p^2t/2m\hbar},\end{aligned} \tag{4.7.10}$$

where we have used the result

$$\int_{-\infty}^{\infty} \mathrm{d}x\, \mathrm{e}^{-zx^2 + \mathrm{i}rx} = (\pi/z)^{1/2}\, \mathrm{e}^{-r^2/4z} \tag{4.7.11}$$

with $z = a^2 + ib$, $a, b, r \in \mathbb{R}$. Therefore

$$|\varphi(\mathbf{p})|^2 = (2\pi\sigma_p^2)^{-3/2}\, \mathrm{e}^{-(\mathbf{p} - \hbar\mathbf{k})^2/2\sigma_p^2} \tag{4.7.12}$$

is a Gaussian whose width $\sigma_p = \langle \hat{p}^2 - \langle \hat{p} \rangle^2 \rangle^{1/2} = \hbar/2\sigma_0$ in each direction is constant in time. We have from (4.7.6) that $\sigma\sigma_p \geqslant \hbar/2$ for all t, the 'minimum uncertainty' being obtained at $t = 0$ (when $\sigma_0\sigma_p = \hbar/2$). We see here an asymmetry in the properties of ψ in its position and momentum space representations. The symmetry is restored below when we consider the probability density function for the *actual* momentum of the particle, rather than for the predicted values of measurements, (4.7.12).

It is sometimes stated that Planck's constant is the smallest quantity of action that may be ascribed a physical meaning in quantum mechanics, but this is not strictly true. Consider for example the case where a Gaussian packet is widely dispersed, $\sigma \gg \sigma_0$. Then from (4.7.6) we have $\hbar \gg 2m\sigma_0^2/t \sim$ (classical action of a free particle that moves a distance σ_0 in time t). There are many other examples where quantities of action associated with the wave or particle dynamics may be smaller than $\hbar$.

4.7.2 Particle motion

From (4.7.8) we find the velocity field

$$\mathbf{v} = \mathbf{u} + (\mathbf{x} - \mathbf{u}t)\hbar^2 t/4m^2\sigma_0^2\sigma^2. \tag{4.7.13}$$

Wherever it is initially placed, the initial velocity of the particle is $\mathbf{u}$. Let us consider the motion in one direction (x) and assume that $u_1 > 0$. Then for points in the front of the packet ($x > u_1 t$) the particle accelerates, but in the rear of the packet ($x < u_1 t$) there is at each time t a point where the effects of the overall motion and dissipation of the packet cancel and we have a stagnation point ($\nabla S = 0$). This reflects the asymmetry in the effective wavelength in the front and rear halves of the packet referred to above.

Substituting (4.7.6) into (4.7.13) a straightforward integration yields for the trajectory

$$\mathbf{x}(t) = \mathbf{u}t + \mathbf{x}_0[1 + (\hbar t/2m\sigma_0^2)^2]^{1/2}, \tag{4.7.14}$$

where $\mathbf{x}_0$ is the initial position. The first term on the right hand side reflects the overall motion, and the second term the spreading, of the packet. The motion with respect to a frame moving with the centre of the packet is thus hyperbolic in each direction (Fig. 4.8). The equation of the asymptotes to the hyperbolae is $\mathbf{x} = \pm\mathbf{x}_0\hbar t/2m\sigma_0^2$ which shows that the further the particle is from the centre at $t = 0$ the larger is its subsequent acceleration or preceding deceleration. Notice that if the particle is initially at the centre of the packet ($\mathbf{x}_0 = 0$) it subsequently stays there and moves uniformly (classical free motion).

To see the origin of these accelerations we evaluate the quantum potential from (4.7.4):

$$Q = (\hbar^2/4m\sigma^2)[3 - (\mathbf{x} - \mathbf{u}t)^2/2\sigma^2] \tag{4.7.15}$$

(see Fig. 4.9 for the case $\mathbf{u} = 0$). Then the quantum force is given by

$$\mathbf{F} = -\nabla Q = (\hbar^2/4m\sigma^4)(\mathbf{x} - \mathbf{u}t). \tag{4.7.16}$$

From this expression it is clear that the further the particle is from the

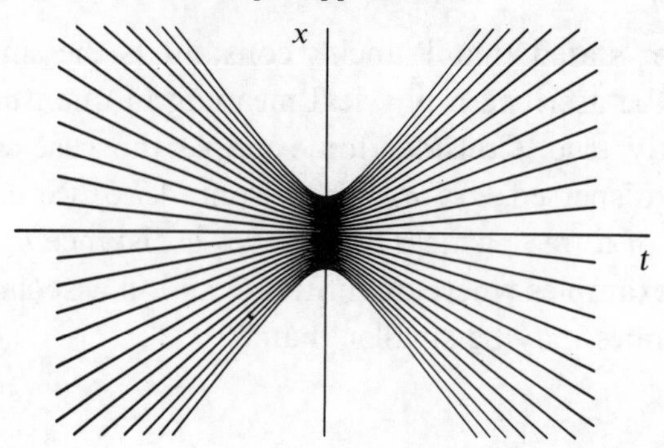

Fig. 4.8 Possible hyperbolic trajectories for a particle in a one-dimensional Gaussian packet at rest ($u = 0$). For $t < 0$ ($t > 0$) the packet is contracting (expanding).

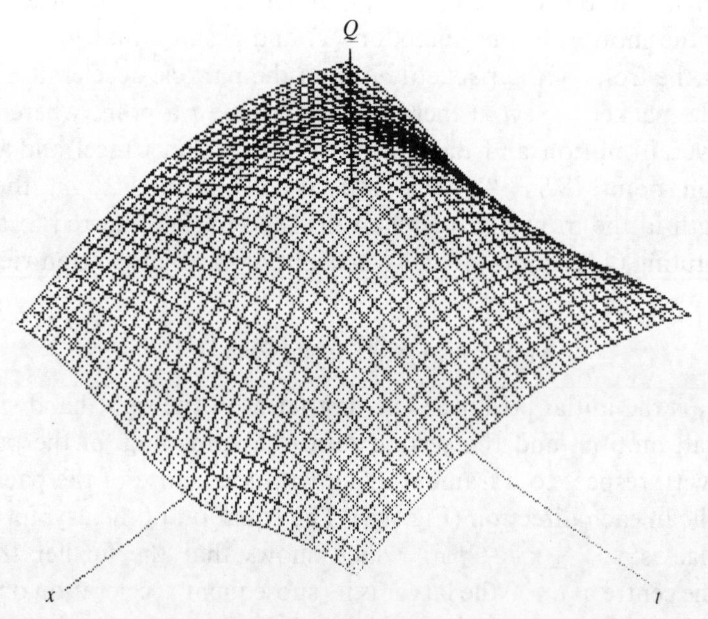

Fig. 4.9 Quantum potential for the Gaussian packet at rest.

centre in each direction (i.e., the larger $x - u_1 t$ etc. are), the greater is the force acting on it. As $t \to \infty$, $\mathbf{F} \to 0$ and the motion becomes uniform and rectilinear (approaching the asymptotes). A constant $-\ t$ section of Q is a parabola corresponding to an inverted oscillator potential. This gives an example where $|Q|$ and $|\mathbf{F}|$ increase in magnitude as the wave amplitude decreases. It is easy to see that the particle energy $(-\partial S/\partial t)$ evaluated along a trajectory is not conserved.

We can use the above results to plot the possible trajectories of a particle passing through a slit system that generates a Gaussian packet (the region $t > 0$ in Fig. 4.8). This provides a simple model of an aperture whose edges are not perfectly sharp. We assume an initial wavefunction of the form

$$\psi_0(x, y) = (2\pi\sigma_0^2)^{-1/4}\, \mathrm{e}^{iky - x^2/4\sigma_0^2} \tag{4.7.17}$$

which describes a Gaussian at rest in the x-direction and a plane wave in the y-direction. Since (4.7.17) is a product of x- and y-dependent functions, and this factorizability is preserved for all t by the free wave equation, it follows that the motions in the two directions are independent (the x-motion is given by (4.7.14) and the y-motion is $y = \hbar kt/m$). The same picture is obtained if we plot x against t or y (apart from a change of scale).

Finally, we consider the probability distribution of particle momentum induced by the position distribution $|\psi|^2$. This is (cf. (3.6.20))

$$g(\mathbf{p}) = \int \mathrm{d}^3x\, R^2(\mathbf{x})\, \delta(\mathbf{p} - \nabla S).$$

To evaluate this expression we substitute for ∇S from (4.7.8) and, using the following property of the δ-function,

$$\int f(x)\, \delta(ax + b)\, \mathrm{d}x = a^{-1} f(-b/a), \tag{4.7.18}$$

and substituting for R^2 from (4.7.4), we find

$$g(\mathbf{p}) = (2\pi\Sigma^2)^{-3/2}\, \mathrm{e}^{-(\mathbf{p} - m\mathbf{u})^2/2\Sigma^2} \tag{4.7.19}$$

where

$$\Sigma = \hbar^2 t/4m\sigma_0^2\sigma = \sigma_p[1 + (2m\sigma_0^2/\hbar t)^2]^{-1/2} \tag{4.7.20}$$

is the root mean square spread in actual momentum in each direction. This is the momentum analogue of the distribution function describing the likely position of a particle at time t. The packet (4.7.19) has an identical structure to R^2 given by (4.7.4) except that it has different spreading properties. Comparing (4.7.20) with (4.7.7) we see that

$$\Sigma = m\, \mathrm{d}\sigma/\mathrm{d}t, \tag{4.7.21}$$

a not unexpected relation given the connection between $\mathbf{p}$ and $\mathbf{x}$ ($\mathbf{p} = m\dot{\mathbf{x}}$). When $t = 0$, $\Sigma = 0$ which expresses the fact pointed out earlier that the initial velocity of each particle in the ensemble is the same ($\mathbf{u}$).

We emphasize that it is Σ that represents the spread in the current momentum over the ensemble, and not σ_p as is commonly assumed. As $t \to \infty$, $\Sigma \to \sigma_p$ and (4.7.19) approaches (4.7.12). The distribution of actual momentum then coincides with the distribution obtained if we perform a measurement of momentum. This is an instance of a general result proved in §8.6.

4.8 Gaussian packet in a uniform field

Let us suppose now that the initial Gaussian packet (4.7.1) is placed in a uniform potential $V = \mathbf{K}\cdot\mathbf{x}$. This might for example be an electric field, or gravity ($V = m\mathbf{g}\cdot\mathbf{x}$). The wavefunction at time t is given by (de Broglie, 1930, p. 190):

$$\psi(\mathbf{x}, t) = (2\pi s_t^2)^{-3/4} \exp\{-(\mathbf{x} - \mathbf{u}t + \mathbf{K}t^2/2m)^2/4s_t\sigma_0 + (im/\hbar)[(\mathbf{x} - \mathbf{K}t/m)\cdot(\mathbf{x} - \tfrac{1}{2}\mathbf{u}t) - \mathbf{K}^2t^3/6m]\} \quad (4.8.1)$$

so that

$$R(\mathbf{x}, t) = (2\pi\sigma^2)^{-3/4}\, \mathrm{e}^{-(\mathbf{x}-\mathbf{u}t+\mathbf{K}t^2/2m)^2/4\sigma^2} \quad (4.8.2)$$

$$S(\mathbf{x}, t) = -(3\hbar/2)\tan^{-1}(\hbar t/2m\sigma_0^2) + (m\mathbf{u} - \mathbf{K}t)\cdot(\mathbf{x} - \tfrac{1}{2}\mathbf{u}t) - \mathbf{K}^2t^3/6m + (\hbar^2 t/8m\sigma_0^2\sigma^2)(\mathbf{x} - \mathbf{u}t + \mathbf{K}t^2/2m)^2. \quad (4.8.3)$$

The centre of the packet accelerates in the direction $\mathbf{K}$. The degree of spreading, as measured by σ, is given by (4.7.6) and is not affected by the external potential.

The trajectories are found by integrating

$$\dot{\mathbf{x}} = \mathbf{u} - \mathbf{K}t/m + \hbar^2 t(\mathbf{x} - \mathbf{u}t + \mathbf{K}t^2/2m)(4m^2\sigma_0^4 + \hbar^2t^2)^{-1}. \quad (4.8.4)$$

Writing $\mathbf{X} = \mathbf{x} - \mathbf{u}t + \mathbf{K}t^2/2m$, (4.8.4) becomes

$$\dot{\mathbf{X}} = \mathbf{X}\frac{\mathrm{d}}{\mathrm{d}t}\log\left(1 + \frac{\hbar^2t^2}{4m^2\sigma_0^4}\right)^{1/2},$$

so that

$$\mathbf{X}(t) = \mathbf{X}(0)(1 + \hbar^2t^2/4m^2\sigma_0^4)^{1/2}$$

and finally

$$\mathbf{x}(t) = \mathbf{u}t - \mathbf{K}t^2/2m + \mathbf{x}_0[1 + (\hbar t/2m\sigma_0^2)^2]^{1/2}. \quad (4.8.5)$$

We see that the motion is a superposition of classical terms and a term due to the spreading of the packet. If the particle lies initially at the centre of the packet ($\mathbf{x}_0 = 0$) it subsequently moves on the classical trajectory in this field of force.

Further insight into the motion is gained from a consideration of the quantum potential. This is easily calculated from (4.8.2) to be

$$Q = (\hbar^2/4m\sigma^2)[3 - (\mathbf{x} - \mathbf{u}t + \mathbf{K}t^2/2m)^2/2\sigma^2] \quad (4.8.6)$$

and hence the total force acting on the particle is given by

$$-\nabla(Q+V)=(\hbar^2/4m\sigma^4)(\mathbf{x}-\mathbf{u}t+\mathbf{K}t^2/2m)-\mathbf{K}. \tag{4.8.7}$$

The quantum force field carries information on the classical force.

Suppose we have a gravitational field. Then equating (4.8.7) to $m\ddot{\mathbf{x}}$ and substituting from (4.8.5) we get for the particle acceleration

$$\ddot{\mathbf{x}}=\hbar^2\mathbf{x}_0/4m^2\sigma_0\sigma^3-\mathbf{g}. \tag{4.8.8}$$

Classically the equation of motion is just $\ddot{\mathbf{x}}=-\mathbf{g}$, which expresses the result that all bodies fall at an equal rate independently of their mass. Quantum mechanically, however, we see from the right hand side of (4.8.8) that the motion in an external gravitational field does depend on the mass. For a system comprising a Gaussian packet freely falling under gravity, all particles do not fall at an equal rate. This is a general feature and is discussed further in §6.8.

4.9 Harmonic oscillator

We examine here two types of motion in the one-dimensional potential $V=\frac{1}{2}m\omega^2x^2$: the stationary states and the coherent packet. The stationary states are given by

$$\psi_n(x,t)=u_n(x)\,\mathrm{e}^{-\mathrm{i}E_nt/\hbar}, \tag{4.9.1}$$

where $u_n(x)$ are real functions (proportional to the Hermite polynomials) and $E_n=(n+\frac{1}{2})\hbar\omega$, $n=0,1,2,\ldots$. The function (4.9.1) implies that outside nodes $\nabla S=0$ and so the particle is at rest:

$$v=0\Rightarrow x=x_0, \tag{4.9.2}$$

where x_0 cannot lie on a node. The easiest way to calculate the quantum potential is to use the Hamilton–Jacobi equation. Since $-\partial S/\partial t=E_n$ and $\partial S/\partial x=0$ we have

$$Q=(n+\tfrac{1}{2})\hbar\omega-\tfrac{1}{2}m\omega^2x^2. \tag{4.9.3}$$

The total energy comprises just quantum and classical potential energy and the motion has nothing in common with the oscillatory behaviour of a classical oscillator (other than that both systems are conservative). Note that the zero-point energy $\frac{1}{2}\hbar\omega$ comes from Q.

To obtain periodic motion of the type executed by a classical oscillator we have to form a wave packet. A nondispersive Gaussian-shaped packet can be constructed by an appropriate superposition of the stationary wave-functions (4.9.1):

$$\psi(x, t) = \sum_{n=0}^{\infty} A_n u_n(x) \, e^{-iE_n t/\hbar}. \tag{4.9.4}$$

If we choose

$$A_n = (m\omega/\hbar)^{n/2} a^n (2^n n!)^{-1/2} \, e^{-m\omega a^2/4\hbar}$$

then the packet is a Gaussian centred around $x = a$ at $t = 0$ with half-width $\sigma_0 = (\hbar/2m\omega)^{1/2}$ (Schiff, 1968, p. 74):

$$\psi(x, t) = (m\omega/\pi\hbar)^{1/4} \exp\{-(m\omega/2\hbar)(x - a\cos\omega t)^2 \\ - (i/2)[\omega t + (m\omega/\hbar)(2xa \sin\omega t - \tfrac{1}{2}a^2 \sin 2\omega t)]\}. \tag{4.9.5}$$

Thus

$$|\psi(x, t)|^2 = |\psi_0(x - a\cos\omega t)|^2 \tag{4.9.6}$$

and the packet oscillates harmonically without change of shape between the points $x = \pm a$.

The phase function is given by

$$S(x, t) = -\tfrac{1}{2}\hbar\omega t - \tfrac{1}{2}m\omega(2xa \sin\omega t - \tfrac{1}{2}a^2 \sin 2\omega t) \tag{4.9.7}$$

and hence the trajectory by the solution of

$$m\dot{x} = \partial S/\partial x = -m\omega a \sin\omega t \tag{4.9.8}$$

which is

$$x(t) = x_0 + a(\cos\omega t - 1), \tag{4.9.9}$$

where x_0 is the initial position. The particle therefore performs a simple harmonic motion of amplitude a about a centre point $x = x_0 - a$, and is at rest with respect to the packet. Over the ensemble, the particles oscillate about different centre points and do not cross; for any two trajectories, $x_1(t) - x_2(t) = x_{10} - x_{20}$ (Fig. 4.10).

This is precisely the motion of a classical oscillator in the potential $V = \frac{1}{2}m\omega^2(x - x_0 + a)^2$, with equation of motion

$$\ddot{x} + \omega^2 x = \omega^2(x_0 - a). \tag{4.9.10}$$

With the initial condition $\dot{x}(0) = 0$, the solution to (4.9.10) is (4.9.9). There is no inconsistency here since the trajectory (4.9.9) is completely independent of $\hbar$.

Unlike the classical oscillator, the particle moving in the oscillating packet is not a conservative system. Evaluated along the trajectory (4.9.9) the particle

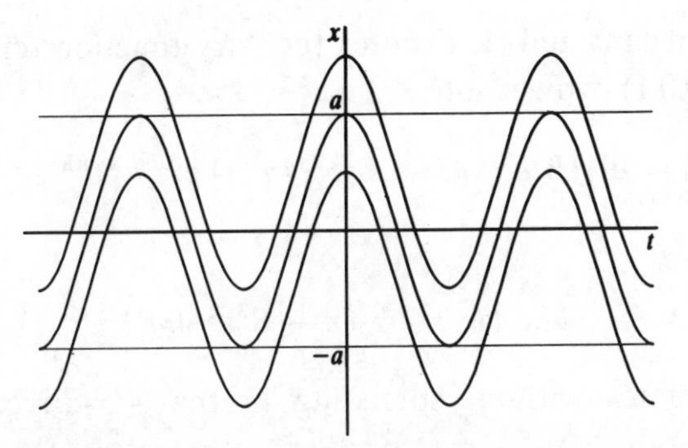

Fig. 4.10 Harmonic oscillator orbits pass beyond the classical amplitude $x = |a|$ to exhibit the single-valuedness of the quantum velocity field.

energy is given by

$$E(x(t), t) = \tfrac{1}{2}\hbar\omega + \tfrac{1}{2}m\omega^2 a^2 + m\omega^2(x_0 - a)\cos\omega t, \tag{4.9.11}$$

which, for $x_0 \neq a$, is a function of t. The reason for this difference is that the effective potential acting on the particle is not of the classical type. From the Hamilton–Jacobi equation we find that

$$Q = -\tfrac{1}{2}m\omega^2(x - a\cos\omega t)^2 + \tfrac{1}{2}\hbar\omega \tag{4.9.12}$$

(a rigid oscillating parabola) and hence the classical potential is cancelled when we form $(Q + V)$. The quantum equation of motion involves a net time-dependent force:

$$\ddot{x} = -\omega^2 a\cos\omega t \tag{4.9.13}$$

which should be compared with (4.9.10). Notice that $Q(x(t), t) =$ constant.

Oscillators in higher dimensions have different properties. For example, in two dimensions the coordinates are dynamic in stationary states (see §12.5.2).

4.10 Nonspreading free packet

It is a remarkable but not widely known fact that there exists a packet-like solution to the free Schrödinger equation which translates as a whole without change of shape. This is the 'Airy packet' (Berry and Balazs, 1979). Apart from the plane wave, this is the only free wavefunction to have the nonspreading property.

At $t = 0$, the wavefunction in one dimension is

$$\psi_0(x) = Ai(Bx/\hbar^{2/3}) \tag{4.10.1}$$

where B is a constant and Ai denotes the Airy function (cf. §6.8). At time t the function (4.10.1) evolves into

$$\psi(x, t) = Ai[(B/\hbar^{2/3})(x - B^3t^2/4m^2)] \, e^{(iB^3t/2m\hbar)(x - B^3t^2/6m^2)}. \tag{4.10.2}$$

Thus

$$|\psi(x, t)|^2 = |\psi_0(x - B^3t^2/4m^2)|^2 \tag{4.10.3}$$

and $|\psi|^2$ propagates without spreading in the $+x$-direction with speed $B^3t/2m^2$.

The function (4.10.2) may be decomposed into plane waves and corresponds to the choice $\varphi(k) = [\hbar^{2/3}/(2\pi)^{1/2}B] \, e^{i\hbar^2k^3/3B^3}$ in (4.6.1). Note that the principle of stationary phase does not apply here since $|\varphi(k)| = \text{constant}$.

Up to a constant, the phase function is given by

$$S(x, t) = (B^3t/2m)(x - B^3t^2/6m^2). \tag{4.10.4}$$

Therefore, solving $m\dot{x} = \partial S/\partial x$ yields for the trajectory

$$x(t) = B^3t^2/4m^2 + x_0, \tag{4.10.5}$$

where x_0 may be chosen to lie anywhere outside the nodes. The particle uniformly accelerates with the packet.

The particle energy is

$$E(x(t), t) = -(B^3/2m)(x_0 - B^3t^2/4m^2) \tag{4.10.6}$$

and so the motion is nonconservative. The quantum potential is easily found from the quantum Hamilton–Jacobi equation:

$$Q = (B^3/2m)(B^3t^2/4m^2 - x) \tag{4.10.7}$$

and so along a trajectory it is constant. The motion is therefore one in which the (quantum) potential energy remains constant while the kinetic energy and hence the total energy vary with time. The constant quantum force is given by (4.10.7): $-Q' = B^3/2m$. This should be compared with a classical potential of the form $V \sim x$. There the potential energy varies to compensate changes in the kinetic energy and the total energy is conserved.

Notice that in this example there is no contradiction with Ehrenfest's theorem which requires the mean position to move uniformly in the absence of classical forces. The reason is that the function (4.10.2) is not square integrable (it has no 'centre') and hence does not satisfy the conditions at infinity required in the proof of Ehrenfest's theorem (cf. §3.8.3).

4.11 Dislocated waves

4.11.1 Edge dislocation

In the hydrodynamic picture the single-valuedness requirement (3.2.10),

$$\oint_C \mathrm{d}S = nh, \qquad n \in \mathbb{Z}, \tag{4.11.1}$$

where $n \neq 0$, is interpreted as implying the existence of quantized vortices in the fluid (Hirschfelder, Goebel & Bruch, 1974b). An alternative way of interpreting (4.11.1) is that the singularities in the phase function (which occur at the nodes of ψ) are *dislocations* in the set of wavefronts, since they have properties very similar to dislocations in crystals.† For example, we may think of the wavefronts of a plane wave as analogous to the lattice of a perfect crystal. If we deform the wave by causing it to have nodes (for example, by the superposition of plane waves having different amplitudes) then (4.11.1) may be interpreted as a Burgers circuit in which the net number of crystal planes (or wavecrests) encountered in one circuit is equal to n. This means that n wavecrests have come to an end within C – we speak of a dislocation of 'strength n'. We have already encountered an example of this in §4.5 where the phase function (4.5.6) corresponds to a multiple edge dislocation (see below) of strength m, the dislocation lying along the z-axis.

To illustrate the method of forming dislocations, which we present here for no reason other than its intrinsic interest, we superpose two plane waves of variable amplitude, each a solution of the free Schrödinger equation. Consider

$$\left.\begin{aligned} \psi_1 &= Akx\, \mathrm{e}^{\mathrm{i}(ky-\omega t)} \\ \psi_2 &= A(ky-2\omega t)\, \mathrm{e}^{\mathrm{i}(ky-\omega t)} \end{aligned}\right\} \tag{4.11.1}$$

where $\omega = \hbar k^2/2m$ and A is a constant. These are evidently plane waves, but not, of course, momentum eigenfunctions (§4.2). Both waves possess nodes, but not dislocations ($n = 0$ in (4.11.1)). Forming $\psi = \psi_1 + \mathrm{i}\psi_2$ we have

$$\psi(x, y, t) = A[kx + \mathrm{i}(ky - 2\omega t)]\, \mathrm{e}^{\mathrm{i}(ky-\omega t)} \tag{4.11.3}$$

so that up to a constant

$$S = \hbar\{\tan^{-1}[(ky - 2\omega t)/kx] + ky - \omega t\}. \tag{4.11.4}$$

The function (4.11.3) has a nodal line parallel to the z-axis which moves

† The analogy between crystal defects and wavefront singularities was first pointed out and systematically discussed by Nye & Berry (1974). The edge and screw dislocated solutions described in this reference and in Berry (1981) and Holland (1987) have been adapted so that they obey the Schrödinger equation.

along the y-axis with speed $u = 2\omega/k = \hbar k/m$. The wave possesses a dislocation of strength unity and this is of the edge-type since the closure failure in performing a circuit around the z-axis in the xy-plane is orthogonal to the dislocation line.

The phase (4.11.4) may be written:

$$S = \hbar\{\tan^{-1}[(y - ut)/x] + ky - \omega t\}. \tag{4.11.5}$$

To find the corresponding trajectories we have to solve:

$$\left.\begin{aligned} \dot{x} &= -(\hbar/m)(y - ut)[x^2 + (y - ut)^2]^{-1} \\ \dot{y} &= u + (\hbar/m)x[x^2 + (y - ut)^2]^{-1} \\ \dot{z} &= 0. \end{aligned}\right\} \tag{4.11.6}$$

The paths evidently lie in a plane parallel to the xy-plane. To solve for x and y we pass to a system of coordinates comoving with the dislocation: $x' = x$, $y' = y - ut$. Then (4.11.6) becomes

$$\dot{x}' = -(\hbar/m)y'(x'^2 + y'^2)^{-1}, \qquad \dot{y}' = (\hbar/m)x'(x'^2 + y'^2)^{-1} \tag{4.11.7}$$

which readily integrates to

$$x' = -a \sin \Omega t, \qquad y' = a \cos \Omega t, \tag{4.11.8}$$

where we have assumed that $x = 0$, $y = a$ at $t = 0$, and $\Omega = \hbar/ma^2$. The orbit is therefore a circle whose centre coincides with the dislocation, performed with an angular speed Ω.

Returning to the coordinates fixed in space we have

$$x = -a \sin \Omega t, \qquad y = ut + a \cos \Omega t. \tag{4.11.9}$$

The motion is a spiral in a plane orthogonal to the z-axis and is depicted in Fig. 4.11. Each time Ωt increases by 2π, x returns to itself and y increases by an amount $2\pi ka^2$.

Comparing with the results of §4.5, we have in (4.11.3) an example of a (nonnormalizable, free) eigenfunction of $\hat{L}_z$, corresponding to the eigenvalue $\hbar$, studied in a reference system (x, y) that is moving relative to that used in §4.5.

4.11.2 Screw dislocation

We now examine a case where the closure failure is parallel to the dislocation so that the latter is of screw-type. Once again this may be generated by a

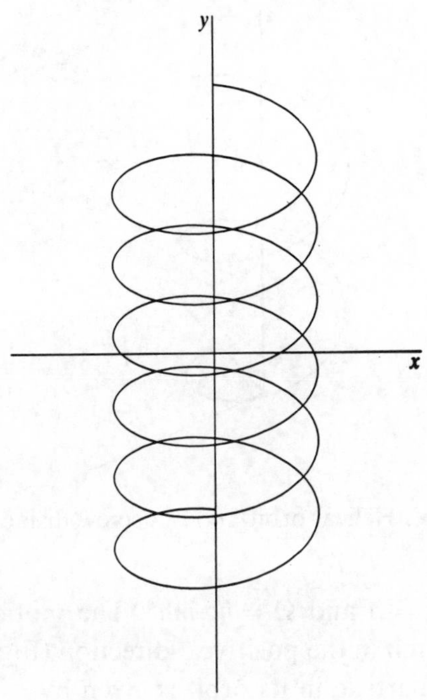

Fig. 4.11 As the edge dislocation moves up the page along the y-axis, the particle circles it at a constant radius.

superposition of variable amplitude plane waves:

$$\psi = Ak(x + \mathrm{i}y)\,\mathrm{e}^{\mathrm{i}(kz - \omega t)}.$$

We shall, in fact, consider a multiple screw dislocation (the analogue of the multiple edge dislocation of §4.5) with wavefunction

$$\psi = Ak^q(x + \mathrm{i}y)^q\,\mathrm{e}^{\mathrm{i}(kz - \omega t)} \tag{4.11.10}$$

where q is a positive integer. This function has a nodal line along the z-axis and its phase is, up to a constant,

$$S(\mathbf{x}, t) = \hbar[q \tan^{-1}(y/x) + kz - \omega t]. \tag{4.11.11}$$

At t = constant the wavefronts form a set of q intertwined helicoids about the z-axis. Integrating round a curve $x^2 + y^2$ = constant we find

$$\oint \nabla S \cdot \mathrm{d}\mathbf{x} = qh, \tag{4.11.12}$$

so that the dislocation is of strength q.

Leaving out the details, the trajectory implied by (4.11.11) is

$$x = -a \sin \Omega t, \qquad y = a \cos \Omega t, \qquad z = \hbar kt/m, \tag{4.11.13}$$

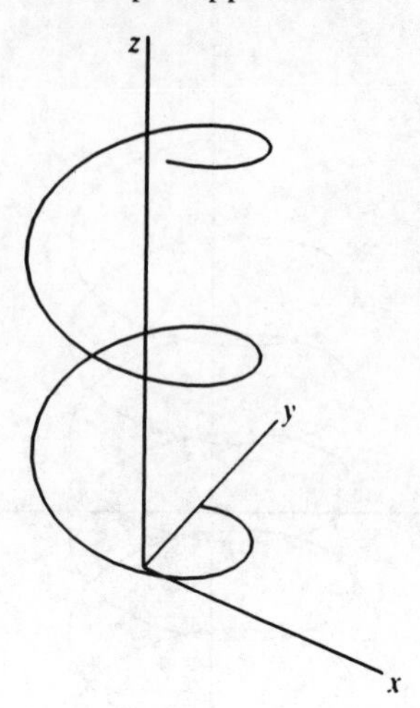

Fig. 4.12 Helical orbit about a screw dislocation.

where $x_0 = z_0 = 0$, $y_0 = a$ and $\Omega = \hbar q/ma^2$. The motion is thus a helix of constant radius and pitch in the positive z-direction (Fig. 4.12). The quantum potential keeping the particle in its orbit is given by

$$Q = -\hbar^2 q^2/2m(x^2 + y^2) \tag{4.11.14}$$

and is a constant of the motion, as is the total energy $\hbar\omega$.

5

Interference and tunnelling

5.1 Interference by division of wavefront. The two-slit experiment

5.1.1 Preliminary remarks

In this chapter we shall be concerned with various types of interference effects exhibited by matter waves. It might be argued that all quantum mechanical phenomena are ultimately a consequence of interference, being governed as they are by a linear wave equation. Of particular interest here is the analogue for material waves of Young's double-slit experiment in optics, a classic problem in the history of quantum theory. Feynman, Leighton & Sands (1965) state that it '... has in it the heart of quantum mechanics. In reality, it contains the *only* mystery'. We shall also treat the process of tunnelling into classically forbidden regions, a phenomenon which also involves interference in an essential way.

To observe the interference of waves issuing from a common source we must split the emerging beam into two spatially separated coherent parts and subsequently recombine them (the superposition of waves from physically independent sources is a many-body problem – see §§8.8 and 12.7). As in optics there are two methods of achieving this: by division of wavefront where, for example, a wave falls on a screen containing two apertures, and by division of amplitude in which a wave is incident on a semi-transparent barrier. Interference effects achieved by the first method, discussed in this section, have been observed for electrons (Jönsson, 1961; Lichte, 1986; Tonomura *et al.*, 1989), neutrons (Zeilinger *et al.*, 1988) and (helium) atoms (Carnal & Mlynek, 1991). The second, conceptually similar, method is discussed in §5.4 and has been demonstrated for neutrons (see §5.4) and (sodium and calcium) atoms (Keith *et al.*, 1991; Kasevich & Chu, 1991; Riehle *et al.*, 1991). In all cases the predictions of quantum mechanics have been confirmed.

The physical phenomenon to be explained is the sequence of patterns depicted in the *frontispiece* for the double-slit diffraction of electrons. This picture shows that *the empirical manifestation of the wave aspect of matter is a statistical effect* resulting from an aggregation of discrete single-particle processes. The pattern does not appear in its entirety all at once and indeed the detection of one electron can occur before the next has left the source. When we speak of 'wave-particle duality' we do not, or cannot, mean that matter manifests itself either as a wave in its classical sense or as a particle (again in its classical sense) depending on the experimental arrangement, and never the two simultaneously. The 'wave' is only *ever* made apparent by observation of particle positions and in this regard has little in common with our classical notion of wave motion. In classical optical or ripple tank experiments one observes a distribution of interference fringes that is continuous in time. Although we may simulate this structure in quantum mechanics by sending a large number of (noninteracting) electrons through the interferometer in a short time, the pattern demonstrating the wave character is basically granular. In this respect the experiment is similar to firing a statistical ensemble of classical particles one at a time through two slits which would likewise gradually build up a pattern, although, of course, the corresponding quantal one is not amenable to a classical explanation. The 'wave-particle duality' seems to be a picturesque way of expressing the fact that in each trial the electron appears to pass through both slits, which is the kind of thing a wave does, because the final pattern is not simply the linear sum of those obtained when only one slit or the other is open.

One interpretation of the experimental results is that the concept of a localized particle is a good one (evidence for which is given by the discrete, localized emission and reception process) but that on the quantum scale the particle follows a spacetime track different from that of a classical particle of the same physical characteristics (mass etc.) moving through the same apparatus (evidence for which comes from the cumulative diffraction-like pattern). This is indeed already suggested by single-slit diffraction where no question arises as to 'which slit' the electron passes through and yet one observes, over a period, the growth of a pattern characteristic of the diffraction of waves. The question then devolves around identifying the new force that escapes detection at the usual classical level but that is evident in these experiments. The latter do not compel us to adopt this kind of interpretation but equally they do not require us to give up our customary notions of objective reality.

To explain the *statistical* emergence of the interference pattern along these lines the causal interpretation postulates that an objective wave is involved in *each* process that ends in a single localized detection event on the observing screen. That is, while a particle propagates from source to screen and passes through one of the slits as in classical physics, it is accompanied by its own wave (the ψ-function) which passes through both slits and in the vicinity of the screen (where the component waves recombine) guides the particle to a certain location. The influence of the wave over many trials is to build up the observed pattern in accordance with the probability distribution $|\psi|^2$. This explanation gives substance to the notion that what is essentially involved is the 'self-interference' of electrons, a concept that is often invoked to indicate that the pattern is built up from a series of single-particle events but whose physical meaning is obscure if ψ is treated as no more than a 'wave of probability'.

Before entering into the technical account we make a general remark concerning the treatment of this type of problem. Especially in descriptions of the two-slit experiment, emphasis is often laid on the necessary intervention of a classically describable apparatus without which, it is claimed, the phenomenon is not completely defined or brought to a close. Thus, it is suggested that the predictions embodied in ψ actually presuppose a classical observing system that is not explicitly included in the theoretical account (i.e., in the Schrödinger equation obeyed by ψ). In contrast, our treatment has the distinct advantage that neglect of the observing apparatus (or parts of it) as an explicit component in the wavefunction is a question not of principle but of formulating valid approximations in physical theory. An exact analysis would include an account of all the atoms and potentials making up the instrument, as well as the electron, described by a gigantic many-body Schrödinger equation. In principle the entire universe should be included. It is a property of the world that to a very good approximation we may not only isolate the electron and laboratory apparatus from the 'rest of the universe' and study it as a unit in itself, but we may obtain a tractable problem by replacing the interaction between the electron and the apparatus by external potentials inserted into the Schrödinger equation for the electron alone. The physical independence of systems is mathematically expressed through the factorizability of the many-body wavefunction (Chap. 7). Actually, specifying the conditions under which bits of matter may be isolated and treated as closed systems is a rather subtle problem. One of the lessons of the EPR analysis (Chap. 11), for example, is that a large separation is not a sufficient condition for the physical independence of material systems, even though this criterion is generally applicable in classical physics (where interparticle potentials fall off with distance). The point is that it is the theory that should tell us the varying conditions under which it is permissible to treat part of an interacting system as 'classical' and 'given'.

In the following we shall assume that this analysis has been carried out and the true electron–slit interaction, which in a more complete analysis is likely to be a quite involved many-body problem, may be approximated by a single body in an external force. We also assume that the electron wave emitted by the source in each trial is identical, although in practice there will be some drift from this ideal and the true distribution is a sharply peaked mixed state. The ultimate arbiter of these assumptions is experiment.

5.1.2 Particle trajectories in the electron interferometer

The description that follows is based on Philippidis *et al.* (1979). Referring to Fig. 5.1, electrons are emitted by a source S_1, pass through two slits B, B' in a barrier P and arrive at a screen S_2. The detection process at the screen S_2 is nontrivial but it plays no causal role in the basic phenomenon of the interference of electron waves and is not discussed here (for remarks on this see Chaps. 8 and 12). In the two-dimensional system of coordinates (x, y) whose origin O is shown, the centres of the slits lie at the points $(0, \pm Y)$. The wave incident on the slits will be taken as plane, $\psi = a\,\mathrm{e}^{\mathrm{i}(k_1x + k_2y)}$ where a is a constant and $u_1 = \hbar k_1/m$ and $u_2 = \hbar k_2/m$ are the x- and y-components

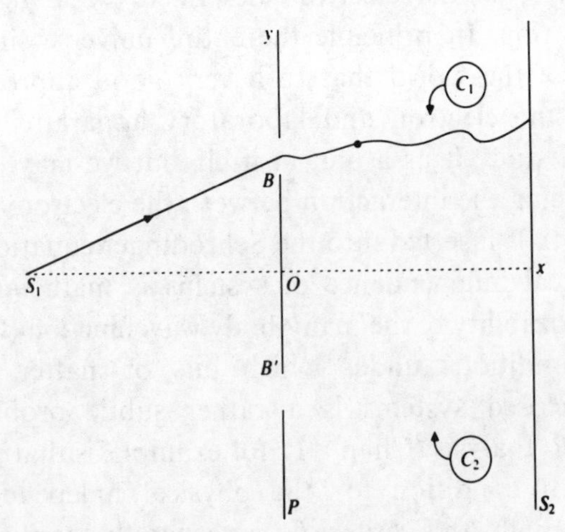

Fig. 5.1 Electrons from a source S_1 pass through slits B, B' in a barrier P and are detected at a screen S_2. A typical trajectory is shown. C_1 and C_2 are counters whose role is described in §5.1.5.

of velocity. To avoid the mathematical complexity of Fresnel diffraction at a sharp-edged slit, we assume the slits have 'soft' edges that generate waves having identical Gaussian profiles in the y-direction. The plane wave in the x-direction is unaffected. The instant at which the packets are formed will be our zero of time. The two waves emerging from the slits are therefore initially (cf. §4.7):

$$\psi_{B0} = a(2\pi\sigma_0^2)^{-1/4}\, e^{-(y-Y)^2/4\sigma_0^2 + i[k_1 x + k_2(y-Y)]} \tag{5.1.1}$$

$$\psi_{B'0} = a(2\pi\sigma_0^2)^{-1/4}\, e^{-(y+Y)^2/4\sigma_0^2 + i[k_1 x - k_2(y+Y)]} \tag{5.1.2}$$

where σ_0 is the half-width of each slit. Wave B is a packet moving in the y-direction with velocity u_2 and a plane wave in the x-direction, and wave B' is similar except it has group velocity $-u_2$. Subsequently the packets move with relative group velocity $2u_2$ and spread into one another. This is how we 'recombine' the beams, although they are not strictly nonoverlapping at $t = 0$ due to the Gaussian tails. The interference comes about from both effects (relative motion and dispersion) or just one of them if the other is negligible.

At time t the total wavefunction at a space point (x, y) is given by

$$\psi(x, y, t) = N[\psi_B(x, y, t) + \psi_{B'}(x, y, t)], \tag{5.1.3}$$

where (see §4.7)

$$\psi_B(x, y, t) = a(2\pi s_t^2)^{-1/4}\, e^{-(y-Y-u_2 t)^2/4\sigma_0 s_t}\, e^{i[k_1 x + k_2(y - Y - \frac{1}{2}u_2 t) - E_1 t/\hbar]} \tag{5.1.4}$$

$$\psi_{B'}(x, y, t) = a(2\pi s_t^2)^{-1/4}\, e^{-(y+Y+u_2 t)^2/4\sigma_0 s_t}\, e^{i[k_1 x - k_2(y + Y + \frac{1}{2}u_2 t) - E_1 t/\hbar]} \tag{5.1.5}$$

with $E_1 = \frac{1}{2}mu_1^2$ and $N = (2 + 2\,e^{-Y^2/2\sigma_0^2})^{-1/2}$ is a normalization constant. These wavefunctions differ slightly from those written down by Philippidis *et al.* (1979). These authors assumed a point source S_1 generating a free Green function which implies an additional spreading term of classical origin in the waves formed at each slit (this is the quantum generalization of the dispersive Gaussian packet described in §2.5.3). For a source placed far from the slits the extra terms are negligible and we obtain essentially our wavefunctions (5.1.4) and (5.1.5).

It is evident that the total wavefunction is factorizable in orthogonal directions:

$$\psi(x, y, t) = f_1(x, t) f_2(y, t) \tag{5.1.6}$$

where f_1 is a plane wave factor. Hence the components of the particle motion in the x- and y-directions are independent and the quantum potential is a function of just y:

$$Q = -\frac{\hbar^2}{2mR}\left(\frac{\partial^2 R}{\partial x^2} + \frac{\partial^2 R}{\partial y^2}\right) = -\frac{\hbar^2}{2mf_2}\frac{\partial^2 f_2}{\partial y^2}. \tag{5.1.7}$$

The trajectory, obtained by integrating $m\dot{\mathbf{x}} = \nabla S$ where S is the phase of (5.1.3) and $\mathbf{x}_0$ must be specified, has a uniform and rectilinear x-component,

$$x = u_1 t \tag{5.1.8}$$

($x_0 = 0$ since all particles pass through the slits at $t = 0$), and a plot of y vs t will look the same as a space plot (y vs x) apart from a difference in scale. To give a flavour of the complexity of the y-motion, we write down the relevant differential equation (for $u_2 = 0$) which must be solved numerically:

$$\begin{aligned}\frac{\mathrm{d}y}{\mathrm{d}t} = F\{&-A\sin[Byt(1 + Ct^2)^{-1}] + Dyt\cos[Byt(1 + Ct^2)^{-1}] \\ &+ \tfrac{1}{2}Dt(y - Y)\,\mathrm{e}^{Ey(1+Ct^2)^{-1}} + \tfrac{1}{2}Dt(y + Y)\,\mathrm{e}^{-Ey(1+Ct^2)^{-1}}\} \\ &\times (1 + Ct^2)^{-1}\{\mathrm{e}^{Ey(1+Ct^2)^{-1}} + \mathrm{e}^{-Ey(1+Ct^2)^{-1}} \\ &+ 2\cos[Byt(1 + Ct^2)^{-1}]\}^{-1},\end{aligned} \tag{5.1.9}$$

where A, B, C, D, E and F are constants and we choose a range of initial positions y_0 in the slits. Notice that $\partial S(y, t)/\partial y = -\partial S(-y, t)/\partial y$ which expresses the symmetry of the velocity field about the x-axis.

To see the nature of the interference pattern we write down the amplitude squared of the total wavefunction:

$$\begin{aligned}R^2(y, t) = {}& a^2 N^2 (2\pi\sigma^2)^{-1/2}\,\mathrm{e}^{-[y^2 + (Y + u_2 t)^2]/2\sigma^2} \\ &\times \{\mathrm{e}^{y(Y+u_2 t)/\sigma^2} + \mathrm{e}^{-y(Y+u_2 t)/\sigma^2} + 2\cos[2k_2 y - (Y + u_2 t)y\hbar t/2m\sigma_0^2\sigma^2]\}.\end{aligned} \tag{5.1.10}$$

The term outside the curly brackets defines the (Gaussian) envelope of the pattern and the terms inside the fringes. The wavefunction has nodes (strict dark fringes) only at the spacetime points defined by

$$t = -Y/u_2, \qquad y = (n + \tfrac{1}{2})\pi/k_2, \tag{5.1.11}$$

where n is an integer and we must have $u_2 < 0$ (the packets approach one another) so that $t > 0$. In the case $u_2 \geqslant 0$, ψ has no strict nodes at all and there is no region of space that a particle may not potentially visit if y_0 is suitably chosen. The inteference structure is evidently a dynamic one that remains symmetric about the x-axis and ultimately evolves into the familiar pattern, as we shall see.

To study the motion in detail we insert values for the parameters characterizing the apparatus of the order of those used in actual experiments (Jönsson, 1961): $u_1 = 1.3 \times 10^{10}$ cm s^{-1}, $u_2 = +1.5 \times 10^4$ cm s^{-1}, $\sigma_0 = 10^{-5}$ cm and $Y = 5 \times 10^{-5}$ cm (with $m = 9.11 \times 10^{-28}$ g and $\hbar = 1.055 \times 10^{-27}$ g cm^2 s^{-1}). The incident kinetic energy is $\sim$45 keV. In the experiment of Tonomura *et al.* (1989) the current corresponds to 10^3 electrons s^{-1}. The trajectories and quantum potential are plotted in the region $|y| \leqslant 1.9 \times 10^{-4}$ cm and $0 < x < 35$ cm which corresponds via (5.1.8) to $0 < t < 2.7 \times 10^{-9}$ s. The distributions at $x = 1.5$ cm and at the screen ($x = 35$ cm) are shown in Fig. 5.2 and were calculated for the above values except that $u_2 = 0$.

It is straightforward to calculate the quantum potential Q from (5.1.10) but nothing is gained by explicitly displaying its complicated form. The quantum potential as viewed from the screen S_2 looking towards the slits is shown in Fig. 5.3 and a 150° azimuthal view from behind the slits in Fig. 5.4. The two parabolic peaks located in the slit plane will be recognized as the quantum potentials corresponding to single-slit Gaussian packets (§4.7). As we move away from the slits along the x-axis high, rapidly varying spikes appear at about 1.5 cm. Comparing with Fig. 5.2(*a*), we have a good example of how Q reflects the form or variation of R rather than its absolute value

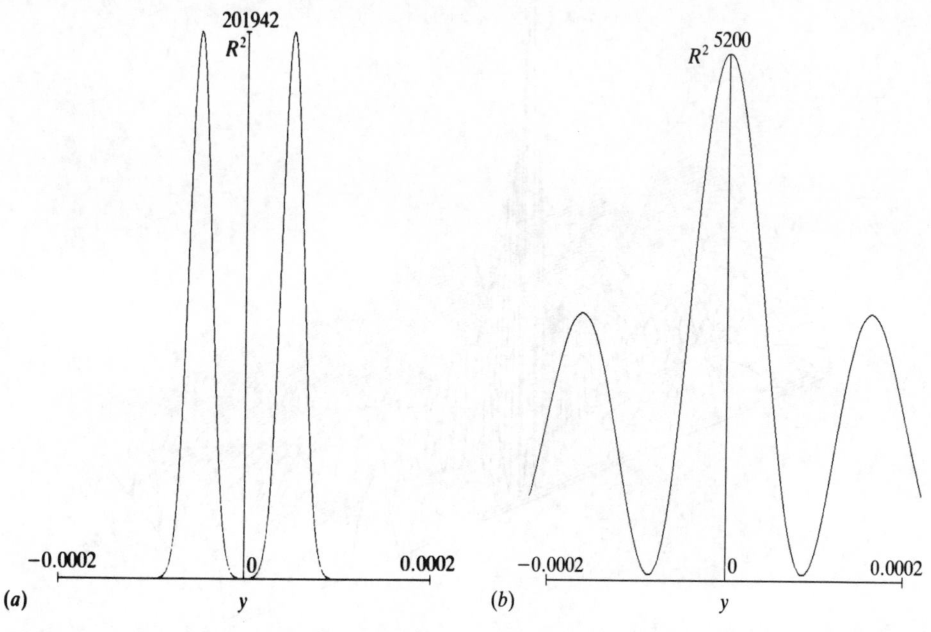

Fig. 5.2 (*a*) Probability distribution at 1.5 cm from two Gaussian slits in an electron interferometer. (*b*) Probability distribution at 35 cm from the slits.

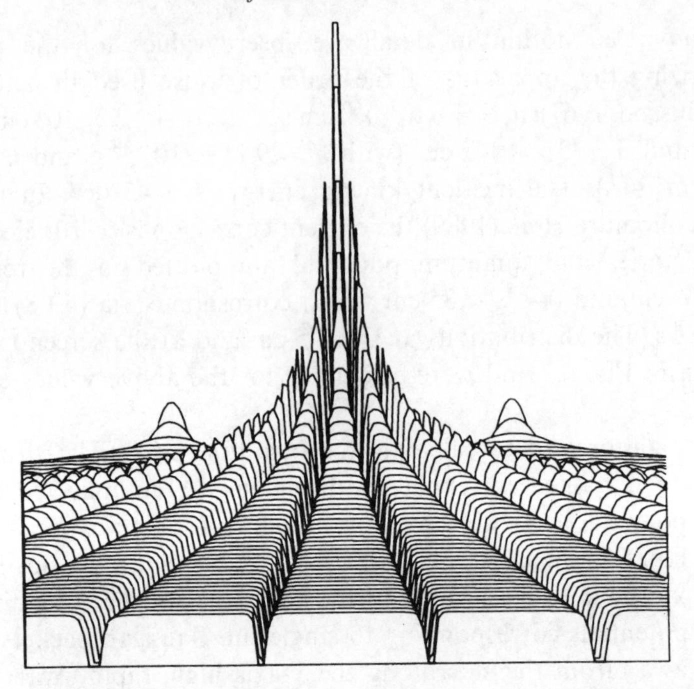

Fig. 5.3 The quantum potential for two Gaussian slits viewed from the screen S_2 (from Philippidis *et al.* (1979)).

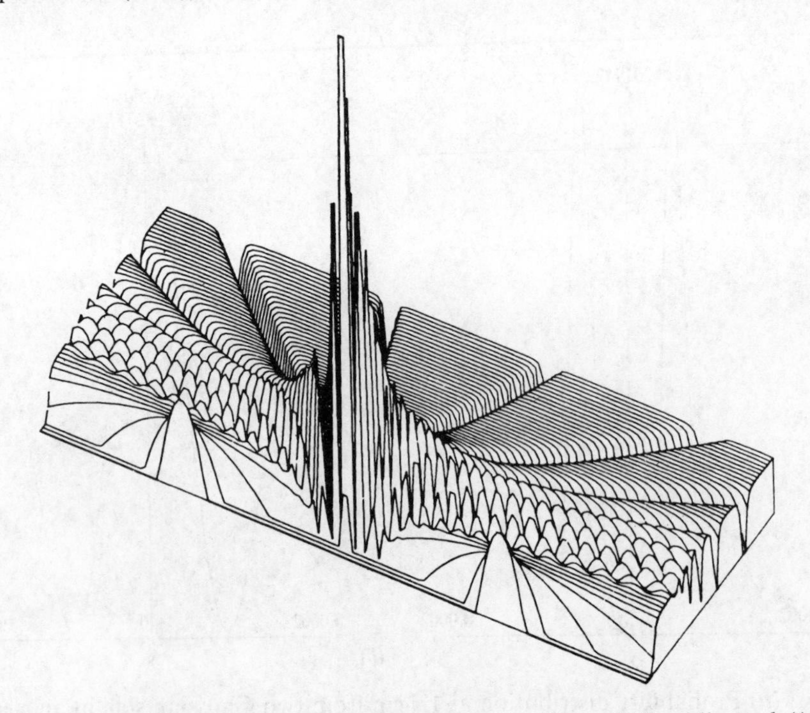

Fig. 5.4 150° azimuthal view of the quantum potential (from Philippidis *et al.* (1979)).

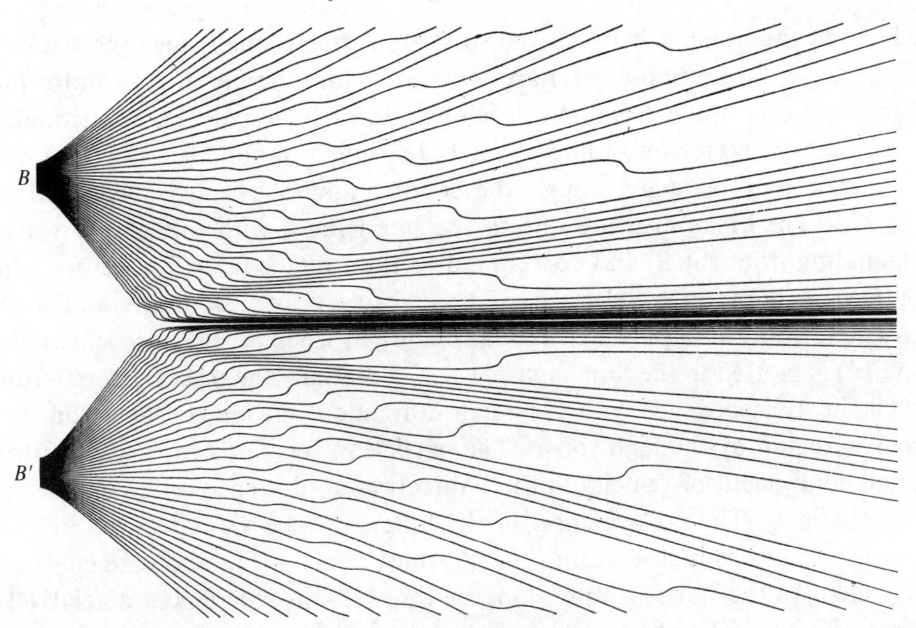

Fig. 5.5 Trajectories for two Gaussian slits with a uniform distribution of initial positions at each slit (from Philippidis *et al.* (1979)).

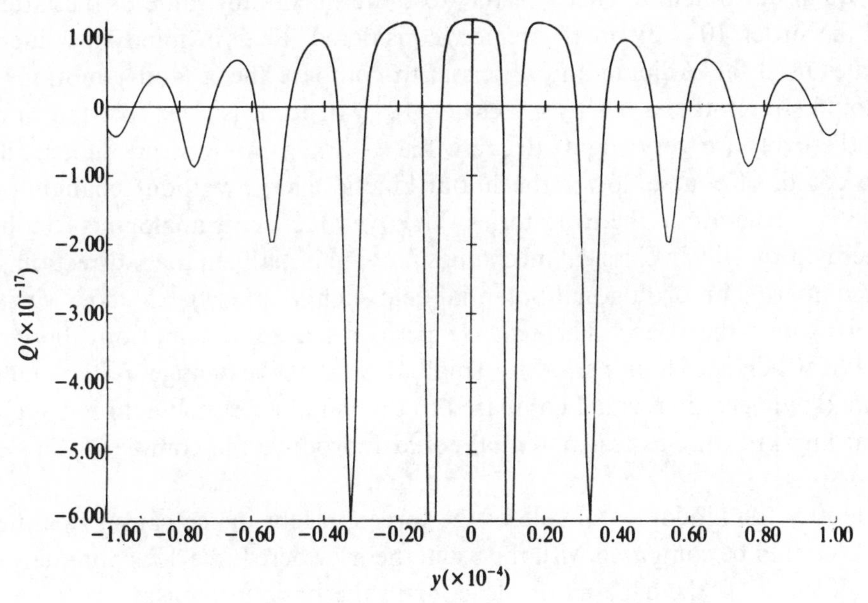

Fig. 5.6 Cross-section of the quantum potential at 18 cm from the slits (from Philippidis *et al.* (1979)).

which, in the domain between the slits, is relatively small since the packets do not yet appreciably overlap. This structure finally decays into the background at about 6 cm and develops into a set of plateaux and troughs. Q is finite in the region of interest, only becoming unbounded as $y \rightarrow \pm\infty$.

A trajectory passing through the centre point of slit B is displayed in Fig. 5.1. The kinks in it are due to the interference with the 'empty wave' emanating from slit B' and correspond to the troughs in Q. To see how the pattern is built up a selection of trajectories drawn from the statistical ensemble (only one of which is realized when a single electron leaves the source) is plotted in the same figure, Fig. 5.5. Initially the tracks diverge from each slit in accordance with Gaussian diffraction but are repelled from the central region by the high spikes. The particle moves fairly uniformly with a small component of velocity in the y-direction until it encounters one of the troughs in Q. The cross-section of the latter may be viewed as a series of potential wells as is clear from Fig. 5.6 which depicts Q at about 18 cm from the slits. As the particle enters one of the wells it experiences a relatively strong force $-\partial Q/\partial y$ first accelerating it, then decelerating it, until it emerges onto another plateau where the force is again weak. The particle is not trapped in a well owing to its large kinetic energy ($\sim$45 keV) compared to its quantum potential energy – the absolute maximum value of the latter is of the order 10^{-4} eV in the region considered. Bear in mind the different scales used for ordinate and abscissa; to compare the x- and y-motions we should stretch the x-axis by a factor $\sim 10^4$ whence it is clear that the ripples in the orbit are very slight. Because the x- and y-motions are independent we could, of course, lower the initial kinetic energy without changing the relative structure of the trajectories. The overall effect is analogous to a high energy projectile in classical mechanics moving initially in the x-direction and encountering an orthogonal potential field such as $V(x, y) = V_0 \cos ky$ which 'corrugates' the space. The particle performs a zig-zag motion about the x-axis which is of long period for small V_0 so that the change in its potential and kinetic energy is small compared to the initial energy. We do not suggest that any known classical potential could reproduce the complexity of Q, of course.

Notice the fundamental role of the force concept in the causal treatment. This should be compared with the usual theory where force does not intervene at all, either at the explanatory level or in the basic formalism.

The result of all this is that at any instant most of the quantal trajectories are running along the plateaux which therefore correspond to the bright fringes. Any cross-section at a suitable distance from the slits will therefore display the expected interference pattern but this is a dynamic structure and particles may cross the troughs (the dark fringes) far from the slits. The envelope of the interference pattern is reflected in the diminishing cross-section of Q as $|y|$ increases.

To compare with the actual experiments we pass to the Fraunhofer limit $x \gg \sigma_0$ or, using (5.1.8), $t \gg \sigma_0/u_1 \sim 10^{-15}$ s. Alternatively we may characterize the limit by supposing that the width σ of each packet is much larger than the initial width σ_0:

$$\hbar t/2m\sigma_0^2 \gg 1. \tag{5.1.12}$$

This is equivalent to $t \gg 2 \times 10^{-10}$ s or $x \gg 3$ cm, a condition that is certainly obeyed at the screen S_2 ($x = 35$ cm) where $\sigma \sim 14\sigma_0$. In this limit the cos term in (5.1.10) becomes

$$2\cos(2myY/\hbar t) = 2\cos(2k_1yY/x). \tag{5.1.13}$$

The pattern then resembles the Fraunhofer limit of two slits producing cylindrical waves (Jönsson, 1961) modulated by a Gaussian envelope. The distance between any two neighbouring maxima is, however, not quite given by the classical formula $\delta y = \lambda x/2Y$ quoted in elementary optics, where $\lambda = 2\pi/k_1$ is the de Broglie wavelength, because of the additional exponential terms in the curly brackets in (5.1.10). The precise positions of the bright fringes may be calculated from $\partial R^2/\partial y = 0$ or by inspection from Fig. 5.2(*b*).

The trajectories only become rectilinear in the Fraunhofer limit if, in addition to (5.1.12), we have $y \gg Y$, for then we find from (5.1.9) $\mathrm{d}y/\mathrm{d}t \to y/t$ (with $u_2 = 0$).

In Fig. 5.5 the distribution of initial positions is uniform and does not reflect the actual Gaussian distribution in each slit. As a result the depicted intensity on the screen S_2 does not accurately reproduce the quantum probability function $|\psi|^2$. A selection of trajectories whose initial positions more closely simulate $|\psi_0|^2$ is shown in Fig. 5.7. A similar channelling of the orbits into bands along the y-axis occurs. Each particle with a given initial position pursues, of course, the same trajectory determined by the wavefunction whatever the initial distribution (the individual is not dependent on the ensemble).

5.1.3 More detailed predictions than are contained in the wavefunction

The classical analogue of the two-slit experiment would proceed as follows. We assign a distribution function to the slit system which reflects our knowledge of the initial position of the particle:

$$\rho_0(y) = \tfrac{1}{2}[f(y - Y) + f(y + Y)], \tag{5.1.14}$$

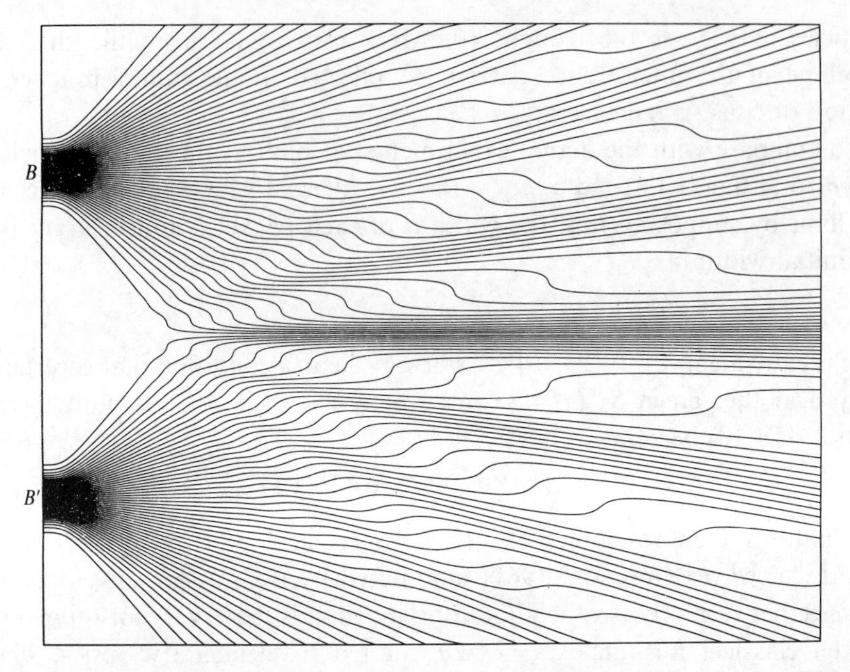

Fig. 5.7 Trajectories for two Gaussian slits with a Gaussian distribution of initial positions at each slit. The probability density is proportional to the number of lines per unit length in the y-direction (from Philippidis *et al.* (1982)).

where $f(y)$ is a normalized function peaked around $y = 0$ and the slits are assumed identical. The probability that the particle passes through slit B, say, is

$$\int_{y-Y/2}^{y+Y/2} \rho_0(y)\, \mathrm{d}y = \tfrac{1}{2}. \tag{5.1.15}$$

The subsequent flow of probability, governed by Liouville's equation (cf. §2.5), is induced by the particle law of motion. From observation of the position on the screen S_2 we can infer which slit the particle passed through. In principle, this may be confirmed by direct observation at the slit without any appreciable disturbance of the orbit. The change in ρ on determination of the trajectory does not in any way affect the distribution of particles on the screen as ρ is a function just of the physical characteristics of the slits and source and not a physical property of the system.

In the quantum theory of motion we can make more detailed statements about the behaviour of individual elements in the ensemble than are contained in the distribution function, of the type possible in classical mechanics, but a new element enters that prevents us from empirically checking these predictions without destroying the very phenomenon we are studying. We cite three examples of predictions pertaining to individual particles.

First and most obvious is that each particle passes through one slit or the other.

Next, the single-valuedness of the wavefunction implies that no trajectory can cross or even intersect the axis of symmetry of the apparatus (the x-axis) because this coincides with the trajectory commencing at $y_0 = 0$ and ending at the peak of $|\psi|^2$ (recall that with Gaussian slits there is a small but finite probability that y_0 lies in the domain $|y| \ll Y$). The physical reason is that the quantum potential repels particles from this line. This feature is peculiar to quantum mechanics and indeed will follow for all types of slits, provided they are identical. It is not due to the symmetry of the device alone for in classical mechanics spreading Gaussian ensembles (§2.5.3) will cross. In the case of sharp slits (so that $y_0 = 0$ is forbidden) we see that the quantal mean does not coincide with any actually realizable motion, for $\langle y \rangle = 0$. In the general quantum case we can then infer that if a particle is detected on S_2 in the region $y > 0$ (<0) it passed through slit $B(B')$. Because the ensemble is a function of the individual we can associate conditional probabilities with subensembles. The probability that the corpuscle passes through slit B is $\frac{1}{2}$, as in (5.1.15), and the probability density that it lands at a point y given that it passed through slit B is $|\psi(x, y > 0, t)|^2$.

The final individual prediction made by the quantum potential model that we mention concerns the influence of the classical potential defining the slit system that is mediated by the wavefunction far from the slit plane. It is clear that the effect of the quantum potential does not fall off with distance and a particle travelling in a region corresponding to one fringe may cross over to a neighbouring fringe, even in the Fraunhofer limit.

These detailed predictions demonstrate that if one wishes to subdivide the phenomena and speak of trajectories inside the interferometer this must be carried out consistently using a notion of trajectory appropriate to the quantum domain and not as has sometimes been done by importing unwarranted classical conceptions, such as the assumption that between slits and screen the orbit is a straight line because this region is apparently free of forces. In fact, we have seen that Q carries the influence of the localized classical potential (the slit plate) to distant points. In a classical treatment we may be led to assign the trajectory to the wrong slit and thus contradict the single-valuedness of the wavefunction.

We emphasize that these predictions concerning individual behaviour do not contradict the statistical ones of quantum mechanics. Their purpose is simply to contribute to the explanation of how the latter come about.

5.1.4 Relative 'wave and particle knowledge'

Because in the causal interpretation the wavefunction is more than a mere repository of our current information about a system, and its law of motion induces that of the particle, it is natural to expect that the insertion of a probe in the apparatus to check the above predictions will modify the wave in such a way that the particle orbit suffers a disturbance. In fact, as we discuss in more detail in Chap. 8, the disturbance involved in identifying precisely the particle track is sufficient to wash out the interference pattern.

One can envisage a less severe disruption of the wave, whereby it is altered by interaction with a probe but in a way that does not completely obliterate the interference effects. It has been proposed that such experiments allow one to glean varying degrees of 'knowledge' of the particle and wave aspects of matter (Wootters & Zurek, 1979). Suppose we simulate the action of an absorbing system inserted into one arm of the interferometer by varying the relative width of the slits. For example, in Fig. 5.1 we may decrease the width σ_0 of slit B', keeping slit B the same. This will be analogous to placing an absorber behind slit B' whose overall effect is roughly to decrease the amplitude of the wave issuing from that slit. The smaller σ_0 is the greater the relative probability that the particle passes through slit B and it has been suggested that this implies we have greater knowledge of which 'path' the particle took through the interferometer, i.e., our 'particle knowledge' grows. In the limit $\sigma_0 = 0$ this knowledge is perfect. On the other hand when the slit widths are equal we have no knowledge of which path the particle took and our 'wave knowledge' is supposed to be maximized.

Before commenting on the view that quantum mechanics is essentially concerned with human knowledge of physical systems we note that some significance has been attributed in the literature to the circumstance that the fringes may remain substantially visible when the 'particle knowledge' is relatively large. This is a simple consequence of the definition of the 'contrast' of an interference pattern. Let $\psi = \psi_B + \psi_{B'}$, where $\psi_B, \psi_{B'}$ are arbitrary waves of amplitude $R_B, R_{B'}$, and write $I = |\psi|^2$. Then the contrast, which has been presented as a measure of our 'wave knowledge', is defined by

$$C = \frac{I_{\max} - I_{\min}}{I_{\max} + I_{\min}} = \frac{2R_B R_{B'}}{R_B^2 + R_{B'}^2} \tag{5.1.16}$$

with $0 \leqslant C \leqslant 1$. As the ratio $R_{B'}/R_B$ decreases, C decreases but at a slower rate. To bring out the implications of this define a complementary quantity characterizing the 'particle knowledge' (Greenberger & Yasin, 1988):

$$P = (R_B^2 - R_{B'}^2)/(R_B^2 + R_{B'}^2). \tag{5.1.17}$$

We have

$$P^2 + C^2 = 1. \tag{5.1.18}$$

As an example, suppose that $R_{B'}^2/R_B^2 = 1/99$, i.e., in every 100 trials it is likely that 99 electrons will pass through slit B. Then $P = 0.98$, indicating almost perfect 'particle knowledge', yet $C = 0.2$ so the fringes are still significantly visible. Because C is defined in terms of I it is a measurable quantity and we can also empirically deduce P up to a sign via (5.1.18). The numerical predictions of quantum mechanics in this case have been confirmed in neutron interferometry (Rauch, 1990; see §5.4) and in optics (Mittelstaedt, Prieur & Schieder, 1987).

The problem with this discussion is that for it to be meaningful we must give credence to the notion that a particle has a 'path' through the interferometer that we could have 'knowledge' of. This concept has been slipped in without an explicit recognition that it is an additional assumption. A correct application of the usual interpretation to the above example would say that there is a relative probability of 0.99 for *finding* an electron behind slit B, not that it is actually there in a fraction 0.99 of cases. It is, in fact, only in the causal interpretation that the 'which path' debate fully makes sense. In this connection one must remember that C and P defined above are primarily expressions reflecting properties of the interfering waves and their variation represents objective changes in the system (note that in general they are functions in space). As secondary properties they yield statistical information. To see this let us apply the definitions (5.1.16) and (5.1.17) to the system of equal Gaussian slits studied in §5.1.2. Inserting (5.1.10) we find

$$C(y, t) = \operatorname{sech}(yY/\sigma^2), \qquad P(y, t) = \tanh(yY/\sigma^2). \tag{5.1.19}$$

These are position-dependent quantities; the fact that the slits are of equal width does not imply maximum contrast ($C = 1$) and minimum 'particle knowledge' ($P = 0$) except at special points. Rather, C and P vary across the interference plane in the following way:

$$\left.\begin{array}{lll} P \to 0, & C \to 1 & \text{as} \quad y \to 0 \\ P \to \pm 1, & C \to 0 & \text{as} \quad y \to \pm\infty. \end{array}\right\} \tag{5.1.20}$$

According to the meanings attached to C and P above we should say that in the centre of the pattern the wave (particle) knowledge is maximized (minimized), which is reasonable, but as we move away towards the edges of the pattern the roles are reversed and the particle knowledge becomes more dominant! We would have to conclude that in the same experiment we have simultaneous and arbitrarily precise knowledge of both aspects of matter, albeit in different regions of space. Similarly, in single-slit diffraction the 'particle knowledge' is maximal and yet the wave aspect is empirically manifest.

The characterization of the wave and particle aspects of matter by (5.1.16) and (5.1.17) does not exhaust all we can say about these concepts and does not reflect the meaning given to them in the causal interpretation. In particular the function P has no connection with the particle orbit. The situation is rather the following: With both slits open we discover properties of the wave by detecting the particle aspect, i.e., by observing the particle probability density. With only one slit open the wave aspect generically manifests itself again in a nonclassical particle probability density. The fact is that quantum mechanics is not simply a theory of our knowledge of physical systems, a kind of generalization of classical statistical mechanics without the trajectories, and diffraction phenomena demonstrate this. The principle of linear superposition represents a way of generating qualitatively new types of behaviour which can never be obtained from distribution functions such as (5.1.14).

It is a simple matter to see in qualitative terms what is actually happening in an interferometer when the size of one of the slits is decreased, through the redistribution in space of the particle trajectories. A change in the relative amplitudes at the slits is not translated in a simple way into alterations in the quantum potential and the superposition of a very small amplitude wave can significantly alter the latter. Although the quantum potential approaches the form associated with single-slit diffraction (trajectories may cross the symmetry axis) it does not do so in step with the slit reduction (for an illustration along these lines see Home & Kaloyerou (1989)). Of course, as remarked above, a proper treatment of the interaction between a probe and the electron is a many-body problem. Einstein proposed as a method for determining which slit the particle passes through mounting the plate containing the slits on a movable support (Jammer, 1974, p. 128). A measurement of the momentum of the plate before and after passage of the particle would, he claimed, allow one to infer which slit had been traversed. A causal version of Einstein's two-slit experiment in which the electron and the slit system are treated as a coupled two-body system has not been developed in detail. It is clear from the results of Chap. 7, for example the two-body oscillator system (§7.4), that the final position of the slits and the point where the electron lands on the screen S_2 at time $t > 0$ sensitively depend on the initial positions of both the electron and the slits and importing classical notions of what the trajectories of either should be in 'free space' may well lead us to an incorrect inference regarding which slit the particle actually travelled through. We expect to recover the usual result that the interaction washes out the interference pattern (appearing in the partial density matrix pertaining to the electron) but understand in detail why this happens through the properties of the quantum potential (cf. §8.8.2).

5.1.5 Delayed-choice experiments

An example of the paradoxes that may arise if one attempts to subdivide the phenomena without adopting a consistent model that describes what an electron is and how it behaves in an interferometer is provided by the 'delayed-choice' class of experiments (Wheeler, 1978). The argument runs as follows. We arrange our two-slit experiment so that included in the apparatus are two counters (C_1 and C_2 in Fig. 5.1). These can be rapidly swung into the two arms of the interferometer between the slits and the detecting screen S_2 in a time much shorter than it takes the electron to traverse this region (this time is defined in terms of the motion of the centre of a wave packet). When in place the clicking of one or the other counter reveals which 'path' the electron took through the interferometer. The point is that the decision on whether to insert the counters (and hence determine 'which path') or to leave them out and let the electron contribute to the interference pattern (the electron takes 'both paths') may be made *after* the electron has 'passed' the slit plane. This suggests, so the argument goes, that the earlier behaviour of the electron (passage through one or both slits) may be influenced by our later decision whether or not to insert C_1 and C_2.

To avoid the paradoxical conclusion apparently flowing from this that the electron somehow traversed both slits *and* just one slit, it is proposed by Wheeler that the past has no existence except as it is recorded in the present. The phenomenon of electron passage 'is not a phenomenon until it is an observed phenomenon'. In other words, confusion will be avoided only if we desist from analysing the functioning of the device.

This conclusion is trivially sidestepped in the quantum potential approach (Bell, 1980; Bohm, Dewdney & Hiley, 1985). The above argument is interpreted simply as evidence that the wrong model of an individual physical system has been employed, or rather that no serious attempt has been made to develop one. The passage of the corpuscle through one slit and the wave through both forms a well-defined time-dependent physical process in itself. What happens subsequently has no bearing on it at all. If both paths through the interferometer are open the particle will respond to the overlapping waves. The detecting plate reveals that this has happened, but does not influence it.

If instead the counters are inserted prior to the overlap of the waves the evolved total wave is different and in detecting the particle the counters again simply reveal this. According to the causal interpretation, *the present merely reveals the past and has no influence upon it*. In this regard at least there is no need to revise our customary conceptions of cause and effect.

For Bohr the form of the experimental apparatus and the content of the experimental results compose an indivisible and unanalysable whole. If certain parameters characterizing the atomic object or apparatus are altered this leads to a new phenomenon whose cause is likewise unanalysable and whose details cannot be compared to the original one. Although we do not accept that conceptual analysis of the double-slit experiment necessarily results in ambiguity and confusion, it seems that something resembling Bohr's perception of the wholeness of physical phenomena reappears in the delicate dependence of the quantum potential and hence particle orbits on all the parameters relevant to the process – slit width and separation, group velocity of packets, particle mass, and so on. An alteration of any parameter, even if small, may lead to a significant change in Q. This indeed leads to a relatively simple intuitive picture of quantum wholeness, as well as a counterexample to the view that Bohr's analysis is implied by quantum mechanics (or vice versa). But apart from emphasizing the unity of the particle and its environment, the causal model has little in common with the Bohrian philosophy.

5.2 The Aharonov–Bohm effect

5.2.1 *Effect of vector potential on charged particle trajectories*

The effect due to Aharonov & Bohm (AB) (1959) concerns the existence of electromagnetic influences on interfering charged particle beams that are confined to spacetime regions containing no electric or magnetic fields. Consider the modified two-slit experiment depicted in Fig. 5.8(a). A beam of electrons emitted by a source S_1 is split into two partial beams that pass on either side of an infinite cylindrical solenoid before being coherently recombined to yield an interference pattern on the screen S_2. A steady current in the solenoid generates a flux given by

$$\Phi = \int \mathbf{B} \cdot d\mathbf{S} = \oint_C \mathbf{A} \cdot d\mathbf{x} \qquad (5.2.1)$$

where C is any loop surrounding the solenoid. The magnetic field $\mathbf{B}$ vanishes outside the coil but the vector potential $\mathbf{A}$, satisfying $\mathbf{B} = \text{curl}\,\mathbf{A}$, must be finite somewhere along the loop in order that (5.2.1) is satisfied, whatever

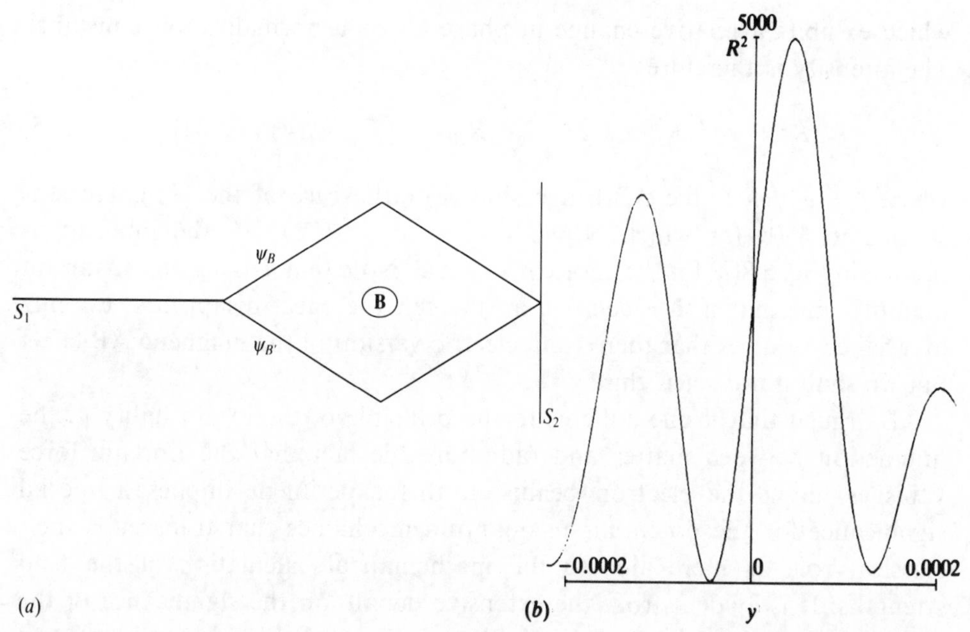

Fig. 5.8 (*a*) Coherent electron beams passing on either side of an inacessible magnetic field **B** experience a relative phase shift that is a periodic function of the flux. (*b*) The fringe shift within the envelope for two Gaussian slits.

gauge is chosen. The solenoid is surrounded by a barrier so that there is no overlap between it and the electron waves. Inserting a specific choice of vector potential corresponding to this set-up into the Schrödinger equation, for example,

$$\mathbf{A} = (\Phi/2\pi r)\hat{\boldsymbol{\theta}} \tag{5.2.2}$$

in cylindrical coordinates, rigorous solution shows that the fringes formed at the screen S_2 are shifted within the envelope of the interference pattern by an amount depending on the flux Φ through the coil.

This may be seen simply as follows. When the magnetic field is switched off, let $\psi_B(\mathbf{x})$ and $\psi_{B'}(\mathbf{x})$ denote the free wavefunctions for the beams passing above and below the solenoid respectively. When the field is switched on, each wave picks up a phase factor,

$$\psi_a \to \psi_a \,\mathrm{e}^{(\mathrm{i}e/c\hbar)\int_a \mathbf{A}\cdot\mathrm{d}\mathbf{x}}, \qquad a = B, B', \tag{5.2.3}$$

where the integral is evaluated along the upper or lower path. It follows that when the beams are recombined, the total wavefunction is

$$\psi = N'[\psi_B + \psi_{B'}\,\mathrm{e}^{\mathrm{i}e\Phi/c\hbar}]\,\mathrm{e}^{(\mathrm{i}e/\hbar c)\int_B \mathbf{A}\cdot\mathrm{d}\mathbf{x}}, \tag{5.2.4}$$

which exhibits a relative change in phase (N' is a normalization constant). The intensity is therefore

$$R^2 = N'^2\{R_B^2 + R_{B'}^2 + 2R_B R_{B'} \cos[(S_B - S_{B'})/\hbar - \delta]\}, \qquad (5.2.5)$$

where $\delta = e\Phi/c\hbar$ is the AB-fringe shift. For the case of the Gaussian slits studied in §5.1, for which $N' = (2 + 2\cos\delta\, \mathrm{e}^{-Y^2/2\sigma_0^2})^{-1/2}$, the intensity is plotted in Fig. 5.8(*b*) for the choice $\delta = -\pi/2$. Note that Φ is a gauge invariant quantity and that if $\Phi = nch/e$, n an integer, the effect disappears. Lorentz invariance requires that there is an electric version of the magnetic AB effect but we shall not discuss this.

AB argued that if one adheres to the principle of the localizability of the interaction between matter and radiation, the fact that the Lorentz force vanishes along the electron beams in this experiment implies a special significance for the potentials in quantum mechanics that transcends their classical role as mere aids to the mathematical calculation of the field strengths. It is evident from the extensive debate on the significance of the AB effect that physicists feel the need to go beyond the description offered by the wavefunction and account for the fringe shift in terms of some physical mechanism (for a discussion of early reactions see Aharonov & Bohm (1961, 1962, 1963). Recent comprehensive reviews are given by Olariu & Popescu (1985) and Peshkin & Tonomura (1989)). But what could this be given that the Lorentz force is zero and **A** is gauge-dependent? Although the quantum mechanical predictions in this case are now experimentally well established (Tonomura *et al.*, 1986 and references therein) the issues raised are still not perceived in the literature as completely resolved. Actually, as we shall see below, in the causal interpretation the AB effect does not introduce any fundamentally new issue of principle not already flowing from the general property of classical potentials that they have nonclassical effects in the quantum domain. Indeed, the bare two-slit experiment itself already poses a similar dilemma. The basic problem in the usual approach is that there is no way to connect the classical potentials with their nonclassical effects via some intelligible model. The quantum potential provides this link.

In the quantum theory of motion the AB effect is explained through the local but indirect action of the vector potential on the particle via the quantum force (Philippidis, Bohm & Kaye, 1982). We recall from §3.11.2 the Hamilton–Jacobi equation and the force law:

$$\partial S/\partial t + \tfrac{1}{2}m\mathbf{v}^2 + eA_0 + Q = 0 \qquad (5.2.6)$$

$$m\,\mathrm{d}\mathbf{v}/\mathrm{d}t = -\nabla Q + \mathbf{F}. \qquad (5.2.7)$$

Here $\mathbf{F}$ is the Lorentz force, and the physical momentum $m\mathbf{v} = \nabla S - (e/c)\mathbf{A}$ and Q are gauge invariant quantities. Even when $\mathbf{F} = 0$ the quantum potential is modified by $\mathbf{A}$. The latter may therefore be expected to cause a redistribution of the trajectories in the two-slit experiment shown in Fig. 5.8(*a*). We may write

$$Q(\mathbf{x}, \Phi) = Q_0(\mathbf{x}) + f(\mathbf{x}, \Phi), \tag{5.2.8}$$

where Q_0 is independent of the flux. The necessity of such an action follows already from the single-valuedness condition,

$$\oint_C m\mathbf{v}\cdot \mathrm{d}\mathbf{x} = nh - (e/c)\Phi, \tag{5.2.9}$$

where C encircles the solenoid since, if $\Phi \neq nhc/e$, uniform motion is not possible. However, from this observation alone one cannot understand the AB effect as the outcome of deviations of an electron trajectory unless one introduces the quantum potential as a local causal agent. This will not conflict with our classical notions since in the classical limit $Q \to 0$ (see Chap. 6) and the vector potential no longer has a vehicle for its action.

The simplest way to gain insight into the AB effect is to apply the phase shift in (5.2.5) to the Gaussian two-slit pattern studied in §5.1 (Philippidis *et al.*, 1982). The quantum potential is modified in quite a complicated way by the phase shift (it is not a simple additive or multiplicative change). The overall effect is to introduce an asymmetry into the spatial structure of Q whereby the set of plateaux and troughs are shifted to one side. A selection of particle trajectories corresponding to a phase shift $\delta = -\pi/2$ is shown in Fig. 5.9. The additional quantum force changes the momenta such that the shifted central fringe is composed mainly of trajectories from the upper slit.

The momentum of the particle is given by (5.1.9) modified by the addition of δ to the arguments of the trigonometric functions. This modification implies that the particles no longer start from rest but have an initial momentum dependent on their position. It also implies (since $\mathrm{d}y/\mathrm{d}t \neq 0$ when $y = 0$) that the trajectories now cross the optical axis. In accordance with Ehrenfest's theorem, the mean momentum $\langle p \rangle$ is constant (since the Lorentz and mean quantum forces vanish) and is proportional to $\sin\delta$. Thus, if $\delta \neq \pi n$, the mean position varies linearly in time:

$$\langle y \rangle = \langle p \rangle t/m. \tag{5.2.10}$$

The physical role of the potentials is implicit in any problem in which we solve the Schrödinger equation for a particle in an electromagnetic field. Even

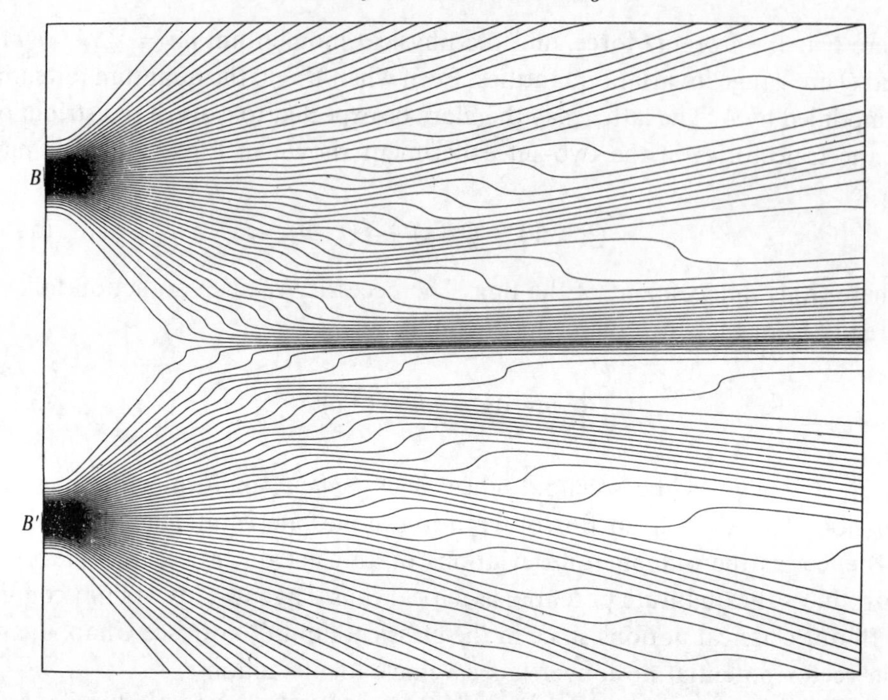

Fig. 5.9 Trajectories for the AB effect issuing from two Gaussian slits (from Philippidis *et al.* (1982)).

when the Lorentz force is finite, it cannot account *on its own* for the behaviour of quantum charged particles for the quantum potential generally carries information on the potentials not present in the field strengths. As a simple example, we may consider the effect of a uniform magnetic field applied to the two-slit AB interference pattern. The net result is that the fringes and the envelope are shifted as a whole in addition to the AB fringe shift, so that in (5.2.5) we should replace y by $(y - a)$ where a depends on the magnetic force (Kobe, 1979). This overall shift clearly cannot be understood as simply due to the nonzero Lorentz force acting on the quantum potential. This example also serves to show that there is no classical explanation for the behaviour of electromagnetic fields in quantum theory because they imply modifications of an effect that is already intrinsically quantum mechanical (interference). This is so even when the Lorentz force is finite, let alone for the pure AB effect, and it therefore seems fruitless to search for a 'classical explanation' of the latter unless one has already classically explained the fundamental phenomenon of interference, something we would argue is not possible.

5.2.2 Connection between the vector and quantum potentials

What constitutes a complete description of electromagnetism in quantum mechanics? An examination of the AB experiment led Wu & Yang (1975) to the following conclusion. The field strengths **E** and **B** provide an incomplete description of electromagnetic effects on the wavefunction since observable results depend on the flux Φ when the electrons can only pass through regions where $\mathbf{E} = \mathbf{B} = 0$. That is, different physical situations in a spacetime domain may be associated with the same fields. On the other hand, **A** and Φ overdescribe electromagnetic processes since one may associate different values of these quantities with the same physical situation (**A** is defined only up to a gauge transformation, and Φ and $\Phi + hnc/e$ are indistinguishable in an interference experiment). What provides a complete description that is neither too much nor too little is the nonintegrable phase factor

$$e^{(ie/\hbar c)\oint \mathbf{A}\cdot d\mathbf{x}}. \tag{5.2.11}$$

This is a gauge invariant quantity and a periodic function of $e\Phi/\hbar c$, and it is indeed through this term that Φ enters $|\psi|^2$ in (5.2.5) (and other observable quantities). Electromagnetism in this view is a gauge invariant manifestation of a nonintegrable phase factor.

The Wu–Yang phase factor characterizes the effect of electromagnetism in quantum mechanics without reference to the quantum state. While it locates uniquely that electromagnetic property which is observable in quantal experiments, it does not throw light on how the results come about through a process involving definite and quantifiable effects on an electron's motion. In contrast, the quantum potential approach suggests that the physical content of electromagnetism in quantum mechanics cannot be understood independently of the quantum state. In fact, the quantum potential is living proof that one may construct a gauge invariant function which embodies local information on **A** beyond that contained in the field strengths, and in this sense it implies a new type of electromagnetic field. But if we take seriously the physical reality of **A** and its ability to influence orbits via Q, how do we counter the usual objection to this notion relating to gauge-dependence, that there are an infinite number of functionally distinct vector potentials compatible with the same quantum potential and with observation? This question may be addressed by enquiring if in classical electromagnetism one may already construct from **A** a local gauge invariant field which provides more detailed information on **A** than is displayed in the Lorentz force. If such a quantity exists, the possibility arises that we could associate it with the physical component of the electromagnetic field, in particular outside the solenoid in the AB effect, and quantify precisely that physical information picked up by Q which is common to all the mathematically inequivalent **A**s.

Given an arbitrary vector potential **A** corresponding to a general problem in classical electrodynamics, the following vector field is invariant with respect to transformations $\mathbf{A} \to \mathbf{A} + \nabla f$ where f is a single-valued scalar function (Prokhorov, 1988; Dubovik & Shabanov, 1989, 1990):

$$\mathbf{G}(\mathbf{x}) = \mathbf{A} - \nabla \partial_i \Delta^{-1} A_i \tag{5.2.12}$$

(sum over $i = 1, 2, 3$). Here

$$\Delta^{-1}\mathbf{A}(\mathbf{x}) = -\int_{\text{all space}} \mathrm{d}^3 y\, \mathbf{A}(\mathbf{y})/4\pi|\mathbf{x} - \mathbf{y}|. \tag{5.2.13}$$

It satisfies the equation $\operatorname{div} \mathbf{G} = 0$. The magnetic field is obtained in the standard way from this new field: $\mathbf{B} = \operatorname{curl} \mathbf{G}$. Clearly, **G** is generally finite in regions where $\mathbf{B} = 0$ but, when **A** is globally a pure gauge function ($\mathbf{A} = \nabla f$), $\mathbf{G} = 0$. The vector **G** therefore has the properties we might expect of a physical field. Its use removes the difficulty in the causal interpretation alluded to above concerning the gauge-dependence of **A** and locates the exact electromagnetic information encoded in Q. That **G** indeed contains sufficient detail to describe the effect of the infinite solenoid on the wavefunction in the AB effect is clear from its form, which is basically a gauge transformation applied to **A**; we know any such **A** will do. In cylindrical coordinates, **G** is given by (5.2.2).

The proposal is then to adopt the vector field **G** as the 'true' physical degree of freedom generated by solutions of Maxwell's equations. Yet at this stage in the development of physical theory **G** is basically confined to the role of a conceptual aid, for in neither classical nor quantum physics can we prove the 'reality' of this field or even require its use in formal calculations. In classical physics observations on the behaviour of an ensemble of charged particles in a spacetime domain only allow us to infer properties of the field strengths. In quantum physics all gauge-equivalent potentials generate the same observable effects and one cannot infer any privileged role for **G**. We understand the physical reason for this: **G** acts only indirectly on charged particles, via Q. But while quantum mechanics cannot be said to provide evidence for the 'reality' of **G**, its introduction makes clearer how ψ comes to be modified when it propagates in regions devoid of Lorentz forces, and shows precisely which attribute of the electromagnetic field it is that Q acts as a vehicle for. The approach suggested here tends to point away from the global formulation of gauge fields of the Wu–Yang type towards a more intuitive local description. Paraphrasing Wu and Yang, we may say that electromagnetism is a gauge-invariant manifestation of the Lorentz force and the quantum force.

5.2.3 Locality and nonlocality

There are several aspects to the problem of locality and nonlocality in quantum mechanics and the word 'nonlocality' especially is used in several senses. The AB effect is often cited as an example of a nonlocal phenomenon because it is felt the fields should have primary significance and they are therefore somehow influencing the electrons outside the domain where they are finite. It is indeed claimed that the gauge-dependence of the potentials prevents any local explanation of the AB effect at all (e.g., Aharonov (1984)) although precisely such an account is given in the quantum potential picture. The latter suggests that one is liable to be led astray when investigating this question by focussing exclusively on the numbers tested by experiment. That observable results may depend only on what are apparently global attributes of a physical process (here Φ) does not mean that the underlying process is irreducibly of this kind. For Φ is just a parameter associated with the source of the field (it may be written in terms of the **B**-field as in (5.2.1)). The ψ-wave locally encounters the vector potential (5.2.2) emanating from the source and picks up in a gauge-invariant way information which includes Φ. This is no different conceptually from the electron wavefunction in the hydrogen-like atom depending on the charge Ze of the nucleus. In both cases the motion of the corpuscle depends on the source parameter (Φ or Ze) via the local mediation of Q. In this view the fact that ψ depends on Φ does not indicate evidence of 'nonlocality'.

If the AB effect is 'nonlocal' because no Lorentz force acts where the electrons exist, we should conclude, to be consistent, that all quantum mechanical effects in domains where classical forces are negligible are 'nonlocal', since clearly the usual local action of the classical forces cannot account for the effects. The basic two-slit experiment itself is of this kind – $|\psi|^2$ varies in spacetime far from the slits and we should therefore speak of the latter as exerting a nonlocal influence. An argument in support of this thesis may well be valid but from the causal viewpoint it is not necessary – the nonlocality is merely apparent and represents a failure to identify the local causal link between the classical potentials and the quantum particles, i.e., Q.

It is certainly true that the dependence of Q on all the factors contributing to a physical problem implies that it locally reflects the whole. However, it seems better to avoid the term 'nonlocality' when describing this global feature and reserve this for a description of certain aspects of time-dependent and many-body processes, as discussed in Chaps. 7 and 11, particularly §11.6.

5.3 Tunnelling through a square barrier

In §4.4 we analysed the plane wave treatment of scattering from a square potential and pointed out that this does not provide a description in which the particle is initially free, i.e., having negligible contact with the scatterer, because the initial wavefunction already carries the imprint of the external potential throughout space. To provide a more realistic time-dependent picture of the process we should start with a wave train initially localized far from the scattering centre. To this end we shall consider the one-dimensional scattering of a Gaussian packet of mean energy E from a square potential barrier $V > 0$ (Fig. 5.10).

The interaction of the packet with the barrier leads eventually to the formation of reflected and transmitted packets of diminished amplitude, perhaps together with a small packet that may persist inside the barrier for some time. According to the causal theory the particle ends up in one of these. We shall consider two cases: $E < V$, where the principal nonclassical feature, compared to a classical ensemble of particles each of energy E, is that a particle may pass inside or beyond the barrier (tunnelling); and $E > V$, where the main nonclassical aspect is that particles may be reflected. In both cases the effects come about from the modification of the total energy of each particle, initially $\approx E$, due to the rapid spacetime fluctuation of the ψ-wave in the vicinity of the barrier. Recall that the total particle energy is

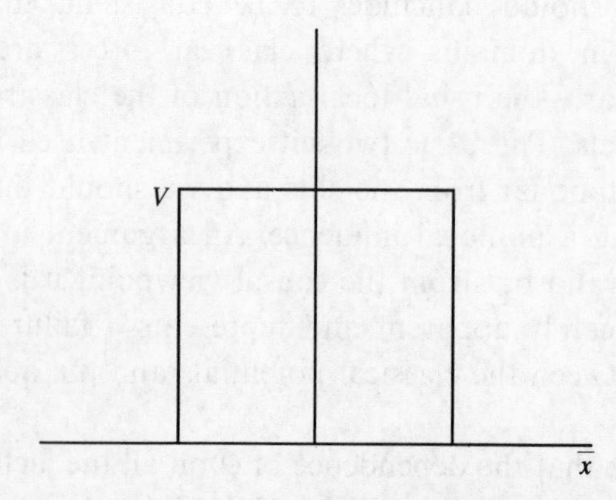

Fig. 5.10 The square potential barrier. The physical potential seen by a particle is $(Q + V)$, not V.

defined by

$$-\partial S/\partial t = (1/2m)(\partial S/\partial x)^2 + V + Q \tag{5.3.1}$$

evaluated along a trajectory $x(t, x_0)$. The effective 'barrier' encountered by the particle, $(V + Q)$, may be higher or lower than V and vary outside the 'true' barrier, and the kinetic energy, $(1/2m)(\partial S/\partial x)^2$, may likewise take on values greater or less than its initial value. Note that the variations in the total potential energy and the kinetic energy need not balance since the total particle energy is not conserved in this time-dependent example. In the case $E < V$ we may find that the effective potential is lowered and the particle enters the barrier, but we can envisage situations where the effective potential is raised ($> V$) and yet tunnelling still occurs because the total energy is sufficiently elevated. All that is necessary is that $-\partial S/\partial t \geqslant V + Q$ in the barrier region. The kinetic energy may, of course, be smaller than $(V + Q)$ and the total and effective potential energies may become negative in some spacetime domains. The causal picture is completed by consideration of the action of the effective force, $-\partial(V + Q)/\partial x$.

The following account is based on Dewdney & Hiley (1982). In a unit system where $\hbar = 1$ and $m = 0.5$ the initial wavefunction is given by

$$\psi_0(x) = A\, e^{-(x-a)^2/4\sigma_0^2}\, e^{ikx}, \tag{5.3.2}$$

which describes a packet centred about the point $x = a$ and moving with group velocity $2k$ towards V, with which it has negligible overlap. The mean energy $E = k^2 + \frac{1}{4}\sigma_0^{-2}$. The time-dependent Schrödinger equation is solved numerically by finite difference techniques described by Goldberg, Schey & Schwartz (1967). A notable point in this treatment is that the parameters are so chosen that the dispersion of the packet is negligible during the scattering process and so does not mask the effects of interest. This amounts to treating (5.3.2) as a 'classical' wavefunction (cf. §6.6.2). All particles in the ensemble start with momentum k and total energy $\approx E \approx k^2$.

The constants are chosen as follows: $a = 0.5$, $\sigma_0 = 0.05/\sqrt{2}$, the barrier width $= 0.064$ and $V = 2(50\pi)^2$. Suppose first that $k = 50\pi$ so the mean energy is half the barrier height. To see how the fate of a particle can depend sensitively on its initial position, a selection of orbits corresponding to a uniform distribution of x_0s (which does not reflect the actual Gaussian distribution) is shown in Fig. 5.11. We emphasize that only one of these is actually realized in each scattering event and that the one chosen does not depend in any way on the other potential starting points. For this choice of energy tunnelling occurs only if x_0 lies in the front of the packet ($x > a$). Particles in the forward part of the front are not significantly affected by the

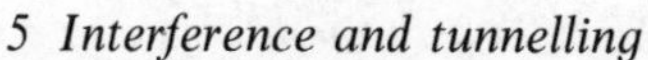

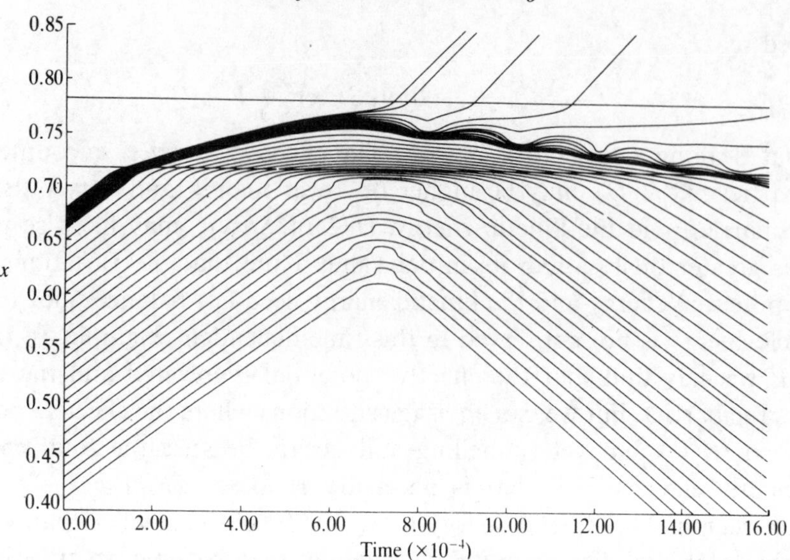

Fig. 5.11 Trajectories associated with a Gaussian packet incident on a barrier of twice the mean energy (from Dewdney & Hiley (1982)).

reflected wave in front of the barrier which is not yet appreciable and maintain substantially uniform (classical) motion. They enter the barrier where they are initially slowed down. Those in the extreme front are eventually accelerated and transmitted to pursue uniform motion again – these are the particles that 'tunnel'. But notice that many other trajectories enter the barrier and subsequently join the reflected beam; they also 'tunnel'. The motion of some of these particles can be quite complicated and to show the generic oscillations extra trajectories for particles trapped in the barrier are plotted.

Particles lying in the rear of the packet ($x < a$) and some in the front never actually reach the barrier. They are brought to rest and turned around when they encounter the effective force induced by the interference of the incident and reflected waves, where Q propagates the influence of V far from where the latter is finite. This is a further nonclassical feature of the model. At points where the effective force is small trajectories congregate and give rise to fringes, a matter-wave analogue of Wiener's optical fringes. The cross-section of Q in the region in front of the barrier is similar to the set of potential wells displayed in Fig. 5.6 for two overlapping packets. It will be noted that the particle starting at the centre of the packet ($x = a = \langle x \rangle(0)$) deviates from the corresponding classical trajectory and is among those reflected before reaching the barrier.

We observe from the distribution of trajectories in Fig. 5.11 that the centres of gravity of the reflected and transmitted packets are displaced in time beyond those of the corresponding classical packets. This reflects the single-valuedness of the wavefunction which implies that the trajectories do not cross, and the time delay is an analogue for matter waves of the optical Goos–Hänchen shift (Hirschfelder *et al.*, 1974a).

It is useful to compare the above explanation of the tunnel effect with those that employ the wavefunction alone. First of all, the evolution of averages of dynamical quantities does not provide insight into the behaviour of ensemble elements – the mean energy for example, remains equal to $E < V$ throughout the process. Secondly, a common line of thought aimed at explaining the tunnel effect, whose existence provides further evidence that physicists are not positivists, is the 'energy-borrowing' argument (Hirschfelder *et al.*, 1974a). This is an attempt to assign a definite value of energy to a system which is not in an energy eigenstate, and values of energy different from the eigenvalue when the system is in an energy eigenstate. One considers the Fourier analysis of the wavefunction in the barrier region:

$$\psi(x, t) = h^{-1/2} \int \varphi(p)\, e^{ipx/\hbar}\, dp. \tag{5.3.3}$$

Each elementary wave, it is suggested, corresponds to a particle having a virtual energy $E' = V + p^2/2m$. There is a finite probability that the particle will 'borrow' energy at the entrance to the barrier and 'ride over' it by transferring from a low to a high energy orbit. The borrowed energy is returned when the particle exits from the barrier and it adopts a low energy trajectory again. The probability of a certain virtual energy is induced by the probability that the corresponding virtual momentum lies between p and $p + dp$, $|\varphi(p)|^2\, dp$.

This is not a satisfactory explanation for several reasons. First, the argument rests on the assumption that the Fourier components refer to momentum values that a particle in the state ψ may actually have rather than those that are found if a measurement is performed. Quantum mechanics only allows the latter interpretation. It is reasonable to attempt to attribute a definite momentum to a particle in an arbitrary state but as shown by the causal interpretation this cannot be one of those associated with an individual plane wave in (5.3.3), as if ψ were simply a mixture of mutually exclusive alternatives, because these waves are interfering. Next, what is the energy borrowed from? This notion borders on the assumption that the ψ-wave is something physically real – can a 'wave of probability' exchange energy with a particle? And what in any case is doing the 'borrowing'? If one has in mind a massive object one should adopt a consistent theory of it and its motion. The causal definition of momentum and energy implies an explanation for

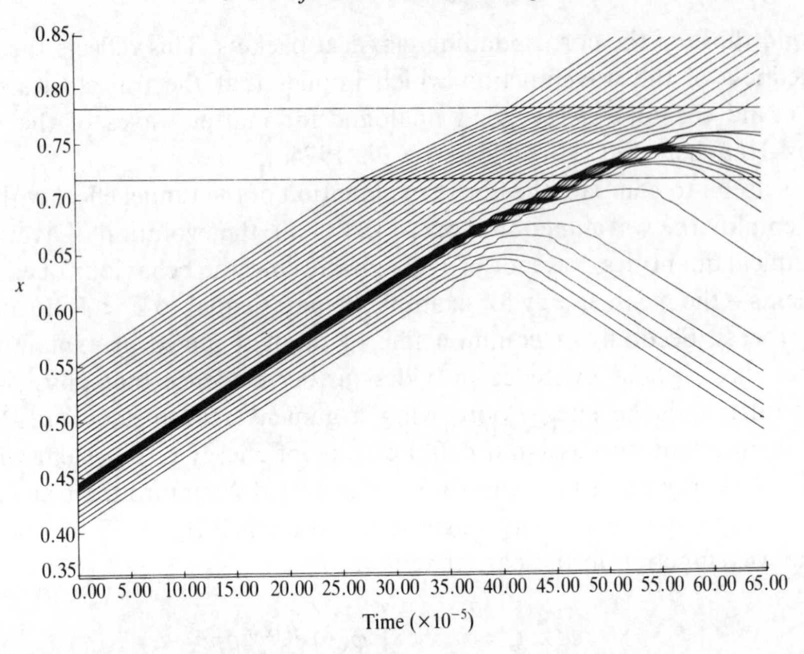

Fig. 5.12 Trajectories incident on a barrier equal to half the mean energy. The two groups of reflected orbits correspond to the formation of two reflected packets (from Dewdney & Hiley (1982)).

tunnelling close in concept to 'energy borrowing' but the details differ considerably. In particular, as noted above, we require the total energy to be greater than the effective potential and not just V, and the energy is a sensitive function of x and t.

We pass now to the case $k = 100\pi$, corresponding to a mean energy twice the barrier strength. We see from Fig. 5.12 that particles in the leading half of the incident packet pass through the barrier substantially like classical particles except that the refraction at the edges, reflecting the lower speed within the barrier, is smooth. For these particles the quantum potential is negligible and this is a good example of the approach to the classical limit (Chap. 6). As regards particles in the rear of the incident packet, some suffer a delay in the barrier and are transmitted and only those in the extreme tail are reflected. The latter form two groups: those that first enter the barrier and those that are turned back without ever reaching it. These groups correspond to the formation of two separate but overlapping reflected packets of small amplitude that are visible in a plot of the probability density (Goldberg *et al.*, 1967).

The above account illustrates the general principles governing the various physical phenomena involving penetration into a classically forbidden domain.

The essential point is that the physical potential seen by the particle is not V but $(V + Q)$. One specific example, a beam splitter, is discussed in §5.4. Another example is the decay of nuclei in which particles must surmount a potential barrier. This example in particular provides a particularly striking illustration of the inadequacy of a description based on the wavefunction alone, for the latter may take a very long time to leak out completely whereas the particle may be detected after a relatively short time. Just when, and whether, the particle passes through the barrier is uniquely determined by its initial position.

The trajectories for the one-dimensional scattering of a Gaussian packet incident on a static barrier have also been given by Leavens (1990a,b), and for a time-modulated barrier by Leavens & Aers (1991). The one-dimensional theory discussed so far could be turned into a theory in two space dimensions for a barrier lying along the y-axis if we assume that the total wave has a y-factor $e^{ik_2 y}$ and we reparametrize the t-axis by $y = \hbar k_2 t/m$. Trajectories for the scattering of a two-dimensional Debye–Picht wave packet from a potential barrier are plotted by Hirschfelder *et al.* (1974a). The two-dimensional character of the packet implies that in the region where the incident and reflected waves interfere the trajectory field displays new properties such as the formation of closed orbits, or vortices encircling nodes associated with wavefront dislocations (cf. §4.11). Similar vortex effects appear in congruences flowing through a two-dimensional duct with a bend (McCullough & Wyatt, 1971; Hirschfelder & Tang, 1976a) and in the scattering of two particles from a spherically symmetric square potential (Hirschfelder & Tang, 1976b). We mention also that stationary state quantal trajectories have been exhibited for repulsive Coulomb scattering (Fröbrich, Lipperheide & Thoma, 1977) and elastic scattering from a hard sphere (Korsch & Möhlencamp, 1978b).

A study of the one-dimensional scattering of a Gaussian packet from a potential well yields results similar to those described above for the barrier (Dewdney & Hiley, 1982). Particles in the leading edge of the packet are transmitted through the well much as classical particles are (the orbits being refracted so that a particle speeds up rather than slows down as in the barrier) and orbits may be trapped for a period inside the well or reflected before reaching it, depending on x_0.

5.4 Interference by division of amplitude

5.4.1 The neutron interferometer

We pass now to the matter-wave analogue of the second class of optical interferometers mentioned in §5.1, those in which the amplitude of a wave is coherently divided by a beam splitter and the resulting spatially separated partial beams are subsequently recombined to yield an interference pattern. It is convenient to describe this device in terms of its most successful practical realization, the perfect crystal neutron interferometer. The advent of this branch of experimental physics in the last 20 years (initiated by Rauch, Treimer & Bonse (1974)) has provided a means of testing many fundamental predictions of quantum mechanics which would be technically very difficult in electron or optical interferometry and would otherwise simply be *gedanken* experiments. We first briefly outline the principles of the instrument (following Greenberger (1983)) and then present a simple model that illustrates how the interference effects emerge from a sequence of individual processes in which the emission of a neutron is continuously and causally connected with its detection.

Referring to Fig. 5.13(*a*), a block of perfect silicon crystal is cut so that three slabs of crystal protrude from a common base. An incoming neutron beam is incident on the first ear where it is split into two equal partial beams by Bragg diffraction from the atomic planes A orthogonal to the crystal face. The emerging beams are incident on the second ear at B and C which acts as a beam splitter for each in the same way. Finally, the two reflected partial beams from the second ear converge on the third ear where they interfere at D. The beams emerging behind the crystal enter the counters C_1 and C_2. They each carry information on both paths through the interferometer. An interference pattern is obtained by inserting a plate inside the device which when rotated presents different thicknesses to the beams and alters their relative phase χ and hence the count rate in either C_1 or C_2 (Fig. 5.13(*b*)). The source is arranged so that only one neutron is in the apparatus at a time. The wave taking the upper (lower) route will be denoted ψ_1 (ψ_2) and the total wavefunction in the interference region is

$$\psi = \psi_1 + \psi_2 \, e^{i\chi}. \tag{5.4.1}$$

The same expression (5.4.1) will be used to denote each outgoing beam entering C_1 and C_2. Thus, for the 'forward' beam entering detector C_2, ψ_1 is the reflected component of the beam traversing path 1 and ψ_2 is the transmitted component of the beam traversing path 2. By symmetry we have $\psi_1 = \psi_2$ in the forward beam in the ideal case of equal reflection and transmission ratios at each slab. This ensures maximum contrast in the interference pattern:

$$|\psi|^2 = 2|\psi_1|^2(1 + \cos \chi). \tag{5.4.2}$$

The layout is obviously similar to the optical Mach–Zehnder interferometer.

Because the crystal planes in the three slabs are perfectly aligned, the

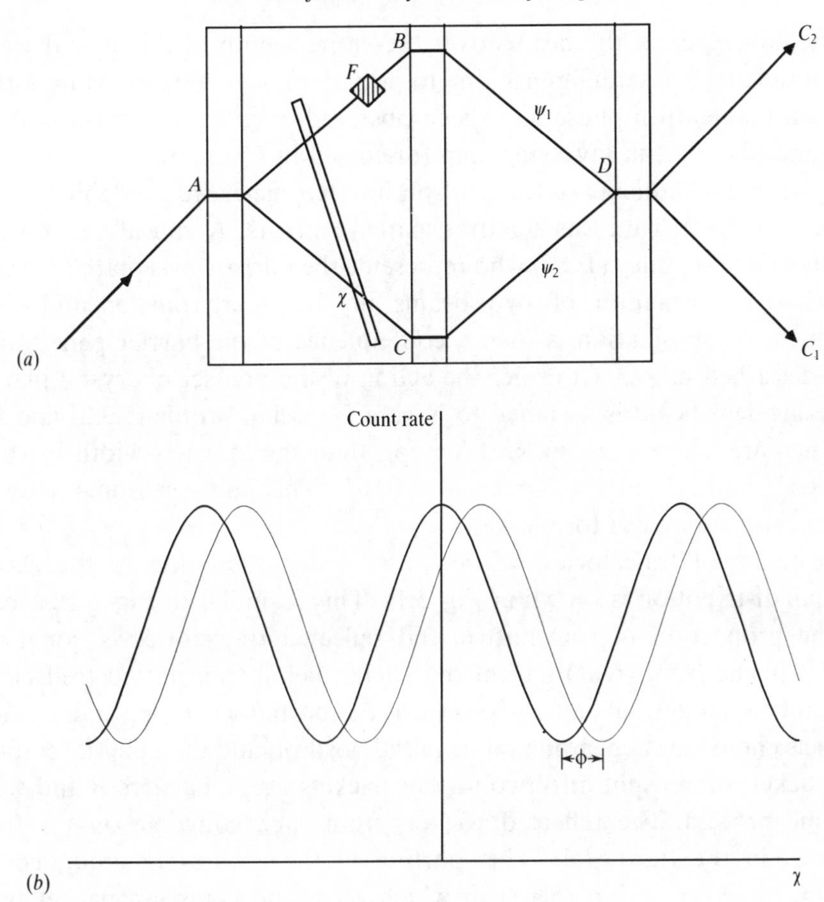

Fig. 5.13 (*a*) The perfect crystal neutron interferometer described in the text. (*b*) The interference pattern displayed in the count rate of C_1 and C_2 obtained by varying the relative phase χ between the split beams. An additional phase shift ϕ due to some external agent is manifested by changes in the count rate for each setting of χ.

neutron beams remain coherent over the length of the device, ~ 10 cm. Moreover, they are sufficiently separated so that one can apply an external agency to one without affecting the other, or at least affecting it to an observably different degree. Thus, a device generically labelled F in Fig. 5.13(*a*) will alter the upper beam in such a way as to vary the count rate in each detector by introducing an additional phase shift ϕ between the two beams. To illustrate the sensitivity of the instrument we cite the first experiment to detect the influence of the earth's gravitational field on quantal phases (Colella, Overhauser & Werner, 1975; see also §6.8). The apparatus is rotated so that the interfering beams are at different heights and hence gravitational potentials. The potential energy difference is a factor 10^7 smaller than the kinetic energy and might have been expected to be negligible. In fact, the coherence of the beams over the entire length of the crystal yields an accumulated phase difference due to gravity of $\phi \sim 100$ rad. Many other effects on the neutron phase have been observed, such as the rotation of the earth, and phenomena involving spin (discussed in Chap. 9).

To give a simple model of the neutron interferometer we may think of the diffracting crystal planes as semitransparent mirrors. A causal version has been given by Dewdney (1985) who represents the mirrors by square potential barriers. The formation of two beams in the interferometer and their subsequent recombination is then a consequence of the barrier penetration effects described in §5.3. Consider the action of the first set of crystal planes, A. The incident beam is assumed to have a Gaussian profile (5.3.2) and the constants are chosen as in §5.3 except that the barrier width $= 0.016$, $V = (50\pi)^2$ and $k = 1.08 \times 50\pi$ so $E = 1.16V$. This choice ensures a transmission ratio very close to one-half.

A selection of trajectories whose initial distribution mimics the actual Gaussian distribution is shown in Fig. 5.14. This is similar to Fig. 5.12 except that the proportion of transmitted and reflected trajectories is equal, all particles in the front (rear) half of the packet being transmitted (reflected). The two emerging Gaussian packets form the partial waves ψ_1 and ψ_2. The corpuscle enters one depending on its initial position and the other is 'empty'. Each packet is then split into two further packets at the barriers B and C by the same process. The reflected packets from each converge on the final barrier D in the third slab. This portion of the process is simulated by assuming a wavefunction (5.4.1) in which ψ_2 is the Gaussian packet used above (centred about $x = 0.5$) and ψ_1 is a Gaussian packet centred about $x = 1$ and having group velocity $-k$.

Trajectories for the case $\chi = 0$ are displayed in Fig. 5.15. The most striking feature is that the interference inside and outside the barrier causes all the orbits in the respective incident packets to be reflected. No trajectories cross, in accordance with the single-valuedness requirement. As in the double-slit experiment, a particle taking the upper route 1 ends up in detector C_2. This contrasts considerably with classical scattering where two incident beams (with $E > V$) would simply pass through the barrier and one another.

That varying the relative phase χ changes the intensities of the beams entering C_1 and C_2 is evident from Fig. 5.16 where $\chi = \pi/2$. There is nearly perfect transmission of particles arriving in wave ψ_2 due to the cancellation of the reflected component of ψ_2 with the transmitted part of ψ_1. A relative phase $\chi = 3\pi/2$ yields the same distribution as Fig. 5.16 except that C_1 and C_2 are interchanged. Clearly, any extra phase shift ϕ will have a similar effect

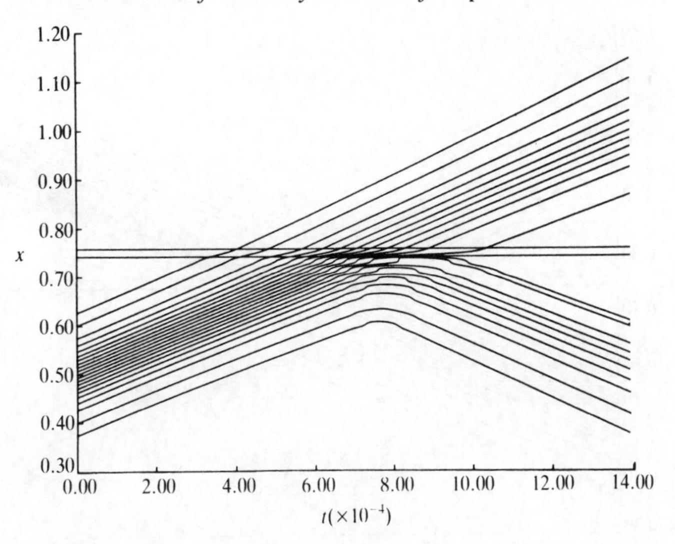

Fig. 5.14 Trajectories having a Gaussian distribution of initial positions scattered by the square potential barrier *A* with a transmission ratio of one-half (from Dewdney, Holland, Kyprianidis and Vigier (1986)).

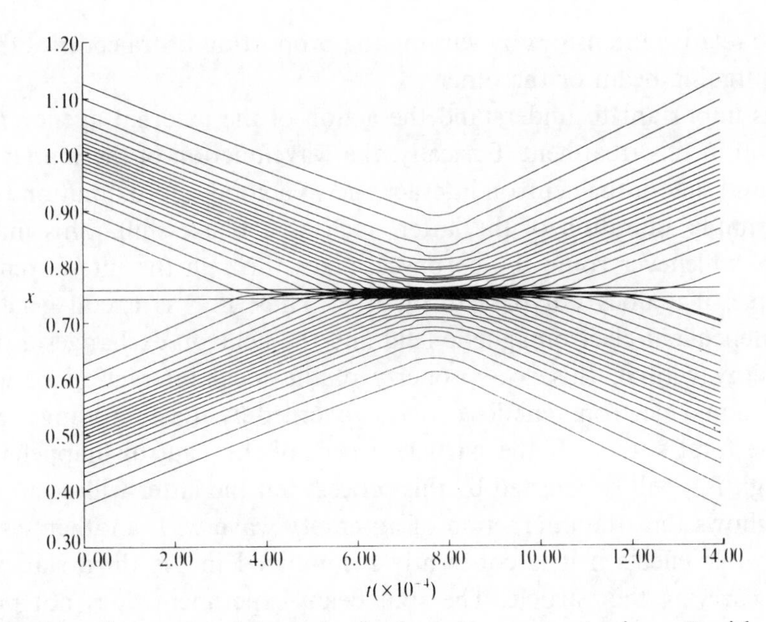

Fig. 5.15 Trajectories scattered from the final semitransparent mirror *D* with $\chi = 0$ (from Dewdney (1985)).

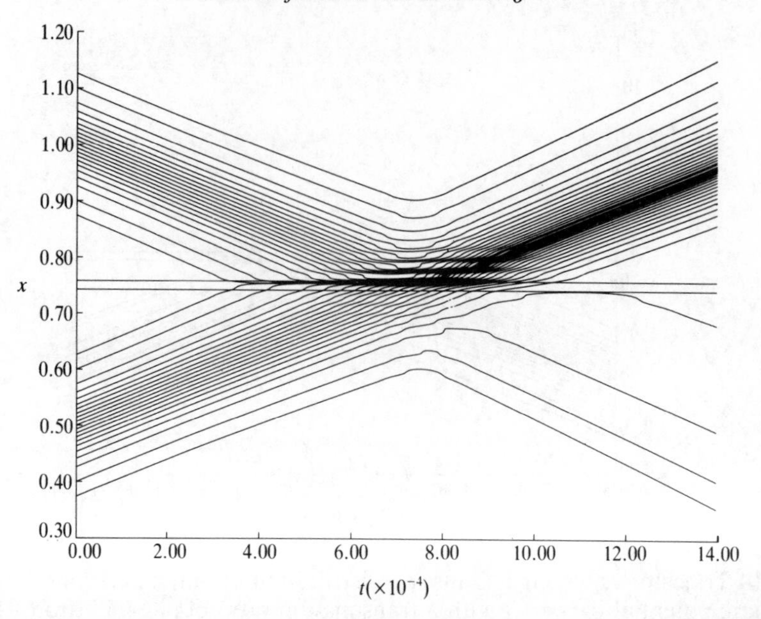

Fig. 5.16 Trajectories scattered from D with $\chi = \pi/2$ (from Dewdney (1985)).

on the relative intensities by varying the proportion of trajectories that enter one outgoing beam or the other.

It is important to understand the action of the external agency F on the neutron in this treatment. Basically, the wavefunction of the matter making up F and that of the neutron interact and to a good approximation this leads to a simple alteration of the latter, such as a phase shift. This interaction occurs whichever route the corpuscle takes through the interferometer, for example, if it does not pass through F. Thus, if F is a coil generating a time-dependent electromagnetic field and the neutron exchanges a quantum of energy with it, this is a property of the local interaction between the neutron and electromagnetic ψ-waves governed by the Schrödinger equation for the total system. If the particle aspect of the neutron happens to pass through F it will be affected by this process but the latter will occur anyway. This shows that the interaction of an empty wave with another system has observable effects if it is coherently recombined in the third slab with the wave carrying the particle. The split beam experiment does not prove the existence of empty waves since we cannot show that the particle went along one path or the other, but they do have explanatory power (see §8.8).

5.4.2 Beam attenuation effects

Suppose that for each setting of the phase shifter χ a large number N of single neutrons associated with the same wavefunction ψ are sent through the interferometer. Varying χ for each N will build up the usual interference figure. Suppose that in a fraction $(1 - a)N$ of trials, $0 < a < 1$, we insert in one arm of the interferometer, say the upper, a perfect absorber which completely blocks the wave ψ_1 (F in Fig. 5.13(a)). Equivalently, the wave ψ_1 propagates unhindered in only a fraction aN of cases, path 2 being always open. Then the interference pattern and its contrast will obviously be diminished. Suppose instead that for all N trials we place in the upper arm a partially absorbing slab with a transmission probability a, and leave the lower arm completely open. Then again, it is obvious that the interference pattern and its contrast will be reduced. Will the degrees of diminution in the two experiments be the same? At first sight we may think this will be so because essentially the same fraction of neutrons, $(1 - a)N/2$, is removed in both cases. In fact, our intuitive (classical) expectation is wrong: quantum mechanics predicts a substantial difference in the two patterns and this has been observed (Rauch & Summhammer, 1984; Summhammer, Rauch & Tuppinger, 1987, 1988).

In the first case, the appropriate description of the forward beam is a mixed state composed of the superposition (5.4.1) with probability a and the wave $\psi_2 \, e^{i\chi}$ with probability $(1 - a)$. The intensity is therefore

$$|\psi|^2 = 2|\psi_1|^2 a(1 + \cos\chi) + |\psi_1|^2(1 - a) \tag{5.4.3}$$

$$= |\psi_1|^2(1 + a + 2a\cos\chi). \tag{5.4.4}$$

In practice, the periodic introduction of the absorber is achieved using a time-dependent chopper (i.e., a rotating disc with one or more slits in it). The wavefunction behind the chopper, $\psi_1(x, t)$, has in general a complicated spacetime dependence and the forward beam should be described by a pure state $\psi_1(x, t) + \psi_2 \, e^{i\chi}$. However, the time-averaged intensity implied by this wavefunction is just the expression (5.4.4) given by an ideal absorber periodically inserted into the upper arm and we shall continue the discussion in terms of the latter.

In the second case, the forward beam is a pure state $a^{1/2}\psi_1 + \psi_2 \, e^{i\chi}$ (neglecting any phase shift due to the static absorber) in all trials and the intensity is

$$|\psi|^2 = |\psi_1|^2(1 + a + 2a^{1/2}\cos\chi). \tag{5.4.5}$$

As expected, the patterns are reduced compared with the free case (5.4.2), but differently: the time-dependent contrast implied by (5.4.4) is a fraction

$a^{1/2}$ of the static contrast implied by (5.4.5). The different outcomes of the two experiments have been interpreted as indicating the differing degrees of wave or particle 'knowledge' that they provide (cf. §5.1.4). Instead we can understand them in the causal interpretation through the objective alteration in the waves and subsequent particle motions brought about by the different characteristics of the two absorption processes, even though both imply the same probability of transmission (Kyprianidis & Vigier, 1987).

To see this, let us consider first the static absorber and simulate its action by a potential barrier. As we have seen in §5.3, an incident packet will be split into a transmitted part of reduced amplitude and reflected and trapped components. If the particle enters either of the latter two packets this will correspond to absorption. Because we cannot tell in advance whether the neutron will be absorbed or transmitted the action of the absorber is sometimes described as 'stochastic' but we see that the actual outcome is deterministically decided by the initial position of the particle. Whether the particle is absorbed or not, (5.4.5) describes a case where there is always a transmitted component $a^{1/2}\psi_1$ which propagates to the third slab and contributes to the interference effect. The total number of particles arriving in the interference region is $\frac{1}{2}aN$(path 1) + $\frac{1}{2}N$(path 2) and they all contribute to the pattern, whatever their route. If the particle is absorbed the waves propagating in the two arms are both 'empty' and they eventually dissipate (they do not contribute to the detection process); there is no 'collapse of the wavefunction' in this model (see Chap. 8).

Let us compare this description with the time-dependent case and suppose first the particle takes open route 2, with probability $\frac{1}{2}$. Then in a fraction $\frac{1}{2}aN$ of cases the empty wave coursing along route 1 will reach the third slab unmodified and the particle will contribute to the usual interference pattern. But in a fraction $\frac{1}{2}(1-a)N$ of trials the empty wave will be blocked and hence the particle will merely add to the background intensity (second term in (5.4.3)). For the particles that traverse route 1, a fraction $\frac{1}{2}aN$ are transmitted and will combine with the empty wave in arm 2 to contribute to the normal pattern. The total number of particles contributing to the interference is therefore aN (first term in (5.4.3)). This is less than the number contributing to interference in the static case and yet the contrast of the pattern formed just by these particles is maximum (unity). The reason why the overall contrast in the time-dependent case is less than in the static case is that the maximum contrast formed by the interfering particles is reduced by the addition of the background intensity.

We see then that while the same total number of particles arrive at the third ear in both experiments, and the probability of transmission along arm 1 is the same, the quite distinct evolution of the wavefunction in the two cases implies different count rates in the detectors.

By changing the characteristics of the chopper, such as speeding it up, one can alter the intensity measured in the detectors and approach that obtained with the static absorber. This is readily understood in terms of the altered structure of the attenuated wave propagating through the chopper.

5.5 Time in quantum theory

5.5.1 Time of transit – a testable prediction beyond quantum mechanics?

The empirical predictions of quantum mechanics flow from a mathematical formalism which nowhere makes use of the assumption that matter comprises a corpuscular aspect pursuing a definite track in spacetime. It follows that the results of experiments designed to test the predictions of the theory, which are also subject to its laws, do not permit us to infer any statement regarding the putative corpuscle – not even its independent existence. The detailed predictions made by the causal interpretation described in previous sections explain how the results of quantal experiments come about but are not tested by them. To draw inferences from experiment regarding the particle track we have to change or augment quantum mechanics in such a way that the testable predictions of the (modified) theory are in some degree functions of the trajectory assumption. The question raised here is whether the de Broglie–Bohm particle law of motion can be made relevant to experiment. We do not contest the correctness of quantum mechanics in the domain where it is unambiguous, testable and confirmed but enquire whether that domain can be enlarged. This becomes a significant issue on the fringes of the formalism where its predictions are unclear but the trajectory interpretation is sharply formulated. An example is the treatment of time in the two theories.

An historical goal of quantum mechanics has been the association of self-adjoint operators with observable quantities. One would like, for example, the outcomes of experiments designed to measure the time of transit of a particle through space to be the eigenvalues of some 'time operator'. But time is not an observable in this sense. In the usual approach to quantum mechanics in which operators act on wavefunctions in a Hilbert space $\mathscr{H}$, time appears merely as a classical parameter labelling the unitary evolution group. It is not possible to introduce a self-adjoint time operator T conjugate to the Hamiltonian H because, in addition to generating time translations, H determines energy levels and so must be bounded from below (Pauli, 1958).

Is the apparently innocuous query 'What is the mean time spent by a particle in a certain region of space?' physically meaningful? The problem is particularly pertinent to tunnelling processes where a variety of proposals have been made for the definition of the time (actual and mean) spent by a particle within a classically forbidden barrier (Hauge & Støvneng, 1989). The competing definitions are flawed in various ways but the debate is significant in indicating how the natural instinct of physicists to describe nature as it is ultimately asserts itself. For the correct mathematical definition of barrier traversal time inevitably raises questions as to the nature of the individual process that we call 'tunnelling'. Although conducted without direct reference to the foundational controversy over the interpretation of quantum mechanics, this is what the discussion really involves. A common theme in these studies is that the definition of mean time, if attempted at all, should be restricted to expressions derived just from the wavefunction, even when the possibility of going beyond the basic formalism is countenanced. An example (in one dimension) is the following (Hauge & Støvneng, 1989, p. 934). Starting from the usual expression for the probability that at time t a particle lies in the domain (x_a, x_b) along the x-axis,

$$pr(x_a, x_b, t) = \int_{x_a}^{x_b} |\psi(x, t)|^2 \, dx, \tag{5.5.1}$$

it is proposed that the mean time spent in this region is

$$\tau = \int_{-\infty}^{\infty} dt \, pr(x_a, x_b, t). \tag{5.5.2}$$

Notice that this expression does not directly depend on the phase of the wavefunction.

From the de Broglie–Bohm point of view the difficulty in giving a generally valid definition of tunnelling time stems from the artificial restrictions placed on the theoretical analysis at the outset. The wavefunction simply does not contain enough information on the process under consideration to give an unambiguous meaning to such times, either individually or on the average. It is not in itself descriptive of the individual tunnelling event precisely because it lacks any feature corresponding to the time at which tunnelling commences. Similar difficulties arise if one attempts to give meaning to the time of arrival concept using ψ alone (Allcock, 1969).

The causal interpretation solves this problem by adding to the wavefunction an entity that can 'take time' to traverse a region, the physical corpuscle. It has been pointed out that this approach gives a clear definition of the time taken for an individual particle to cross a barrier, and hence the mean time over an ensemble as well (Hirschfelder *et al.*, 1974a; Spiller *et al.*, 1990; Leavens, 1990a,b; Leavens & Aers, 1991), but it is evident that it presents an unambiguous answer to all questions pertaining to transit times in all physical problems. Many such times may be isolated – in tunnelling, there are the time of arrival at the barrier region, the dwell time in the barrier for particles eventually transmitted or reflected, and so on. These are all special cases of the precise time attributed to any track linking one space point with another as a concomitant of the law of motion $m\dot{x} = \partial S/\partial x$ (for convenience we work in one dimension). Specifying the initial position x_0, the formula for the orbit $x = x(x_0, t)$ may be inverted (algebraically or by inspection of a graph) to yield the time taken for a particle guided by the wave ψ to travel from x_0 to x: $t = t(x_0, x)$. For example, a glance at Fig. 5.11 tells us how long particles with a range of initial positions spend inside a barrier when a packet is scattered. For those states for which $\partial S/\partial x$ is independent of t (including the stationary states) we may write for the traversal time

$$t = m \int_{x_0}^{x} \mathrm{d}x\, (\partial S/\partial x)^{-1}. \tag{5.5.3}$$

In the general case a particle may visit the same region more than once and $t = t(x_0, x)$ is a multivalued function (e.g., an oscillator packet). The mean transit time over the ensemble connected with ψ may be deduced from that of the individual ensemble elements by assigning probabilities to them, and not simply written down as in (5.5.2) for an ensemble whose individual laws are unknown or even have no meaning. Different sorts of probabilities and averages may be defined corresponding to the different types of transit times connected with various subensembles.

Fig. 5.17 shows a single trajectory, connected with some arbitrary initial wavefunction $\psi_0(x)$, displaying the types of motions that might be expected in an ensemble crossing a domain bounded by the points (x_a, x_b). The latter

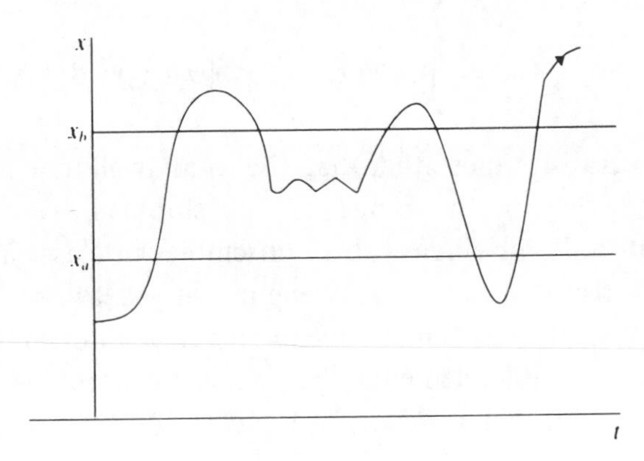

Fig. 5.17 A trajectory displaying generic features.

may coincide with a barrier for example. For a given x_0 we can invert the formula $x = x(x_0, t)$ to find the times $t_i(x_0, x_a, x_b)$, $i = 1, \ldots, N$, at which $x = x_a$ or x_b. The particle may spend a range of periods in the domain and the total time is

$$t(x_0, x_a, x_b) = (t_2 - t_1) + (t_4 - t_3) + \cdots + (t_N - t_{N-1}). \qquad (5.5.4)$$

In principle we can now invert (5.5.4) to give the set of initial positions $x_0 = x_0(t, x_a, x_b)$ which correspond to a fixed dwell time t.

Our aim is to derive an expression for the probability $P(t, x_a, x_b)\,\mathrm{d}t$ that a particle spends a time in the interval $(t, t + \mathrm{d}t)$ in the segment (x_a, x_b). Fix attention on those x_0s which correspond to a transit time t. If none exist then $P(t, x_a, x_b) = 0$. If some do then the required probability of dwell time t will be equated to the probability of the corresponding set of x_0s:

$$P(t, x_a, x_b)\,\mathrm{d}t = \sum_{x_0 \text{ corr. to } t} |\psi_0[x_0(t, x_a, x_b)]|^2\,\mathrm{d}x_0. \qquad (5.5.5)$$

Writing $\mathrm{d}x_0 = (\partial x_0/\partial t)\,\mathrm{d}t$ we have for the probability per unit time

$$P(t, x_a, x_b) = \sum_{x_0 \text{ corr. to } t} |\psi_0[x_0(t, x_a, x_b)]|^2[\partial x_0(t, x_a, x_b)/\partial t] \qquad (5.5.6)$$

with

$$\int_0^\infty P(t, x_a, x_b)\,\mathrm{d}t = \int_{-\infty}^\infty |\psi_0(x)|^2\,\mathrm{d}x = 1. \qquad (5.5.7)$$

From these formulae we readily deduce two expressions for the mean time spent by the ensemble in the domain:

$$\langle t\rangle(x_a, x_b) = \int_0^\infty tP(t, x_a, x_b)\,\mathrm{d}t \qquad (5.5.8)$$

$$= \int_{-\infty}^\infty t(x_0, x_a, x_b)|\psi_0(x_0)|^2\,\mathrm{d}x_0. \qquad (5.5.9)$$

Probabilities for transit times other than the total dwell time may be obtained as in (5.5.6) by varying the range of x_0s summed over. The formulae (5.5.3)–(5.5.9) all hold for single-valued ensembles in classical physics.

The time of transit $t = t(x_0, x_a, x_b)$ depends, in general, on x_0, but in many cases it does not. If it does not the mean traversal time coincides with that of an individual ensemble element. The simplest example is the plane wave $\psi = \mathrm{e}^{ikx}$, $k = mv/\hbar$, for which $P(t, x_a, x_b) = \delta[t - (x_b - x_a)/v]$ and

$$\langle t\rangle(x_a, x_b) = (x_b - x_a)/v. \qquad (5.5.10)$$

Similar results are obtained for the plane wave treatment of the potential step where it will be observed from Fig. 4.3 that all trajectories take the same time to traverse any space slice. A simple example where the transit time depends on x_0 is the free Gaussian packet at rest.

The individual transit times implied by the causal interpretation for all types of transport processes are clearly dependent on the particle law of motion (e.g., (5.5.3)), and so are the probability density $P(t)$ and the mean time. It is possible that the numbers calculated for the latter could in some cases be obtained from expressions not involving the causal law of motion, such as in (5.5.2). In general, though, (5.5.8) yields numbers not given by other methods and the possibility arises of experimentally distinguishing the various proposals. This is not a question of offering more detailed predictions in order to interpret an unambiguous formalism, but of giving precise meaning to a quantity, $\langle t \rangle$, which on physical grounds we expect should be definable but about which orthodox quantum mechanics is rather coy. The *formalism* of the quantum theory of motion can extend in a nonarbitrary and general way the quantum formalism in an area where the latter is ill-defined. Under the guise of expounding an 'interpretation' of quantum mechanics, the de Broglie–Bohm approach potentially extends the predictive power of the theory, as well as providing insight. Although empirical confirmation of these predictions would not prove the 'reality' of the particle trajectory, it would provide strong circumstantial evidence in its favour, being a test of the particle law of motion.

We present this as a possibility to explore. In attributing definite times to events the causal interpretation does not start from or require connection with a time operator. The implication is that the historical problem of time in quantum theory may have been artificially created by the demand that all physical observables be associated with self-adjoint operators. It is therefore possible that the quantum theory of measurement (Chap. 8) should be modified to include a treatment of observables not directly connected with such operators, but we shall not pursue this line of enquiry.

5.5.2 Age

The notion of time employed above is an essentially classical one, external to the system. One can envisage a different concept of 'inner time' or 'age' of a physical system, perhaps relevant to thermodynamical processes rather than the dynamical ones just discussed, and we conclude with some remarks on this.

One way to overcome the objection against the existence of a time operator cited earlier, the dual role of the Hamiltonian, is to discuss time evolution in terms of the Liouville–von Neumann equation (Prigogine, 1980; Misra & Prigogine, 1983; see also Holland, 1988c):

$$i\hbar\, \partial\rho/\partial t = L\rho \tag{5.5.11}$$

where ρ is the density matrix and $L = H \times 1 - 1 \times H$ is the Liouville 'superoperator' acting in the product space $\mathscr{H} \otimes \mathscr{H}$ in which ρ is treated as a 'supervector'. H continues to determine energy levels but now L generates the time evolution. Because L is not required to be semibounded it becomes possible to introduce for systems whose Hamiltonian has an absolutely continuous spectrum extending from 0 to ∞ a conjugate self-adjoint time superoperator T obeying the commutation rule:

$$[T, L] = i\hbar. \tag{5.5.12}$$

Usually it is assumed that one can attribute a definite 'age' to a physical system only if the density matrix is an eigenfunction of T, in accordance with the views of Bohr (1932) regarding the complementary character of thermodynamics and dynamics of which (5.5.12) is supposed to be an expression. But if we follow the method of the causal interpretation (§3.12) it becomes possible to attribute a definite age to arbitrary states, as we shall now see.

We work in the energy representation in which the density matrix has components $\rho(E, E')$ and pass to new indices $\rho(\Omega, \omega)$ where $\Omega = \frac{1}{2}(E + E')$, $\omega = E - E'$. Then $L = \omega$ and $T = i\hbar\, \partial/\partial\omega$. The solution of (5.5.11) is

$$\rho(\Omega, \omega, t) = \rho_0(\Omega, \omega)\, e^{-i\omega t/\hbar}. \tag{5.5.13}$$

The eigenfunctions of T are a special case of (5.5.13) in which ρ_0 is a function of Ω alone and the corresponding eigenvalues are t, the external time parameter. For a general state, the mean age is easily seen to satisfy

$$\frac{\langle\rho, T\rho\rangle_t}{\langle\rho, \rho\rangle_t} = \frac{\langle\rho, T\rho\rangle_0}{\langle\rho, \rho\rangle_0} + t. \tag{5.5.14}$$

That is, the mean age keeps step with the external time. For an eigenstate the right hand side of (5.5.14) reduces to just t.

Yet age and external time are not the same thing and we show this by giving a formula for the age associated with any state. We define this as the 'local expectation value' of the operator T in the energy representation:

$$\mathscr{A}(\Omega, \omega, t) = \mathrm{Re}\, \rho^*(\Omega, \omega, t)(T\rho)(\Omega, \omega, t)/|\rho(\Omega, \omega, t)|^2. \tag{5.5.15}$$

Treating the density matrix as a supervector in the higher space, we express it in polar form:

$$\rho(\Omega, \omega, t) = R(\Omega, \omega)\, e^{iS(\Omega,\omega,t)/\hbar}, \tag{5.5.16}$$

where R and S are real functions. Substituting into (5.5.15) and using (5.5.13) yields

$$\begin{aligned} \mathscr{A}(\Omega, \omega, t) &= -\partial S(\Omega, \omega, t)/\partial\omega \\ &= t - \partial S_0(\Omega, \omega)/\partial\omega \end{aligned} \tag{5.5.17}$$

where S_0 is the phase of ρ_0.

We thus arrive at a notion of time that represents an intrinsic property of matter in the sense that it depends on the state of the system, as well as on the external time. It reduces to the latter for eigenstates. Eq. (5.5.17) expresses the difference between the two types of time: systems may have different ages at the same instant t, or the same age at different t. This seems to be in accord with our intuitive notions regarding the relation between time as measured by radio-carbon dating and time measured by a periodic process. We present this as an illustration of how the method of the causal interpretation provides a point of departure for new lines of enquiry but we will not pursue it further.

6

The classical limit

6.1 Conceptual and formal problems

Many of the problems surrounding the interpretation of quantum mechanics stem from the absence of a sufficiently general and physically transparent analysis of the connection with the domain in which classical concepts are assumed to apply. It is the purpose of this chapter to describe with the aid of a few examples how the quantum potential approach provides a context within which the conceptual and formal issues raised by this problem may be clarified. A key question is to explain how the state or context dependence which is so characteristic of the quantum domain turns into state independence at a certain level, characterized as 'classical'.

To illustrate one aspect of the conceptual problem we recall that in Bohr's interpretation the validity of classical concepts is already presupposed since, it is suggested, it is only in terms of these that one can unambiguously communicate the results of experiments in the quantum domain. Thus, classical physics must be considered as prior to quantum mechanics and the latter is a generalization of the former in that it provides a new set of laws governing the application of classical concepts.[†] According to this view, any procedure by which classical mechanics is recovered from quantum mechanics as a mathematical limit can only be a demonstration of consistency with the already postulated epistemological relation between the two theories and not as a 'derivation' of classical mechanics from quantum mechanics (George, Prigogine & Rosenfeld, 1972). Yet in spite of this admonition the discussion of the problem is usually carried on as if quantum mechanics is the more fundamental theory from which classical mechanics emerges when certain parameters naturally occurring in the theory are varied (in much the same

[†] It is noted that the procedure of 'quantization' whereby one generates a quantum theory by replacing Poisson brackets with commutators also presupposes that classical mechanics is correct in a certain domain, and indeed the classical potential energy appears in the wave equation. Similarly, the path integral method starts from the validity of the classical expression for the action function.

way as classical physics is supposed to be a special case of relativity when the velocity is small compared to the speed of light). As a result a great deal of confusion still surrounds this question and there is not even agreement on what constitutes a universal mathematical criterion to characterize the classical limit.

The most widely quoted criterion for the classical limit is $\hbar \to 0$ (vanishing Planck's constant). Other related limits that are routinely employed are high quantum numbers ($n \to \infty$), short de Broglie wavelength, large mass ($m \to \infty$) and the 'macroscopic level', or combinations of these. At first sight these requirements seem reasonable in that $\hbar$ does not appear in the classical laws of motion, and there is no evidence of discrete energy levels or interference phenomena in the everyday macroworld of classical experience. Indeed, these features, in particular the finiteness of $\hbar$, are supposed to characterize the quantum domain. However, there are several problems connected with a limiting programme that concerns itself just with the variation of parameters.

To begin with, it seems to be a legitimate question to ask what happens if Planck's constant takes on a value different (smaller) from the one it actually has. But in a universal theory we should be able to account for the classical level, where $\hbar$ is not relevant, without letting $\hbar$ deviate from its real value. In any case, the notion that $\hbar$ is 'small' has no absolute meaning because its value depends on the system of units (Berry, 1991). For both these reasons one should arrange that $\hbar$ is compared with quantities having the same dimension (of action) involving parameters and functions that naturally take on a range of values (such as m and n and the action). In general, this is how we shall construe the expression '$\hbar \to 0$' in what follows, although we shall occasionally interpret it in absolute terms. Similar remarks apply to the variations of other quantities. If there were no other problems with this process we could in this way account for classical physics in a world in which $\hbar$ is, in fact, finite.

But the question of whether or not classical mechanics follows when certain parameters are varied depends very much on the nature of the equations containing the parameters that we expect to have classical analogues. The wavefunction, in order to satisfy an equation in which $\hbar$, for instance, appears in the differential coefficients, must itself depend on $\hbar$: $\psi = \psi(x, t; \hbar, n, m, \ldots)$. It is not at all obvious *a priori* that limits such as $\hbar \to 0$ will result in the classical equations, even if they are performed with due regard to the need to compare only like quantities. In fact they often do not. The intervention of the quantum state implies that the problem of connecting quantum and classical physics is considerably more involved than the corresponding Newtonian limit of special relativistic mechanics.

To illustrate these points we write down the Schrödinger equation

$$\mathrm{i}\hbar\frac{\partial\psi}{\partial t} = -\frac{\hbar^2}{2m}\nabla^2\psi + V\psi \tag{6.1.1}$$

and make our usual substitution $\psi = R\,\mathrm{e}^{\mathrm{i}S/\hbar}$ to get

$$\frac{\partial S}{\partial t} + \frac{1}{2m}(\nabla S)^2 - \frac{\hbar^2}{2m}\frac{\nabla^2 R}{R} + V = 0 \tag{6.1.2}$$

$$\frac{\partial R^2}{\partial t} + \nabla\cdot\left(\frac{R^2\nabla S}{m}\right) = 0. \tag{6.1.3}$$

The crudest method of obtaining the classical limit (and this is done in some of the best textbooks) would be to claim that the only term involving $\hbar$ in these last two equations is the quantum potential in (6.1.2) and that this vanishes when $\hbar \to 0$ to leave an equation resembling the classical Hamilton–Jacobi equation:

$$\frac{\partial S}{\partial t} + \frac{1}{2m}(\nabla S)^2 + V = 0. \tag{6.1.4}$$

The fallacy in the argument is obvious: R (and S) generically depends on $\hbar$ and hence in general $Q \nrightarrow 0$ as $\hbar \to 0$. We could after all have set $\hbar = 0$ in (6.1.1) and obtained a nonsensical result. Indeed, the circumstance that the limiting quantum potential is not always negligible in comparison with other relevant energies remains valid in other limits too (such as $n \to \infty$). Moreover, the various conceivable limiting processes that may be applied in a given physical problem are not always mathematically equivalent (for examples of the inequivalence of $\hbar \to 0$ and $n \to \infty$ see Liboff (1975, 1984)).

Clearly, in order to recover (6.1.4) the quantum potential term must somehow disappear. The closest classical analogue of quantum mechanics in the position representation is the statistical mechanics of a single-valued ensemble in the Hamilton–Jacobi language. In the process of taking the limit the typical quantum effects should disappear and ψ must take the form $R\,\mathrm{e}^{\mathrm{i}S/\hbar}$ where R and S are independent of $\hbar$ in the relevant spacetime domain. This presumably is the end the method of variation of parameters is tacitly striving to achieve. But, while a means to that end, this procedure does not in itself *characterize* the goal because it ignores the role of the quantum state. There is an infinite family of physically acceptable functionally distinct solutions to (6.1.1) in the given potential V. To demand that all of these should imply the classical law when the constants on which they depend take on particular relative values would surely be too stringent a requirement, and would, in fact, exclude many states displaying characteristic quantum properties (such as stationary states). As we shall see, some states have no classical limiit because we can never obtain a vanishing quantum potential. We need then some other criterion, involving the wavefunction and universally applicable, in order to decide when we are approaching the classical limit. The obvious candidate is the behaviour of the quantum potential itself, as we shall argue in due course.

In this connection it is important to realize that it is not sufficient when studying the problem of the classical limit to focus attention on what appears to be a key aspect of the quantum description, say energy levels, devise a scheme whereby classical (continuous) energies are obtained in the high quantum number limit, and then claim that, since the correct classical law has been obtained from the quantum formula, one has entered the classical regime. A similar objection may be raised against claims of having achieved the classical limit when close agreement is obtained between classical and quantum probability densities or cross-sections. All features relevant to the problem must be examined, and the basic mathematical problem is to obtain the classical *law of motion*. Unless this is found no conclusions concerning the classical limit can be drawn from the limiting behaviour of other quantities such as energy levels, probabilities or cross-sections. If we do not obtain motions typical of the classical domain in the given potential, we cannot claim to have recovered classical mechanics from quantum mechanics (even though other quantities such as energy levels may be quasi-classical).

It is when this facet of the problem is taken into account that the inadequacy of characterizing the classical limit by, say, high quantum numbers becomes evident. As an example, consider the hydrogen-like atom (§4.5). When $n \to \infty$, E_n becomes quasi-continuous (in the sense that adding or subtracting one quantum does not appreciably alter E_n). Yet the electron will persist in nonclassical circular motion in a surface coplanar with the equatorial plane, due to the action of Q. The limit of high quantum numbers does not always imply the classical limit because the quantum potential may remain finite.

Before we continue this discussion of the formal relation between classical and quantum mechanics, of one being the limit of the other, we must raise and confront a basic conceptual problem that besets the usual approach to this question. For what lies behind the various limiting processes investigated in the literature is a desire to understand the properties of matter as it actually is, independently of observation, to see the connection with the more familiar world of relatively autonomous objects in motion. Yet the '$\mathbf{x}$' appearing in the argument of the Schrödinger wavefunction is usually supposed to represent only the potential locations of a particle, one of which is realized with a probability $|\psi|^2$ if a measurement is performed to determine the position. It does not refer to a current material location. The classical analogue of this system is an ensemble of point particles of mass m each deterministically pursuing a trajectory $\mathbf{x}(t)$, with a certain probability of occupying a particular region of space at each instant. How can one pass from a theory in which ψ merely represents statistical knowledge of the state of a system to one in which matter has substance and form independently of our knowledge of it? That is, even if one contrives to obtain (6.1.4) in some limit, why are we justified in identifying this with the Hamilton–Jacobi equation describing the propagation of the S-function associated with an ensemble of precisely defined trajectories whose law of motion is given by $\dot{\mathbf{x}} = \nabla S/m$? Or in treating the probability of being as a limit of the probability of finding? The answer is that one is not justified in doing these things; one cannot logically deduce a model of substantial matter and its motion from an algorithm which has no theory of matter and motion in it at all (i.e., which makes no statements as to what matter *is*). In order to connect quantum and classical dynamics smoothly in the usual approach the law $\dot{\mathbf{x}} = \nabla S/m$ and the notion of a precise initial position $\mathbf{x}_0$ of a material object are slipped in as additional postulates. Indeed these postulates, which cannot be derived from the wavefunction, are made in a domain where *quantum mechanics is valid* (for (6.1.4) is here an instance of the Schrödinger equation).

An obvious further question therefore presents itself: if we are prepared to accept the objective existence of matter in this special case of quantum mechanics so as to ensure consistency with the classical theory, why do we not countenance it at an earlier stage in the approximation procedure, or even in the exact theory before any approximation is made?

In fact, particle trajectories are routinely plotted in studies of the semiclassical approximation to quantum mechanics. The semiclassical trajectory can only be given a consistent interpretation if we admit Q as a causal agent, for it is defined as a solution to the law of motion $\dot{\mathbf{x}} = \nabla S/m$ where S is the phase of the semiclassical (i.e., nonclassical) wavefunction (cf. §6.4). Since it is in the nature of an approximation to have ill-defined boundaries of application it indeed appears arbitrary to accept at one extreme of the semiclassical theory the meaningfulness of the trajectory concept and at the other to deny it. The only consistent position seems to be to adopt the trajectory assumption from the outset, in the exact theory. One could still accept the formal consistency of the trajectory concept in the exact theory but claim that it does not correspond to anything physically 'real' since the orbit cannot be 'measured' (we cannot determine position and momentum to arbitrary accuracy at the same time). But then one foregoes the possibility of ever connecting classical mechanics with quantum mechanics.

Conceptual problems of this kind do not arise in the causal interpretation, because it starts from an unambiguous theory of material motion in which probabilities refer to actual states. It does not presuppose a 'classical level' but rather treats reality as a whole as basically quantum mechanical in nature and seeks to deduce classical mechanics both formally and informally (Bohm & Hiley, 1987; Holland & Kyprianidis, 1988); it is thus more in harmony with the way in which most physicists think of the relation between the two theories (whether it is legitimate to conceive of the classical theory as entirely contained within quantum theory is examined in §6.2). Conceptually, the classical domain is that case where the wave component of matter is passive and exerts no influence on the corpuscular component, i.e. the state of the particle is independent of the state of the field. Mathematically this is characterized by the state-dependent condition $Q \to 0$ (see §6.2) which encapsulates within a single universal criterion the condition of validity of the various limiting processes that have been proposed hitherto. In this limit particle effects such as quantized energies, interference and so on, for which Q is responsible, will be absent and the effect of a finite $\hbar$ will no longer be evident (even though $\hbar$ remains a fixed number). The fact that the variation of parameters does not always result in classical behaviour because ψ depends on them is incorporated through the nonvanishing of the quantum potential.

It might be thought that all interpretations would agree at least on the correct mathematical criterion for this limit, given the apparent necessity of recovering (6.1.4). This does not seem to be so however. The conventional interpretation gives a nonclassical interpretation to some classical wave-functions (solutions of (6.1.3) and (6.1.4)) so the same function may be associated with both an objective classical system and a quantum system (§6.3).

Finally, note that we use the phrase 'classical limit' in the sense of 'classical-like behaviour'. This does not necessarily imply that the systems under consideration are macroscopic. An electron may under certain circumstances behave like a classical point particle, as in processes adequately treated using classical electrodynamics, and in other situations quantally. One has to take into account not only the system in itself but how it interacts with other systems. On the other hand, quantum mechanics potentially describes a situation where typical quantum effects such as interference or tunnelling are displayed by macroscopic systems (the '$\mathbf{x}$' in the Schrödinger equation may be the centre of mass (cm) coordinate of a many-body system). The quantum theory of motion has the advantage that it makes no essential distinction between the macro- and micro-levels; it applies to reality as a whole and does not arbitrarily cut off parts of the world and deem them unanalysable.

Summarizing, there are two logically distinct problems to be considered:

(1) characterizing and achieving the classical domain as a mathematical limit of quantum mechanics;
(2) making the conceptual connection between the classical determinist theory of matter and motion and the quantum statistical theory of observation.

In this chapter we discuss a selection of the relevant questions which illustrate the value of the quantum potential picture in clarifying these problems. It is proposed that the mathematical and conceptual benefits of this approach render it superior to any of the other partial treatments of the problem.

6.2 A state-dependent criterion for the classical limit

In passing to the classical domain, the fundamental relations we want to recover from the quantum theory are the law of particle motion:

$$\mathrm{d}\mathbf{p}/\mathrm{d}t = -\nabla V|_{\mathbf{x}=\mathbf{x}(t)} \tag{6.2.1}$$

and the law of energy change:

$$\mathrm{d}E/\mathrm{d}t = \partial V/\partial t|_{\mathbf{x}=\mathbf{x}(t)} \tag{6.2.2}$$

for each particle starting from a specified point $\mathbf{x}_0$ in an ensemble of identical particles. These equations give a complete description of the dynamical behaviour of a classical particle in the potential V. The ψ-function of classical mechanics (§2.6) has a purely subjective significance and does not enter into the definition of the physical state of the system (which is completely specified by $\mathbf{x}(t)$) or the equation of motion. We have argued in §6.1 that the usual interpretation of quantum mechanics cannot logically deduce material structures obeying eqs. (6.2.1) and (6.2.2); there is nothing in it corresponding to $\mathbf{x}_0$ and the particle trajectory is simply introduced as an additional postulate in a certain domain. On the other hand, no such problem arises in the quantum theory of motion and comparison of these relations with the analogous equations (3.9.36) and (3.9.37) shows that the relation between classical and quantum mechanics is mathematically characterized in this approach by the behaviour of the quantum potential Q. The quantum dynamics will coincide with, or approach, the classical dynamics when the quantum force ($-\nabla Q$) and power ($\partial Q/\partial t$) become negligible in comparison with the classical force and power, i.e., when Q is effectively constant along the particle orbit. Requiring that the entire ensemble of particles associated with the same ψ-function should exhibit classical behaviour in the prescribed potential V evidently implies that Q should be constant throughout the regions of space and time accessible to the particles.

To formulate the criterion for the classical limit we need to take note of one further requirement. In quantum mechanics the zero of the energy scale is the (arbitrary) zero of the classical potential energy V (e.g., for a Coulomb potential the zero is at $r \to \infty$, for a harmonic oscillator at $r = 0$). We should therefore require that in regions where $V \to 0$ we also have $Q \to 0$ in the classical limit. But we have said that in this limit $Q \to$ constant throughout the relevant region of spacetime (including where $V \neq 0$), so we conclude that the condition

$$Q \to 0 \tag{6.2.3}$$

characterizes the regime in which the classical limit is being approached. A constant but nonzero quantum potential would signify a quantum mechanical contribution to the total energy (e.g., zero-point energy) and a nonclassical situation, even though the particle motions may be identical in both the classical and quantum cases. (More generally, the motions will be the same if Q is purely time-dependent. Again, however, this is a nonclassical case; see §6.3.)

It is possible that a subensemble may display classical behaviour, while the whole ensemble does not, if the condition (6.2.3) is obeyed only in a restricted region of the space potentially accessible to the particles. In this case quantum and classical mechanics may be distinguished globally since mean values computed in the two theories may not coincide (see §6.7).

Inserting (6.2.3) into the Schrödinger equation, or equivalently the quantum Hamilton–Jacobi and continuity equations, we obtain the nonlinear classical wave equation (2.6.7), or equivalently the classical Hamilton–Jacobi and continuity equations:

$$\partial S/\partial t + (1/2m)(\nabla S)^2 + V = 0 \tag{6.2.4}$$

$$\partial R^2/\partial t + \nabla \cdot (R^2 \nabla S/m) = 0. \tag{6.2.5}$$

Eqs. (6.2.1) and (6.2.2) follow from (6.2.4) by applying the operators ∇ and $\partial/\partial t$ in turn and writing $\dot{\mathbf{x}} = \nabla S/m$. Eqs. (6.2.4) and (6.2.5) show that, when it is defined, *the classical limit of the quantum mechanical one-body pure state problem is a particular classical statistical mechanics of a single body in the Hamilton–Jacobi language*, i.e., a field theory describing a single-valued family of classical trajectories as tracks orthogonal to the surfaces $S =$ constant in which the position probability distribution uniquely fixes the distribution in momentum. Note that, since (6.2.4) is a limit of (6.1.2), S is not, in general, a complete integral (it does not depend explicitly on a set of nonadditive constants). Conceptually, we enter a domain where ψ ceases to be part of actuality and merely becomes a repository of our knowledge concerning actuality (i.e., $\mathbf{x}(t)$). There will be no vestige of quantized energies, interference and so on in the motions of particles.

Eq. (6.2.3) expresses a condition on the quantum state. It is by no means the case that one can achieve this limit for all states corresponding to a given physical problem. It is convenient to classify the solutions to the Schrödinger equation into classes which do possess a classical limit and those which do not:

(*a*) $Q = 0$ exactly for all $\mathbf{x}$ and t. In this case ψ is a simultaneous exact solution of the classical and quantum wave equations. For examples see §6.3.

(*b*) $Q \to 0$ as the result of a variation of parameters on which ψ depends, or the behaviour of V, or on passing to certain regions of space and time, or all three. For examples see §§6.4–6.8.

(*c*) Neither (*a*) nor (*b*) is satisfied, whatever the values of parameters and the potential and throughout the spacetime regions of interest. For examples see §§6.4–6.8.

The necessary and sufficient condition for the classical limit is embodied in (6.2.3), which may be thought of as a Correspondence Principle.

An important point is that it is not sufficient in obtaining the classical limit that the absolute value of the quantum potential, $|Q|$, be small in comparison with the other energies appearing in the Hamilton–Jacobi equation. In §5.1 we saw that $|Q|$ is practically negligible compared with the kinetic energy of an electron in the two-slit experiment, and yet of course the channelling of the trajectories which yields the characteristic interference pattern is entirely due to the action of Q. The particle responds to the spatio-temporal *variation* of Q (i.e., force) rather than its numerical value. This, of course, is the case in classical mechanics where the absolute magnitude of a potential has no bearing on the motion. It is necessary that Q be numerically negligible ($Q \ll$ other energies in the Hamilton–Jacobi equation) and slowly varying (at least its first derivatives should be negligible: $\nabla Q \ll \nabla V$) and the arrow in (6.2.3) is intended to denote this. When finite, the quantum force is often independent of $\hbar$ in ultra-quantum cases such as stationary states when it balances the classical force (cf. §§6.6.1, 6.8). It is not surprising therefore that the condition $\hbar \ll$ (classical action) is not sufficient for the classical limit, although we find that it is necessary in the examples we give that do have a limit. Note also that Q may be slowly varying when the amplitude R is a slowly varying spatial function (so that $\nabla^2 R \approx$ constant), but we may obtain the same result for Q even when R is a rapidly varying function because the

R in the denominator of Q can cancel the variation in $\nabla^2 R$ (for an example see §6.5.1).

The possibilities (*a*) and (*b*) establish the conditions under which a solution to the Schrödinger equation implies motion in an ensemble of particles indistinguishable from the motion of a classical ensemble moving in the same external potential V. This does not mean that classical mechanics, as a theory of matter and motion, has thereby been deduced in its entirety from quantum mechanics. It merely makes clear when the two theories imply the same objective motions. There are two further questions to be addressed here, beyond the problem of logical consistency discussed in §6.1 which is not of course a problem in the causal interpretation.

First, the coincidence of the motions in the given potential does not imply that classical and quantum particles will behave similarly in interactions with other systems. For example, if we have a quantum state in which an electron is pursuing a quasi-classical trajectory (e.g., a wave packet) and we perform a momentum measurement we will generally obtain a distribution of results rather than the single momentum value that the particle had prior to measurement as in classical mechanics (cf. Chap. 8). This is connected with the fact that the criterion (6.2.3) pertains to one representation of quantum mechanics, the position representation. In order to achieve the classical result we would have to further require that $Q \to 0$ for the combination of system plus measuring apparatus.

Second, just as there are solutions to the Schrödinger equation with no classical limit ('quantum systems with no classical analogue'; case (*c*) above), so one cannot exclude *a priori* the possibility that there exists a class of solutions to the classical equations of motion which does not correspond to the limit of some class of quantum solutions ('classical systems with no quantum analogue'). We recall from Chap. 2 that classical trajectory fields in the potential V are generically multivalued and we obviously cannot obtain such ensembles as a limit of the quantum mechanical pure state. It might be thought that this may be achieved using a quantum mixed state but this is not guaranteed since the pure states making up the mixture, while not unique, are individually conserved whereas classical ones are not (cf. §2.6). If the classical limit of a quantum pure state exists it will correspond to a single-valued family. Even if we restrict attention to the latter, we are not aware of a theorem establishing that for all physically acceptable potentials V there exist quantal wavefunctions which imply either exactly or in some limit the ensemble of motions associated with all of the single-valued classical S-functions corresponding to V, where these exist. This is because one generally performs the limiting process on particular solutions of the

Schrödinger equation and not the general solution in the potential V. Weaker requirements would be that we should recover at least one classical S-function or a subset of a classical family from the quantum ensemble. That (6.2.3) is always fulfilled in at least a limited domain for the physically interesting potentials seems reasonable and is indeed confirmed by our examples, but we still lack a theorem to this effect.

Moreover, even when a classical limit does exist and a classical S-function is obtained from the quantum mechanical phase, the coupling of the quantum R and S fields implies that the distribution function R^2 found in the limit will be but one of the infinite variety of conceivable classical distribution functions which generally correspond to the relevant classical Hamilton–Jacobi function. There may exist of course several distinct quantum phases which have the same classical S-function as a limit. Then we would expect to obtain a range of R^2s in the limit. Whether we may obtain in this way all the acceptable classical distribution functions corresponding to the classical S is not known.

To illustrate what we have in mind consider scattering from a potential step or barrier for a classical ensemble of energy $E < V$ and a quantum ensemble of mean energy $< V$. Assume we have some initially free classical packet function $\psi_0 = R_0 \, e^{iS_0/\hbar}$ where R_0, S_0 are independent of $\hbar$ (e.g., a Gaussian). The initial phase space distribution is $f_0(\mathbf{x}, \mathbf{p}) = R_0^2(\mathbf{x}) \, \delta(\mathbf{p} - \nabla S_0)$. As we saw in §2.6.2 the classical Liouville evolution implies that the initial pure state develops into a mixed state which expresses the fact that the incident and reflected orbits cross. In contrast, the quantum pure case remains pure and the paths deviate from classical straight lines in order to maintain a single-valued congruence (cf. Fig. 5.11). The only way one could hope to get coincidence of the two types of ensemble for all t would be if the ensemble contains essentially one element, i.e., if R_0^2 approaches a δ-function (for an example where this works see §6.6.1). But whereas a classical δ-distribution in x and p retains its integrity and remains a pure state, a free quantal δ-function corresponds to equiprobability of momentum and rapidly spreads over space. Thus, there is no limit in which we can obtain the classical ensemble from a quantum one for all t, although the ensembles initially coincide.

It appears reasonable to conceive of classical mechanics as a special case of quantum mechanics in the sense that the latter exhibits new elements ($\hbar$ and Q) not anticipated in the former. But the possibility that the classical theory admits more general types of ensembles than can ever be reached from the limit of quantum ensembles, because the latter correspond to a specific type of linear wave equation and satisfy special conditions such as being built from single-valued conserved pure states, suggests that the two statistical theories

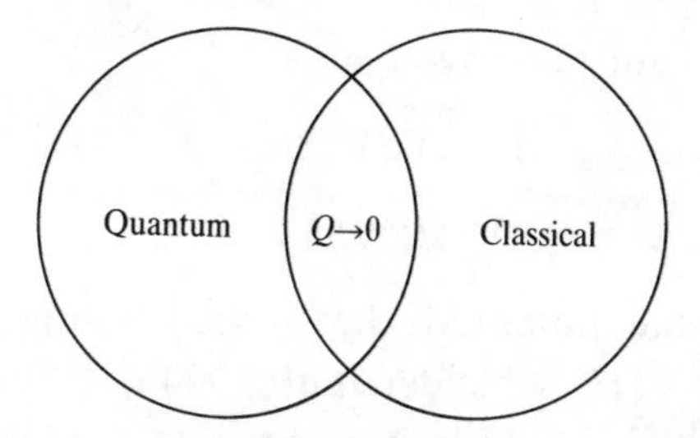

Fig. 6.1 The connection between quantum and classical statistical ensembles.

are disparate while having a common domain of application. The intersection is characterized by $Q \to 0$ in the quantum theory (Fig. 6.1). There is a well-defined conceptual and formal connection between the classical and quantum domains but as regards ensemble theory they merely intersect rather than one being contained in the other.

6.3 Identical quantum and classical motions

Under what circumstances will the ensemble of quantum mechanical motions coincide exactly with some classical ensemble in the same potential V? This will occur if the quantum potential is purely time-dependent or constant throughout the region of interest:

$$\nabla^2 R = \kappa R, \tag{6.3.1}$$

where $\kappa = \kappa(t)$ is real. (For complete equivalence of the quantum and classical cases we require $\kappa = 0$, as argued in §6.2. $\kappa \neq 0$ corresponds to a quantum contribution to the energy but the momenta are unaffected.) This equation must be satisfied in addition to the classical Hamilton–Jacobi and continuity equations (6.2.4) and (6.2.5). But since the three equations together involve only two unknowns (R and S) we see that, in general, there will be no solution (there is, of course, no requirement in classical statistical mechanics that a distribution function should obey (6.3.1)). Instead, we may treat V as an unknown function and deduce from the three equations those potentials for which simultaneous solutions of the quantum and classical wave equations do exist.

This may be achieved as follows. Solve (6.3.1) for $R(\mathbf{x}, t)$. Insert this into the continuity equation (6.2.5) and solve for ∇S and hence S. Finally, V is determined from (6.2.4). Note that the integration constants appearing in this process may be time-dependent.

In the case of stationary states, where $R = R(\mathbf{x})$, $S = S_0(\mathbf{x}) - Et$ and (6.2.4) and (6.2.5) reduce to (putting now $\kappa = 0$)

$$E = (\nabla S)^2/2m + V \tag{6.3.2}$$

$$\nabla \cdot (R^2 \nabla S) = 0, \tag{6.3.3}$$

some of the acceptable potentials have been enumerated (Rosen 1964a; Berkowitz & Skiff, 1972; Hirschfelder *et al.*, 1974a). It turns out that the more familiar potentials which imply bound states (e.g., Coulomb, harmonic) do not lie in this set. This is an instance of the general result that most stationary states of physical interest have no classical limit however we might try to obtain it, in the sense that (6.2.3) will never be fulfilled in some domain. Moreover, V, in general, depends on E so one cannot obtain a complete set of eigensolutions all of which behave classically. An exception occurs when $R =$ constant and

$$S = \mathbf{p} \cdot \mathbf{x} - Et, \tag{6.3.4}$$

where $\mathbf{p} =$ constant and $E = \mathbf{p}^2/2m$. Then $V = 0$ and ψ is a monochromatic plane wave (product of momentum eigenfunctions). Another example is the spherical wave (§4.2).

The general solution of the classical Hamilton–Jacobi and conservation equations when S takes the form (6.3.4) is

$$\psi(\mathbf{x}, t) = R_0(\mathbf{x} - \mathbf{v}t)\, \mathrm{e}^{\mathrm{i}(\mathbf{p} \cdot \mathbf{x} - Et)/\hbar}, \tag{6.3.5}$$

where $\mathbf{v} = \mathbf{p}/m$ and $R_0^2(\mathbf{x})$ is the initial distribution. Except when $R_0 =$ constant this is not an exact solution of the Schrödinger equation ($Q \neq 0$) but it can be obtained as the limit of a quantal wavefunction (see §6.6.2).

The existence of simultaneous classical and quantum wavefunctions highlights one of the interpretative idiosyncrasies of the usual version of quantum mechanics. For example, the plane wave momentum eigenfunction having phase (6.3.4) satisfies the classical wave equation and describes in that case an ensemble of uniform and rectilinear trajectories. Yet in the quantum context it is given a highly nonclassical interpretation as describing a state of definite momentum but completely unknown or 'fluctuating' or 'unreal' position (without, moreover, making clear what physical entity it is that has a definite momentum). This view is perhaps supportable to the extent that the wavefunction (or any ψ-function) does not in itself describe anything resembling a definite track. But then neither does it in the classical case. The point is: how in a universal physical theory can the same S-function be connected with a definite particle motion in one case and in the other nothing at all? Under what circumstances or limit does it become legitimate to treat this solution as describing an objective classical ensemble?

A class of nonstationary simultaneous classical and quantum solutions, again with $Q = 0$, is obtained when $R = R(t)$. Among these are the Green functions for the constant, linear and harmonic potentials. In the free case, the Green function is given by

$$G(\mathbf{x}, t; \mathbf{x}_0, t_0) = [m/ih(t - t_0)]^{3/2} \, e^{im(\mathbf{x}-\mathbf{x}_0)^2/2\hbar(t-t_0)}. \tag{6.3.6}$$

It has the property

$$\lim_{t \to t_0} G(\mathbf{x}, t; \mathbf{x}_0, t_0) = \delta(\mathbf{x} - \mathbf{x}_0) \tag{6.3.7}$$

and so describes an isotropic source of particles emanating from the point $\mathbf{x}_0$ (§2.3). The phase of (6.3.6) is the classical action and the motion is uniform and orthogonal to the expanding spherical wavefront. In describing an ensemble for which $\mathbf{x}_0$ is fixed and $\mathbf{p}$ arbitrary and uniformly distributed ($g(\mathbf{p})$ = constant), the free Green function is dual or complementary to the plane wave which describes an ensemble of fixed $\mathbf{p}$ but arbitrary $\mathbf{x}_0$, again uniformly distributed. Evaluating $-\partial S/\partial t$ along a trajectory shows that the energy is a constant; we have an example of how energy may be conserved for all members of an ensemble in a nonstationary state (cf. §4.1).

6.4 Semiclassical wave mechanics

6.4.1 The WKB approximation

In this section we shall be concerned with some of the problems connected with semiclassical approximations to quantum mechanics. We begin with the WKB approximation. The idea behind this is that reflection and hence interference tend to occur at places where the potential V varies sharply. Thus it might be expected that for obstacles with a sufficiently slowly varying spatial structure no appreciable reflection will occur and consequently one enters the classical regime. In fact, this expectation is not always borne out; we shall see that the conditions of validity of the WKB approximation are not sufficient to obtain classical motion. Moreover, they are not even necessary. The basic drawback with the WKB method is that it attempts to formulate the classical limit in terms of properties of the external potential and the de Broglie wavelength which do not make reference to the quantum state of the system.

To show the connection with the quantum potential it suffices to work in one dimension and consider stationary states of energy $E > V$. In the WKB approach one notes that ψ is a function of $\hbar$ and assumes that ϕ, where $\psi = e^{i\phi/\hbar}$, may be asymptotically expanded as a polynomial in $\hbar$ (Wentzel,

1926; Kramers, 1926; Brillouin, 1926; Bohm, 1951, Chap. 12; Schiff, 1968, p. 270):

$$\phi = \phi_0 + \hbar\phi_1 + \hbar^2\phi_2 + \hbar^3\phi_3 + \cdots. \tag{6.4.1}$$

Inserting this expression into the Schrödinger equation

$$-\frac{\hbar^2}{2m}\frac{\partial^2\psi}{\partial x^2} + V\psi = E\psi \tag{6.4.2}$$

and equating powers of $\hbar$ yields expressions for each ϕ_i in terms of $(E - V)$:

$$\phi_0 = \pm \int [2m(E - V)]^{1/2}\,\mathrm{d}x, \qquad \phi_1 = (\mathrm{i}/4)\log[2m(E - V)] \tag{6.4.3}$$

and so on. The first two terms in the expansion of ϕ are roughly comparable and both must be retained in the approximate wavefunction. The criteria which allow us to neglect the terms in the expansion beyond ϕ_0 and ϕ_1 are easily found:

$$m\hbar|V'|/[2m(E - V)]^{3/2} \ll 1, \tag{6.4.4}$$

$$m\hbar^2|V''|/[2m(E - V)]^2 \ll 1, \tag{6.4.5}$$

$$m\hbar^3|V'''|/[2m(E - V)]^{5/2} \ll 1 \tag{6.4.6}$$

and so on for higher derivatives of V ($V' = \mathrm{d}V/\mathrm{d}x$ etc.). These are evidently conditions on the smoothness of variation of the external potential and the magnitude of the (reduced) de Broglie wavelength:

$$\lambda\!\!\!^{-} = \hbar/[2m(E - V)]^{1/2}. \tag{6.4.7}$$

What (6.4.4) means is that the fractional change in $\lambda\!\!\!^{-}$ within the distance of a wavelength is small compared with unity. We may write (6.4.4) in the form

$$|\nabla\lambda\!\!\!^{-}| \ll 1. \tag{6.4.8}$$

Thus, when V is a slowly varying function of x and $(E - V)$ is not too small, we expect from (6.4.3) that a good approximation to the wavefunction will be given by one of the following two expressions:

$$\psi^{\pm}_{\mathrm{WKB}}(x, t) = A_{\pm}[2m(E - V)]^{-1/4}\exp\left((\mathrm{i}/\hbar)\left\{\pm\int^x [2m(E - V)]^{1/2}\,\mathrm{d}x - Et\right\}\right) \tag{6.4.9}$$

where $A_{\pm}$ is a constant. These 'classical wavefunctions' are called the WKB solutions. They satisfy exactly the classical equations (6.3.2) and (6.3.3) and approximately the quantum equation (6.4.2).

We shall denote by Q_{WKB} the quantum potential evaluated from the amplitude of the function (6.4.9):

$$Q_{\mathrm{WKB}} = -(\hbar^2/2m)(E - V)^{1/4}[\mathrm{d}^2(E - V)^{-1/4}/\mathrm{d}x^2]. \qquad (6.4.10)$$

Then the WKB conditions (6.4.4)–(6.4.6) are seen to be consistent with the classical nature of the solutions (6.4.9) since from them we can deduce that

$$|Q_{\mathrm{WKB}}| \ll E - V \qquad (6.4.11)$$

and

$$|Q'_{\mathrm{WKB}}| \ll |V'|. \qquad (6.4.12)$$

These relations state that the quantum potential is negligible in comparison with $(E - V)$, the classical kinetic energy, and that the quantum force has negligible effect in comparison with the classical force (in fact,

$$\tfrac{1}{4}|Q_{\mathrm{WKB}}|(E - V)^{-1} = |\hbar^2 \phi_3|$$

so the requirement $\hbar^2|\phi_3| \ll 1$ directly implies (6.4.11)). They are thus in accord with our criterion (6.2.3).

We observe, however, that the WKB conditions (6.4.4)–(6.4.6) are not sufficient to guarantee that the quantum potential associated with an *arbitrary* solution of the Schrödinger equation (6.4.2) varies slowly within the space of a de Broglie wavelength, and hence on their own they do not constitute universal criteria for when we are approaching the classical limit. This is easily demonstrated by simple examples; for instance, the superposition of the two WKB wavefunctions (6.4.9),

$$\psi = N(\psi^{+}_{\mathrm{WKB}} + \psi^{-}_{\mathrm{WKB}}), \qquad (6.4.13)$$

where N is a normalization constant, satisfies the Schrödinger equation and not the classical wave equation. Such a solution might be obtained by the reflection of ψ^{+}_{WKB} from a distant turning point. Inserting the explicit expressions (6.4.9) and assuming that $A_+ = A_- = A$ is real, the amplitude of (6.4.13) is given by

$$R = 2NA[2m(E - V)]^{-1/4}\left|\cos\left\{\int [2m(E - V)]^{1/2}\,\mathrm{d}x/\hbar\right\}\right|, \qquad (6.4.14)$$

whose sinusoidal variation may become rapid in a region far from the classical turning points ($E - V(x) \gg 0$) and when the quantum number is high (E large). The distribution of nodes implied by (6.4.14) indicates that the motion cannot be that of a classical particle as generated by either of (6.4.9) individually. Indeed, the spatial part of the phase of (6.4.13) is constant and the particle is at rest. And the quantum potential is not negligible; on the contrary, the Hamilton–Jacobi equation yields $Q = E - V$ which, by the requirements of the approximation we are using, is large. It will be noted that for the function (6.4.13) ∇Q is independent of $\hbar$ and will not vanish if $\hbar \to 0$, in all cases. We shall meet examples of this sort in §§6.6 and 6.8.

But are not the conditions of slowly varying potential and high quantum number, if not sufficient, at least necessary to guarantee a negligible quantum potential? In fact they are not; we may obtain classical behaviour, at least for a subensemble, in the region of a sharply changing potential and this is illustrated by the example of a packet incident on a potential barrier where the mean energy $E > V$. Referring to Fig. 5.12 it will be observed that particles in the leading edge of the packet pass through the barrier and, apart from the effects of spreading, are refracted very much like classical particles of energy $E > V$, for they move in a region where no appreciable reflected wave has yet formed (recall from §4.4 that we could not obtain this result using an incident plane wave since the reflected wave is already present over all space). As E is increased the agreement with classical mechanics improves for more members of the ensemble. Indeed, if the WKB conditions were necessary for the classical limit we would never be able to obtain as a special case of quantum mechanics the motion of a classical particle encountering a potential step (assuming this exists; see §6.2). This example also serves to show that smoothing the potential so that it varies slowly within the space of a de Broglie wavelength does not always lead to a vanishing quantum potential in regions where the amplitude is small (Dewdney, 1988).

The origin of the problems with the WKB conditions as criteria for the classical limit is the initial expansion assumption (6.4.1). It is fairly obvious that not all physically valid wavefunctions can be expressed in this way and once again we see that a proper characterization of the classical limit must involve the quantum state. The value of λ has no direct bearing on how Q is varying.

6.4.2 Semiclassical wavefunctions

The circumstance that generic classical ensembles exhibit multivalued trajectory fields is exploited in a technique for building up a semiclassical approximation to the quantal wavefunction (Berry & Mount, 1972; Korsch & Möhlenkamp, 1978a). One associates with each trajectory passing through a space point $\mathbf{x}$ an elementary WKB wavefunction. In three dimensions now this is

$$\psi_\mu(\mathbf{x}) = \rho_\mu^{1/2}(\mathbf{x})\, e^{iS_\mu(\mathbf{x})/\hbar}, \tag{6.4.15}$$

where μ labels the trajectory, $S_\mu(\mathbf{x}) = \int^{\mathbf{x}} [2m(E - V)]^{1/2}\, ds_\mu$ is the classical action and ρ_μ is a solution of the classical equation of continuity corresponding to the trajectory field generated by S_μ (we ignore an overall time factor in what follows). In classical mechanics one may in certain circumstances describe the statistics of a multivalued velocity field by a mixed state composed of the single-valued pure states (6.4.15) (Chap. 2). As an approximation to the quantum wave, one instead *linearly superposes* the elementary semiclassical wavefunctions to obtain the 'global' semiclassical wavefunction:

$$\begin{aligned}\psi(\mathbf{x}) &= \sum_\mu \psi_\mu(\mathbf{x}) \\ &= \sum_\mu \rho_\mu^{1/2}(\mathbf{x})\, \mathrm{e}^{\mathrm{i}(S_\mu(\mathbf{x})/\hbar - M_\mu \pi/2)}. \end{aligned} \tag{6.4.16}$$

The sum over μ includes not only the classical trajectories traversing the point $\mathbf{x}$ but also classically forbidden complex-valued paths (for which S_μ is a complex solution to the classical Hamilton–Jacobi equation; see the above references and Knoll & Schaeffer (1976)). In this way one obtains a finite amplitude at points where classical orbits do not pass and ψ would otherwise be zero. The integers M_μ reflect the topology of the trajectory field, i.e., the caustics which occur where $\rho_\mu \to \infty$.

The expression (6.4.16) has something of a heuristic character in that it cannot be rigorously deduced from the Schrödinger equation except in some special cases. It breaks down at caustics and, for fixed $\hbar$, for long times (Berry, 1980). Of course, it can be simply postulated and, evaluating the trajectories from $\dot{\mathbf{x}} = \nabla S/m$, compared with the exact quantum treatment where this is available. In e.g. repulsive Coulomb scattering, the results are quite close (Korsch & Möhlenkamp, 1978b). To the extent that it is valid, we evidently leave the domain of classical mechanics in forming such a combination of classical wavefunctions; (6.4.16) is not a solution of the classical wave equation and exhibits interference effects typical of quantum mechanics (a one-dimensional example is provided by (6.4.14)). This is natural since it is the purpose of the method to study quantal deviations from classical behaviour. And because the classical quantities ρ_μ, S_μ are independent of $\hbar$ it clearly possesses the property that the quantum potential does not generally vanish when $\hbar \to 0$. As a simple example, suppose there are only two classical contributions of equal density, $\rho_1 = \rho_2$. Then computing the quantum potential outside the nodes we obtain

$$\lim_{\hbar \to 0} Q = (m/8)(\mathbf{v}_1 - \mathbf{v}_2)^2, \tag{6.4.17}$$

where $\mathbf{v}_1$, $\mathbf{v}_2$ are the classical velocities. This generalizes slightly the example (6.4.14).

The method described so far is a means to construct a semiclassical quantum state from known classical building blocks. It is of interest to examine the converse problem where the form (6.4.16) has been successfully deduced from the exact quantum result in the 'semiclassical limit', $\hbar \to 0$. For simplicity we suppose there are just two classical contributions. A good example is the limiting behaviour of a stationary state attractive Coulomb scattering wavefunction where ρ_1, S_1 describe the 'in' beam (the segments of all classical trajectories prior to crossing the z-axis) and ρ_2, S_2 the 'out' beam (the segments of all classical trajectories after crossing the z-axis) (Rowe, 1987). In what sense is the derived expression

$$\psi = \rho_1^{1/2}\,\mathrm{e}^{\mathrm{i}(S_1/\hbar - M_1\pi/2)} + \rho_2^{1/2}\,\mathrm{e}^{\mathrm{i}(S_2/\hbar - M_2\pi/2)} \tag{6.4.18}$$

connected with the classical description? The probability density is given by

$$R^2 = \rho = \rho_1 + \rho_2 + 2(\rho_1\rho_2)^{1/2}\cos\delta \tag{6.4.19}$$

where $\delta = (S_1 - S_2)/\hbar - (M_1 - M_2)\pi/2$. In the corresponding classical case the probability density is simply the mixture of the two possibilities:

$$\rho = \rho_1 + \rho_2. \tag{6.4.20}$$

Eq. (6.4.20) may be recovered from (6.4.19) in two ways: if ρ_1, ρ_2 do not overlap, or if the interference term varies rapidly within a characteristic distance, say a de Broglie wavelength, and averages to zero in this domain. In the general semiclassical case the partial waves do overlap (since the method describes points where orbits cross) and one must invoke the second argument. Although this is open to the objection that the *mathematical* act of averaging a probability does not modify its real structure as determined by the actual ensemble (see also §6.6.1), we accept it for the sake of argument. But even if one attains mathematical identity of the densities on the average we cannot claim to have recovered the classical motion. Consider the velocity field

$$\mathbf{v} = \rho^{-1}[\rho_1\mathbf{v}_1 + \rho_2\mathbf{v}_2 + (\rho_1\rho_2)^{1/2}(\mathbf{v}_1 + \mathbf{v}_2)\cos\delta]. \tag{6.4.21}$$

The derived trajectories obviously have little in common with those implied by the classical fields $\mathbf{v}_1$ and $\mathbf{v}_2$. Suppose though that we invoke the argument that smoothing the velocity field over a small region destroys the interference terms, which we provisionally accept as reasonable. We then find an expression for the velocity which is the one expected from the classical mixture:

$$\mathbf{v} = (\rho_1 + \rho_2)^{-1}(\rho_1\mathbf{v}_1 + \rho_2\mathbf{v}_2). \tag{6.4.22}$$

Have we thereby obtained the ensemble of classical motions? The answer is that we have not: the classical expression (6.4.22) is derived *from* the individual velocity fields; at each point the velocity of a particle is either $\mathbf{v}_1$ or $\mathbf{v}_2$. In contrast, all we can deduce from (6.4.22) as derived from the quantum wavefunction is that the velocity of an ensemble element passing the point $\mathbf{x}$ is $\mathbf{v}$, not $\mathbf{v}_1$ or $\mathbf{v}_2$.

Thus, it is erroneous to claim, even in the case that we smear out the interference, that the wavefunction (6.4.16) 'corresponds' to families of classical orbits, each family being generated by S_μ. This is true only in a rather indirect sense. The claim is even less warranted if we retain the interference terms. This function is indeed associated with an ensemble of trajectories but these are not those of classical mechanics because of the nontrivial nature of the quantum potential in this case. One can indeed never derive from a limit of the single-valued quantal state function the multivalued trajectory fields of classical mechanics, for the latter are described by a mixed state and not a linear superposition.

Further evidence for this conclusion is that the total semiclassical function can be linearly decomposed into sets of functions in an infinite number of ways; eq. (6.4.16) is but one choice. In quantum mechanics it is always the total wave that has physical significance and whose properties must be evaluated and not the particular partial sum into which it may conveniently be expanded.

A key point in this discussion is that even if ρ_1, ρ_2 do not overlap, so that $\mathbf{v} = \mathbf{v}_1$ $(\mathbf{v}_2)$ in regions where ρ_1 (ρ_2) is finite, *it is only in the causal interpretation* that we can claim to have recovered the classical ensemble. In the usual interpretation ρ refers to the probability distribution for the outcomes of position measurements, not to the objective likely location of a material particle as in classical mechanics. Time and again we find this lacuna in the logical development of connecting quantum to classical mechanics, of deriving a classical mixture from a quantum linear superposition. It is the crux of the so-called 'measurement problem' (Chap. 8). The problem is solved in the quantum theory of motion because we have a particle which is definitely in either ρ_1 or ρ_2 with velocity $\mathbf{v}_1$ or $\mathbf{v}_2$, just as in classical statistical mechanics.

More generally, the semiclassical limit of quantum solutions will continue to display all the subsidiary requirements of quantum mechanics, that ψ be single-valued, finite and continuous. In this connection we note that a single-valued quantal field of rays generated by a pure state cannot be focussed to form a caustic (locus of points touched by neighbouring trajectories). Indeed, such focussing cannot be achieved by any judicious choice of mixed state for, although trajectories may cross in this case, the total density is always finite.

6.4.3 Is classical mechanics the short-wave limit of wave mechanics?

Is it legitimate to characterize the classical limit of quantum mechanics as akin to the geometrical optics limit of wave optics? In optics we mean two things: passage to the short-wave regime and introduction of the ray concept which facilitates the solution of optical problems by geometry. In the quantum theory of motion the geometrical structure of rays is already defined in the exact (Schrödinger) wave theory and this is not a characteristic of just classical mechanics. What about the other requirement, that of short wavelength? We have already indicated above that the shortness of λ compared to the distance over which V varies significantly does not generally ensure classical behaviour. But then λ is defined in terms of the momentum of a classical particle in the potential V and perhaps a more appropriate requirement is that the state-dependent wavelength we have introduced (§§3.3.11, 4.1),

$$\lambda_Q = h/[2m(E - V - Q)]^{1/2}, \tag{6.4.23}$$

should be comparatively short. This definition certainly yields a more accurate reflection of the actual state of affairs. For example, in a region far from the classical turning points (where $E - V = 0$) in which λ is small and slowly varying, we may have a quantum turning point ($E - V - Q = 0$) where $\lambda_Q \to \infty$ (cf. high quantum number limit of stationary states in §§6.5, 6.6). This may be taken as an indication of the nonclassical nature of the solution in that domain, and we might propose that a better criterion for the classical limit is that $\nabla ƛ_Q \ll 1$. In fact, there does not seem to be any reason why this should be so. All that happens when $Q \to 0$ is that $\lambda_Q \to \lambda$ and $\nabla ƛ_Q \to \nabla ƛ$. In a world in which $\hbar$ is a fixed parameter λ remains well defined but no interference effects persist in the limit that would demonstrate its finiteness. There are residues of quantum mechanics which are not manifest at the classical scale but they must remain well defined if we are to connect the two domains in a consistent manner.

Notice that it is not sufficient merely that $\lambda_Q \to \lambda$ in passing to the classical regime. This is illustrated by the treatment of the double-slit experiment in §5.1 where although Q is numerically negligible compared to the total energy so that $\lambda_Q \sim \lambda$, ∇Q and hence $\nabla ƛ_Q$ are not small everywhere since Q varies rapidly (the troughs interspersed among the plateaux). This explains why λ is correctly connected with the experimental results (fringe separation in the Fraunhofer limit) although it is defined in terms of a classical Hamilton–Jacobi wave which does not account for the formation of the interference pattern.

It is sometimes stated that it would be difficult to observe diffraction patterns for macrosystems due to the smallness of the de Broglie wavelength, but this reasoning does not seem to be cogent. The real practical problem is to isolate the system sufficiently from its environment so that the wavefunction may be split and recombined in a manner that preserves coherence.

6.5 The particle in a box

6.5.1 Stationary states

We consider the one-dimensional problem of a particle of mass m trapped between impenetrable walls placed at $x = 0$ and $x = a$. The external potential is given by

$$V = \begin{cases} 0, & 0 < x < a \\ \infty, & x \leqslant 0,\ x \geqslant a. \end{cases} \tag{6.5.1}$$

Applying the boundary conditions $\psi = 0$ at $x = 0$ and $x = a$, the normalized stationary state wavefunctions take the form

$$\psi_n(x, t) = (2/a)^{1/2} \sin(k_n x)\, \mathrm{e}^{-\mathrm{i}E_n t/\hbar} \tag{6.5.2}$$

inside the box and vanish outside. Here $k_n = n\pi/a$, n is a nonzero integer (henceforth assumed positive) and $E_n = \hbar^2 k_n^2/2m$. Eq. (6.5.2) is an example of the interference of two plane (classical) waves each associated with uniform motion having momentum $\pm p_n = \pm\hbar k_n$ in opposite directions and corresponds to a special case of the state studied in §4.3. This solution, although mathematically trivial and physically rather unrealistic, contains within it most of the essential points relating to the difficulty of connecting classical and quantum mechanics. It is the example employed by Einstein in his criticism of the causal interpretation.

The amplitude and phase of the function (6.5.2) are given by

$$R(x, t) = (2/a)^{1/2}|\sin k_n x|, \qquad 0 \leqslant x \leqslant a, \tag{6.5.3}$$

and in the segments $(r - 1)a/n < x < ra/n$

$$S(x, t) = \begin{cases} -E_n t, & r = 1, 3, \ldots, n-1\ (n\ \text{even})\ n\ (n\ \text{odd}) \\ -E_n t + \pi\hbar, & r = 2, 4, \ldots, n\ (n\ \text{even})\ n-1\ (n\ \text{odd}) \\ \text{undefined}, & x = ra/n,\ 0 \leqslant r \leqslant n. \end{cases} \tag{6.5.4}$$

Apart from the nodes at $x = 0$ and a required by the boundary conditions, the function (6.5.2) has a set of nodes ($(n + 1)$ in all) at the points $x_r = ar/n$, $0 < r < n$. These are separated by a distance $x_{r+1} - x_r = a/n$ and so become more densely packed as n increases.

It is evident from (6.5.4) that $\partial S/\partial x = 0$ everywhere except at the nodes where it varies infinitely fast and is undefined. It follows that the particle is at rest:

$$v = (1/m)(\partial S/\partial x) = 0 \Rightarrow x = x_0 \tag{6.5.5}$$

where x_0 is any nonnodal point. The kinetic energy is zero and all the energy resides in quantum potential energy:

$$E_n = Q = p_n^2/2m = \hbar^2\pi^2 n^2/2ma^2. \tag{6.5.6}$$

This has the same form as the (kinetic) energy associated with a plane wave of momentum p_n but is of quite different origin. In this completely static situation the total effective potential, a constant, evidently prevents the particle from passing through nodes if it initially lies outside such points.

The distribution of positions x_0 is determined by R^2, from (6.5.3). The corresponding actual momentum distribution is given by

$$\begin{aligned} g(p) &= \int_0^a R^2(x)\,\delta(p - \partial S/\partial x)\,\mathrm{d}x \\ &= \delta(p) \end{aligned} \tag{6.5.7}$$

because $\partial S/\partial x = 0$ when $R^2 \neq 0$ and R^2 is normalized. The result (6.5.7) obviously reflects (6.5.5). The phase space distribution function is therefore

$$f(x, p, t) = (2/a)\sin^2(k_n x)\,\delta(p). \tag{6.5.8}$$

6.5.2 Classical ensembles

The state (6.5.2) provides an example of how classical solutions (plane waves) that are combined in classical mechanics in an incoherent mixture of mutually exclusive options are linearly added in quantum mechanics to produce qualitatively new types of solutions and motions. For an ensemble of quantum and classical particles of total energy E_n in a box of zero potential each quantum particle has zero momentum and is at rest while each classical particle has two possible values of its momentum, $\pm p_n = \pm(2mE_n)^{1/2}$, and performs an oscillation of period $2a/p_n$ (see Fig. 6.2). Is there any connection between these two descriptions of motion? To aid our discussion below of this problem it will be useful to have at hand the solution of the general classical statistical problem, obtained by solving Liouville's equation in the box potential (6.5.1) (for further details see Born (1955), Born & Hooton (1956) and Born & Ludwig (1958)).

We start with an arbitrary distribution function $f_0(x, p)$ on the phase space (x, p) which is finite inside the box and zero elsewhere. The distribution

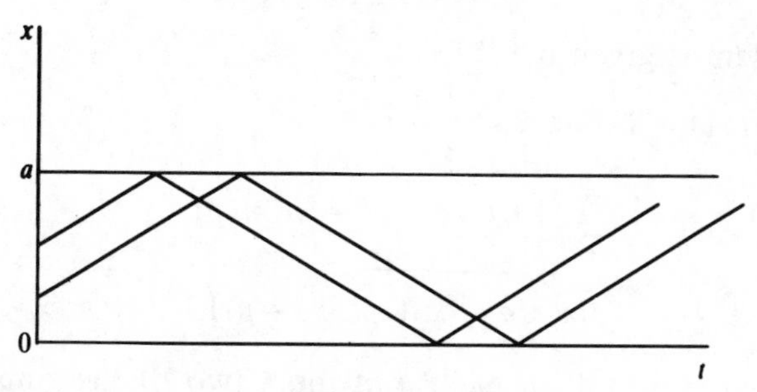

Fig. 6.2 Two possible trajectories of a classical particle in a box. Each point is associated with two momentum values.

function at time t must satisfy the following periodicity conditions in order that after each reflection at the walls the probability density at any point x has the same value:

$$\left.\begin{aligned} f(x, p, t) &= f(x, -p, t - 2mx/p) \\ &= f(x, p, t + 2ma/p), \qquad 0 < x < a. \end{aligned}\right\} \tag{6.5.9}$$

These are simple consequences of the law $\mathrm{d}f/\mathrm{d}t = 0$. To find a function having these properties we imagine that the perfectly reflecting walls are absent and take their effect into account by imposing the conditions (6.5.9) on a solution to the *free* Liouville equation. The general solution to the latter along the entire x-axis is

$$F(x, p, t) = F_0(x - pt/m, p), \tag{6.5.10}$$

where F_0 is an arbitrary function of x and p. This function will obey the relations (6.5.9) if

$$F_0(x, p) = F_0(x + 2a, p) = F_0(-x, -p) \tag{6.5.11}$$

for all x. To connect such a periodic function with the actual initial function $f_0(x, p)$ we define

$$F_0(x, p) = \sum_{r=-\infty}^{\infty} [f_0(2ra + x, p) + f_0(2ra - x, -p)]. \tag{6.5.12}$$

This expression satisfies the conditions (6.5.11) and reduces to f_0 in the interval $0 < x < a$, corresponding to the first $r = 0$ term in the infinite sum. For each value of p the remaining terms in the sum describe a set of mirror image virtual distribution functions regularly spaced along the x-axis whose collective uniform movements in one or other direction simulate in the domain $0 < x < a$ and for all time the sought-for solution of Liouville's equation in this potential. That is, using (6.5.10), the solution to the statistical evolution problem is given by

$$f(x, p, t) = F(0 < x < a, p, t)$$
$$= \sum_{r=-\infty}^{\infty} [f_0(2ra + x - pt/m, p)$$
$$+ f_0(2ra - x + pt/m, -p)], \qquad 0 < x < a. \quad (6.5.13)$$

At each instant $t > 0$ and for each p at most two of the summands in the sum will contribute to the distribution function in the interval $0 < x < a$. The function F is normalized in this domain if f_0 is:

$$\int_{-\infty}^{\infty} \mathrm{d}x \int_{-\infty}^{\infty} \mathrm{d}p\, f(x, p, t) = \int_0^a \mathrm{d}x \int_{-\infty}^{\infty} \mathrm{d}p\, F(x, p, t) = 1. \quad (6.5.14)$$

Notice that while the ensemble elements individually perform periodic motions, the dependence on p of the period $2a/p$ of the phase space density in (6.5.9) implies that the marginal position probability density,

$$P(x, t) = \int f(x, p, t)\, \mathrm{d}p, \quad (6.5.15)$$

is not generally periodic. This means that initially localized functions $P_0(x)$ may reach statistical equilibrium for long times:

$$P(x, t \to \infty) = 1/a. \quad (6.5.16)$$

We are obviously free, in principle, to choose f_0 to be whatever we like. As a pertinent example, suppose that all the particles in the ensemble have the same initial momentum p_n and an initial position density given by the quantum mechanical expression $|\psi_n|^2$:

$$f_0(x, p) = (2/a) \sin^2(k_n x)[\theta(x) - \theta(x - a)]\, \delta(p - p_n), \quad (6.5.17)$$

where $\theta(x) = 1$ $(x > 0)$ and 0 $(x < 0)$. This is a classical pure state (as defined in §2.5) corresponding to a classical wavefunction with phase $S = k_n x - E_n t$. At time t we have by substituting (6.5.17) into (6.5.15):

$$P(x, t) = (2/a)\Big\{\sin^2[k_n(x - p_n t/m)] \sum_{r=-\infty}^{\infty} [\theta(2ra + x - p_n t/m)$$
$$- \theta((2r - 1)a + x - p_n t/m)] + \sin^2[k_n(x + p_n t/m)]$$
$$\times \sum_{r=-\infty}^{\infty} [\theta(2ra - x - p_n t/m) - \theta((2r + 1)a - x - p_n t/m)]\Big\}.$$
$$(6.5.18)$$

As shown in §2.5 this is a conserved density. It displays the expected flow of probability following the underlying to-and-fro motion so that particles do not cross nodes. The segments of the trajectories moving in the positive x-direction are associated with one nonconserved pure state and those moving in the opposite direction with another. f turns into a mixed state but returns to its initial pure form after time $2am/p_n$. It is not therefore of the type (6.5.16). Similar results are obtained if we start with an initial mixed state

$$f_0(x, p) = P_0(x)\tfrac{1}{2}[\delta(p + p_n) + \delta(p - p_n)]. \tag{6.5.19}$$

6.5.3 Einstein's critique

Einstein used this example in two ways: to argue that the usual interpretation is deficient in that it does not ascribe to matter objective properties in a situation where it would reasonably be expected to have them, and to suggest that the causal interpretation is of no help in resolving this issue for, while it does attribute definite properties to matter, these lead to results that are apparently inconsistent with what he regarded as the classical limit. He argued as follows (Einstein (1953). See also the presentation of his view in Born (1971), in particular the letters of Pauli. For replies see Bohm (1953c), de Broglie (1953b, 1956, p. 131) and Bohm & Hiley (1985)).

Quantum mechanics is a universal physical theory and should, in principle, apply to macroscopic objects. In particular, if we have a sphere of mass m, diameter 1 mm and of constant energy trapped between perfectly reflecting walls 1 m apart, then (6.5.2) should represent a possible description of its physical state when we pass to the 'macroscopic limit' of this function (x is associated with the cm coordinate and it is assumed that the internal degrees of freedom may be neglected). Einstein characterized the macroscopic limit by the requirement that the de Broglie wavelength be small in comparison with the box width: $2\pi/k_n \ll a$. This is equivalent to $n \gg 2$, the limit of high quantum numbers, and to $p_n a \gg h$, which states that the quantum of action is negligible compared with the action of a classical particle of momentum p_n. Yet, in this limit, the solution (6.5.2) does not attribute to the particle definite properties, such as quasi-localization around a definite point, which we are sure it should have in such cases, independently of measurement or observers. Quantum mechanics does not, therefore, provide a complete description of the individual system.

As regards the causal interpretation, Einstein's objection rested on the assumption that in the macrolimit we should recover from the guidance formula the classical motion, and yet $v = 0$ remains rigorously true independently of the value of the quantum number. His feeling was therefore that the treatment of the individual process by the causal interpretation could not be an accurate description of reality.

Note that Einstein's objections do not relate in any way to the issue of determinism (cf. §1.4). He was not opposed to the statistical postulate – on the contrary, he considered it essential. Referring to the particle in a box example, Einstein wrote to Born: 'I have written a little nursery song about physics, which has startled Bohm and de Broglie a little. It is meant to demonstrate the indispensability of your statistical interpretation of quantum mechanics . . .' (Born, 1971, p. 199). He seeks rather to demonstrate that quantum mechanics cannot account for the actual individual facts of experience. According to Einstein, the theory's statements refer only to an ensemble of similarly prepared systems, and not to the properties of an individual. The statistical character of the theory is a consequence of the incomplete description it provides.

In reply, we may say first of all that we agree with Einstein's contention that quantum mechanics in its conventional formulation cannot lead to classical mechanics in the 'macroscopic' limit. Indeed, we reinforce Einstein's point: it is not possible to deduce the classical theory of matter from *any* solution of the Schrödinger equation in any limit, even those which are fairly well localized (packets) and which to a good approximation remain so in the course of time. The pure theory of linear fields whose dynamics quantum mechanics describes must be supplemented by a physical postulate (of the kind made in the causal interpretation). One may attempt to get round this by making a literal identification of a packet solution with the particle, but apart from a few special cases one is always ultimately faced with the fact of dispersion (see §6.6.2). In fact, the usual interpretation of a wave packet is that if one performs a measurement to determine the position of a body, it will be found with greatest likelihood in a definite, small portion of space. This postulate does not show, and does not claim to show, that a material object is actually 'there' with the property of definite coordinates independently of the measurement, as in classical mechanics. One cannot therefore use the conventional interpretation to deduce the classical theory which is based on the fundamental assumption that matter has substance and possesses attributes such as a fairly definite location at each moment independently of whether or not it is 'observed'.

On the other hand, we part company with Einstein in his assumption that classical mechanics should emerge in the 'macroscopic' limit for *all* valid quantum mechanical states. This does not seem to be a well-founded requirement, and indeed the case under discussion provides a counterexample.

Einstein's treatment of this question is connected with his belief that while the claim that ψ provides a complete description of the individual is untenable, the ensemble interpretation (in which ψ is admitted to be incomplete, giving statistical information on an ensemble as in classical statistical mechanics) provides a satisfactory resolution of the problem in that the statistical predictions of (6.5.2) agree with those of classical mechanics in the high quantum number limit. This is indeed true in the case of momentum measurements but it becomes obvious that the statistical interpretation does not generally imply the same results as classical mechanics if we enquire into the outcome of position measurements in the two theories. The wavefunction (6.5.2) possesses a distribution of nodes which become more numerous in the 'macro'-limit. A measurement of position would – if the conditions for establishing (6.5.2) were technically met, an extraordinarily difficult task – reveal that the particle is never found at certain points, a circumstance quite incompatible with classical to-and-fro uniform motion. Note that the non-classical nature of the solution is not that it exhibits nodes – we encountered a classical ensemble with this property in (6.5.18). It is that the quantum nodal structure is *static*, corresponding in the causal interpretation to the particle being at rest with finite energy.

It is easy to see that if one carried out a momentum measurement on a particle in the state (6.5.2) one would obtain in the high n limit essentially just the classical values $\pm p_n$, although this is not so for low quantum numbers. Because of the confinement, ψ is not simply a superposition of e^{ik_nx} and e^{-ik_nx} with equal amplitudes for all x. Rather, the Fourier analysis of (6.5.2) yields a continuous distribution of observed momentum values:

$$|\varphi_n(p)|^2 = \left(\frac{\hbar}{8\pi^2 a}\right)\left|\frac{-\exp[-i(p_n+p)a/\hbar]}{p_n+p} - \frac{\exp[i(p_n-p)a/\hbar]}{p_n-p} + \frac{2p_n}{p_n^2-p^2}\right|^2. \quad (6.5.20)$$

But for high quantum numbers this distribution is a sum of two effectively nonoverlapping packets peaked around the classical momenta $\pm p_n$. And

$$\lim_{n\to\infty} |\varphi_n(p)|^2 = \tfrac{1}{2}[\delta(p+p_n) + \delta(p-p_n)] \quad (6.5.21)$$

as for a classical ensemble.

Since the quantum predictions do not converge on the classical ones in the case of high n position measurements, the implication of Einstein's reasoning is that quantum mechanics itself must fail in this instance, rather than any particular interpretation of it (Bohm & Hiley, 1985). But there is no particular reason to doubt that if we can sufficiently isolate a system it could exist in such a single-frequency state for any value of the quantum number. The causal treatment of this problem is consistent with this possibility and indeed highlights the fact that the state (6.5.2) has no classical analogue. The equation of motion (6.5.5) is independent of the quantum number and although the energy becomes quasi-continuous the quantum potential remains finite. Clearly, the state-independent characterization of the classical regime in terms of the relative value of the de Broglie wavelength does not satisfactorily reflect the phase function or distinguish different physical states. The de Broglie wavelength reflects the periodicity of the probability density R^2 (whose argument is the classical phase or Hamilton–Jacobi function) and not the structure of the actual phase (6.5.4) of the total wavefunction.

Note that even if agreement were found between the classical and quantal probability densities in some problem, the ensemble interpretation does not imply the law of motion of a classical object since it is completely silent on the issue of what laws govern the underlying ensemble elements, in all circumstances.

How can the state of rest implied by the quantum theory of motion be compatible with the finite classical values obtained in a measurement of momentum? To see this we shall consider the causal version of such a measurement carried out by the 'time-of-flight' method which involves suddenly removing the confining walls and detecting the particle at a point x at a time $t \to \infty$ (the theoretical justification of this method is given in §8.6. See also §11.6). The number mx/t is the value of the momentum immediately preceding the position measurement and the position distribution reflects the momentum distribution (6.5.20). At time t after removal of the walls the wavefunction is

$$\psi_n(x, t) = \left(\frac{h^{1/2}}{2\pi}\right) \int_{-\infty}^{\infty} \varphi(\hbar k)\, \mathrm{e}^{\mathrm{i}(kx - \hbar k^2 t/2m)}\, \mathrm{d}k \tag{6.5.22}$$

for all values of the quantum number (for an alternative form of (6.5.22) see (11.6.19)). Two identical separating packets begin to form. The particle, starting from rest, will gradually acquire kinetic energy as the wave spreads and its quantum potential energy is liberated. The δ-distribution (6.5.7) continuously evolves into the function (6.5.20) as $t \to \infty$ and the particle eventually ends up at a distant point on one side or other of the box with an actual momentum calculated from the gradient of the phase of the function (6.5.22) that is uniquely determined by its initial position. The results of a momentum measurement are thus causally but not directly related to the actual momentum of the system (zero) prior to the interaction (see Chap. 8 for the general theory). We see that in the limit $n \to \infty$ we have that $g(p)$ will evolve into (6.5.21) and hence the final actual momentum of the particle is $\pm p_n$, as expected.

The prediction of the causal interpretation that the actual momentum is zero in the stationary state (6.5.2) is therefore consistent with the fact that a momentum measurement would reveal the classical to-and-fro momenta in the high quantum number limit since the quantum state and hence the quantum potential are transformed by the process.

One can recover motion approaching that of the classical system without applying an external agent by superposing a sufficient number of eigenfunctions (6.5.2) to form a packet:

$$\psi(x, t) = \sum_n c_n \psi_n(x, t). \tag{6.5.23}$$

The eventual spreading of such a function rules out any attempt at identifying the wave itself with a material object. Interestingly, the loss in coherence is only temporary; each energy level is an integer multiple of E_1, the ground state energy, and hence after a time h/E_1 the packet will regain its initial form. Hence a spreading packet will never attain a uniform distribution in the long time limit. This conclusion remains valid for any mixed state. Thus, no quantum ensemble in a box can ever reach statistical equilibrium.

To conclude, the quantum theory of motion resolves Einstein's conceptual objection to the conventional interpretation by assuming from the outset that physical systems comprise a corpuscular aspect. It also makes evident that quantum states do not necessarily approach classical states when $\hbar$ is negligible compared to the action. This answers the implications of Einstein's argument concerning the limiting validity of quantum mechanics and also his criticism of the causal interpretation.

6.6 Further examples of the classical limit

6.6.1 Harmonic oscillator

This section is devoted to further investigation of the problems encountered when the classical limit is characterized in terms of the variation of parameters on which the wavefunction depends.

We begin with the one-dimensional harmonic oscillator, a system displaying properties similar to those found for the nonharmonically oscillating particle in a box (§6.5). In classical mechanics, the trajectory of a particle with potential energy $V = \frac{1}{2}m\omega^2 x^2$ is given by

$$x(t) = x_0 \cos \omega t + (\dot{x}_0/\omega) \sin \omega t. \tag{6.6.1}$$

This may be generated by a variety of Hamilton–Jacobi functions describing different distributions of initial momenta. One choice, which at first sight should be the closest analogue of the quantal stationary state, is obtained by separating the t-variable:

$$S(x, t) = -\tfrac{1}{2}m\omega[a^2 \sin^{-1}(x/a) + x(a^2 - x^2)^{1/2}] - Et, \qquad x < |a|, \quad (6.6.2)$$

where $E = \frac{1}{2}m\omega^2 a^2$ is the total energy, and a is the amplitude of the motion, for all particles in the ensemble. Assuming that the distribution function is time-independent, the solution of the probability conservation law uniquely implied by (6.6.2) is

$$\rho(x) = \begin{cases} 1/\pi(a^2 - x^2)^{1/2}, & x < |a| \\ 0, & x > |a|. \end{cases} \quad (6.6.3)$$

This is the distribution function usually quoted for the classical oscillator but, of course, there are other possibilities. A different probability function corresponding to the S-function (6.6.2) is obtained if we suppose that ρ is space- and time-dependent and initially assumes the form of a δ-distribution, $\rho_0 = \delta(x - a)$, describing a particle whose initial position is known precisely to be $x_0 = a$. The initial velocity is $\dot{x}_0 = 0$. Then, as shown in §2.5,

$$\rho(x, t) = \delta(x - a\cos\omega t), \quad (6.6.4)$$

which is finite only in the domain $x < |a|$.

We pass now to the quantum oscillator of the same frequency and mass and examine its classical limit (the reader should refer to §4.9 for the derivations of formulae quoted here). Whereas in the classical case the underlying particle motion is always oscillatory (cf. (6.6.1)) whatever the S- or ρ-functions, here the motion is a function of the quantum state. For the nth stationary state the particle, lying between the nodes, is at rest for all t. The total energy $E_n = (n + \frac{1}{2})\hbar\omega$ is stored as quantum and classical potential energy. As with the particle in a box, the limit of high quantum numbers (or any limit) will never result in the classical oscillation although the energy becomes quasi-continuous. This is confirmed by consideration of the nth eigenfunction which becomes in the limit $n \gg 1$ (Morse & Feshbach, 1953, p. 1643):

$$u_n \approx \begin{cases} (2/\pi)^{1/2}(a_n^2 - x^2)^{-1/4} \\ \quad \times \cos[\frac{1}{2}\beta x(a_n^2 - x^2)^{1/2} + n\sin^{-1}(x/a_n) - \frac{1}{2}\pi n], & x < |a_n|, \quad (6.6.5) \\ n!^{1/2}(\beta/\pi)^{1/4}(\beta x^2)^{n/2}\,\mathrm{e}^{-\beta x^2/2}, & x > |a_n|, \quad (6.6.6) \end{cases}$$

where $E_n = \frac{1}{2}m\omega^2 a_n^2$ so that a_n is the amplitude of a classical oscillator of energy E_n, and $\beta = m\omega/\hbar$. Using these definitions we can reexpress the condition of high quantum numbers as the requirement that $\hbar$ is small relative to the action of a classical oscillator of energy E_n: $\hbar \ll m\omega a_n^2$. The wavefunction is real for all x and the part (6.6.5) is nothing more than a superposition (6.4.13) of WKB wavefunctions with the phase function given by (6.6.2). The probability density $|u_n|^2$ for $x < |a_n|$ is the classical distribution (6.6.3) modulated by an oscillatory function displaying a set of nodes that become more densely packed as n increases. The state of rest implied by the causal interpretation in the high quantum number limit is evidently consistent with this structure while finite oscillation is not. The quantal density is not infinite at the classical turning points $x = \pm a_n$ and it is not zero for $x > |a_n|$ (whereas (6.6.3) is both of these things) (for further discussion of the connection between the classical and quantal densities see Leubner, Alber & Schupfer (1988) and for another example of this sort see Home & Sengupta (1984)).

The following argument is sometimes invoked at this point to extract classical mechanics from the quantum functions. First of all, to obtain the classical probability (6.6.3) one 'averages' $|u_n|^2$ in the domain $x < |a_n|$ over an interval $\mathrm{d}x$ containing many oscillations and replaces the $\cos^2$ term by $\frac{1}{2}$. The result will be denoted $\overline{|u_n|^2}$. Next, to obtain the classical motion, one equates the probability of a particle being in the interval $\mathrm{d}x$, $\overline{|u_n|^2}\,\mathrm{d}x$, to the fraction of the semiperiod that the particle spends in this interval, $\mathrm{d}t/\frac{1}{2}T$ $(T = 2\pi/\omega)$. This yields

$$\mathrm{d}x/\mathrm{d}t = \omega/\pi\overline{|u_n|^2} = (2/m)^{1/2}(E_n - \tfrac{1}{2}m\omega^2x^2)^{1/2} \qquad (6.6.7)$$

which is the velocity of a classical oscillator of energy E_n.

This procedure is open to two objections. Firstly, the fact that *mathematically* the classical distribution is the local average of $|u_n|^2$ does not have any bearing on the *actual* state of affairs – the fine structure of nodes remains intact and the particle remains at rest. To transform $|u_n|^2$ physically into the classical function requires interaction with a further system. This may well lead to finite motion but the particle is then no longer a harmonic oscillator in a stationary state. Secondly, it will be noted that the probability of finding a particular outcome if a measurement of position is performed, which is how $|u_n|^2$ is usually interpreted, has been surreptitiously replaced by the classical notion of the probability that a particle *is* located in an interval $\mathrm{d}x$ at an instant. The causal formulation provides a justification for the latter interpretation but then we must work directly with the quantal wavefunction itself.

We conclude that the harmonic oscillator stationary bound states have no classical limit. One can multiply this sort of example indefinitely. In Coulomb bound stationary states (§4.5) the particle remains at rest in the large n limit rather than following a Kepler orbit. Another example, that of the linear potential, is given in §6.8.

These nonclassical features persist if we superpose a few neighbouring eigenfunctions in the high n limit (Cabrera & Kiwi, 1987). In order to achieve something approaching the classical oscillation of frequency ω we must wipe out the nodes by superposing many eigenfunctions to form a packet. Several types of harmonic oscillator packet may be constructed (such as the squeezed states whose width oscillates in time) but we shall consider just the coherent state described in §4.9, characterized by its constant width.

We recall that the quantum potential associated with a packet whose centre oscillates with frequency ω between the points $x = \pm a$ is

$$Q = \tfrac{1}{2}\hbar\omega - \tfrac{1}{2}m\omega^2(x - a\cos\omega t)^2 \tag{6.6.8}$$

and hence the quantum force is

$$-\partial Q/\partial x = m\omega^2(x - a\cos\omega t). \tag{6.6.9}$$

The total effective force, $-m\omega^2 a\cos\omega t$, which is of the type generated by an electric field applied to a charged particle, induces the orbit

$$x(t) = x_0 + a(\cos\omega t - 1). \tag{6.6.10}$$

All the particles in the ensemble connected with ψ start from rest. If the particle lies at the centre of the packet ($x_0 = a$) its motion is identical to that of a classical particle in the same potential V. However, while in this case the quantum force (6.6.9) vanishes, and the total energy is conserved, the energy still contains a zero-point contribution $\frac{1}{2}\hbar\omega$. In interactions with other systems this nonclassical energy may be released and is indicative that even for this special choice of initial conditions we are not in the classical regime. The motion of particles lying outside the centre of the packet is oscillatory of frequency ω but not about the classical centre point $x = 0$; Q and $-\partial Q/\partial x$ are finite in these cases.

To recover classical behaviour for the entire ensemble we pass to the limit in which the width of the packet is negligible in comparison with the amplitude of the oscillation: $a \gg \Delta x$. Since $(\Delta x)^2 = \hbar/2m\omega$ we may write

$$m\omega^2 a^2 \gg \tfrac{1}{2}\hbar\omega, \tag{6.6.11}$$

which may be interpreted as the condition that $\hbar$ is small relative to the classical action. This limit is equivalent to constructing the packet from a superposition of energy eigenstates peaked around a high quantum number n defined by $n\hbar\omega = \frac{1}{2}m\omega^2 a^2$ (Schiff, 1968, p. 76). We find that the Gaussian distribution $|\psi(x, t)|^2 \to \delta(x - a\cos\omega t)$ so that all the particles in the ensemble are very near the centre of the packet ($x_0 = a$). For each member of the ensemble we see from (6.6.9) that the quantum force vanishes and from (6.6.8) that $Q \approx \frac{1}{2}\hbar\omega$, the zero-point energy. But the total kinetic and classical potential energy $\frac{1}{2}m\omega^2 a^2 \sin^2 \omega t + \frac{1}{2}m\omega^2 x^2 \approx \frac{1}{2}m\omega^2 a^2$ so we see from (6.6.11) that Q is negligible. Therefore, in what is effectively the high quantum number limit, we do, indeed, recover the classical system for this particular quantum state. The quantum phase, which up to an additive constant is given by

$$S = -\tfrac{1}{2}\hbar\omega t - m\omega a x \sin \omega t + \tfrac{1}{4}m\omega a^2 \sin 2\omega t, \qquad (6.6.12)$$

reduces to the classical function (6.6.2) apart from the zero-point contribution but this is negligible compared with E. The relevant classical distribution function is (6.6.4) (and not (6.6.3)).

Ironically, the classical conserved system is derived not from the stationary state but from the time-dependent coherent packet solution. Notice that the classical ensemble obtained is single-valued (since it comprises essentially only one element). To generate the generic multivalued classical trajectory field one may take the limit of a quantum mechanical mixed state composed of coherent packets oscillating with the same amplitude and frequency but having a range of initial positions. This is to be distinguished from the linear superposition of coherent states centred around distinct points which displays interference effects and may have no classical limit (Holland & Kyprianidis, 1988).

6.6.2 Free Gaussian packet

It is sometimes claimed that although free packets eventually dissipate (an exception being given in §4.10), they may approximate classical force-free motion if the spreading is slow compared to characteristic experimental times. As a mathematical result this is easily demonstrated for the Gaussian packet although it is only in the causal interpretation that the physical conclusion may be drawn, and it has perhaps not been appreciated that it is not the nondispersive character of the solution that is the crucial component in the derivation.

Referring to §4.7 we recall that a particle at the centre of the three-dimensional Gaussian wavefunction (4.7.2) always pursues a classical orbit. We now consider the limiting behaviour of the whole ensemble for all times t for which the increase in width $(\sigma^2 - \sigma_0^2)^{1/2}$ is negligible in comparison with the initial width σ_0. Thus

$$\hbar t/2m\sigma_0^2 \ll 1. \qquad (6.6.13)$$

This may evidently be reexpressed as the condition that $\hbar$ is small relative to the action of a classical particle of mass m that moves a distance σ_0 in the time t. From (4.7.4) and (4.7.5) we see that the amplitude and phase functions respectively tend to

$$R \to (2\pi\sigma_0^2)^{-3/4}\, \mathrm{e}^{-(\mathbf{x}-\mathbf{u}t)^2/4\sigma_0^2}, \tag{6.6.14}$$

$$S \to m\mathbf{u}\cdot\mathbf{x} - Et, \tag{6.6.15}$$

where $E = \frac{1}{2}m\mathbf{u}^2$. The quantum potential (4.7.15) and quantum force (4.7.16) both approach zero. The classical limit of the quantum Gaussian ensemble is therefore a set of particles all moving uniformly with velocity $\mathbf{u}$, described by a plane Hamilton–Jacobi function, and having a nondispersive Gaussian distribution (cf. §2.5). As $t \to \infty$ the quantal spreading will eventually reassert itself but this may be staved off if the mass is chosen large enough. Note that the wavelength $\lambda_Q \to \lambda = h/m|\mathbf{u}|$ which is not necessarily small compared to σ_0.

This limit also yields the vanishing of the interference effects in the two-slit experiment performed with Gaussian packets (§5.1) if the packets have negligible relative motion. This has been explicitly portrayed in plots of the quantum potential when a value of $\hbar$ smaller than its real one is inserted, a procedure whose effect is similar to increasing the mass or slit-width (Philippidis, 1980).

In the case above the classical limit of a quantal system involves a nondispersive packet. But we have not entered the classical regime *because* of this. This is to confuse a localized probability with a localized object. Rather, the classical nature of the solution derives from the motion of the underlying ensemble being generated in the limit by a particular solution of the classical Hamilton–Jacobi equation. Although the emergence of the latter is often correlated with the emergence of a nondispersive density this is not always so (an extreme example is the Green function, §6.3). There is after all nothing specially classical about a localized probability distribution; classical probabilities generically spread over spatial regions (cf. §§2.5, 6.5.2) and need not remain peaked around a single trajectory. If it is required that in the classical limit the position density is so peaked, how could one ever recover from quantum mechanics classical ensembles which do not behave in this way? In the causal interpretation the trajectory concept is valid for all types of waves and is not associated with the formation of narrow packets.

An example where we cannot obtain a classical limit from an interference pattern is provided by two Gaussian packets with classical broadening (cf. §2.5.3). In the quantum case we have extra dispersion. Then in the limit where the quantal dispersion is negligible we will still have classical dispersion which classically would lead to overlapping packets and the crossing of trajectories. In the quantum case we have interference when the packets overlap, which prevents orbits crossing.

Quasi-coherent states may be constructed for many types of atomic and other potentials (Nieto & Simmons, 1979; Nieto, 1980; Gutschick & Nieto, 1980; Gerry & Kiefer, 1988; Dačić Gaeta & Stroud, 1990; Heller, 1991) and atomic wave packets have been produced in the laboratory (Yeazell & Stroud, 1988; ten Wolde *et al.*, 1988). These might be dubbed prototypes of a 'classical atom', subject to our remarks above. The notion that an 'electron' conceived as a classical particle traverses the quasi-classical orbit cannot, of course, be derived from the wavefunction alone.

We emphasize that the interaction of a classical packet with an external potential proceeds along quite different paths according to the law of motion. If this is incident for instance on a potential barrier we may obtain tunnelling effects from Schrödinger's law but simple reflection from Newton's.

6.7 Mean values and the Heisenberg relations

6.7.1 Operators and mean values

Having established a criterion which enables us to pass continuously from the quantum to the classical equations of motion, one might ask whether we can arrive at the classical limit of the remaining aspects of the quantum formalism – such as the operator structure, mean values and Heisenberg's relations – in a similar fashion and thus bring under the wing of the limiting programme of the causal theory the entire quantal edifice. We shall see that this is possible, although consideration of the limiting behaviour of the quantum potential alone is not always a sufficient criterion.

In the causal interpretation the physical content of operators is understood through the quantities they determine when applied to wavefunctions via what we have termed the 'local expectation value' (cf. §3.5). Starting with a classical function on phase space, $A(\mathbf{x}, \mathbf{p})$, the substitution $\mathbf{p} \rightarrow -i\hbar\nabla$ is equivalent to writing $\mathbf{p} = \nabla S$ and modifying A by the introduction of new quantum terms. Neglect of these additional terms, which may be justified on the basis of a variety of limiting processes, leads us back to the classical phase space expression in the Hamilton–Jacobi theory.

The limit of mean values follows directly from these considerations. According to classical statistical mechanics (in the Hamilton–Jacobi formalism) the mean momentum and energy of a free particle, over an ensemble, are given by

$$\langle \mathbf{p} \rangle = \int R^2 \nabla S \, \mathrm{d}^3x \tag{6.7.1}$$

$$\langle \mathbf{p}^2/2m \rangle = \int R^2 [(\nabla S)^2/2m] \, \mathrm{d}^3x \tag{6.7.2}$$

respectively, where R^2 is the probability density (R being the amplitude of the classical wavefunction). In quantum mechanics we expect the expression (6.7.1) for the mean momentum to be the same, but (6.7.2) will be modified since a (classically) free particle has quantum potential energy. That is, the mean energy of a (classically) free quantum mechanical system will be given by:

$$\int R^2[(\nabla S)^2/2m + Q]\,\mathrm{d}^3x. \qquad (6.7.3)$$

Now it is hardly surprising that (6.7.1) and (6.7.3) are just the usual quantum mechanical expectation values for the operators $\hat{\mathbf{p}} = -i\hbar\nabla$ and $\hat{\mathbf{p}}^2/2m$, respectively. And so we see that the quantum mechanical definition of averages, $\langle\hat{A}\rangle = \langle\psi, \hat{A}\psi\rangle$ where $\hat{A}$ is some nonmultiplicative operator, has in these cases a simple physical meaning: it is equivalent to taking an ensemble average according to the classical prescription, but taking care to include contributions of purely quantum mechanical origin.

Following our remarks above, this result can be easily extended to the case where $\hat{A}$ is an arbitrary polynomial function of $\hat{\mathbf{x}}$ and $\hat{\mathbf{p}}$. Should polynomials in $\hat{\mathbf{p}}$ higher than the second order be relevant we will find contributions to the mean values of such operators analogous to Q, in addition to the classical terms (for $\hat{\mathbf{p}}^n$ the latter is $(\nabla S)^n$, etc.). And once again the reason for the appearance of the nonclassical contributions to the ensemble average comes about from the properties of the ψ-field, and the mean value defined in this way is mathematically equivalent to the usual definition of $\hat{A}$ given above.

The classical limit of quantum mechanical expectation values is then readily established: this simply follows when we are able to neglect the extra contributions to the ensemble average which are characteristic of quantum mechanics. This condition is evidently weaker than requiring that we neglect the extra *local* contribution to the expectation value (in (6.7.3) this is Q).

Note there will be instances where the additional characteristic quantum terms will not be negligible when Q is. Thus, while the particle motion is indistinguishable from that of a classical ensemble in the same potential V, other properties will be distinguishable. These will show up in mean values and may indicate that, in interactions with other systems (especially measurements, cf. Chap. 8), the system of interest will not behave like a classical object.

6.7.2 *Ehrenfest's theorem*

At this point it is appropriate to examine the significance of Ehrenfest's theorem (§3.8.3) since this is often invoked in discussions on the classical limit.

Recall that the mean position and momentum satisfy the system of equations

$$m\, \mathrm{d}\langle \mathbf{x} \rangle/\mathrm{d}t = \langle \mathbf{p} \rangle, \tag{6.7.4}$$

$$\mathrm{d}\langle \mathbf{p} \rangle/\mathrm{d}t = \langle \mathbf{F}(\mathbf{x}) \rangle, \tag{6.7.5}$$

where $\mathbf{F} = -\nabla V$, for all sufficiently localized wavefunctions. These equations resemble the classical equations of motion but only in special cases do they imply motion identical to the orbit of a classical particle, or the mean over an ensemble, in the same force field. Indeed, in neither the classical nor the quantum cases does the mean motion generally follow one of the individual orbits in the respective ensembles (e.g., the two-slit experiment with identical sharp slits; in both cases $\langle \mathbf{x} \rangle$ moves along the symmetry axis where no particle can go).

The mean quantum (and classical) motion will coincide rigorously with that of a classical particle when the following condition is satisfied (Messiah, 1961, p. 217):

$$\langle \mathbf{F}(\mathbf{x}) \rangle = \mathbf{F}(\langle \mathbf{x} \rangle). \tag{6.7.6}$$

This relation is obeyed by polynomial potentials of order no higher than the second (which includes constant, linear and harmonic oscillator potentials). For other force fields the dependence of $\langle \mathbf{F} \rangle$ on the quantum state implies that the evolution of $\langle \mathbf{x} \rangle$ will not generally resemble that of an individual classical particle (unless $\mathbf{F}$ is practically constant over the entire domain where ψ is appreciable). An example is provided by the one-dimensional scattering of a packet incident on a potential step V located at $x = 0$. The force is given by $F = -V\delta(x)$ and inserting this in (6.7.5) yields

$$\mathrm{d}\langle p \rangle/\mathrm{d}t = -V|\psi(0, t)|^2. \tag{6.7.7}$$

When the packet is far from the step and free ($t \to -\infty$) the right hand side of (6.7.7) vanishes and the mean motion is uniform, as expected for a classical free particle. However, in the vicinity of the step the incident wavefunction has a complicated time dependence and (6.7.7) is no longer zero. In this region $\langle x \rangle$ departs from the classical orbit. A similar result is found for the particle in a box where the mean motion of a packet departs from the classical oscillation as soon as the reflected wave is appreciable (Brandt & Dahmen, 1989, p. 34). Indeed, if a classical particle, a quantum particle, the classical mean and the quantal mean coincide at $t = 0$, all four will, in general, part company for $t > 0$.

We conclude that in general the mean motion deduced from quantum mechanics will not reproduce that of a classical particle in the same potential. But even in those special cases (6.7.6) where the two motions do coincide there are two reasons why one still cannot claim to have deduced the classical limit. First, viewed from the conventional interpretation, Ehrenfest's theorem does not describe the mean *of* anything because ψ does not refer to the likely current state of a particle (there is no underlying objective ensemble). That a mathematical point traverses a classical orbit does not imply the existence of a substantive material system pursuing that orbit. To assume the latter is essentially to invoke the causal interpretation, but if one accepts there is in reality a trajectory it is immediately obvious, and this is our second point, that Ehrenfest does not describe it. The actual motion of the individual ensemble members may be brought about by a finite quantum force and so have little in common with classical behaviour – vastly different ensembles can generate the same mean. Note that these points remain valid for macro-objects whose motion must be consistent with the predictions embodied in $|\psi|^2$.

Hence Ehrenfest's theorem cannot be portrayed in itself as an example of the 'derivation' of classical mechanics. Limiting procedures applied to Ehrenfest's equations can lead to classical formulae (Hepp, 1974) but these do not provide a basis for a general theory of the classical limit of quantum mechanics.

6.7.3 Heisenberg's relations

The inequalities due to Heisenberg (1927) were fundamental in establishing the dominance of the view that the classical conception of material systems had to be replaced by a more nebulous view of physical reality. We shall examine the validity of this claim in §8.5. Here we shall be concerned with the mathematical origin of the inequalities and their connection with classical mechanics.

Heisenberg's relations are usually presented as deductions from the commutation relations:

$$[\hat{x}_i, \hat{p}_j] = \mathrm{i}\hbar\delta_{ij}, \qquad i, j = 1, 2, 3. \tag{6.7.8}$$

Defining

$$\left.\begin{aligned}(\Delta\hat{x}_i)^2 &= \langle\hat{x}_i^2\rangle - \langle\hat{x}_i\rangle^2\\ (\Delta\hat{p}_i)^2 &= \langle\hat{p}_i^2\rangle - \langle\hat{p}_i\rangle^2\end{aligned}\right\} \tag{6.7.9}$$

where in rectangular Cartesian coordinates, with $\psi = R\,\mathrm{e}^{\mathrm{i}S/\hbar}$,

$$\langle\hat{x}_i^2\rangle = \int x_i^2|\psi|^2\,\mathrm{d}^3x = \int R^2x_i^2\,\mathrm{d}^3x \tag{6.7.10}$$

$$\langle\hat{p}_i^2\rangle = -\hbar^2\int\psi^*\,\partial_i^2\psi\,\mathrm{d}^3x = \int R^2(\partial_iS)^2\,\mathrm{d}^3x - \hbar^2\int R^2(\partial_i^2R/R)\,\mathrm{d}^3x, \tag{6.7.11}$$

the Heisenberg relations are

$$(\Delta\hat{x}_i)^2(\Delta\hat{p}_j)^2 \geqslant (\hbar/2)^2\delta_{ij}. \tag{6.7.12}$$

One usually proves (6.7.12) by applying the Schwarz inequality:

$$\int |f|^2\,\mathrm{d}^3x \int |g|^2\,\mathrm{d}^3x \geqslant \left|\int f^*g\,\mathrm{d}^3x\right|^2, \tag{6.7.13}$$

where f, g are complex functions and it is assumed that the integrals converge. In one dimension, the proof proceeds (Bohm, 1951, p. 205) by letting $f = x\psi$, $g = \hat{p}\psi$, so that the right hand side of (6.7.13) becomes $|\int \psi^* x\hat{p}\psi\,\mathrm{d}x|^2$, and writing $x\hat{p} = \frac{1}{2}(x\hat{p} + \hat{p}x) + \frac{1}{2}i\hbar$. To locate the origin of the inequality we shall demonstrate (6.7.12) in a different way which shows that the essential result on which it rests is independent of Planck's constant.

We start by proving an inequality that is valid for any normalized function $R(x)$ which vanishes sufficiently fast at infinity:

$$-\int (x_i - \langle x_i\rangle)^2 R^2\,\mathrm{d}^3x \int R\,\partial_j^2 R\,\mathrm{d}^3x$$

$$= \int (x_i - \langle x_i\rangle)^2 R^2\,\mathrm{d}^3x \int (\partial_j R)^2\,\mathrm{d}^3x \quad \text{(by parts)} \tag{6.7.14}$$

$$\geqslant \left|\int (x_i - \langle x_i\rangle)\tfrac{1}{2}\partial_j R^2\,\mathrm{d}^3x\right|^2 \tag{6.7.15}$$

$$= \tfrac{1}{4}\delta_{ij}, \qquad i, j = 1, 2, 3, \tag{6.7.16}$$

where in going from lines (6.7.14) to (6.7.15) we have used (6.7.13) with $f = (x_i - \langle x_i\rangle)R$ and $g = \partial_j R$ and from (6.7.15) to (6.7.16) the normalization of R^2 and $\partial_j\langle x_i\rangle = 0$. The proof tacitly assumes that the integrals $(\Delta x_i)^2$ and $\int (\partial_j R)^2\,\mathrm{d}^3x$ are finite; if they are not, the demonstration breaks down. We may write the inequality thus

$$(\Delta x_i)^2 \int (\partial_j R)^2\,\mathrm{d}^3x \geqslant \tfrac{1}{4}\delta_{ij}, \qquad i, j = 1, 2, 3. \tag{6.7.17}$$

We also have a second inequality, which is obvious. Define

$$(\Delta p_i)^2_{\mathrm{cl}} = \int R^2(\partial_i S)^2\,\mathrm{d}^3x - \left(\int R^2\,\partial_i S\,\mathrm{d}^3x\right)^2. \tag{6.7.18}$$

This is the mean square deviation of the ith component of momentum over the ensemble connected with ψ. The subscript 'cl' indicates that it has the same form as the classical expression although, of course, ψ obeys the

Schrödinger equation so R and S generally depend on $\hbar$. Then

$$\Delta x_i (\Delta p_j)_{\text{cl}} \geqslant 0. \tag{6.7.19}$$

Observe that

$$(\Delta \hat{p}_i)^2 = (\Delta p_i)^2_{\text{cl}} + \hbar^2 \int (\partial_i R)^2 \, \mathrm{d}^3 x. \tag{6.7.20}$$

Now both inequalities, (6.7.17) and (6.7.19), hold in quantum mechanics and in the classical statistical mechanics of a single-valued ensemble. Suppose in the classical case that we multiply (6.7.17) by an arbitrary constant α^2, where α has the dimension of action, and add the result, which now has the dimension of (momentum)^2, to the inequality (6.7.19). Then we deduce, in classical ensemble theory, the inequality (6.7.12) with $\hbar$ replaced by α. Clearly, exactly the same argument applies if we treat (6.7.17) and (6.7.19) as pertaining to a quantum ensemble and write $\alpha = \hbar$. This is indeed how we derive the Heisenberg theorem (6.7.12) from our two inequalities. The point of this is that there is nothing especially quantum mechanical about the inequality (6.7.12) *per se*. It may be derived in the classical theory but there it has no particular significance because R^2 has a purely statistical meaning and (6.7.17) does not relate in any way to the particle dynamics. In quantum mechanics in contrast (6.7.17) is a statistical relation that reflects also the underlying dynamics and is potent when combined with (6.7.19) for that reason, since R defines the force as well as the distribution.

What we wish to propose then is that in the classical limit as carried out according to our programme ($Q \to 0$ in a world where $\hbar$ remains finite) the relations (6.7.12) continue to hold unmodified (in the sense that the right hand side is nonzero) but that the decoupling of the R and S fields implies that they have a purely formal significance. This is reminiscent of the case of the de Broglie wavelength which likewise remains mathematically well-defined in the limit but is not relevant to the physical dynamics (§6.4.3). This position is supportable only to the extent that the system continues to behave classically in interactions with other systems. If it does not, such as in quantum measurements, then the tacitly valid relations will become relevant to the dynamical description.

Consider the Gaussian packet studied in §6.6.2. We have in general

$$(\Delta p_x)^2_{\text{cl}} = (\hbar^2 t / 4 m \sigma_0^2 \sigma)^2, \tag{6.7.21}$$

which depends on $\hbar$ but vanishes if the spreading may be neglected. This is intuitively reasonable since all the particles in the ensemble will have the same momentum. Thus, since $\Delta x = \sigma$, we have $\Delta x (\Delta p_x)_{\text{cl}} \to 0$. This yields

$$(\Delta \hat{x})^2(\Delta \hat{p}_x)^2 \to (\Delta \hat{x})^2\hbar^2 \int (\partial R/\partial x)^2 \, d^3x = \hbar^2/4. \tag{6.7.22}$$

Cancelling $\hbar^2$ we recover (6.7.17) with an equals sign and with R independent of $\hbar$. Similar relations are obtained for the y- and z-component relations. The classical Gaussian corresponds to a 'minimum uncertainty' packet for all time.

Our argument does not therefore provide support for the view that in the classical limit $\Delta\hat{x}_i\,\Delta\hat{p}_i$ should either tend to zero or be very large compared to $\hbar$.

The physical interpretation of $(\Delta\hat{p}_i)^2$ is considered in §8.5 where it is identified with a component of the total stress tensor of the ψ-field. In one dimension (x) we can establish a connection with the quantum potential for the inequality (6.7.17) implies that

$$(\Delta x)^2\langle Q\rangle \geqslant \hbar^2/8m, \tag{6.7.23}$$

where $\langle Q\rangle$ denotes the mean quantum potential over the ensemble. This result clearly establishes the connection with dynamics. If Q is strictly zero for all x and t (cf. §6.3) then the proof of (6.7.23) breaks down. In three dimensions the proof may remain valid for states in which $Q = 0$.

6.8 The effect of gravity on quantum systems

6.8.1 The principle of equivalence in quantum mechanics

We pass now to an analysis of the extent to which the classical equivalence of gravitation and inertia remains valid in quantum mechanics, and show how the equal rate of fall of all bodies independent of their mass is not always guaranteed in the WKB limit of high quantum numbers (Holland, 1989).

The causal interpretation is especially useful in the context of the equivalence principle, since it enables one to formulate a possible quantum analogue in terms of concepts similar to those employed in the classical case, *viz.* the motion of a particle acted on by an external field. We can thus clearly establish whether or not it remains valid in the quantum domain. Actually, it is necessary to distinguish different aspects of the equivalence principle, since not all its components require the trajectory concept.

To begin with, we discriminate between three statements which are equivalent according to classical physics:

(a) inertial mass = passive gravitational mass: $m = m_g$.

(*b*) All sufficiently small test bodies fall with an equal acceleration independently of their mass or constitution in a gravitational field: $\ddot{x} = -g$.

(*c*) With respect to the mechanical motion of particles, a state of rest in a sufficiently weak, homogeneous gravitational field is physically indistinguishable from a state of uniform acceleration in a gravity-free space.

These three assertions, classically equivalent statements of the weak equivalence principle (WEP), are logically distinct, and it is important to separate them clearly in order to discuss quantum analogues of WEP.

The Einstein equivalence principle (EEP) comprises WEP (*a*) and (*b*) and extends (*c*) so as to include all the nongravitational laws of physics. It provides a way of incorporating the effect of gravitation on a system by performing a coordinate transformation. Notice that the coordinate transformation envisaged in WEP (*c*) or EEP is independent of the characteristics of the systems acted upon (in particular, their mass).

How do the various parts of WEP and of EEP carry over into the quantum theory? Let us consider this question first of all from the point of view of the orthodox interpretation of quantum theory and then examine it using the causal interpretation.

If we suppose that the most complete possible information concerning individual quantum systems is contained in the wavefunction, then the problem reduces to an examination of the properties of the Schrödinger equation. In this context, those aspects of classical equivalence that refer to trajectories have no meaning (WEP (*b*) and (*c*)). However, WEP (*b*) *might* be replaced by some principle such as the following, denoted WEP (*q*):

> The results of experiments in an external potential comprising just a (sufficiently weak, homogeneous) gravitational field, as determined by the wavefunction, are independent of the mass of the system.

This statement is manifestly violated in quantum mechanics. All interference phenomena are strongly mass-dependent, in particular those associated with gravitational fields (Greenberger, 1968, 1983).

On the other hand, the extension of WEP (*a*) to quantum mechanics is unobjectionable. The reasonableness of inserting the classical potential $V = mgx$ into the Schrödinger equation has been checked and confirmed by gravity-induced interference experiments (Colella *et al.*, 1975).† In addition, that part of EEP which relates to the local equivalence of gravitation and acceleration remains valid if we apply it to the Schrödinger equation treated

† There is theoretical evidence from finite-temperature quantum field theory that inertial and gravitational masses are not equal but the discrepancy is too small to be detected with present techniques – see Donoghue & Holstein (1987).

as a field equation, into which gravity enters in the guise of the potential just stated. This, in fact, is the natural quantum analogue of WEP (*c*).

To see this, suppose x is the coordinate of an inertial system, and let us transform to a rigid uniformly accelerating system with coordinates $x' = x - \frac{1}{2}gt^2$ and $t' = t$. Then the Schrödinger equation with an arbitrary potential $V(x, t)$ is transformed into the equation

$$i\hbar \frac{\partial u(x', t')}{\partial t'} = \left[\frac{-\hbar^2}{2m}\frac{\partial^2}{\partial x'^2} + V'(x', t') + mgx'\right] u(x', t'), \qquad (6.8.1)$$

with $V'(x', t') = V(x, t)$ if the old and new wavefunctions are related by the following unitary transformation (Greenberger & Overhauser, 1979):

$$\psi(x, t) = e^{i\phi(x', t')}u(x', t'), \qquad \phi = (m/\hbar)(gt'x' + \tfrac{1}{6}g^2t'^3). \qquad (6.8.2)$$

The Schrödinger equation in the accelerated frame (6.8.1) will be observed to have the same form as that for a state of rest in an external gravitational field, and so the statistical predictions in the two frames are identical. (A test confirming the predictions of the Schrödinger equation in a noninertial frame is reported by Bonse & Wroblewski (1983).) It might appear strange that the natural quantum analogue (WEP (*q*)) of WEP (*b*) is violated, and yet one cannot locally distinguish gravitation from uniform acceleration, given that in classical physics one tends to think of EEP as a stronger version of WEP. The explanation for this paradox lies in the nature of the unitary transformation (6.8.2). The transformation into the accelerating frame is not independent of the details of the system under consideration – it is a function of the mass. In this regard (6.8.2) is quite unlike the transformation law of tensors in classical physics. However, although the physical consequences of quantum mechanics in an accelerating frame or in a gravitational field at rest are mass-dependent, the same physics is nevertheless obtained in both frames. In short, there is no contradiction between the violation of WEP (*q*) and the continued validity of that part of EEP which refers to the local equivalence of gravitation and acceleration provided we admit mass-dependent transformations.

We shall now consider this question from the point of view of the individual particle motions which go to make up the statistical ensemble, according to the quantum potential approach. In the gravitational potential $V = mgx$ the quantum version of Newton's second law is:

$$\ddot{x} = -g - (1/m)\, \partial Q/\partial x|_{\mathbf{x}=\mathbf{x}(t)}. \qquad (6.8.3)$$

This equation enables us to formulate precisely quantum analogues of WEP (*b*) and (*c*) in terms similar to those of classical physics. It is immediately clear that the assumption of the equality of the inertial and passive gravitational masses (WEP (*a*)) does not imply that all bodies fall with an equal acceleration in a given gravitational field, due to the intervention of the mass-dependent quantum force term (apart from the $(1/m)$ coefficient, Q depends on m). WEP (*b*) therefore breaks down in the quantum domain for generic solutions of the Schrödinger equation. It thus has meaning in the causal interpretation to postulate WEP (*b*) as potentially viable, and to see clearly why it is false. This provides a dynamical explanation for why WEP (*q*) does not hold.

In general Q also depends on g and so provides a vehicle for the nonclassical influence of gravity on a particle (e.g., (6.8.11) below). While the de Broglie waves may be thought of as in 'free fall' the particles guided by them are most certainly not.

The quantum analogue of WEP (*c*) for the trajectory would state that (6.8.3) takes on the same form in a gravitational field as in a uniformly accelerating frame. Since (6.8.3) is a consequence of the Schrödinger equation which possesses this symmetry, it follows that WEP (*c*) indeed remains valid in this case. Gravitation is thus locally equivalent to acceleration not only at the level of the statistical predictions of quantum theory but also at the level of the individual processes that go to build up the statistical ensemble.

We find though that, in spite of this conclusion, bodies do not, in general, fall at an equal rate in a gravitational field independently of their mass. We shall now show that this result persists in the domain where we might expect to recover the classical symmetry, that of high quantum numbers, if we make a special choice of states in this limit.

6.8.2 Breakdown of weak equivalence (b) in the WKB limit

We suppose that a particle of mass m is confined to the region $x > 0$ and is subject to the potential

$$V = \begin{cases} mgx, & x > 0 \\ \infty, & x \leqslant 0. \end{cases} \tag{6.8.4}$$

This may be due to the earth's gravity or to uniform acceleration in a space free of forces. While classically the particle will continuously accelerate and bounce, quantally a wide range of motions is feasible depending on the choice of wavefunction, as we have seen. We have introduced the boundary at $x = 0$ in order to generate discrete energy levels.

The energy eigenfunctions that are finite for all x and vanish as $x \to \infty$ are

(Landau & Lifschitz, 1958, §22; Langhoff, 1971):

$$\psi_E(x, t) = A Ai(\xi)\, e^{-iEt/\hbar}, \qquad \xi = (2m^2/g\hbar^2)^{1/3}(x - E/mg), \qquad (6.8.5)$$

where A is a constant and $Ai(\xi)$ is the Airy function. This solution is not square-integrable. The free solution in the inertial frame corresponding to the stationary Airy function in the potential (6.8.4) is given by the unitary transformation (6.8.2) and is just the nonstationary Airy function studied in §4.10. The boundary condition $\psi_E(0, t) = 0$ fixes the energy levels in terms of the zeroes of the Airy function. We can obtain an approximate expression for these in the WKB limit.

Let $H = E/mg$ be the maximum height attained by an equivalent classical particle of energy E, and restrict attention to the classically accessible region $0 < x < H$, where x is not too close to the discontinuity in the potential at $x = 0$. The WKB condition (6.4.4) becomes $|H - x| \gg (\hbar^2/8m^2g)^{1/3}$. In this limit (6.8.5) has the asymptotic form

$$\psi_E(x, t) = A|\xi|^{-1/4} \sin[\tfrac{2}{3}|\xi|^{3/2} + \tfrac{1}{4}\pi]\, e^{-iEt/\hbar}, \qquad \text{large negative } \xi. \qquad (6.8.6)$$

The boundary condition gives $\sin[(m/3g\hbar)(2E/m)^{3/2} + \frac{1}{4}\pi] = 0$ which has the solutions

$$E_n = (9\pi^2 mg^2\hbar^2/8)^{1/3}(n - \tfrac{1}{4})^{2/3}, \qquad n = 1, 2, \ldots. \qquad (6.8.7)$$

The conditions of the approximation imply that $n \gg 1$, i.e., the state is highly excited and 'classical' according to common versions of the Correspondence Principle. Indeed, (6.8.6) is the 'semiclassical' wavefunction, being related to the WKB solutions

$$\psi^{\pm}_{\mathrm{WKB}}(x, t) = A_{\pm}(-\xi)^{-1/4}\, e^{i(\pm\frac{2}{3}(-\xi)^{3/2} - Et/\hbar)} \qquad (6.8.8)$$

as follows (we choose $A_{\pm} = \frac{1}{2}A\, e^{\mp i\pi/4}$):

$$\psi_n = \psi^{+}_{\mathrm{WKB}} + \psi^{-}_{\mathrm{WKB}}, \qquad n \gg 1. \qquad (6.8.9)$$

Needless to say, in view of our previous remarks in this chapter, the state (6.8.6) actually implies ultra-quantum rather than classical behaviour. To begin with, for each n, ψ_n has a set of nodes at the points (apart from $x = 0$)

$$x_{n'} = (9\pi^2\hbar^2/8m^2g)^{1/3}[(n - \tfrac{1}{4})^{2/3} - (n' - \tfrac{1}{4})^{2/3}] \qquad (6.8.10)$$

where n' is an integer with $1 \ll n' < n$. Notice that the fringe spacing, $x_{n'} - x_{n'+1}$, is independent of n, i.e., it is the same for all levels, a point noted in a similar context by Berry (1982), but that it depends on the mass. Such a static distribution of nodes is quite incompatible with the constant, continuous acceleration of a classical object.

The motion of the ensemble of particles associated with the wave (6.8.6) according to the causal interpretation is, however, compatible with the existence of nodes. Since the space-dependent part of the wavefunction is real, $\nabla S = 0$ between the nodes and thus each particle is at rest. The energy balance in the system is given by the Hamilton–Jacobi equation which reduces to

$$E_n = V + Q = mgx + (E_n - mgx). \tag{6.8.11}$$

All the energy is stored in (classical plus quantum) potential energy. Classically, the particle is only at rest at a turning point (where $E = V(x)$), and we are far from such points in our asymptotic approximation. Indeed, there is no limit in which (6.8.6) leads to finite motion of a particle.

To study further this question, suppose we fix the position of the classical turning point, H_n. Then from (6.8.7) it is clear that $n \propto m$. In the macroscopic limit the quantum number increases and from (6.8.10) the fringe separation decreases. Let us put in some numbers. For a particle of mass 1 g in the earth's gravity ($g \sim 10^3$ cm s^{-2}) and $H_n = 100$ cm we find for the number of the state $n \sim 10^{32}$ and the distance between nodes $\sim 10^{-19}$ cm. This corresponds to a de Broglie wavelength $\lambda \sim 10^{-30}$ cm. The magnitude of these quantities is sometimes cited (e.g., ter Haar (1964, p. 88)) as indicating the enormous precision with which the motion of macroscopic bodies follows the laws of classical mechanics. We would cite it instead as evidence that the state in question is simply not relevant to the classical domain. Of course, these results are simply further instances of the conclusions already drawn from our study of the stationary states of a particle in a box and a harmonic oscillator. As remarked for the latter (§6.6.1), one can average away the fine nodal structure either mathematically or by physical interaction with a further system, but quantum mechanics nonetheless predicts that states such as (6.8.6) can exist and according to our treatment they have no classical analogue.

Let us examine the implications of these results for WEP (*b*). Obviously, for the trajectories calculated from the classical solutions (6.8.8) it is valid. For the wavefunction (6.8.6) on the other hand, the quantum force exactly balances the gravitational force, and so $\ddot{x} = 0$ for all particles, independently of their mass. Thus, this solution is formally consistent with WEP (*b*) in the sense that all particles are affected in the same way, even though this is not a state of constant finite acceleration. Nevertheless, if an experiment were performed to determine the position of the particle a set of fringes would be observed whose separation depends on the mass.

To find a solution for which the actual particle motion does vary with mass in the high quantum number limit, we can superpose two solutions of the form (6.8.6) corresponding to energies n and n', say. The particle is no longer stationary; the quantum potential, total energy, and momentum vary in time with frequency $(E_n - E_{n'})/\hbar$, and this frequency, moreover, depends on the mass (since $E \propto m^{1/3}$). It is therefore possible to construct superpositions of high quantum number states of a particle in a gravitational field for which the motion depends sensitively on the mass. The classical symmetry is therefore not recovered. Of course, the physical preparation of states such as (6.8.6) and their superposition is naturally very difficult, since they are fragile against perturbations.

As with all problems pertaining to the classical limit, the relevant question is the behaviour of the quantum state. We have found states for which the universality of free fall is invalid in the domain in which it is usually thought to apply. To recover the classical accelerating particle we must construct a packet by integrating over a range of energies.

As an example we cite the freely falling Gaussian packet, with the barrier at $x = 0$ removed (§4.8). The acceleration of a particle in such a packet is given by (4.8.8):

$$\ddot{x} = -g + (x_0\hbar^2/4m^2\sigma_0^4)[1 + (\hbar t/2m\sigma_0^2)^2]^{-3/2}, \tag{6.8.12}$$

where x_0 is the initial position. The following features of the motion may be deduced from this law. For a particle in the front (rear) of the packet, $x_0 > 0$ (<0), the deceleration is less (greater) than g because the force due to dispersion opposes (enhances) that due to gravity. It is interesting to see to what extent Aristotle's law of inertia (that heavier bodies fall faster) is valid. Consider a fixed instant t and let $x_0 > 0$. Then it is obvious from (6.8.12) that the acceleration of a particle of mass $M > m$ may be greater or less than that of m depending on the size of the mass ratio. Thus it may happen that heavier objects fall more *slowly*. A particle at the centre of the packet ($x_0 = 0$) will always pursue the classical trajectory. To obtain the classical limit for the entire ensemble we pass to the case where the spreading may be neglected (§6.6.2) and the quantum force becomes negligible.

The deviations from the universality of free fall described above for particles in highly excited states in a gravitational field have apparently not been subjected to experimental verification. Where the free fall of electrons, neutrons and molecules has been tested the equivalence principle (WEP (*b*)) has been confirmed (Misner *et al.*, 1973, p. 13; Koester, 1976). Yet one could envisage constructing a mass-analyser in which a mixture of (noninteracting) particles of different mass but the same initial wavefunction (yielding an identical position distribution and initial velocity, e.g., a Gaussian packet at rest), which classically cannot be distinguished by gravitational means alone, will be discriminated by quantum effects.

Our theme has been that gravitation and uniform acceleration are locally equivalent in their effects on de Broglie waves. When we consider quantum systems that are otherwise free, the classical symmetry which identifies the equivalence of gravity and inertia with the universality of free fall of mechanical systems is lost. The dynamical origin of the mass-dependent effects may be understood as the outcome of the quantum force acting on particles guided by Schrödinger waves, whose statistical effect is to reproduce the distributions determined by experiment.

6.9 Remarks on the path integral approach

Apart from the algebraic approach of Heisenberg and the wave-theoretical one of Schrödinger, there is a third independent route to quantum mechanics due to Feynman (Feynman, 1948; Feynman & Hibbs, 1965). This rests on the trajectory concept and so may be expected to have some connection with the causal formulation. We present here some remarks concerning this relationship.

Recall that the transition amplitude (Green function or propagator) $G(\mathbf{x}, t; \mathbf{x}_0, t_0)$ connecting two spacetime points is the wavefunction describing a point source at time $t = t_0$:

$$\lim_{t \to t_0} G(\mathbf{x}, t; \mathbf{x}_0, t_0) = \delta(\mathbf{x} - \mathbf{x}_0) \tag{6.9.1}$$

(an example was given in §6.3). It is the quantum generalization of the classical concept of attaching an action function to an entire spacetime path. Feynman provides a technique for computing G from the classical Lagrangian $L(\mathbf{x}, \dot{\mathbf{x}}, t) = \frac{1}{2}m\dot{\mathbf{x}}^2 - V(\mathbf{x}, t)$. One considers all the paths $\mathbf{x}(t)$ that may link the two points (see Fig. 6.3) and associates with each an amplitude, or elementary wave, $e^{(i/\hbar)S[\mathbf{x}(t)]}$, where $S[\mathbf{x}(t)]$ is the classical action function,

$$S[\mathbf{x}(t)] = \int_{\mathbf{x}_0, t_0}^{\mathbf{x}, t} L \, dt, \tag{6.9.2}$$

evaluated along the path $\mathbf{x}(t)$. These tracks are therefore called 'classical paths' although only one (or at most a subset) will minimize the integral (6.9.2). The actual classical path, if it exists, will be denoted $\bar{\mathbf{x}}(t)$ and $S[\bar{\mathbf{x}}(t)]$ will be denoted S_{cl}. Feynman attaches equal weights to each of the paths and constructs the propagator by linearly superposing the elementary amplitudes:

$$G(\mathbf{x}; t; \mathbf{x}_0, t_0) = N \int e^{(i/\hbar)S[\mathbf{x}(t)]} \, d[\mathbf{x}(t)]. \tag{6.9.3}$$

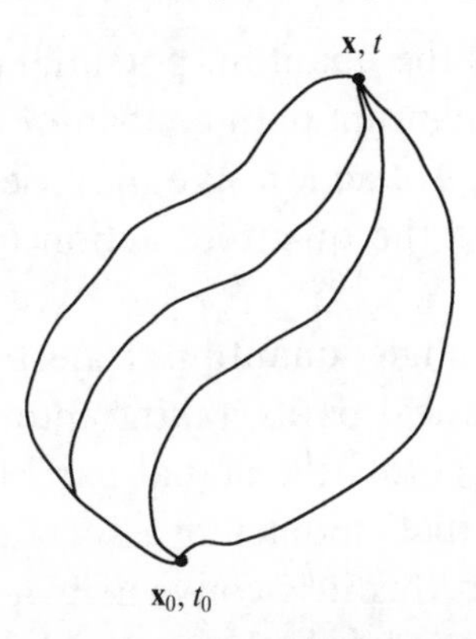

Fig. 6.3 Possible paths connecting two spacetime points. If they exist, at least one is the actual classical path (minimum classical action) and one is the actual quantum path (minimum quantum action).

Here N is a normalization constant and the notation implies that one sums (integrates) over all the paths $\mathbf{x}(t)$ linking the two points.

It is readily verified that the expression (6.9.3) satisfies the Schrödinger equation in the variables $\mathbf{x}, t$ and the complex conjugate wave equation in the variables $\mathbf{x}_0, t_0$ (for $\mathbf{x} \neq \mathbf{x}_0, t \neq t_0$) and hence that, for an arbitrary initial wavefunction $\psi_0(\mathbf{x})$, the solution at time t is given by

$$\psi(\mathbf{x}, t) = \int G(\mathbf{x}, t; \mathbf{x}_0, t_0)\psi_0(\mathbf{x}_0)\,\mathrm{d}^3x_0 \tag{6.9.4}$$

(Huygens' principle).

The path concept introduced above arises at a different level of theory from that of the quantum theory of motion where the trajectory is employed alongside the wave equation rather than as a means of deriving it. Nevertheless, there is a connection with Feynman's method and to see what it is we first observe that of all the paths envisaged as linking two spacetime points, one of them will be the actual trajectory pursued by the quantum particle according to the guidance formula $\dot{\mathbf{x}} = \nabla S/m$, where S is the phase of the Green function ($G = R\,\mathrm{e}^{\mathrm{i}S/\hbar}$). We assume that the pair of points in question are indeed joined by some physical track; as in classical mechanics, not all points can be so connected. For example, in the two-slit experiment a path starting in one of the slits and crossing the axis of symmetry of the apparatus is possible for Feynman but forbidden for us (cf. §5.1). We shall refer to the causal trajectory, when it exists, as the 'quantum path'. It is that path for which the *quantum* action

$$S(\mathbf{x}, t; \mathbf{x}_0, t_0) = \int_{\mathbf{x}_0,t_0}^{\mathbf{x},t} (\tfrac{1}{2}m\dot{\mathbf{x}}^2 - V - Q)\,\mathrm{d}t \tag{6.9.5}$$

is an extremal, where Q is the quantum potential constructed from R. In the path integral (6.9.3) the quantum path contributes, as do all paths, via an, in general, nonextremal *classical* action. We may view the Feynman procedure as a method for obtaining the quantum action from the set of all classical actions.

Correspondingly, the single quantum trajectory may be treated as a 'superposition' of the classical paths, taking into account their interference. This means that we can express the actual particle momentum at the point $(\mathbf{x}, t)$ in terms of the 'virtual' momenta associated with each of the paths crossing that point. To do this it is convenient to write (6.9.3) as a discrete sum, $G = N \sum_i e^{iS_i/\hbar}$, where i labels the path. Then, at the point $(\mathbf{x}, t)$,

$$\nabla S = N^2 R^{-2} \left\{ \sum_i \nabla S_i + \sum_{i \neq j} \nabla S_i \cos[(S_i - S_j)/\hbar] \right\}. \tag{6.9.6}$$

This formula expresses the momentum of the quantum particle as a sum of the 'classical' average of the momenta ∇S_i attached to each path and terms due to interference between the elementary waves. In general, the momentum ∇S_i corresponding to the classical action evaluated along the quantum path i is not equal to ∇S.

Now it is apparent that the set of paths potentially available to a quantum mechanical particle according to the causal interpretation is but a subset of all Feynman paths. The quantum path between two points is unique once the initial momentum is fixed, whereas there are an infinite number of Feynman paths having the same initial gradient. The latter are not therefore 'physically real' and the causal interpretation does not apparently throw any light on the problem of why Feynman's technique works. Formula (6.9.3) is analogous to the representation of a wave as a Fourier sum, and (6.9.6) to the corresponding decomposition of the momentum into a sum of the average momentum connected with the plane waves and terms arising from their interference. There, likewise, the individual plane waves have only a mathematical significance. It has been shown that if one performs a sequence of measurements designed to determine whether a particle moves along a particular spacetime path, the particle follows with probability unity the path that is being checked (Aharonov & Vardi, 1980). But, as shown in detail in Chap. 8, such a process involves an interaction of the system of interest with further systems which brings about a transformation of the original wave-function. It does not reveal what the trajectory would have been had one not performed the measurement and hence does not demonstrate the independent physical reality of a Feynman path.

According to Feynman we approach the classical limit when the classical actions in the sum (6.9.3) are very large in comparison with $\hbar$. For then the action functions vary rapidly with position and neighbouring ones will tend to cancel. An exception occurs for the stationary action S_{cl} which by definition is unchanged by small variations in the path. Thus

$$\lim_{\hbar \to 0} G = A\, e^{iS_{cl}/\hbar} \tag{6.9.7}$$

where A is a slowly varying function of position. Clearly, the quantum potential constructed from A is negligible and $\nabla S \approx \nabla S_{cl}$. Feynman's criterion for the classical limit is thus correct but, of course, it only applies to the Green function. For an arbitrary wavefunction (6.9.4) substitution of (6.9.7) does not yield the classical limit, for this function still displays interference. The universal criterion remains $Q \to 0$.

Finally, we note an interesting picture of the quantum path implied by the path integral method. For an infinitesimal time interval ε, the propagator is just the classical wavefunction:

$$G(\mathbf{x}, t_0 + \varepsilon; \mathbf{x}_0, t_0) = A(\varepsilon)\, e^{(i/\hbar)S_{cl}(\mathbf{x}, t_0 + \varepsilon; \mathbf{x}_0, t_0)}. \tag{6.9.8}$$

In this interval the particle pursues a classical trajectory in the prescribed external potential. But a finite path may be decomposed into many such infinitesimal steps, the net propagator being obtained by successive applications of Huygens' construction (6.9.4). It follows that the quantum path, in particular, may be decomposed into a sequence of segments along each of which the classical action is a minimum (Fig. 6.4). The 'jump' onto a new

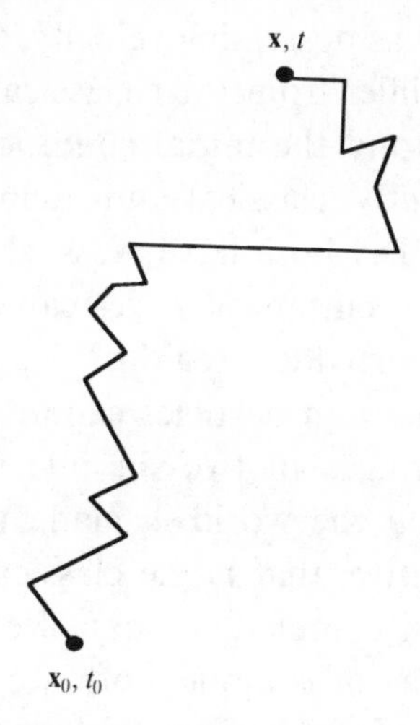

Fig. 6.4 A quantum trajectory may be decomposed into a sequence of elementary classical extremals which smoothly fit together in the limit $\varepsilon \to 0$.

track at the end of each segment is caused by the interference of the elementary waves vibrating there.

6.10 What is the connection between the quantum world and the classical world?

6.10.1 The principle of linear superposition

The classical world view is based on the objective existence of distinct material systems occupying definite portions of space at each instant. Although the precise control of initial conditions is impossible in practice, the classical theory provides a complete theoretical deterministic description of individual events in which statistics is secondary to dynamics and arises from ignorance of the actual facts rather than irreducible contingency. Whatever the merits of this conception, and it is certainly open to objection, we accept its validity in a certain domain of experience and enquire if quantum mechanics can be made organically consistent with it, rather than simply accepting it as 'given' in some ultimate Bohrian sense.

The classical and quantum worlds are connected in the causal interpretation by virtue of an unambiguous relation between the primitive notions of 'state' in the two theories. The quantum state is defined by $\psi(\mathbf{x}, t)$ and $\mathbf{x}(t)$ which evolve as a unit in a deterministic manner. When the former has no influence on the latter the classical and quantum states coincide. It is on the basis of this connection that we can establish clear criteria for the coincidence of all other aspects of the two theories – equations of motion, mean values, particle properties, and so on. Both theories permit a complete deterministic description of individuals in their respective domains of applicability. The quantum particles may be conceived as possessing velocity, energy, angular momentum etc., but these quantities differ from their classical counterparts in that they depend on something beyond the initial conditions and external force: the quantum state. Consequently, classical and quantum particles differ not in their intrinsic properties (they both have mass, charge and so on) but in the character of their motion: quantum particles can go where classical particles cannot (into 'classically forbidden regions', e.g., tunnelling) and classical particles can go where quantum particles cannot (into 'quantally forbidden regions', e.g., the dark fringes in two-slit interference). According to the quantum theory of motion the world is made up of waves and particles. Where the waves are inactive, that is the classical domain. The description of states is as theoretically complete as that of classical mechanics but it is not exhausted by $\mathbf{x}(t)$ alone or ψ alone; both are needed.

Note the implications of this. If we wish to treat classical systems as special cases of quantum ones the primitive type of wave embodied in the

Hamilton–Jacobi equation is not merely fictitious but physically real, albeit ineffective. All matter in this view has a wave aspect. The quantum potential characterizes whether this aspect will be manifested in motion.

The conventional interpretation has no definition of the state of a system independently of a specification of the experimental arrangement employed to produce or investigate that state. The ψ-function merely embodies knowledge gleaned from experiment. Matter acquires 'position' only when one performs certain laboratory operations. It is for this reason that one cannot logically deduce the classical theory of matter from the pure quantum mechanical formalism as a limiting case. A theory of substantial matter cannot emerge from a statistical theory of observation.

This circumstance is a powerful argument in favour of the causal interpretation. Most physicists would surely accept the assertion that a billiard ball has a fairly well-defined location at each instant, even though we cannot control it with absolute precision. They do not believe this to be so only when the ball is illuminated and observed to have a quasi-definite position. Now if quantum mechanics is a universal physical theory it should be possible to assign to the billiard ball a wavefunction. If we assume with the conventional view that ψ provides an exhaustive physical description of the individual system, here a billiard ball, then it is legitimate to ask which feature or features of ψ it is that describe the fact of localization of the ball independently of observation. Clearly, there are no such features. According to the orthodox view ψ merely tells us the relative probabilities of the ball being found to have this or that position, not that it *has* a position and what that position is. This is so regardless of the form of ψ. The fact that the centre of a packet moves along a well-defined orbit as if it were a particle of mass m does not demonstrate that there *is* such a particle pursuing that orbit. It is only in the causal interpretation that we can consistently claim that the classical-like motion of a packet when dispersion may be neglected is, in fact, the approach to the classical limit, since one starts by assuming the particle trajectory.

Therefore, on its own, ψ does not provide a complete description of a real matter of fact for the individual case. Yet we feel that a billiard ball does indeed have a quasi-definite location. This leads to the conclusion that if we wish to retain the classical notion of 'reality' in the classical domain *and* we require that quantum mechanics should be applicable in this domain, we must supplement the wavefunction by further variables which specify location. But if we accept the incompleteness of the ψ-description and the necessity of a further postulate relating to position in the case of a billiard ball, why are we so sure that ψ exhausts the physical description of an electron? After all, the microscopic and macroscopic levels are closely connected and it is a matter of convenience how we distinguish them. The classical and quantum levels are too closely interwoven for the notion of an unambiguous 'cut' between them to be tenable. This indeed is recognized in Bohr's approach, for he applied quantum mechanics to shutters – surely 'classical objects'. The ambiguity between the classical and quantum levels is apparent also in the semiclassical approximation which applies the trajectory concept in a non-classical domain and would lose all meaning if we maintained the position that 'quantum matter' is something fundamentally distinct from 'classical matter'.

An important conclusion of these considerations is that the independent reality of the instruments by which we come to know the world (lab. bench, electron source, photographic plate, eyes, brains) cannot be accounted for by our fundamental theory of matter if we start from the assumption that ψ is in itself the most complete description of an individual that is in principle possible. The only reasonable assumption is that matter possesses objective properties at all levels and under all circumstances. This point is taken up again in Chap. 8.

From this conceptual vantage point we have appraised some of the mathematical procedures proposed in the literature for arriving at the classical formalism from the quantum one. Our theme has been that while this end may naturally be achieved by the relative variation of the parameters on which ψ depends, the success of the mission should be judged by a state-dependent criterion that is consistent with the continued finiteness of $\hbar$.† The obvious universal criterion for gauging when quantum states approach classical states is to investigate the behaviour of the quantum potential. Notice that one's acceptance of the mathematical condition $Q \to 0$ as the definition of the classical limit is closely connected with one's view of the conceptual status of quantum mechanics. The usual view which claims that physical quantities ('observables') are only well defined when the system is in an eigenstate would give a highly nonclassical interpretation to states for which $Q = 0$.

On first acquaintance with quantum mechanics one gains the impression that its formal apparatus – linear noncommuting operators acting on functions in Hilbert space – has no parallel in classical physics. The mathematics, it seems, reflects the fundamentally new physical content of a theory which must account for an apparently discrete and nonlocal world

† In view of the parameter-dependence of ψ, it would be interesting to know if there are counterexamples to the assertion that nonrelativistic quantum mechanics is obtained when we let $c \to \infty$ in the relativistic theory.

governed by laws of intrinsic contingency. On closer inspection this feeling dissipates somewhat when it is realized that many aspects of the formalism are anticipated by, or at least consistent with, classical conceptions. Thus the conceptual structure of classical mechanics is not incompatible with the use of Hilbert space operators and probability amplitudes and these notions are valuable in formulating certain aspects of the theory (for examples see Koopman (1931), Schiller (1962a,b), Prigogine (1962) and George & Prigogine (1979)). But more importantly, the presentation of classical and quantum mechanics as instances of a generalized Hamilton–Jacobi theory suggests that where they basically differ as theories of matter and motion is that the quantum theory admits a principle of linear superposition. It is this that is responsible for the emergence of a new kind of force and the nonclassical feature of the context dependence of particle motion. The 'wave equation' of classical mechanics is nonlinear (cf. §2.6):

$$i\hbar\frac{\partial\psi}{\partial t} = \left(\frac{-\hbar^2}{2m}\nabla^2 + V - Q\right)\psi, \tag{6.10.1}$$

where Q is the quantum potential. If we enquire how one may modify classical mechanics so as to obtain a linear equation in ψ, the simplest (but not unique) possibility is to add $Q\psi$ to the right hand side of (6.10.1). Planck's constant then enters for dimensional reasons in order that we may define Q.

Hilbert space is a structure naturally suited to a theory which applies a linear superposition principle to continuous functions. Having obtained a linear vector space, one may interpret the physical content of the Hermitian operators acting in it in terms of classical expressions for the corresponding quantities and modifications thereto (cf. §3.5). Conversely, the use of linear operators is an indirect expression of the necessity of modifying the classical definition of physical properties to accommodate the superposition principle.

It is sometimes claimed that the circumstance that $\psi(\mathbf{x})$ depends on only half the phase space variables used in classical mechanics is an expression of 'wave-particle duality' and indicative of the impossibility of a trajectory interpretation. But we see rather that the Hilbert space theory follows from a linearization of a field-theoretic formulation of classical statistical mechanics, where similarly the field ψ (i.e., ρ, S) is a function of just $\mathbf{x}$. According to the causal interpretation indeterminacy is not a consequence of the finiteness of Planck's constant but arises for the same reason as in classical mechanics (lack of information concerning the real situation which is in itself well defined). The impossibility of controlling the initial particle coordinates without disturbing the wavefunction is a consequence of interactions being governed by the linear wave equation. *The role of Planck's constant is to allow the linearization of the classical wave equation.* The novelty of quantum mechanics lies not just in the finiteness of $\hbar$ but in the specific combination of this with the principle of linear superposition of wavefunctions that is embodied in the quantum potential. If for whatever reason the latter may be neglected Planck's constant will play no role in the dynamics, but it remains finite. (Naturally one can envisage a theory admitting linear superposition of functions that does not involve $\hbar$, classical electromagnetic theory being a case in point. Actually, Maxwell's equations correspond to a kind of 'first quantization' – see §12.6.)

In passing from the nonlinear to the linear wave equation more stringent conditions are imposed on ψ (that it be continuous, finite and single-valued) than are required in the classical case, as befits the mathematical representation of a real physical field. Ironically, it is the classical wavefunction, which is associated with a definite conception of matter, that has a purely probabilistic and descriptive significance.

When we think of classical physics as conceptually and mathematically contained within quantum physics we are right in the sense that the latter contains more than the former ($\hbar$ or its vehicle, Q). But because the ψ-function retains mathematical characteristics in the limit that are not requirements in classical theory, we may also think of quantum mechanics as a special case of classical ensemble theory. Taking into account also the possibility that we may not be able to arrange $Q \to 0$ for *any* Schrödinger solution propagating in some potential V, it seems that classical and quantum mechanics are, in some instances, disparate theories.

We conclude that the quantum world and the classical world are both aspects of a single, undivided universe. One continuously passes from one regime to the other by varying the effectiveness of the quantum potential. This is the Correspondence Principle of the quantum theory of motion.

6.10.2 Remarks on chaos

A fundamental feature of the quantum potential theory is the extremely sensitive dependence of quantal motions on the initial conditions. A useful indicator of the degree of instability is the equation for the relative acceleration of neighbouring trajectories (Misner *et al.*, 1973, p. 272). Consider a family of trajectories $\mathbf{x} = \mathbf{x}(t, \mathbf{x}_0)$ where $\mathbf{x}_0$ labels which trajectory and t a point on a trajectory. Define the ith component of the separation vector as the matrix function $n_{ij}(t, \mathbf{x}_0) = \partial x^i/\partial x_0^j|_t$ for each $j = 1, 2, 3$. Then

$$m\left.\frac{\partial^2 n_{ij}}{\partial t^2}\right|_{\mathbf{x}_0} + \left.\frac{\partial^2 (Q+V)}{\partial x^i\,\partial x^k}\right|_{\mathbf{x}=\mathbf{x}(t)} n_{kj} = 0. \qquad (6.10.2)$$

The relative acceleration (the first term) is brought about by the 'tide-producing' relative force (the second term). Infinitesimal differences in initial coordinates can be rapidly amplified so that in the long run the motion is unpredictable, although deterministic. This resembles in some ways the chaotic regime that it is now recognized many classical systems evolve into (Lighthill, 1986) and we conclude with some tentative remarks on the relation between the two types of instability.

In the conventional approach to quantum mechanics one obviously cannot define the analogue of classical chaos in terms of the behaviour of particle orbits. One deals instead with the evolution of the mean values of operators or the structure of spectra or nodes which in the absence of a theory of matter are essentially mathematical concepts rather than objective properties. A significant feature of these studies is that the chaotic effects in some classical systems tend to be suppressed in the corresponding quantum systems (Blümel & Smilansky, 1990). Because classically chaotic systems will presumably be the limit of some quantum ones, Berry (1987) makes the following definition: 'Quantum Chaology is the study of semiclassical, but non-classical, behaviour characteristic of systems whose classical motion exhibits chaos.' This is reasonable since semiclassical wavefunctions are built from classical ones (§6.4.2). An advantage of the quantum theory of motion is that it can go beyond this definition and propose conditions for the emergence of quantum chaos in the same terms as used in classical mechanics, but without reference to whether the classical counterparts exhibit chaos. More importantly, we can speak clearly of what it is physically that becomes chaotic (particle orbits).

This may throw light on a possible dynamical origin of the quantal suppression of classical chaos as stemming from the single-valuedness of the wavefunction. We have often emphasized the essential difference between quantum and classical ensembles, that an initially single-valued quantal congruence connected with a pure state ψ will remain single-valued for all t. The motion in phase space $(\mathbf{x}, \mathbf{p})$ is confined to the moving subspace $\mathbf{p} = \nabla S(\mathbf{x}, t)$ whereas the classical phase point may wander at will so that the mapping into configuration space yields clusters of orbits that generically cross. Although the quantum mechanical system may not possess constants of the motion (in the causal sense that, for example, the total energy including Q is not conserved), the global structure of the trajectories is stringently organized by the requirement of forming a single-valued congruence (something approaching the classical complexity may perhaps be achieved using mixed states). And naturally, if the quantum state is suitably chosen, the orbits may have no features at all in common with their classical counterparts so the latter cannot be reached by any mathematical limiting process applied to the wavefunction. As an example, consider a classically chaotic system such as the two-dimensional stadium potential. It is found that the stationary states of the corresponding quantum mechanical problem display irregularities in their nodal patterns and probability densities (McDonald & Kaufman, 1979; Gutzwiller, 1990, Chap. 15) and yet the space parts of these states are real so the quantum system is, in fact, highly regular in that all particles in the ensemble are at rest!

The instability of the quantum ensemble derives, in general, from the complexity of the quantum potential. In contrast classical instabilities occur for relatively simple potentials (e.g., bounded planes of various shapes). In neither case is the complex behaviour a consequence of our ignorance of what are, in fact, well-defined initial coordinates, i.e., that our knowledge is at best coarse-grained, or due to an inadequacy in the law of motion. It reflects a basic dynamical instability inherent in the latter, or rather in (6.10.2). Classifying the quantum paths as closed (stable and unstable), integrable, ergodic and so on, and defining analogues of Kolmogorov entropy etc., is largely uncharted territory.

7

Many-body systems

7.1 General properties of many-body systems

7.1.1 Definition of individual system

So far we have studied the properties of single-body systems in prescribed external potentials. An individual physical system comprises a wave and a particle, both evolving in three-dimensional Euclidean space. Of course, isolating a system and representing the influence of other matter by an 'external potential' is an approximation to the real state of affairs in that we neglect the source of the potential. As a fundamental theory of matter quantum mechanics should apply to a closed many-body system (and ultimately to the universe as a whole) and reduce to a theory of systems of a few degrees of freedom as a special case under conditions where it is legitimate to neglect the 'rest of the universe'. The extension of the quantum theory of motion to many-body systems is straightforward, although it displays some striking features not evident in the one-body case. To begin with, we define an *individual n-body system* as comprising:

(*a*) A wavefunction $\psi = \psi(\mathbf{x}_1, \ldots, \mathbf{x}_n, t)$ defined in a $3n$-dimensional configuration space in which $\mathbf{x}_1, \ldots, \mathbf{x}_n$ provide a set of rectangular Cartesian coordinates.

(*b*) A set of n point particles pursuing trajectories $\mathbf{x}_i(t)$, $i = 1, \ldots, n$, in three-dimensional Euclidean space. A single configuration space trajectory is equivalent to n particle trajectories in Euclidean space.

When we speak of a 'many-body system' we mean then a *single* wavefunction together with a set of particles. There is, in general, no wave associated just with each particle individually (see §7.2 for the special case when this does happen). Notice that we have introduced the notion that an *individual physical system resides in a multidimensional (configuration) space.* While the particles each move in 3-space, the guiding wave is, in general, irreducibly defined in $3n$-space. Since we conceive of the wave as a physical influence on the particles,[†] *we ascribe to configuration space as much physical reality as we do to three-dimensional Euclidean space in the one-body theory*. This is a radical step, but it is a necessary one if we wish to extend the quantum theory of motion consistently to embrace many-body problems.

This definition should be compared with the usual approach. In the two-body case, for example, one forms the direct product of the Hilbert spaces $\mathscr{H}$ associated with each single-body system and defines the state $|\psi\rangle$ to be a vector in $\mathscr{H} \otimes \mathscr{H}$. Our wavefunction above is then a representative of this vector: $\psi(\mathbf{x}_1, \mathbf{x}_2) = \langle \mathbf{x}_1, \mathbf{x}_2 | \psi \rangle$. But in the general, entangled (see §7.2), case, one still tends to speak of 'two systems' as if they can be distinguished and have some identity or separation (e.g., the two electrons in the hydrogen molecule; see §7.6) even though all we conventionally have available to describe the composite is $|\psi\rangle$ which implies there is just *one* 'system' in configuration space. As we shall see, it is a feature of the quantum theory of motion that it makes acceptable and consistent the informal pictures physicists tend to slip into in this context by introducing, beyond $|\psi\rangle$, two particles. It is the combination of the three entities that constitutes the 'two-body system'. With this proviso, we shall continue to use the conventional terminology.

7.1.2 Equations of motion

Most of the formulae and general results set out in Chap. 3 for the one-body theory remain valid when $\mathbf{x}$ is replaced by $\mathbf{x}_1, \ldots, \mathbf{x}_n$, so we need only give an outline of the theory of the wave and particle equations of motion.

The equation of motion of the wavefunction is the Schrödinger equation:

$$i\hbar \frac{\partial \psi}{\partial t} = \left[\sum_{i=1}^{n} (-\hbar^2/2m_i)\nabla_i^2 + V(\mathbf{x}_1, \ldots, \mathbf{x}_n, t) \right] \psi, \tag{7.1.1}$$

where m_i is the mass of the ith particle and V is the classical potential energy which includes interparticle and external potentials. The value of ψ at time t is uniquely determined once we specify its value at $t = 0$: $\psi_0(\mathbf{x}_1, \ldots, \mathbf{x}_n) = \psi(\mathbf{x}_1, \ldots, \mathbf{x}_n, 0)$.

The quantum theory based on (7.1.1) can be reformulated in terms of the real configuration space functions $R(\mathbf{x}_1, \ldots, \mathbf{x}_n, t)$ and $S(\mathbf{x}_1, \ldots, \mathbf{x}_n, t)$ where $\psi = R\,e^{iS/\hbar}$. Substituting this expression into (7.1.1) and separating into real

[†] As real as Faraday's electromagnetic field in that it brings about observable changes in the configurations of matter.

and imaginary parts yields two field equations:

$$\partial S/\partial t + \sum_{i=1}^{n} (\nabla_i S)^2/2m_i + Q + V = 0, \tag{7.1.2}$$

$$\partial R^2/\partial t + \sum_{i=1}^{n} \nabla_i \cdot (R^2 \nabla_i S/m_i) = 0, \tag{7.1.3}$$

where

$$Q = \sum_{i=1}^{n} -\hbar^2 \nabla_i^2 R/2m_i R \tag{7.1.4}$$

which will be interpreted below as the many-body quantum potential energy. These two field equations have the same content as (7.1.1) if we impose on R and S the boundary and continuity conditions etc. corresponding to those satisfied by ψ. In particular, the requirement that ψ be a single-valued function may be expressed as

$$\sum_{i=1}^{n} \oint_C \nabla_i S \cdot d\mathbf{x}_i = nh, \qquad n \in \mathbb{Z}, \tag{7.1.5}$$

where C is a circuit in configuration space passing through regions where $\psi \neq 0$. One may interpret (7.1.5) as stating that n wavefronts $S =$ constant come to an end somewhere within C. The functions S and $\nabla_i S$, $i = 1, \ldots, n$, are undefined in nodal regions (where $\psi = 0$).

In a straightforward generalization of the one-body theory we now treat (7.1.2) as a generalized Hamilton–Jacobi equation for a system of n particles, with a set of momentum fields

$$\mathbf{p}_i(\mathbf{x}_1, \ldots, \mathbf{x}_n, t) = \nabla_i S(\mathbf{x}_1, \ldots, \mathbf{x}_n, t), \qquad i = 1, \ldots, n, \tag{7.1.6}$$

and velocity fields $\mathbf{v}_i = \mathbf{p}_i/m_i$. The particle trajectories are the solutions $\mathbf{x}_i(t)$ to the following system of n simultaneous differential equations

$$\begin{aligned}\frac{d\mathbf{x}_i}{dt} &= \mathbf{v}_i(\mathbf{x}_1(t), \ldots, \mathbf{x}_n(t), t)\\ &= (1/m_i)\nabla_i S(\mathbf{x}_1, \ldots, \mathbf{x}_n, t)|_{\mathbf{x}_j = \mathbf{x}_j(t)}, \qquad i, j = 1, \ldots, n.\end{aligned} \tag{7.1.7}$$

In order to solve for any one trajectory $\mathbf{x}_i(t)$ we have to specify the initial positions $\mathbf{x}_{io}$ of *all* the particles.

That Q may be treated in respect of each particle motion as an external potential on the same footing as V is justified as follows. Applying the operator ∇_i to (7.1.2) and rearranging yields

$$\left[\partial/\partial t + \sum_j (1/m_j)\nabla_j S \cdot \nabla_j\right] \nabla_i S = -\nabla_i (Q + V). \tag{7.1.8}$$

Identifying $\dot{\mathbf{x}}_i$ with $\nabla_i S/m_i$, (7.1.8) becomes

$$m\ddot{\mathbf{x}}_i = -\nabla_i(Q + V)|_{\mathbf{x}_j=\mathbf{x}_j(t)}, \qquad i, j = 1, \ldots, n. \tag{7.1.9}$$

Note that $\equiv \mathrm{d}/\mathrm{d}t = \partial/\partial t + \sum_i \dot{\mathbf{x}}_i \cdot \nabla_i$ is the time rate of change with respect to a moving point in configuration space.

Substituting (7.1.7) into (7.1.2) the total energy of the particles is given by

$$\begin{aligned} E &= -\partial S/\partial t|_{\mathbf{x}_j=\mathbf{x}_j(t)} \\ &= \sum_i \tfrac{1}{2} m_i \dot{\mathbf{x}}_i^2 + Q(\mathbf{x}_1(t), \ldots, \mathbf{x}_n(t), t) + V(\mathbf{x}_1(t), \ldots, \mathbf{x}_n(t), t). \end{aligned} \tag{7.1.10}$$

This is not, in general, a sum of (kinetic plus potential) energies of each individual particle, even in the case where V is an additive function ($V = \sum_i V_i(\mathbf{x}_i)$), due to the irreducible character of Q.

Because the set of functions $\nabla_i S$, $i = 1, \ldots, n$, forms a single-valued function of position in configuration space, the system-point configuration space trajectories corresponding to different choices of $\mathbf{x}_{i0}$, $i = 1, \ldots, n$, do not cross. However, since the individual particle trajectories $\mathbf{x}_i(t)$ all lie in the same three-dimensional space, the n projections in 3-space of a single configuration space orbit will, in general, cross. Note that the configuration space trajectories may not pass through nodes (where $\nabla_i S$ is undefined) and hence neither may the individual paths. The total effective potential, $(V + Q)$, will always act so as to ensure that the system point avoids nodes (independently of the probability hypothesis introduced below).

Thus far we have described the dynamics of a single physical system on the assumption that the initial conditions ψ_0 (or R_0 and S_0) and $\mathbf{x}_{i0}$ are precisely known to us. We now relax this assumption and pass to the case of the statistical mechanics of waves and particles in which these conditions are only imprecisely known. Thus, just as was done in Chap. 3, we introduce the concept of a representative ensemble, each element of which is an individual many-body system as defined in §7.1 having one of the set of initial conditions available to the actual system of interest. The frequency with which these elements occur in the ensemble is proportional to the probability that the system of interest has the initial conditions possessed by those elements. The following postulate will be made concerning the distribution of particle positions:

Postulate. The probability that at time t particle 1 lies in the volume element d^3x_1 around the point $\mathbf{x}_1$, particle 2 lies in the volume element

d^3x_2 around the point $\mathbf{x}_2$ etc. is given by

$$R^2(\mathbf{x}_1, \ldots, \mathbf{x}_n)\, d^3x_1 \cdots d^3x_n \qquad (7.1.11)$$

(it is assumed that $R^2 = |\psi|^2$ is a normalizable function).

If (7.1.11) is true at any one time, the conservation law (7.1.3) will ensure its validity at all times. As usual, the probability of *finding* the system in a certain state is a special case of the probability of *being*, to which the postulate refers. And as usual, the assumption that the particle distribution is determined by a function that is also a physical field does not introduce any inconsistency into the theory; the ensemble is a function of the individual, as in classical statistical mechanics.

This deals with the uncertainty in our knowledge of the particle coordinates, given the wavefunction. If the precise form of the latter is unknown we must introduce a density matrix, but we shall not discuss this further at the moment.

7.1.3 Novel properties of the many-body system

The novel features of the quantum theory of motion described in §§3.3, 3.4 are carried over into the many-body theory, but there are some additional characteristics to which we now call attention (Belinfante, 1973; Bohm, 1987).

(i) State dependence

It was emphasized in the one-body theory that the quantum potential acting on the particle is not a preassigned function of the coordinates but is determined by the *quantum state* of the system. Therefore, an infinite set of possible quantum potentials is associated with the same physical situation and Schrödinger equation. This property takes on a further significance in the many-body case for the following reason. In classical physics, an interparticle potential $V(\mathbf{x}_i, \mathbf{x}_j)$ connecting two particles with coordinates $\mathbf{x}_i, \mathbf{x}_j$ is preassigned, i.e., it is uniquely specified just by the nature of the particles under consideration (e.g., if they are charged then V may be a Coulomb interaction). In quantum physics, the interparticle quantum potential $Q(\mathbf{x}_i, \mathbf{x}_j)$ associated with a two-body system is not a uniquely specified function of the coordinates. *The interaction potential is determined by something* (ψ) *external to* (*although dependent on the characteristics of*) *the particles under consideration* (e.g., if they are charged any of an infinite set of possible Qs can be chosen corresponding to different choices of solution to the wave equation). The fact that Q is not a preassigned function, that it is not unique for a given physical problem, is a new feature of the quantum theory of motion and is not anticipated in classical physics. We may say that whereas in classical dynamics the whole is the sum of the parts and their interactions, in quantum mechanics the whole is prior to the parts (particles) and its properties cannot be explained by a kind of superposition of the properties of the parts.

(ii) *Nonlocal connection*

The circumstance that the configuration space wavefunction depends on a single evolution parameter t implies that the state of the n particles is specified at a common time, and that there is a nonlocal connection between them brought about by the classical and quantum potentials. This means the following. First of all, it is evident from the equation of motion (7.1.7) that the instantaneous motion of any one particle depends on the coordinates of all the other particles *at the same time*. A different choice of initial position of one particle implies a different subsequent motion of all the particles. Such correlations in the individual motions have a significant consequence. For, if we disturb part of the system in a localized region of three-dimensional space, the configuration space wave as a whole will respond and consequently all the particles making up the system will be affected *instantaneously*.

Such nonlocal connections are not foreign to classical physics, but there is an additional element here. It is characteristic of classical forces that they fall off with interparticle distance so that at large enough separations particles can be considered to be effectively independent. In this case V becomes a sum of functions each depending on the coordinates of only one particle: $V = \sum_i V_i(\mathbf{x}_i)$. Then $-\nabla_i V = -\nabla_i V_i$. In contrast, the contribution to the total force acting on the ith particle coming from the quantum potential, $-\nabla_i Q$, does not necessarily fall off with distance and indeed the forces between particles may become stronger, even though $|\psi|$ may decrease in this limit. This is because Q depends on the form of ψ rather than on its absolute value. The equation of motion (7.1.9) of the ith particle in a many-body system in the limit of large separations is therefore generically

$$m_i \ddot{\mathbf{x}}_i = -[\nabla_i Q(\mathbf{x}_1, \ldots, \mathbf{x}_n) + \nabla_i V_i(\mathbf{x}_i)]|_{\mathbf{x}_j = \mathbf{x}_j(t)} \tag{7.1.12}$$

and so is a function of all the other particle coordinates.

The three features brought out here – the dependence of each particle orbit on all the others, the response of the whole to localized disturbances, the extension of actions to large interparticle distances – will be referred to as 'nonlocal connection'. Nonlocality is a generic feature of quantum many-body systems and the conditions under which it can be generally expected to occur are discussed in §7.2 and further in relation to the EPR problem in Chap. 11.

7.1.4 Identical but distinguishable particles

We shall define a set of particles to be 'identical' if the intrinsic parameters associated with them (mass, charge, magnetic moment etc.) are the same. In classical physics we would incorporate this feature by requiring that in a symmetric potential $V(\mathbf{x}_1, \ldots, \mathbf{x}_n)$ the Hamilton–Jacobi function $S(\mathbf{x}_1, \ldots, \mathbf{x}_n)$ satisfies symmetry conditions so that the interchange of any two sets of coordinates $\mathbf{x}_i$, $\mathbf{x}_j$ does not change the momentum and energy fields $\nabla_i S$ and $-\partial S/\partial t$. This does not present any conceptual difficulty because, although we cannot distinguish the particles by their intrinsic properties, they are distinguished or labelled by the continuity of their trajectories.

Correspondingly, in the case of the classical statistical mechanics of a single-valued ensemble, we require that the distribution function $R^2(\mathbf{x}_1, \ldots, \mathbf{x}_n)$ be invariant under arbitrary permutations of its arguments.

We may therefore characterize the notion of identity in classical physics by the following symmetry conditions:

$$R(\mathbf{x}_1, \ldots, \mathbf{x}_i, \ldots, \mathbf{x}_j, \ldots, \mathbf{x}_n) = R(\mathbf{x}_1, \ldots, \mathbf{x}_j, \ldots, \mathbf{x}_i, \ldots, \mathbf{x}_n) \tag{7.1.13}$$

$$\left.\begin{aligned} \nabla_k S(\mathbf{x}_1, \ldots, \mathbf{x}_i, \ldots, \mathbf{x}_j, \ldots, \mathbf{x}_n) &= \nabla_k S(\mathbf{x}_1, \ldots, \mathbf{x}_j, \ldots, \mathbf{x}_i, \ldots, \mathbf{x}_n) \\ \partial S(\mathbf{x}_1, \ldots, \mathbf{x}_i, \ldots, \mathbf{x}_j, \ldots, \mathbf{x}_n)/\partial t &= \partial S(\mathbf{x}_1, \ldots, \mathbf{x}_j, \ldots, \mathbf{x}_i, \ldots, \mathbf{x}_n)/\partial t \end{aligned}\right\} \tag{7.1.14}$$

for all $i, j, k = 1, \ldots, n$.

We now propose to adopt precisely the same characterization of identity in the quantum theory. It is easy to see that the conditions (7.1.13), (7.1.14) are consistent with the usual requirement of quantum mechanics that the wavefunction be symmetric or antisymmetric with respect to an arbitrary permutation of its arguments:

$$\psi(\mathbf{x}_1, \ldots, \mathbf{x}_i, \ldots, \mathbf{x}_j, \ldots, \mathbf{x}_n) = \pm\psi(\mathbf{x}_1, \ldots, \mathbf{x}_j, \ldots, \mathbf{x}_i, \ldots, \mathbf{x}_n) \tag{7.1.15}$$

for all $i, j = 1, \ldots, n$. Note that it is a subtle problem to formulate the precise conditions to be imposed on ψ in order to derive the result (7.1.15) – it does not follow from the single requirement that $|\psi|^2$ be permutation invariant (Girardeau, 1965; Leinaas & Myrheim, 1977). It would be interesting to know how far the property of phase symmetry, quite natural from the perspective of the quantum theory of motion, must be assumed.

In the usual formulation of the quantum theory the imposition of symmetry requirements on the wavefunction of a set of identical particles is considered to render the particles 'indistinguishable' in some absolute sense, for there is nothing available to label them individually. In contrast, the situation in the quantum theory of motion is the same as in classical mechanics. The set of identical particles, guided by a wave which has had all identifying or labelling marks removed (by symmetrization or antisymmetrization), are nevertheless distinguishable by their individual histories. The symmetrization or antisymmetrization of the wavefunction has nothing to do with 'indistinguishability', but, in fact, implies *the introduction of forces between the particles* making up the system, which bring about correlations in their motion (see §7.2.3).[†]

Labelling the particles by their trajectories does not contradict the independence of observable properties from those labels. The difference between the classical and quantum cases is that (as with all effects characteristic of the quantum domain) in the latter the symmetrization is made manifest in the correlated particle motions by the quantum potential determined by ψ whereas in the former the only forces acting are due to the classical potential V, which is independent of the Hamilton–Jacobi function. It is the combination of symmetrization (a classical requirement) and the principle of superposition that leads to nonclassical effects (§7.2.3). In particular, the structure of Q for a set of bosons (symmetric ψ; we neglect spin here) is distinct from that for a set of fermions (antisymmetric ψ) and the correlations in the particle motions are correspondingly different.

For fermions, the antisymmetry of $\psi(\mathbf{x}_1, \ldots, \mathbf{x}_n)$ implies that $\psi = 0$ if any two or more sets of coordinates are equal. Since the configuration space path cannot pass through nodes, it follows that two or more fermions cannot occupy the same point in 3-space at the same time (Pauli's exclusion principle). The total potential will always act in accordance with this requirement. Notice that we have not invoked probability to arrive at this conclusion.

Starting from an arbitrary entangled state and symmetrizing or antisymmetrizing, will two bosons tend to come closer to one another than two fermions with the same intrinsic properties and initial positions? It does not seem possible to prove a general theorem in this direction, and, in fact, our intuitive expectation that this will be so is not borne out by the detailed study of examples (see §7.4). The only general result we can state is that the 3-space trajectories of bosons can cross while, as we have just said, those of fermions cannot.

[†] Einstein (1925) had an inkling of this when he wrote that the differences between Maxwell–Boltzmann and Bose–Einstein statistics 'express indirectly a certain hypothesis on a mutual influence of the molecules which for the time being is of a quite mysterious nature' (quoted by Pais (1982, p. 430)).

7.1.5 Conservation laws

In an obvious generalization of the one-body theory §(3.9) one can define field energy, momentum etc. densities in configuration space and set up field conservation laws. For example, the energy density is given by

$$\mathscr{H} = R^2\left\{\sum_i [(\nabla_i S)^2/2m_i + \hbar^2(\nabla_i R)^2/2m_i] + V\right\}. \tag{7.1.16}$$

The total field energy is equal to the expectation value of the Hamiltonian operator

$$\int \mathscr{H}\, \mathrm{d}^3x_1 \cdots \mathrm{d}^3x_n = \langle \hat{H} \rangle \tag{7.1.17}$$

and is conserved if the external potential is time-independent.

Let us consider the particle conservation laws. From (7.1.2) or (7.1.10) we obtain

$$\frac{\mathrm{d}E}{\mathrm{d}t} = \frac{\partial}{\partial t}(Q + V)|_{\mathbf{x}_j = \mathbf{x}_j(t)}, \qquad j = 1, \ldots, n, \tag{7.1.18}$$

and from (7.1.9), by summing,

$$\sum_{i=1}^{n} \frac{\mathrm{d}\mathbf{p}_i}{\mathrm{d}t} = -\sum_{i=1}^{n} \nabla_i (Q + V)|_{\mathbf{x}_j = \mathbf{x}_j(t)}. \tag{7.1.19}$$

Let us suppose that the system is classically closed in the sense that V is purely an interparticle potential ($V = V(\mathbf{x}_i - \mathbf{x}_j)$) and is time-independent ($\partial V/\partial t = 0$). Then the mean energy and mean total momentum, averaged over an ensemble, are conserved:

$$\mathrm{d}\langle E \rangle/\mathrm{d}t = \langle \partial V/\partial t \rangle = 0. \tag{7.1.20}$$

$$d\sum_i \langle \mathbf{p}_i \rangle/\mathrm{d}t = -\left\langle \sum_i \nabla_i V \right\rangle = 0. \tag{7.1.21}$$

The energy and total momentum of each element (many-body system) of the ensemble are not necessarily conserved, however. From (7.1.18) and (7.1.19) we obtain

$$\mathrm{d}E/\mathrm{d}t = \partial Q/\partial t|_{\mathbf{x}_j = \mathbf{x}_j(t)} \tag{7.1.22}$$

$$\sum_i \mathrm{d}\mathbf{p}_i/\mathrm{d}t = -\sum_i \nabla_i Q|_{\mathbf{x}_j = \mathbf{x}_j(t)} \tag{7.1.23}$$

because the conditions on V are not necessarily translated into similar conditions on Q (if suitable solutions of the wave equation are chosen).

A classically closed system of particles may not be closed quantum mechanically. Since (7.1.23) states that there is generically a resultant force acting on the many-body system, we see that Newton's third law may fail in quantum mechanics where classically it is strictly valid.

It is usually claimed that energy and momentum are conserved in individual quantum mechanical processes involving time-independent external potentials, and not just on the average. In general, as we have seen, this will not be true as the quantum potential can put energy and momentum into a system. Do we then predict a violation of the conservation laws? The answer is 'no' in the particular instances where one tests for conservation by performing measurements of energy and momentum, which is the case to which the usual claim refers. We have had occasion to remark several times before that, according to the quantum theory of motion, the state of affairs obtaining after a measurement of some quantity is quite distinct from, though causally and continuously connected to, that which exists prior to the measurement (see Chap. 8). In the present case we find that although in the course of an arbitrary process (in a classical time-independent potential) the energy and momentum continuously vary in time, the total values of these quantities will nevertheless be found to be conserved if we measure them at the beginning and end of the process. The reason is easy to see. Consider the energy changes involved in the collision of two systems (§7.5). In order to test the energy conservation law, we must know the initial energies of the colliding systems (i.e., we must perform energy measurements). This means that the systems are each in stationary states, with energies E_1 and E_2, say. Let us suppose that the outcome of the collision is that the systems are again in stationary states (or we perform further energy measurements to bring this about), with energies E_1' and E_2' respectively. Then the initial total energy, $E_1 + E_2$, is equal to $\langle \hat{H} \rangle$ and, since this is constant throughout the process (because $\partial \hat{H}/\partial t = 0$), we have that

$$E_1 + E_2 = E_1' + E_2', \tag{7.1.24}$$

the final total energy. It is this relation between energy eigenvalues that constitutes the usual quantum mechanical law of energy conservation in individual processes. In contrast, if the initial state of one of the systems is a packet made up of a range of energy eigenfunctions, then the total energy of the systems prior to the (time-independent) interaction will not in general equal the final energy, due to the effect of the quantum potential. But, we emphasize, the latter case is not the one relevant to an experimental check of the validity of the law of conservation. Similar remarks apply to momentum and angular momentum conservation.

There are, of course, cases where energy and momentum are conserved even though the wavefunction is not an eigenfunction of the corresponding operator. Thus, energy will be conserved whenever the phase function has the form

$$S(\mathbf{x}_1, \ldots, \mathbf{x}_n, t) = S_0(\mathbf{x}_1, \ldots, \mathbf{x}_n) - Et \tag{7.1.25}$$

and $\partial Q/\partial t = 0$. The stationary states form that subclass of the states satisfying (7.1.25) for which $R(\mathbf{x}_1, \ldots, \mathbf{x}_n)$ is independent of time. For an example where momentum is conserved, although the wavefunction is not an eigenfunction of the total momentum operator, see §7.4.

7.1.6 Empty waves

Suppose the wavefunction is a superposition of two waves: $\psi(\mathbf{x}_1, \ldots, \mathbf{x}_n) = \psi_1(\mathbf{x}_1, \ldots, \mathbf{x}_n) + \psi_2(\mathbf{x}_1, \ldots, \mathbf{x}_n)$. The amplitude is given as usual by

$$R^2 = R_1^2 + R_2^2 + 2R_1R_2\cos[(S_1 - S_2)/\hbar]. \tag{7.1.26}$$

Suppose that the functions ψ_1 and ψ_2 have no common support (no points in common) in configuration space. Then the system point is located in one of them, say ψ_1, and the other (ψ_2) is an 'empty wave'. Although it has associated energy, momentum etc. densities, ψ_2 has no influence on the particle motions. Then (7.1.26) reduces to

$$R^2 \approx R_1^2 + R_2^2 \tag{7.1.27}$$

and ψ is equivalent to a classical mixture. If ψ_1 and ψ_2 subsequently overlap, so that R_1 and R_2 are appreciable in the same region, the wave ψ_2 actively influences the particles and the interference term in (7.1.26) is finite.

Examples of the interference of configuration space wavefunctions are presented in §§7.3–7.6.

7.2 Factorizability, entanglement and correlations

7.2.1 Factorization of the wavefunction

To study the conditions under which we can expect to find correlations in the particle motions, and those under which they are independent, we restrict attention to a two-body system with wavefunction $\psi(\mathbf{x}_1, \mathbf{x}_2)$ (t is suppressed). Consider the case where this function is a product of functions associated respectively with each of the particles:

$$\psi(\mathbf{x}_1, \mathbf{x}_2) = \psi_A(\mathbf{x}_1)\psi_B(\mathbf{x}_2). \tag{7.2.1}$$

We shall call such a wavefunction 'strictly factorizable' or just 'factorizable'.

Eq. (7.2.1) expresses the physical independence of the two particles (even though the functions $\psi_A(\mathbf{x})$, $\psi_B(\mathbf{x})$ may be finite in overlapping domains in Euclidean 3-space). It is easy to see that the phase and amplitude functions are given by

$$S(\mathbf{x}_1, \mathbf{x}_2) = S_A(\mathbf{x}_1) + S_B(\mathbf{x}_2) \tag{7.2.2}$$

$$R(\mathbf{x}_1, \mathbf{x}_2) = R_A(\mathbf{x}_1)R_B(\mathbf{x}_2). \tag{7.2.3}$$

Clearly, from (7.1.7),

$$m_1\dot{\mathbf{x}}_1 = \nabla_1 S_A|_{\mathbf{x}_1=\mathbf{x}_1(t)}, \qquad m_2\dot{\mathbf{x}}_2 = \nabla_2 S_B|_{\mathbf{x}_2=\mathbf{x}_2(t)} \tag{7.2.4}$$

and

$$\begin{aligned} Q = Q_A(\mathbf{x}_1) + Q_B(\mathbf{x}_2), \qquad Q_A &= -\hbar^2\nabla_1^2 R_A/2m_1R_A, \\ Q_B &= -\hbar^2\nabla_2^2 R_B/2m_2R_B. \end{aligned} \tag{7.2.5}$$

The total energy of each particle is well defined; for example

$$E_A = -\partial S_A/\partial t = (\nabla_1 S_A)^2/2m_1 + Q_A + V_A \tag{7.2.6}$$

and the total energy of the system is $E = E_A + E_B$. The motion of either particle is independent of the location of the other and there is no nonlocal correlation. Particle 1(2) is associated with just $\psi_A(\psi_B)$ which satisfies its own Schrödinger equation. If there is no classical interaction the form (7.2.1) will be preserved by the wave equation for all time. In short, *factorizability implies physical independence*. Moreover, it implies that the particles are statistically independent. We shall say that the particles obey Maxwell–Boltzmann (MB) statistics.

Note that the conditions (7.2.2) and (7.2.3) are sufficient for independence but not necessary. Wavefunctions exist whose phases take the form (7.2.2) but whose amplitudes do not factorize. Then (7.2.4) remains valid and, from the Hamilton–Jacobi equation (7.1.2), $Q + V = f_A(\mathbf{x}_1) + f_B(\mathbf{x}_2)$. The individual orbits depend only on their respective initial conditions but the functions S_A and S_B may depend on parameters associated with both particles (for an example see §7.3).

Consider now the case where the wavefunction may be expressed as a sum of factorizable wavefunctions:

$$\psi(\mathbf{x}_1, \mathbf{x}_2) = N[\psi_A(\mathbf{x}_1)\psi_B(\mathbf{x}_2) + \psi_C(\mathbf{x}_1)\psi_D(\mathbf{x}_2)], \tag{7.2.7}$$

where N is a normalization constant. Such solutions may be constructed when there is no classical interaction between the particles. Suppose that the summands in (7.2.7) do not overlap, by which we mean that they are separated by regions of configuration space that are practicallly nodal. It is sufficient that either ψ_A and ψ_C, or ψ_B and ψ_D, have no common support. Such a function will be called 'effectively factorizable' because it behaves as if the wave is either

$$\psi = \psi_A(\mathbf{x}_1)\psi_B(\mathbf{x}_2) \quad \text{or} \quad \psi_C(\mathbf{x}_1)\psi_D(\mathbf{x}_2) \tag{7.2.8}$$

i.e., as a mixture (in which in a sequence of trials one or other of the summands is realized with a certain probability). The amplitude squared of (7.2.7),

$$R^2 = N^2\{R_A^2 R_B^2 + R_C^2 R_D^2 + 2R_A R_B R_C R_D \cos[(S_A + S_B - S_C - S_D)/\hbar]\} \tag{7.2.9}$$

reduces to (cf. (7.1.27))

$$R^2 \approx N^2(R_A^2 R_B^2 + R_C^2 R_D^2). \tag{7.2.10}$$

The system point is located in one or other of the regions in configuration space where ψ is finite and, assuming the component waves do not overlap at any time, we may ignore the empty wave and treat the system as simply two independent particles.

The notion of effective factorizability may be generalized. If $\eta_i(\mathbf{x}_2)$ form a complete set of functions, any configuration space function may be expressed as

$$\psi(\mathbf{x}_1, \mathbf{x}_2) = \sum_i \xi_i(\mathbf{x}_1)\eta_i(\mathbf{x}_2) \tag{7.2.11}$$

where $\xi_i(\mathbf{x}_1)$ are a set of expansion coefficients. If the summands do not overlap, ψ is equivalent to a mixture and, although they are potentially active, we may drop from attention all the summands in which the system point does not lie.

If in the course of time the particles classically interact, the Schrödinger evolution will turn a function (7.2.1) into (7.2.7) or generally (7.2.11) where the summands do overlap; the wavefunction is 'nonfactorizable' or 'entangled' (Schrödinger, 1935a, 1936). For (7.2.7) the condition for this is that

$$\psi_A \cap \psi_C \neq \varnothing \quad \text{and} \quad \psi_B \cap \psi_D \neq \varnothing. \tag{7.2.12}$$

The interference term in (7.2.9) becomes finite and the motions nonlocally correlated. It now has no meaning to say that particle 1(2) is associated just with wave $\psi_A(\psi_B)$ or $\psi_C(\psi_D)$ – both particles are guided by one wave, $\psi(\mathbf{x}_1, \mathbf{x}_2)$. While the particles individually have well-defined kinetic energies, the potential energy and hence the total energy are properties of the system as a whole. In addition, the probability distribution is no longer factorizable. Bearing in mind that the primary property of ψ is the influence it exerts on the particles, we deduce that the particles are *statistically correlated because they are physically connected.*

While strict factorization is a property common to all representations, *effective* factorizability is a representation-dependent concept. It is fairly obvious, for example, that if we transform to the momentum representation of the function (7.2.7) the Fourier coefficient will not generally be effectively factorizable in the two sets of momentum coordinates. This does not contradict our assertion that particles in effectively factorizable states are independent because the Fourier coefficient pertains to a situation where a measurement of the momentum is carried out and this requires an external interaction which couples the two initially nonoverlapping summands and thus destroys the effective factorizability. Thus, the true physical independence of two systems is maintained under all circumstances only when the wavefunction remains strictly factorized. When it is not, there will always be circumstances where the disturbance of one particle will be translated into effects on the other particle. The condition for nonlocality is thus nonfactorizability (for further discussion of this point see Chap. 11, especially §11.1.2).

We can mathematically test a wavefunction for 'entanglement' by use of differential and integral criteria applied to continuous matrices (Takabayasi, 1954; Holland, 1986). The generalization of the notions of factorizability and entanglement to a system of n bodies is obvious.

7.2.2 Factorization in other coordinates

If we think of the function $\psi(\mathbf{x}_1, \mathbf{x}_2)$ as a continuous matrix, then the degree of finiteness of the off-diagonal elements provides a measure of the correlations in the motions of the particles. Between the two extremes of no correlations (factorizability or maximum nondiagonality) and maximum correlations (nonfactorizability or diagonality, e.g., the EPR case) there is a wide range of degrees of correlation. A particularly important point is that the off-diagonal elements may be a function of interparticle distance and decrease at large distances.

We can get an alternative perspective on correlations if we pass to another system of configuration space coordinates in which the wavefunction naturally factorizes due to the symmetry of the problem. Suppose this occurs in cm and difference coordinates:

$$\mathbf{X} = (m_1\mathbf{x}_1 + m_2\mathbf{x}_2)/(m_1 + m_2), \qquad \mathbf{r} = \mathbf{x}_1 - \mathbf{x}_2 \tag{7.2.13}$$

so that

$$\psi(\mathbf{X}, \mathbf{r}) = \psi_1(\mathbf{X})\psi_2(\mathbf{r}). \tag{7.2.14}$$

This form is preserved if the potential is of the form $V = V(\mathbf{r})$, for example.

Then

$$\left.\begin{aligned} S(\mathbf{X}, \mathbf{r}) &= S_1(\mathbf{X}) + S_2(\mathbf{r}) \\ R(\mathbf{X}, \mathbf{r}) &= R_1(\mathbf{X})R_2(\mathbf{r}). \end{aligned}\right\} \tag{7.2.15}$$

If we think of two fictitious 'particles' with coordinates $\mathbf{X}$ and $\mathbf{r}$, then these points will move independently. The physical particles with coordinates $\mathbf{x}_1$ and $\mathbf{x}_2$ will of course pursue correlated motions in the state (7.2.14).

7.2.3 *Identical particles*

For a generic two-body system with wavefunction $\psi(\mathbf{x}_1, \mathbf{x}_2)$ the quantum potential exerts a force between the particles, and this may extend to large interparticle distances. If the particles are identical, the symmetrization of the wavefunction introduces further forces between the particles in addition to those generically present in any entangled state. This is easy to see, for if the wavefunction is given by

$$\Psi_\pm(\mathbf{x}_1, \mathbf{x}_2) = N_\pm[\psi(\mathbf{x}_1, \mathbf{x}_2) \pm \psi(\mathbf{x}_2, \mathbf{x}_1)] \tag{7.2.16}$$

then

$$\begin{aligned} |\Psi_\pm|^2 = N_\pm^2[&|\psi(\mathbf{x}_1, \mathbf{x}_2)|^2 + |\psi(\mathbf{x}_2, \mathbf{x}_1)|^2 \\ &\pm \psi^*(\mathbf{x}_1, \mathbf{x}_2)\psi(\mathbf{x}_2, \mathbf{x}_1) \pm \psi^*(\mathbf{x}_2, \mathbf{x}_1)\psi(\mathbf{x}_1, \mathbf{x}_2)] \end{aligned} \tag{7.2.17}$$

and the quantum potential constructed from $R_\pm$ includes contributions from the interference terms. Although in the quantum theory of motion the particles are distinguishable in all cases, in the conventional view they will be so only if the functions $\psi(\mathbf{x}_1, \mathbf{x}_2)$ and $\psi(\mathbf{x}_2, \mathbf{x}_1)$ do not overlap, a situation depicted in Fig. 7.1. In that case (7.2.16) is effectively equal to one or other of the summands, with the system point lying in one of them, and behaves like any entangled state.

Suppose now that two identical particles are far apart and each is associated with a packet function (1 with ψ_A, 2 with ψ_B so that $\psi_\mathrm{A} \cap \psi_\mathrm{B} = \varnothing$). At this stage they are conventionally distinguishable and the wavefunction is effectively (7.2.1). If now the packets approach one another (without classically interacting) and overlap, so that

$$(\langle\mathbf{x}_1\rangle - \langle\mathbf{x}_2\rangle)^2 \leqslant \sigma_\mathrm{A}^2 + \sigma_\mathrm{B}^2$$

where $\sigma_\mathrm{A}, \sigma_\mathrm{B}$ are the packet widths, then, according to the usual view, the particles are indistinguishable in that we cannot tell to which packet each belongs. Correlations in the subsequent behaviour of the particles then arise from our 'ignorance'. Actually, even from the conventional point of view this cannot be the whole story since it would imply an incoherent mixture of

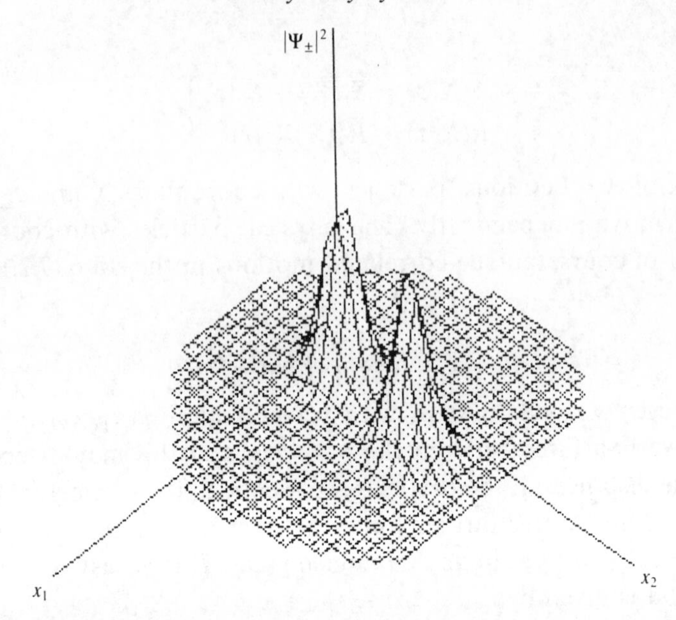

Fig. 7.1 The amplitude squared of the wavefunction of identical particles in the case where these are conventionally treated as distinguishable. Only two of the six configuration space dimensions are shown.

possible wavefunctions whereas expectation values depend on the interference of the summands. For us the correlations come about from the introduction of new forces which, as we have said, are due to the physical interference of the summands making up the symmetrized wavefunction. The latter is given by

$$\psi_{\pm}(\mathbf{x}_1, \mathbf{x}_2) = N_{\pm}[\psi_{\mathrm{A}}(\mathbf{x}_1)\psi_{\mathrm{B}}(\mathbf{x}_2) + \mathrm{e}^{\mathrm{i}\varphi}\,\psi_{\mathrm{A}}(\mathbf{x}_2)\psi_{\mathrm{B}}(\mathbf{x}_1)], \tag{7.2.18}$$

where $\varphi = 0$ for Bose–Einstein (BE) statistics and $\varphi = \pi$ for Fermi–Dirac (FD) statistics, and the condition (7.2.12) that the summands overlap becomes:

$$\psi_{\mathrm{A}} \cap \psi_{\mathrm{B}} \neq \varnothing. \tag{7.2.19}$$

Yet, from a physical point of view, where does the second term in (7.2.18) come from? One cannot pass from (7.2.1) to (7.2.18) as part of the evolution of the overlapping packets described by the Schrödinger equation. We conclude that the assumption that the wavefunction of two distantly separated identical particles is factorizable is, in fact, incorrect. The actual state is really always (7.2.18) which, when the functions ψ_{A} and ψ_{B} do not overlap, is effectively factorizable and so physically equivalent to the product state. Does this mean that we can never isolate a particle from all the other particles in the universe of the same species and ascribe to it its own wavefunction?

Strictly speaking, no, if the symmetrization postulate is exact. This then implies a very strong form of state-dependence. However, in many situations in which particles are distantly separated, classical interactions may be neglected and the component parts of a symmetrized wavefunction will not overlap; ψ will then be effectively factorizable. It is legitimate in that case to treat the many-body system as a collection of independent systems obeying MB statistics.

7.2.4 The separation of two particles

In a discussion on locality and nonlocality clarity is enhanced if the notions of the distance between two systems in 3-space, and the time at which an event occurs, have meanings that can be unambiguously defined. In the usual approach one can, strictly speaking, only discuss the probability of finding a particular distance between two particles if one makes position measurements. However, one may attempt to give the notion of separation a meaning independently of observation using the properties of the wavefunction. Thus, if a two-body system has wavefunction (7.2.1) and ψ_A and ψ_B are packet functions, then we might roughly define the distance between systems 1 and 2 as the distance between the centres of the packets. However, this is not an unambiguous definition if the packets overlap, and it cannot be extended to an arbitrary entangled state where the particles cannot be ascribed individual wavefunctions. Indeed, using the wave alone one has no concept at all of a material system occupying a definite portion of space since it is in the nature of waves to spread out over a region with ill-defined boundaries.

In the quantum theory of motion the separation of two particles has a precise meaning for all quantum states. At any instant t, the distance between two particles pursuing tracks $\mathbf{x}_1(t)$ and $\mathbf{x}_2(t)$ is defined by

$$d = |\mathbf{x}_1(t) - \mathbf{x}_2(t)|. \tag{7.2.20}$$

This generalizes the notion of distance introduced in the one-body theory which enabled us, for example, to give a meaning to the position of a particle in an interferometer.

This definition allows us to give a clear meaning to the notion of the *extension of matter*. We may retain the classical definition of a rigid body as a collection of mass points which remain at constant distance from one another. As in classical mechanics, the number of independent degrees of freedom is six – the position of one point in the body, and the orientation of a set of body axes, relative to axes fixed in space. This is the material system described by the quantum mechanics of rigid bodies, where the wavefunction depends on six independent parameters. For further discussion of the rigid rotator see Chap. 10.

7.2.5 Particle properties

It was proposed in §3.5 that quantum mechanical operators could be connected with the physical properties of a particle via the 'local expectation value' of the operator. Extending this method to the many-body case introduces a new element. Consider a two-body system and an Hermitian operator $\hat{A}(\hat{\mathbf{x}}_1, \hat{\mathbf{p}}_1)$ pertaining just to particle 1. Can we connect this with some physical characteristic of the particle? We immediately see that for an entangled state the local expectation value

$$A = \mathrm{Re}\,\psi^*(\mathbf{x}_1, \mathbf{x}_2)\hat{A}(\hat{\mathbf{x}}_1, \hat{\mathbf{p}}_1)\psi(\mathbf{x}_1, \mathbf{x}_2)/|\psi(\mathbf{x}_1, \mathbf{x}_2)|^2 \tag{7.2.21}$$

is, in general, a function of both $\mathbf{x}_1$ and $\mathbf{x}_2$, and not just of $\mathbf{x}_1$ alone. We have already encountered this situation with the momentum, for when $\hat{A} = \hat{\mathbf{p}}_1$, (7.2.21) is equal to $\nabla_1 S(\mathbf{x}_1, \mathbf{x}_2)$. Nevertheless, in this case (7.2.21) is unequivocally connected with just particle 1 via (7.1.7) (although the trajectory $\mathbf{x}_1(t)$ is correlated with particle 2). Similarly, the operators $\hat{\mathbf{x}}_1$ and $\hat{\mathbf{x}}_1 \times \hat{\mathbf{p}}_1$ are connected with properties of just particle 1 via (7.2.21) and (7.1.7).

There are other operators, however, ostensibly pertaining to just particle 1, whose local expectation value is irreducibly a function of both sets of coordinates and has no natural interpretation as a property of particle 1 alone. For example, substituting $\hat{A} = \hat{\mathbf{p}}_1^2/2m_1$, the kinetic energy operator, into (7.2.21) yields

$$A = (\nabla_1 S)^2/2m_1 - \hbar^2\nabla_1^2 R/2m_1 R. \tag{7.2.22}$$

Evaluating this expression along a (configuration space) trajectory $\mathbf{x}_1 = \mathbf{x}_1(t)$, $\mathbf{x}_2 = \mathbf{x}_2(t)$, we see that the first term is the particle 1 kinetic energy $\frac{1}{2}m_1\dot{\mathbf{x}}_1^2$, but the second term is a function of both particles (part of the quantum potential energy of the whole system). Insofar as (7.2.21) provides a reasonable way of interpreting the physical meaning of operators, it is clear that they generally have a global significance that belies their local functional dependence.

In the case of factorizable wavefunctions, (7.2.21) always describes a property of particle 1 alone as it coincides with our definition of particle properties in the single-body theory.

In general, we cannot ascribe to the particles their own quantum potential energy – Q is the quantum potential energy of the whole system and cannot be subdivided. It follows that the total energy $(-\partial S/\partial t)$ is likewise not simply a sum of kinetic and potential energies of each particle. This is not a feature peculiar to quantum mechanics since two classical particles connected by an interaction potential $V(\mathbf{x}_1 - \mathbf{x}_2)$ have the same property.

7.2.6 Classical limit

How do we reconcile the quantal account of many-body systems in which the whole, described by $\psi(\mathbf{x}_1, \ldots, \mathbf{x}_n)$, is prior to and independent of the parts $(\mathbf{x}_1(t), \ldots, \mathbf{x}_n(t))$, with the domain of everyday experience in which it is apparently legitimate to treat matter as a collection of autonomous objects whose properties are independent of the context, and which is largely accounted for by classical mechanical notions? There are two principal ways in which we may pass from the quantum to the classical paradigm:

(1) When the many-body quantum potential and force are negligible compared to their classical counterparts, in an obvious generalization of the criteria of Chap. 6. Then we obtain a statistical ensemble of interacting many-particle classical systems, connected with a single-valued velocity field and described in the Hamilton–Jacobi language. In the case of identical particles the characteristic effects of wavefunction symmetrization are no longer translated into dynamical effects.
(2) When the wavefunction is strictly or effectively factorizable, which are sufficient conditions for the physical independence of systems. We may then apply the usual limiting procedure of (1) and Chap. 6 to the factorized subwholes.

In both cases the state of the whole will be defined in the limit by the state of the particles and their classical interactions. In particular, distantly separated bodies may be treated as independent.

The possibilities (1) and (2) may apply in a variety of circumstances. As described in detail in §7.5 and Chap. 8, the effective factorization of a wavefunction not held together by classical potentials tends to come about whenever a system is perturbed by its thermodynamic environment.

The limit considered here is not to be confused with the 'macroscopic' level. Quantum mechanics is generally required to explain the macroscopic properties of an object that as a unit is effectively factorized from its environment and behaves according to classical laws. For example, the rigidity of a billiard ball is explained at the atomic and molecular levels by the forces exerted by the quantum potential (§§4.5 and 7.6). There is, however, a much stronger sense in which quantum mechanics potentially applies at the macroscopic scale, for there is nothing other than technical difficulties to rule out the possibility of observing coherent superpositions of many-body ($\sim 10^{23}$) wavefunctions. This is still an open question (Leggett, 1980, 1987) but the existence of such effects could be easily encompassed in the causal scheme and does not imply convoluted philosophical conundrums concerning the nature of macroscopic reality.

7.3 Two-particle interferometry

7.3.1 General results

To illustrate the meaning of entanglement, we shall first consider a two-body wavefunction built out of four one-body packet wavefunctions, $\psi_A(\mathbf{x}_1)$, $\psi_C(\mathbf{x}_1)$, $\psi_B(\mathbf{x}_2)$ and $\psi_D(\mathbf{x}_2)$ whose distribution in three-dimensional Euclidean space is depicted in Fig. 7.2. ψ_A and ψ_C satisfy the same Schrödinger equation, as do ψ_B and ψ_D. Initially there is negligible overlap between the packets and corpuscle 1 will be located within either ψ_A or ψ_C and corpuscle 2 within ψ_B or ψ_D. Let us suppose that the state of the total two-body system is given by (7.2.7),

$$\psi(\mathbf{x}_1, \mathbf{x}_2) = N[\psi_A(\mathbf{x}_1)\psi_B(\mathbf{x}_2) + e^{i\varphi/\hbar}\,\psi_C(\mathbf{x}_1)\psi_D(\mathbf{x}_2)], \qquad (7.3.1)$$

where φ is a constant (we do not consider the problem of how physically to construct a source of particles in this state). This cannot, of course, be depicted in 3-space but must be represented in configuration space. The initial form of (7.3.1) corresponds to two humps, similar to Fig. 7.1. The form of (7.3.1) implies that the initial positions are correlated; since the system point lies in only one of the summands, particle 1 lies within ψ_A and 2 within ψ_B *or* 1 is within ψ_C and 2 in ψ_D.

We now assume that the packets ψ_A and ψ_C move towards one another and overlap, and similarly for the packets ψ_B and ψ_D, the two pairs of packets remaining well separated from one another. The particles are classically noninteracting so the expression (7.3.1) will be preserved by the Schrödinger evolution. Will interference fringes be observed in each of the disjoint regions accessible to particles 1 and 2? To answer this we note that all the statistical predictions for experiments pertaining to particle 1, for example, are contained

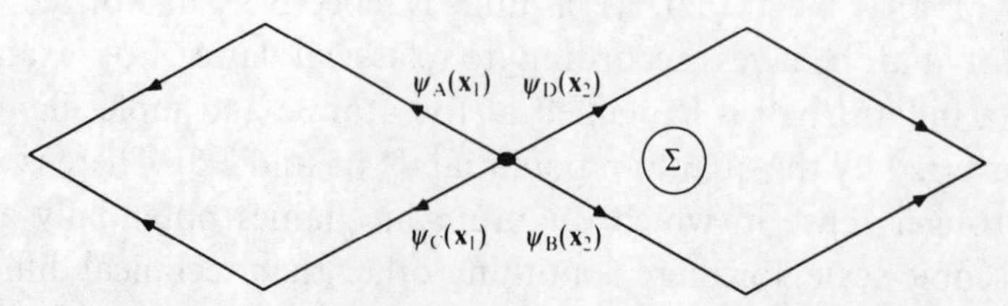

Fig. 7.2 The distribution in 3-space of four packet functions associated with two particles $\mathbf{x}_1$ and $\mathbf{x}_2$. The arrows indicate the directions of motion of the packets.

in the reduced density matrix

$$\rho_1(\mathbf{x}_1, \mathbf{x}_1') = \int \psi(\mathbf{x}_1, \mathbf{x}_2)\psi^*(\mathbf{x}_1', \mathbf{x}_2)\, \mathrm{d}^3x_2$$

$$= N^2[\psi_A(\mathbf{x}_1)\psi_A^*(\mathbf{x}_1') + \psi_C(\mathbf{x}_1)\psi_C^*(\mathbf{x}_1') + c\psi_A(\mathbf{x}_1)\psi_C^*(\mathbf{x}_1')$$

$$+ c^*\psi_C(\mathbf{x}_1)\psi_A^*(\mathbf{x}_1')], \qquad (7.3.2)$$

where

$$c = \mathrm{e}^{-\mathrm{i}\varphi/\hbar} \int \psi_B(\mathbf{x}_2)\psi_D^*(\mathbf{x}_2)\, \mathrm{d}^3x_2 \qquad (7.3.3)$$

is a complex constant and we have assumed that ψ_B and ψ_D are normalized. Writing the one-body wavefunctions in polar form and $c = |c|\,\mathrm{e}^{\mathrm{i}(s-\varphi)/\hbar}$, the probability density of particle 1 regardless of the position of particle 2 is given by

$$\rho_1(\mathbf{x}_1, \mathbf{x}_1) = N^2(R_A^2 + R_C^2 + 2R_A R_C|c| \cos[(S_A - S_C + s - \varphi)/\hbar]. \qquad (7.3.4)$$

We see that, as long as $|c| \neq 0$, an interference pattern will indeed be observed in the region of overlap of the particle 1 packets. The contrast

$$C = 2R_A R_C|c|/(R_A^2 + R_C^2) \qquad (7.3.5)$$

is proportional to $|c|$ so the interference image is suppressed if $|c| < 1$ (C gives a measure of the 'orthogonality' of the particle 2 packets). Similar results are obtained for the probability density of particle 2 regardless of particle 1, where the number analogous to c above is given by $c' = \int \psi_A(\mathbf{x}_1)\psi_C^*(\mathbf{x}_1)\, \mathrm{d}^3x_1$.

To bring out the new feature implied by the entangled state (7.3.1) in this context, we write down the joint probability density:

$$|\psi(\mathbf{x}_1, \mathbf{x}_2)|^2 = N^2\{R_A^2 R_B^2 + R_C^2 R_D^2$$

$$+ 2R_A R_B R_C R_D \cos[(S_A + S_B - S_C - S_D - \varphi)/\hbar]\}. \qquad (7.3.6)$$

The interference term in (7.3.6) is finite only if R_A and R_C appreciably overlap, and similarly for R_B and R_D (conditions which are satisfied here). If we suppose now that the functions ψ_A and ψ_C, and the functions ψ_B and ψ_D, are orthogonal, then $c = c' = 0$ and we see from (7.3.5) that in this case no interference will be observed in regard to each of the particles considered in isolation. Nevertheless, the interference term in (7.3.6) is finite and manifests itself through *correlations* in the particle motions. The fringes are a property of configuration space and not of 3-space; we have here an example of a 'two-particle interferometer' (Horne, Shimony & Zeilinger, 1989).

As an illustration of the effects implied by such a device, suppose that particle 1 is electrically neutral while particle 2 is charged, and that a solenoid of the type used in the Aharonov–Bohm effect (§5.2) is placed in a region Σ (see Fig. 7.2) so that the packets ψ_B and ψ_D pass on either side of it and pick up a relative phase φ. Then the individual particle probability densities will continue to show no interference effects, but the cos term in (7.3.6) will be finite and contain a phase shift equal to the AB-phase shift brought about by the solenoid between the particle 2 packets. As a result the statistical correlations in the particle motions will be altered. The trajectory of the neutral particle 1 will be nonlocally affected by the electromagnetic potential via the influence of the two-body quantum potential constructed from $R(\mathbf{x}_1, \mathbf{x}_2)$ (the same will be true if particle 1 is charged of course – see Silverman (1990)). This is a distant effect of an electromagnetic field similar to that occurring in the spin version of the EPR problem (Chap. 11).

7.3.2 *Plane waves*

As an idealization of the case treated above let us assume that the component waves may be taken to be approximately plane and that the total wave is a stationary state:

$$\psi(\mathbf{x}_1, \mathbf{x}_2) = N(e^{i(\mathbf{k}_A\cdot\mathbf{x}_1 + \mathbf{k}_B\cdot\mathbf{x}_2)} + e^{i\varphi}\, e^{i(\mathbf{k}_C\cdot\mathbf{x}_1 + \mathbf{k}_D\cdot\mathbf{x}_2)})\, e^{-iEt/\hbar}, \qquad (7.3.7)$$

where the total energy is given by

$$E = \hbar^2\mathbf{k}_A^2/2m_1 + \hbar^2\mathbf{k}_B^2/2m_2 = \hbar^2\mathbf{k}_C^2/2m_1 + \hbar^2\mathbf{k}_D^2/2m_2 \qquad (7.3.8)$$

in order that (7.3.7) obeys the wave equation. We have explicitly included a phase factor φ which may represent a phase shift between the waves ψ_B and ψ_D, for example, brought about by some external potential. The function (7.3.7) is a superposition of two equal-amplitude six-dimensional plane waves.

The amplitude and phase are given by:

$$R = 2N|\cos\theta|, \qquad \theta = [(\mathbf{k}_A - \mathbf{k}_C)\cdot\mathbf{x}_1 + (\mathbf{k}_B - \mathbf{k}_D)\cdot\mathbf{x}_2 - \varphi]/2, \qquad (7.3.9)$$

$$S = (\hbar/2)[(\mathbf{k}_A + \mathbf{k}_C)\cdot\mathbf{x}_1 + (\mathbf{k}_B + \mathbf{k}_D)\cdot\mathbf{x}_2 + \varphi] - Et + \eta\hbar, \qquad (7.3.10)$$

where $\eta = 0$ for $-\pi/2 < \theta < \pi/2$ and $\eta = \pi$ for $\pi/2 < \theta < 3\pi/2$. The function S is undefined at the zeroes of (7.3.9) and undergoes a discontinuous jump there. The nodes lie on the set of planes

$$(\mathbf{k}_A - \mathbf{k}_C)\cdot\mathbf{x}_1 + (\mathbf{k}_B - \mathbf{k}_D)\cdot\mathbf{x}_2 - \varphi = (2n+1)\pi \tag{7.3.11}$$

where n is an integer.

The trajectories found by solving (7.1.7) are straight lines each depending only on the respective initial conditions:

$$\left.\begin{aligned} \mathbf{x}_1(t) &= (\hbar/2m_1)(\mathbf{k}_A + \mathbf{k}_C)t + \mathbf{x}_{10}, \\ \mathbf{x}_2(t) &= (\hbar/2m_2)(\mathbf{k}_B + \mathbf{k}_D)t + \mathbf{x}_{20}. \end{aligned}\right\} \tag{7.3.12}$$

The particle motions are therefore independent but the speeds depend on parameters connected with the system as a whole (the amplitude is variable and nonfactorizable while the phase is an additive function of $\mathbf{x}_1$ and $\mathbf{x}_2$). Combining (7.3.8) and (7.3.12) yields that the system trajectory lies in the plane

$$(\mathbf{k}_A - \mathbf{k}_C)\cdot\mathbf{x}_1 + (\mathbf{k}_B - \mathbf{k}_D)\cdot\mathbf{x}_2 = c, \tag{7.3.13}$$

where c is a constant determined by the initial positions. Comparing with (7.3.11) we see that this surface is coplanar with the nodal surfaces and so we confirm that, if the system point initially lies outside a nodal region, it will remain so for all time.

The quantum potential evaluated from (7.3.9) is a constant:

$$Q = -\hbar^2(\mathbf{k}_A - \mathbf{k}_C)^2/8m_1 - \hbar^2(\mathbf{k}_B - \mathbf{k}_D)^2/8m_2. \tag{7.3.14}$$

Adding to this the total kinetic energy of the particles,

$$(\nabla_1 S)^2/2m_1 + (\nabla_2 S)^2/2m_2 = \hbar^2(\mathbf{k}_A + \mathbf{k}_C)^2/8m_1 + \hbar^2(\mathbf{k}_B + \mathbf{k}_D)^2/8m_2, \tag{7.3.15}$$

we recover the total energy (7.3.8).

The partial probability densities do not exhibit fringes. The total density evidently does but in this idealization the trajectories are only trivially correlated in that their speeds are connected through (7.3.8) which ensures they cannot pass through nodes. The phase shift φ fixes the location of the nodes. (For a treatment of two-particle interferometry using wave packets see Lam & Dewdney (1990).)

A pair of identical particles is described by the special case $\mathbf{k}_A = \mathbf{k}_D$, $\mathbf{k}_B = \mathbf{k}_C$, and $m_1 = m_2$ with $\varphi = 0\ (\pi)$ for BE (FD) statistics. The particles have equal speeds and the only difference between the BE and FD cases is that the interference pattern is shifted by $\pi/2$. The uniform and rectilinear motion is identical in form to that obtained with a factorizable wavefunction (MB statistics).

7.4 Two harmonic oscillators

7.4.1 The three statistics

To illustrate in a more illuminating way the differences in the motions induced by waves obeying the three different statistics (MB, BE and FD), we examine a system of two classically noninteracting one-dimensional harmonic oscillators of equal mass and frequency (Dewdney, Kyprianidis & Vigier, 1984; Dewdney & Holland, 1988; Kyprianidis, 1988b). The Schrödinger equation is

$$i\hbar\frac{\partial\psi(x_1, x_2, t)}{\partial t} = \left[-\frac{\hbar^2}{2m}\frac{\partial^2}{\partial x_1^2} - \frac{\hbar^2}{2m}\frac{\partial^2}{\partial x_2^2} + \tfrac{1}{2}m\omega^2 x_1^2 + \tfrac{1}{2}m\omega^2 x_2^2\right]\psi(x_1, x_2, t). \tag{7.4.1}$$

Since the external potential is an additive function of x_1 and x_2 we can construct solutions out of two one-particle wavefunctions, which we take to be nondispersive packets oscillating between the points $x = \pm a$ (§4.9). Let

$$\begin{aligned}\psi_A(x, t) = (m\omega/\pi\hbar)^{1/4} \exp\{&-(m\omega/2\hbar)(x - a\cos\omega t)^2 \\ &- (i/2)[\omega t + (m\omega/\hbar)(2xa\sin\omega t - \tfrac{1}{2}a^2\sin 2\omega t)]\}\end{aligned} \tag{7.4.2}$$

be the packet initially centred about $x = a$, and $\psi_B(x)$ be the packet initially centred about $x = -a$ (replace a by $-a$ in (7.4.2)). The width of each packet is given by

$$\Delta x = (\hbar/2m\omega)^{1/2}. \tag{7.4.3}$$

The three normalized wavefunctions to be considered are as follows:

$$\psi_{MB}(x_1, x_2) = \psi_A(x_1)\psi_B(x_2), \tag{7.4.4}$$

$$\psi_{BE} = N_{BE}[\psi_A(x_1)\psi_B(x_2) + \psi_B(x_1)\psi_A(x_2)], \tag{7.4.5}$$

$$\psi_{FD} = N_{FD}[\psi_A(x_1)\psi_B(x_2) - \psi_B(x_1)\psi_A(x_2)], \tag{7.4.6}$$

where the normalization constants are given by

$$\left.\begin{aligned}N_{BE} &= [2(1 + e^{-2m\omega a^2/\hbar})]^{-1/2} \\ N_{FD} &= [2(1 - e^{-2m\omega a^2/\hbar})]^{-1/2}.\end{aligned}\right\} \tag{7.4.7}$$

These will be treated in turn. The width (7.4.3) is arbitrary and ψ_A and ψ_B may overlap initially if a is not too large. The functions (7.4.5) and (7.4.6) describe a system of identical particles and (7.4.4) does as well when $\psi_A = \psi_B$ (a degenerate case of (7.4.5)).

Maxwell–Boltzmann

According to §7.2 the factorizable function (7.4.4) implies two independent one-body simple harmonic motions as described in §4.9. Particle 1 lies in packet ψ_A and particle 2 in ψ_B. The trajectories are given by

$$\left.\begin{aligned} x_1(t) &= x_{10} + a(\cos \omega t - 1) \\ x_2(t) &= x_{20} - a(\cos \omega t - 1). \end{aligned}\right\} \tag{7.4.8}$$

Denoting the separation of the particles by $u = x_1 - x_2$ we have

$$u(t) = u_0 - 4a \sin^2 \tfrac{1}{2}\omega t. \tag{7.4.9}$$

A selection of trajectories is shown in Fig. 7.3 (where it is obvious that the choice of initial position of each particle has no influence on the motion of the other particle) and $u(t)$ in Fig. 7.4. The quantum potential decomposes into a sum of functions of x_1 and x_2:

$$Q(x_1, x_2, t) = \hbar\omega - \tfrac{1}{2}m\omega^2(x_1 - a \cos \omega t)^2 - \tfrac{1}{2}m\omega^2(x_2 + a \cos \omega t)^2. \tag{7.4.10}$$

Bose–Einstein

The initial wavefunction is built from four one-body packets whose spatial distribution is shown in Fig. 7.5. The corresponding configuration space function comprises two two-dimensional Gaussian packets and is depicted in

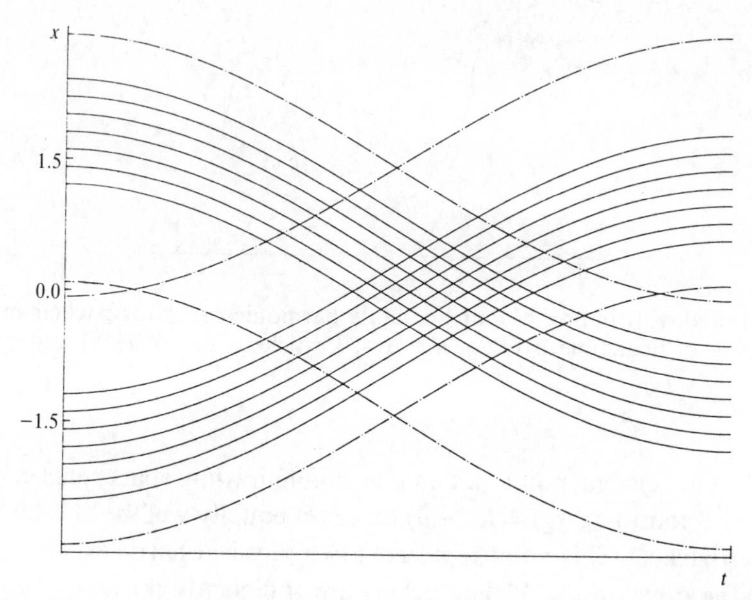

Fig. 7.3 Particle trajectories for two harmonic oscillators with a factorizable wave-function (Maxwell–Boltzmann statistics (solid lines)). The interrupted lines show the variances of the individual wave packets. The initial position of particle 1(2) is only relevant to the motion of 1(2) (from Dewdney & Holland (1988)).

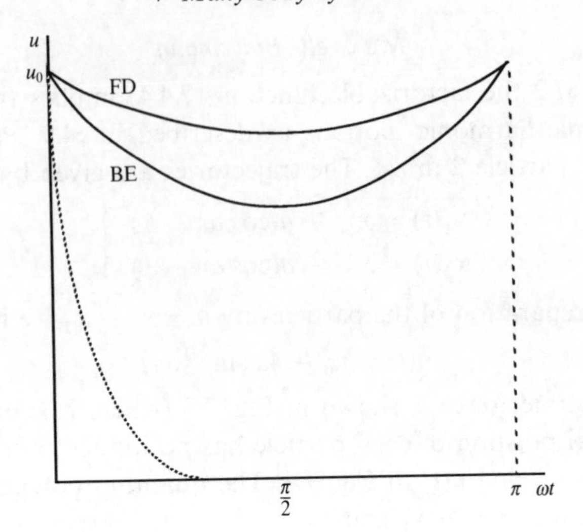

Fig. 7.4 The relative particle separation $u = x_1 - x_2$ for two harmonic oscillators obeying Bose–Einstein (BE) and Fermi–Dirac (FD) statistics in the limit $|au| \ll (\Delta x)^2$. The Maxwell–Boltzmann case is represented by the dashed line (adapted from Kyprianidis (1988b)).

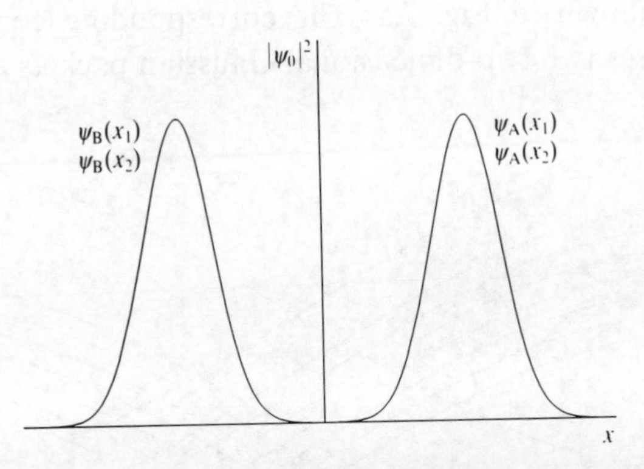

Fig. 7.5 Initial distribution of four one-body harmonic oscillator packets in 3-space (in the case of negligible overlap, $a \gg \Delta x$). Particle 1 lies in $\psi_A(x)$ and particle 2 in $\psi_B(x)$.

Fig. 7.6. The system point lies in the configuration space packet centred around the point $(x_1, x_2) = (a, -a)$ (it could equally well be chosen to lie in the other packet). This point represents two physical particles, one in packet ψ_A and the other in ψ_B. If the packets are sufficiently far apart (so that ψ_A and ψ_B do not appreciably overlap) the wavefunction is effectively factorizable and the particles move independently (as in MB statistics). The motions become correlated as the packets approach one another (in Figs. 7.5 and 7.6)

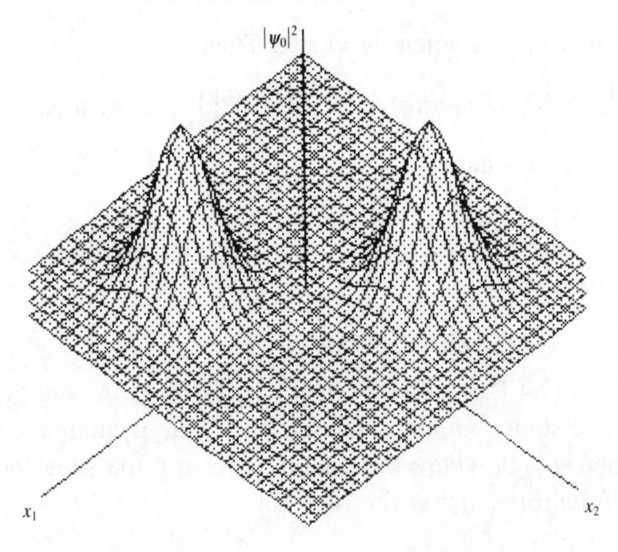

Fig. 7.6 The amplitude squared of the initial wavefunction of identical harmonic oscillators consists of two two-dimensional Gaussian packets in configuration space (in the case of negligible overlap, $a \gg \Delta x$). The MB case corresponds to just one two-dimensional packet.

and interfere, which gives rise to fringes, and eventually become independent again as the packets separate.

Substituting (7.4.2) into (7.4.5) gives

$$\begin{aligned}\psi_{\text{BE}} = N_{\text{BE}}(m\omega/\pi\hbar)^{1/2} \exp(-\{i[\omega t - (m\omega a^2/2\hbar)\sin 2\omega t] \\ + (m\omega a^2/\hbar)\cos^2 \omega t + (m\omega/2\hbar)(x_1^2 + x_2^2)\}) \\ \times \{\exp[(m\omega au/\hbar)\cos \omega t - (im\omega au/\hbar)\sin \omega t] \\ + \exp[-(m\omega au/\hbar)\cos \omega t + (im\omega au/\hbar)\sin \omega t]\}.\end{aligned} \tag{7.4.11}$$

The square of the amplitude is therefore

$$\begin{aligned}R_{\text{BE}}^2 = N_{\text{BE}}^2(m\omega/\pi\hbar)\exp[-(m\omega/\hbar)(x_1^2 + x_2^2) - (2m\omega a^2/\hbar)\cos^2 \omega t] \\ \times \{\exp[(2m\omega au/\hbar)\cos \omega t] + \exp[-(2m\omega au/\hbar)\cos \omega t] \\ + 2\cos[(2m\omega au/\hbar)\sin \omega t]\}.\end{aligned} \tag{7.4.12}$$

If $a \gg \Delta x$ and $|u| \gg \Delta x$ then at $t = 0$ this is a superposition of nonoverlapping packets and, as we have said, the motion is essentially MB-like. If not, then the initial motions are correlated.

The function (7.4.12) evidently possesses a set of nodes. To get an idea of their distribution, consider the case where the packets perfectly overlap

($R_A = R_B$), which occurs when $\omega t = \pi/2$. Then

$$R_{BE}^2 = N_{BE}^2(2m\omega/\pi\hbar)\,e^{-(m\omega/\hbar)(x_1^2+x_2^2)}[1 + \cos(2m\omega au/\hbar)] \qquad (7.4.13)$$

and the nodes occur where

$$2m\omega au = (2n+1)\pi\hbar, \qquad n \in \mathbb{Z} \qquad (7.4.14)$$

and the maxima where

$$2m\omega au = 2n\pi. \qquad (7.4.15)$$

Eqs. (7.4.14) and (7.4.15) define a set of parallel lines in configuration space, separated by a distance $\pi\hbar/m\omega a$. In particular, the probability density has a maximum when $u = 0$, where the particles occupy the same point in space.

The phase function satisfies the relation

$$\tan[(S + \hbar\omega t - \tfrac{1}{2}m\omega a^2 \sin 2\omega t)/\hbar] = -\tanh[(m\omega au/\hbar)\cos\omega t]\tan[(m\omega au/\hbar)\sin\omega t] \qquad (7.4.16)$$

outside the nodes. It is evident that the phase depends only on the relative coordinate $u = x_1 - x_2$, and not on the centre of mass coordinate $\frac{1}{2}(x_1 + x_2)$. Hence, adding the equations of motion

$$m\dot{x}_1 = \partial S(u)/\partial x_1, \qquad m\dot{x}_2 = \partial S(u)/\partial x_2, \qquad (7.4.17)$$

it follows that $\dot{x}_1 + \dot{x}_2 = 0$ and

$$x_1(t) + x_2(t) = x_{10} + x_{20} \qquad (7.4.18)$$

so that the cm is at rest (which we might guess from the symmetry of the problem). Note that the cm momentum is conserved even though the state (7.4.11) is not an eigenfunction of the total momentum operator $\hat{P} = -i\hbar(\partial/\partial x_1 + \partial/\partial x_2)$, and $\hat{P}$ does not commute with the Hamiltonian operator in (7.4.1) (this is so in the MB and FD cases also).

Subtracting eqs. (7.4.17) we find for the relative coordinate

$$\begin{aligned}\dot{u} &= (2/m)\,\partial S/\partial u, \qquad \partial/\partial u = \tfrac{1}{2}(\partial/\partial x_1 - \partial/\partial x_2)\\ &= -2\omega a\{\cos\omega t \sin[(2m\omega au/\hbar)\sin\omega t] + \sin\omega t \sinh[(2m\omega au/\hbar)\cos\omega t]\}\\ &\quad \times \{\cosh[(2m\omega au/\hbar)\cos\omega t] + \cos[(2m\omega au/\hbar)\sin\omega t]\}^{-1}. \qquad (7.4.19)\end{aligned}$$

This is not readily integrated analytically but we may get a rough idea of the solution if we assume that the particles are close together in comparison with the width of the packet (7.4.3) ($|u| \ll \Delta x$) and that the amplitude of the

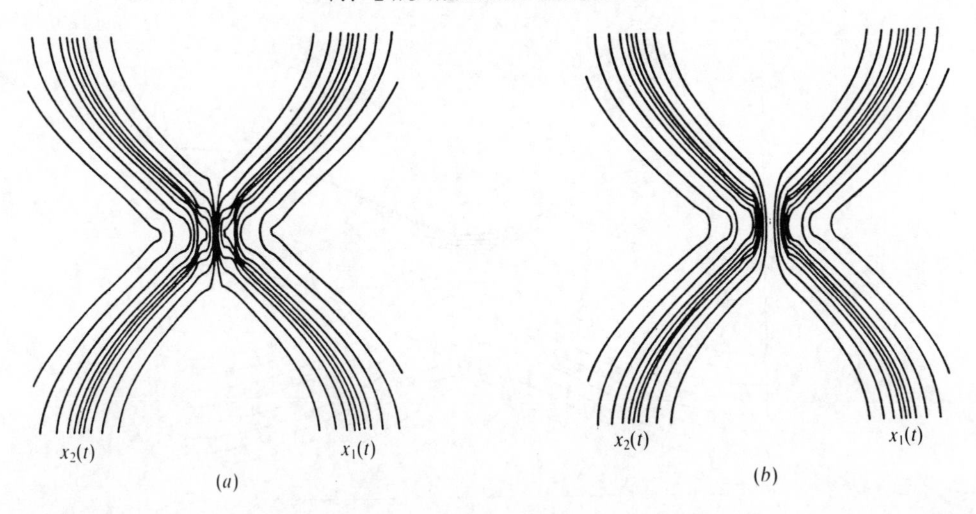

Fig. 7.7 Two-particle trajectories $x_1(t)$ and $x_2(t)$ with initial positions $x_2(0) = -x_1(0)$ concentrated around the packet maxima: (*a*) Bose–Einstein, and (*b*) Fermi–Dirac (from Vigier, Dewdney, Holland & Kyprianidis (1987)).

oscillation is much smaller than Δx ($a \ll \Delta x$). Then $|2m\omega a u/\hbar| \ll 1$ and (7.4.19) reduces approximately to

$$\dot{u} \approx -(2m\omega^2 a^2 u/\hbar) \sin 2\omega t. \qquad (7.4.20)$$

This has the solution

$$u(t) \approx u_0[1 - (a/\Delta x)^2 \sin^2 \omega t], \qquad (7.4.21)$$

where $u_0 = x_{10} - x_{20}$ (see Fig. 7.4). Using (7.4.18), the individual trajectories are therefore given in this approximation by

$$\left.\begin{aligned} x_1(t) &= \tfrac{1}{2}(x_{10} + x_{20}) + \tfrac{1}{2}(x_{10} - x_{20})[1 - (a/\Delta x)^2 \sin^2 \omega t], \\ x_2(t) &= \tfrac{1}{2}(x_{10} + x_{20}) - \tfrac{1}{2}(x_{10} - x_{20})[1 - (a/\Delta x)^2 \sin^2 \omega t]. \end{aligned}\right\} \qquad (7.4.22)$$

It is clear from these formulae that while the trajectories bunch, they do not cross ($u \neq 0$ if $u_0 \neq 0$). The point of closest approach occurs when $t = \pi/2\omega$ (perfect overlap of packets).

The trajectories resulting from the numerical solution of the exact differential equation (7.4.19), in the case where $x_{10} + x_{20} = 0$, are shown in Fig. 7.7(*a*). We observe the initial MB-like motion until the particles enter the region where the packets appreciably overlap and the fringes are formed, and the trajectories bunch around the point $x = 0$. The correlations are shown in a different way in Fig. 7.8,† where x_{10} is chosen to be the same for each pair

† In Figs. 7.8–7.10 the initial packets partly overlap but the initial positions are chosen to lie in nonoverlapping regions so that the initial motion is MB-like.

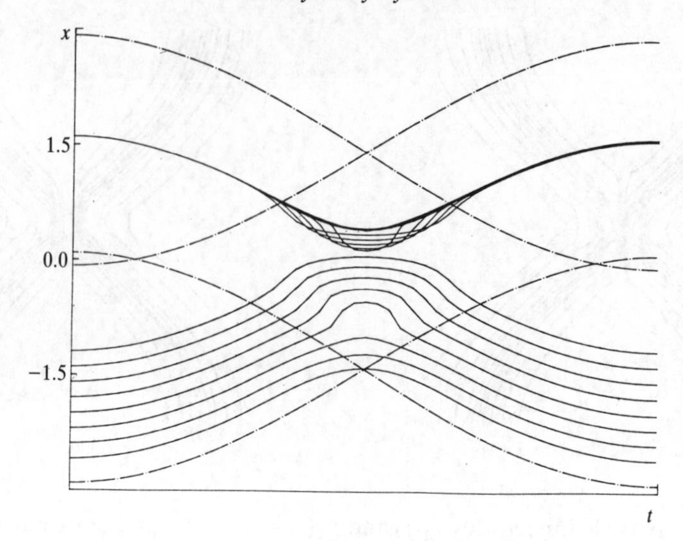

Fig. 7.8 Correlated pairs of particle trajectories for two harmonic oscillators, symmetric wavefunction. Initial position of particle 1 is the same for a range of initial positions of particle 2. The solid lines represent particle trajectories and the interrupted lines variances of the wave packets (from Dewdney & Holland (1988)).

of trajectories while x_{20} is varied from trial to trial. To see more clearly which are the correlated pairs, the region of overlap is shown in more detail in Fig. 7.9 (the numbers on the right hand side indicate the order of the trajectories at the centre).

To complete this discussion on correlations we write down the general form of the two-body quantum potential which may be evaluated from (7.4.12):

$$Q(x_1, x_2, t) = \hbar\omega - \tfrac{1}{2}m\omega^2(x_1^2 + x_2^2) + f(u, t), \qquad (7.4.23)$$

where $f(u, t)$ is a complicated function of u and t which we do not give. Since the effective potential $(Q + V)$, with $V = \tfrac{1}{2}m\omega^2(x_1^2 + x_2^2)$, is independent of the cm coordinate we confirm the conservation of the cm momentum. From $(Q + V)$ we can readily evaluate the total force acting on each particle. These are mathematically complex functions but they simplify somewhat in two cases: when $t = 0$, and in the overlap region ($\omega t = \pi/2$). We have

$$F_1 = -\partial(Q + V)/\partial x_1$$

$$= \begin{cases} -m\omega^2 a\{\tanh(m\omega a u/\hbar) + (m\omega a u/\hbar)[\cosh(m\omega a u/\hbar)]^{-2}\}, & t = 0 \\ m\omega^2 a\{\tan(m\omega a u/\hbar) + (m\omega a u/\hbar)[\cos(m\omega a u/\hbar)]^{-2}\}, & \omega t = \pi/2 \end{cases} \qquad (7.4.24)$$

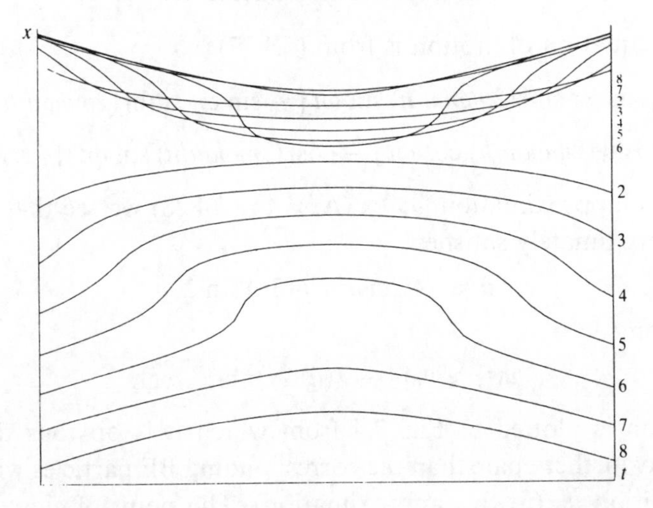

Fig. 7.9 Correlated pairs of trajectories for two harmonic oscillators, symmetric wavefunction, in the region of overlap. The numbering of the upper trajectories gives the vertical ordering at the centre of the plot (from Dewdney & Holland (1988)).

and

$$F_2 = -F_1. \tag{7.4.25}$$

These expressions take the following forms in the approximation $|m\omega au/\hbar| \ll 1$:

$$F_1 \approx \begin{cases} -2m^2\omega^3 a^2 u/\hbar, & t = 0 \\ 2m^2\omega^3 a^2 u/\hbar, & \omega t = \pi/2. \end{cases} \tag{7.4.26}$$

These formulae may be confirmed by evaluating $\ddot{u}$ from (7.4.20) and putting in turn $t = 0$, $t = \pi/2\omega$. The forces are such as to bring the particles together, but they become repulsive in the overlap region.

Fermi–Dirac

The analysis for FD statistics may be carried out in the same way. The initial wavefunction is shown in Figs. 7.5 and 7.6 but now we have an interference structure different from the BE case; the expression for R_{FD}^2 is given by (7.4.12) and (7.4.13) with N_{BE} replaced by N_{FD} and a minus sign in front of the cos term. Then the nodes and maxima are interchanged with respect to the BE distribution and the probability density is zero when $u = 0$.

The phase function satisfies the relation

$$\begin{aligned} &\tan[(S + \hbar\omega t - \tfrac{1}{2}m\omega a^2 \sin 2\omega t)/\hbar] \\ &\qquad = -\coth[(m\omega au/\hbar)\cos\omega t]\tan[(m\omega au/\hbar)\sin\omega t]. \end{aligned} \tag{7.4.27}$$

The trajectories are again solutions of (7.4.17) and obey (7.4.18). The relative coordinate equation of motion is from (7.4.27)

$$\dot{u} = 2\omega a\{\cos\omega t \sin[(2m\omega au/\hbar)\sin\omega t] - \sin\omega t \sinh[(2m\omega au/\hbar)\cos\omega t]\} \times \{\cosh[(2m\omega au/\hbar)\cos\omega t] - \cos[(2m\omega au/\hbar)\sin\omega t]\}^{-1}. \qquad (7.4.28)$$

Applying the approximation $|au| \ll (\Delta x)^2$ to (7.4.28) we see that the relative motion approximately satisfies

$$\dot{u} \approx -(2m\omega^2 a^2 u/3\hbar)\sin 2\omega t \qquad (7.4.29)$$

with solution

$$u(t) \approx u_0[1 - \tfrac{1}{3}(a/\Delta x)^2 \sin^2 \omega t]. \qquad (7.4.30)$$

This function is plotted in Fig. 7.4 from which it is obvious that the FD particles stay further apart than the corresponding BE particles with the same initial condition u_0 (in this approximation). The point of closest approach occurs at time $t = \pi/2\omega$ (perfect overlap).

The trajectories resulting from the numerical solution of (7.4.28) in the case $x_{10} + x_{20} = 0$ are depicted in Fig. 7.7(*b*). The interchange of the FD and BE interference fringes is evident. An illustration of the correlations when x_{10} is chosen to be the same for each pair of trajectories while x_{20} is varied is shown in Fig. 7.10. Comparing with Fig. 7.8 we observe that the FD particles are further apart in the overlap region than the corresponding BE ones.

The two-body quantum potential takes the form (7.4.23) where now $f(u, t)$

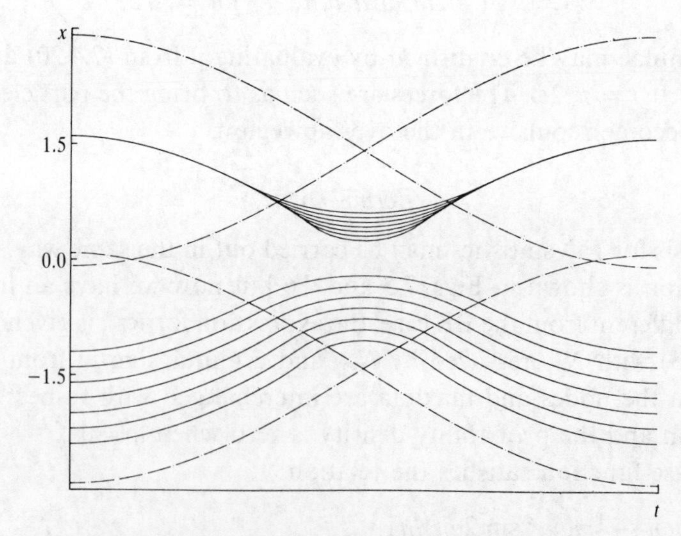

Fig. 7.10 Correlated pairs of particle trajectories (solid lines) for two harmonic oscillators, antisymmetric wavefunction. Initial position of particle 1 is chosen to be the same for a range of initial positions of particle 2. The interrupted lines show the variances of the individual wave packets (from Dewdney & Holland (1988)).

is a different function of its arguments. The forces acting initially and in the region of perfect overlap are as follows:

$$F_1 = \begin{cases} m\omega^2 a\{-\coth(m\omega a u/\hbar) + (m\omega a u/\hbar)[\sinh(m\omega a u/\hbar)]^{-2}\}, & t = 0 \\ m\omega^2 a\{-\cot(m\omega a u/\hbar) + (m\omega a u/\hbar)[\sin(m\omega a u/\hbar)]^{-2}\}, & \omega t = \pi/2 \end{cases} \tag{7.4.31}$$

while (7.4.25) remains valid. In the limiting case $|m\omega a u/\hbar| \ll 1$ we find

$$F_1 \approx \begin{cases} -2m^2\omega^3 a^2 u/3\hbar, & t = 0 \\ 2m^2\omega^3 a^2 u/3\hbar, & \omega t = \pi/2. \end{cases} \tag{7.4.32}$$

While being initially attracted, the particles are repelled in the overlap region (but with a force *less* than in the BE case (7.4.26)).

7.4.2 General remarks. The exclusion principle

It is by no means true that for given initial conditions a pair of trajectories always come closer in the BE case than in the FD case. This is evident from Fig. 7.7 where BE particles near the centre of the packets ψ_A, ψ_B may be repelled while the FD particles are attracted, and in Figs. 7.8 and 7.10 where particles in the outer part of ψ_B are repelled more in the BE than in the FD case. One cannot therefore prove a theorem to the effect that, for specified initial conditions, BE particles tend to be attracted and FD tend to be repelled throughout each individual process – a slight change in u_0 can give rise to quite distinct subsequent relative motions. The only general result we have is that when $u \to 0$, so that $R_{FD} \to 0$, the effective potential $(Q + V)$ prevents the FD particles from occupying the same point in space at an instant. The situation in this regard is the usual one when we encounter nodes.

On the other hand, evaluation of the *average* separations does yield results that confirm our expectations. The mean separation $\langle u \rangle = 0$ for each of the wavefunctions (7.4.4), (7.4.5) and (7.4.6), while the mean square separations at each instant are given by

$$\langle (x_1 - x_2)^2 \rangle_{MB} = (\Delta x)^2 + 4a^2 \cos^2 \omega t, \tag{7.4.33}$$

$$\langle (x_1 - x_2)^2 \rangle_{BE} = (\Delta x)^2 + 4a^2[\cos^2 \omega t - (e^{2m\omega a^2/\hbar} + 1)^{-1}], \tag{7.4.34}$$

$$\langle (x_1 - x_2)^2 \rangle_{FD} = (\Delta x)^2 + 4a^2[\cos^2 \omega t + (e^{2m\omega a^2/\hbar} - 1)^{-1}]. \tag{7.4.35}$$

The third term on the right hand side of each of (7.4.34) and (7.4.35) corresponds to interference between the summands in (7.4.5) and (7.4.6). Clearly, the mean square separation is less (more) in the BE (FD) case than in the MB case, but these results naturally do not capture any of the detail revealed by a study of the individual processes that make up the ensemble.

In some cases the mean square separations will coincide. Firstly, suppose that the amplitude of the packet motion is much larger than the width of the packet: $a \gg (\hbar/2m\omega)^{1/2}$. Then the interference terms in (7.4.34) and (7.4.35) are negligible (since the packets overlap for only a small fraction of the period) and we recover (7.4.33). Secondly, the BE and MB mean squares agree if ψ_A and ψ_B initially coincide ($a = 0$) (the FD case has no meaning here).

To conclude, the exclusion principle is incorporated in the quantum theory of motion in that particles cannot pass through nodes. But for small separations, the particles in the example studied here are also repelled in the BE case and for larger separations it may be found that FD particles come closer than BE ones with identical initial positions. The detailed structure of the motions implied by nonfactorizable wavefunctions is therefore subtle and no general statement to the effect that FD particles are repelled, and BE particles are attracted, in comparison with the MB case, can be supported.

7.5 Theory of quantum jumps

7.5.1 The collision of an electron with a hydrogen atom

In this section and the next we shall see how one may give an objective description of two specific but typical many-body processes: the transition between stationary states of an atom (including ionization) and the formation of a diatomic molecule. These processes may occur fortuitously or as part of specially designed experiments. In both cases, however, our account gives them an objective significance which in no way relies on the intervention of an 'observer' or a 'measuring apparatus' (Bohm, 1952a; Belinfante, 1973, p. 99).

To begin with we consider the transition of a hydrogen atom from the ground state to an excited state and show how the discrete change in energy may be understood as the outcome of a basically continuous process. A proper treatment of transitions is a many-body problem since a given system cannot make a transition unless there is another system available to give up or receive energy (in simpler treatments an applied classical field can induce transitions but there is no explanation of the discrete character of the energy change). Let us suppose that the transition takes place as a result of an inelastic collision between an incident electron and the hydrogen atom (this will give a simplified treatment of the Franck–Hertz experiment). The precise amount of change in energy is uniquely determined by the initial positions of the incident and atomic electrons.

Initially the atomic electron (with coordinate $\mathbf{x}$) is in the ground state $\psi_0(\mathbf{x})\,e^{-iE_0t/\hbar}$ and the bombarding particle (with coordinate $\mathbf{y}$) is a free wave packet

$$F(\mathbf{y}, t) = \int f(\mathbf{k} - \mathbf{K})\, \mathrm{e}^{\mathrm{i}(\mathbf{k}\cdot\mathbf{y} - \hbar\mathbf{k}^2 t/2m)}\, \mathrm{d}^3 k,$$

where $f(\mathbf{k} - \mathbf{K})$ is peaked around $\mathbf{K}$. The centre of the packet moves along the path $\mathbf{y} = \hbar\mathbf{K}t/m$ towards the atom. The atomic electron is at rest and has total energy E_0 (cf. §4.5) and the incident electron has a variable energy which is approximately $\hbar^2\mathbf{K}^2/2m$.

While the incident packet and the atom are separated in space and do not overlap the total wavefunction is a product:

$$\Psi_i(\mathbf{x}, \mathbf{y}, t) = \psi_0(\mathbf{x})\, \mathrm{e}^{-\mathrm{i}E_0 t/\hbar} F(\mathbf{y}, t).$$

The $\mathbf{x}$- and $\mathbf{y}$-electron motions are at this stage independent. When the packet reaches the neighbourhood of the atom, the two electrons interact with one another and with the nucleus via Coulomb potentials (we ignore exchange effects). We can expand the wavefunction at this stage in terms of a complete set of atomic eigenfunctions $\psi_n(\mathbf{x})$:

$$\Psi(\mathbf{x}, \mathbf{y}, t) = \Psi_i(\mathbf{x}, \mathbf{y}, t) + \left(\sum_n + \int\right)\psi_n(\mathbf{x})\, \mathrm{e}^{-\mathrm{i}E_n t/\hbar} f_n(\mathbf{y}, t), \qquad (7.5.1)$$

where $f_n(\mathbf{y}, t)$ are a set of expansion coefficients. This wavefunction contains contributions from the excited states of the atom ($n = 0$ corresponds to the ground state). The system point, initially lying in the region of configuration space where Ψ_i is appreciable, will now be influenced by the interference of Ψ_i and the scattered waves $\psi_n(\mathbf{x})f_n(\mathbf{y})$ in the region where these appreciably overlap (so that Ψ is nonfactorizable). As a result the $\mathbf{x}$- and $\mathbf{y}$-motions become closely correlated and, in general, rather complex. At this stage the total energy of each electron alone cannot be independently defined. Their momenta are highly variable and may differ considerably from the initial values, or the values associated with the excited states.

The asymptotic form of (7.5.1) is (Mott & Massey, 1965, p. 136):

$$\Psi = \Psi_i + \left(\sum_n + \int\right)\psi_n(\mathbf{x})\, \mathrm{e}^{-\mathrm{i}E_n t/\hbar} \int f(\mathbf{k}_n - \mathbf{K}_n)$$
$$\times\, r^{-1} \mathrm{e}^{\mathrm{i}(\mathbf{k}_n\cdot\mathbf{y} - \hbar\mathbf{k}_n^2 t/2m)} g(\theta, \phi, \mathbf{k}_n)\, \mathrm{d}^3 k_n \qquad (7.5.2)$$

with $\mathbf{y} = (r, \theta, \phi)$ and

$$\hbar^2\mathbf{k}_n^2/2m + E_n = \hbar^2\mathbf{K}^2/2m + E_0 \qquad (7.5.3)$$

for each n. It is assumed that $\hbar^2\mathbf{K}^2/2m + E_0 > E_n$, i.e., that the incident electron is sufficiently energetic to excite the nth state of the atom for some

$n \neq 0$. The function (7.5.2) is a sum of outgoing packets, each of group velocity $\hbar \mathbf{K}_n/m$, and each is correlated with an atomic eigenfunction $\psi_n(\mathbf{x})$. Eventually these packets will cease to overlap in **y**-space and attain a classically describable separation. The wavefunction Ψ as a whole will then be a superposition of nonoverlapping configuration space functions into one of which the representative point will move. If the regions between the packets are nodal then the point cannot pass through them because the tangent to a trajectory is undefined. If these regions are not strictly nodal then $|\Psi|^2$ will be very small there and so there is negligible probability of the system point passing from one packet to another. In either case, we see that the system remains in one outgoing packet and thus we explain how a definite outcome is achieved. The remaining configuration space packets are 'empty waves' and, as regards the motion of the particles, may be disregarded (it is assumed that the process is fast enough for spreading to be ignored). In other words, Ψ is statistically equivalent to a mixture. If the system enters the nth summand in (7.5.2) the wavefunction is therefore effectively

$$A_n \Psi_n(\mathbf{x}, \mathbf{y}, t) = \psi_n(\mathbf{x})\, \mathrm{e}^{-\mathrm{i}E_n t/\hbar} \int f(\mathbf{k}_n - \mathbf{K}_n) \times r^{-1}\, \mathrm{e}^{\mathrm{i}(\mathbf{k}_n \cdot \mathbf{y} - \hbar \mathbf{k}_n^2 t/2m)} g(\theta, \phi, \mathbf{k}_n)\, \mathrm{d}^3 k_n. \qquad (7.5.4)$$

Here we have included a real renormalization constant A_n which we may do since multiplication by a constant does not affect any physically significant quantity. The wavefunction being thus effectively factorizable into a function of **x** times a function of **y**, the atomic electron and the bombarding particle are again independent and their total energies are individually well defined. The atom is in the nth stationary state and has energy E_n and the outgoing particle is free and has a kinetic energy which is roughly $\hbar^2 \mathbf{K}_n^2/2m$ (so that it is fairly constant before and after the collision).

The atom has therefore absorbed a discrete energy $E_n - E_0$ and 'jumped' to a new stationary state. If the process of interaction occurs in a sufficiently short time it will appear to be discontinuous. Yet it is the outcome of an entirely continuous (but rapid) process. The atomic electron is initially at rest and finally in uniform circular motion, and between these states it has an unstable but well-defined motion. The actual packet eventually joined by the system point is uniquely determined by the initial positions $\mathbf{x}_0$ and $\mathbf{y}_0$ and the initial wavefunction.

Of course, in practice, we do not precisely know $\mathbf{x}_0$ and $\mathbf{y}_0$ and small differences in them can result in quite different outcomes (i.e., the atom may be left in the n'th state where $n' \neq n$). The probability density is as usual given by $|\Psi(\mathbf{x}, \mathbf{y}, t)|^2$. For the purposes of calculating expectation values, the 'state' of the two-body system after the collision is described by the density matrix

$$\rho(\mathbf{x}, \mathbf{y}; \mathbf{x}', \mathbf{y}') = \sum_n A_n^2 \Psi_n(\mathbf{x}, \mathbf{y}) \Psi_n^*(\mathbf{x}', \mathbf{y}'), \tag{7.5.5}$$

where A_n^2 is the probability that the atom is left in the nth eigenstate and the colliding particle is in the nth packet. Since $\psi_n(\mathbf{x})$ is normalized, we have from (7.5.4) that

$$A_n^2 = \int \left| \int f(\mathbf{k}_n - \mathbf{K}_n) r^{-1} \, \mathrm{e}^{\mathrm{i}(\mathbf{k}_n \cdot \mathbf{y} - \hbar \mathbf{k}_n^2 t/2m)} g(\theta, \phi, \mathbf{k}_n) \, \mathrm{d}^3 k_n \right|^2 \mathrm{d}^3 y, \tag{7.5.6}$$

where the integration is carried out over the region of space occupied by the nth outgoing packet.

What is to prevent the empty waves subsequently being brought back together by some method that does not affect the atom, overlapping with the wave that the system point actually entered and interfering with it, and thus destroying the possibility of a definite outcome (so that the atom is not in a stationary state although it will have, of course, a well-defined motion)? Such a scenario is possible, but it will be seen to be overwhelmingly unlikely if we take into account that in any real system the outgoing particle will soon interact with other systems, and in particular with systems comprising many degrees of freedom (such as the walls of a container). This indeed is necessary if we are to 'measure' the energy in order to determine (by inference) the final state of the atom, a process that involves an amplification to the macroscopic scale (see Chap. 8). Let us denote the coordinates of the further system by z. Then when the $\mathbf{y}$-particle couples with this new system, the total wavefunction (7.5.2) evolves into

$$\Psi(\mathbf{x}, \mathbf{y}, z, t) = \Psi_i + \left(\sum_n + \int \right) \psi_n(\mathbf{x}) \, \mathrm{e}^{-\mathrm{i}E_n t/\hbar} f_n(\mathbf{y}, z, t). \tag{7.5.7}$$

Now clearly, unless the packets overlap not only in $\mathbf{x}$- and $\mathbf{y}$-space but also for *all* the coordinates z, this function is equivalent to an incoherent mixture. Given the enormous number of degrees of freedom generally included in z, we conclude that for all practical purposes the probability that the functions $\psi_n(\mathbf{x})$ will subsequently interfere with one another, even though they may overlap, is negligible.

Our treatment of the scattering process contains the basic elements of a causal theory of measurement (Chap. 8). This involves a series of causally connected states: two initially independent systems come into contact, mutually transform one another, separate, and one of the systems undergoes an irreversible change making subsequent interference practically impossible. It is important to emphasize that we have given an entirely objective account of the process of transition between stationary states in terms of the actual determinist behaviour of waves and particles. We have not invoked any 'collapse' hypothesis and probability only enters because the precise initial particle positions, although well defined, are unknown to us.

We observe that the above treatment also includes the possibility of ionization, where the atomic electron is ejected from the atom with an energy E lying in the continuous spectrum. For this to occur we would require the incident energy of the bombarding particle to satisfy the condition

$$\hbar^2\mathbf{K}^2/2m + E_0 > E. \tag{7.5.8}$$

The Franck–Hertz experiment is considered to provide evidence for the conservation of energy in individual processes. If the initial energy of the colliding electron is sufficiently controlled, then (7.5.3) expresses a relation between the actual initial and final energies of the two particles and hence the conservation of energy. In general, however, the initial and final energy of the **y**-particle is not quite constant and includes quantum potential energy.

7.5.2 *Pauli's objection*

The example studied in this section is conceptually the same as that used by Pauli in his criticism of de Broglie's pilot-wave theory (the inelastic scattering of a particle off a rigid rotator – for details of the actual example used by Pauli see Fermi (1926), *Electrons et Photons* (1928), de Broglie (1956) and Jammer (1974, p. 111)). Pauli argued that the initial wavefunction should be a product of the atomic ground state wavefunction with an incident plane wave ((7.5.2) with $f(\mathbf{k} - \mathbf{K}) = \delta(\mathbf{k} - \mathbf{K})$). Repeating the steps given above with this new wavefunction, one sees from (7.5.2) that the asymptotic wavefunction will not evolve into a set of nonoverlapping parts, each corresponding to a single atomic wavefunction, but rather into a single configuration space wave in which the atomic eigenfunctions continue to interfere. As a result the **x**- and **y**-motions remain correlated and no definite result seems to be obtained. Since it is known from experiment that both the atom and outgoing electron do eventually obtain definite energies, Pauli concluded that the pilot-wave theory is untenable.

There are two answers to this objection (Bohm, 1952b, App. B; de Broglie, 1956; Belinfante, 1973, p. 96). First of all, *whatever the interpretation* of quantum mechanics, the use of a plane wave is an excessive idealization never realized in practice. Although plane waves are convenient abstractions for theoreticians, the incident wave will in reality be of finite extent. Then the scattered waves are finite and Pauli's conclusion will not follow, in the quantum theory of motion or in any other approach to quantum theory. Secondly, it may indeed be the case that the incident packet spreads out over such a large region that the outgoing packets always overlap to some extent, and so an unambiguous outcome is not obtained. But this simply means that we have not satisfied the conditions necessary for a precise determination of the energy. If a further apparatus were introduced in order to effect an energy measurement, then we would find that the configuration space wavefunction breaks up into a set of nonoverlapping packets of the kind that we have described in this section. And again, this will be so regardless of the interpretation we may put on the process.

7.6 The formation of a molecule

As a further illustration of a typical quantum phenomenon treated as an objective process, we shall show that the force between two neutral atoms which causes them to form a stable molecule (attractive at large distances, repulsive at small) is due to the quantum potential. No known classical force can explain the covalent bond so we are dealing here with a purely quantum effect (due to exchange degeneracy). It is sufficient to treat the simplest example of a covalent bond, that of the formation of a hydrogen molecule from two hydrogen atoms in the ground state (Heitler & London, 1927; Pitzer, 1954, Chap. 8), and to work in the approximation of first-order perturbation theory. In order to avoid calculating the quantum potential of the two-electron interaction from the perturbed two-body wavefunction, which is difficult, we find it using the standard calculation for the perturbed energy level. The latter has a simple relationship with Q via the Hamilton–Jacobi equation for the electrons (see the Appendix).

The time-independent wave equation for a system of two hydrogen atoms is (Schiff, 1968, p. 447):

$$[-(\hbar^2/2m)(\nabla_1^2 + \nabla_2^2) - (\hbar^2/2M)(\nabla_A^2 + \nabla_B^2) + V]\Psi = E\Psi, \quad (7.6.1)$$

where

$$V = -\frac{e^2}{r_{A1}} - \frac{e^2}{r_{B2}} - \frac{e^2}{r_{A2}} - \frac{e^2}{r_{B1}} + \frac{e^2}{r_{12}} + \frac{e^2}{r}, \quad (7.6.2)$$

$\Psi = \Psi(\mathbf{x}_1, \mathbf{x}_2, \mathbf{r}_A, \mathbf{r}_B)$, $\mathbf{x}_i$, $i = 1, 2$, are the coordinates of the electrons, $\mathbf{r}_i$, $i = \text{A, B}$, are the coordinates of the nuclei, m is the electron mass, M is the

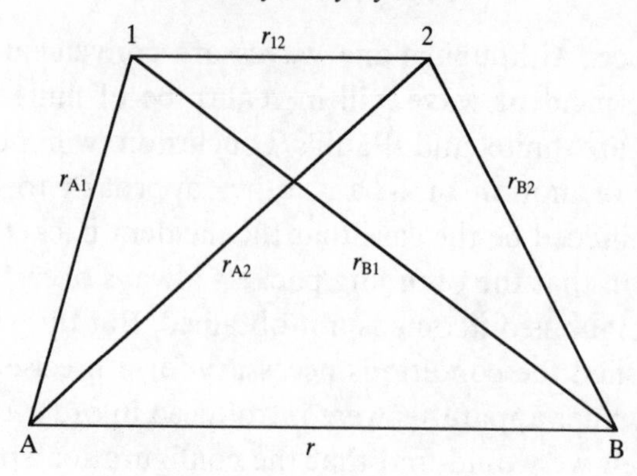

Fig. 7.11 The definition of the distances used in the text.

nucleon mass, and the other distances are defined in Fig. 7.11. Bearing in mind the large disparity between m and M, we assume that the nuclei may be treated as fixed in space and their motion neglected. We write

$$\Psi(\mathbf{x}_1, \mathbf{x}_2, \mathbf{r}_A, \mathbf{r}_B) = \psi_{\mathbf{r}_i}(\mathbf{x}_1, \mathbf{x}_2)w(\mathbf{r}_A, \mathbf{r}_B), \tag{7.6.3}$$

where the function $\psi_{\mathbf{r}_i}(\mathbf{x}_1, \mathbf{x}_2)$ satisfies the equation

$$[-(\hbar^2/2m)(\nabla_1^2 + \nabla_2^2) + V]\psi_{\mathbf{r}_i}(\mathbf{x}_1, \mathbf{x}_2) = E(\mathbf{r}_A, \mathbf{r}_B)\psi_{\mathbf{r}_i}(\mathbf{x}_1, \mathbf{x}_2). \tag{7.6.4}$$

(The assumption made here is not only that the nucleonic kinetic energies may be neglected in relation to the electronic kinetic energies, but also that part of the quantum potential is negligible.) The nuclear coordinates appear as parameters in the two-body electron equation (7.6.4). Substituting (7.6.3) into (7.6.1) and using (7.6.4), one finds that the energy eigenvalue $E(\mathbf{r}_A, \mathbf{r}_B)$ of the electronic motion is approximately the potential energy of interaction of the nuclei. We shall treat (7.6.4) as the perturbed wave equation.

Consider two hydrogen atoms initially far apart and gradually brought closer together. The electrons are in the ground state with energy E^0 (cf. §4.5):

$$\left.\begin{aligned} \psi_A(\mathbf{x}_1, t) &= (\pi a^3)^{-1/2}\,\mathrm{e}^{-r_{A1}/a}\,\mathrm{e}^{-iE^0t/\hbar} \\ \psi_B(\mathbf{x}_2, t) &= (\pi a^3)^{-1/2}\,\mathrm{e}^{-r_{B2}/a}\,\mathrm{e}^{-iE^0t/\hbar}. \end{aligned}\right\} \tag{7.6.5}$$

The identity of the particles requires that we symmetrize or antisymmetrize the spatial wavefunction (§7.2.3):

$$\psi_\pm(\mathbf{x}_1, \mathbf{x}_2) = N_\pm[\psi_A(\mathbf{x}_1)\psi_B(\mathbf{x}_2) \pm \psi_A(\mathbf{x}_2)\psi_B(\mathbf{x}_1)], \tag{7.6.6}$$

where $N_\pm$ is a normalization constant. Noting that electrons are fermions, and taking into account the electron spin, the set of possible totally antisymmetric wavefunctions is as follows:

$$\psi_+(\mathbf{x}_1, \mathbf{x}_2)(1/\sqrt{2})[u_+(1)u_-(2) - u_-(1)u_+(2)] \quad (7.6.7)$$

$$\psi_-(\mathbf{x}_1, \mathbf{x}_2)\begin{Bmatrix} (1/\sqrt{2})[u_+(1)u_-(2) + u_-(1)u_+(2)] \\ u_+(1)u_+(2) \\ u_-(1)u_-(2) \end{Bmatrix} \quad (7.6.8)$$

where $u_+(1)$ means particle 1 is spin up etc. (Chap. 9). The spin will play no part in our considerations below and we shall henceforth ignore it. Because of their connection with the spin functions, ψ_+ will be called the singlet state and ψ_- the triplet state.

The function (7.6.6) is our unperturbed wavefunction. Since the spatial part is real, the electrons are at rest ($\nabla_1 S = \nabla_2 S = 0$). The summands in (7.6.6) satisfy different wave equations. For the first term the unperturbed Hamiltonian is

$$\hat{H}^0 = -(\hbar^2/2m)(\nabla_1^2 + \nabla_2^2) - \frac{e^2}{r_{\mathrm{A}1}} - \frac{e^2}{r_{\mathrm{B}2}} \quad (7.6.9)$$

and the perturbing term is

$$\lambda V = -\frac{e^2}{r_{\mathrm{B}1}} - \frac{e^2}{r_{\mathrm{A}2}} + \frac{e^2}{r_{12}} + \frac{e^2}{r} \quad (7.6.10)$$

while for the second term (7.6.9) and (7.6.10) are modified by interchanging the labels 1 and 2. More generally this will be so for the summands of wavefunctions of the type

$$\psi_{nn'}(\mathbf{x}_1, \mathbf{x}_2) = \psi_{\mathrm{A}n}(\mathbf{x}_1)\psi_{\mathrm{B}n'}(\mathbf{x}_2) \pm \psi_{\mathrm{A}n}(\mathbf{x}_2)\psi_{\mathrm{B}n'}(\mathbf{x}_1), \quad (7.6.11)$$

where $\psi_{\mathrm{A}n}(\mathbf{x}_1)$ is the nth eigenstate of electron 1 associated with nucleus A etc. (n includes the three quantum numbers (n, l, m); cf. §4.5). In order to apply perturbation theory we should therefore insert one or other of the λVs in evaluating the matrix element (cf. (A5))

$$V_{nn',ss'} = \int \psi^*_{nn'} \lambda V \psi_{ss'} \, \mathrm{d}^3x_1 \, \mathrm{d}^3x_2 \quad (7.6.12)$$

according as λV multiplies one or other of the summands (Heitler, 1956, p. 127). It turns out that this is equivalent to always using (7.6.10) in (7.6.12) since the integral is symmetric with respect to the interchange of 1 with 2.

The perturbed energy level depends on just the internuclear distance r.

The result is

$$E_{\pm}(r) = \int \psi_{\pm}^* \hat{H} \psi_{\pm} \, d^3x_1 \, d^3x_2, \qquad \hat{H} = \hat{H}^0 + \lambda V$$
$$= 2E^0 + (C \pm X)(1 \pm \Delta)^{-1}. \tag{7.6.13}$$

The first term on the right hand side of (7.6.13) ($2E^0$) is the total energy of two noninteracting electrons, and the remainder is due to the interaction. Here $N_{\pm}^{-2} = 1 \pm \Delta(r)$,

$$C(r) = \int |\psi_A(\mathbf{x}_1)|^2 |\psi_B(\mathbf{x}_2)|^2 \left(\frac{e^2}{r} + \frac{e^2}{r_{12}} - \frac{e^2}{r_{A2}} - \frac{e^2}{r_{B1}} \right) d^3x_1 \, d^3x_2$$

represents the Coulomb interaction of the 'charge cloud' around A with nucleus B, of that around B with nucleus A, of the two charge clouds, and of the two nuclei, and

$$X(r) = \int \psi_A^*(\mathbf{x}_1)\psi_B(\mathbf{x}_1)\psi_A(\mathbf{x}_2)\psi_B^*(\mathbf{x}_2) \left(\frac{e^2}{r} + \frac{e^2}{r_{12}} - \frac{e^2}{r_{A2}} - \frac{e^2}{r_{B1}} \right) d^3x_1 \, d^3x_2$$

is the exchange integral which arises from the interference of the summands. It is the latter which is numerically the most significant term in explaining the formation of a bond. The behaviour of $E_{\pm}(r)$ for all r is sketched in Fig. 7.12. The energy of the triplet state, $E_-(r)$, increases as the atoms approach one another and is a repulsive potential function at all distances. In contrast, the singlet state energy $E_+(r)$ decreases and has a stable minimum at a separation of the order of the Bohr radius a. For smaller r it becomes strongly repulsive.

The Hamilton–Jacobi equation (7.1.2) for the two-electron system derived

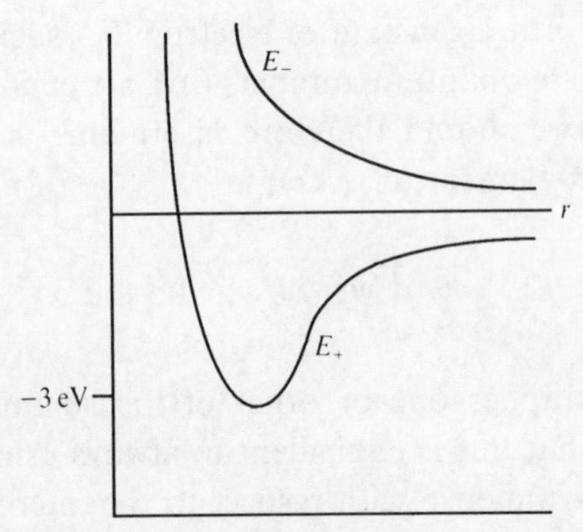

Fig. 7.12 The internuclear potentials in the singlet (E_+) and triplet (E_-) cases.

from (7.6.4) reduces to

$$E_{\pm}(r) = Q_{\pm} + V, \tag{7.6.14}$$

where $Q_{\pm} = Q_{\pm}(\mathbf{x}_1, \mathbf{x}_2, r)$ is the quantum potential constructed from the perturbed wavefunction, since to this order of approximation the kinetic energy of the electrons may be neglected (see Appendix). The r-dependence of the total potential energy of the electrons depicted in Fig. 7.12 is therefore due to $Q_{\pm}$ which cancels the $(\mathbf{x}_1, \mathbf{x}_2)$-coordinate-dependent part of $V(\mathbf{x}_1, \mathbf{x}_2, r)$ to leave just a function of r. In particular, Q_+ is responsible for $E_+(r)$ becoming negative for a certain range of r. Recalling that $E_{\pm}(r)$ may be treated as the internuclear potential, we have from (7.6.14) that the effective internuclear force is given by

$$-\frac{\mathrm{d}E_{\pm}(r)}{\mathrm{d}r} = -\frac{\partial Q_{\pm}}{\partial r} + \frac{e^2}{r^2}. \tag{7.6.15}$$

According to this treatment, *the existence of an internuclear potential, which implies the attraction of neutral atoms and the formation of a stable bond, is due to the quantum potential energy of interaction of the electrons in the singlet state.* This conclusion will evidently remain valid in a more accurate treatment.

It follows from (7.6.14) that the total force acting on each electron, $-\partial(Q_{\pm} + V)/\partial\mathbf{x}_i$, $i = 1, 2$, is zero, so the particles are in uniform motion. We can gain an idea of their likely position by considering the probability density implied by (7.6.6) and (7.6.5):

$$|\psi_{\pm}|^2 = N_{\pm}^2(\pi a^3)^{-2}(\mathrm{e}^{-2(r_{\mathrm{A}1}+r_{\mathrm{B}2})/a} + \mathrm{e}^{-2(r_{\mathrm{A}2}+r_{\mathrm{B}1})/a} \pm 2\,\mathrm{e}^{-(r_{\mathrm{A}1}+r_{\mathrm{B}2}+r_{\mathrm{A}2}+r_{\mathrm{B}1})/a}). \tag{7.6.16}$$

The interference term is appreciable only in the region between the nuclei. In the singlet case (ψ_+), (7.6.16) is maximum in this region (and practically zero in the triplet case (ψ_-)).

We have seen how the identity of particles implies the existence of a nonclassical attractive force between uncharged atoms. The same principle underlies the formation of more general molecular patterns, which are thereby explained by the properties of the quantum potential.

7.7 Other approaches to a causal interpretation

One might wonder whether it is possible to circumvent the use of multi-dimensional waves and associate fields propagating in Euclidean 3-space with each particle in a many-body system. Two possible approaches to this problem will be briefly described here.

One method is to define the 'state' of the ith particle as the reduced density matrix formed by taking the trace of the (pure state) density matrix of the whole many-body system over all indices except the ith:

$$\rho_i(\mathbf{x}_i, \mathbf{x}_i') = \int \psi(\mathbf{x}_1, \ldots, \mathbf{x}_{i-1}, \mathbf{x}_i, \mathbf{x}_{i+1}, \ldots, \mathbf{x}_n) \times \psi^*(\mathbf{x}_1, \ldots, \mathbf{x}_{i-1}, \mathbf{x}_i', \mathbf{x}_{i+1}, \ldots, \mathbf{x}_n)\, \mathrm{d}^3x_1 \cdots \widehat{\mathrm{d}^3x_i} \cdots \mathrm{d}^3x_n. \quad (7.7.1)$$

This is a reasonable definition insofar as knowledge of ρ_i is sufficient to calculate the expectation values of all operators pertaining to the ith system alone. We may then define a momentum field

$$\mathbf{p}_i(\mathbf{x}_i) = (\hbar/\rho_i(\mathbf{x}_i, \mathbf{x}_i)) \operatorname{Im} \nabla_{\mathbf{x}_i} \rho_i(\mathbf{x}_i, \mathbf{x}_i')|_{\mathbf{x}_i = \mathbf{x}_i'}$$

$$= \int R^2 \nabla_i S\, \mathrm{d}^3x_1 \cdots \widehat{\mathrm{d}^3x_i} \cdots \mathrm{d}^3x_n \bigg/ \int R^2\, \mathrm{d}^3x_1 \cdots \widehat{\mathrm{d}^3x_i} \cdots \mathrm{d}^3x_n \quad (7.7.2)$$

and a density $R_i^2(\mathbf{x}_i) = \rho_i(\mathbf{x}_i, \mathbf{x}_i)$. These identifications are justified by the fact that one may derive the conservation law

$$\frac{\partial R_i^2}{\partial t} + \nabla_i \cdot (R_i^2 \mathbf{p}_i/m_i) = 0 \quad (7.7.3)$$

by integrating (7.1.3) with respect to the variables $\mathbf{x}_1, \ldots, \mathbf{x}_{i-1}, \mathbf{x}_{i+1}, \ldots, \mathbf{x}_n$. Note that $\mathbf{p}_i$ is not a gradient field.

We might then propose that the wave component of the state of the ith particle is specified by the physical fields R_i^2 (which in addition would be a probability density) and $\mathbf{p}_i$ (from which we would derive trajectories by integrating $\dot{\mathbf{x}}_i = \mathbf{p}_i/m$ given $\mathbf{x}_{i0}$). This definition is consistent in the case of factorizable wavefunctions for then R_i^2 and $\mathbf{p}_i$ reduce to the usual expressions for a single body. However, in the general entangled case this approach is unsatisfactory because (7.7.1) involves a significant loss of information concerning the behaviour of the ith particle in relation to the remaining particles, and, in fact, leads to inconsistencies.

To begin with, the same partial density matrix may be associated at an instant with different wavefunctions and hence, while its value at any subsequent instant is well defined, it is not unique. More seriously, the trajectory of the ith particle calculated from $\mathbf{p}_i$ will not, in general, coincide with that calculated from the many-body wavefunction via (7.1.7), since the latter depends on the initial positions of all the particles. In this connection we see that, when there is no classical interaction, (7.7.1) (and hence $\mathbf{p}_i$) is insensitive to external disturbances applied to any of the particles other than the ith (§11.3). Yet $\mathbf{x}_i(t)$ evaluated from (7.1.7) is indeed altered by such perturbations.

We see that the attempt to associate fields with each particle in a many-body system in this way does not provide a complete and accurate description of the behaviour of the particles because it takes no account of interparticle correlations.

A second possible approach to a causal formulation in three dimensions starts from the observation that in the classical mechanics of many-body systems it is not necessary to apply Hamilton–Jacobi theory in configuration space. Instead, one may introduce a set of Hamilton–Jacobi functions, one for each particle and satisfying its own equation, into which the coordinates of all the other particles enter as parameters, their trajectories being assumed known. If one introduces a quantum-potential-type term into each of these equations, the possibility arises of treating a many-body system as a set of mutually interacting waves in 3-space, one wave being associated with each particle (de Broglie, 1953a, pp. 79 & 86; Freistadt, 1957). However, the basic equations describing the evolution of these waves depend essentially on the trajectory concept and so there is a question of compatibility with the Schrödinger equation which, of course, does not. Moreover, since the Hamilton–Jacobi functions depend on all the particle coordinates, it seems that the feature of nonlocality present in the causal interpretation based directly on the Schrödinger equation will still occur here and it is not clear what has been gained. We will not discuss this approach further.

We conclude that a complete and accurate account of the motions of particles moving in accordance with the laws of quantum mechanics must be directly connected with multidimensional waves dynamically evolving in configuration space.

Appendix: The connection between Q and stationary perturbation theory

For the purposes of the following demonstration it is sufficient to restrict attention to the one-body problem. The unperturbed time-independent wave equation will be written as

$$H^0\psi_n^0 = E_n^0\psi_n^0, \qquad H^0 = -(\hbar^2/2m)\nabla^2 + V^0(\mathbf{x}) \tag{A1}$$

and the perturbed wave equation as

$$H\psi_n = E_n\psi_n, \qquad H = H^0 + \lambda V, \tag{A2}$$

where λ is a small real constant. The time-independent eigenfunctions ψ_n^0 and ψ_n are normalized. To first order in λ we have the following relations between the perturbed and unperturbed energy levels and eigenfunctions (Schiff, 1968, p. 245):

$$E_n = E_n^0 + \lambda V_n = \int \psi_n^{0*} H \psi_n^0 \, \mathrm{d}^3 x, \tag{A3}$$

$$\psi_n = \psi_n^0 + \lambda \sum_{s \neq n} \psi_s^0 V_{sn}/(E_n^0 - E_s^0), \tag{A4}$$

where

$$V_{ns} = \int \psi_n^{0*} V \psi_s^0 \, \mathrm{d}^3 x \tag{A5}$$

and $V_n \equiv V_{nn}$.

Let us consider how the particle motion is affected by the perturbation. Writing $\psi_n^0 = R_n^0 \, \mathrm{e}^{\mathrm{i}S_n^0/\hbar}$ and $\psi_n = R_n \, \mathrm{e}^{\mathrm{i}S_n/\hbar}$ we find from (A1) and (A2) two Hamilton–Jacobi equations:

$$E_n^0 = (\nabla S_n^0)^2/2m + Q_n^0 + V^0, \qquad Q_n^0 = -\hbar^2 \nabla^2 R_n^0/2mR_n^0 \tag{A6}$$

$$E_n = (\nabla S_n)^2/2m + Q_n + V^0 + \lambda V, \qquad Q_n = -\hbar^2 \nabla^2 R_n/2mR_n. \tag{A7}$$

These are the expressions for the total unperturbed and perturbed particle energies respectively. The connection between the perturbed and unperturbed momentum fields may be evaluated from (A4):

$$\begin{aligned} \nabla S_n = \nabla S_n^0 + \lambda (R_n^0)^{-2} \Big((\hbar/2\mathrm{i}) \sum_{s \neq n} & \{ [\psi_n^{0*} \nabla \psi_s^0 - (\nabla \psi_n^{0*}) \psi_s^0] V_{sn} \\ & + [(\nabla \psi_n^0) \psi_s^{0*} - \psi_n^0 \nabla \psi_s^{0*}] V_{ns} \} / (E_n^0 - E_s^0) \\ & - \nabla S_n^0 \sum_{s \neq n} (\psi_n^{0*} \psi_s^0 V_{sn} + \psi_n^0 \psi_s^{0*} V_{ns})/(E_n^0 - E_s^0) \Big), \end{aligned} \tag{A8}$$

where we have used the following relation between the probability densities:

$$R_n^2 = (R_n^0)^2 + \lambda \sum_{s \neq n} (\psi_n^{0*} \psi_s^0 V_{sn} + \psi_n^0 \psi_s^{0*} V_{ns})/(E_n^0 - E_s^0). \tag{A9}$$

Given E_n from (A3), (A7) implies a formula for the sum of the kinetic and quantum potential energies of the particle in the perturbed state. Consider the case where the nth unperturbed eigenfunction ψ_n^0 is real (ψ_s^0, $s \neq n$, may be complex) so that $\nabla S_n^0 = 0$ (outside nodes). Then by (A8), ∇S_n is of order λ so that the perturbed kinetic energy $(\nabla S_n)^2/2m$ is negligible to this order of approximation. Eq. (A7) then reduces to

$$E_n = Q_n + V^0 + \lambda V. \tag{A10}$$

The perturbed energy level is therefore determined by the quantum potential and the total classical potential. Conversely, given E_n, (A10) provides a formula for Q_n. The total force acting on the particle is zero and so the motion is uniform.

8

Theory of experiments

8.1 Measurement in classical physics

Measurements in classical physics are the means by which we come to know the current state of a mechanical system without appreciably disturbing it. The laws of classical mechanics imply that we may interact with the system in such a way that this ideal may be approached arbitrarily closely, at least in principle (i.e., in theory, ignoring practical problems). The aim of this chapter is to examine how far we must revise this programme in the quantum theory of measurement.

Did the classical physicist believe in the objective existence of material systems with well-defined position and momentum just because these quantities could be empirically determined with arbitrary precision at the same time? It seems she made a stronger assumption than this; that two successive observations of an apple, first in a tree and then on the ground, are *connected* to one another by virtue of the fact that the apple actually *has* a definite position and momentum at each instant, whether it is observed or not. Thus, the programme of classical physics was not merely to predict the results of future experiments by inserting data from previous ones into the laws of motion, but to show how the observations are a function of a causally connected sequence of unique processes undergone by the system of interest, which possesses properties independently of measurements. Measurement was then something of a secondary affair and indeed, prior to quantum mechanics, the theory of measurement as a distinct branch of theoretical physics did not exist.

It is interesting to note that if, as was the case, it is believed that classical physics (Newton's laws and electromagnetism) is universal, it is a logical necessity in a complete treatment of observation that the observer must be included in the theory as an object in her own right. Nevertheless, we can break the chain of inference because it is legitimate to neglect the effect of the apparatus (and any further apparatuses observing it) on the object under investigation. There is, in particular, no need to invoke the consciousness of the perceiving subject as an active participant – reading a meter just reveals what has already happened and has no influence on the course of phenomena. It is necessary to make these rather obvious remarks for later comparison with the quantum theory.

We can illustrate the essential characteristics of a classical measurement by means of the following simple model (our quantum treatment in §8.3 will be closely related to this). Suppose we wish to measure the value of a physical quantity $A(x, p_x)$ which is a function of the coordinate x of a particle of mass m and the canonical momentum p_x (we work in one dimension for simplicity). We shall do this by letting the system x interact with an apparatus 'particle' of mass M having coordinate y and canonical momentum p_y, the interaction Hamiltonian being

$$H = gA(x, p_x)p_y, \tag{8.1.1}$$

where g is a coupling constant. It is assumed that H is impulsive and of sufficient strength that the individual kinetic and potential energies of the interacting particles may be neglected during the period of interaction. The Hamiltonian (8.1.1) has the property that the value of A is not changed by the evolution it generates, since

$$\mathrm{d}A/\mathrm{d}t = \{A, H\} = 0, \tag{8.1.2}$$

where $\{\ ,\ \}$ are Poisson brackets. Likewise, p_y will not change. The idea is that the coordinate y will become correlated with system x in such a way that observing it will reveal by inference the value of A at the instant of the measurement.

To see that this indeed happens we substitute (8.1.1) into Hamilton's canonical equations:

$$\dot{x} = \partial H/\partial p_x = g(\partial A/\partial p_x)p_y, \tag{8.1.3}$$

$$\dot{y} = \partial H/\partial p_y = gA, \tag{8.1.4}$$

$$\dot{p}_x = -\partial H/\partial x = -g(\partial A/\partial x)p_y, \tag{8.1.5}$$

$$\dot{p}_y = -\partial H/\partial y = 0. \tag{8.1.6}$$

These may be readily integrated once we know A and x_0, y_0, p_{x0} and p_{y0}, the initial values of the four canonical coordinates. It will be observed from (8.1.4) that the coordinate y (meter needle) is indeed correlated with A. The strength and duration of the interaction are chosen so as to allow y to be distinguished from y_0. In general the position x and canonical momentum p_x of the system of interest are also transformed by this process, as is a conjugate quantity $B(x, p_x)$ satisfying $\{A, B\} \neq 0$, but as we have said, A itself is invariant.

We now assert that the disturbances in the various x-quantities may be made arbitrarily small. From (8.1.6) we obtain $p_y = p_{y0}$ and substituting in (8.1.3) and (8.1.5) gives $\dot{x} = \dot{p}_x \approx 0$ if p_{y0} is chosen sufficiently small (but still finite). Integrating (8.1.3)–(8.1.6) we therefore obtain, in this limit,

$$x = x_0, \quad y = y_0 + gAT, \quad p_x = p_{x0}, \quad p_y = p_{y0}, \tag{8.1.7}$$

where T is the period of the impulse. Hence, we can infer by observation of y the value of A immediately prior to the measurement with negligible disturbance of any particle property. Position and momentum measurements correspond to inserting $A = x$ or p_x in (8.1.7):

$$x = x_0, \quad y = y_0 + gx_0T, \quad p_x = p_{x0}, \quad p_y = p_{y0}, \tag{8.1.8}$$

$$x = x_0, \quad y = y_0 + gp_{x0}T, \quad p_x = p_{x0}, \quad p_y = p_{y0}. \tag{8.1.9}$$

Pursuing this method of measurement, it is clear that we may simultaneously measure the position and momentum of a classical particle to arbitrary accuracy, i.e., with negligible disturbance of either. To do this we let the system x interact simultaneously with two apparatuses, having coordinates y and z and masses M and M' respectively, via an impulsive interaction Hamiltonian

$$H = gxp_y + g'p_xp_z, \tag{8.1.10}$$

where g and g' are coupling constants. Applying Hamilton's equations to (8.1.10) as above and taking the limit of negligible p_{y0} and p_{z0}, we obtain

$$\left.\begin{aligned} x = x_0, \quad y = y_0 + gx_0T, \quad z = z_0 + g'p_{x0}T, \\ p_x = p_{x0}, \quad p_y = p_{y0}, \quad p_z = p_{z0}. \end{aligned}\right\} \tag{8.1.11}$$

Observation of y and z reveals the current values of x and p_x respectively, and the latter are effectively unaltered.

To compare with our quantum treatment later on we now introduce a statistical element and consider a Gibbs ensemble of identical interacting x- and y-systems which differ only in their initial phase space coordinates. We suppose that the apparatus has initial distribution function $f_2(y, p_y)$ whose width Δy in y-space is much smaller than the differences between the readings corresponding to distinct values of A, and that the object variables are initially distributed according to the density $f_1(x, p_x)$. The joint phase space density $f(x, p_x, y, p_y)$ evolves according to Liouville's equation:

$$\partial f/\partial t + \{H, f\} = 0 \tag{8.1.12}$$

where H is given by (8.1.1). To simplify matters let us consider a momentum measurement, $A = p_x$. Then (8.1.12) becomes

$$\frac{\partial f}{\partial t} + gp_y\frac{\partial f}{\partial x} + gp_x\frac{\partial f}{\partial y} = 0, \tag{8.1.13}$$

which has the general solution

$$f = f_0(x - gp_yT, p_x, y - gp_xT, p_y) \tag{8.1.14}$$

apart from an additive function of p_x and p_y. Then, since the initial distribution function is factorizable,

$$f_0(x, p_x, y, p_y) = f_1(x, p_x)f_2(y, p_y), \tag{8.1.15}$$

we obtain in the limit (8.1.9),

$$f(x, p_x, y, p_y, T) = f_1(x, p_x)f_2(y - gp_xT, p_y). \tag{8.1.16}$$

The distribution function of system x is unaffected whereas that of the apparatus is correlated with p_x and has shifted as a unit. In an ensemble of p_x-measurements we obtain a range of y-results from which we may infer the partial distribution function $\rho(p_x) = \int f_1(x, p_x)\,\mathrm{d}x$. The rms scatter in the values obtained, Δp_x, represents the spread of the true momentum values prior to the measurement. We may express (8.1.16) as a superposition of nonoverlapping 'probability packets' in phase space:

$$f(x, p_x, y, p_y, T) = \int \mathrm{d}p'_x f_1(x, p'_x)\delta(p_x - p'_x)f_2(y - gp'_xT, p_y). \tag{8.1.17}$$

On making an observation of y, we may infer in each individual case the true value of momentum, p'_x, and subsequently work with a new distribution function obtained by deleting all the summands in (8.1.17) that do not correspond to the actual result and renormalizing. Then

$$f_1(x, p_x) \to \delta(p_x - p'_x)f_1(x, p'_x)[\rho(p'_x)]^{-1}. \tag{8.1.18}$$

The original distribution function has therefore 'collapsed' as a result of our observation. This does not, of course, represent any objective change in the system x.

These considerations are easily extended to more general cases. In a simultaneous measurement of x and p_x, f_1 will be replaced by $\delta(x - x')\delta(p_x - p'_x)$ in each trial. If the observation of y by which we gain information on x is less than fully precise, we should replace f_1 by a distribution function reflecting our less than perfect knowledge of y, and hence x.

Finally, we note that to compare like with like, we should consider a classical ensemble associated with a single-valued momentum field that may be ascribed a classical wavefunction as described in Chap. 2. This adds nothing essential to the above discussion, being simply a special case. For a position measurement, the Hamilton–Jacobi equation is

$$\frac{\partial S}{\partial t} + gx\frac{\partial S}{\partial y} = 0, \tag{8.1.19}$$

where $S = S(x, y, t)$. For the laws of motion of the observed system and the apparatus we obtain respectively $\dot{x} = (1/m)(\partial S/\partial x) \approx 0$ and

$$\dot{y} = (1/M)(\partial S/\partial y) + gx \approx gx$$

due to the conditions of approximation of the problem, in accord with (8.1.7) and (8.1.8). The solution of (8.1.19) together with that for the conservation law yields the following classical wavefunction when the interaction terminates:

$$\Psi(x, y, T) = \psi(x)\phi_0(y - gxT), \tag{8.1.20}$$

where $\psi(x)$ is the initial classical wavefunction of the system and $\phi_0(y)$ that of the apparatus. Precisely this formula results from a quantum impulsive measurement of position (§8.4).

8.2 The measurement problem in quantum theory

The classical ideal of passive measurements which simply reveal a preexisting reality cannot be sustained when we come to a quantum treatment of this problem. In crude terms, one may say that at the quantum level the probe becomes as significant as the probed so one cannot 'calculate away' its influence to leave pure information regarding preexisting properties of an object. Indeed, it is commonplace in this discussion to question whether a quantum system 'has' properties at all beyond the specification of experimental devices employed to observe it. It has proved extraordinarily difficult to explain and account for this holistic aspect of quantum processes. This is not simply a question of the finite value of Planck's constant, although this certainly contributes to fixing the scale at which the quantum aspects become relevant. It is also a question of the specific properties of the equation (Schrödinger) that is assumed to govern all physical interactions, especially its linearity which is the feature responsible for generating irreducible couplings between objects and apparatuses. Because these interactions entail transformations of the object that only in special cases reveal the values of properties without altering them, the continued use of the word 'measurement' in this context is liable to fuel misconceptions (it has been suggested that it be banned altogether from physics (Bell, 1990a)). Certainly, the subject has been made opaque by an unfortunate and inconsistent use of language. The phrases 'observable' and 'measuring an operator', for instance, do not convey an accurate picture of measuring processes which follow in our treatment (§8.3) (although to maintain continuity with previous authors we shall often slip into the jargon). Before coming to this, we review in this section some of the difficulties that are posed by this most notorious of quantal puzzles. We start with Bohr and then go on to consider von Neumann's contribution.

The interpretation of Bohr (1948) avoids analysing the details of the interaction between an object and an apparatus by declaring the two to form an indivisible and unanalysable whole. Bohr considers that a condition for the well-defined application of the formalism is that the whole experimental arrangement, described in classical terms, should be specified in advance. That is, the ψ-function in a sense tacitly involves the specification of a classically-describable apparatus which is not explicitly included in the formalism. A distinction should be made in principle between measuring instruments and objects under investigation, but these do not have equal status: the description of the instruments is based on spacetime pictures but the objects are nonvisualizable and should not be assigned conventional physical attributes. Yet, Bohr says, one cannot separate the behaviour of the objects from their interaction with the instruments.

Now, simply assuming that there are systems around that can play the role of classically describable apparatuses hardly seems satisfactory in a fundamental physical theory. Physical instruments are composed of atoms undergoing complicated many-body and thermodynamic transformations and are, surely, suitable candidates for quantum-theoretical investigation. In fact, Bohr admits that the construction and functioning of any apparatus will depend on properties of materials which are themselves subject to quantal laws. He requires though that there is a limit in which one may disregard the fine quantal structure of the device and invoke classical concepts to describe it. He characterizes this limit as the case where the instruments are sufficiently heavy compared with the atomic objects under investigation.

Two objections relevant to the subject of this chapter may be raised against Bohr's theses. First, if we accept that the experimental device may itself be made the subject of quantum mechanics, consistency in Bohr's position would require that a further classically-functioning instrument is postulated that can provide the conditions for an unambiguous application of the formalism to the initial device. Then we rapidly get into an infinite regress in which the universe as a whole should, in principle, be subject to quantal laws and yet there must be an 'external' observer available to define the conditions under which the theory may be applied to it, thus contradicting the notion that the universe embraces 'all that is'. Second, in admitting the necessity of invoking a limiting operation to ensure the consistency of his interpretation, Bohr seems to be tacitly accepting the need for some kind of account of the object–apparatus interaction that would tell us when it is legitimate to apply classical concepts to one and not the other. That is, he makes the existence of the classical measuring instrument, which serves to specify the conditions under which the phenomena appear, depend on ill-defined limiting procedures. We saw in Chap. 6, for example, that a large relative mass is not generally sufficient to guarantee the classical behaviour of a quantum system. And what in any case are these massive 'atomic objects'? Bohr claims it is meaningless to attribute position and momentum to them, but they are apparently sufficiently well defined to have a mass which can be compared with the masses of other systems. The attempt to devise a consistent approach to these two problems brings us to von Neumann.

We recall that the empirical content of quantum mechanics relates both to the outcomes of individual measurements, in which it is predicted that dynamical variables are found to be eigenvalues of self-adjoint operators, and to an ensemble of similarly prepared individual measurements, in which the eigenvalues are predicted to be distributed according to a specified probability law. Putting Bohr aside, what happens if we do try to apply the usual formalism to account for the actions by which we test these predictions by treating them as special but typical many-body processes? In fact, the attempt to give a *physical* treatment of the measuring process in this way rapidly flounders if it is assumed that the state vector furnishes a complete description of individuals and that this description just comprises information on the statistical frequency of the outcomes of the very processes it is desired to explain. For it is necessary to account for properties of matter that are not included in this type of statistical, predictive description.

To see this, let us follow the procedure of von Neumann (1955) and divide the world into two parts: an observed system and a system that may 'observe', say a meter. This is closer to how most physicists think of practical measurements than Bohr's approach. To perform its role the apparatus must combine quantum and classical properties. It must behave quantum mechanically when interacting with the quantum mechanical object of interest, but the meter needle must be classically describable in that we should be able to assign to it a fairly well-defined initial coordinate y_0 which is differentiated at the macroscopic scale (i.e., the scale of humans) from its location y when the measurement is complete. Then the outcomes registered by the apparatus will allow us to infer something about a property of the object. And, of course, a classical instrument will have a definite continuously variable position at all stages of the measuring process, after it commences and before it terminates. We meet again the problem of logical consistency raised in Chap. 6: the wavefunction we attribute to the apparatus does not embrace a description of all the properties we require it to have, namely that it is an objective material system possessing autonomous properties such as occupying a definite portion of space. The assumption that the apparatus wavefunction is an eigenstate of the position operator $\hat{y}$, or sharply peaked about a classical orbit, does not address this difficulty (see below).

The problem just stated is the first component of the measurement problem. Actually, it is not the aspect that is usually emphasized and so, leaving it aside for a moment, we pass to an examination of what may be considered the second component of this problem (see London & Bauer (1939), d'Espagnat (1976), Wigner (1983a,b), Redhead (1989)).

Suppose the system under investigation is initially in a pure state $|\psi_i\rangle \in \mathcal{H}_1$ which is an eigenstate of an Hermitian operator $\hat{A}$ pertaining to that system, corresponding to eigenvalue a_i (for simplicity the spectrum is assumed to be discrete and nondegenerate). Let the initial state of the apparatus which is to 'measure' the dynamical variable corresponding to $\hat{A}$ be $|\phi_0\rangle \in \mathcal{H}_2$. This is one of an orthonormal set of eigenstates $|\phi_i\rangle$ of the position operator $\hat{y}$ corresponding to macroscopically distinct eigenvalues y_i. Then the initial state of the composite system is an element of the direct product space $\mathcal{H}_1 \otimes \mathcal{H}_2$:

$$|\Psi_0\rangle = |\psi_i\rangle \otimes |\phi_0\rangle. \tag{8.2.1}$$

In this space the operators take the form $\hat{A} \otimes 1$ and $1 \otimes \hat{y}$. It is now required that the two systems interact in such a way as to cause the total state to evolve unitarily into

$$|\Psi\rangle = |\psi_i\rangle \otimes |\phi_i\rangle. \tag{8.2.2}$$

Such a transformation belongs to the class of ideal measurements. The state of the apparatus has become correlated with the initial object eigenstate while the latter is left undisturbed. 'Observing' the pointer reading therefore enables us to infer the preexisting state of the object under investigation and hence the eigenvalue. Apart from the difficulty pointed out above of giving objective meaning to the notion of a meter reading, and the corresponding problem pertaining to the object of what it is that the eigenvalue a_i is a property of, this process is unproblematic, provided of course a suitable interaction Hamiltonian exists.

Of course, the initial object state will not generally be an eigenfunction. Generalizing, suppose that the initial state of the system is an arbitrary linear superposition of the eigenstates of $\hat{A}$ so that the initial composite state is

$$|\Psi_0\rangle = \sum_i c_i |\psi_i\rangle \otimes |\phi_0\rangle \tag{8.2.3}$$

with $\sum_i |c_i|^2 = 1$. Then by the linearity of the Schrödinger evolution we have from (8.2.1) and (8.2.2):

$$|\Psi\rangle = \sum_i c_i |\psi_i\rangle \otimes |\phi_i\rangle. \tag{8.2.4}$$

Again, the apparatus has become correlated with the observed system. But the state (8.2.4) does *not* describe an outcome in which the joint system is associated with the state (8.2.2), and the dynamical variable has eigenvalue a_i, with probability $|c_i|^2$; rather, it is a linear superposition of such states. Neither the apparatus nor the object are left in eigenstates of their respective operators.

To see what has happened, consider the statistical state of the apparatus alone. This may be obtained from the density operator associated with the pure state (8.2.4),

$$\begin{aligned} \rho_{\text{pure}} &= |\Psi\rangle\langle\Psi| \\ &= \sum_{i,j} c_i c_j^* |\psi_i\rangle\langle\psi_j| \otimes |\phi_i\rangle\langle\phi_j|, \end{aligned} \tag{8.2.5}$$

by tracing over the object variables to give the partial density operator

$$\begin{aligned} \rho_{\text{app}} &= \sum_k \langle\psi_k|\rho_{\text{pure}}|\psi_k\rangle \\ &= \sum_i |c_i|^2 |\phi_i\rangle\langle\phi_i|. \end{aligned} \tag{8.2.6}$$

This describes an improper mixture in which all the pure states into which ρ_{app} is decomposed 'coexist'; the apparatus is spread over a range of macroscopically distinct eigenstates. Yet it is surely a fact of experience that the apparatus variable has a unique value. Somehow, then, the outcome of the measurement *should* be that in each trial the final composite state evolves into one of the summands in (8.2.4):

$$|\Psi\rangle \to |\psi_i\rangle \otimes |\phi_i\rangle \tag{8.2.7}$$

with probability $|c_i|^2$. Were this to happen, the apparatus would be left in a definite state that is correlated with the object state and we could infer the latter. Over an ensemble, the transition (8.2.7) induces the transformation of the pure state density operator (8.2.5) into a *proper* mixture of mutually exclusive alternatives:

$$\rho_{\text{pure}} \to \rho_{\text{mixed}} = \sum_i |c_i|^2 |\psi_i\rangle\langle\psi_i| \otimes |\phi_i\rangle\langle\phi_i|. \tag{8.2.8}$$

Accounting for the transition (8.2.7), or equivalently (8.2.8), is what is usually called the measurement problem of quantum mechanics. Note though that it is additional to the first component of the problem cited above.

Now it is evident that the transformation (8.2.7) can never be generated by a unitary Schrödinger evolution applied to the unique initial state vector (8.2.3) since, having specified the Hamiltonian, this evolution results continuously, deterministically and with certainty in a unique future state vector (i.e., (8.2.4)) and not in a set of unpredictable distinct alternatives. Thus, except in the special case $|\psi_0\rangle = |\psi_i\rangle$, the usual quantum formalism cannot account for the evolution (8.2.7), at least not on its own. This impossibility is often expressed through the observation that the Schrödinger equation cannot map a pure state into a mixture but note that we have to achieve a *proper* mixture, so killing the interference terms in (8.2.5) does not solve the problem (see below).

Suppose we tried to bring about a definite state by coupling the correlated composite system to a further apparatus that could 'observe' the first apparatus. Clearly, all this would achieve is a new linear superposition of macroscopically different states of three coupled systems. One then rapidly gets into an infinite regress if further apparatuses are invoked. To break the chain of inference, von Neumann proposed that at some point Schrödinger's evolution must be suspended and replaced by a different one, i.e., (8.2.7), that is discontinuous, noncausal and nonunitary. The proposal that this new transformation, not described by quantum mechanics, actually occurs in nature is termed the 'projection postulate' or the hypothesis of 'wavefunction collapse'. It forms the core of von Neumann's theory of measurement.

There are several obvious difficulties with this hypothesis. For a start, we are not told when the collapse is supposed to occur, how long it takes or what brings it about. Nor are we told how many systems should be coupled to the object of interest in the chain of inference before the collapse is invoked. The types of processes we have in mind when discussing idealizations of experiments are not very different from those that may be expected to occur naturally all the time when physical systems interact with their environments. If we cannot account for the measuring process by applying the usual many-body Schrödinger theory, this implies a massive incompleteness in the quantum mechanical treatment of general natural processes, since the collapse must be constantly occurring. One may avoid this conclusion by adopting the proposal of von Neumann and others that the collapse takes place only when the human observer intervenes and becomes aware of the pointer reading. But is it reasonable to invoke consciousness at this level of physics when it is not necessary to include it in the formal description of any of the processes to which the quantum theory is usually applied?

A celebrated example of the difficulties this last suggestion leads to is due to Schrödinger (1935b). A coupled system composed of a radioactive atom and a macroscopic body (a cat) is confined to a closed box. As the atom decays, the system evolves into a linear superposition of states corresponding to a live cat and a dead cat. According to von Neumann's idea, the wholly passive act of opening the box and looking is supposed to determine which alternative is actually realized (a live cat or a dead cat).

Reducing the physical content of quantum mechanics to random acts by human brains is objectionable on several grounds but we shall concentrate on just one: the collapse postulate, wherever it is deemed to intervene, does not account in the first place for the objective existence of cats, independently of their possible coupling to a quantum mechanical system. Even if we could devise some mathematical means to bring about the stochastic evolution (8.2.7), such as modifying the Schrödinger equation, the definite state so obtained will still only be a wavefunction that in itself exhibits no feature that may be identified with the reality of a definite pointer position, or the centre of mass of a cat. Such an assignation of definite variables constitutes an assumption that goes beyond quantum mechanics. If it is regarded as reasonable to attribute classical variables to classical objects, the impossibility of precisely defining the difference between macro- and micro-levels implies we should attribute such variables to quantum objects too. Thus we arrive at the de Broglie–Bohm theory. But if this possibility is seriously entertained, it is also possible that we can avoid the collapse hypothesis altogether.

Von Neumann's postulate appears not so much as a resolution of the measurement problem as a statement of what one might aim to achieve in such a resolution. It is significant that the perceived necessity of this postulate is itself an admission that quantum mechanics is incomplete and must be supplemented by further hypotheses. What is remarkable is that the seemingly far more natural conclusion has not been widely drawn that the inability of the pure wavefunction formalism to account for definite measurement outcomes is evidence that it should be supplemented by variables that more completely specify the state of an individual system. As we shall see in §8.3, this allows one to assign a privileged role to one of the summands in (8.2.4) whilst retaining the usual Schrödinger evolution.

To bring out what might be expected of a solution to the measurement problem, we conclude this section by enumerating four ways in which the density operator resulting from quantum mechanics, (8.2.5), differs from the desired one, (8.2.8).

(1) Coherence

The two density operators are mathematically and hence statistically distinguished by the presence of interference terms in (8.2.5). In the basis indicated in which $\hat{A}$ and $\hat{y}$ are diagonal, ρ_{pure} has finite off-diagonal elements $c_i c_j^*$, $i \neq j$. This implies differences in the expectation values of certain operators pertaining to the joint system (although not in the expectation values of operators pertaining to either system alone, such as $\hat{A}$ or $\hat{y}$).

(2) Proper vs improper mixtures

The two density operators do not differ simply in their coherence properties. Suppose it can be arranged that the off-diagonal elements of (8.2.5) are negligible, for example by coupling the composite to its thermodynamic environment and treating (8.2.5) as a partial density operator obtained from the new, larger density operator (Zurek, 1982; Halliwell, 1989). Then (8.2.5) and (8.2.8) effectively coincide and the expectation values of any operator calculated using the two operators agree. But the so-derived mixture is improper; the pure components into which it is decomposed are simultaneously present, as if ρ_{pure} is akin to a state in its own right. In contrast, the proper mixture (8.2.8) represents a case where the state of the composite is pure (just one of the components) but unknown. In this case the density operators are statistically equivalent but physically inequivalent. The decoherence of ρ_{pure} does not therefore in itself put the system in a definite state. The summands must be distinguished in some way.

(3) Preferred apparatus basis

The partial density operator of the apparatus derived by tracing over the object variables in (8.2.8) coincides with the expression (8.2.6) obtained from ρ_{pure}. As follows from point (2), the former is a proper mixture while the latter is improper. Given the total improper density operator, which is all the quantum formalism provides us with, we may expand it in an infinite number of ways into distinct pure states. In contrast, the proper apparatus mixed state is expressed in terms of a unique physically preferred basis. Why should one expansion be singled out in the improper case?

(4) Autonomous properties of matter

The desired proper density operator (8.2.8) must ultimately describe a mixture of states representing systems having actual pointer readings. The wavefunctions in ρ_{pure} do not include these parameters in the description they provide. If nothing is added to the usual single interpretative attribute of the wavefunction that it determines the statistical frequency of measurement outcomes, arranging for the apparatus state to be peaked about a classical orbit does not address this problem. For all such localization tells us is that a particle is most likely to be found in the vicinity of the peak of the packet if a position measurement is performed. This is not logically connected with the assertion that a system *is* in this region. To assert that means that the usual interpretation of the state vector has been tacitly altered, or supplemented.

A successful resolution of the measurement problem does not require achieving absolute identity of ρ_{pure} and ρ_{mixed} under all circumstances. The most important physical problem is point (4). We will see in §8.3 that when that difficulty is removed the other problems outlined above may also be circumvented.

8.3 Causal theory of ideal measurements

8.3.1 General remarks

In §8.1 we saw that in classical mechanics individual or simultaneous measurements of arbitrary precision may, in principle, be carried out to reveal the current values of canonically conjugate quantities associated with a mechanical system. This is because it is possible to invent interactions that transform measuring devices with negligible disturbance of the system of interest. Given this information we may predict the future behaviour of the system and retrodict its past, assuming of course, we have knowledge of the forces to which it is subject. On the face of it, the quantal analysis of the measurement problem outlined in §8.2 has little in common with the classical treatment. Indeed the two treatments cannot be compared if one discusses the quantal problem purely in terms of the wavefunction for, as we have said, there is no way to attribute the definite outcomes of experiments to the properties of material systems since the systems themselves are undefined. This is true as much for the apparatus as it is for the observed system. The 'observables' of conventional quantum theory are not actually (directly) observable at all; their values are inferred from the values of 'true' observables, the additional positions of classical particles making up the apparatus which the wavefunction does not describe and that at some stage must be postulated.

In contrast, the analysis of the quantum mechanical measurement process that follows shares several key features with the classical theory. We assume the universal validity of the dynamical equations of the theory, i.e., Schrödinger's equation and the particle law of motion, throughout the process, just as

Newton's laws apply throughout the classical measurement. We also make a distinction between the observed system and the observing apparatus but this division is not absolute and both systems have equal ontological status and are subject to the same physical laws. Moreover, the human observer plays no essential role in the physics she observes. Most importantly, we share the property that the true 'observables' of the theory, the things that immediately present themselves in experiments, are the *positions of particles*, particularly the position of the apparatus pointer. Our treatment should lay to rest the misapprehension that the 'hidden variables' (i.e., the positions of particles) are 'metaphysical' and 'unobservable'. The hidden variables *are* the observables and the system and apparatus always have definite positions, whatever the quantum state may be.

In our approach measurement is a partial investigation into the kinematical properties of a particle (such as its position or momentum) that may be inferred from the observation, or registration, of the position of some other system (the apparatus) with which the system of interest becomes correlated. We aim to describe this as a typical many-body interaction process that is special only insofar as the interaction is required to have the property that the system under investigation is left in a particular state, an eigenfunction of an Hermitian operator (the 'operator measured') while the apparatus is left in a state whose subsequent behaviour in no way influences the system of interest. It is this effect of the interaction that signals the key difference between the classical and quantum cases: in order that we may extract from the apparatus variable unambiguous information on the system of interest, the wavefunction of the latter must, in general, undergo an irreducible and unpredictable transformation so that the values of the relevant physical properties of the particle that it determines are no longer those obtaining prior to the interaction. As a result, measurements do not and cannot, in general, reveal the current values of physical quantities. A principal goal is to account for this fact through the usual continuous, causal and unitary evolution of quantum mechanics without resorting to the projection postulate.

It is emphasized that while in the following great emphasis is laid on the dynamics of the wave, it is ultimately properties of the particle that the process is intended to 'measure', just as in classical physics. This allows a continuity of interpretation between the quantum and classical descriptions. The extent to which we may empirically determine the characteristics of the wave component of a material system is discussed in §8.7.

Since we are analysing a slice drawn from the entire history of causally connected processes into which the system enters, the initial state $(\psi_0(\mathbf{x}), \mathbf{x}_0)$ is the result of some previous 'state preparation' process and is not generally known in advance. In order to test the statistical predictions of quantum mechanics we need to know ψ_0 in order to compute the transition probabilities $|c_i|^2 = |\langle\psi_0|\psi_i\rangle|^2$, where ψ_i are the eigenfunctions of the relevant operator. In principle, an arbitrary ψ_0 may be prepared by applying to some other wavefunction a set of external classical potentials (Lamb, 1969; similar techniques may be devised to 'measure' arbitrary operators). To simplify matters we assume this has been done and the system is in a known pure state ψ_0. The generalization of the theory to a mixed state is straightforward and does not introduce anything essentially new. The distribution of the initial position $\mathbf{x}_0$ is therefore given by $|\psi_0|^2$ and we cannot control this more precisely without disturbing ψ_0.

A quantal wavefunction is also assigned to the classically-describable apparatus. This will obey the usual conditions of single-valuedness etc. and so will describe an ensemble of a special kind. Assigning the apparatus a narrow packet function has nothing to do with the definiteness of the pointer position (this is definite whatever our knowledge of it); it is necessary rather in order that we may notice a change in the meter variable and hence infer from it something about the observed system, just as in classical statistical mechanics.

We consider the case of ideal measurements, those that do not destroy the system and are reproducible in the sense that immediate repetition yields the same result. The outcome of the measurement, the location of the apparatus variable, allows us to infer the final system wavefunction, which is the initial wavefunction for all subsequent interactions, and hence the final actual value of the physical quantity that corresponds to the operator measured. We do not, in general, know the final particle position, unless it is the position that happens to be measured.

The period of measurement may be naturally divided into two stages:

(1) A *state preparation* of a certain kind in which the wavefunction of the system under investigation becomes correlated with the wavefunction of the observing apparatus and evolves into an eigenfunction of an Hermitian operator.
(2) An irreversible act of amplification which allows one indelibly to *register* the outcome and infer the value of the physical property of the particle corresponding to the operator.

The effect of stage (1) is analogous to the action of a prism on white light which brings about the spatial separation of the monochromatic waves making up the light, the intensity of each of the colours being proportional to the square of the coefficients in the Fourier decomposition of the incident wave. This stage may or may not be a many-body problem – the position of the particle may itself be an apparatus variable (as in the Stern–Gerlach experiment, cf. §9.5). It is, however, in principle, reversible. The second, irreversible, stage is a many-body problem. The whole treatment closely follows von Neumann's approach, but supplementing this with a consideration of the actual trajectories pursued by the particles making up the system and apparatus removes the problems flowing from the collapse hypothesis.

Accounts of the causal theory describing the particle dynamics are given by Bohm (1952b), Freistadt (1957), Belinfante (1973) and Bohm & Hiley (1984, 1988). Although this constitutes the core of the theory, in our treatment we also consider the evolution of the actual values of the physical quantity that corresponds in the causal interpretation to an Hermitian operator. This serves to bring out completely the physical significance of the measurement, which does not follow solely from a consideration of the particle position.

8.3.2 Stage 1: state preparation

Let $\psi(\mathbf{x}, t)$ be the wavefunction associated with a one-body system. We wish to deduce some information about the particle variable $A(\mathbf{x}, t)$ that is associated with an operator $\hat{A}(\hat{\mathbf{x}}, \hat{\mathbf{p}})$ via the local expectation value (cf. §3.5):

$$A(\mathbf{x}, t) = \operatorname{Re} \psi^*(\mathbf{x}, t)(\hat{A}\psi)(\mathbf{x}, t)/|\psi(\mathbf{x}, t)|^2 \tag{8.3.1}$$

(we assume this has one component). Evaluating (8.3.1) along a particle trajectory $\mathbf{x} = \mathbf{x}(t, \mathbf{x}_0)$ yields the true values of the physical quantity for each element of the ensemble associated with ψ. Initially, the actual value for an element will be

$$A_0(\mathbf{x}_0) = \operatorname{Re} \psi_0^*(\mathbf{x})(\hat{A}\psi_0)(\mathbf{x})/|\psi_0(\mathbf{x})|^2|_{\mathbf{x}=\mathbf{x}_0}. \tag{8.3.2}$$

In practice $\hat{A}$ is envisaged as being one of only a handful of physically important operators (such as a component of position, momentum or angular momentum, or energy) but it is convenient to keep the discussion general.† For simplicity it is assumed that $\hat{A}$ is nondegenerate.

The system interacts with an apparatus whose initial wavefunction $\phi_0(y)$ is a packet, the one-dimensional coordinate $y(t, y_0)$ defining the continuously variable location of the meter needle. The interaction is assumed to be impulsive so that the independent evolution of either system during it may be neglected. A suitable interaction Hamiltonian is the quantized version of (8.1.1):

$$H = g\hat{A}\hat{p}_y, \tag{8.3.3}$$

† The class of physical variables associated with Hermitian operators does not exhaust the quantities we may wish to determine empirically. Mass (inferred from position in a mass spectrograph), wavelength (inferred from fringe spacing in an interference experiment) and time are examples of quantities that do not correspond to the eigenvalues of such operators.

where g is a coupling constant and $\hat{p}_y$ is the momentum operator conjugate to $\hat{y}$. (In our treatment of the Stern–Gerlach experiment (§9.5) the relevant apparatus operator will be $\hat{y}$ rather than $\hat{p}_y$.) Neglect of the kinetic energy operators implies that the kinetic and quantum potential energies of the two systems are negligible while the interaction lasts. Since the operator $\hat{A}$ commutes with H it is usually stated that this interaction is one in which one measures the 'observable' $\hat{A}$ without changing its value (cf. the discussion in §3.9.3). In fact, as we shall see, the evolution generated by H fundamentally alters the prior actual value (8.3.2) except in the special case that ψ_0 is an eigenfunction of $\hat{A}$ and hence A_0 is an eigenvalue. The significance of H for us is rather that it ensures the measurement is reproducible.

The interacting systems are initially independent and so the initial total wavefunction is factorizable:

$$\Psi_0(\mathbf{x}, y) = \psi_0(\mathbf{x})\phi_0(y). \tag{8.3.4}$$

During the interaction the Schrödinger equation is

$$i\hbar \frac{\partial \Psi(\mathbf{x}, y, t)}{\partial t} = -i\hbar g \hat{A} \frac{\partial \Psi(\mathbf{x}, y, t)}{\partial y} \tag{8.3.5}$$

and the wavefunction becomes entangled in the four-dimensional configuration space. To solve (8.3.5), we expand Ψ in terms of a complete set of eigenfunctions $\psi_a(\mathbf{x})$ of the operator $\hat{A}$, where $\hat{A}\psi_a = a\psi_a$ and a is an eigenvalue:

$$\Psi(\mathbf{x}, y, t) = \sum_a f_a(y, t)\psi_a(\mathbf{x}). \tag{8.3.6}$$

The sum includes a possible contribution from a continuous part of the spectrum. Substituting (8.3.6) into (8.3.5) and using the orthonormality of the eigenfunctions, we find the following equation for the coefficients:

$$\frac{\partial f_a(y, t)}{\partial t} = -ga \frac{\partial f_a(y, t)}{\partial y}. \tag{8.3.7}$$

This has the solution

$$f_a(y, T) = f_{a0}(y - gaT), \tag{8.3.8}$$

where $f_{a0}(y)$ are the initial values and T is the period of the impulse. Expanding the initial wavefunction as $\psi_0(\mathbf{x}) = \sum_a c_a \psi_a(\mathbf{x})$ where c_a are constants, we find by equating (8.3.4) to (8.3.6) at $t = 0$ that $f_{a0}(y) = c_a \phi_0(y)$. Substituting (8.3.8), (8.3.6) therefore yields for the wavefunction at the termination of the interaction

$$\Psi(\mathbf{x}, y, T) = \sum_a c_a \psi_a(\mathbf{x}) \phi_0(y - gaT). \tag{8.3.9}$$

The subsequent evolution proceeds according to the free Hamiltonian.

It is evident that the wavefunction is now nonfactorizable and the system point $(\mathbf{x}(t), y(t))$ performs a complicated motion during the interaction. Note that in this period the equation of motion of the system point is not given by the usual expression $(\dot{\mathbf{x}}, \dot{y}) = (m^{-1}\nabla S, M^{-1}\, \partial S/\partial y)$, where S is the phase of Ψ, because the interaction Hamiltonian (8.3.3) is momentum-dependent (recall the guidance formula (3.11.18) in an external electromagnetic field) and by the conditions of the impulse approximation the phase terms are negligible. In any case, the y-coordinate has become correlated with the eigenvalues a. The centres of the packets ϕ_0 move along the tracks

$$y_a = gat, \qquad 0 < t < T \tag{8.3.10}$$

so that when the interaction terminates the separation of two packets corresponding to neighbouring eigenvalues a, $a + \delta a$ is given by

$$\delta y_a = gT\delta a. \tag{8.3.11}$$

It is required that the strength and duration of the impulse, together with the subsequent free evolution, are such that the packets ϕ_0 corresponding to different eigenvalues have no appreciable overlap and so are orthogonal. The condition for this is that the width of the packet $\Delta y \ll \delta y_a$. If this outcome is not achieved then the process described so far does not form part of a complete measurement – the measurement is rather 'incomplete' (see §8.5). It is assumed that the packets ϕ_0 eventually attain a classically describable separation. Notice that although they have separated in space the packets all have essentially the same momentum. If in (8.3.3) we had coupled to $\hat{y}$ instead of $\hat{p}_y$, we would have found at the end of the impulse that the packets have different momenta but the same location (cf. §9.5).

Assuming the condition for a complete measurement is satisfied, the wavefunction splits up into a set of nonoverlapping configuration space functions (even though the $\psi_a(\mathbf{x})$s may overlap). Referring to Fig. 8.1, the system point enters a region where one of these functions is finite so that, assuming the summands do not subsequently overlap (see below), the wavefunction (8.3.9) may be effectively replaced as regards the particle dynamics by just one of the summands:

$$\Psi \to c_a \psi_a(\mathbf{x}) \phi_0(y - gaT). \tag{8.3.12}$$

The system point will henceforth remain in the domain where the function

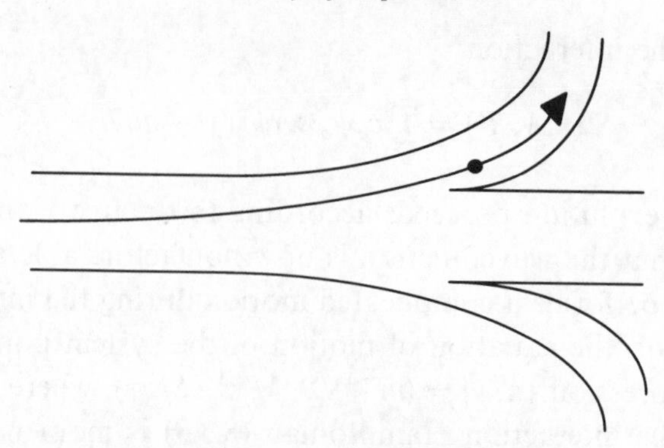

Fig. 8.1 The measurement interaction causes the configuration space wavefunction to split up into a set of nonoverlapping packets. The system point (the dot) enters one of them, the rest being empty.

(8.3.12) is finite, for the various summands are separated by regions in which $\Psi = 0$ where the orbit cannot pass (because the phase is undefined). If Ψ is not strictly zero between the summands there will be an overwhelmingly small probability of the system crossing these regions, as shown in §8.3.4. The fate of the remaining empty waves is discussed below.

We find out the configuration space function that is actually realized in (8.3.12), labelled by a, by observation of the value of y, say $y = Y$. The final wavefunction of the observed system is obtained by inserting $y = Y$ in (8.3.9), or equivalently (8.3.12). But note that the process embodied in (8.3.12) occurs regardless of whether we are aware that $y = Y$. This latter observation (looking at the meter) can be treated by a further quantum mechanical interaction as an ideal measurement of the operator $\hat{y}$ (§8.4), or just by applying the classical theory. But however one describes it, this further process simply reveals what has already happened and has no causal role in it. Thus, from the reading for y we may unambiguously infer that the $\mathbf{x}$-system is associated with a wavefunction $\psi_a(\mathbf{x})$, an eigenfunction of the operator 'measured'. The effective factorization of the final wavefunction implies that the subsequent $\mathbf{x}$- and y-motions are independent. In particular, the wavefunction of the observed system may henceforth be taken to be $\psi_a(\mathbf{x})$.

To complete our account of this stage of the measurement we must consider the evolution of the actual value (8.3.1) of the physical quantity corresponding to $\hat{A}$. During the interaction this becomes a configuration space function:

$$A(\mathbf{x}, y, t) = \mathrm{Re}\, \Psi^*(\mathbf{x}, y, t)(\hat{A}\Psi)(\mathbf{x}, y, t)/|\Psi(\mathbf{x}, y, t)|^2. \qquad (8.3.13)$$

The initial value (8.3.2) for an ensemble element thus becomes a function of the apparatus coordinate y. For some quantities, such as position or momentum, A retains its meaning as a well-defined property of the observed system. For others, such as energy, A cannot be attributed to the observed system alone. However, when the interaction is complete and (8.3.9) is a superposition of nonoverlapping functions, substitution into (8.3.13) shows that in the region where the ath summand is finite (cf. §7.2.5),

$$A \to a = \operatorname{Re} \psi_a^*(\mathbf{x})(\hat{A}\psi_a)(\mathbf{x})/|\psi_a(\mathbf{x})|^2. \tag{8.3.14}$$

Thus, a different value of A is associated with each of the outgoing packets and this is a constant throughout the region where each packet is finite. A is once again a property of just the observed system and the value obtained in each trial in an ensemble, an eigenvalue a, depends on which outgoing packet the system point enters. We may therefore infer from observation of y that if the system point lies in the ath packet, the actual value $A = a$. Notice that all ensemble elements that are ultimately guided by ψ_a have the same constant value $A = a$ regardless of their position.

Whatever the initial true value (8.3.2) may have been, it has been deterministically and continuously transformed into an eigenvalue. Which result will be found is unpredictable and depends critically on the initial values $\mathbf{x}_0$, y_0, ψ_0 and ϕ_0. Only if $\psi_0 = \psi_a$ do we find $A_0 = A = a$ whatever the initial positions $\mathbf{x}_0$, y_0.

It is then obvious that if we immediately repeat the same impulsive operation on a system that has been mapped from an arbitrary initial state ψ_0 to ψ_a, the same result $A = a$ will be obtained for all elements of the ensemble associated with ψ_a.

We have thus explained what it means to say that a quantum 'observable' 'has' the value a on completing a measurement. This has been done by connecting the mathematical operator $\hat{A}$ with a physical property of the particle via the continuously variable quantity $A(\mathbf{x}, t)$ which may take on the value a. In the usual approach $\hat{A}$ is connected with a only via the eigenvalue equation, not by direct equality.

We also shed light on why the outcomes of measurements are the eigenvalues of Hermitian operators. Evidently these are real numbers. But the main point is that they are also spacetime constants. Suppose, for example, we observe the meter variable y while the packets still overlap, i.e., before the measurement is complete. Then the value of A is well defined but (configuration-) spacetime-dependent and so we cannot ascertain its value as we do not know the position of the particle. When the conditions for a complete measurement are met, we can infer A without needing to know $\mathbf{x}$. Eigenfunctions thus play the special role of uniquely specifying the relevant particle property independently of the particle location.

The definite outcome is obtained from the usual Schrödinger evolution and does not require the sudden intervention of an unexplained 'collapse'. Our knowledge of the state of the system of interest changes because of the objective transformation (8.3.12) of the physical wave. Wave packet collapse puts it the other way round: the change in the wave (collapse) occurs when our knowledge changes. In our approach no collapse, in the sense that the unrealized summands in (8.3.9) cease to 'exist', actually occurs. When a detector clicks the wavefunction does not 'collapse' from all over space to a point, it is simply that only part of it is now relevant. For this reason the causal theory may offer a useful aid in the development of a relativistic theory of measurement where the instantaneous collapse hypothesis would be difficult to reconcile with the principles of relativity. And because by assumption the apparatus coordinates are always well defined there is no need to invoke further systems that can put the observer into a definite state. We are therefore able to avoid an infinite and inexplicable regress.

We see that far from this process having passively revealed the value of a system variable with negligible external perturbation, the observed and observing systems have mutually transformed one another. The final value of A, a, is causally but not directly related to its initial value (8.3.2). The actual outcome obtained is uniquely determined by ψ_0, ϕ_0 and the initial positions $\mathbf{x}_0$, y_0 and is highly sensitive to small changes in the initial values. One cannot infer from observation of y the final position $\mathbf{x}$ of the observed system (unless the measurement happens to be that of position) and hence one cannot work backwards using the law of motion to deduce the actual premeasurement value (8.3.2).

The above description may be straightforwardly extended to include the simultaneous measurement of a set of commuting operators $\hat{A}_1, \ldots, \hat{A}_n$. One introduces a set of independent meter coordinates $y_1, \ldots, y_n$ and applies an interaction Hamiltonian that is a linear sum of n terms of the type (8.3.3). Expanding Ψ in a set of simultaneous eigenfunctions of the commuting operators, the analysis goes through as before. Observation of the final meter readings, $y_i = Y_i$, allows us to infer the final state of the object, i.e., a simultaneous eigenfunction, and hence the values of each of the particle properties corresponding to the operators. The problem of noncommuting operators is discussed in §§8.4 and 8.5.

8.3.3 Derivation of Born's statistical postulate

According to Born's (1926) postulate, when a system is described by a wavefunction $\psi_0(\mathbf{x}) = \sum_a c_a \psi_a(\mathbf{x})$, the probability of finding the result a when a measurement is performed is given by $|c_a|^2$. In our approach a statistical element enters because we are unable to control the initial positions $\mathbf{x}_0$ and y_0 but only know that they are distributed with densities $|\psi_0|^2$ and $|\phi_0|^2$ respectively. To show how Born's prescription regarding the statistical frequency of measurement outcomes follows in the causal theory, we must consider an ensemble of the individual deterministic processes described above, each performed with an identical initial wavefunction (8.3.4) but with a range of initial system positions.

During the interaction (8.3.3) the total probability is conserved:

$$(\mathrm{d}/\mathrm{d}t)\int |\Psi|^2\,\mathrm{d}^3x\,\mathrm{d}y = 0. \qquad (8.3.15)$$

When the packets cease to overlap, the configuration space probability density is given by

$$|\Psi(\mathbf{x}, t, T)|^2 \approx \sum_a |c_a|^2|\psi_a(\mathbf{x})|^2|\phi_0(y - gaT)|^2 \qquad (8.3.16)$$

since the interference terms are negligible. The probability that the system point lies in the volume element $\mathrm{d}^3x\,\mathrm{d}y$ about a point $(\mathbf{x}, y)$ in the domain where the ath summand is appreciable is therefore

$$P_a\,\mathrm{d}^3x\,\mathrm{d}y = |c_a|^2|\psi_a(\mathbf{x})|^2|\phi_0(y - gaT)|^2\,\mathrm{d}^3x\,\mathrm{d}y. \qquad (8.3.17)$$

Hence, the total probability that $\mathbf{x}$ lies within $\psi_a(\mathbf{x})$ and y within $\phi_0(y - gaT)$ is given by

$$\int P_a\,\mathrm{d}^3x\,\mathrm{d}y = |c_a|^2, \qquad (8.3.18)$$

where the integral extends over all configuration space. We deduce that the probability of the outcome (8.3.12), in which the true value $A = a$, is given by the Born formula (8.3.18). The probabilistic significance of the set of numbers $|c_a|^2$ is a consequence of the assumption that $|\psi|^2$ describes the probability of presence of a particle at an actual location in space and no separate postulate is necessary. Born's statistical postulate is therefore a consequence of the causal theory (the converse is not true).

This result may be seen in another way by computing the probability density of the true values A:

$$g(A) = \int |\Psi(\mathbf{x}, y)|^2\delta[A - \mathrm{Re}\,\Psi^*\hat{A}\Psi/|\Psi|^2]\,\mathrm{d}^3x\,\mathrm{d}y. \qquad (8.3.19)$$

Substituting (8.3.9) and passing to the case where the packets do not overlap, we have from (8.3.14) and (8.3.16)

$$g(A) \to \sum_a |c_a|^2 \int |\psi_a(\mathbf{x})|^2 |\phi_0(y - gaT)|^2 \delta(A - a)\, \mathrm{d}^3x\, \mathrm{d}y$$
$$= |c_A|^2. \tag{8.3.20}$$

Again, we recover the statistical formula of quantum mechanics.

Notice that the numbers $|c_a|^2$ only have the statistical meaning they have just been given when the wave $\psi_0(\mathbf{x})$ is physically disrupted in the way that occurs in a measurement. These numbers do not refer to the probability that the particle *has* the value $A = a$ when the state function is $\psi_0(\mathbf{x})$. The true value of A is in general quite different, given by (8.3.2).

Although the measurement changes in a fundamental way the prior actual value of the observed quantity in each individual trial, the mean value of this quantity over the ensemble is preserved and is therefore given by the weighted sum of the eigenvalues:

$$\langle \hat{A} \rangle = \int \psi_0^*(\mathbf{x})(\hat{A}\psi_0)(\mathbf{x})\, \mathrm{d}^3x$$
$$= \sum_a a|c_a|^2. \tag{8.3.21}$$

Thus, some information concerning the prior state of the system is revealed by the measurement, albeit statistically. The same is true for the mean of any function of $\hat{A}$, since this commutes with the interaction Hamiltonian.

Granted that the system point remains in one of the summands, we may renormalize the actual wavefunction realized since the particle dynamics does not depend on its absolute value. Carrying out this purely mathematical rescaling, we have as the outcome of the ideal measurement that the observed system undergoes the mapping

$$\left.\begin{aligned} \psi_0(\mathbf{x}) &\to \psi_a(\mathbf{x}) \\ A_0(\mathbf{x}_0) &\to a \end{aligned}\right\} \tag{8.3.22}$$

and hence $g(A) = \delta(A - a)$ from then on.

We are now in a position to clarify how the difficulties concerning the density matrix summarized in points (1)–(4) in §8.2 are overcome in the causal theory.

The pure state density matrix of the composite system when the measurement interaction ceases is, from (8.3.9),

$$\rho_{\text{pure}}(\mathbf{x}, y, \mathbf{x}', y') = \sum_{a,a'} c_a c_{a'}^* \psi_a(\mathbf{x})\psi_{a'}^*(\mathbf{x}')\phi_0(y - gaT)\phi_0^*(y' - ga'T). \tag{8.2.23}$$

This includes off-diagonal interference terms, whether the configuration space packets overlap or not. However, as soon as the packets cease to overlap, and *when we take into account the particle*, we see from (8.3.12) that it is legitimate to represent the statistical state of the composite as a *proper* mixture:

$$\rho_{\text{mixed}}(\mathbf{x}, y, \mathbf{x}', y') = \sum_a |c_a|^2 \psi_a(\mathbf{x})\psi_a^*(\mathbf{x}')\phi_0(y - gaT)\phi_0^*(y' - gaT). \quad (8.3.24)$$

Here, each pure state is realized with a probability $|c_a|^2$ in a sequence of trials.

We see that in the causal theory of measurement the preferred pointer basis is picked out by the location of the system point. From a mathematical point of view we may expand Ψ in any basis we choose but the physically relevant one is that in which each summand is an isolated lump in configuration space.

A further significant feature of the causal theory is that it does not require decoherence of the density matrix. Because of this, the density matrices (8.3.23) and (8.3.24) will be statistically indistinguishable only so long as the coexisting packets contributing to the off-diagonal terms in (8.3.23) are not connected in some way. Thus, if the summands should subsequently overlap (so that the apparatus states are not orthogonal) the expectation values of operators pertaining to the object or to the apparatus will be different in the two cases. Or, if the summands do not overlap, one can, in principle, distinguish the two density matrices by performing measurements on an operator pertaining to the composite that has finite matrix elements linking the disjoint configuration space packets (this requires coupling to a further apparatus).

But, while the two density matrices are, in principle, distinguishable, this does not affect or cast doubt on our account of measurements when they *are* indistinguishable, i.e., when the pure states do not overlap and we measure an operator pertaining to the object alone. The definiteness of the final outcome is a property of the definiteness of the pointer location under all circumstances (i.e., whatever the quantum state). This is the crucial point, and not that Ψ is composed of a set of disjoint packets. The latter is simply a necessary condition that allows us to ascertain from the always well-defined pointer reading unambiguous information on an object property, and is not itself the condition for definiteness. This is an advantage of the causal approach as it implies a simple interpretation of incomplete measurements (§8.5) without any essential change in the conceptual structure of the theory.

Having made these remarks, we note that for all practical purposes it is in any case impossible to distinguish the two density matrices, as we now see.

8.3.4 Stage 2: registration

The process described so far is reversible. We can, in principle, apply a suitable interaction Hamiltonian to bring the outgoing packets back together so that they significantly overlap. If this is done the apparatus and object coordinates will still be well defined but correlated in such a way that we can no longer infer from y precise statements regarding the state of $\mathbf{x}$. To account for the fact that real measurements lead to permanent results which do allow the desired inferences to be drawn, we simply observe that, in practice, the apparatus coordinate y represents, or is coupled to, a very large number of degrees of freedom making up the experimental instrument. Then, although it is, in principle, possible that the y-packet the system point enters will be disturbed by interference with the other packets, the likelihood of this happening is negligible. To distinguish the pure and mixed density matrices pertaining to this enlarged composite system would require measuring an operator that couples all these degrees of freedom, an operation that is likewise all but impossible. (We emphasize though that should it prove possible to detect the empty packets, this circumstance is comfortably included in the causal description.)

To see this we may suppose that the variable y represents a range of apparatus coordinates but to be explicit we introduce a separate set of variables $z_1, \ldots, z_n$, $n \sim 10^{23}$, and assume they may be assigned a wavefunction which is denoted initially $\xi_0(z_1, \ldots, z_n)$. The total initial wavefunction is therefore

$$\Psi_0(\mathbf{x}, y, z_1, \ldots, z_n) = \psi_0(\mathbf{x})\phi_0(y)\xi_0(z_1, \ldots, z_n). \tag{8.3.25}$$

After the interaction the further coordinates will be correlated with the y-coordinate:

$$\Psi(\mathbf{x}, y, z_1, \ldots, z_n, T) = \sum_a c_a\psi_a(\mathbf{x})\phi_0(y - gaT)\xi_a(z_1, \ldots, z_n). \tag{8.3.26}$$

This is a linear superposition of nonoverlapping functions in the total configuration space. Once the system point has entered a domain where one of the summands is finite it will stay there, for the probability that the functions will subsequently overlap is overwhelmingly low, even if the ψ_as or ϕ_0s overlap. Hence we see that to all intents and purposes the reading of y will yield a permanent record of the state of the system $\mathbf{x}$ at the termination of the impulse. The packets into which the system point does not enter can have no sensible effect on the subsequent motion. The question whether they can have any observable effect at all is discussed in §8.8.

The complexity of the interaction gives it the appearance of being irreversible but the underlying dynamical law bringing about the wavefunction (8.3.26) remains the reversible Schrödinger equation. That the process so described is not actually irreversible may be considered a weak point in our demonstration. Yet according to our treatment, the general problem of irreversibility in physics, namely the inadequacy of attributing such transformations to our ignorance of the details of a highly complex set of individually reversible motions, is not more acute in the measuring process than in any other branch of the subject. As we have presented it, the theory of measurement may be understood without first answering fundamental questions regarding the physical nature of irreversibility.

To complete the story, we should, in principle, include in our model of measurement explicit reference to the light used to illuminate the meter, the eye of the observer the light is reflected into, the observer's brain, other observers who become aware of the results, and so on. The variables connected with all these systems may be formally included in the theory by extending the set $z_1, \ldots, z_n$. From this it is immediately clear that it is legitimate to make the assumption of classical physics that, while strictly speaking these further processes must be included in a complete account, they have no influence on the problem of primary interest (deriving information on an atomic object) and hence can be left out. In particular, registration by a human observer is quite irrelevant.

In this connection we see how the problem posed by Schrödinger's example of a cat coupled to a radioactive atom (§8.2) may be very simply resolved. The set-up may be thought of as a device for performing an unsharp measurement of position (see §8.5). 'Cat-alive' corresponds to a meter reading telling us that the α-particle is still in the source, and 'cat-dead' corresponds to a meter reading indicating that the α-particle has left the source. Our observation of the meter, i.e., looking at the cat, has nothing to do with the objective process in which the α-wave leaks out of the nucleus and begins to interact with its environment. Whether the particle has emerged prior to the act of observation is determined simply by its initial position. Rather than posing a paradox, this example illustrates that all measurements entail amplifying to the macroscale linear superpositions at the microlevel.

How much should be explicitly included in the description depends on the nature of the problem. A rule of thumb is (Bell, 1981): 'Put sufficiently much into the quantum system that the inclusion of more would not significantly alter practical predictions'. A good example of the application of this dictum that has been discussed in detail is the formation of a cloud chamber track (Mott, 1929b; Heisenberg, 1930; Bell, 1981): It makes no essential difference whether one treats the α-particle as the object and the molecules to be ionized as part of the observing apparatus, or includes the latter in an enlarged object that is subject to a further observing system, and so on.

To conclude this highly idealized account of typical experiments, we emphasize the role of the particle in resolving the measurement problem. It is the assumption of a corpuscle which transforms quantum mechanics into a theory of matter having substance and form. The pure wave dynamics described by Schrödinger's equation does not yield any account of which result is actually realized in an individual measurement operation. The wavefunction collapse hypothesis only gains physical content if actual coordinates for the collapsed system are posited. Since the point at which these are introduced in the chain of connected physical systems is arbitrary, the only consistent assumption is that they are well defined all along. And if that assertion is admitted, the collapse hypothesis becomes superfluous and the universal validity of the Schrödinger equation in all natural processes may be maintained. One may harbour an aesthetic aversion to the proliferation of empty configuration space packets in the causal approach but this seems a small price to pay for the gain in coherence and internal consistency in the theory of measurement. And the empty waves do have explanatory power, as we saw in the two-slit experiment (§5.1). The idea is, after all, not so different from the behaviour of an electromagnetic wave which is distributed throughout space but is tangible only in the vicinity of a charged particle. We shall see in §8.8 that no inconsistencies or contradictions follow from the concept of empty wave packets.

This measurement theory hinges on the assumption that true physical observables are the positions of particles, i.e., that all experiments in the real world ultimately reduce to the determination of position. As far as we are aware this is completely in accord with actual laboratory practice. Everything that is regarded as 'observable' about the object system in the conventional approach is in the causal theory inferred from position measurements. In this connection we note that the separation of the empty packets occurs in the space of the preferred pointer basis, position configuration space, and not in, e.g., momentum configuration space (cf. §7.2). If it proved possible to perform measurements that do not eventually reduce to position determinations (location of meter needle, distribution of dots in computer print-out, . . .) the foregoing theory would naturally have to be modified.

8.4 Position, momentum and the commutation relations

8.4.1 Single and joint measurements

The theory developed so far may be applied in a straightforward way to ideal measurements of position and momentum. We have seen how in the general theory the measurement entails a disturbance of the initial object wavefunction so that the actual value found for the particle property measured is not the preexisting value (unless ψ_0 happens to be an eigenfunction of the relevant operator), although the two are continuously and causally connected. In fact, there is one important exception to this rule: ideal measurements of position. These hold a unique significance in the theory for, while the initial arbitrary wavefunction is transformed, it condenses around the current position of the particle and so we are able, in principle, to infer the premeasurement position as defined by the causal interpretation (in the usual approach the notion of a position independent of measurement has no meaning).

To see this, we employ the interaction Hamiltonian (8.3.3) where now

$$H = g\hat{x}\hat{p}_y. \tag{8.4.1}$$

Restricting attention to one dimension, the total wavefunction when the impulse ceases is given by

$$\Psi(x, y, T) = \psi_0(x)\phi_0(y - gxT) \tag{8.4.2}$$

$$= \int \mathrm{d}x'\, \psi_0(x')\delta(x - x')\phi_0(y - gx'T). \tag{8.4.3}$$

Inserting the observed apparatus coordinate $y = Y$ into (8.4.2) and renormalizing yields the initial postmeasurement object wavefunction. If in (8.4.3) the apparatus packets do not overlap, this yields

$$\psi_0(x) \to \delta(x - x'), \tag{8.4.4}$$

where x' is the result of the position measurement. The probability density of the outcomes is, of course, $|\psi_0(x')|^2$. That this is indeed the premeasurement position of the particle may be seen from the law of motion during the impulse. This is just that of the classical theory studied at the end of §8.1: $\dot{x} \approx 0$, $\dot{y} \approx gx$. Thus

$$x(T) = x_0, \qquad y(T) = y_0 + gx_0T. \tag{8.4.5}$$

The momentum of a particle in the state (8.4.4) is evidently unknown. The state will be recognized as the initial form of the Green function and this is the wavefunction that subsequently guides the particle. This indicates a basic difference with the classical theory where the transformation of the distribution function to a δ-function has no influence on the particle motion.

In contrast, the value of momentum found in an ideal measurement is generally an artefact of the measuring process, unless the initial wavefunction is an eigenstate. The interaction Hamiltonian in this case becomes

$$H = g\hat{p}_x\hat{p}_y. \tag{8.4.6}$$

The final total wavefunction is

$$\Psi(x, y, T) = h^{-1/2} \int \mathrm{d}p\, \varphi(p)\, \mathrm{e}^{\mathrm{i}px/\hbar}\, \phi_0(y - gpT). \tag{8.4.7}$$

If the conditions for an ideal measurement are met we obtain

$$\psi_0(x) \to \mathrm{e}^{\mathrm{i}px/\hbar}, \tag{8.4.8}$$

a momentum eigenfunction. Thus $\partial S_0(x_0)/\partial x \to p$ and the initial and final true momenta may have different values. As shown in (8.3.20), the true probability density $g(p) \to |\varphi(p)|^2$. The postmeasurement initial position of the particle guided by the wave (8.4.8) is well defined but unknown.

It is clear that the physical conditions required to effect a position measurement, implying evolution of the wavefunction into a δ-function, are incompatible with those necessary to bring about a momentum measurement because the relevant operators do not have simultaneous eigenfunctions. This is the operational content of the commutation relation $[\hat{x}, \hat{p}_x] = \mathrm{i}\hbar$ and is how 'complementarity' appears in this model. The requirement that the object be left in an eigenstate of an Hermitian operator therefore implies the impossibility of a simultaneous measurement of position and momentum to arbitrary accuracy. This conclusion follows without direct reference to Heisenberg's relations (cf. §8.5) and is logically independent of the probabilistic aspect of quantum theory, since it refers to each individual experiment.

So what happens if we do try to measure position and momentum together? To investigate this problem we employ the treatment of Arthurs & Kelly (1965) (for a different approach see She & Heffner (1966)) and consider an impulsive interaction in which the Hamiltonian is the quantized version of the classical expression (8.1.10):

$$H = g(x\hat{p}_y + \hat{p}_x\hat{p}_z). \tag{8.4.9}$$

Here x is the coordinate of the observed system and y and z are the coordinates of two independent meters corresponding to commuting operators

$\hat{y}$ and $\hat{z}$. The initial total wavefunction is

$$\Psi_0(x, y, z) = \psi_0(x)\phi_0(y)\xi_0(z) \tag{8.4.10}$$

and we shall assume that the apparatuses are described by normalized Gaussian states

$$\left.\begin{aligned} \phi_0(y) &= (2/\pi b)^{1/4}\,\mathrm{e}^{-y^2/b} \\ \xi_0(z) &= (2b/\pi)^{1/4}\,\mathrm{e}^{-bz^2}, \end{aligned}\right\} \tag{8.4.11}$$

where b is a constant. Setting $\hbar = 1$, the Schrödinger equation is

$$\frac{\partial \Psi}{\partial t} = g\left(-x\frac{\partial \Psi}{\partial y} + \mathrm{i}\frac{\partial^2 \Psi}{\partial x\,\partial z}\right). \tag{8.4.12}$$

Fourier transforming on the variable z, the solution to (8.4.12) subject to (8.4.10) and (8.4.11) is

$$\Psi(x, y, z, T) = (8\pi^3 b)^{-1/4}\int_{-\infty}^{\infty}\psi_0(x - gTw) \times \phi_0(y - gTx + \tfrac{1}{2}wT^2g^2)\,\mathrm{e}^{-w^2/4b + \mathrm{i}wz}\,\mathrm{d}w. \tag{8.4.13}$$

To simplify the algebra we use units in which $gT = 1$. Then the wavefunction (8.4.13) becomes

$$\Psi(x, y, z) = f(y, z)(\pi b)^{-1/4}\,\mathrm{e}^{-(x-y)^2/2b + \mathrm{i}xz}, \tag{8.4.14}$$

where

$$f(y, z) = (4\pi^3 b)^{-1/4}\int_{-\infty}^{\infty}\psi_0(x)\,\mathrm{e}^{-(x-y)^2/2b - \mathrm{i}xz}\,\mathrm{d}x. \tag{8.4.15}$$

The three interacting systems have evidently become correlated. For example, the joint probability density of the two meters is nonfactorizable:

$$\int_{-\infty}^{\infty}|\Psi(x, y, z)|^2\,\mathrm{d}x = |f(y, z)|^2. \tag{8.4.16}$$

To obtain the object state resulting from this interaction we insert the observed meter values $y = Y$ and $z = Z$ into (8.4.14) and renormalize:

$$\psi(x) = \frac{\Psi(x, Y, Z)}{|f(Y, Z)|} = (\pi b)^{-1/4}\,\mathrm{e}^{-(x-Y)^2/2b + \mathrm{i}xZ} \tag{8.4.17}$$

apart from a constant phase factor.

We have thus generated a wavefunction, a coherent state, which in a sense lies between the extremes of a position eigenfunction ($b \to 0$) and a momentum eigenfunction ($b \to \infty$). As may be expected, from the usual point of view the outcome of the joint measurement, i.e., the meter readings, does not allow one to infer values for the particle position or momentum. Rather, the measurement has produced a state that it is claimed displays uncertainty in the position and momentum variables which, for the function (8.4.17), are the minimum possible compatible with the Heisenberg inequality (see §8.5). In the causal interpretation it is also true that we cannot deduce the particle position, but for the state (8.4.17) we do know that the momentum for all ensemble elements is Z. However, checking this prediction requires a further measurement of the momentum variable and if this is carried out we will find a scatter in the results compatible with the Heisenberg relation, as shown in §8.5.

What the 'joint measurement' just described really amounts to then is a state preparation. All measurements have a preparatory character, but for unambiguous inferences to be drawn they must be of a more specific type. The mutual disturbance of the three systems transforms the object wavefunction but prevents its evolution into an eigenfunction of $\hat{x}$ or $\hat{p}_x$. Each meter reading corresponds to a range of object eigenfunctions. The overall effect is similar to an incomplete measurement of a single variable, as we shall see in §8.5. If we wish, the process can be related to the usual axiom that the outcome of a measurement is an eigenfunction of an operator, by observing (Glauber, 1963b) that the coherent state (8.4.17) is an eigenfunction of the non-Hermitian annihilation operator $a = (b/2)^{1/2}(\hat{x}b^{-1} + \mathrm{i}\hat{p}_x)$ corresponding to the complex eigenvalue $(b/2)^{1/2}(Yb^{-1} + \mathrm{i}Z)$ (this operator does not commute with the Hamiltonian (8.4.9)).

The above examples demonstrate that we cannot empirically prove the hypothesis that a quantum system comprises a corpuscular aspect possessing simultaneously well-defined position and momentum variables by experiments compatible with the axioms of quantum measurement. But conversely, and this is equally important in view of the historical claims that have been made (cf. §8.5), it cannot be proved by experiments that are designed to test and that confirm the quantum rules that the concept of a corpuscle is inadmissible. In particular, the attempt to perform a joint measurement does not entail this conclusion. Such a notion can be ruled out only by fiat, such as the assertion by Heisenberg (1927) that a physical concept only has meaning if a means of experimentally determining it is specified. This may be a valid point of view but there is no decisive evidence in its favour (the evidence of experience is equally compatible with a trajectory assumption). It is surely impossible to formulate a theory purely in terms of 'observable' quantities and indeed quantum theory does not do this. All sorts of tacit assumptions on the continuity of space etc. are invoked which according to the theory itself cannot be subjected to empirical verification. Heisenberg's programme excludes the possibility that what is now unobservable may become observable as physical theory develops. In the final analysis the point of concepts is to render intelligible the essence behind the phenomena, and it is in this capacity that they should be judged.

8.4.2 Prediction and retrodiction

What has just been shown to be impossible is a simultaneous determination of the position and momentum of a particle which yields sufficient data to predict the future orbit. This does not rule out an effectively simultaneous measurement of position and momentum at an instant, because in the causal interpretation the actual momentum is well defined for all quantum states and so we may, for instance, infer momentum from a position measurement. That is, if the wavefunction is known, so that the momentum $\mathbf{p} = \nabla S$ at each point is given, a precision position measurement implies we possess precise values for both variables at the same time. Another way of achieving this is by the so-called 'time-of-flight' method, described in §8.6. The idea may be illustrated by the following simple example (Ballentine, 1970). Referring to Fig. 8.2, a plane wave $e^{ipx/\hbar}$ is incident on a screen containing a slit which produces a cylindrical wave. This wave propagates to a detection plane containing a set of detectors that can register the arrival of a particle. According to the causal law of motion the incident momentum is p, in the x-direction. On passing through the slit the energy of the particle remains $E = p^2/2m$ and the particle changes direction to join a new rectilinear orbit under the guidance of the cylindrical wave (cf. §4.2). Detecting the particle

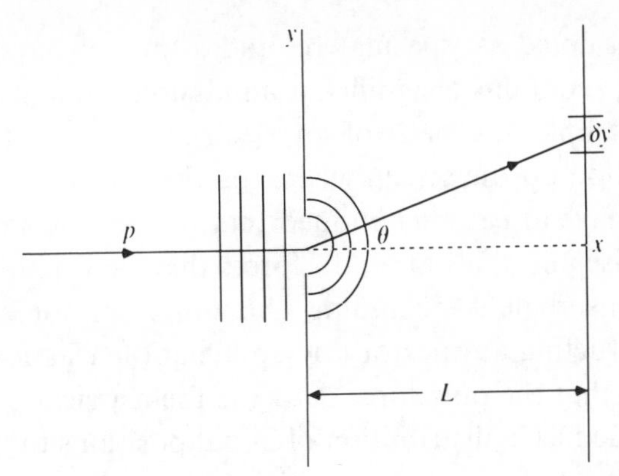

Fig. 8.2 A device for simultaneously measuring the y-component of position and momentum.

at the point y on the screen allows us to infer that the y-component of the momentum immediately prior to the detection is given by $p_y = p \sin \theta$, $\theta = \tan^{-1}(y/L)$. The error δy in the y-measurement may be made arbitrarily small and the error δp_y in the momentum decreases as $L \to \infty$.

While we may discover by such means both the position and momentum coordinates in a single direction to arbitrary accuracy, this information does not allow us to predict the subsequent trajectory. The interaction of the wave with the bank of detectors obviously disrupts it so that in learning the current position we do not know the corresponding momentum implied by the newly created wavefunction. On the other hand, what we can do using such techniques is retrodict the trajectory of the particle from the point of detection. This, of course, is possible only if we possess knowledge of all the forces that previously acted on the system, including both the classical and quantum potentials. Assuming this information is available, we may say that in the causal interpretation *we can know more about the past of a quantum particle than about its future*, a conclusion also arrived at by Aharonov & Albert (1987).

The possibility of performing simultaneous arbitrarily sharp measurements of conjugate variables which refer to the past was recognized by Heisenberg (1930, p. 20). He wrote:

> ... if the velocity of the electron is at first known and the position then exactly measured, the position for times previous to the measurement may be calculated ... but this knowledge of the past is of a purely speculative character, since it can never (because of the unknown change in momentum caused by the position measurement) be used as an initial condition in any calculation of the future progress of the electron and thus cannot be subjected to experimental verification. It is a matter of personal belief whether such a calculation concerning the past history of the electron can be ascribed any physical reality or not.

Now, even if considered as 'speculative' and a 'matter of personal belief', it seems this passage contains a significant admission. First, it accepts that one can trace, for example, the path of an electron through an interferometer prior to its detection on the screen where interference is observed. But more importantly, we have already mentioned that in order to retrodict an orbit, one requires precise information on the forces that were acting on the particle prior to the measurement, including the quantum force. If Heisenberg envisages reconstructing a trajectory by applying the classical law of motion (i.e., by assuming that the only force acting is the classical one) then over an ensemble he will deduce a distribution of initial positions that contradicts the one implied by the probability distribution of quantum mechanics. Thus, if one takes seriously Heisenberg's argument that it is indeed possible to retrodict paths, one cannot escape accepting the postulates of the causal interpretation and its law of motion, or risk inconsistencies.

8.4.3 The Wigner function

There have been many attempts to introduce phase-space-type structures into quantum mechanics. The most famous is due to Wigner (1932) who introduced the function

$$F(x, p) = (1/h) \int_{-\infty}^{\infty} \psi^*(x + \tfrac{1}{2}z)\psi(x - \tfrac{1}{2}z)\, \mathrm{e}^{\mathrm{i}pz/\hbar}\, \mathrm{d}z, \qquad (8.4.18)$$

where $\psi(x)$ obeys the Schrödinger equation (we work in one dimension). The function $F(x, p)$ has many properties that we might expect of a distribution function defined on the 'phase space' whose points are labelled by the independent variables x and p. We shall make some remarks on whether this has any connection with the phase space of the quantum theory of motion whose points are labelled by x and $p = \partial S/\partial x$, with distribution function

$$f(x, p) = R^2(x)\delta(p - \partial S/\partial x). \qquad (8.4.19)$$

For our purposes, the most noteworthy properties of the Wigner function are that its marginal densities are the usual quantum mechanical position and momentum distributions,

$$\int_{-\infty}^{\infty} F(x, p)\, \mathrm{d}p = |\psi(x)|^2 \qquad (8.4.20)$$

$$\int_{-\infty}^{\infty} F(x, p)\, \mathrm{d}x = |\varphi(p)|^2, \qquad (8.4.21)$$

and that it satisfies an equation (induced by the Schrödinger equation) that looks like a generalization of the classical Liouville equation (for other properties see Hillery *et al.* (1984)). F cannot be interpreted as a probability, however, because it may take on negative values. For this reason it is called a quasi-distribution function.

Now, given that the quantum Liouville equation is a kind of generalization of the classical one, the question has been posed (Lee & Scully, 1983; Holland *et al.*, 1986; Lee, 1990) whether it is possible to maintain in the Wigner formalism the classical notion that the (x, p) variables represent the current state of a mechanical system, rather than accepting this as an abstract space having no connection with the actual properties of matter. Actually, conventional presentations of the Wigner theory often speak as if this is some kind of space of states, albeit one in which the coordinates cannot be specified more precisely than is allowed by the Heisenberg relations $\Delta x\, \Delta p \geqslant \hbar/2$ (the variances are computed using F). If the idea of a precise specification were viable, it would presumably amount to a different representation of the motion implied by the usual causal interpretation in which the classical trajectory is maintained in the quantum generalization of Hamilton–Jacobi theory. In support of this approach we may cite the result that for polynomial potentials of order no higher than the second, the equation of motion of the Wigner function actually coincides with the classical Liouville equation. For example, in the case of the harmonic oscillator we have for both the classical and Wigner distribution functions

$$\frac{\partial F}{\partial t} + \frac{p}{m}\frac{\partial F}{\partial x} - m\omega^2 x \frac{\partial F}{\partial p} = 0. \tag{8.4.22}$$

Does this imply that it is legitimate to think in terms of an ensemble of quantum particles each pursuing the elliptic phase space orbit of a classical oscillator of energy

$$E' = p^2/2m + \tfrac{1}{2}m\omega^2 x^2? \tag{8.4.23}$$

It seems not, for the following reason.

The coincidence of the Wigner equation of motion with the classical statistical law in the case of the oscillator indicates not the identity of the two theories but that the quantum equation does not convey all the information contained in the Schrödinger equation from which it is derived. To see this, let us consider stationary states of energy E (so $\partial F/\partial t = 0$). Then, translating the Schrödinger eigenvalue problem into the Wigner language, we obtain the following relation in addition to (8.4.22) (Dahl, 1983; Carruthers & Zachariasen, 1983):

$$E = \frac{p^2}{2m} + \tfrac{1}{2}m\omega^2 x^2 - \frac{\hbar^2}{8mF}\frac{\partial^2 F}{\partial x^2} - \frac{m\omega^2\hbar^2}{8F}\frac{\partial^2 F}{\partial p^2}. \tag{8.4.24}$$

This expression for the energy is evidently different from the classical formula (8.4.23). The third term on the right hand side is purely quantum mechanical in origin and will be recognized as a quantum potential constructed from the Wigner function. The fourth term, also nonclassical, involves an additional contribution from the external potential. This relation thus appears to exclude a classical particle law of motion. If such a law is to make sense we would have to suppose that F is a 'physical' field capable of exerting forces on a particle. This would be consistent with the classical case in that the classical limit (8.4.23) is obtained when the quantum terms in (8.4.24) are negligible (following our discussion in Chap. 6, this limit is not generally achieved when $\hbar \to 0$, for F itself depends on $\hbar$).

But it is doubtful whether any such approach can really be sustained. It is evident from the marginals (8.4.20) and (8.4.21) that the numbers x and p represent the outcomes of mutually exclusive measurements, yet one is trying to combine them to define a space of points that is intended to represent a single state of a system. Clearly, we cannot put a particle in such a state. That a point in the Wigner phase space corresponds to the result of a measurement, or rather to two distinct measurements, and not to the current state of a system, also casts doubt on the view that the space is 'cellular' or 'fuzzy', the limit of resolution being fixed by Heisenberg's inequality. Underlying this latter idea is the suggestion that $|\varphi(p)|^2$ represents the distribution of the *actual* momentum of a particle, a view that is criticized on several occasions in this book (cf. §8.5).

Despite its important formal role, the Wigner formalism does not therefore appear to provide a suitable language for a causal representation of quantum phenomena. The problem of constructing a phase space that refers to current states and carries a genuine probability distribution is solved in the causal interpretation, via (8.4.19).

8.5 Heisenberg's relations

The historical debate concerning the possibility of a simultaneous application of spacetime coordination and the law of causality has centred around the interpretation of Heisenberg's relations:

$$\Delta\hat{x}_i \, \Delta\hat{p}_j \geqslant (\hbar/2)\delta_{ij} \tag{8.5.1}$$

(for the notation and proof of the relations see §6.7.3). The formula (8.5.1) is but one among many inequalities that may be derived from the position–momentum commutation relations and it is not necessarily always the most appropriate for the uses to which it has been put. Here we shall consider its significance according to the quantum theory of motion. Bearing in mind the discussion in §3.8, it appears that three distinct interpretations can be supported, relating to:

(1) the structure of the wave;
(2) the properties of an ensemble of particles; and
(3) our knowledge of the current behaviour of a single particle.

These points of view will be considered in turn.

First of all, it is well known in the classical theory of waves that the width of a wave train in physical space, and the spread in the wave numbers in the Fourier space, are connected by the relation $\Delta x_i \, \Delta k_i \sim 1$, $i = 1, 2, 3$ (cf. §4.6). Applying the de Broglie relation $\mathbf{p} = \hbar\mathbf{k}$ yields (8.5.1) and, interpreted simply as a property of the ψ-wave, this is unproblematic.

Next, we may connect (8.5.1) with the ensemble of particles associated with ψ by proposing that it asserts a limitation on the scatter in the results of a statistical ensemble of identical experiments. This interpretation, which was proposed in the 1930s by Popper (1982; see also Ballentine (1970)), follows from the usual axioms of quantum mechanics regarding measurements. It means the following.

Suppose that the initial wavefunction is $\psi_0(\mathbf{x})$ so the particles have an initial distribution $R_0^2(\mathbf{x})$. Consider the ith particle in the ensemble and at time t measure its x-coordinate, $x_i(t)$. Repeat the same process many times (with the same initial wavefunction, apparatus, time t etc.). After n trials evaluate from these position measurements the average and mean square positions:

$$\langle x \rangle_n = (1/n) \sum_{i=1}^{n} x_i(t)$$

$$(\Delta x)_n^2 = (1/n) \sum_{i=1}^{n} [x_i(t) - \langle x \rangle_n]^2$$

(we may have $x_i = x_j$, $i \neq j$). Starting from the same initial wavefunction, perform now a series of momentum measurements in the x-direction. Let the ith particle in the ensemble be found to have momentum $p_{xi}(t)$ at time t. From these data we evaluate the average and mean square momenta:

$$\langle p_x \rangle_n = (1/n) \sum_{i=1}^{n} p_{xi}(t)$$

$$(\Delta p_x)_n^2 = (1/n) \sum_{i=1}^{n} [p_{xi}(t) - \langle p_x \rangle_n]^2.$$

(The x_is and p_{xi}s in these formulae are, of course, what the actual values of position and momentum evolve into according to the quantum theory of motion.) Let $n \to \infty$. Then defining $\Delta x = \lim_{n\to\infty} (\Delta x)_n$ and $\Delta p_x = \lim_{n\to\infty} (\Delta p_x)_n$ we should find (8.5.1). This is the operational meaning of (8.5.1), i.e., the way one would test it experimentally.

Notice that each of the trials in the ensemble of measurements is a *precision* experiment – the particle is found in an element d^3x or d^3p in position or momentum space and the errors δx, δp_x in the respective measurements are assumed to be very small in comparison with the variances (as they must be in order that the latter have a meaning). Moreover, no reference has been made to the performance of *simultaneous* measurements of $\hat{x}$ and $\hat{p}_x$ on a single system.

The interpretation just described clearly has nothing to do with the issue of whether a particle may be attributed simultaneously well-defined position and momentum coordinates. Yet it is precisely this problem that Heisenberg and subsequent authors claim may be addressed by an application of (8.5.1) to individual cases. For example, it is asserted that the attempt to localize a particle inevitably introduces an unpredictable and uncontrollable uncertainty into its momentum and that this disturbance is quantified by (8.5.1). We now examine how far the causal interpretation provides justification for such arguments.

In previous sections we considered ideal measurements in which the final wavefunction of the system observed is a δ-function in the space of the operator measured. We consider now an imprecise or unsharp measurement where the probe is blunt and can resolve only a range of eigenfunctions. This provides a more realistic model of actual experiments which in the case of continuous spectra cannot, of course, yield a precise result. The measurement interaction will still transform the system ψ-function but not into an eigenfunction and so we cannot infer from the apparatus y-coordinate the precise value of the quantity 'measured' (although this is, in fact, well defined), but can only locate it within a certain range. There is then at the termination of the measurement an 'uncertainty' in our knowledge of the true value of the property of interest, and also in the degree of precision with which we can predict the outcome of a future measurement (as we shall see, the two uncertainties do not, in general, coincide).

As an example of such a process we may consider the impulsive interaction studied in §8.3 but where the process is incomplete in the sense that the outgoing apparatus packets $\phi_0(y - gaT)$ overlap one another and $\phi_0(y)$. Then

$$\Psi_0(\mathbf{x}, y) = \psi_0(\mathbf{x})\phi_0(y)$$

$$\rightarrow \Psi(\mathbf{x}, y, T) = \sum_a c_a \psi_a(\mathbf{x})\phi_0(y - gaT), \tag{8.5.2}$$

which is an entangled function. To find the object wavefunction we insert the observed apparatus coordinate $y = Y$ (obtained by performing an ideal measurement of $\hat{y}$) into (8.5.2) and renormalize. This yields a function of just $\mathbf{x}$:

$$\psi(\mathbf{x}) = \left(\int |\Psi(\mathbf{x}, Y, T)|^2 \, \mathrm{d}^3x\right)^{-1/2} \Psi(\mathbf{x}, Y, T). \tag{8.5.3}$$

Given the details of the interaction, observation of y thus enables us to ascertain the final object function $\psi(\mathbf{x})$ which is, in general, different from $\psi_0(\mathbf{x})$ and not an eigenfunction (unless $\psi_0 = \psi_a$). As a more general possibility we may consider nonideal interactions in which an initial eigenfunction is transformed: $\psi_a(\mathbf{x})\phi_0(y) \to \psi_a'(\mathbf{x})\phi_a(y)$. (The theory of unsharp measurements can be made rigorous by introducing the notion of unsharp observables; see, for instance, Davies (1976) and Busch & Lahti (1984).)

The actual outcome Y is unpredictable and uncontrollable and depends on $\mathbf{x}_0$, y_0 and Ψ_0. Knowing $\psi(\mathbf{x})$, we may compute $A(\mathbf{x}, t)$ for the ensemble of particles associated with ψ but, being spacetime-dependent, we do not know which value of A is the one actually realized because we do not know the particle position (in the special case $\psi = \psi_a$ we know $A = a$ regardless of $\mathbf{x}_0$, as seen in §8.3). From $\psi(\mathbf{x})$ we can compute $\Delta\hat{A}$ and the root mean square value of other quantities.

There is no reason *a priori* why the root mean square deviation from the mean of an operator, or of some other function, is an appropriate measure of the latitude in our knowledge of a physical quantity. Other definitions are certainly conceivable (see below). However, for our purposes here we shall assume that such a measure is appropriate. Also, we shall henceforth restrict attention to measurements of position and momentum in the x-direction.

As regards particle properties, the root mean square deviation in the x-coordinate has two interpretations. First, as already stated, it provides a measure of the degree to which we may predict the outcome of a precision measurement of position, through the scatter in an ensemble. Next, it represents our knowledge of the likely deviation from the mean of the position of an individual particle in the ensemble connected with $\psi(\mathbf{x})$, as a result of the unsharp measurement just described.

Now, the conclusion conventionally drawn from the study of an interaction that determines position with a precision $\Delta\hat{x}$ is that this introduces a corresponding uncontrollable uncertainty in the momentum, given by $\Delta\hat{p}_x \geqslant \hbar/2\,\Delta\hat{x}$. For example, this argument has been applied to a system with definite momentum (i.e., a momentum eigenstate) that is incident on a barrier containing a slit of width $\Delta\hat{x}$ (Heisenberg, 1930, p. 23). From consideration of such examples it has been concluded that no meaning can be given to the notion of the spacetime track pursued by a particle because the attempt to specify ever more closely the position coordinate ($\Delta\hat{x} \to 0$) entails a corresponding loss in determination of the momentum. The meaning of this argument is not very clear. Are we told that we cannot possess precise simultaneous knowledge of the values of conjugate variables that are in themselves well defined (an out-of-focus photograph)? Or that there is a fuzziness in their very definition (a sharp photo of fog)? Or even that they simply have no meaning? But, however the argument may be construed, our principal concern is this: the argument of Heisenberg and his followers rests explicitly or implicitly upon the assumption that the root mean square deviation $\Delta\hat{p}_x$ indeed furnishes a measure of the uncertainty in our knowledge of the *current* momentum of a particle that accompanies an imprecise position or momentum determination. We now show that this is an unjustified assumption and hence the traditional argument against the trajectory concept falls (see also de Broglie (1963, p. 58)).

This we do by examining what meaning may be attributed to $\Delta\hat{p}_x$ in the quantum theory of motion, apart from its interpretation as a measure of the dispersion in the results of precision momentum measurements already discussed. Recall that in this approach the actual momentum of a particle is unknown only because the position is. From the distribution of true momentum in the state $\psi = R\,\mathrm{e}^{\mathrm{i}S/\hbar}$,

$$g(\mathbf{p}) = \int R^2(\mathbf{x})\delta[\mathbf{p} - \nabla S(\mathbf{x})]\,\mathrm{d}^3x, \tag{8.5.4}$$

we find that the mean square deviation from the mean of the x-component of the actual momentum is given by

$$\begin{aligned}(\Delta p_x)^2_{\mathrm{cl}} &= \int p_x^2 g(\mathbf{p})\,\mathrm{d}^3p - \langle p_x\rangle^2 \\ &= \int R^2(\mathbf{x})(\partial S/\partial x)^2\,\mathrm{d}^3x - \langle p_x\rangle^2,\end{aligned} \tag{8.5.5}$$

where

$$\begin{aligned}\langle p_x\rangle &= \int p_x g(\mathbf{p})\,\mathrm{d}^3p \\ &= \int R^2(\mathbf{x})(\partial S/\partial x)\,\mathrm{d}^3x = \langle \hat{p}_x\rangle.\end{aligned} \tag{8.5.6}$$

The subscript 'cl' is used to indicate that this is the same expression one would write down for a single-valued classical ensemble, although, of course, ψ satisfies the Schrödinger equation. Thus, at any instant, (8.5.5) gives a measure for the uncertainty in our knowledge of the x-component of the true momentum of a particle corresponding to the uncertainty Δx in our knowledge of its current x-coordinate. We have the following inequality:

$$\Delta x(\Delta p_x)_{\mathrm{cl}} \geqslant 0. \tag{8.5.7}$$

It is obvious that there is no particular reciprocal limit on the precision with which position and momentum may be known or specified, according to this definition of uncertainty.

The $\Delta\hat{p}_x$ of Heisenberg and (8.5.5) are connected by the following relation:

$$(\Delta\hat{p}_x)^2 = (\Delta p_x)_{\text{cl}}^2 - \hbar^2 \int R(\partial^2 R/\partial x^2)\, \mathrm{d}^3x. \tag{8.5.8}$$

As was shown in §6.7.3, it is the product of Δx with the second term in (8.5.8) that produces the inequality (8.5.1). We emphasize that (8.5.1) and (8.5.7) both hold at each instant. But, assuming that the standard deviation is a reasonable measure of our uncertainty in the current momentum, this is given by $(\Delta p_x)_{\text{cl}}$, and *not* by $\Delta\hat{p}_x$ which refers to a different physical quantity. To see the significance of the latter, we write down the stress tensor of the Schrödinger field in terms of the R and S fields (cf. §3.9):

$$T_{ij} = (1/m)R^2\, \partial_i S\, \partial_j S + \theta_{ij} \tag{8.5.9}$$

where

$$\theta_{ij} = (\hbar^2/2m)\{2\, \partial_i R\, \partial_j R - \delta_{ij}[R\nabla^2 R + (\nabla R)^2]\}. \tag{8.5.10}$$

Clearly, the first term on the right hand side of (8.5.9) represents the classical stress tensor density of a free ensemble of particles and (8.5.10) contains the quantum effects. The total stress tensor is then easily shown to be

$$\int T_{ij}\, \mathrm{d}^3x = (1/m) \int R^2\, \partial_i S\, \partial_j S\, \mathrm{d}^3x - (\hbar^2/m) \int R\, \partial_{ij} R\, \mathrm{d}^3x \tag{8.5.11}$$

by parts. Comparing (8.5.11) with (8.5.5) and (8.5.8) we find

$$(\Delta\hat{p}_x)^2 = (\Delta p_x)_{\text{cl}}^2 + m \int \theta_{11}\, \mathrm{d}^3x. \tag{8.5.12}$$

The $\Delta\hat{p}_x$ of Heisenberg thus differs from our expression for the current uncertainty by a term proportional to the 11-component of the quantum contribution to the total stress tensor (equivalent to the mean 11-component of the stress per particle). This term reflects the quantum contribution to the dynamics of a particle (for a different interpretation see Cohen (1990)). Indeed, there is a close relationship between the quantum stress and the quantum potential, as is evident from the following local and global relations:

$$\sum_j \partial_j \theta_{ij} = R^2\, \partial_i Q \tag{8.5.13}$$

$$\frac{1}{2}\sum_i \int \theta_{ii}\, \mathrm{d}^3x = \int R^2 Q\, \mathrm{d}^3x. \tag{8.5.14}$$

Yet we also know that at the instant $\psi(\mathbf{x})$ is formed we may write

$$(\Delta\hat{p}_x)^2 = \int (p_x - \langle\hat{p}_x\rangle)^2 |\varphi(\mathbf{p})|^2 \, \mathrm{d}^3 p. \tag{8.5.15}$$

It is this way of expressing $\Delta\hat{p}_x$ that is apt to mislead one into thinking that it represents the uncertainty in the current momentum. But the very common tendency to assume that this is indeed a valid interpretation is tantamount to the proposal that $|\varphi(\mathbf{p})|^2$ refers to the distribution in the actual momentum in an ensemble, a meaning it cannot have. Representing $\psi(\mathbf{x})$ as a linear sum of momentum eigenfunctions does not imply that the ensemble is a mixture of particles having momentum $\mathbf{p}$ (eigenvalue of $\hat{\mathbf{p}}$), with relative frequency $|\varphi|^2$, because of interference between the Fourier components. The actual momentum is defined by ∇S and may be totally different from that associated with any of the component plane waves.

What $\Delta\hat{p}_x$ does describe is a measure of the accuracy with which we can predict the outcome of a subsequent precision momentum measurement for a particle in the state $\psi(\mathbf{x})$. This follows because $|\varphi(\mathbf{p})|^2$ gives the probability of the outcome p_x, and (8.5.15) is preserved by the measurement interaction. But checking this entails a transformation of the system ($\psi(\mathbf{x}) \to \mathrm{e}^{\mathrm{i}p_x x/\hbar}$). This explains why it is not $(\Delta p_x)_{\mathrm{cl}}$ that describes the scatter in measurement results and why our claim that this quantity yields the current uncertainty is not inconsistent with experiment.

To summarize, given a wavefunction $\psi(\mathbf{x})$ we may compute Δx, $(\Delta p_x)_{\mathrm{cl}}$ and $\Delta\hat{p}_x$. The quantity $\Delta\hat{p}_x$ may be interpreted as giving information on the current mean value of a particle property but this is not the momentum – it is another quantity, (8.5.12). It obeys with Δx the relation (8.5.1). The current uncertainty in momentum is given by $(\Delta p_x)_{\mathrm{cl}}$ which obeys the relation (8.5.7). $\Delta\hat{p}_x$ also gives a measure of the precision with which we can predict the results of future momentum measurements. Referring to (8.5.12), we may trace the origin of the statistical correlations between position and momentum measurements to the distribution of stresses in the ψ-field. They come about because the field that guides each particle in an ensemble also enters into the definition of mean values.

In view of these remarks, it is apparent that the phrase 'uncertainty relations' is an inappropriate description of the formula (8.5.1). Even less appropriate is the epithet 'indeterminacy relations', as if (8.5.1) expresses an indeterminateness in the very properties of matter. In fact, Heisenberg's inequalities have no bearing at all on the issue of whether matter may be attributed objective properties such as simultaneously well-defined position and momentum variables.

We have taken for granted above that the standard deviation provides a good measure of the uncertainty in a physical quantity. In the paradigmatic case of the Gaussian packet this is a reasonable assumption but it is not in general. In fact, it has been pointed out that the Δx and Δp_x employed by Bohr and Heisenberg in their discussions of various *gedanken* experiments designed to illustrate complementarity are generally *not* the root mean square values of the position and momentum operators in that the values quoted for these quantities conflict with those calculated for the standard deviations (Uffink & Hilgevoord, 1985; Hilgevoord & Uffink, 1988). This is so for the simplest of examples, such as diffraction at a single sharp slit (Beck & Nussenzveig, 1958), and reflects the semiquantitative approach to these questions adopted by Bohr. The assertion by Heisenberg (1930, p. 20) that the retrodicted value of $\Delta x \cdot \Delta p_x$ can be less than $\hbar/2$ also suggests that the uncertainties he has in mind may not be root mean square values if we note that the mathematical theorem (8.5.1) is valid at all instants. Other more suitable rigorous definitions of uncertainty based on different notions of the width of a function are described in the cited references. These lead to an improved mathematical formulation of the uncertainty inequality, but do not provide support for the historical claims associated with Heisenberg's relations.

8.6 Time-of-flight measurements

There are several ways in which we may interact with a system in order to derive information about it, beyond the impulsive measurements described in §8.3. Here we shall consider how momentum may be measured by inference from a position measurement. This will apply to systems that are initially confined and then allowed to evolve freely for a considerable time before the position measurement is effected. A typical application would be to the measurement of the momentum of a particle in a box (§6.5) by a sudden removal of the confining walls. Another example is the determination of electron momentum in a hydrogen atom by rapid ionization (an experiment along these lines has been carried out and confirms the quantal probability distribution; see McCarthy & Weigold (1983)).

That this method provides a legitimate way of measuring momentum, in the sense that it leads to the same statistical distribution as a more direct (e.g., impulsive) method, may be seen as follows (Feynman & Hibbs, 1965, p. 96; Park & Margenau, 1968). It suffices to work in one dimension. Suppose the system has initial wavefunction $\psi_0(x)$ which is finite in the interval $(-b, b)$ and negligible elsewhere, and let $\varphi_0(p)$ denote the wavefunction in momentum space. ψ_0 is allowed to propagate freely and after a time $t \to \infty$ one measures the position x of the particle and forms the number $p = mx/t$. Then the distribution of the numbers p over an ensemble of trials follows that of a momentum measurement at $t = 0$. That is, the probability of finding that the momentum lies in the element $\mathrm{d}p$ around the point p at $t = 0$ is equal to the probability that the position lies in the element $(t/m)\,\mathrm{d}p$ around the point tp/m in the limit $t \to \infty$:

$$|\varphi_0(p)|^2\,\mathrm{d}p = \lim_{t\to\infty} |\psi(x, t)|^2\,\mathrm{d}x|_{x=pt/m}. \qquad (8.6.1)$$

To prove this, we write the wavefunction at time $t > 0$ in the form

$$\psi(x, t) = \int_{-\infty}^{\infty} K(x, t; y, 0)\psi_0(y)\,\mathrm{d}y \qquad (8.6.2)$$

where

$$K(x, t; y, 0) = (m/iht)^{1/2}\,\mathrm{e}^{im(x-y)^2/2\hbar t} \qquad (8.6.3)$$

is the free particle kernel. Substituting $x = pt/m$ yields

$$\psi(pt/m, t) = (m/iht)^{1/2}\,\mathrm{e}^{ip^2t/2\hbar m} \int_{-\infty}^{\infty} \mathrm{e}^{-ipy/\hbar + imy^2/2\hbar t}\psi_0(y)\,\mathrm{d}y. \qquad (8.6.4)$$

Taking the limit $t \to \infty$, we see that the integral in (8.6.4) reduces to

$$h^{1/2}\varphi_0(p) = \int_{-\infty}^{\infty} \mathrm{e}^{-ipy/\hbar}\psi_0(y)\,\mathrm{d}y \qquad (8.6.5)$$

since the factor $\mathrm{e}^{imy^2/2\hbar t} \to 1$ for $y \in (-b, b)$ and $\psi_0(y) \approx 0$ outside this region. Therefore, writing $\mathrm{d}x = (t/m)\,\mathrm{d}p$, we obtain from (8.6.4) the result (8.6.1).

Now, in general, the observation of a classically-free particle at the point $x = pt/m$ at time t, if it started near $x = 0$ at $t = 0$, does not imply that its actual momentum at the instant of detection is given by p. The instantaneous momentum at the point is given by

$$\partial S(x, t)/\partial x|_{x=pt/m} \neq p. \qquad (8.6.6)$$

According to the quantum theory of motion, Newton's first law is valid in quantum mechanics only when the resultant quantum force vanishes as well as the resultant classical force; the (classically-free) motion between the two given points is generically nonuniform. What then is the significance of the value p obtained in the above method of 'momentum measurement'? The answer is that, in the limit, the actual momentum of the particle does, in fact, coincide with p. To see this, we look at the limiting behaviour of the gradient of the phase of the function (8.6.4). A short calculation yields:

$$\begin{aligned}\lim_{t\to\infty} \partial S(x, t)/\partial x|_{x=pt/m} &= \lim_{t\to\infty} \frac{m}{t}\frac{\partial S(pt/m, t)}{\partial p}\\ &= \lim_{t\to\infty} \frac{m}{t}\left[\frac{pt}{m} + \frac{\partial s_0(p)}{\partial p}\right]\\ &= p. \end{aligned} \tag{8.6.7}$$

where $s_0(p)$ is the (time-independent) phase of $\varphi_0(p)$. Thus, the trajectory in this limit becomes uniform: $x(t) \to pt/m$. It may be checked from (8.6.4) that the limiting quantum potential and quantum force are zero.

Substituting (8.6.1) and (8.6.7) into the formula for the probability density of actual momentum,

$$g(p, t) = \int |\psi(x, t)|^2 \delta[p - \partial S(x, t)/\partial x] \, \mathrm{d}x, \tag{8.6.8}$$

we find that the limit of the true momentum distribution coincides with the usual quantal one:

$$\lim_{t\to\infty} g(p, t) = |\varphi_0(p)|^2. \tag{8.6.9}$$

Thus, for a function of compact support and in the long-time limit, we may infer from observation of its position the actual momentum of a particle at the moment of detection. The value obtained is uniquely fixed by the initial position x_0 in the packet ψ_0. The distribution of momentum values so found is that implied by quantum mechanics. Note that this method does not provide a joint measurement of position and momentum to arbitrary accuracy in the classical sense because the detection process transforms the wavefunction so that immediately after it we do not know the momentum. We cannot therefore use the information gained to predict the future trajectory. On the other hand, we can retrodict the trajectory in the neighbourhood of the detection point and we may trace back further (towards $t = 0$) if we know the quantum force. This illustrates how we may know more about the past than the future (§8.4.2).

The objection may be raised that although the probability densities in the time-of-flight and more 'direct' methods of momentum measurement agree, the results obtained when the different methods are applied to the same particle may not. That is, there is no justification for the claim that the value mx/t would result from a 'genuine' momentum measurement (de Muynck, Janssen & Santman, 1979). Translated into our language this means that in an individual case characterized by given ψ_0 and x_0, the eventual actual momentum found will depend on the method of measurement used, although the statistical distributions over an ensemble will be the same. This is clearly so if we recall that the outcome of the 'direct' method depends on the initial values of the apparatus coordinates as well as x_0. Yet, whatever method is employed, all momentum measurements entail a continuous transformation of the initial actual momentum into the one finally 'measured' ($\partial S_0/\partial x \to p$). In general, no method reveals the premeasurement value. Since quantum mechanics only predicts the distribution of outcomes in an ensemble of similarly prepared (identical ψ_0) experiments, it is hard to see why the value of momentum obtained in the 'direct' method should be deemed more 'real' than the time-of-flight value. Insofar as the time-of-flight technique yields the correct distribution $|\varphi_0(p)|^2$, it seems to be a legitimate method of momentum measurement, if we bear in mind that measurements are transformations of systems and do not passively reveal preexisting values.

8.7 Measuring the actual momentum and the wavefunction

So far we have looked at conventional momentum measurements in which the initial true momentum, ∇S_0, is transformed into an eigenvalue. We now raise the question of how one might measure the true momentum at a spacetime point itself. As we shall see shortly, this problem is closely connected with another of great interest, that of 'measuring the wavefunction'. It was noted previously that the initial wavefunction ψ_0 in a measuring process is not generally known in advance although we can, in principle, prepare any given function by a suitable application of external potentials (Lamb, 1969). Can we devise an experimental technique to check that the wavefunction is indeed ψ_0? The transformation of ψ_0 that accompanies a typical measurement of a particle property allows us to infer in each individual case the final quantum state, i.e., an eigenfunction of the relevant operator, but does not reveal the premeasurement state of the wave. Indeed, specification of ψ_0 requires an infinite set of numbers and we would hardly expect to determine it in a single act. What we can do to 'measure' ψ_0 is use its statistical interpretation to reconstruct the function from the relative frequencies with which it is left in other (eigen)states as a result of measurements. That is, we perform a very large ensemble of experiments relating to a range of dynamical variables using the same ψ_0 each time. It might be expected that the statistical information gleaned from measurements pertaining to a single variable is not generally sufficient for this purpose as it furnishes only the transition probabilities $|c_a|^2$ and not the phases of the c_as. Which variables will give sufficient information?

It is at this point that we see the connection with measurements of the actual momentum, for an arbitrary function $\psi(\mathbf{x}, t) = R\,e^{iS/\hbar}$ will be determined up to a constant phase factor if we can measure R^2, ∇S and $\partial S/\partial t$ at each spacetime point. R^2 may be obtained from a sequence of position measurements and the last two functions, the local particle momentum and energy according to the causal interpretation, may be found as follows.

Consider once again the continuity and Hamilton–Jacobi equations derived from Schrödinger's equation:

$$\frac{\partial R^2}{\partial t} + \nabla \cdot \left(\frac{R^2 \nabla S}{m} \right) = 0, \qquad (8.7.1)$$

$$\frac{\partial S}{\partial t} + \frac{(\nabla S)^2}{2m} + Q + V = 0. \qquad (8.7.2)$$

In one dimension we may reconstruct ψ using these relations as follows (Kemble, 1937, p. 71). By repeated position measurements at a range of space points at different times we may determine R^2 and $\partial R^2/\partial t$. At nonnodal points the local momentum field is then empirically determined using (8.7.1) as follows:

$$\partial S(x, t)/\partial x = -[m/R^2(x, t)] \int_{-\infty}^{x} [\partial R^2(x', t)/\partial t]\,\mathrm{d}x' \qquad (8.7.3)$$

provided $R \to 0$ as $x \to \pm\infty$. Next, from R^2 we can compute the quantum potential Q. Substitution of this and (8.7.3) into (8.7.2) then yields empirical values for $\partial S/\partial t$. Hence, we can, in principle, use this method to deduce $\psi(x, t)$, up to a constant phase factor.

This method will not work in more than one dimension (Gale, Guth & Trammell, 1968). To see this, consider the stationary wavefunction

$$\psi(\mathbf{x}) = f(r) P_{lm}(\theta)\,e^{im\phi}. \qquad (8.7.4)$$

Then $\partial R^2/\partial t = 0$ and R^2 is independent of the sign of m. Thus we cannot infer ψ just from position measurements. This example also serves to show that the additional empirical determination of $|\varphi(\mathbf{p})|^2$ (momentum distribution) still yields insufficient information to reconstruct ψ since this quantity is insensitive to the sign of m as well.

To effect a measurement of ∇S in the general case we may proceed as follows (Gale *et al.*, 1968; Royer, 1989). The idea is to modify the function ψ in the vicinity of a point in such a way that the mean momentum operator is just ∇S, evaluated at that point. Consider a small volume ω containing a point $\mathbf{x}_0$. We perform a series of position measurements to obtain $R^2(\mathbf{x}_0)$.

Each time the particle is found in ω we measure its momentum (in the conventional way). The wavefunction after the position measurement will be

$$\psi'(\mathbf{x}) = N f_\omega(\mathbf{x})\psi(\mathbf{x}), \tag{8.7.5}$$

where N is a normalization constant and $f_\omega(\mathbf{x})$ is a real function that vanishes unless $\mathbf{x}$ is in ω. If ω is sufficiently small that we may write $S(\mathbf{x}) = S(\mathbf{x}_0) + (\mathbf{x} - \mathbf{x}_0)\cdot\nabla S(\mathbf{x}_0)$ in ω, we have

$$\langle\psi'|\hat{\mathbf{p}}|\psi'\rangle = \nabla S(\mathbf{x}_0). \tag{8.7.6}$$

Thus, determination of the mean over an ensemble of momentum measurements for a particle lying in ω yields in the limit $\omega \to 0$ the actual particle momentum at the point $\mathbf{x}_0$. Repetition at all spacetime points then allows us to reconstruct the amplitude and momentum fields. Having obtained ∇S, $\partial S/\partial t$ follows from (8.7.2).

Naturally, the assumption that a source can generate identical ψ_0s in each trial is too restrictive an idealization; techniques to measure the density matrix are discussed in the above references.

The statistical determination of the quantum state described here involves inventing techniques to measure what we have defined as the actual particle momentum. The momentum is found by observations carried out on an ensemble of particles and hence this does not constitute a simultaneous measurement of $\mathbf{x}$ and $\mathbf{p}$ in an individual case. Neither does it prove the correctness of the guidance formula. Still, that $\mathbf{p}$ is, in principle, an observable quantity is not without interest and we might wonder whether this could be construed as evidence in favour of its 'reality'. Notice that our reconstruction of ψ proceeds by determining its influence on particles. It is thus indirectly measured in much the same way that one obtains information on fields of force in classical physics.

8.8 On the possibility of testing the hypotheses of the quantum theory of motion

8.8.1 Can we detect empty waves?

The fundamental physical assumption of the de Broglie–Bohm theory is the objective (co)existence of the quantum wave and the particle it guides. In this section we shall consider two ways in which one might attempt to demonstrate empirically the validity of this hypothesis: by studying the propagation of the wave into regions far beyond the location of the corpuscle (this subsection), and by monitoring the passage of the corpuscle without appreciably disturbing the coherence properties of the wave (the next one). In both cases we conclude that it is not, in fact, possible to invent experiments compatible with the current quantum formalism whose results would allow one to infer support for the causal assumptions. But this circumstance does not provide evidence against the causal interpretation; paradoxically, our arguments strengthen the conventional assertions of impossibility by clarifying their physical meaning.

We saw in §8.3 how the measurement process generally involves the generation of a plethora of 'empty wave packets', wavefunctions which are finite in regions of configuration space far from where the system point is located, and having negligible overlap with the directly relevant guiding wave. Indeed, inasmuch as measurements are just special but typical many-body interactions, we expect that empty waves are routinely created in naturally occurring physical processes. Whilst the complexity of the configuration space dynamics implies that there is negligible probability of the empty packets subsequently overlapping the packet actually containing the particle of interest and observably influencing it, the question we raise here is whether we can detect the influence of a wave we know to be empty on the measurable properties of *another* wave-particle composite with which it may interact. It has been suggested that this is possible (Croca, 1987). If so, it would provide empirical evidence in favour of the de Broglie–Bohm theory (by demonstrating the physical reality of the wave) and hence against the wavefunction collapse hypothesis. An argument in support of this possibility is that the empty waves, while devoid of corpuscles, are certainly not devoid of energy-momentum which is propagated throughout space. It has indeed been claimed that empty waves are, in principle, detectable but that this is technically difficult because the energy-momentum they carry is negligible relative to that of the particle (Selleri, 1982). We shall now show that application of the usual many-body quantum theory to this problem forbids such detection.

Suppose an initial packet $\psi(x)$ containing a particle with coordinates $x(t)$ is split into two packets $\psi_1(x)$ and $\psi_2(x)$ that subsequently separate so that eventually they do not appreciably overlap (Fig. 8.3). Suppose that ψ_1 can interact with a detector having wavefunction $\varphi(z)$ and coordinate $z(t)$ that can measure the position x, and ψ_2 interacts with some other system having wavefunction $\phi(y)$ and actual location $y(t)$. Initially, the total wavefunction is

$$\Psi_0(x, y, z) = [\psi_1(x) + \psi_2(x)]\phi(y)\varphi(z). \qquad (8.8.1)$$

The two interactions occur in disjoint regions of configuration space and after they have commenced, the wavefunction is nonfactorizable:

$$\Psi(x, y, z) = \alpha(x, z)\phi(y) + \beta(x, y)\varphi(z), \qquad (8.8.2)$$

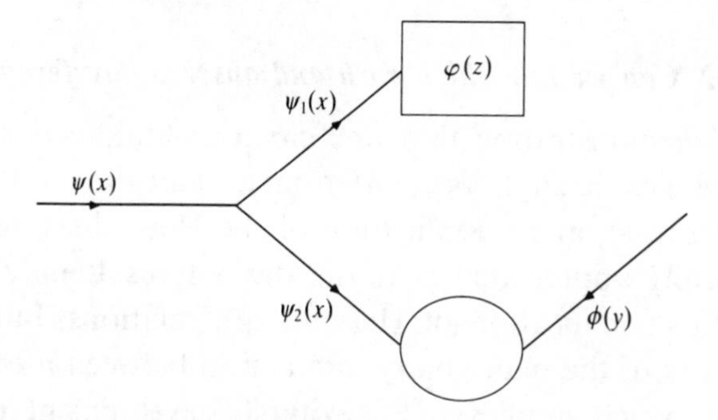

Fig. 8.3 If the particle x is detected, the packet $\psi_2(x)$ is empty. Its interaction with any other system (y) cannot be detected.

where α and β are each entangled in their respective configuration spaces. If the corpuscle x lies in ψ_1 and is detected, we know ψ_2 is an empty wave. Can we distinguish this notion from the proposal of wave packet collapse according to which ψ_2 no longer 'exists' due to the change in our knowledge? Since the two summands in (8.8.2) do not overlap, the system point is in one of them. But we know $x(t)$ is in the first summand so we conclude that $y(t)$ cannot be in the second. Hence, $\psi_2(x)$ can have no observable effect on the behaviour of the particle y and we cannot prove our hypothesis of empty waves.

The point is that empty packets only interact with other *empty* packets – the particles of the other systems are in the same configuration space packet as the particle of interest. All the empty packets do indeed interact with one another and mutually modify their behaviour but this is unobservable since no particles are involved.

This feature of the configuration space dynamics is consistent with experience. If empty waves could really alter the measurable properties of systems, experiments would be constantly perturbed by background noise caused by extraneous ψ-fields and it would be hard to justify the assumption of relative autonomy in which it is legitimate to isolate segments of matter and ignore the rest of the universe.

We also see that the magnitude of the energy-momentum content of the ψ-wave is not actually relevant to the question of detectability. All we ever observe in the outcomes of experiments are the properties of particles. This is the only way we can derive empirical information regarding the wave-function. Our argument here is of a general character and not dependent on the nature of the systems described by $\psi(x)$. This may, for example, be an electromagnetic state function, where x represents the coordinates of the field (cf. Chap. 12).

8.8.2 Can we tell which path and observe interference?

The traditional demonstration that one cannot simultaneously observe the path a particle takes through an interferometer and the interference pattern it contributes to rests on an application of the Heisenberg relations. It has not been generally appreciated that the desired result may be arrived at without any direct invocation of Heisenberg's relations, but follows from general properties of the many-body interaction between a coherent system and a detector which implies the inevitable development of correlations between the two. This seems to provide a deeper understanding of the problem in that applications of Heisenberg's relations have tended to rely on arguments drawn from classical optics whose connection with the quantum formalism is not always clear. We shall see that even if the coherence of the interfering beams is not destroyed by the external path-measuring agency, interference will still not be observed in situations where we determine the path.

Actually, the usual approach to this problem, however it proceeds, is deficient in one fundamental respect: it seeks to define the conditions under which one can or cannot have definite knowledge about the 'path' an atomic object takes while working with a pure wavefunction formalism within which such a notion is meaningless in all circumstances. To discuss this problem consistently means admitting that ψ provides an incomplete description of individual events and must be supplemented by variables pertaining to a theoretically well-defined path. This, of course, is precisely what the causal interpretation provides and it is our purpose here to put on a secure conceptual footing the proof of the impossibility of simultaneous observation of an orbit and an interference image ('simultaneous' means that in an individual case a particle traverses a known path and contributes to an interference pattern generated by the overlap of two coherent waves – the pattern itself is built up over a statistical ensemble as usual). In the process we also show that the trajectory assumption is not disproved by the traditional argument – on the contrary, it serves to justify the latter. Notice that the wave aspect of matter is still relevant when we possess 'which-path' information and the notion that 'wave-particle duality' refers to complementary concepts does not apply in our model.

Consider a normalized wave packet $\psi(x)$ incident on a beam splitter which produces two packets ψ_1 and ψ_2 that propagate along the arms of an interferometer (Fig. 8.4). A device designed to measure the position of the particle x is situated in the upper arm. The meter needle of this device has initial coordinate y_0 lying in a normalized packet state $\phi_0(y)$. It is assumed

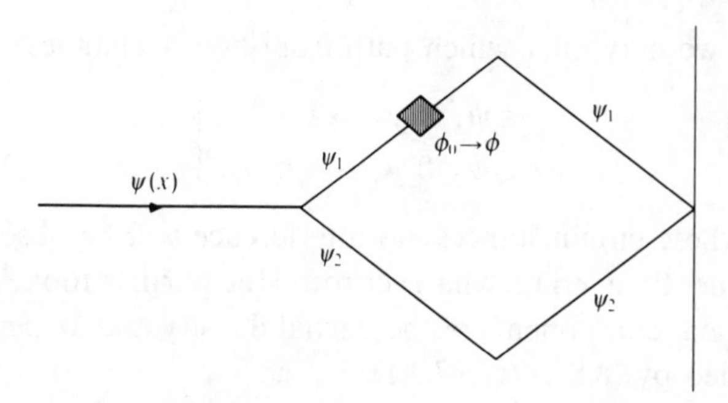

Fig. 8.4 Path detection in an interferometer.

that $\psi_1 \cap \psi_2 = 0$ in order that the observing apparatus interacts only with ψ_1. We can determine the route of the particle by performing a sharp position measurement. However, a 'which-path' determination only requires a weaker measurement which locates the particle in the region of space where ψ_1 is finite. Hence, we assume that the interaction with the observing device does not appreciably alter ψ_1. The aim is to maximize the interference in the x-variable and minimize that between the initial and final states of the y-variable. We shall show that these are mutually incompatible requirements.

At the exit to the beam splitter the total wavefunction is

$$\Psi_0(x, y) = [\psi_1(x) + \psi_2(x)]\phi_0(y) \tag{8.8.3}$$

$$\rightarrow f(x, y) + \psi_2(x)\phi_0(y) \qquad \text{during interaction}$$

$$\rightarrow \Psi(x, y) = \psi_1(x)\phi(y) + \psi_2(x)\phi_0(y) \tag{8.8.4}$$

after the interaction, where $\phi(y)$ is the final state of the meter and where, as we have said, ψ_1 is essentially unaltered. This kind of transformation can be realized in practice if the device interacts with a variable pertaining to the particle (e.g., spin) other than the one displaying interference (x). An example is a sequence of spin flips, spin up to down to up again, induced by the passage of a spin $\frac{1}{2}$ particle through two maser cavities (Scully, Englert & Schwinger, 1989; Scully & Walther, 1989; see also Scully, Englert & Walther, 1991). If ψ_1 and ψ_2 still do not overlap, the configuration space summands in (8.8.4) will not overlap and the system point (x, y) lies in one of them. However, in order to say unambiguously which one from observation of y, we must require that ϕ and ϕ_0 are disjoint ($\phi_0 \cap \phi = 0$), that is the initial and final apparatus states must be orthogonal. (In the above example, this may be achieved if the electromagnetic field in each maser cavity is a number state that changes by one quantum when the spin is flipped.) The paths through the interferometer are then correlated with distinguishable states of the meter and we may infer which path from how y changes:

$$\left.\begin{array}{l} x \in \psi_1 \text{ iff } y_0 \to y \in \phi \\ x \in \psi_2 \text{ iff } y_0 \to y \in \phi_0. \end{array}\right\} \tag{8.8.5}$$

But, under these circumstances, no interference will be observed when ψ_1 and ψ_2 subsequently overlap, whatever route the particle took. This is evident from the diagonal components of the partial density matrix pertaining to the system x implied by (8.8.4) (cf. §7.3.1):

$$I(x) = |\psi_1(x)|^2 + |\psi_2(x)|^2 + [c\psi_2^*(x)\psi_1(x) + \text{cc}] \tag{8.8.6}$$

where

$$c = \int \phi_0^*(y)\phi(y)\,\mathrm{d}y. \tag{8.8.7}$$

Clearly, if $c = 0$ the interference terms vanish. We conclude that a determination of the path a particle takes through an interferometer is incompatible with the observation of interference.

One can envisage various set-ups to investigate this question, such as those discussed in the Einstein–Bohr debate in which path-determination is achieved by interacting directly with the variable displaying interference. But however the path is determined, the general principle at work that causes the pattern to be washed out is clear: it is the creation of entanglement between two initially independent systems which implies that the coherence properties of one system (x) come to depend on the coherence of the other (y). The nub of the issue is that in order to perform its role and unambiguously reveal a change, the initial and final states of the observing device must be orthogonal. Between the extremes of $\phi = \phi_0$ (maximum contrast) and $\phi \cap \phi_0 = 0$ (no interference) there is a continuous range of diminutions in the contrast of the pattern depending on the extent of the overlap of ϕ_0 and ϕ. The mere fact that there is an exchange of energy or momentum between the system and apparatus in these cases is not sufficient to yield which-path information, except when $\phi \cap \phi_0 = 0$.

Our argument against the possibility of a simultaneous observation of path and interference is evidently of a general character and not dependent on the nature of the interacting systems. For example, y may represent the coordinates of an electromagnetic field and x an atomic beam. Notice also that we arrive at this conclusion without invoking the collapse of the wavefunction.

But accepting that we cannot gain information on the path while retaining coherence, can we not at least claim that experiments may be designed to demonstrate that *one* path was, in fact, traversed, even though we do not know which? This possibility has been discussed in the context of a modification of Fig. 8.4 in which a second device is inserted, in the lower arm. Let the initial wavefunction of this further device be $\xi_0(z)$ where the meter needle has coordinate z. We assume that the effect of the two devices, one in each arm, is to alter ψ_1 and ψ_2 in the same way so as not to disturb their mutual coherence:

$$[\psi_1(x) + \psi_2(x)]\phi_0(y)\xi_0(z) \to (\psi_1'(x) + \psi_2'(x))\phi_0(y)\xi_0(z) \qquad (8.8.8)$$

assuming that the states of the devices are essentially unchanged. A practical realization of such a set-up is a neutron interferometer (cf. §§5.4, 9.4) in which time-dependent spin-flipping coils are inserted in each arm (Badurek, Rauch & Tuppinger, 1986). A quantum of energy is exchanged between the electromagnetic fields and the neutron waves when the spin is flipped and the states of the fields are not appreciably altered if they are coherent. It has been suggested that the indivisibility of the energy exchange located in the coils implies that it is associated with the particle aspect of matter (Vigier, Dewdney, Holland & Kyprianidis, 1987). Indeed the energy change suffered by the neutron waves can be measured simultaneously with the observation of interference (Rauch & Vigier, 1990). Can we conclude that such experiments provide at least indirect evidence for the particle trajectory assumption?

The answer seems to be negative, on simple logical grounds. In order to infer from the results of experiments some fact regarding a trajectory, the theory tested in the experiment should depend in some way on the concept of an orbit. The quantum formalism does not. The time-dependent interactions embodied in (8.8.4) and (8.8.8) take place regardless of which path is traversed by the corpuscle postulated by the causal interpretation, or indeed of whether there is a corpuscle at all. The localized, discrete energy exchanges are properties of the interactions of quantum waves governed by the Schrödinger equation and do not provide evidence for the existence of trajectories.

It seems that a direct physical demonstration of the hypotheses of the quantum theory of motion is not possible using just experimental techniques compatible with the current formalism. The role of the more detailed individual predictions concerning particle behaviour that go beyond the statistical predictions contained in the wavefunction is primarily explanatory. But the sharpness of the causal formulation and its open character imply that this is not a closed question. We may entertain the possibility of changing some of the quantal and causal rules which, as discussed by Bohm (1952a,b), would imply testable predictions different from those of the current quantum theory. However, we have at present no clues as to which rules are suspect and which changes would be legitimate. The approach advocated in §5.5 was that we should attempt to employ the causal interpretation to extend the quantum formalism into new domains where it is currently ambiguous, rather than change axioms which as far as we can tell are well established. One possible area for such an application is the definition of transit times.

9

Spin $\frac{1}{2}$: the Pauli theory

9.1 Introduction

The claim that it is impossible to construct a realistic physical model of quantum matter is often presented as definitively and finally proved in the case of systems with spin $\frac{1}{2}$, or internal angular momentum of magnitude $\hbar/2$, which exhibit a characteristic 'nonclassical two-valuedness'. Nevertheless, the extension of the causal interpretation to embrace nonrelativistic spin $\frac{1}{2}$ systems was successfully carried through in the 1950s (Bohm, Schiller & Tiomno, 1955; Bohm & Schiller, 1955; Takabayasi, 1955; Takabayasi & Vigier, 1957) and developed more recently (Dewdney, Holland & Kyprianidis, 1986, 1987a,b; Dewdney, Holland, Kyprianidis & Vigier, 1988). The idea is to assume that the spinor wave represents a new type of physical field propagating in spacetime that exerts an influence on a corpuscle moving within it. The additional information conveyed by the spinor beyond that contained in the spinless Schrödinger wave implies that the state of the particle is not exhausted by specifying its position and momentum, but that it is endowed with further properties such as an internal angular momentum.

The equation governing the dynamics of the spinor wavefunction is the Pauli equation:

$$i\hbar\frac{\partial\psi}{\partial t} = \hat{H}\psi, \tag{9.1.1}$$

where the Hamiltonian operator is given by

$$\hat{H} = -(\hbar^2/2m)[\nabla - (\mathrm{i}e/\hbar c)\mathbf{A}]^2 + \mu\mathbf{B}\cdot\boldsymbol{\sigma} + eA_0 + V. \tag{9.1.2}$$

Here e, m and μ are respectively the charge, mass and magnetic moment of the particle, A_0 and $\mathbf{A}$ are the electromagnetic potentials, $\mathbf{B} = \nabla \times \mathbf{A}$ is the magnetic field, V is any other external scalar potential, and $\boldsymbol{\sigma}$ is a vector whose components are the 2×2 Pauli matrices. The Hamiltonian (9.1.2) is therefore a 2×2 matrix operator and the wavefunction $\psi(\mathbf{x}, t)$ has two complex components. We shall label the spin indices by $a, b, \ldots$ and write $\psi = (\psi^a)$, $a = 1, 2$. The components of the spinor are not simply two Schrödinger functions placed side by side in a column matrix but rather together form a single entity.

The causal formulation starts by observing that (9.1.1) implies the existence of a conserved current

$$\mathbf{j} = (\hbar/2mi)[\psi^\dagger \nabla\psi - (\nabla\psi^\dagger)\psi] - (e/mc)\mathbf{A}\psi^\dagger\psi, \tag{9.1.3}$$

where $\psi^\dagger$ is the Hermitian adjoint spinor. In a fairly obvious generalization of the Schrödinger theory, one associates an ensemble of particles with the same spinor wave and defines the velocity of a particle as $\mathbf{v} = \mathbf{j}/\rho$ where $\rho = \psi^\dagger\psi$. Specifying $\mathbf{x}_0$ we can chart the motion of each particle in the ensemble. Probability enters as usual only because the actual $\mathbf{x}_0$ realized is unknown, and is given by ρ. The detailed development of the theory to include the new spin degrees of freedom in the model of the particle is the subject of this chapter.

This method of treating spin is consistent but not completely satisfactory and its drawbacks will be discussed (§9.7) and resolved (Chap. 10) later. Its principal virtue is to show that spin-related phenomena may indeed be understood on the basis of a continuous, spacetime model and hence that this most novel of quantal characteristics may be brought within the scope of the causal interpretation.

9.2 Spinors and orientation

9.2.1 The triad implied by a spinor

Our aim is to reformulate the wave equation (9.1.1) mathematically in a way that implies a natural physical interpretation. To this end, our first task is to find a way of representing visually the information encoded in a spinor. It might be thought that a natural procedure to follow in attempting to develop a causal formulation of spin would be to start by expressing each of the complex components of a Pauli spinor in the form

$$\psi^a = R^a \, \mathrm{e}^{\mathrm{i}S^a/\hbar}, \qquad a = 1, 2, \tag{9.2.1}$$

where R^a, S^a are real functions, in view of the success of the polar representation of the wavefunction in the Schrödinger case. Although we will have occasion to use this form, a more useful pseudo-polar decomposition which brings out clearly how the formal theory constitutes a generalization of the spin 0

(Schrödinger) theory is the following:

$$\psi^a = R\,e^{iS/\hbar}\varphi^a. \tag{9.2.2}$$

Here R and S are real amplitude and phase functions respectively and $\varphi^a = r^a\,e^{is^a/\hbar}$ is a two-component object satisfying the conditions

$$\varphi^\dagger\varphi = r^\dagger r = 1, \qquad \tfrac{1}{2}(s^1 + s^2) = 0 \tag{9.2.3}$$

so that it has two real degrees of freedom. This means that S is the average phase of ψ (and is not a scalar quantity under spinor transformations). The information relating specifically to spin $\frac{1}{2}$ is carried by φ.

To see the meaning of the terms in the decomposition (9.2.2), we shall interpret the spinor as defining a state of rotation through the mutually orthogonal triad of Euclidean vectors that it determines (Payne, 1952; Kramers, 1957; Cartan, 1966. For the algebraic background see Holland, 1988b).

Unless otherwise stated, we shall assume the summation convention on repeated spinor $(a, b, \ldots)$ and tensor $(i, j, k, \ldots)$ indices and in accordance with common practice we shall often suppress the spinor indices.

We first note that we may form from a spinor ψ and its Hermitian conjugate $\psi^\dagger = (\psi^{a*})$, a real vector $\mathbf{S} = S_i\mathbf{e}_i$ whose components are given by

$$S_i = \psi^\dagger\sigma_i\psi \tag{9.2.4}$$

where $\mathbf{e}_i$, $i = 1, 2, 3$, form an orthonormal set of space axes:

$$\mathbf{e}_i\cdot\mathbf{e}_j = \delta_{ij}, \qquad i, j = 1, 2, 3. \tag{9.2.5}$$

We shall assume the following representation for the Pauli matrices:

$$({\sigma_1}^a{}_b) = \begin{pmatrix} 0 & 1 \\ 1 & 0 \end{pmatrix}, \qquad ({\sigma_2}^a{}_b) = \begin{pmatrix} 0 & -i \\ i & 0 \end{pmatrix}, \qquad ({\sigma_3}^a{}_b) = \begin{pmatrix} 1 & 0 \\ 0 & -1 \end{pmatrix}. \tag{9.2.6}$$

It is easy to show that the magnitude of $\mathbf{S}$ is

$$\rho \equiv R^2 = (S_iS_i)^{1/2} = (|\psi^1|^2 + |\psi^2|^2)^{1/2}. \tag{9.2.7}$$

Note that $\mathbf{S} \neq 0$ if $\rho \neq 0$.

We may generate two other vectors $\mathbf{M} = M_i\mathbf{e}_i$, $\mathbf{N} = N_i\mathbf{e}_i$ by forming a bilinear combination of ψ and the dual spinor $\bar{\psi}$ defined by $\bar{\psi}^1 = -\psi^2$, $\bar{\psi}^2 = \psi^1$. The (real) components of these vectors are given by the real and imaginary parts of the complex vector $\bar{\psi}\sigma_i\psi = \mathrm{M}_i + \mathrm{i}N_i$. Using certain properties of the Pauli matrices it is easy to show that this vector is isotropic (so that $\mathbf{M}\cdot\mathbf{N} = 0$ and $|\mathbf{M}| = |\mathbf{N}| = \rho$, where ρ is given by (9.2.7)) and orthogonal to **S** ($\mathbf{M}\cdot\mathbf{S} = \mathbf{N}\cdot\mathbf{S} = 0$). The set of unit vectors

$$\mathbf{e}_1' = \mathbf{M}/\rho, \qquad \mathbf{e}_2' = \mathbf{N}/\rho, \qquad \mathbf{e}_3' = S/\rho \tag{9.2.8}$$

thus defines an orthonormal triad satisfying

$$\mathbf{e}_i'\cdot\mathbf{e}_j' = \delta_{ij}, \qquad i, j = 1, 2, 3. \tag{9.2.9}$$

A Pauli spinor therefore determines a system of vectors which defines a Cartesian reference frame. Later we shall associate the system $\mathbf{e}_i'$ with the body axes of a rigid body oriented with respect to a set of fixed space axes $\mathbf{e}_i$.

The nine components S_i, M_i and N_i are restricted by five conditions, i.e., (9.2.9), which leaves four real degrees of freedom to describe the spinor (three orientation parameters and one length).

To see how the system $\mathbf{e}_i'$ specifies a state of rotation, suppose that the space axes are in standard orientation, with components $(\mathbf{e}_i)_j = \delta_{ij}$. Denoting the components of $\mathbf{e}_i'$ with respect to $\mathbf{e}_i$ by

$$(\mathbf{e}_i')_j = a_{ij}, \tag{9.2.10}$$

we have

$$\mathbf{e}_i' = a_{ij}\mathbf{e}_j, \tag{9.2.11}$$

where the matrix $a = (a_{ij})$ satisfies from (9.2.9) $a_{ik}a_{jk} = \delta_{ij}$ or $a^{\mathrm{T}}a = aa^{\mathrm{T}} = I$, where I is the unit 3×3 matrix. The real orthogonal matrices a leave invariant the norm $(x_i x_i)^{1/2}$ of any vector $\mathbf{x} = x_i\mathbf{e}_i$ and, if we impose the condition of no reflections, $\det a = 1$, they generate the group $SO(3)$ of proper Euclidean rotations.

The scalar ρ in (9.2.7) is an invariant of SO(3) and of the group of 2×2 unimodular unitary matrices, $SU(2)$. With respect to the latter ψ transforms as

$$\psi' = U\psi, \tag{9.2.12}$$

where

$$UU^\dagger = U^\dagger U = I, \qquad \det U = 1. \tag{9.2.13}$$

This is the invariance group of the Pauli equation (9.2.1).

Given U, we can construct a unique corresponding $SO(3)$ transformation a by the formula

$$U\sigma_i U^{-1} = a_{ij}\sigma_j. \tag{9.2.14}$$

The correspondence between U and a is not, however, one-to-one. By inverting (9.2.14) we may deduce U given a, but only up to a sign. We have

(Hestenes, 1971)

$$U = \pm[(1 + \mathrm{Tr}\, a)I - \mathrm{i}a_{ij}\varepsilon_{ijk}\sigma_k]/2(1 + \mathrm{Tr}\, a)^{1/2}, \tag{9.2.15}$$

where ε_{ijk} is the completely antisymmetric symbol, with $\varepsilon_{123} = 1$. The groups $SU(2)$ and $SO(3)$ are therefore 2–1 homomorphic.

A point to note with regard to this homomorphism is that the order of multiplication of successive corresponding rotations in the two groups is reversed. That is, if U_1 corresponds to a Euclidean rotation a_1, and U_2 to a_2, then

$$U_1 U_2 \leftrightarrow a_2 a_1. \tag{9.2.16}$$

A matrix satisfying (9.2.13) may be expressed in the form

$$(U^a{}_b) = \begin{pmatrix} \alpha & -\beta^* \\ \beta & \alpha^* \end{pmatrix}, \qquad |\alpha|^2 + |\beta|^2 = 1, \qquad \alpha, \beta \in \mathbb{C}. \tag{9.2.17}$$

Just as the rows of the matrix a define via (9.2.10) the components of $\mathbf{e}'_i$ with respect to $\mathbf{e}_i$ (and the columns the components of $\mathbf{e}_i$ with respect to $\mathbf{e}'_i$), so here in an analogous way the rows and columns of (9.2.17) may be interpreted as the components of unit spinors with respect to some standard basis in spin space. The unit spinor with components α, β is analogous, for example, to the vector $\mathbf{e}'_1$ which results from an arbitrary Euclidean rotation. We shall use this fact below where we derive an expression for a spinor in explicit geometric terms.

To put some geometrical content into the above results, we note that the orientation of a rigid body relative to some standard state may be uniquely specified by giving a unit vector $\mathbf{n}$ along the axis of rotation, and the angle of rotation Φ about the axis. Given the matrix $a(\Phi, \mathbf{n})$ describing this rotation, it is easy from (9.2.15) to construct the corresponding spinor transformation:

$$U(\Phi, \mathbf{n}) = \pm[\cos(\Phi/2)I + \mathrm{i}\sin(\Phi/2)\mathbf{n}\cdot\boldsymbol{\sigma}]. \tag{9.2.18}$$

The appearance of a half-angle in (9.2.18) implies that a rotation of 2π about some axis $\mathbf{n}$ in real space, which, of course, is the identity transformation, is represented in spin space by a transformation that reverses sign:

$$U(2\pi, \mathbf{n}) = -U(0, \mathbf{n}) = -I. \tag{9.2.19}$$

Applying (9.2.19) to a spinor, we thus find that it transforms in the following way under a Euclidean rotation of $2\pi n$, n an integer:

$$\psi \to (-1)^n \psi \tag{9.2.20}$$

i.e., it only returns to itself under a rotation which is an even multiple of 2π, and it reverses sign under odd multiples of 2π.

9.2.2 Eulerian representation

It will prove useful to represent the spinor in terms of the Euler angles which specify the orientation of the triad with respect to the fixed set of Cartesian axes in Euclidean space.

The Euler angles (ϕ, θ, χ) are defined by a sequence of three successive rigid rotations of the space axes, and we shall adopt the following convention (Fig. 9.1). Starting from the right handed triad $\mathbf{e}_1, \mathbf{e}_2, \mathbf{e}_3$, we rotate clockwise (as viewed along the negative of the axis of rotation) by an angle ϕ about the $\mathbf{e}_3$-axis to a first intermediate set of axes $\mathbf{f}_1, \mathbf{f}_2, \mathbf{e}_3$. We then rotate clockwise by an angle θ about the axis $\mathbf{f}_1$ to a second intermediate set of axes $\mathbf{f}_1, \mathbf{f}'_2, \mathbf{e}'_3$. Finally, we rotate clockwise through an angle χ about the axis $\mathbf{e}'_3$ to obtain the system $\mathbf{e}'_1, \mathbf{e}'_2, \mathbf{e}'_3$.

The rotation matrices for the three steps are

$$\left.\begin{aligned} a(\phi) &= \begin{pmatrix} \cos\phi & -\sin\phi & 0 \\ \sin\phi & \cos\phi & 0 \\ 0 & 0 & 1 \end{pmatrix}, \quad a(\theta) = \begin{pmatrix} 1 & 0 & 0 \\ 0 & \cos\theta & -\sin\theta \\ 0 & \sin\theta & \cos\theta \end{pmatrix}, \\ a(\chi) &= \begin{pmatrix} \cos\chi & -\sin\chi & 0 \\ \sin\chi & \cos\chi & 0 \\ 0 & 0 & 1 \end{pmatrix} \end{aligned}\right\} \tag{9.2.21}$$

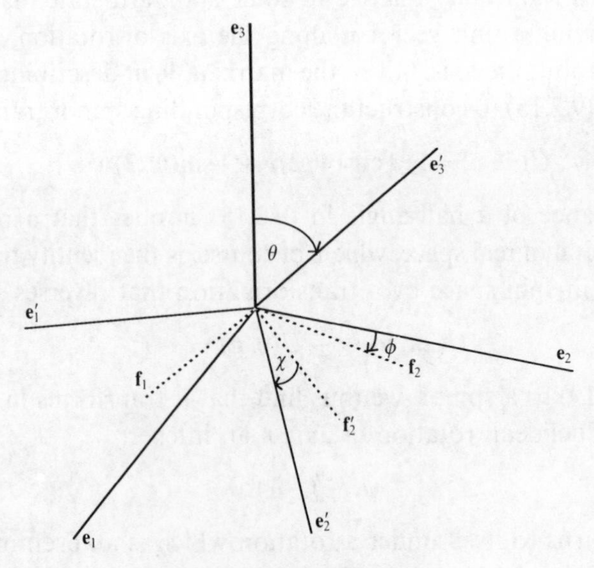

Fig. 9.1 The Euler angles.

and the matrix for the resultant transformation is

$$(a_{ij}) = a(\phi, \theta, \chi) = a(\chi)a(\theta)a(\phi)$$

$$= \begin{pmatrix} \cos\chi\cos\phi - \sin\chi\cos\theta\sin\phi & -\cos\chi\sin\phi - \sin\chi\cos\theta\cos\phi & \sin\chi\sin\theta \\ \sin\chi\cos\phi + \cos\chi\cos\theta\sin\phi & -\sin\chi\sin\phi + \cos\chi\cos\theta\cos\phi & -\cos\chi\sin\theta \\ \sin\theta\sin\phi & \sin\theta\cos\phi & \cos\theta \end{pmatrix} \quad (9.2.22)$$

with

$$0 \leqslant \phi \leqslant 2\pi, \qquad 0 \leqslant \theta \leqslant \pi, \qquad 0 \leqslant \chi \leqslant 2\pi.$$

The matrices corresponding to (9.2.21) in the spinor representation are from (9.2.18) and (9.2.6):

$$U(\phi) = \cos(\phi/2) + \mathrm{i}\sin(\phi/2)\sigma_3 = \begin{pmatrix} \mathrm{e}^{\mathrm{i}\phi/2} & 0 \\ 0 & \mathrm{e}^{-\mathrm{i}\phi/2} \end{pmatrix}$$

$$U(\theta) = \cos(\theta/2) + \mathrm{i}\sin(\theta/2)\sigma_1 = \begin{pmatrix} \cos(\theta/2) & \mathrm{i}\sin(\theta/2) \\ \mathrm{i}\sin(\theta/2) & \cos(\theta/2) \end{pmatrix}$$

$$U(\chi) = \cos(\chi/2) + \mathrm{i}\sin(\chi/2)\sigma_3 = \begin{pmatrix} \mathrm{e}^{\mathrm{i}\chi/2} & 0 \\ 0 & \mathrm{e}^{-\mathrm{i}\chi/2} \end{pmatrix}$$

and hence the resultant transformation homomorphic with (9.2.22) is, bearing in mind (9.2.16),

$$\begin{aligned} (U^a{}_b) &= U(\phi, \theta, \chi) = U(\phi)U(\theta)U(\chi) \\ &= \cos\left(\frac{\theta}{2}\right)\cos\left(\frac{\phi+\chi}{2}\right) + \mathrm{i}\sin\left(\frac{\theta}{2}\right)\cos\left(\frac{\phi-\chi}{2}\right)\sigma_1 \\ &\quad + \mathrm{i}\sin\left(\frac{\theta}{2}\right)\sin\left(\frac{\chi-\phi}{2}\right)\sigma_2 + \mathrm{i}\cos\left(\frac{\theta}{2}\right)\sin\left(\frac{\phi+\chi}{2}\right)\sigma_3 \\ &= \begin{pmatrix} \mathrm{e}^{\mathrm{i}(\chi+\phi)/2}\cos\left(\frac{\theta}{2}\right) & \mathrm{i}\,\mathrm{e}^{\mathrm{i}(\phi-\chi)/2}\sin\left(\frac{\theta}{2}\right) \\ \mathrm{i}\,\mathrm{e}^{\mathrm{i}(\chi-\phi)/2}\sin\left(\frac{\theta}{2}\right) & \mathrm{e}^{-\mathrm{i}(\chi+\phi)/2}\cos\left(\frac{\theta}{2}\right) \end{pmatrix}. \end{aligned} \quad (9.2.23)$$

The rows of (9.2.22) are the Eulerian representation of the components of the triad of vectors $\mathbf{e}'_i$ whose magnitude is unity (cf. (9.2.10)). For vectors of an arbitrary length ρ given by (9.2.7), we thus have

$$\left.\begin{aligned} M_i &= \rho a_{1i} \quad \text{or} \quad \mathbf{M} = \rho\mathbf{e}'_1 \\ N_i &= \rho a_{2i} \qquad\quad\; \mathbf{N} = \rho\mathbf{e}'_2 \\ S_i &= \rho a_{3i} \qquad\quad\; \mathbf{S} = \rho\mathbf{e}'_3 \end{aligned}\right\} \quad (9.2.24)$$

Similarly, comparing (9.2.17) with (9.2.23) we see that the first column of (9.2.23) gives the Eulerian representation of the components of a unit spinor which defines a state of rotation in spin space. For a spinor of arbitrary magnitude $(\psi^\dagger\psi)^{1/2} = \rho^{1/2} = R$ we therefore find

$$\psi = R\,\mathrm{e}^{\mathrm{i}\chi/2}\begin{pmatrix} \cos\left(\dfrac{\theta}{2}\right)\mathrm{e}^{\mathrm{i}\phi/2} \\ \mathrm{i}\sin\left(\dfrac{\theta}{2}\right)\mathrm{e}^{-\mathrm{i}\phi/2} \end{pmatrix}. \tag{9.2.25}$$

Comparing with (9.2.2) we have

$$\varphi = \mathrm{e}^{-\mathrm{i}\pi/4}\begin{pmatrix} \cos\left(\dfrac{\theta}{2}\right)\mathrm{e}^{\mathrm{i}\phi/2} \\ \mathrm{i}\sin\left(\dfrac{\theta}{2}\right)\mathrm{e}^{-\mathrm{i}\phi/2} \end{pmatrix}. \tag{9.2.26}$$

Referring to Fig. 9.2, the relation between the above geometrical representations of the four degrees of freedom in a Pauli spinor may be summarized as follows:

Average phase $= \frac{1}{2}(\chi + \pi/2)$ = half the angle of rotation of (**M**, **N**) as a unit about **S**.
Relative phase $= \pi/2 - \phi$ = azimuthal angle of **S** with respect to the $\mathbf{e}_1$-axis.
Total magnitude $= R$ = square root of the length of **S**.
Relative magnitude $= \tan(\theta/2)$ = tan (half the polar angle of **S**).

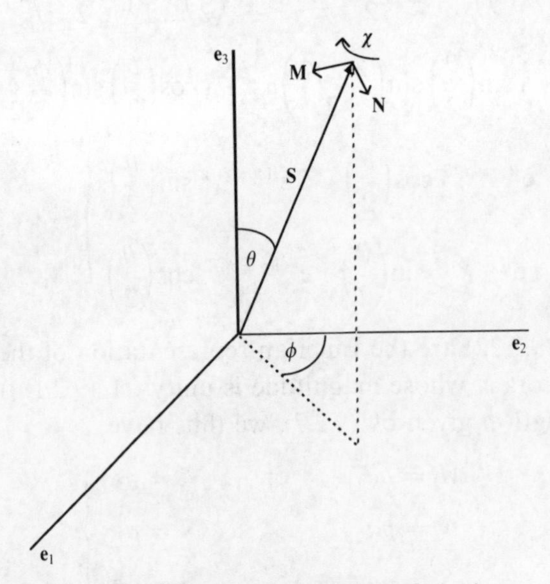

Fig. 9.2 Euclidean representation of a spinor.

This also indicates the meaning of the phases in (9.2.1): $S^1/\hbar = (\chi + \phi)/2$, $S^2/\hbar = (\chi - \phi + \pi)/2$.

The most important property of a spinor, expressed in (9.2.20), is not exhibited in the above model since the sign of ψ does not appear as an independent degree of freedom. Information on the sign is carried by the phase $S/\hbar$ but the angle χ changes by $2\pi n$ when $S/\hbar$ changes by πn, n an integer. It might be felt that this feature of the spinor cannot be experimentally observed since physical observables are quadratic functions of spinor components. However, *relative to another state* whose sign stays fixed, a reversal of sign can be empirically relevant (Aharonov & Susskind, 1967; Bernstein, 1967). This is possible because the spinor enters physics as a field that can be coherently split into two disjoint parts which may pick up a relative phase and amplitude before being recombined. If we consider the coherent superposition of two spinors ψ_1 and ψ_2,

$$\psi = \psi_1 + c\psi_2, \qquad c \in \mathbb{C}, \tag{9.2.27}$$

then the vectors associated with ψ depend sensitively on the value of c. If in particular $c = \pm 1$, which corresponds to considering the different signs of ψ_2 relative to the state ψ_1, we shall see that a body guided by ψ behaves differently in the two cases (§9.4). The 4π-symmetry of spinors was first observed in neutron interferometry (Rauch *et al.*, 1975, Werner *et al.*, 1975). An analogy may be drawn between this property of spin $\frac{1}{2}$ systems and the well-known physical inequivalence of identical objects in Euclidean space that are rotated by odd or even multiples of 2π relative to a fixed environment (Misner *et al.*, 1973, Chap. 41; Bolker, 1973).

The above model also leads to a simple picture of a gauge transformation of the electromagnetic potentials in (9.1.2) (cf. §3.11.2). If the gauge function is denoted λ, the spinor is multiplied by a factor $e^{ie\lambda/\hbar}$ which corresponds to the transformation $\chi \to \chi + 2e\lambda/\hbar$. This induces a rotation of the vectors $\mathbf{M}, \mathbf{N}$ as a unit about the vector $\mathbf{S}$.

9.3 Physical content of the Pauli equation

9.3.1 Physical model

We pass now to the case of a spinor field, $\psi = \psi(\mathbf{x}, t)$. The configuration space of the system is $\mathbb{R}^3 \times SU(2)$ and ψ is an element of the Hilbert space defined on this manifold. Using the results of §9.2, we interpret the additional content in the field beyond the amplitude and phase of the Schrödinger wavefunction through the field of frames $\mathbf{e}'_i(\mathbf{x}, t)$ that it determines, which now vary from point to point, or equivalently through the spacetime-dependent set of Euler angles: $\phi(\mathbf{x}, t)$, $\theta(\mathbf{x}, t)$, $\chi(\mathbf{x}, t)$. Further, we postulate that embedded in the wave is a corpuscle of mass m. Corresponding to the new attributes of the wave, we suppose the particle has sufficient structure that it may be attributed an orientation defined by the rotation of a rigid set of body axes relative to fixed space axes. It is assumed that if the centre of mass of the body lies at the point $\mathbf{x}$ at time t, the body axes are defined by the frame $\mathbf{e}'_i(\mathbf{x}, t)$, $i = 1, 2, 3$, implied by the spinor field. As the field dynamically evolves, so the axes rotate in a corresponding fashion and the centre of mass moves.

The total angular momentum of a body now consists of its orbital angular momentum associated with the centre of mass translation and a continuously variable internal or intrinsic angular momentum (spin) characterized by the triad or Euler angles. Note that whilst the surrounding spinor wave depends on properties of the particle such as its mass, there is no direct physical reaction of the particle on the wave.

There is then naturally (i.e., independently of probability) associated with the wave a fictitious ensemble of spinning bodies, the motion of each being determined by its location in the field (like iron filings oriented in a magnetic field). Only one body actually accompanies each wavefunction. The centre of mass motion of the ensemble is derived from the velocity field implied by (9.1.3) and we may impose the consistent subsidiary condition that the spatial distribution of the ensemble is given by $\rho = \psi^\dagger\psi$.

We treat the triad of vectors as the principal axes of the body, but other than the fact that it has an orientation the structure of the body turns out not to be relevant. This is because the Pauli theory is not sensitive to any more precise details; for example, the moment of inertia does not enter the theory as a parameter (see Chap. 10). The causal theory of the Pauli equation is, therefore, consistent with a wide range of possible underlying structures. In particular this includes rigid bodies since the theory neglects relativistic effects which, of course, would exclude perfect rigidity on the grounds of causality (no instantaneous signalling). For a body the size of the classical electron radius ($\sim 10^{-13}$ cm) the velocity of rotation of the periphery exceeds the speed of light, for a magnetic moment $\hbar e/2mc$.

For convenience we may think of the body as a point particle with an attached frame. Note that in this theory we do not modify the kind of model we might give for a dipole in classical mechanics. As we shall see, the novelty lies in the forces and torques the body is subject to, compared with a classical rotator.

The Pauli theory does have one feature, however, that is not generally true for rigid body motion: the spin vector (i.e., the Euclidean vector which defines the axis of rotation) always lies along a principal axis. To see this, we must first justify our definition of the spin vector.

According to quantum mechanics, the expectation value of the spin observable in a state $\psi(\mathbf{x}, t)$ is

$$\langle \hat{\mathbf{s}} \rangle = \langle (\hbar/2)\boldsymbol{\sigma} \rangle = (\hbar/2) \int_{-\infty}^{\infty} \psi^\dagger \boldsymbol{\sigma} \psi \, \mathrm{d}^3x \tag{9.3.1}$$

with ψ normalized:

$$\int_{-\infty}^{\infty} \psi^\dagger \psi \, \mathrm{d}^3x = 1. \tag{9.3.2}$$

This suggests that the vector $(\hbar/2)$ $\mathbf{S}$ with $\mathbf{S}$ defined by (9.2.4) is a local spin density and hence that a suitable candidate for the spin vector field associated with the ensemble of particles is (cf. §3.12.2)

$$\left.\begin{aligned} \mathbf{s} = (\hbar/2\rho)\mathbf{S} &= (\hbar/2)\varphi^\dagger \boldsymbol{\sigma} \varphi \\ &= (\hbar/2)(\sin\theta \sin\phi, \sin\theta\cos\phi, \cos\theta), \end{aligned}\right\} \tag{9.3.3}$$

$$(\mathbf{s}\cdot\mathbf{s})^{1/2} = \hbar/2. \tag{9.3.4}$$

The spin vector of an individual particle is (9.3.3) evaluated along the trajectory: $\mathbf{s} = \mathbf{s}(\mathbf{x}(t), t)$. This is justified by the fact that just this vector couples to a magnetic field in the Hamilton–Jacobi equation (see below) in the expected way and, moreover, that $s_z \to \pm(\hbar/2)$, i.e., the eigenvalues of $\hat{s}_z$, when a measurement of $\hat{s}_z$ is performed (§9.5). The condition (9.3.4) implies that the body cannot be at rest.

The unit vector $\mathbf{S}/\rho$ is known in the literature as the 'polarization' vector and this is how we shall think of $\mathbf{s}$. Note that such a quantity, which is a local expectation value rather than a global one, sits rather uncomfortably in the conventional interpretation where strictly speaking it has no particular significance because the notion that there is an actual individual process underlying the statistical predictions of the formalism is meaningless.

Suppose that in the standard orientation the principal axes coincide with the space axes $\mathbf{e}_i$. After a rotation, the principal axes coincide with the body axes $\mathbf{e}'_i$. Since the spin vector (9.3.3) is collinear with one of the body axes ($\mathbf{e}'_3$), it follows that in the Pauli theory the body always rotates about one of its principal axes. A more realistic theory, for which this result only holds on the average and which is sensitive to more details of the model (e.g., moment of inertia) is described in the next chapter.

To conclude, the four real degrees of freedom in the spinor field may be represented by the set of variables $(\rho, S, \mathbf{s})$ or $(\rho, \chi, \phi, \theta)$ where each quantity is a spacetime function.

9.3.2 Equations of motion

For convenience we start by writing down again the Pauli equation (9.1.1):

$$i\hbar\frac{\partial\psi}{\partial t} = \{-(\hbar^2/2m)(\nabla - (ie/\hbar c)\mathbf{A})^2 + \mu\mathbf{B}\cdot\boldsymbol{\sigma} + eA_0 + V\}\psi \qquad (9.3.5)$$

or in component form:

$$i\hbar\frac{\partial\psi^a}{\partial t} = \{[-(\hbar^2/2m)(\nabla - (ie/\hbar c)\mathbf{A})^2 + eA_0 + V]\delta^a{}_b + \mu B_i\sigma_i{}^a{}_b\}\psi^b,$$

where e, m and μ are respectively the charge, mass, and magnetic moment, $\mathbf{A}(\mathbf{x}, t)$ is the external vector potential, $\mathbf{B} = \nabla \times \mathbf{A}$, $A_0(\mathbf{x}, t)$ is the external electric potential, $V(\mathbf{x}, t)$ represents any other external scalar potentials, c is the speed of light and $\hbar$ is Planck's constant divided by 2π. Both the translational and spin motion of a charged particle (such as an electron, with $\mu = e\hbar/2mc$) are affected by an external electromagnetic field, whereas only the spin couples to a magnetic field for an uncharged particle (e.g., a neutron).

Our aim is to reformulate the wave equation in terms of relations describing the translational and orientational behaviour of the ensemble of bodies. On substituting our expressions (9.2.2) or (9.2.25) for the spinor in (9.3.5) we expect to obtain four real coupled equations of motion for the evolution of the field variables (ρ, S, $\mathbf{s}$) or (ρ, ϕ, θ, χ). The more elegant way of obtaining the field equations is via a canonical formalism but here we shall employ a more direct 'sledgehammer' method.

First, to find two of the equations, contract (9.3.5) with $\psi^\dagger$ and separate into real and imaginary parts. The result is a Hamilton–Jacobi-type equation for S and a conservation equation for ρ:

$$\frac{\partial S}{\partial t} - i\hbar\varphi^\dagger\frac{\partial\varphi}{\partial t} + \tfrac{1}{2}m\mathbf{v}^2 + Q + Q_s + \left(\frac{2\mu}{\hbar}\right)\mathbf{B}\cdot\mathbf{s} + eA_0 + V = 0 \qquad (9.3.6)$$

and

$$\partial\rho/\partial t + \nabla\cdot(\rho\mathbf{v}) = 0. \qquad (9.3.7)$$

Here we may give a variety of forms for the various terms. Thus

$$\left.\begin{aligned}\mathbf{v} &= (\hbar/2mi\rho)[\psi^\dagger\nabla\psi - (\nabla\psi^\dagger)\psi] - (e/mc)\mathbf{A}\\ &= (1/m)(\nabla S - i\hbar\varphi^\dagger\nabla\varphi) - (e/mc)\mathbf{A}\\ &= (\hbar/2m)(\nabla\chi + \cos\theta\nabla\phi) - (e/mc)\mathbf{A}\end{aligned}\right\} \qquad (9.3.8)$$

is the velocity field derived from the Pauli current (9.1.3),

$$Q = -\hbar^2\nabla^2 R/2mR \qquad (9.3.9)$$

is the quantum potential associated with aspects of the motion not depending explicitly on the spin, and

$$Q_s = (1/2m)\, \partial_i s_j\, \partial_i s_j = (\hbar^2/8m)[(\nabla\theta)^2 + \sin^2\theta(\nabla\phi)^2] \qquad (9.3.10)$$

is a spin-dependent addition to the quantum potential.

Eqs. (9.3.6) and (9.3.7) obviously constitute the generalization to the spin case of the translational Hamilton–Jacobi and conservation equations of the Schrödinger theory (Schönberg, 1954; Schiller, 1962c). Indeed they reduce to the latter when $\phi =$ constant, $\theta =$ constant (i.e., φ or $\mathbf{s} =$ constant) and $\mu = 0$. Thus, if we think of R and S as analogous to the amplitude and phase of the Schrödinger wavefunction, then terms involving $\mathbf{s}$, or ϕ and θ, or φ, represent purely spin-dependent effects.

It seems reasonable to suppose that the particle energy in the ensemble is given by the field

$$E(\mathbf{x}, t) = -\frac{\partial S}{\partial t} + i\hbar\varphi^\dagger \frac{\partial \varphi}{\partial t} = -\frac{\hbar}{2}\left(\frac{\partial \chi}{\partial t} + \cos\theta \frac{\partial \phi}{\partial t}\right) \qquad (9.3.11)$$

which contains a spin-dependent contribution similar to the velocity (9.3.8).

Introducing now the postulate of the causal interpretation that the particle centre of mass has a well-defined translational motion given by $\mathbf{x} = \mathbf{x}(t)$, this may be found by solving

$$d\mathbf{x}/dt = \mathbf{v}(\mathbf{x}, t)|_{\mathbf{x}=\mathbf{x}(t)} \qquad (9.3.12)$$

given the initial position $\mathbf{x}_0 = \mathbf{x}(0)$. Clearly, the trajectories will, in general, differ from those of the Schrödinger theory, *even when there are no external potential energies* (so that each component of the spinor field satisfies its own free Schrödinger equation). This will occur when the wavefunction is not an eigenstate of the spin operator in some direction, i.e., when the wave is a superposition of eigenstates so that the degrees of freedom associated with the spin are functions of position and time. In general, then, the trajectories are not orthogonal to the surfaces $S =$ constant. This is the analogue of the result that the trajectory of a free particle in the Schrödinger case is not rectilinear if the wavefunction is a superposition of states of different momentum, e.g., a wave packet. To see this in more detail we shall evaluate the force law implied by (9.3.6) by taking the gradient of that equation. In order to write this succinctly we require the remaining two independent equations of motion implied by the Pauli equation, describing the evolution of $\mathbf{s}$, or ϕ and θ.

These may be found most efficiently by first deriving the equation of motion of the spin vector $\mathbf{s}$. Contracting (9.3.5) with $(\psi^\dagger\boldsymbol{\sigma})$ yields a vector equation.

Subtracting from this the Hermitian conjugate equation then gives an expression for $\partial \mathbf{s}/\partial t$:

$$\partial \mathbf{s}/\partial t + \mathbf{v}\cdot\nabla\mathbf{s} = \mathbf{T} + (2\mu/\hbar)\mathbf{B} \times \mathbf{s}, \tag{9.3.13}$$

where

$$\mathbf{T} = (1/m\rho)\mathbf{s} \times \partial_i(\rho\, \partial_i \mathbf{s}) \tag{9.3.14}$$

(for the method of proof see Holland (1988b)). In component form, the vector product $(\mathbf{B} \times \mathbf{s})_k = \varepsilon_{ijk} B_i s_j$. Identifying $\mathbf{v}$ with $\dot{\mathbf{x}}$, the field equation (9.3.13) describes the precession of the spin vector as one moves with one of the particles in the ensemble:

$$\frac{\mathrm{d}\mathbf{s}}{\mathrm{d}t} = (\mathbf{T} + (2\mu/\hbar)\mathbf{B} \times \mathbf{s})|_{\mathbf{x}=\mathbf{x}(t)} \tag{9.3.15}$$

where $\mathrm{d}/\mathrm{d}t = \partial/\partial t + \dot{\mathbf{x}}\cdot\nabla$ is the total time derivative with respect to a point moving with the particle.

Notice that even when the external field is zero ($\mathbf{B} = 0$) or when the particle has no magnetic moment ($\mu = 0$), the spin vector will still precess due to the action of the first term on the right hand side, the quantum torque (9.3.14). Such a situation arises when the wave is a superposition of spin eigenstates, as pointed out above for the trajectories. On the other hand, when the quantum torque vanishes ($\mathbf{T} = 0$) (9.3.15) reduces to the classical precession of a spin vector about an applied magnetic field. In general (9.3.15) may be written in classical form

$$\frac{\mathrm{d}\mathbf{s}}{\mathrm{d}t} = \frac{2\mu}{\hbar}\mathbf{B}_{\text{eff}} \times \mathbf{s} \tag{9.3.16}$$

where

$$\mathbf{B}_{\text{eff}} = \mathbf{B} - \frac{\hbar}{2\mu m\rho}\,\partial_i(\rho\, \partial_i \mathbf{s})$$

is an effective magnetic field.

It is a simple matter now to substitute (9.3.3) into (9.3.15) to find the time rate of change of ϕ and θ:

$$\frac{\hbar}{2}\frac{\mathrm{d}\phi}{\mathrm{d}t} = \frac{\hbar^2}{4m\rho \sin\theta}[-\partial_i(\rho\, \partial_i \theta) + \rho \cos\theta \sin\theta (\nabla\phi)^2] + \frac{\mu}{\sin\theta}\mathbf{B}\cdot(\cos\theta \sin\phi, \cos\theta\cos\phi, -\sin\theta) \tag{9.3.17}$$

$$\frac{\hbar}{2}\frac{\mathrm{d}\theta}{\mathrm{d}t} = \frac{\hbar^2}{4m\rho\sin\theta}\partial_i(\rho \sin^2\theta\, \partial_i\phi) - \frac{2\mu}{\hbar\sin\theta}(\mathbf{B} \times \mathbf{s})_3. \tag{9.3.18}$$

Returning to the Hamilton–Jacobi equation (9.3.6), taking its gradient, rearranging and substituting for $\mathrm{d}\mathbf{s}/\mathrm{d}t$ from (9.3.15) we obtain the generalization of Newton's force law for quantum systems of spin $\frac{1}{2}$:

$$m\frac{\mathrm{d}v_i}{\mathrm{d}t} = -\partial_i(V+Q) + F_i - \frac{2\mu}{\hbar}(\partial_i B_j)s_j - \frac{1}{m\rho}\partial_k(\rho\,\partial_i s_j\,\partial_k s_j), \quad (9.3.19)$$

where F_i is the Lorentz force. Clearly, in the absence of external fields, the particle is subject to the usual quantum force $(-\partial_i Q)$ together with a spin-dependent addition (the last term on the right hand side of (9.3.19)).

To summarize, we have transcribed the Pauli equation (9.3.5) into a mathematically equivalent set of four field equations (9.3.6), (9.3.7) and (9.3.13) (or (9.3.17) and (9.3.18)). These must be supplemented by conditions on the new variables corresponding to those imposed on ψ (continuity etc.). The equations also describe particle dynamics; we have associated the vector field $\mathbf{v}$ with the translational motion of the centre of mass of an ensemble of particles via (9.3.12) and the vector $\mathbf{s}$ with the internal rotation. A completely deterministic system of field and particle is obtained if we supplement the initial spinor wavefunction $\psi_0 = \psi(\mathbf{x}, 0)$ with the initial position of the particle $\mathbf{x}_0 = \mathbf{x}(0)$ and solve (9.3.12) for $\mathbf{x} = \mathbf{x}(t, \mathbf{x}_0)$. The initial velocity is, of course, fixed by (9.3.8) and the initial spin by (9.3.3) once ψ_0 and $\mathbf{x}_0$ are specified. The consistency of the statistical interpretation of ρ is ensured by (9.3.7) which implies that the normalization condition (9.3.2) is preserved in time. In practice it is easier to solve the wave equation in its linear form (9.3.5) and treat the variables ρ, S and $\mathbf{s}$ as derived quantities.

Note that the unit vectors of the triad $\mathbf{e}'_i$, which provide a covariant specification of a state of rotation, all satisfy the same equation of evolution. Thus, starting from the unit vector $\mathbf{e}'_3 = (2/\hbar)\mathbf{s}$, it is easy to show from (9.3.15) that

$$\mathrm{d}\mathbf{e}'_i/\mathrm{d}t = (\hbar/2m\rho)\mathbf{e}'_i \times \partial_j(\rho\,\partial_j\mathbf{s}) + (2\mu/\hbar)\mathbf{B} \times \mathbf{e}'_i, \qquad i = 1, 2, 3. \quad (9.3.20)$$

It follows from (9.3.20) that

$$\mathrm{d}(\mathbf{e}'_i \cdot \mathbf{e}'_j)/\mathrm{d}t = 0 \qquad \text{for all } i, j = 1, 2, 3 \quad (9.3.21)$$

so that the orthonormality of the triad is preserved by the Pauli equation.

We have remarked earlier that we recover the Schrödinger theory from the present theory when $\mu = 0$ and when the spin degrees of freedom are decoupled from position, i.e., when

$$\psi^a(\mathbf{x}, t) = \Psi(\mathbf{x}, t)\Phi^a, \quad (9.3.22)$$

where Φ is a constant unit spinor (so that $\phi = \text{constant}$, $\theta = \text{constant}$ or equivalently $\mathbf{s}$ is a constant vector). All the entities of the Pauli theory reduce to their Schrödinger counterparts when (9.3.22) holds. This result relies, however, on the assumption that the velocity (9.3.8) of the particle has been correctly identified. This particular expression for $\mathbf{v}$ has been postulated because of its similarity in form with the definition of velocity in the Schrödinger theory. It is not self-evident that (9.3.8) is the correct choice in the Pauli theory though since, given the extra degrees of freedom that are available over and above those present in the Schrödinger case, other definitions are possible (Landau & Lifschitz, 1958, p. 486).

For example, as an analogue to the Gordon decomposition of the Dirac current in the relativistic quantum mechanics of spin $\frac{1}{2}$ particles (§12.2), we might propose that the Pauli current (9.1.3) is the convection part of a total current which also comprises a spin part:

$$j_{\text{tot}\,i} = j_i + (1/m)\varepsilon_{ijk}\,\partial_j(\rho s_k). \tag{9.3.23}$$

This is a viable definition of current in that (from (9.3.7)) it satisfies a conservation equation

$$\partial\rho/\partial t + \nabla\cdot\mathbf{j}_{\text{tot}} = 0 \tag{9.3.24}$$

due to the antisymmetry of ε_{ijk}. One could therefore define the velocity as $\mathbf{v}_{\text{tot}} = \mathbf{j}_{\text{tot}}/\rho$. The resulting theory does *not* reduce to the Schrödinger theory under the assumption (9.3.22) however (Gurtler & Hestenes, 1975). In view of this we shall continue to use the definition (9.3.8) in the following, but it should be borne in mind that the resulting causal theory does not follow uniquely from the Pauli equation.

Finally, we consider the stationary states in which the wavefunction takes the form $\psi(\mathbf{x}, t) = \psi_0(\mathbf{x})\,\mathrm{e}^{-\mathrm{i}Et/\hbar}$. In terms of the model of Fig. 9.2, this describes a case where at each point the vectors $\mathbf{M}$ and $\mathbf{N}$ linearly rotate about $\mathbf{S}$. The total energy (9.3.11) is equal to E, as expected. For hydrogen atom stationary states, it turns out that the spin does not modify the orbit obtained in the spinless case (§4.5), *viz.* a circle in a plane normal to the z-axis (Bohm & Schiller, 1955). The spin vector lies at a constant angle with respect to the z-axis, and precesses uniformly about it.

9.3.3 Is there a classical analogue of spin?

As mathematical objects spinors are not peculiar to quantum mechanics. Their use in characterizing the state of rotation of a rigid body in classical dynamics, for example, is well known (Goldstein, 1980), the connection being essentially that outlined in §9.2. In relation to this, we note that in circumstances where we can neglect the total quantum potential $(Q + Q_s)$ and the quantum torque $\mathbf{T}$ in (9.3.6) and (9.3.13), we obtain the translational Hamilton–Jacobi equation and equation of spin precession corresponding to an ensemble of classical dipoles. Eq. (9.3.7) describes their statistical mechanics. Indeed, we could start from these three equations (without the quantum terms), formally introduce a spinor field and work backwards to translate them into a single nonlinear wave equation, as we did for the classical mechanics of an ensemble of spinless particles in §2.6 (Schiller, 1962c). It follows that the existence of spin $\frac{1}{2}$ states in quantum theory does not signal difficulties of attributing 'reality' to the systems considered. Rather, spin $\frac{1}{2}$ is a property of the wave and in passing from the quantum to the classical domain the basic physical structure is left intact. The discreteness revealed in spin measurements is a special case of a continuous theory (§9.5).

As a simple example of the approach to the classical limit, we may consider the case where ψ factorizes as in (9.3.22) so that the spin plays no role in the dynamics and the particle behaves as a Schrödinger system. Then choosing a Schrödinger solution with a well-defined classical limit, we obtain the classical limit of the spin $\frac{1}{2}$ system. As emphasized in Chap. 6, the interaction of such a system with another object will generally be highly nonclassical. Note that the classical limit may be achieved for systems of fixed internal angular momentum and it is not necessary to pass to the case of infinite, or zero, spin.

We conclude that the classical analogue of the systems governed by the Pauli equation is an ensemble of charged dipoles and one passes continuously between the two regimes by varying the effectiveness of the quantum potential and torque. The spinning object does not 'disappear' in the limit, it simply evolves differently. None of the difficulties usually connected with this question therefore arise.

9.4 The superposition law for spinors

9.4.1 A quantum spin flipper

The first application we shall make of the preceding theory is a consideration of the effects of a free spinor wave on a particle's motion and orientation, for the case where two coherent waves overlap and interfere. There is no one-to-one correspondence between the operations of addition for spinors and for the vectors generated by them and the physical superposition of two spinor waves ψ_1, ψ_2 can imply complicated effects on vectors in Euclidean space. For example, the spin vector associated with $(\psi_1 + \psi_2)$ may possess properties that the spin vectors associated with the individual waves do not have, since in

$$(\psi_1 + \psi_2)^\dagger \boldsymbol{\sigma}(\psi_1 + \psi_2) = \psi_1^\dagger \boldsymbol{\sigma}\psi_1 + \psi_2^\dagger \boldsymbol{\sigma}\psi_2 + \psi_1^\dagger \boldsymbol{\sigma}\psi_2 + \psi_2^\dagger \boldsymbol{\sigma}\psi_1 \quad (9.4.1)$$

the cross terms may be nontrivial.

We first explain what we mean by 'spin up' and 'spin down'. Let us denote the eigenfunctions of the operator σ_z by $u_\pm$:

$$u_+ = \begin{pmatrix} 1 \\ 0 \end{pmatrix}, \qquad u_- = \begin{pmatrix} 0 \\ 1 \end{pmatrix}. \quad (9.4.2)$$

Then from (9.3.3) the spin vectors corresponding to these states are respectively

$$\mathbf{s}_+ = (\hbar/2)(0, 0, 1), \qquad \mathbf{s}_- = (\hbar/2)(0, 0, -1). \quad (9.4.3)$$

Hence, the state u_+ (u_-) corresponds to the case where the spin vector is pointing in the $+$ ($-$) z-direction and thus describes 'spin up (down)' in that direction. Note that the orthogonality of the spinors u_+, u_- in spin space translates into oppositely pointing vectors in Euclidean space, and that the x- and y-components of the vectors are not undefined but actually zero. A particle in a state proportional to u_+ will be said to be 'polarized' in the z-direction.

It is clear from (9.4.1) that the resultant spin vector of a superposition of spin up and down eigenstates has a component in a plane orthogonal to that of the spin vectors associated with the constituent spinors.

Suppose we have a beam of spin $\frac{1}{2}$ particles polarized in the z-direction ($\mathbf{e}_3$) and incident in the y-direction ($\mathbf{e}_2$) on a beam splitter. The two emerging beams separate in the x-direction ($\mathbf{e}_1$) as they move along the y-axis, one of the beams being subjected to a device which is capable of altering its spin, and are subsequently recombined so that they overlap by some means that does not disturb their individual spin states. It is this latter stage of the process in which we are interested. At any one time a particle is in one of the subbeams, the other being empty. We are only concerned with motion in the x-direction and so shall suppress the y-motion although all three components of $\mathbf{s}$ are important. Each component beam is in a spin eigenstate and is assumed to have the profile of a Gaussian packet. At $t = 0$ the packets are centred about the points $x = \pm a$ on the x-axis with initial half-width $\sigma_0 < a$. They then move towards one another with speed u, overlap about the point $x = 0$, pass through one another and separate (Fig. 9.3). Each summand and the total wave satisfy the free Pauli equation throughout and we shall work in the approximation where the dispersion of the packets may be neglected (§6.6.2).

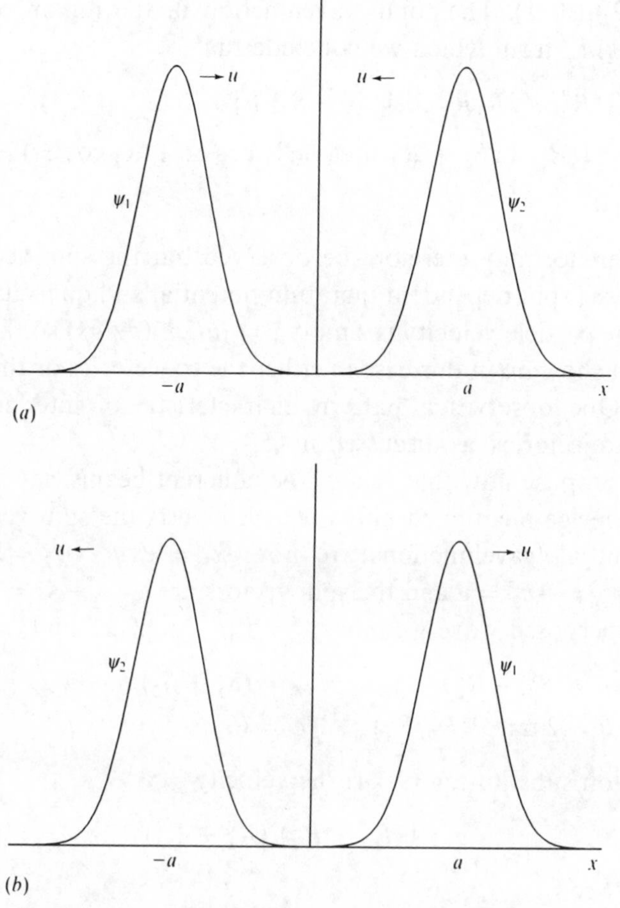

Fig. 9.3 One-dimensional motion of packets (*a*) at $t = 0$, (*b*) at $t = 2a/u$. A particle starting in ψ_1 (ψ_2) ends in ψ_2 (ψ_1).

At time t the beams have a profile given by the normalized functions $\Psi_1(\mathbf{x}, t)$, $\Psi_2(\mathbf{x}, t)$ with amplitudes and phases

$$\left.\begin{aligned} R_1 &= (2\pi\sigma_0^2)^{-1/4}\,\mathrm{e}^{-(x+a-ut)^2/4\sigma_0^2}, \\ R_2 &= (2\pi\sigma_0^2)^{-1/4}\,\mathrm{e}^{-(x-a+ut)^2/4\sigma_0^2}, \\ S_1/\hbar &= k(x + a - \tfrac{1}{2}ut), \\ S_2/\hbar &= -k(x - a + \tfrac{1}{2}ut), \end{aligned}\right\} \tag{9.4.4}$$

where $u = \hbar k/m$. We shall treat three types of superposition.

(i) In the first case we suppose that both beams are in the same spin state, a spin up eigenstate of σ_z. The wavefunction for each beam is then $\psi_1 = \Psi_1 u_+$, $\psi_2 = \Psi_2 u_+$ so that initially $v_1 = -v_2 = u$ and the two spin vectors are

$\mathbf{s}_1 = \mathbf{s}_2 = (\hbar/2)(0, 0, 1)$. The total wavefunction in the region of overlap is $\psi = (\Psi_1 + \Psi_2)u_+$ from which we conclude that

$$\left.\begin{aligned} \rho &= R_1^2 + R_2^2 + 2R_1R_2\cos[(S_1 - S_2)/\hbar] \\ \chi/2 &= \tan^{-1}[(R_1 \sin S_1 + R_2 \sin S_2)(R_1 \cos S_1 + R_2 \cos S_2)^{-1}] \\ \phi &= \theta = 0. \end{aligned}\right\} \quad (9.4.5)$$

Spatial interference may therefore be observed but the spin vector remains constant and the spin-dependent quantum potential and quantum torque are both zero. The particle velocity is simply $v = (\hbar/2m)(\partial\chi/\partial x)$ so that the effects of spin do not show up in this case in either the trajectories or the spin vector orientation. One observes a pattern characteristic of interference in the grouping of trajectories, as discussed in §5.1.

(ii) Let us suppose now that one of the coherent beams, say 2, has passed through the device mentioned above which inverts the spin vector prior to $t = 0$. The initial wavefunctions are now $\psi_1 = \Psi_1 u_+$, $\psi_2 = \Psi_2 u_-$ which implies that $v_1 = -v_2 = u$ and the spin vectors are $\mathbf{s}_1 = -\mathbf{s}_2 = (\hbar/2)(0, 0, 1)$.

For the superposed wave we have $\psi = \Psi_1 u_+ + \Psi_2 u_-$ and

$$\left.\begin{aligned} \rho &= R_1^2 + R_2^2, & \chi &= (S_1 + S_2)/\hbar - \pi/2, \\ \theta &= 2\tan^{-1}(R_2/R_1), & \phi &= (S_1 - S_2)/\hbar + \pi/2, \end{aligned}\right\} \quad (9.4.6)$$

which yields on substituting (9.4.4) the velocity

$$v = u(R_1^2 - R_2^2)/(R_1^2 + R_2^2) \quad (9.4.7)$$

and the spin vector

$$\mathbf{s} = [\hbar/2(R_1^2 + R_2^2)]\{2R_1R_2\cos[(S_1 - S_2)/\hbar], \\ -2R_1R_2\sin[(S_1 - S_2)/\hbar], R_1^2 - R_2^2\} \quad (9.4.8)$$

with $S_1 - S_2 = 2mux$.

On comparing (9.4.6) with (9.4.5) we see that in this case there is no spatial interference but whereas classically the ensemble of trajectories associated with each packet crosses the axis of symmetry ($x = 0$), here they are prevented in accordance with the single-valuedness of the velocity field. What happens is that the particles are reflected from the region around $x = 0$ by the quantum force. This is evident from (9.4.7). If at $t = 0$ the packets have negligible overlap and the particle lies in packet 1, then $v = u$. At $t = a/u$, we have $R_1 = R_2$ so $v = 0$ and the particle is brought to rest. Subsequently it reverses its motion so that at $t = 2a/u$, $v = -u$. Similar behaviour is obtained for a particle starting in packet 2.

In the process the spin vector (9.4.8) varies along the orbit. A particle which enters the region of overlap with spin up ($\theta = 0$) has its spin vector continuously rotated so that it emerges with spin down ($\theta = \pi$), and conversely for initially spin down particles. We thus have a device for flipping the spin just by action of the quantum torque.

Notice that when $R_1 = R_2$, i.e., $\theta = \pi/2$, **s** lies entirely in the x–y plane. From (9.4.4) this occurs when $x = 0$ (for all t) or $t = a/u$ (for all x).

This example thus illustrates the spin state superposition law which in certain regions of superposition implies that a spin vector lies in a plane orthogonal to that of the constituent spin vectors, and that reflection and spin flipping may be induced by purely quantum mechanical means.

(iii) Finally, we consider the case where one of the beams has had its spin vector completely rotated (i.e., by 2π) around an axis lying in the x–y plane prior to $t = 0$, so that the spinor associated with the beam is rotated by π, i.e., its direction in spin space is reversed. Then $\psi_1 = \Psi_1 u_+$, $\psi_2 = -\Psi_2 u_+$ so that the total wave is $\psi = (\Psi_1 - \Psi_2)u_+$. This situation is identical to that of case (i) (spatial interference, no spin effects) except that the phase of ψ_2 has been shifted by π, which yields

$$\rho = R_1^2 + R_2^2 - 2R_1 R_2 \cos[(S_1 - S_2)/\hbar] \tag{9.4.9}$$

and so the maxima and minima are interchanged and the motion is altered accordingly. A further shift of π will bring us back to case (i).

This illustrates the physical inequivalence of 2π and 4π rotations, discussed in §9.2.

9.4.2 Application to neutron interferometry

Experiments to test the superposition law for spinors have been performed with neutron interferometers (Summhammer *et al.*, 1982; Badurek *et al.*, 1983). In §5.4 it was shown how the interference of two beams in the interferometer in the final set of crystal planes may be modelled by a system of two wave packets incident from either side on a square potential barrier. Here we shall extend this theory using spinor wave packets to give a simple account of the final stage of the spin superposition experiments (Dewdney *et al.*, 1987a).

The particular result that has been verified in the experiments is the prediction that the superposition of spin up and spin down beams results in a spin vector lying in the x–y plane, as discussed in §9.4.1. The essential details of the apparatus used to demonstrate this are shown in Fig. 9.4. A neutron beam polarized in the z-direction (defined by a magnetic guide field $B = B_z \mathbf{e}_3$ into the paper, introduced to minimize depolarization effects) enters the

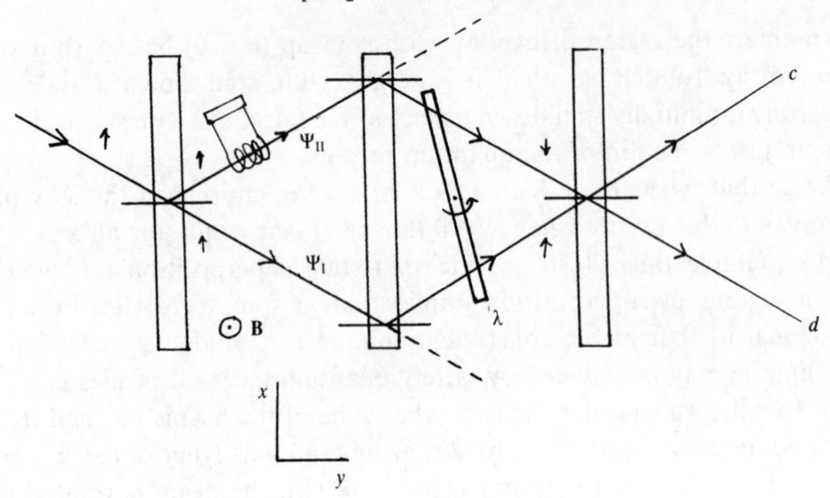

Fig. 9.4 Neutron interferometer with a spin rotating coil and phase shifter (from Dewdney *et al.* (1987*a*)).

device and is split into an upper (II) and a lower (I) component. The upper beam is passed through a dc spin-flipping device, consisting of a coil whose homogeneous magnetic field $\mathbf{B}'$ couples with the magnetic moment and rotates the spin vector about the y-axis. Denoting the upper and lower beam spinors by

$$\psi_{II} = \begin{pmatrix} \Psi_{II} \\ 0 \end{pmatrix}, \qquad \psi_{I} = \begin{pmatrix} \Psi_{I} \\ 0 \end{pmatrix} \tag{9.4.10}$$

respectively, the effect of the magnetic field $\mathbf{B}' = B'\mathbf{e}_2$ is a unitary transformation (9.2.18): $\psi_{II} \to [\cos(\alpha/2)I + \mathrm{i}\sin(\alpha/2)\sigma_y]\psi_{II}$. Here the angle of rotation is, from the Pauli equation (9.3.5), $\alpha = -(2\mu/\hbar)\int B'\,\mathrm{d}t$ where the integration is carried out over the period T that the neutron spends in the field. For a suitable choice of B' and T, the angle may be fixed at $\alpha = -\pi$ so that

$$\psi_{II} \to -\mathrm{i}\sigma_y\psi_{II} = \begin{pmatrix} 0 \\ \Psi_{II} \end{pmatrix}. \tag{9.4.11}$$

The phase of each beam may be altered by a sheet placed in the interferometer which may be rotated so that it presents different thicknesses to each beam. This has the effect of introducing a relative scalar phase shift λ.

The two beams with opposite spin and this phase difference coherently converge on the final set of crystal planes where each beam is split once more. Consequently the emerging forward and deviated beams both contain a coherent superposition of components of opposite spin.

The final crystal planes may be treated as a semitransparent mirror. Representing this by a square potential V, the Pauli Hamiltonian operator in the region of superposition is

$$\hat{H} = -\frac{\hbar^2}{2m}\nabla^2 + V \tag{9.4.12}$$

and the particle motion and spin orientation evolve according to the theory of §9.3.2 under this assumption. In (9.4.12) we have neglected the term $\mu B_z \sigma_z$ arising from the background guide field.

The initial spinor (i.e., just before the final semitransparent mirror) is a superposition of the spin up and down beams:

$$\psi_0 = \begin{pmatrix} \Psi_{I0} \\ \Psi_{II0}\,\mathrm{e}^{i\lambda} \end{pmatrix} \tag{9.4.13}$$

with the packet functions

$$\Psi_{I0} = (2\pi\sigma_0^2)^{-1/4}\,\mathrm{e}^{[-(x-0.5)^2/4\sigma_0^2]+ikx} = R_{10}\,\mathrm{e}^{iS_{10}/\hbar}$$

$$\Psi_{II0} = (2\pi\sigma_0^2)^{-1/4}\,\mathrm{e}^{[-(x-1)^2/4\sigma_0^2]-i(kx+\Phi)} = R_{20}\,\mathrm{e}^{iS_{20}/\hbar}$$

where Φ is a constant factor introduced to symmetrize the two wavefunctions with respect to the potential V. The initial spin vectors are from (9.4.10) and (9.4.11), $\mathbf{s}_{I0} = (\hbar/2)(0, 0, 1)$, $\mathbf{s}_{II0} = (\hbar/2)(0, 0, -1)$. At time t the Euler parameters of the wave that develops from (9.4.13) are given by

$$\left.\begin{aligned} \theta &= 2\tan^{-1}(R_2/R_1) \\ \chi &= \frac{S_1 + S_2}{\hbar} + \frac{\pi}{2} + \lambda \\ \phi &= \frac{S_1 - S_2}{\hbar} - \frac{\pi}{2} - \lambda \end{aligned}\right\} \tag{9.4.14}$$

and the spin vector in the region of overlap is given by (9.4.8), apart from the extra phase difference λ (as functions of space and time the spin vector and trajectories here differ from those of §9.4.1(ii) due to the potential barrier).

Solving the Pauli equation numerically, a set of trajectories whose range of initial positions simulates the actual Gaussian distribution in each incident beam is displayed in Fig. 9.5 (with units as in §5.4). The horizontal lines indicate the position of the square potential. Notice that all the trajectories originating in the upper (lower) beam inside the interferometer enter the upper (lower) beam outside. This leads to an equal count rate in the two detectors, placed beyond the interferometer, independent of the relative phase.

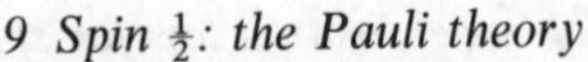

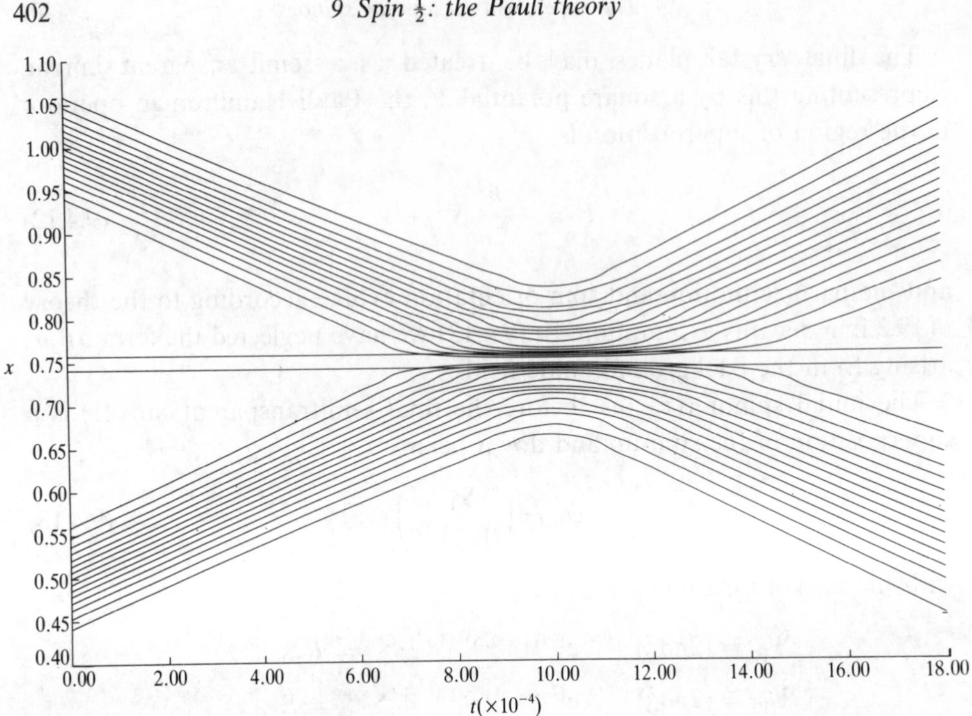

Fig. 9.5 Neutron trajectories at the last semitransparent surface (from Dewdney *et al.*, 1987a).

This is to be expected since the probability density

$$\rho = |\Psi_{\text{I}}|^2 + |\Psi_{\text{II}}|^2 \tag{9.4.15}$$

is independent of λ. This is in contrast with the superposition of two spin up partial beams for which the density would be a function of λ (cf. §5.4).

The spin orientation fields are shown in Figs. 9.6 and 9.7. In the emerging beams the spin vector lies in the x–y plane ($\theta = \pi/2$). Mentally superposing the trajectories on these figures allows us to see the way in which the spin orientation changes in a continuous manner during the scattering from the potential, depending on the route the particle takes (upper beam, spin down, top left hand corner; lower beam, spin up, lower left hand corner). Changing the relative phase λ does not affect the polar angle but does, as is shown in Fig. 9.8, cause the final direction of the spin vector in the x–y plane to be rotated, as we expect from (9.4.14). Notice that once the spin vector has been rotated into the x–y plane in each emerging beam, it stays there. This is in contrast to the case of §9.4.1(ii) where the spin vectors are rotated back into the z-direction after the packets leave the region of overlap. The difference is of course brought about by the potential barrier; each emerging

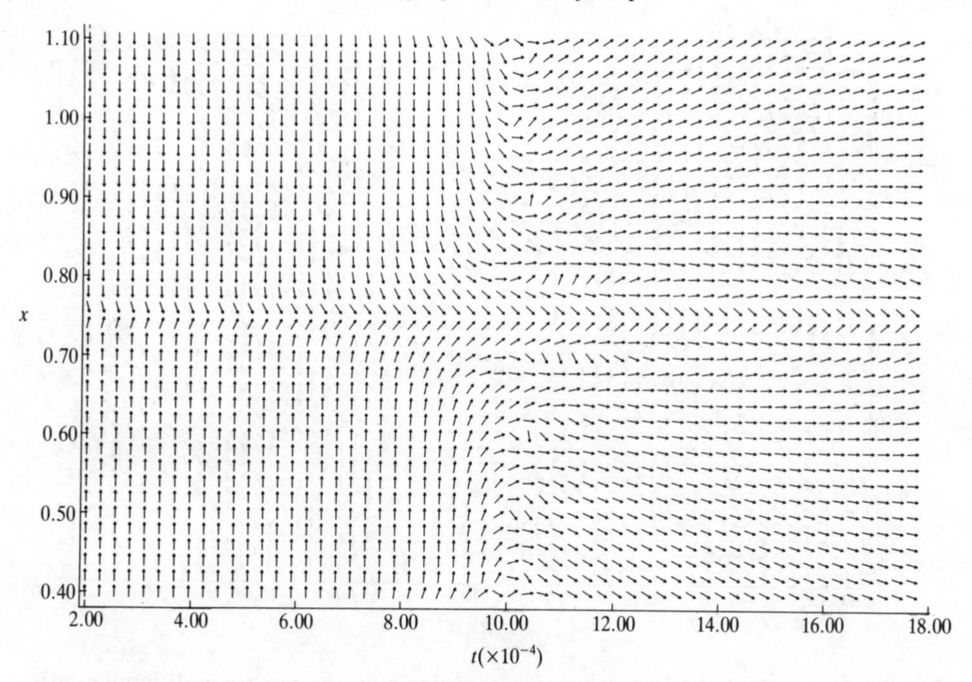

Fig. 9.6 The polar orientation angle of the spin vector, $\theta(x, t)$, plotted on a mesh of points associated with Fig. 9.5 (from Dewdney *et al.* (1987a)).

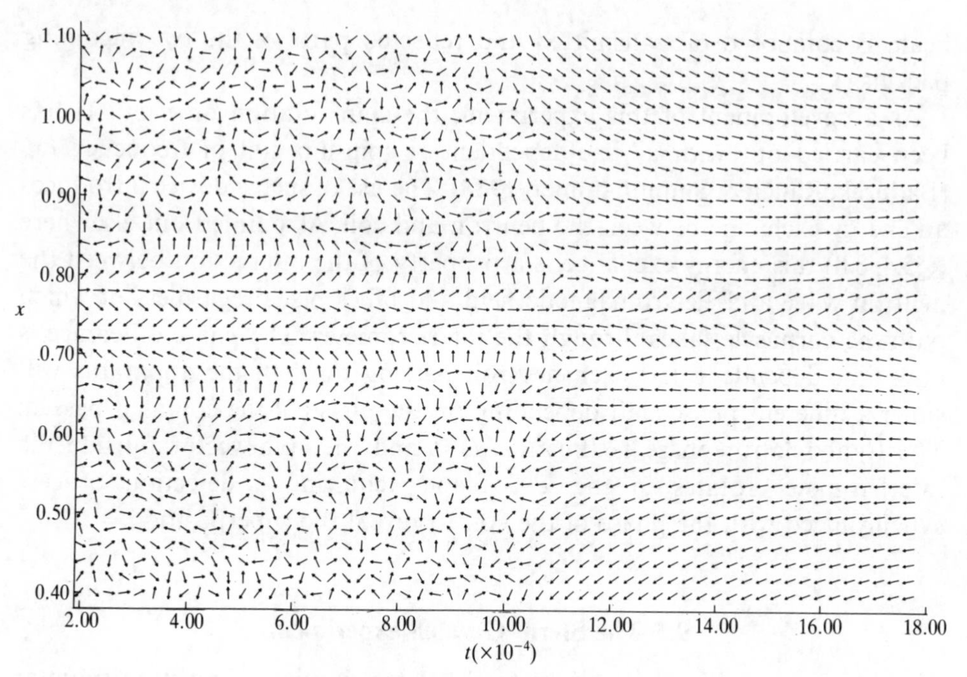

Fig. 9.7 The azimuthal orientation angle of the spin vector, $\phi(x, t)$, associated with Fig. 9.5 (from Dewdney *et al.* (1987a)).

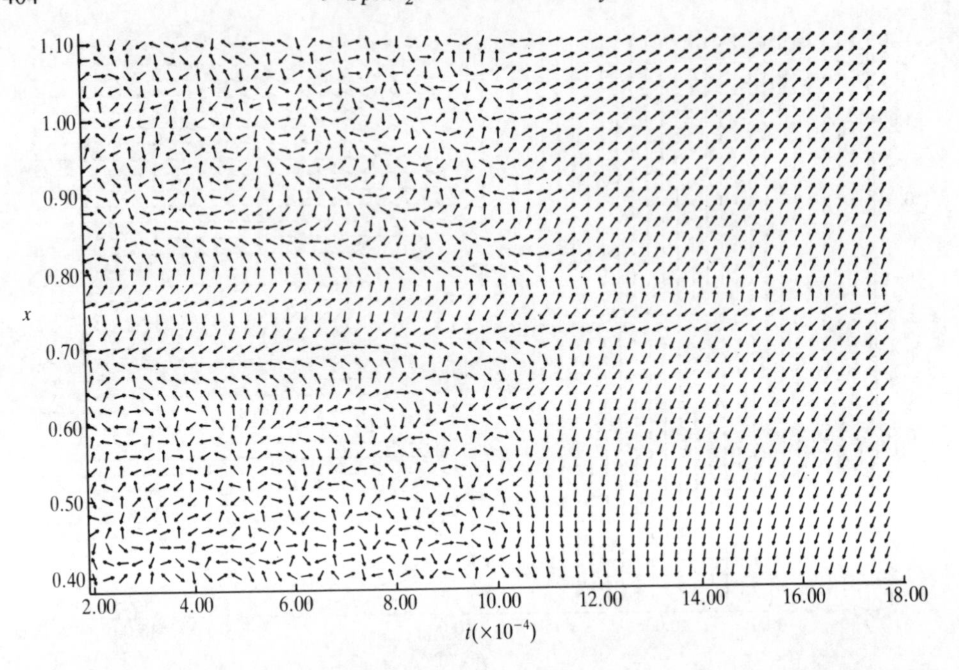

Fig. 9.8 The azimuthal orientation angle of the spin vector with an additional relative phase λ, introduced by rotating the phase plate (from Dewdney *et al.* (1987a)).

beam is composed of transmitted and reflected parts of the two incoming packets.

As a development of this experiment, the spin rotation on beam II has been carried out with a time-dependent rf spin-flip coil of frequency ω_{rf} (Badurek, Rauch & Summhammer, 1983). The use of such a device introduces an energy exchange between the neutron and coil. As pointed out elsewhere (§§5.4, 8.8) this energy exchange is a property of the interaction between the neutron wave and electromagnetic field and takes place regardless of which path the corpuscle takes through the interferometer. The phase difference is now time-dependent and each neutron passing through the apparatus will suffer a different phase shift depending on the phase of the coil as it passed. The total relative phase is thus $\lambda - \omega_{\text{rf}}t$ and the spin vector outside the interferometer rotates in the x–y plane, requiring a detection process synchronized with the phase of the coil to reveal the superposition.

9.5 The Stern–Gerlach experiment

All experiments ultimately reduce to the determination of position (pointer reading, splash of light on a screen, arm on a clock). The purpose of the

Stern–Gerlach experiment is to establish a coupling, or correlation, between the angular momentum of a system and its position so that the value of the latter allows us to infer the value of the former. In this section we shall apply the causal interpretation of spin $\frac{1}{2}$ particles to analyse in detail the functioning of a Stern–Gerlach device, with the aim of explaining what is meant by the phrase 'measuring the spin observable'. This will bring out in a special case the general theory of measurement discussed in Chap. 8 (Dewdney, Holland & Kyprianidis, 1986).

The usual account of the measurement of spin in some direction, say z, proceeds in terms of the notion of a projection of the wavefunction into an eigenstate of the spin operator $\hat{s}_z = (\hbar/2)\sigma_z$. The spin becomes definite in this direction with magnitude (eigenvalue) $\pm \hbar/2$. The spin operators $\hat{s}_x$, $\hat{s}_y$ in the two perpendicular directions do not commute with $\hat{s}_z$ or with each other:

$$[\hat{s}_x, \hat{s}_z] = -2i\hbar\hat{s}_y, \qquad [\hat{s}_y, \hat{s}_z] = 2i\hbar\hat{s}_x, \qquad [\hat{s}_x, \hat{s}_y] = 2i\hbar\hat{s}_z. \qquad (9.5.1)$$

When $\hat{s}_z$ is definite, the components of $\hat{\mathbf{s}}$ in the x- and y-directions are completely undetermined and are to be thought of as 'randomly fluctuating' between $\pm\hbar/2$. Thus before measurement the spin is a 'potentiality', and yet afterwards it is still indefinite in two directions. As regards the result which is actually realized in any given experiment, the most that one can do is to state the probabilities of finding spin 'up' and 'down'.

There is an evident lack of clarity as to what all this means physically in each individual experiment, in particular how the apparent discreteness (up or down, nothing between) comes about. In contrast, the spin vector of the causal interpretation, which is always well defined and continuously variable, evolves by the action of the quantum torque to take up the values predicted by quantum mechanics for the eigenvalues of the spin operator being 'measured', whilst the particle moves continuously to a position in space that allows us to infer which spin state it is actually in (for a spin $\frac{1}{2}$ particle, two well-separated spatial regions). The particle property under investigation is therefore $\mathbf{s}$, the 'local expectation value' of $\hat{\mathbf{s}}$. Which outcome is obtained follows uniquely from the initial position $\mathbf{x}_0$.

The conventional accounts of the Stern–Gerlach experiment as given for example in the texts of Bohm (1951, Chap. 22) and Böhm (1986, Chap. 13) are readily adapted to the causal interpretation. The idealized experimental set up is shown in Fig. 9.9. A beam of electrically neutral spin $\frac{1}{2}$ atoms having a magnetic moment (with an intensity so low that only one is present in the apparatus at any one time) moves in the y-direction between the poles of a magnet which has a gradient in the z-direction. The beam is polarized by which we mean the initial spin vector is arbitrary but constant in spacetime

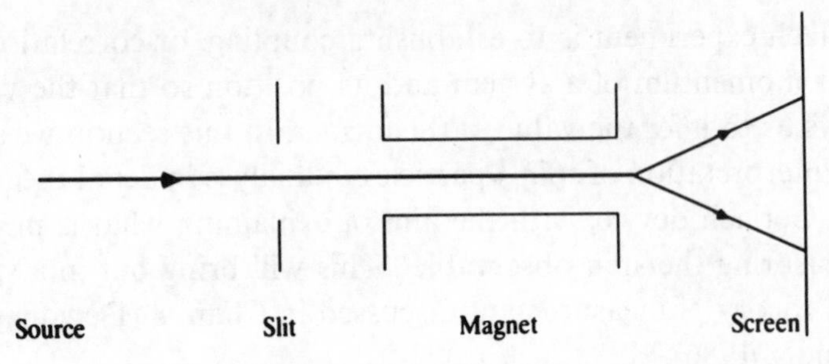

Fig. 9.9 Idealized Stern–Gerlach apparatus.

and the same for all ensemble elements. The case of measurement on an initially unpolarized beam corresponds to a case where the spin vector has a range of directions over the ensemble and adds nothing essentially new. The beam is split by the magnetic field into two parts which separate in the z-direction and are incident on a screen where two spots are formed (this is not quite what one observes in practice). In order to achieve an unambiguous result it is necessary to have some idea of the initial particle position and so the beam passes through a slit which creates a wave packet in the z-direction before entering the magnet. To bring out clearly what is essential in the operation of the device, we shall make the following simplifying assumptions:

(i) We suppose that the interaction with the magnetic field is impulsive, i.e., the particle spends a short time T in the field whose effect is so great that the kinetic energy terms in the Pauli Hamiltonian operator may be neglected during the interaction (the value of T necessary for an unambiguous result will be estimated below).

(ii) We shall only take into account the magnetic field in the z-direction, and will assume that $\mathbf{B}$ may be Taylor expanded about $z = 0$ with the terms of order z^2 and above being negligible. That is,

$$B_x = B_y = 0, \qquad B_z = B_0 + zB'_0 \tag{9.5.2}$$

where B_0 is the uniform field and $B'_0 = \partial B_z/\partial z|_{z=0}$ is the field gradient. Such a vector field clearly does not satisfy Maxwell's equations (since div $\mathbf{B} \neq 0$) and physically B_x is necessarily nonzero. Neglect of this component is justified by Böhm (1986, Chap. 13).

According to Bohr, inclusion of this component induces a Lorentz force on a charged particle which tends to blur the spots on the screen and thus prevents the measurement of the spin of, e.g., a free electron (Mott, 1929a; Mott & Massey, 1965, p. 214; Wheeler & Zurek, 1983, p. 699).

(iii) We shall only explicitly describe motion in the z-direction. In particular the y-motion is not affected by the magnetic field. It is therefore constant throughout the experiment and will be suppressed.

(iv) The impulsive action of **B** is assumed to be large enough to mask the spread of the wave packets.
(v) The entrance slit will be taken to be Gaussian.

(More general treatments of the formal theory are given by Scully, Lamb & Barut (1987), Garbaczewski (1988) and Busch & Schroeck (1989).)

As we have said, observing the position of the particle allows us to infer the value of the spin. The 'observed system' is therefore the spin variable, and the coordinates of the 'observing apparatus' comprise the particle position, the magnet and the detecting screen. The part of the apparatus of principal concern is the particle position. The magnet coordinates will be ignored and the detecting screen may be included by a further coupling, as described in §8.3.

After passing through the entrance slit, but before entering the field, the initial wavefunction represents a state where the position and spin are independent. Expanding in terms of a complete set of spin eigenfunctions of the spin operator in the z-direction, we have

$$\psi_0(z) = \psi(z, 0) = f(z)(c_+u_+ + c_-u_-), \tag{9.5.3}$$

where $f(z)$ is a real normalized packet function, and c_+, c_- are complex constant coefficients which define the direction of polarization of the beam and the probabilities of the outcomes of the experiment ($|c_+|^2, |c_-|^2$). They satisfy $|c_+|^2 + |c_-|^2 = 1$, which ensures that

$$\int_{-\infty}^{\infty} \psi_0^\dagger \psi_0 \, \mathrm{d}z = 1. \tag{9.5.4}$$

The initial velocity is then $v_0 = 0$ (in the z-direction) and, writing

$$c_+ = |c_+| \, \mathrm{e}^{\mathrm{i}s_+/\hbar}, \qquad c_- = |c_-| \, \mathrm{e}^{\mathrm{i}s_-/\hbar}, \tag{9.5.5}$$

the initial orientation is from (9.2.25)

$$\theta_0 = \cos^{-1}(|c_+|^2 - |c_-|^2), \qquad \phi_0 = \frac{s_+ - s_-}{\hbar} + \frac{\pi}{2}, \qquad \chi_0 = \frac{s_+ + s_-}{\hbar} - \frac{\pi}{2}. \tag{9.5.6}$$

To find the effect of the impulsive field we have to solve, using (9.3.5),

$$\mathrm{i}\hbar \frac{\partial \psi}{\partial t} = \mu(B_0 + zB_0')\sigma_z \psi. \tag{9.5.7}$$

Writing $\psi = \psi_+u_+ + \psi_-u_-$, (9.5.7) is equivalent to two equations

$$i\hbar \frac{\partial \psi_+}{\partial t} = \mu(B_0 + zB_0')\psi_+$$

$$i\hbar \frac{\partial \psi}{\partial t} = -\mu(B_0 + zB_0')\psi_-.$$

Clearly each spinor component evolves independently. The solution is

$$\psi_\pm = f(z)c_\pm \, e^{\mp i\mu(B_0 + zB_0')T/\hbar},$$

where we have used the initial condition (9.5.3) and T is the period of interaction. The solution at the exit of the magnet is therefore

$$\psi(z, T) = f(z)[c_+ \, e^{-i(\Delta + z\Delta')}u_+ + c_- \, e^{i(\Delta + z\Delta')}u_-], \tag{9.5.8}$$

where $\Delta = \mu B_0 T/\hbar$, $\Delta' = \mu B_0' T/\hbar$.

Notice that while the particle is in the field its kinetic energy and the quantum potential and quantum torque (which are all determined by the kinetic energy operator in the Pauli equation) are negligible. However, on leaving the field these terms become relevant as the particle motion and spin evolve according to the free Pauli equation

$$i\hbar \frac{\partial \psi}{\partial t} = -\frac{\hbar^2}{2m}\frac{\partial^2 \psi}{\partial z^2}. \tag{9.5.9}$$

Again, each spinor component evolves independently. To solve (9.5.9) subject to the initial condition (9.5.8), it is convenient first to Fourier analyse the initial packet:

$$f(z) = \frac{1}{(2\pi)^{1/2}} \int_{-\infty}^{\infty} g(k) \, e^{ikz} \, dk, \tag{9.5.10}$$

where $g(k)$ is a real normalized packet in momentum space centred around $k = 0$. Now the general solution to (9.5.9) is

$$\psi(z, t) = \int_{-\infty}^{\infty} [\psi_{+k}(z) \, e^{-iE_k^+ t/\hbar} u_+ + \psi_{-k}(z) \, e^{-iE_k^- t/\hbar} u_-] \, dk, \tag{9.5.11}$$

where $\psi_{\pm k}(z)$ are a complete set of energy eigenfunctions for each spinor component:

$$\frac{d^2 \psi_{\pm k}}{dz^2} + \frac{2mE_k^\pm}{\hbar^2}\psi_{\pm k} = 0. \tag{9.5.12}$$

But from (9.5.11), (9.5.8) and (9.5.10)

$$\psi(z, 0) = \frac{1}{(2\pi)^{1/2}} \int_{-\infty}^{\infty} g(k)\, e^{ikz}\, dk(c_+ \, e^{-i(\Delta + z\Delta')} u_+ + c_- \, e^{i(\Delta + z\Delta')} u_-)$$

$$= \int_{-\infty}^{\infty} [\psi_{+k}(z) u_+ + \psi_{-k}(z) u_-]\, dk,$$

which implies that

$$\psi_{\pm k}(z) = \frac{1}{(2\pi)^{1/2}} g(k)\, e^{ikz} c_{\pm}\, e^{\mp i(\Delta + z\Delta')}.$$

Substituting this expression into (9.5.12) yields

$$E_k^{\pm} = \hbar^2 (k \mp \Delta')^2 / 2m.$$

We therefore find that at time t after the particle leaves the magnet, the kth Fourier component of the initial packet picks up a factor $e^{-i\hbar(k \mp \Delta')^2 t/2m}$ and the general time-dependent solution is

$$\psi(z, t) = \frac{1}{(2\pi)^{1/2}} \int_{-\infty}^{\infty} g(k)(c_+ \, e^{i[-\Delta + (k - \Delta')z - (\hbar/2m)(k - \Delta')^2 t]} u_+$$

$$+ c_- \, e^{i[\Delta + (k + \Delta')z - (\hbar/2m)(k + \Delta')^2 t]} u_-)\, dk. \qquad (9.5.13)$$

At this stage it is clear that the wavefunction splits up into two packets, associated with the spin up and spin down basis functions, which separate in the z-direction. To see this, note that the centre of each packet occurs where the phase has an extremum, i.e., where

$$\frac{\partial}{\partial k} \left[\mp\Delta + (k \mp \Delta')z - \frac{\hbar}{2m}(k \mp \Delta')^2 t \right]_{k=0} = 0,$$

which implies that

$$z = \mp ut, \qquad u = \hbar\Delta'/m, \qquad (9.5.14)$$

for the motion of the centres of the packets associated with u_+, u_- respectively. It is only necessary to ensure that the inhomogeneous field and the duration of interaction are sufficiently large that a classical scale separation is subsequently achieved ($t > 0$) for the packets, with respect to which their natural spreading is negligible: $|u|t \gg \sigma$, the packet half-width.

The condition for this is that the momentum imparted to each packet by the magnetic field should be much greater than the initial half-width σ_p (in momentum space) of the packet on entering the magnet, since for large t, σ approaches $\sigma_p t/m$ (§4.7.1). Thus, $m|u| \gg \sigma_p$, or from (9.5.14) and $\sigma_p = \hbar/2\sigma_0$

where σ_0 is the initial half-width, $-B_0'T \gg \hbar/2\mu\sigma_0$. Note that we assume $B_0' < 0$ so that spin up (down) is correlated with upward (downward) motion in space.

The final possible states of the apparatus (including the packet coordinates) are therefore nonoverlapping in configuration space and in Euclidean space. The particle enters one of the packets and, assuming that they are not brought together again and interfere, the packet which the particle does not enter can be dropped from further attention. Moreover, which packet it enters is determined solely by its initial position at the entrance slit, given the values of c_+ and c_-. The measurement is complete when a spot appears on the screen which indicates which packet the particle joined on leaving the magnet. This irreversible stage of the process merely tells us what has already objectively happened, and there is no need to invoke the 'wave packet collapse' hypothesis.

To see this in detail it simply remains to determine the wavefunction explicitly by evaluating the integral (9.5.13), using our assumption that the entrance slit is Gaussian. Substituting $g(k) = (2\sigma_0^2/\pi)^{1/4} \mathrm{e}^{-k^2\sigma_0^2}$ into (9.5.13), performing the integration and rearranging, we deduce (cf. §4.7.1)

$$\psi(z, t) = (2\pi s_t^2)^{-1/4}(c_+ \mathrm{e}^{-(z+ut)^2/4\sigma_0 s_t}\, \mathrm{e}^{-\mathrm{i}[\Delta + (z+\frac{1}{2}ut)\Delta']}u_+ + c_- \mathrm{e}^{-(z-ut)^2/4\sigma_0 s_t}\, \mathrm{e}^{\mathrm{i}[\Delta + (z-\frac{1}{2}ut)\Delta']}u_-), \tag{9.5.15}$$

where $u = \hbar\Delta'/m$ is the speed of the centre of each packet, $\Delta = \mu B_0 T/\hbar$ and $\Delta' = \mu B_0' T/\hbar$. The normalization (9.5.4) is, of course, preserved.

It is convenient to write

$$\psi(z, t) = R_+ \mathrm{e}^{\mathrm{i}S_+/\hbar} c_+ u_+ + R_- \mathrm{e}^{\mathrm{i}S_-/\hbar} c_- u_-, \tag{9.5.16}$$

where

$$\left.\begin{aligned} R_\pm &= (2\pi\sigma^2)^{-1/4}\, \mathrm{e}^{-(z\pm ut)^2/4\sigma^2}, \\ \frac{S_\pm}{\hbar} &= \mp[\Delta + (z + \tfrac{1}{2}ut)\Delta'] - \tfrac{1}{2}\tan^{-1}\left(\frac{\hbar t}{2m\sigma_0^2}\right) + \frac{\hbar t(z \pm ut)^2}{8m\sigma^2\sigma_0^2}. \end{aligned}\right\} \tag{9.5.17}$$

Noting (9.5.5), a comparison of (9.2.25) with (9.5.16) gives

$$\left.\begin{aligned} \rho &= R_+^2|c_+|^2 + R_-^2|c_-|^2 \\ \chi &= (1/\hbar)(S_+ + S_- + s_+ + s_-) - \pi/2 \\ \theta &= 2\tan^{-1}(R_-|c_-|/R_+|c_+|) \\ \phi &= (1/\hbar)(S_+ - S_- + s_+ - s_-) + \pi/2 \end{aligned}\right\} \tag{9.5.18}$$

in terms of which the velocity is given as usual by

$$v = \frac{\hbar}{2m}\left(\frac{\partial \chi}{\partial z} + \cos\theta \frac{\partial \phi}{\partial z}\right) \tag{9.5.19}$$

and the spin vector by (9.3.3).

We shall describe the process beyond the magnet for which the initial wavefunction is (9.5.8). The orientation at this moment is given by

$$\theta_T = \theta_0, \qquad \phi_T = \phi_0 - 2(\Delta + z\Delta'), \qquad \chi_T = \chi_0, \tag{9.5.20}$$

where θ_0, ϕ_0, χ_0 are given in (9.5.6). Thus, although the particle position has not changed during the period in the magnet, the impulse has given each particle in the ensemble a z-velocity $v_T = -u\cos\theta_0$.

In the examples we give below of the possible motions implied by substituting (9.5.17) into (9.5.18) and (9.5.19) it will be assumed that c_+ and c_- are real and nonnegative so that $s_+ = s_- = 0$ in the above formulae. Units are chosen as in §5.4. It is clear from (9.5.18) that there is no spatial interference so we expect the trajectories to be smoothly curved.

When $c_- = 0$ ($c_+ = 0$) all the trajectories move upward (downward) to reach the upper (lower) portion of the detecting screen, and the polar angle θ maintains a constant value of 0 (π). The initial state is already an eigenstate of σ_z and the particle enters the upper (lower) beam regardless of its initial position.

Fig. 9.10 shows the trajectories for the choice $c_+^2 = 0.75$, $c_-^2 = 0.25$, which corresponds to an initial orientation $\theta_0 = 60°$, $\phi_0 = \pi/2$, $\chi_0 = -\pi/2$ and $v_T = -u/2$. Evidently which packet the particle enters depends only on its initial position within the wave packet for the given values of c_+, c_-. Although the initial position is independent of the initial orientation, the subsequent translational motion is determined by the total quantum force acting on the particle in (9.3.19), which depends on the internal degrees of freedom. It will be observed in Fig. 9.10 how the usual probabilities of the outcomes up or down ($|c_+|^2$ and $|c_-|^2$) are reflected in the relative proportions of trajectories entering the two packets. This is evident also from the expectation value of position which for the state (9.5.15) is given by

$$\langle z \rangle = \int_{-\infty}^{\infty} z\psi^\dagger\psi \, dz = -ut(|c_+|^2 - |c_-|^2) = -v_T t. \tag{9.5.21}$$

Fig. 9.11 depicts the field $\theta(z, t)$, Fig. 9.12 the field $\phi(z, t)$ and Fig. 9.13 shows for a few trajectories how the information of Figs. 9.10 and 9.11 is combined.

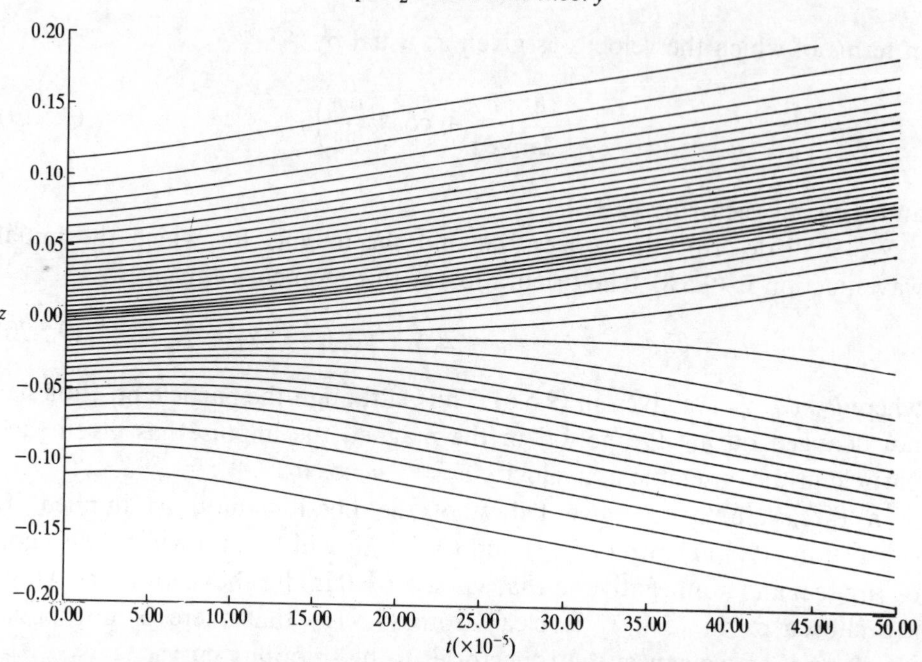

Fig. 9.10 The trajectories for a Gaussian distribution of initial positions beyond the exit of a Stern–Gerlach magnet, with $c_+^2 = 0.75$, $c_-^2 = 0.25$ (from Dewdney, Holland & Kyprianidis (1986)).

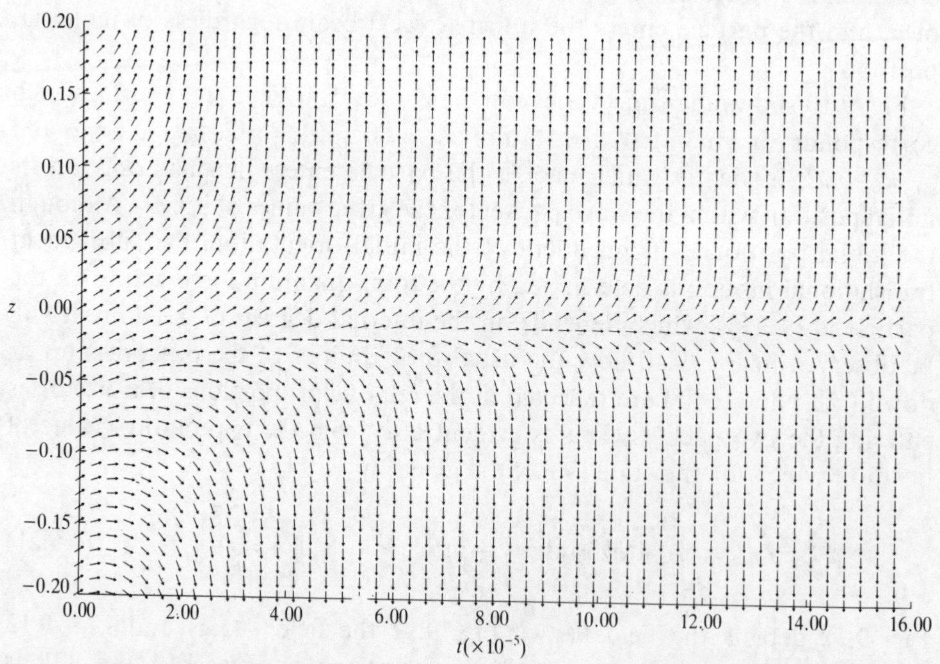

Fig. 9.11 The field of orientations $\theta(z, t)$ associated with Fig. 9.10 (from Dewdney, Holland & Kyprianidis (1986)).

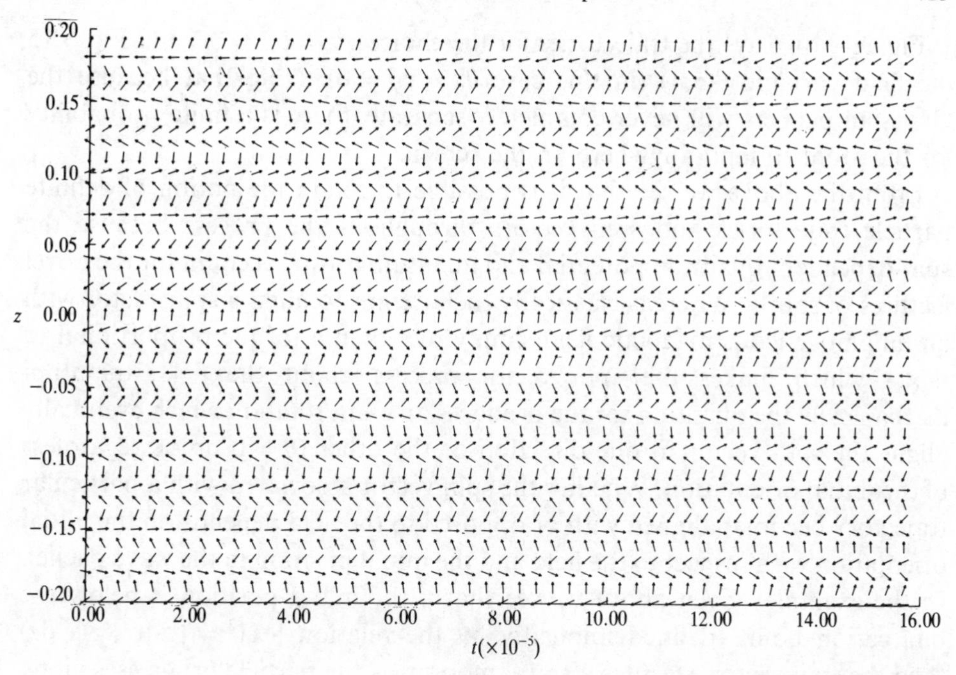

Fig. 9.12 The field of orientations $\phi(z, t)$ associated with Fig. 9.10 (from Dewdney, Holland & Kyprianidis (1986)).

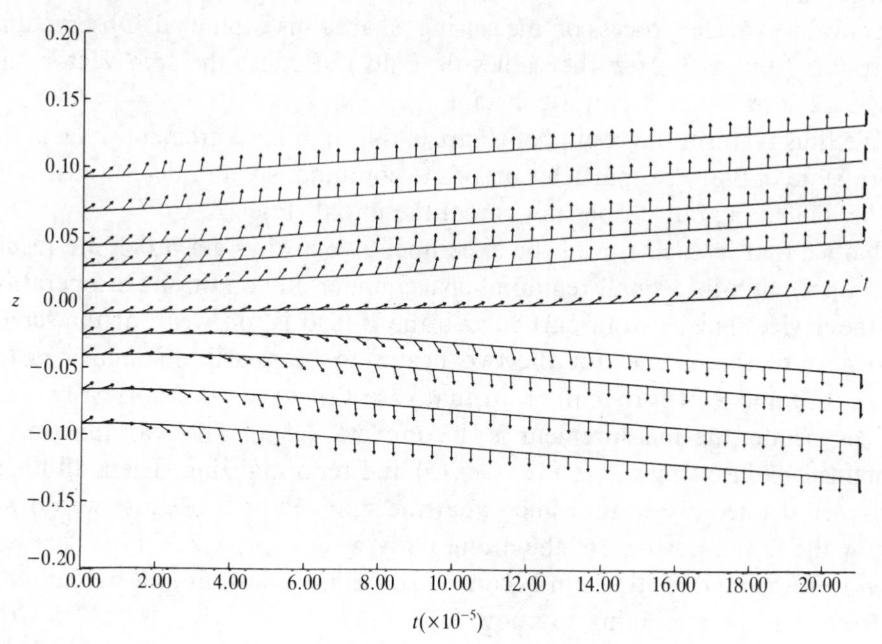

Fig. 9.13 Trajectories and orientations $\theta(z, t)$ associated with Figs. 9.10 and 9.11 (from Dewdney, Holland and Kyprianidis (1986)).

Fig. 9.14 displays the trajectories for the choice $c_+^2 = c_-^2 = 0.5$ and Fig. 9.15 the field variable $\theta(z, t)$ (in this case $\theta_0 = v_T = 0$). Note that because the Gaussian packets will never completely separate, there is a finite probability for the atom to land anywhere on the screen.

From the above we see that it is possible to retain the notion of definite particle trajectories and spin vectors throughout the process because the spin-dependent quantum potential and quantum torque account for the novel features of spin $\frac{1}{2}$. Whereas classically the attempt to align a spin vector with an inhomogeneous magnetic field simply results in a torque proportional to $\mathbf{B} \times \mathbf{s}$ which makes the spin vector precess about the field, quantum mechanically the quantum torque produces an extra rotation which eventually aligns the spin vector in one direction or the other ($\theta = 0$ or π), regardless of the initial orientation. Whether the spin vector associated with a particular trajectory becomes aligned with or opposed to the field depends on the initial orientation with respect to the field and the initial position in the wave packet. In the examples given above we see that if the initial position is below the bifurcation point (defined implicitly by the relation $|c_+|^2 = \int_z^\infty |\psi_0(z)|^2 \, dz$) then the spin vector executes a spiral motion, as the particle progresses along its downward path, towards the value $\theta = \pi$ with $\mathbf{s} = (\hbar/2)(0, 0, -1)$. Similarly, if the initial position is above the bifurcation point the spin vector spirals to the value $\theta = 0$ with $\mathbf{s} = (\hbar/2)(0, 0, 1)$. The discrete possible outcomes (eigenvalues) of the process of 'measuring' $\hat{s}_z$ are thus explained. Observe that once θ is fixed as 0 or π the values of ϕ do not affect the spin vector and Fig. 9.12 shows that ϕ is not constant.

We thus reaffirm the results of Chap. 8: the 'spin measurement' relates the prior state of the wave-particle composite continuously and deterministically to its final state, but it does not reveal the initial state.

Notice that in each run of the experiment we always get a definite result, i.e., a detection in a small region of space, under all conditions of operation of the device. This is so in particular if the **B**-field is too weak or the screen too near to the magnet for the two beams to be clearly distinguished (cf. Figs. 9.10 and 9.14 for small t). In that case the wave does not evolve into an eigenstate; the measurement is 'incomplete' (§8.5). The wavefunction is obtained by inserting $z = Z$ into (9.5.15) and renormalizing. Then, although it is well defined, we cannot infer what the spin vector is because we do not know the initial position of the atom. Only when the packets are effectively nonoverlapping does the spin become a spacetime constant and we can infer it from Z without needing to know $\mathbf{x}_0$.

It is clear that a further measurement of $\hat{s}_x$, say, on the upper beam emerging from a Stern–Gerlach magnet oriented in the z-direction corresponds

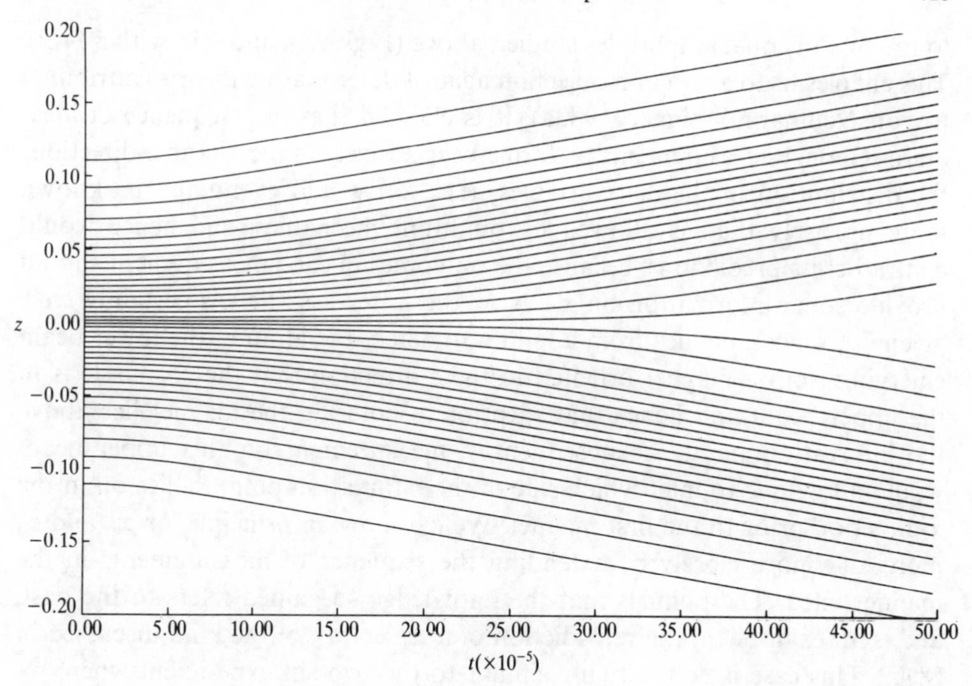

Fig. 9.14 Trajectories during spin measurement with $c_+^2 = c_-^2 = 0.5$ (from Dewdney, Holland & Kyprianidis (1986)).

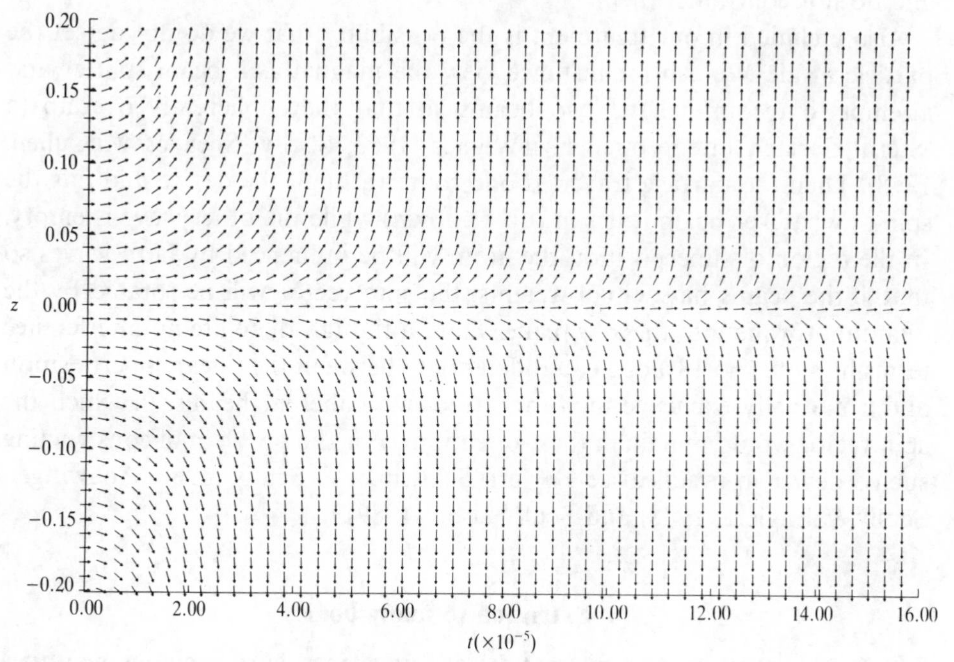

Fig. 9.15 Field of orientations $\theta(z, t)$ associated with Fig. 9.14 (from Dewdney, Holland & Kyprianidis (1986)).

to the case of equal amplitudes studied above (Figs. 9.14 and 9.15 with $z \to x$). This enables us to answer an objection against deterministic theories attributed to von Neumann (Wigner, 1983a). It is claimed that in a sequence of ideal Stern–Gerlach measurements performed successively in the x- and z-directions the distribution of 'hidden variables' (i.e., $\mathbf{x}_0$) would eventually be known more precisely than is allowed by quantum mechanics, and hence would contradict its predictions, because the outcomes of each measurement should provide some information on $\mathbf{x}_0$. Now the latter assertion is certainly true; given ψ_0, we can deduce from a failure to detect the atom in the lower beam emerging from a magnet oriented in the z-direction that the corpuscle is in the upper beam and hence that z_0 prior to entering the magnet lies above the bifurcation point. A subsequent $\hat{s}_x$-measurement on this upper beam would allow one to infer which side of the bifurcation point x_0 lies on in the x-direction prior to the first magnet. We can now, in principle, fix x_0 and z_0 more and more closely by extending the sequence of measurements in the manner cited. The point is that the knowledge so gained refers to the past, and is an example of the retrodiction of a trajectory of the kind discussed in §8.4.2. This case is conceptually similar to the two-slit experiment where we may infer which slit a particle passed through from its position on the detecting screen. Such inferences are consistent with the quantal predictions and do not contradict them.

Also included in our treatment is the possibility that we do not detect the particle at all. We can for instance view the magnet as a beam splitter and attempt to recombine the two beams so that they coherently overlap (a 'Stern–Gerlach interferometer') (Wigner, 1963; Scully, Shea & McCullen, 1978). Then the particle traverses one route through the device and has the spin of whatever beam it lies in (up or down) while the other wave is empty. In the region of superposition the atom will be influenced by both waves so that, if the beams have equal weights, the spin vector will be rotated by the quantum torque into the x–y plane. Thus, although there are no interference terms in ρ (cf. (9.5.18)) one can differentiate between the linear superposition and a mutually exclusive mixture of the two coherent beams, in which the spin vector would remain in the z-direction. (For the problem of constructing such a device in practice see Englert, Schwinger & Scully (1988), Schwinger, Scully & Englert (1988) and Scully *et al.* (1989).)

9.6 Extension to many bodies

The theory developed so far may be extended in a fairly obvious way to a many-body system. It is found that quantities analogous to those of the one-body theory may be defined (e.g., spin vectors) but that they have quite distinct properties. A detailed examination of this problem, including proofs, may be found in Holland (1988b) and we shall confine ourselves here to stating some pertinent results.

It suffices to consider a two-body system whose translational configuration space is $\mathbb{R}^3 \times \mathbb{R}^3$. The two-body wavefunction $\psi^{ab}(\mathbf{x}_1, \mathbf{x}_2)$ carries two spin indices. If it factorizes, $\psi^{ab}(\mathbf{x}_1, \mathbf{x}_2) = \psi_1^a(\mathbf{x}_1)\psi_2^b(\mathbf{x}_2)$, the following theory reduces to two copies of the one-body theory.

The configuration space dependence of ψ implies that we cannot interpret the wavefunction as a wave propagating in three-dimensional Euclidean space as was possible in the one-body case, but we can nevertheless retain the notion of individual well-defined (but correlated) particle trajectories in three-dimensional space, and endow these moving particles with further properties such as spin angular momentum. As we have said, it is the proposal of the causal interpretation that we should ascribe as much physical reality to waves propagating in multidimensional spaces as we do to ones in Euclidean space, and that they manifest themselves indirectly through the quantum potential and quantum torque by the influence they exert on the motions and physical properties of the correlated particles.

Our physical model is a system of two small rigid bodies, or even point particles, embedded in the configuration space field $\psi^{ab}(\mathbf{x}_1, \mathbf{x}_2, t)$. For clarity we shall suppose that the particles have magnetic moments μ_1 and μ_2 but are electrically neutral and only subject to magnetic fields. The two-body Pauli equation is:

$$i\hbar \frac{\partial \psi^{ab}}{\partial t} = \left\{\left(-\frac{\hbar^2}{2m_1}\nabla_1^2 - \frac{\hbar^2}{2m_2}\nabla_2^2\right)\delta^a{}_c\delta^b{}_d + \mu_1 B_{1i} e_{1i}{}^{ab}{}_{cd} + \mu_2 B_{2i} e_{2i}{}^{ab}{}_{cd}\right\}\psi^{cd}. \quad (9.6.1)$$

Here $e_{1i}{}^{ab}{}_{a'b'} = \sigma_i{}^a{}_{a'}\delta^b{}_{b'}$, $e_{2i}{}^{ab}{}_{a'b'} = \delta^a{}_{a'}\sigma_i{}^b{}_{b'}$, where σ_i, $i = 1, 2, 3$, are Pauli matrices.

It is convenient to express ψ in pseudo-polar form

$$\psi^{ab} = R\, e^{iS/\hbar}\, \phi^{ab} \quad (9.6.2)$$

where $R^2 = \rho = \psi^{ab*}\psi^{ab}$, S is the average phase and ϕ^{ab} satisfies the condition $\phi^{ab*}\phi^{ab} = 1$ and has zero average phase. The functions R and S are analogous to the amplitude and phase of the spin 0 two-body theory and the information on spin is carried by ϕ. The content of the latter is represented by the following tensor quantities. Two spin vectors are defined by

$$\left.\begin{aligned} \mathbf{s}_1 &= (\hbar/2\rho)\psi^{ab*}\boldsymbol{\sigma}^a{}_{a'}\psi^{a'b} \\ \mathbf{s}_2 &= (\hbar/2\rho)\psi^{ab*}\boldsymbol{\sigma}^b{}_{b'}\psi^{ab'}. \end{aligned}\right\} \quad (9.6.3)$$

This definition is justified on several grounds. First of all, these vectors are 'local averages' of the spin operators $\hat{\mathbf{s}}_1 = (\hbar/2)\mathbf{e}_1$ and $\hat{\mathbf{s}}_2 = (\hbar/2)\mathbf{e}_2$ since, for example,

$$\langle \hat{\mathbf{s}}_1 \rangle = (\hbar/2) \int \psi^\dagger \mathbf{e}_1 \psi \, \mathrm{d}^3x_1 \, \mathrm{d}^3x_2 \bigg/ \int \psi^\dagger \psi \, \mathrm{d}^3x_1 \, \mathrm{d}^3x_2. \tag{9.6.4}$$

Next, it is just these vectors which couple to the magnetic fields in the Hamilton–Jacobi equation (9.6.8) derived from the wave equation. Moreover, in a spin measurement of a two-body system, the magnitudes of these vectors eventually take on the values $\pm\hbar/2$ (Chap. 11). Finally, when the wavefunction factorizes, the vectors (9.6.3) become single-body spin vectors (9.3.3) for each particle. In the general case, the spin vectors defined here are functions of both sets of coordinates.

Notice that unlike the one-body case $|\mathbf{s}_1|$ and $|\mathbf{s}_2|$ are continuously variable and therefore not of fixed magnitude:

$$|\mathbf{s}_1| = |\mathbf{s}_2| = (\hbar/2\rho)(2\Omega - \rho^2)^{1/2}, \qquad \Omega = \psi^{ab*}\psi^{cb}\psi^{cd*}\psi^{ad}. \tag{9.6.5}$$

As a result, they may be zero when $\rho \neq 0$, something that is impossible for the one-body spin vector.

In addition, we define a local spin correlation tensor:

$$s_{ij} = (\hbar/2\rho)\psi^{ab*}\sigma_i{}^a{}_{a'}\sigma_j{}^b{}_{b'}\psi^{a'b'}. \tag{9.6.6}$$

This is motivated by the definition of the spin correlation function which in a given state ψ is

$$P_\psi(\mathbf{a}, \mathbf{b}) = \frac{\langle(\hat{\mathbf{s}}_1 \cdot \mathbf{a})(\hat{\mathbf{s}}_2 \cdot \mathbf{b})\rangle - \langle \hat{\mathbf{s}}_1 \cdot \mathbf{a}\rangle\langle \hat{\mathbf{s}}_2 \cdot \mathbf{b}\rangle}{[\langle(\hat{\mathbf{s}}_1 \cdot \mathbf{a})^2\rangle\langle(\hat{\mathbf{s}}_2 \cdot \mathbf{b})^2\rangle]^{1/2}}, \tag{9.6.7}$$

where $\langle \ \rangle$ is given by (9.6.4) and $\mathbf{a}, \mathbf{b}$ are two unit vectors. It is a natural definition since in the factorizable case $s_{ij} = (2/\hbar)s_{1i}s_{2j}$, which implies no correlations between the spin vectors.

Together with the phase S, the 16 real tensor quantities ρ, s_{1i}, s_{2i} and s_{ij} (which are interconnected by identities so that only seven of them are independent) cover all the degrees of freedom in the wavefunction and help to explain its meaning.

We now transcribe the wave equation (9.6.1) into a system of equations connecting the variables just introduced.

Contracting (9.6.1) with $\psi^\dagger$ and taking the real and imaginary parts we obtain

$$\frac{\partial S}{\partial t} - i\hbar\phi^\dagger \frac{\partial \phi}{\partial t} + \tfrac{1}{2}m_1\mathbf{v}_1^2 + \tfrac{1}{2}m_2\mathbf{v}_2^2$$
$$+ Q_1 + Q_2 + Q_{1s} + Q_{2s} + \frac{2\mu_1}{\hbar}\mathbf{B}_1\cdot\mathbf{s}_1 + \frac{2\mu_2}{\hbar}\mathbf{B}_2\cdot\mathbf{s}_2 = 0 \quad (9.6.8)$$

and

$$\frac{\partial \rho}{\partial t} + \nabla_1\cdot(\rho\mathbf{v}_1) + \nabla_2\cdot(\rho\mathbf{v}_2) = 0 \quad (9.6.9)$$

respectively, where

$$\mathbf{v}_\mu = \frac{-i\hbar}{2\rho m_\mu}\psi^\dagger\overleftrightarrow{\nabla}_\mu\psi = \left(\frac{1}{m_\mu}\right)(\nabla_\mu S - i\hbar\phi^\dagger\nabla_\mu\phi), \qquad \mu = 1, 2, \quad (9.6.10)$$

are velocity fields,

$$Q_\mu = \frac{-\hbar^2}{2m_\mu}\frac{\nabla_\mu^2 R}{R}, \qquad \mu = 1, 2, \quad (9.6.11)$$

are the usual quantum potentials which arise in the spinless two-body problem, and

$$Q_{\mu s} = \frac{\hbar^2}{2m_\mu}[\partial_{\mu i}\phi^\dagger\,\partial_{ui}\phi + (\phi^\dagger\nabla_\mu\phi)^2], \qquad \mu = 1, 2,$$

are spin-dependent additions. This may be written as

$$Q_{\mu s} = \frac{1}{4m_\mu}[\partial_{\mu i}s_{jk}\,\partial_{\mu i}s_{jk} + \partial_{\mu i}s_{1j}\,\partial_{\mu i}s_{1j} + \partial_{\mu i}s_{2j}\,\partial_{\mu i}s_{2j}]. \quad (9.6.12)$$

Contracting (9.6.1) now with $\psi^\dagger e_{1i}$ and $\psi^\dagger e_{2i}$ and taking in each case the imaginary part we find the equations of evolution of the two spin-vector fields:

$$\frac{\partial \mathbf{s}_\nu}{\partial t} + (\mathbf{v}_1\cdot\nabla_1 + \mathbf{v}_2\cdot\nabla_2)\mathbf{s}_\nu = \mathbf{T}_\nu + \frac{2\mu_\nu}{\hbar}\mathbf{B}_\nu \times \mathbf{s}_\nu, \qquad \nu = 1, 2, \quad (9.6.13)$$

where $\mathbf{T}_\nu$, $\nu = 1, 2$, are quantum torques given by

$$\left.\begin{aligned} T_{1k} &= \frac{1}{2\rho m_1}\varepsilon_{ijk}[s_{1i}\,\partial_{1l}(\rho\,\partial_{1l}s_{1j}) + s_{ir}\,\partial_{1l}(\rho\,\partial_{1l}s_{jr})] \\ &\quad + \frac{1}{2\rho m_2}\varepsilon_{ijk}[s_{1i}\,\partial_{2l}(\rho\,\partial_{2l}s_{1j}) + s_{ir}\,\partial_{2l}(\rho\,\partial_{2l}s_{jr})], \\ T_{2k} &= [1 \leftrightarrow 2]. \end{aligned}\right\} \quad (9.6.14)$$

Finally, contracting (9.6.1) with $\psi^\dagger e_{1i}e_{2j}$ and taking the imaginary part we get the equation of motion of the spin correlation tensor:

$$\frac{\partial s_{kt}}{\partial t} + (v_{1i}\,\partial_{1i} + v_{2i}\,\partial_{2i})s_{kt} = \frac{2\mu_1}{\hbar}\,\varepsilon_{klr}B_{1l}s_{rt} + \frac{2\mu_2}{\hbar}\,\varepsilon_{tlr}B_{2l}s_{kr}$$
$$+ \frac{1}{2m_1\rho}\{\varepsilon_{ijk}[s_{1i}\,\partial_{1l}(\rho\,\partial_{1l}s_{jt}) + s_{it}\,\partial_{1l}(\rho\,\partial_{1l}s_{1j})]$$
$$+ \varepsilon_{ijt}[s_{ki}\,\partial_{1l}(\rho\,\partial_{1l}s_{2j}) + s_{2i}\,\partial_{1l}(\rho\,\partial_{1l}s_{kj})]\}$$
$$+ \frac{1}{2m_2\rho}\{\varepsilon_{ijk}[s_{1i}\,\partial_{2l}(\rho\,\partial_{2l}s_{jt}) + s_{it}\,\partial_{2l}(\rho\,\partial_{2l}s_{1j})]$$
$$+ \varepsilon_{ijt}[s_{ki}\,\partial_{2l}(\rho\,\partial_{2l}s_{2j}) + s_{2i}\,\partial_{2l}(\rho\,\partial_{2l}s_{kj})]\}. \quad (9.6.15)$$

The new equations allow one to interpret the meaning of the quantum formalism if a further postulate is made: that the integral curves of the velocity fields (9.6.10) are the trajectories of the (centres of mass) of the two rigid bodies:

$$\mathbf{v}_\mu = \mathrm{d}\mathbf{x}_\mu/\mathrm{d}t, \qquad \mu = 1, 2. \quad (9.6.16)$$

We can then associate the vectors $\mathbf{s}_1$ and $\mathbf{s}_2$ with spin angular momenta of the particles and bring (9.6.13) to the form of equations of precession:

$$\frac{\mathrm{d}\mathbf{s}_\nu}{\mathrm{d}t} = \mathbf{T}_\nu + (2\mu_\nu/\hbar)\mathbf{B}_\nu \times \mathbf{s}_\nu, \qquad \nu = 1, 2. \quad (9.6.17)$$

One might object that this transcription of the wave equation into a mathematically equivalent form (i.e., eqs. (9.6.8), (9.6.9), (9.6.13) and (9.6.15)) makes the single linear equation (9.6.1) look terribly complicated. This is true. Yet it has the merit of indicating the physical significance of the wave equation in two ways. First, when the complicated bits may be neglected we obtain the classical equations of a pair of spinning dipoles, including in (9.6.9) their statistical mechanics. Second, the complications express the very subtle character of the motions implied by quantum mechanics, as will be seen in the application we make of this theory in Chap. 11. The point is that the mathematically convenient version of the theory, (9.6.1), does not directly connect with the physical interpretation. A superior approach to the many-spin problem is described in the next chapter.

9.7 Minimalist approach and problems with the Pauli theory

In §3.2.5 we noted that there is a 'minimalist' approach to the causal interpretation in which one eschews attributing any kinematical properties to a quantum system other than its position within the guiding wave. This view has been advocated in the case of the spin $\frac{1}{2}$ theory described in this chapter (Bell, 1966, 1981, 1982; Bohm & Hiley, 1989). The proposal is that one adopts the velocity field (9.3.8) and defines trajectories as solutions to the guidance equation $\dot{\mathbf{x}} = \mathbf{v}$, but that no other properties should be attributed to the particle such as a spin vector. This approach is subject to the same criticisms we levelled against its Schrödinger counterpart. We have seen, for example, that when the specifically quantum contributions in (9.3.6) are negligible we obtain the translational Hamilton–Jacobi equation corresponding to an ensemble of classical dipoles. If this indeed represents the classical limit, why should the particle intrinsic angular momentum cease to be well defined when one passes to the quantum domain? An additional objection relevant to the spin case is that the velocity field (9.3.8) is not sensitive to all the physically significant information conveyed by the spinor field that is pertinent to a causal picture of a process. The additional information is contained in the spin vector.

As a simple example of what we mean, consider an initial spinor wavefunction in which the space and spin parts are decoupled:

$$\psi_0(\mathbf{x}) = f(\mathbf{x})(c_+ u_+ + c_- u_-)$$

where $u_\pm$ are eigenfunctions of σ_z. Suppose the system is subject to an impulsive uniform magnetic field B_0 in the z-direction for a time T. Then, when the field is switched off, we have (cf. (9.5.8))

$$\psi(\mathbf{x}, T) = f(\mathbf{x})(c_+ \,\mathrm{e}^{-\mathrm{i}\Delta} u_+ + c_- \,\mathrm{e}^{\mathrm{i}\Delta} u_-), \tag{9.7.1}$$

where $\Delta = \mu B_0 T/\hbar$. It is immediately obvious from this expression that the interaction with the external field has left the space factor and hence the velocity of the particle unaltered; the trajectory is insensitive to this change in the quantum state of the system. Yet clearly, the statistics of subsequent measurements performed on (9.7.1) will be influenced by the transformation. The meaning of the latter is exhibited not in the orbit but in the spin vector which has undergone a rotation about the z-axis. Since it is the purpose of the causal interpretation to provide insight into all aspects of quantum phenomena, we conclude that the minimalist approach is not adequate to this task.

A more striking example of how minimalism does not provide a complete explanation is the EPR two-body problem. We will see in §11.3 that the trajectory of a particle may be unaffected by an interaction undergone by a distant particle with which it is coupled, and yet the spins of the particles exhibit a strong correlation. These correlations would be quite mysterious unless we were also able to define and investigate the change in the spin vector of the unmeasured particle.

There is nevertheless a valid criticism that may be raised against the theory of this chapter. This has to do with the manner in which particle properties have been defined as local functions of space and time through quadratic combinations of spinor components. What we are doing is summing over spin indices and hence, from the point of view of spin space, these functions are *averages*. To put the method in context, we may think of the configuration space of the system (continuous translational and discrete spin indices) as analogous to that of a two-body system. Then the procedure followed is akin to tracing over the coordinates of one body. The resulting theory of the other body retains some information on the whole but details on the correlations between the bodies are missing (cf. §7.7). Likewise in the spin case, the insight provided into typical spin phenomena by consideration of the orbit and spin vector is adequate as we have seen, but there is the hint of a more detailed theory lurking behind the scenes. Following the method outlined in §3.12.2 we might try to define particle properties as 'local' functions in the configuration space by bracketing operators with spinor components and not summing over the internal indices. It is immediately obvious, however, that this would not lead to a meaningful theory. It was pointed out in §3.12.2 that the causal interpretation can best be developed if one works in a coordinate representation of the underlying quantum theory. But in the typical textbook treatment of spin $\frac{1}{2}$ we have followed here, the translational and orientational coordinates do not appear on an equal footing. Rather, we mix two representations: the position representation for the linear momentum operators, and the angular momentum representation for the angular momentum operators. This suggests that the Pauli theory does not employ a representation of the quantum theory of angular momentum appropriate to a complete causal formulation, and that the latter may be obtained if the quantum theory is developed in a representation in which the angular momentum operators are realized as differential operators. Then we could give meaning to fully local entities in a new configuration space of the system where the discrete spin indices are replaced by continuous Euler angles.

An additional problem with the approach based on spinors is that there does not seem to be any obvious way of introducing Euler angles to describe orientation in the theory for more than one body, since triads of orthogonal vectors which could represent a state of rotation for each particle in a many-body system do not naturally arise (Holland, 1988b).

Apart from questions of technique in formulating the causal interpretation, the points just made also relate to a difficulty in the Pauli equation itself.

Insofar as this makes no reference to the moment of inertia of a body, it may be considered to be defective if treated as a basic theory of spinning bodies. Indeed, it is compatible with a wide class of underlying models, including point particles.

The difficulties mentioned are overcome in the model of the next chapter.

10

Spin $\frac{1}{2}$: the rigid rotator

10.1 Classical dynamics of rigid bodies

The difficulties in formulating the theory of spin $\frac{1}{2}$ systems outlined at the end of the last chapter may be avoided if we start from a classical theory of rotation in which the translational and orientational coordinates enter on a symmetrical footing, and pass to the quantum theory via the standard canonical quantization procedure. The result is not simply the Pauli theory written in a different matrix representation but a more general physical theory which explicitly incorporates in the wave equation the extended character of the bodies considered, through moment-of-inertia-dependent terms. As expected, the spin $\frac{1}{2}$ character of the wave is manifested through the dependence of the body motion on the total quantum state via quantum potential and torque terms. Spin is not usually approached from this angle in basic quantum mechanics texts so it is worthwhile to build up the theory from scratch (for a more rigorous treatment see Normand (1980) and for further details: Bopp & Haag (1950), Rosen (1951, 1984), van Winter (1954), Dankel (1970), Holland (1988a,b)).

We begin by reviewing some basic results in the classical canonical theory of rigid body motion. Our treatment is confined to a rigid homogeneous sphere with moment of inertia I about any axis passing through its centre, where the centre of mass is located. The sphere carries a rigid charge and a magnetic moment. By Chasle's theorem an arbitrary displacement of a rigid body is composed of a translation together with a rotation about a line fixed in the body.

As was shown in §9.2, the orientation of a rigid body may be expressed through the inclination of a set of axes $\mathbf{e}'_i$ fixed in the moving body with respect to a set of space axes $\mathbf{e}_i$, via the Euler angles. We shall place the origin of the frame $\mathbf{e}'_i$ at the centre of mass. For reasons which will become clear later we denote the Euler angles here by (α, β, γ), which correspond respectively to (θ, ϕ, χ) as defined in Fig. 9.1. Sometimes it will be convenient to write $\boldsymbol{\alpha} = (\alpha_i) = (\alpha, \beta, \gamma)$ where $i = 1, 2, 3$ is, of course, not a vector index. The configuration space of the system is therefore a six-dimensional manifold $\mathbb{R}^3 \times SO(3)$ comprising the centre of mass Cartesian coordinates x_i and the orientation parameters α_i. We denote the centre of mass velocity by $\mathbf{v} = \dot{\mathbf{x}}$.

With reference to Fig. 9.1, the angular velocity vector is given by $\boldsymbol{\omega} = -\dot{\alpha}\mathbf{f}_1 - \dot{\beta}\mathbf{e}_3 + \dot{\gamma}\mathbf{e}'_3$. We can write down the components of $\boldsymbol{\omega}$ with respect to either the space or body axes (assuming the summation convention):

$$\boldsymbol{\omega} = \omega_i \mathbf{e}_i = \omega'_i \mathbf{e}'_i \tag{10.1.1}$$

and, using

$$\mathbf{f}_1 = \cos\beta\,\mathbf{e}_1 - \sin\beta\,\mathbf{e}_2, \qquad \mathbf{e}'_3 = \sin\alpha\sin\beta\,\mathbf{e}_1 + \sin\alpha\cos\beta\,\mathbf{e}_2 + \cos\alpha\,\mathbf{e}_3, \tag{10.1.2}$$

these are given by

$$\left.\begin{aligned}
\omega_1 &= -\dot{\alpha}\cos\beta - \dot{\gamma}\sin\alpha\sin\beta \\
\omega_2 &= \dot{\alpha}\sin\beta - \dot{\gamma}\sin\alpha\cos\beta \\
\omega_3 &= -\dot{\beta} - \dot{\gamma}\cos\alpha \\
\omega'_1 &= -\dot{\alpha}\cos\gamma - \dot{\beta}\sin\alpha\sin\gamma \\
\omega'_2 &= -\dot{\alpha}\sin\gamma + \dot{\beta}\sin\alpha\cos\gamma \\
\omega'_3 &= -\dot{\gamma} - \dot{\beta}\cos\alpha.
\end{aligned}\right\} \tag{10.1.3}$$

The magnitude of $\boldsymbol{\omega}$, $|\boldsymbol{\omega}|$, is a scalar and is, of course, the same in both reference systems:

$$|\boldsymbol{\omega}|^2 = \dot{\alpha}^2 + \dot{\beta}^2 + \dot{\gamma}^2 + 2\dot{\beta}\dot{\gamma}\cos\alpha. \tag{10.1.4}$$

The components of the angular momentum with respect to the two sets of axes are $I\omega_i$ and $\mathrm{I}\omega'_i$, $i = 1, 2, 3$.

Let us suppose that the body has mass m and charge e and moves in an external electromagnetic field with potentials $A_0, \mathbf{A}$, with magnetic field $\mathbf{B} = \nabla \times \mathbf{A}$, and that it has a potential energy V as a result of coupling to any other external scalar potential. All the potentials are functions of $\mathbf{x}$ and t only (although we could include the possibility that V depends on all six coordinates, for example when the body is placed in an external gravitational field). The magnetic moment vector is proportional to the angular momentum (i.e., $\mu' I\boldsymbol{\omega}$, where μ' is a scalar constant) and so the magnetic moment has a potential energy given by $\mu' I\boldsymbol{\omega}\cdot\mathbf{B}$. The Lagrangian is therefore

$$\begin{aligned}
L &= L(\mathbf{x}, \mathbf{v}, \boldsymbol{\alpha}, \dot{\boldsymbol{\alpha}}, t) \\
&= \tfrac{1}{2}m\mathbf{v}^2 + (e/c)\mathbf{A}\cdot\mathbf{v} - eA_0 - V + \tfrac{1}{2}I\omega^2 - \mu' I\mathbf{B}\cdot\boldsymbol{\omega}.
\end{aligned} \tag{10.1.5}$$

The components of the canonical momenta with respect to $\mathbf{e}_i$ are then given by

$$p_i = \partial L/\partial v_i = mv_i + (e/c)A_i \tag{10.1.6}$$

and the canonical angular momenta p_{α_i} conjugate to α_i by

$$\left.\begin{aligned} p_\alpha &= \partial L/\partial\dot{\alpha} = I\dot{\alpha} + \mu' I\mathbf{B}\cdot(\cos\beta, -\sin\beta, 0),\\ p_\beta &= \partial L/\partial\dot{\beta} = I(\dot{\beta} + \dot{\gamma}\cos\alpha) + \mu' I B_3,\\ p_\gamma &= \partial L/\partial\dot{\gamma} = I(\dot{\gamma} + \dot{\beta}\cos\alpha) + \mu' I\mathbf{B}\cdot(\sin\alpha\sin\beta, \sin\alpha\cos\beta, \cos\alpha). \end{aligned}\right\} \tag{10.1.7}$$

Inverting these relations yields

$$\left.\begin{aligned} I\dot{\alpha} &= p_\alpha - \mu' I\mathbf{B}\cdot(\cos\beta, -\sin\beta, 0),\\ I\dot{\beta} &= (1/\sin^2\alpha)[p_\beta - \cos\alpha\, p_\gamma + \mu' I\mathbf{B}\cdot\sin\alpha(\cos\alpha\sin\beta, \cos\alpha\cos\beta, -\sin\alpha)],\\ I\dot{\gamma} &= (1/\sin^2\alpha)[p_\gamma - \cos\alpha\, p_\beta - \mu' I\mathbf{B}\cdot(\sin\alpha\sin\beta, \sin\alpha\cos\beta, 0)]. \end{aligned}\right\} \tag{10.1.8}$$

Substituting (10.1.8) into (10.1.3) we find the two sets of angular momentum components in terms of the canonical angular momenta:

$$\left.\begin{aligned} I\omega_i &= M_i + \mu' I B_i\\ I\omega'_i &= M'_i + \mu' I B'_i, \end{aligned}\right\} \tag{10.1.9}$$

where

$$\left.\begin{aligned} M_1 &= -\cos\beta\, p_\alpha + \sin\beta\cot\alpha\, p_\beta - \sin\beta\,\mathrm{cosec}\,\alpha\, p_\gamma\\ M_2 &= \sin\beta\, p_\alpha + \cos\beta\cot\alpha\, p_\beta - \cos\beta\,\mathrm{cosec}\,\alpha\, p_\gamma\\ M_3 &= -p_\beta\\ M'_1 &= -\cos\gamma\, p_\alpha - \sin\gamma\,\mathrm{cosec}\,\alpha\, p_\beta + \sin\gamma\cot\alpha\, p_\gamma\\ M'_2 &= -\sin\gamma\, p_\alpha + \cos\gamma\,\mathrm{cosec}\,\alpha\, p_\beta - \cos\gamma\cot\alpha\, p_\gamma\\ M'_3 &= -p_\gamma \end{aligned}\right\} \tag{10.1.10}$$

and $B'_i = \mathbf{B}\cdot\mathbf{e}'_i$ are the components of the magnetic field with respect to the moving frame.

Performing a Legendre transformation in the usual way yields the Hamiltonian

$$\begin{aligned} H &= H(x_i, p_i, \alpha_i, p_{\alpha_i})\\ &= p_i v_i + p_{\alpha_i}\dot{\alpha}_i - L\\ &= (1/2m)[\mathbf{p} - (e/c)\mathbf{A}]^2 + (1/2I)(\mathbf{M} + \mu' I\mathbf{B})^2 + eA_0 + V. \end{aligned} \tag{10.1.11}$$

In evaluating the last line we have used the easily proved result

$$p_{\alpha_i}\dot{\alpha}_i = M_i\omega_i \tag{10.1.12}$$

and have substituted for v_i and ω_i from (10.1.6) and (10.1.9) respectively.

We define the Poisson brackets (PBs) of two functions A and B of the canonical variables $x_i, \alpha_i, p_i, p_{\alpha_i}$ by

$$\{A, B\} = \frac{\partial A}{\partial x_i}\frac{\partial B}{\partial p_i} + \frac{\partial A}{\partial \alpha_i}\frac{\partial B}{\partial p_{\alpha_i}} - \frac{\partial A}{\partial p_i}\frac{\partial B}{\partial x_i} - \frac{\partial A}{\partial p_{\alpha_i}}\frac{\partial B}{\partial \alpha_i}.$$

The basic PBs are then

$$\left.\begin{aligned} \{x_i, p_j\} = \delta_{ij}, \qquad \{\alpha_i, p_{\alpha_j}\} = \delta_{ij}, \\ \{x_i, \alpha_j\} = \{x_i, p_{\alpha_j}\} = \{\alpha_i, p_j\} = \{p_i, p_{\alpha_j}\} = 0 \end{aligned}\right\} \tag{10.1.13}$$

for all i, j and from (10.1.10) one easily deduces

$$\left.\begin{aligned} \{M_i, M_j\} = \varepsilon_{ijk} M_k, \qquad \{M'_i, M'_j\} = -\varepsilon_{ijk} M'_k \\ \{M_i, M'_j\} = 0, \qquad \text{all } i, j \end{aligned}\right\} \tag{10.1.14}$$

for the components of the free angular momentum vector.

The translational equation of motion and equation of spin precession follow from the Hamiltonian equations in PB form. To begin with, from $\dot{x}_i = \{x_i, H\}$ and $\dot{\alpha}_i = \{\alpha_i, H\}$ we obtain the relations (10.1.6) and (10.1.7). Furthermore, $\dot{p}_i = \{p_i, H\}$ yields the equation of motion

$$m\,\mathrm{d}\mathbf{v}/\mathrm{d}t = \mathbf{F} - \nabla V - \mu' I \nabla(\boldsymbol{\omega}\cdot\mathbf{B}), \tag{10.1.15}$$

where $\mathbf{F} = e\mathbf{E} + (e/c)\mathbf{v} \times \mathbf{B}$ is the Lorentz force with $\mathbf{E} = -\nabla A_0 - (1/c)(\partial\mathbf{A}/\partial t)$, $\mathbf{B} = \nabla \times \mathbf{A}$. Finally, $\dot{p}_{\alpha_i} = \{p_{\alpha_i}, H\}$ implies

$$I\,\mathrm{d}\boldsymbol{\omega}/\mathrm{d}t = -\mu' I \boldsymbol{\omega} \times \mathbf{B} + \mu' I\,(\mathrm{d}\mathbf{B}/\mathrm{d}t). \tag{10.1.16}$$

For later comparison with the analogous quantum case, we write down the Hamilton–Jacobi equation corresponding to an ensemble of rigid rotators, derived from (10.1.11):

$$(\partial S/\partial t) + (1/2m)(\nabla S - (e/c)\mathbf{A})^2 + (1/2I)(\hat{\boldsymbol{\lambda}} S + \mu' I\mathbf{B})^2 + eA_0 + V = 0 \tag{10.1.17}$$

where S depends on x_i, α_i and t and $\hat{\boldsymbol{\lambda}}$ is defined in (10.2.9). The classical theory of rotators corresponding to the classical limit of the Pauli quantum theory (§9.3.3) follows from the above theory when one averages over the variables α_i, as will be seen in §10.4.

10.2 The quantum rigid rotator

In order to quantize the rigid rotator we proceed as usual and replace the canonical classical variables by operators $\hat{x}_i, \hat{p}_i, \hat{\alpha}_i, \hat{p}_{\alpha_i}$. As a result, the domain of definition of the wavefunction comprises translational and orientational parameters (Euler angles), and one obtains a scheme describing both integral and half-integral spin. The PBs are replaced by commutators: $\{\ \} \rightarrow -(\mathrm{i}/\hbar)[\ ]$ so that (10.1.13) become

$$\left.\begin{array}{c}[\hat{x}_i, \hat{p}_j] = \mathrm{i}\hbar\delta_{ij}, \qquad [\hat{\alpha}_i, \hat{p}_{\alpha_j}] = \mathrm{i}\hbar\delta_{ij}, \\ [\hat{x}_i, \hat{\alpha}_j] = [\hat{x}_i, \hat{p}_{\alpha_j}] = [\hat{\alpha}_i, \hat{p}_j] = [\hat{p}_i, \hat{p}_{\alpha_j}] = 0\end{array}\right\} \tag{10.2.1}$$

for all i, j. In a representation in which $\hat{x}_i$ and $\hat{\alpha}_i$ are diagonal, relations (10.2.1) are satisfied by

$$p_i = -\mathrm{i}\hbar\,\partial/\partial x_i, \qquad \hat{p}_{\alpha_i} = -\mathrm{i}\hbar\,\partial/\partial\alpha_i \tag{10.2.2}$$

and the components of the free angular momentum vector (10.1.10) become the operators

$$\left.\begin{array}{l}\hat{M}_1 = \mathrm{i}\hbar(\cos\beta\,\partial_\alpha - \sin\beta\cot\alpha\,\partial_\beta + \sin\beta\operatorname{cosec}\alpha\,\partial_\gamma) \\ \hat{M}_2 = \mathrm{i}\hbar(-\sin\beta\,\partial_\alpha - \cos\beta\cot\alpha\,\partial_\beta + \cos\beta\operatorname{cosec}\alpha\,\partial_\gamma) \\ \hat{M}_3 = \mathrm{i}\hbar\,\partial_\beta \\ \hat{M}'_1 = \mathrm{i}\hbar(\cos\gamma\,\partial_\alpha + \sin\gamma\operatorname{cosec}\alpha\,\partial_\beta - \sin\gamma\cot\alpha\,\partial_\gamma) \\ \hat{M}'_2 = \mathrm{i}\hbar(\sin\gamma\,\partial_\alpha - \cos\gamma\operatorname{cosec}\alpha\,\partial_\beta + \cos\gamma\cot\alpha\,\partial_\gamma) \\ \hat{M}'_3 = \mathrm{i}\hbar\,\partial_\gamma\end{array}\right\} \tag{10.2.3}$$

with respect to the space and body axes respectively.

The components $\hat{M}_i$ are the infinitesimal generators of the rotation group and span the Lie algebra of $SO(3)$, and the components $\hat{M}'_i$ satisfy the 'anomalous' commutation relations (cf. (10.1.14)):

$$\left.\begin{array}{c}[\hat{M}_i, \hat{M}_j] = \mathrm{i}\hbar\varepsilon_{ijk}\hat{M}_k, \qquad [\hat{M}'_i, \hat{M}'_j] = -\mathrm{i}\hbar\varepsilon_{ijk}\hat{M}'_k \\ [\hat{M}_i, \hat{M}'_j] = 0 \qquad \text{for all } i, j.\end{array}\right\} \tag{10.2.4}$$

The generator of an arbitrary finite clockwise rotation is given by the sequence of operators

$$D(\alpha, \beta, \gamma) = \mathrm{e}^{-\mathrm{i}\gamma\hat{M}'_3/\hbar}\,\mathrm{e}^{-\mathrm{i}\alpha(\cos\beta\hat{M}_1 - \sin\beta\hat{M}_2)/\hbar}\,\mathrm{e}^{-\mathrm{i}\beta\hat{M}_3/\hbar}$$

(cf. the definition of the Euler angles in §9.2.2). This expression may be reduced to

$$D(\alpha, \beta, \gamma) = \mathrm{e}^{-\mathrm{i}\beta\hat{M}_3/\hbar}\,\mathrm{e}^{-\mathrm{i}\alpha\hat{M}_1/\hbar}\,\mathrm{e}^{-\mathrm{i}\gamma\hat{M}_3/\hbar}. \tag{10.2.5}$$

Starting from the Hamiltonian (10.1.11) and quantizing as above, we obtain the Schrödinger equation for a rigid rotating sphere of mass m, moment of inertia I, charge e and magnetic moment μ':

$$\left.\begin{array}{c}\mathrm{i}\hbar\,(\partial\psi/\partial t) = \hat{H}\psi, \\ \hat{H} = (1/2m)[\hat{\mathbf{p}} - (e/c)\mathbf{A}]^2 + (1/2I)(\hat{\mathbf{M}} + \mu' I\mathbf{B})^2 + eA_0 + V,\end{array}\right\} \tag{10.2.6}$$

where A_0, $\mathbf{A}$ and V are unaffected by quantization. Here the wavefunction $\psi = \psi(\mathbf{x}, \alpha, \beta, \gamma, t) \in L^2[\mathbb{R}^3 \times SU(2)]$ i.e., it lies in the Hilbert space associated with the manifold $\mathbb{R}^3 \times SU(2)$. Notice that we anticipate the result derived below that ψ may correspond to half-integral angular momentum values by writing the universal covering group of $SO(3)$, $SU(2)$.

The inner product of two states ϕ, ψ is defined in this Hilbert space by the positive definite expression

$$\langle \phi | \psi \rangle = \int \phi^* \psi \, \mathrm{d}\Omega \, \mathrm{d}^3 x, \tag{10.2.7}$$

where $\mathrm{d}\Omega = \sin\alpha \, \mathrm{d}\alpha \, \mathrm{d}\beta \, \mathrm{d}\gamma$ is the measure on $SU(2)$ ($0 \leqslant \alpha \leqslant \pi$, $0 \leqslant \beta \leqslant 2\pi$, $0 \leqslant \gamma \leqslant 4\pi$). The wavefunction will be assumed to be normalized:

$$\int |\psi|^2 \, \mathrm{d}\Omega \, \mathrm{d}^3 x = 1. \tag{10.2.8}$$

We shall interpret $|\psi|^2 \, \mathrm{d}\Omega \, \mathrm{d}^3 x$ as the probability that at time t a particle has a position and orientation lying in the volume element $\mathrm{d}\Omega \, \mathrm{d}^3 x$ about the point $\mathbf{x}$ and inclination (α, β, γ). The angular momentum operators are Hermitian with respect to the inner product (10.2.7) because the functions they operate on are periodic (cf. Carruthers & Nieto, 1968).

It is convenient to introduce the real operators

$$\nabla = \hat{\mathbf{p}}/(-\mathrm{i}\hbar), \qquad \hat{\boldsymbol{\lambda}} = \hat{\mathbf{M}}/(-\mathrm{i}\hbar) \tag{10.2.9}$$

and write the Hamiltonian operator (10.2.6) as

$$\hat{H} = -\frac{\hbar^2}{2m}\left[\nabla - \left(\frac{\mathrm{i}e}{c\hbar}\right)\mathbf{A}\right]^2 - \frac{\hbar^2}{2I}\left[\hat{\boldsymbol{\lambda}} + \left(\frac{\mathrm{i}\mu' I}{\hbar}\right)\mathbf{B}\right]^2 + eA_0 + V. \tag{10.2.10}$$

In the case of no external potentials the Schrödinger equation reduces to

$$\mathrm{i}\hbar \frac{\partial \psi}{\partial t} = \left(-\frac{\hbar^2}{2m}\nabla^2 + \frac{1}{2I}\hat{\mathbf{M}}^2\right)\psi, \tag{10.2.11}$$

where

$$\hat{\mathbf{M}}^2 = -\hbar^2 \operatorname{cosec}\alpha \left[\frac{\partial}{\partial\alpha}\left(\sin\alpha \frac{\partial}{\partial\alpha}\right) + \operatorname{cosec}\alpha\left(\frac{\partial^2}{\partial\beta^2} + \frac{\partial^2}{\partial\gamma^2} - 2\cos\alpha \frac{\partial^2}{\partial\beta\,\partial\gamma}\right)\right]. \tag{10.2.12}$$

In order to specify the dependence of ψ on (α, β, γ), we first seek simultaneous eigenfunctions $D^s_{ab}(\alpha, \beta, \gamma)$ of the three mutually commuting operators $\hat{\mathbf{M}}^2$, $\hat{M}_3$ and $\hat{M}'_3$:

$$\left.\begin{aligned}\hat{\mathbf{M}}^2 D^s_{ab} &= s(s+1)\hbar^2 D^s_{ab}\\ \hat{M}_3 D^s_{ab} &= a\hbar D^s_{ab}\\ \hat{M}'_3 D^s_{ab} &= b\hbar D^s_{ab}.\end{aligned}\right\} \tag{10.2.13}$$

The functions D^s_{ab} are labelled by the total angular momentum quantum number $s = 0, \frac{1}{2}, 1, \frac{3}{2}, \ldots$ and the angular momentum quantum numbers $a = -s, \ldots, s$ and $b = -s, \ldots, s$ in the z-direction with respect to the space and body axes respectively. The restrictions that s must be integral or half-integral and on the range of values that a and b may take follow purely from the commutation relations (10.2.4).

We shall briefly sketch a method by which one may derive the functions D^s_{ab}. First of all, we introduce the two sets of raising and lowering operators $\hat{M}_1 \pm \mathrm{i}\hat{M}_2$ and $\hat{M}'_1 \mp \mathrm{i}\hat{M}'_2$. We have from (10.2.4)

$$\hat{\mathbf{M}}^2 = (\hat{M}_1 - \mathrm{i}\hat{M}_2)(\hat{M}_1 + \mathrm{i}\hat{M}_2) + \hbar M_3 + \hat{M}_3^2$$

and hence from (10.2.13)

$$(\hat{M}_1 - \mathrm{i}\hat{M}_2)(\hat{M}_1 + \mathrm{i}\hat{M}_2)D^s_{ab} = \hbar^2(s-a)(s+a+1)D^s_{ab}. \tag{10.2.14}$$

Since, in addition, from (10.2.4)

$$\hat{M}_3(\hat{M}_1 + \mathrm{i}\hat{M}_2)D^s_{ab} = (a+1)\hbar(\hat{M}_1 + \mathrm{i}\hat{M}_2)D^s_{ab},$$

so that

$$(\hat{M}_1 + \mathrm{i}\hat{M}_2)D^s_{ab} = C^s_{ab}D^s_{a+1,b},$$

where C^s_{ab} is a complex constant, we find from (10.2.14) that

$$(\hat{M}_1 \pm \mathrm{i}\hat{M}_2)D^s_{ab} = \hbar[(s \mp a)(s \pm a + 1)]^{1/2}D^s_{a\pm 1,b}, \tag{10.2.15}$$

where the phase of C^s_{ab} is suitably chosen. Similarly,

$$(\hat{M}'_1 \mp \mathrm{i}\hat{M}'_2)D^s_{ab} = \hbar[(s \mp b)(s \pm b + 1)]^{1/2}D^s_{a,b\pm 1}. \tag{10.2.16}$$

Substituting (10.2.3) into the last two equations of (10.2.13) yields

$$D^s_{ab}(\alpha, \beta, \gamma) = \mathrm{e}^{-\mathrm{i}(a\beta + b\gamma)}\, d^s_{ab}(\alpha). \tag{10.2.17}$$

Substituting this, in turn, into (10.2.15) implies the following differential equation for $d^s_{ab}(\alpha)$:

$$\frac{\mathrm{d}d^s_{ab}}{\mathrm{d}\alpha} \pm b \operatorname{cosec}\alpha\, d^s_{ab} \mp a \cot\alpha\, d^s_{ab} = -\mathrm{i}[(s \mp a)(s \pm a + 1)]^{1/2}d^s_{a\pm 1,b}.$$

When $a = \pm s$ this is easily solved to give

$$d^s_{\pm sb}(\alpha) = A_{sb}\sin^s\alpha\,(\tan\alpha/2)^{\mp b} \tag{10.2.18}$$

where A_{sb} is a constant. To find $d^s_{ab}(\alpha)$ for all $a = -s, \ldots, s$ we simply apply the raising and lowering operators $\hat{M}_1 \pm \mathrm{i}\hat{M}_2$ to (10.2.18). The constants A_{sb} in (10.2.18) are chosen so that the functions $D^s_{ab}(\alpha, \beta, \gamma)$ satisfy (10.2.15) and (10.2.16) and the orthonormality relations

$$\int \mathrm{d}\Omega\, D^{s'*}_{a'b'}(\alpha, \beta, \gamma) D^s_{ab}(\alpha, \beta, \gamma) = \delta_{ss'}\delta_{aa'}\delta_{bb'}. \tag{10.2.19}$$

The total set of functions $D^s_{ab}(\alpha, \beta, \gamma)$ is given by substituting $d^s_{ab}(\alpha)$ into (10.2.17).

In addition to the relations (10.2.19), the simultaneous eigenfunctions obey closure relations:

$$\sum_{s,a,b} D^{s*}_{ab}(\boldsymbol{\alpha}) D^s_{ab}(\boldsymbol{\alpha}') = \operatorname{cosec} \alpha\, \delta_{\alpha\alpha'}\delta_{\beta\beta'}\delta_{\gamma\gamma'}. \tag{10.2.20}$$

They thus form a complete orthonormal set and we may expand an arbitrary function in the Hilbert space as follows:

$$\psi(\mathbf{x}, \alpha, \beta, \gamma, t) = \sum_{s,a,b} \psi^{ab}_s(\mathbf{x}, t) D^s_{ab}(\alpha, \beta, \gamma). \tag{10.2.21}$$

The functions $D^s_{ab}(\boldsymbol{\alpha})$ are the transformation matrices linking the angular momentum representation spanned by the kets $|s, a, b\rangle$ and the angular coordinate representation spanned by the kets $|\boldsymbol{\alpha}\rangle$: $D^s_{ab}(\boldsymbol{\alpha}) = \langle \boldsymbol{\alpha} | s, a, b\rangle$.

The solution (10.2.21) describes a state of indefinite total spin angular momentum in the conventional sense that it is not an eigenstate of $\hat{\mathbf{M}}^2$. For each value of s, the $(2s + 1)^2$ functions D^s_{ab} span the $\mathscr{D}^s$ representation of the rotation group. An arbitrary function in the Hilbert space of functions of fixed s, i.e., an eigenfunction of $\hat{\mathbf{M}}^2$, may be written

$$\psi(\mathbf{x}, \alpha, \beta, \gamma, t) = \sum_{a,b=-s}^{s} \psi^{ab}(\mathbf{x}, t) D^s_{ab}(\alpha, \beta, \gamma). \tag{10.2.22}$$

The wave equation (10.2.6) clearly preserves a solution corresponding to a given value of s. If we wish we may also confine attention to a particular eigenstate of $\hat{M}_3$ or $\hat{M}'_3$. If we fix b, say, then an arbitrary simultaneous eigenfunction of $\hat{\mathbf{M}}^2$ and $\hat{M}'_3$ (but not of $\hat{M}_3$) may be written

$$\psi(\mathbf{x}, \alpha, \beta, \gamma, t) = \sum_{a=-s}^{s} \psi^a(\mathbf{x}, t) D^s_{ab}(\alpha, \beta, \gamma). \tag{10.2.23}$$

Again, the wave equation preserves such solutions (i.e., it does not mix eigenfunctions corresponding to different values of b). The coefficients ψ^a, $a = -s, \ldots, s$, form the $(2s + 1)$-components of a spinor.

Let us now come to the case that interests us here, spin ½. Putting $s = \frac{1}{2}$ in (10.2.18) and substituting into (10.2.17) we obtain the following four angular momentum basis functions:

$$\left.\begin{aligned} u_+ &= D^{\frac{1}{2}}_{\frac{1}{2}\frac{1}{2}} = (2\sqrt{2}\,\pi)^{-1}\cos(\alpha/2)\,e^{-i(\gamma+\beta)/2} \\ u_- &= D^{\frac{1}{2}}_{-\frac{1}{2}\frac{1}{2}} = -i(2\sqrt{2}\,\pi)^{-1}\sin(\alpha/2)\,e^{i(-\gamma+\beta)/2} \\ v_+ &= D^{\frac{1}{2}}_{\frac{1}{2}-\frac{1}{2}} = -i(2\sqrt{2}\,\pi)^{-1}\sin(\alpha/2)\,e^{i(\gamma-\beta)/2} \\ v_- &= D^{\frac{1}{2}}_{-\frac{1}{2}-\frac{1}{2}} = (2\sqrt{2}\,\pi)^{-1}\cos(\alpha/2)\,e^{i(\gamma+\beta)/2}. \end{aligned}\right\} \tag{10.2.24}$$

A spin ½ wavefunction with b fixed as $b = \frac{1}{2}$, say, is then by (10.2.23)

$$\psi(\mathbf{x}, \alpha, \beta, \gamma, t) = \psi_+(\mathbf{x}, t)u_+(\alpha, \beta, \gamma) + \psi_-(\mathbf{x}, t)u_-(\alpha, \beta, \gamma), \tag{10.2.25}$$

where we have written $\psi^1 = \psi_+$, $\psi^2 = \psi_-$.

The existence of such solutions shows that spin ½ is a property independent of the requirement of relativistic invariance.

Writing $\hat{M}^{\pm} = \hat{M}_1 \pm i\hat{M}_2$, we have from (10.2.15):

$$\hat{M}^+u_+ = \hat{M}^-u_- = 0, \qquad \hat{M}^+u_- = \hbar u_+, \qquad \hat{M}^-u_+ = \hbar u_-. \tag{10.2.26}$$

10.3 Causal formulation

10.3.1 One body

The causal interpretation of the quantum theory of rigid rotators proceeds from the usual supposition that associated with the wave $\psi = \psi(\mathbf{x}, \boldsymbol{\alpha}, t)$ there is a particle whose external and internal motions are determined by the space and time development of ψ. The particle is assumed to have the characteristics ascribed to a classical rigid sphere, i.e., simultaneously well-defined position ($\mathbf{x}(t)$) and momentum variables associated with the translation of the centre of mass, and simultaneously well-defined orientation ($\boldsymbol{\alpha}(t)$) and angular momentum variables associated with the rotational motion of the body about a line fixed in it. The action of the wave on the particle of course induces motions which are highly nonclassical in nature, as we shall see later. Because it is defined in the configuration space $\mathbb{R}^3 \times SU(2)$, we cannot ascribe to ψ a direct physical significance in Euclidean space. The situation in this respect is similar to that obtaining in the many-body theory (Chap. 7) but, as in that case, this facet does not prevent us from discussing the motion of a particle in ordinary space. It does imply, however, that the internal and external motions of the particle are likely to be very closely coupled in the case of an arbitrary wavefunction. As usual, each wavefunction is associated with an infinite fictitious ensemble of corpuscles distinguished by their initial co-ordinates, independently of the probability hypothesis.

First of all, we aim to rewrite the Schrödinger equation in terms of velocity and angular velocity fields in configuration space. Since the wavefunction does not carry any free indices, we may perform our usual trick of expressing it in polar form:

$$\psi = R\,\mathrm{e}^{\mathrm{i}S/\hbar} \tag{10.3.1}$$

where $R = R(\mathbf{x}, \alpha, \beta, \gamma, t)$ and $S = S(\mathbf{x}, \alpha, \beta, \gamma, t)$ are real amplitude and phase functions, respectively, on the manifold $\mathbb{R}^3 \times SU(2)$. Substituting (10.3.1) into the Schrödinger equation with (10.2.10) and separating into real and imaginary parts yields as usual a Hamilton–Jacobi-type equation

$$\partial S/\partial t + \tfrac{1}{2}m\mathbf{v}^2 + \tfrac{1}{2}I\boldsymbol{\omega}^2 + Q + Q_s + eA_0 + V = 0 \tag{10.3.2}$$

and a continuity equation in configuration space:

$$\partial\rho/\partial t + \nabla\cdot(\rho\mathbf{v}) + \hat{\boldsymbol{\lambda}}\cdot(\rho\boldsymbol{\omega}) = 0. \tag{10.3.3}$$

Here

$$\left.\begin{aligned} \mathbf{v} &= (\hbar/2mi\rho)[\psi^*\nabla\psi - (\nabla\psi^*)\psi] - (e/mc)\mathbf{A} \\ &= (1/m)(\nabla S - (e/c)\mathbf{A}) \end{aligned}\right\} \tag{10.3.4}$$

is the velocity field corresponding to the translation of the centres of mass of an ensemble of rigid spheres, $\rho = R^2 = |\psi|^2$ is the probability density,

$$\left.\begin{aligned} \boldsymbol{\omega} &= (1/I\rho)\,\mathrm{Im}(\psi^*\hat{\mathbf{M}}\psi) + \mu'\mathbf{B} \\ &= (1/I)(\hat{\boldsymbol{\lambda}}S + \mu' I B) \end{aligned}\right\} \tag{10.3.5}$$

is the angular velocity field, and $(Q + Q_s)$ is the total quantum potential, where

$$\left.\begin{aligned} Q &= -\frac{\hbar^2}{2m}\frac{\nabla^2 R}{R} \\ Q_s &= \frac{1}{2I}\frac{\hat{\mathbf{M}}^2 R}{R} = \frac{-\hbar^2}{2I}\frac{\hat{\boldsymbol{\lambda}}^2 R}{R}. \end{aligned}\right\} \tag{10.3.6}$$

Eq. (10.3.5) defines the spin, or angular momentum, vector field $I\boldsymbol{\omega}$ whose components with respect to the space and body axes are given by:

$$\left.\begin{aligned} I\omega_i &= M_i + \mu' I B_i \\ I\omega_i' &= M_i + \mu' I B_i', \end{aligned}\right\} \tag{10.3.7}$$

where

$$\left.\begin{aligned} M_1 &= -\cos\beta\,\partial S/\partial\alpha + \sin\beta\cot\alpha\,\partial S/\partial\beta - \sin\beta\operatorname{cosec}\alpha\,\partial S/\partial\gamma \\ M_2 &= \quad\sin\beta\,\partial S/\partial\alpha + \cos\beta\cot\alpha\,\partial S/\partial\beta - \cos\beta\operatorname{cosec}\alpha\,\partial S/\partial\gamma \\ M_3 &= -\partial S/\partial\beta \\ M'_1 &= -\cos\gamma\,\partial S/\partial\alpha - \sin\gamma\operatorname{cosec}\alpha\,\partial S/\partial\beta + \sin\gamma\cot\alpha\,\partial S/\partial\gamma \\ M'_2 &= -\sin\gamma\,\partial S/\partial\alpha + \cos\gamma\operatorname{cosec}\alpha\,\partial S/\partial\beta - \cos\gamma\cot\alpha\,\partial S/\partial\gamma \\ M'_3 &= -\partial S/\partial\gamma. \end{aligned}\right\} \quad (10.3.8)$$

Eqs. (10.3.4) and (10.3.7) with (10.3.8) will be recognized as nothing more than the corresponding classical expressions (10.1.6) and (10.1.9) with p_i and p_{α_i} replaced by $\partial S/\partial x_i$ and $\partial S/\partial \alpha_i$ respectively. The difference here, of course, is that the function $S(\mathbf{x}, \boldsymbol{\alpha}, t)$ satisfies the Hamilton–Jacobi equation (10.3.2), with quantum potential contributions. The energy field of the ensemble, given by

$$E(\mathbf{x}, \boldsymbol{\alpha}, t) = -\partial S/\partial t, \quad (10.3.9)$$

is thus composed of kinetic energy terms of translation and rotation and potential energy terms arising from external potentials and the quantum potential derived from the quantum state.

We have transcribed the Schrödinger equation into (10.3.2) and (10.3.3). As usual, we must supplement the latter with conditions on R and S corresponding to those imposed on ψ. Notice that (10.3.3) preserves the normalization (10.2.8) of ψ.

Chasle's theorem (that an arbitrary motion of a rigid body is composed of a translation together with a rotation about a line fixed in the body) remains valid in quantum mechanics since its proof depends only on the formal properties of an orthogonal rotation matrix. Notice though that these two facets of the motion are coupled in a fundamental way in the above equations. However, if the wavefunction factorizes in the form

$$\psi = \psi_1(\mathbf{x}, t)\psi_2(\boldsymbol{\alpha}, t) \quad (10.3.10)$$

we obtain a separation of the motion into independent translational and rotational components. For then, writing $\psi_j = R_j\,\mathrm{e}^{\mathrm{i}S_j/\hbar}$, $j = 1, 2$, we obtain from (10.3.6)

$$\left.\begin{aligned} Q &= -\frac{\hbar^2}{2m}\frac{\nabla^2 R_1}{R_1} = Q(\mathbf{x}, t) \\ Q_s &= -\frac{\hbar^2}{2I}\frac{\hat{\boldsymbol{\lambda}}^2 R_2}{R_2} = Q_s(\boldsymbol{\alpha}, t) \end{aligned}\right\} \quad (10.3.11)$$

and from (10.3.4) and (10.3.5)

$$\left.\begin{aligned}\mathbf{v} &= (1/m)(\nabla S_1 - (e/c)\mathbf{A}) = \mathbf{v}(\mathbf{x}, t)\\ \boldsymbol{\omega} &= (1/I)(\hat{\boldsymbol{\lambda}} S_2 + \mu' I\mathbf{B})\end{aligned}\right\} \tag{10.3.12}$$

the latter being a function of only $\boldsymbol{\alpha}$, if $\mathbf{B}$ = constant. The condition (10.3.10) thus represents a state where the internal and external motions are decoupled.

Returning to the general case, both the translational and the angular velocities (10.3.4) and (10.3.5) depend on all six coordinates and the time. In order to find the trajectories and the evolution of the orientation we have to solve for $\mathbf{x}(t)$ from

$$\mathrm{d}\mathbf{x}/\mathrm{d}t = \mathbf{v}(\mathbf{x}, \boldsymbol{\alpha}, t)|_{\mathbf{x}=\mathbf{x}(t), \boldsymbol{\alpha}=\boldsymbol{\alpha}(t)} \tag{10.3.13}$$

and for $\boldsymbol{\alpha}(t)$ by expressing $\boldsymbol{\omega}$ in terms of $\mathrm{d}\boldsymbol{\alpha}/\mathrm{d}t$ from (10.1.8) with p_{α_i} replaced by $\partial S/\partial\alpha_i$:

$$\left.\begin{aligned}I\dot{\alpha} &= \partial S/\partial\alpha - \mu' I\mathbf{B}\cdot(\cos\beta, -\sin\beta, 0)\\ I\dot{\beta} &= \operatorname{cosec}^2\alpha[\partial S/\partial\beta - \cos\alpha\,\partial S/\partial\gamma + \mu' I\mathbf{B}\cdot\sin\alpha(\cos\alpha\sin\beta,\\ &\qquad \cos\alpha\cos\beta, -\sin\alpha)]\\ I\dot{\gamma} &= \operatorname{cosec}^2\alpha[\partial S/\partial\gamma - \cos\alpha\,\partial S/\partial\beta - \mu' I\mathbf{B}\cdot(\sin\alpha\sin\beta,\\ &\qquad \sin\alpha\cos\beta, 0)]\end{aligned}\right\} \tag{10.3.14}$$

where the right hand sides are evaluated along the system trajectory. The system (10.3.13) and (10.3.14) constitutes a set of six simultaneous first-order differential equations whose solution uniquely follows once we specify the initial wavefunction $\psi_0(\mathbf{x}, \boldsymbol{\alpha})$, solve the Schrödinger equation to find the function S, and specify the initial position $\mathbf{x}_0 = \mathbf{x}(0)$ and orientation $\boldsymbol{\alpha}_0 = \boldsymbol{\alpha}(0)$. We have then

$$\left.\begin{aligned}\mathbf{x} &= \mathbf{x}(t, \mathbf{x}_0, \boldsymbol{\alpha}_0),\\ \boldsymbol{\alpha} &= \boldsymbol{\alpha}(t, \mathbf{x}_0, \boldsymbol{\alpha}_0).\end{aligned}\right\} \tag{10.3.15}$$

Recall that in the Pauli theory the evolution of the Euler angles along a trajectory is uniquely fixed by specifying the initial wavefunction and the initial position. Here we need to know $\boldsymbol{\alpha}_0$ also.

As in the Pauli theory, the spin-orbit coupling just described implies that the trajectories will differ from those calculated from the spinless Schrödinger equation, even in the free case (10.2.11). To see this we derive the equation of motion of the centre of mass by taking the spatial gradient of (10.3.2). The result is

$$m\frac{\mathrm{d}\mathbf{v}}{\mathrm{d}t} = -\nabla(Q + Q_s + V) + \mathbf{F} - \mu' I\omega_i\nabla B_i, \tag{10.3.16}$$

where $\mathbf{F}$ is the Lorentz force and

$$\mathrm{d}/\mathrm{d}t = \partial/\partial t + \dot{\mathbf{x}}\cdot\nabla + \dot{\alpha}_i\,\partial/\partial\alpha_i. \tag{10.3.17}$$

Here we have used the easily proved result

$$\boldsymbol{\omega}\cdot\hat{\boldsymbol{\lambda}} = \dot{\alpha}_i\,\partial/\partial\alpha_i = \dot{\alpha}\frac{\partial}{\partial\alpha} + \dot{\beta}\frac{\partial}{\partial\beta} + \dot{\gamma}\frac{\partial}{\partial\gamma}, \tag{10.3.18}$$

which follows from the definitions (10.1.3), (10.2.3) and (10.2.9).

If the particle is free a spin-dependent quantum force $-\nabla(Q + Q_s)$ acts. The potential Q_s is expected to contribute to this force when the wavefunction does not factorize in the form (10.3.10).

Similarly, a quantum torque acts on the spin vector causing it to precess in addition to any precession arising from external fields. The precession of the spin vector (10.3.7) may be found by applying the operator $\hat{\boldsymbol{\lambda}}$ to the Hamilton–Jacobi equation (10.3.2). After some rearrangements and use of the commutation relations

$$[\hat{\lambda}_i, \hat{\lambda}_j] = -\varepsilon_{ijk}\hat{\lambda}_k, \tag{10.3.19}$$

which follow from (10.2.4), we find

$$I\,\mathrm{d}\boldsymbol{\omega}/\mathrm{d}t = -\hat{\boldsymbol{\lambda}}(Q + Q_s) - \mu' I\boldsymbol{\omega}\times\mathbf{B} + \mu' I\,\mathrm{d}\mathbf{B}/\mathrm{d}t, \tag{10.3.20}$$

where $\mathrm{d}/\mathrm{d}t$ is given by (10.3.17). A quantum torque

$$\mathbf{T} = -\hat{\boldsymbol{\lambda}}(Q + Q_s) \tag{10.3.21}$$

is present which rotates the spin vector even when the terms involving $\mathbf{B}$, which are classically the sole agents of precession, are absent.

The discussion so far in this section has made no reference to the magnitude of the spin of the particle. Indeed one of the advantages of the theory is that it is applicable to systems of arbitrary spin. Now we shall consider spin $\frac{1}{2}$. We can be more specific about the form of $\mathbf{v}$ and $\boldsymbol{\omega}$ in this case, since the functional dependence of the phase function S on the Euler angles is fixed in the wavefunction (10.2.25). Later (§10.4) we shall see clearly how $\mathbf{v}$ and $\boldsymbol{\omega}$ differ from their Pauli counterparts. As a first step in that calculation let us write $\psi_\pm = R_\pm\,\mathrm{e}^{\mathrm{i}S_\pm/\hbar}$ in(10.2.25). With $\psi = R\,\mathrm{e}^{\mathrm{i}S/\hbar}$ and substituting the expressions (10.2.24) for $u_\pm$, we obtain

$$\tan\left(\frac{S}{\hbar}\right) = \frac{R_+\cos\frac{\alpha}{2}\sin\left[-\frac{(\gamma+\beta)}{2} + \frac{S_+}{\hbar}\right] - R_-\sin\frac{\alpha}{2}\cos\left[\frac{\beta-\gamma}{2} + \frac{S_-}{\hbar}\right]}{R_+\cos\frac{\alpha}{2}\cos\left[-\frac{(\gamma+\beta)}{2} + \frac{S_+}{\hbar}\right] + R_-\sin\frac{\alpha}{2}\sin\left[\frac{\beta-\gamma}{2} + \frac{S_-}{\hbar}\right]}, \tag{10.3.22}$$

$$R^2 = (8\pi^2)^{-1}\left[\cos^2\frac{\alpha}{2}R_+^2 + \sin^2\frac{\alpha}{2}R_-^2 + R_+R_- \sin\alpha\sin\left(\beta + \frac{S_- - S_+}{\hbar}\right)\right]. \tag{10.3.23}$$

By simple differentiation and neglecting the magnetic field terms, the rate of change of the Euler angles, (10.3.14), is given by

$$\left.\begin{aligned} I\dot{\alpha} &= -\hbar(4\pi R)^{-2}R_+R_- \cos[\beta + (S_- - S_+)/\hbar] \\ I\dot{\beta} &= (\hbar/2)\operatorname{cosec}\alpha(4\pi R)^{-2}\{(R_-^2 - R_+^2)\sin\alpha \\ &\qquad + 2R_+R_- \cos\alpha\sin[\beta + (S_- - S_+)/\hbar]\} \\ I\dot{\gamma} &= -(\hbar/2)\operatorname{cosec}\alpha(4\pi R)^{-2}\{(R_+^2 + R_-^2)\sin\alpha \\ &\qquad + 2R_+R_- \sin[\beta + (S_- - S_+)/\hbar]\}. \end{aligned}\right\} \tag{10.3.24}$$

We also easily find the components of the angular momentum with respect to the space axes by substituting (10.3.8) into (10.3.7), with

$$\left.\begin{aligned} M_1 &= \hbar(4\pi R)^{-2}\{R_+R_- \cos[(S_- - S_+)/\hbar] + \tfrac{1}{2}(R_+^2 + R_-^2)\sin\beta\cos\alpha\}, \\ M_2 &= \hbar(4\pi R)^{-2}\{R_+R_- \sin[(S_- - S_+)/\hbar] + \tfrac{1}{2}(R_+^2 + R_-^2)\cos\beta\cos\alpha\}, \\ M_3 &= \hbar(4\pi R)^{-2}\left(R_+^2\cos^2\frac{\alpha}{2} - R_-^2\sin^2\frac{\alpha}{2}\right). \end{aligned}\right\} \tag{10.3.25}$$

Finally, a straightforward calculation yields the following expression for the spin $\frac{1}{2}$ velocity field:

$$\begin{aligned} \mathbf{v} = \left(\frac{1}{2m}\right)\left(\frac{1}{2\pi R}\right)^2 \Bigg\{ & R_+^2\cos^2\frac{\alpha}{2}\nabla S_+ + R_-^2\sin^2\frac{\alpha}{2}\nabla S_- \\ & + \tfrac{1}{2}\sin\alpha\sin\left[\beta + \frac{(S_- - S_+)}{\hbar}\right]R_+R_-\nabla(S_+ + S_-) \\ & + \left(\frac{\hbar}{2}\right)\sin\alpha\cos\left[\beta + \frac{(S_- - S_+)}{\hbar}\right] \\ & \times (R_-\nabla R_+ - R_+\nabla R_-)\Bigg\} - \left(\frac{e}{mc}\right)\mathbf{A}. \end{aligned} \tag{10.3.26}$$

It will be noted that the angular momentum (10.3.25) and the velocity (10.3.26) are independent of the angle γ. This is a consequence of restricting ourselves to the $b = \frac{1}{2}$ eigenstate of $\hat{M}_3'$.

Notice that the particle functions have 2π periodicity but as explained in Chap. 9 they are sensitive to the 4π-symmetry of the wavefunction because of the linear superposition principle.

As remarked also in Chap. 9, the novelty of spin $\frac{1}{2}$ lies not in the structure of the material object described by the theory (a rigid top whose orientational configuration space is $SO(3)$) but in its motion (guided by a wave whose spin configuration space is $SU(2)$). The spin $\frac{1}{2}$ rotator thus has a well-defined classical analogue and as usual one approaches it when the quantum potential, force and torque are relatively negligible.

10.3.2 Many bodies

The generalization of this theory to a many-spin $\frac{1}{2}$ system is straightforward and offers no surprises. In the two-body case one has a set of cm coordinates $\mathbf{x}_\mu$ and a set of Euler angles $\boldsymbol{\alpha}_\mu$ for each particle, $\mu = 1, 2$. The latter specify the orientation of the two sets of body axes with respect to a set of space axes. For fixed eigenvalues of the z-components of the angular momentum operators with respect to both sets of body axes, the wavefunction may be expanded as

$$\psi(\mathbf{x}_1, \mathbf{x}_2, \boldsymbol{\alpha}_1, \boldsymbol{\alpha}_2, t) = \sum_{a,b=1}^{2} \psi^{ab}(\mathbf{x}_1, \mathbf{x}_2, t) u_{1a}(\boldsymbol{\alpha}_1) u_{2b}(\boldsymbol{\alpha}_2), \quad (10.3.27)$$

where u_{1a} and u_{2a} are sets of functions (10.2.24). For a free system this satisfies the equation

$$i\hbar \frac{\partial \psi}{\partial t} = \left(-\frac{\hbar^2}{2m_1} \nabla_1^2 - \frac{\hbar^2}{2m_2} \nabla_2^2 + \frac{1}{2I_1} \hat{\mathbf{M}}_1^2 + \frac{1}{2I_2} \hat{\mathbf{M}}_2^2 \right) \psi, \quad (10.3.28)$$

where the particles have masses m_1, m_2 and moments of inertia I_1, I_2 respectively. Writing $\psi = R\, e^{iS/\hbar}$ one may then develop a causal interpretation in strict analogy to the one-body case, with formulae such as $\mathbf{v}_\mu = (1/m_\mu)\nabla_\mu S$, $\mu = 1, 2$, bearing in mind that the phase, the quantum potential and the quantum torque depend on the parameters associated with both particles. In addition to the velocities and spin vectors, one may also define in analogy to the Pauli case a spin correlation tensor:

$$M_{ij} = \mathrm{Re}(\psi^* \hat{M}_{1i} \hat{M}_{2j} \psi / \psi^* \psi). \quad (10.3.29)$$

Applications of this theory are made in §§10.6 and 11.4.

10.4 Relation with the Pauli theory

10.4.1 Wave equation

So far we have worked in the angular coordinate representation in which the Euler angle operators are diagonal and represented by real numbers. To see the connection with the quantum theory of spin $\frac{1}{2}$ as expressed through the Pauli equation we must transform to the angular momentum representation in which the angular momentum operators $\hat{\mathbf{M}}^2$, $\hat{M}_3$, $\hat{M}'_3$ are diagonal, and perform a suitable limit to a point particle. The causal interpretation of the Pauli theory studied in Chap. 9 arises by averaging the various configuration space fields in the extended theory over the internal Euler angle coordinates, and taking the limit. This formation of mean values just expresses in a different representation the method used in Chap. 9 in which we generated the probability density, velocity and spin vector fields by summing over the internal spin indices. We shall now look at the connection between the two theories in more detail.

The first point to note which will prove of some significance in §10.6.2 is that the restriction to $s = \frac{1}{2}$ eigenstates entails a corresponding restriction on the angular momentum operators. Thus, when acting on spin $\frac{1}{2}$ states (an arbitrary superposition of the four basis functions (10.2.24)), it is easy to show that the angular momentum operators satisfy the defining relations of a Clifford algebra:

$$\hat{M}_i\hat{M}_j + \hat{M}_j\hat{M}_i = 2(\hbar/2)^2\delta_{ij}, \tag{10.4.1}$$

$$\hat{M}'_i\hat{M}'_j + \hat{M}'_j\hat{M}'_i = 2(\hbar/2)^2\delta_{ij}, \tag{10.4.2}$$

in addition to the commutation relations (10.2.4). It is interesting that anticommutation relations between differential operators follow as a consequence of the canonical quantization procedure which replaces PBs by commutation relations. Conversely, if we impose the Clifford algebra relations on $\hat{M}_i$,$\hat{M}'_i$ we are automatically restricted to the spin $\frac{1}{2}$ subspace. We see that the physical content of an anticommutative algebra is expressed in the dynamics of matter by the quantum potential. When the latter is negligible the motion is not sensitive to the algebra, although this structure remains well-defined.

Given (10.4.1), it is not surprising to learn that the matrix representations of the operators $\hat{M}_i$ with respect to the basis states (10.2.24) are ($\hbar/2$ times) the Pauli matrices in the representation (9.2.6):

$$\begin{aligned}(\hat{M}_i)^{a'}{}_a{}^{b'}{}_b &= \int D^{1/2*}_{a'b'}(\alpha, \beta, \gamma)\hat{M}_i D^{1/2}_{ab}(\alpha, \beta, \gamma)\,\mathrm{d}\Omega \\ &= (\hbar/2)\sigma_i{}^{a'}{}_a\delta^{b'}{}_b.\end{aligned} \tag{10.4.3}$$

This result has been derived using the raising and lowering relations (10.2.15) and the normalization condition (10.2.19). The operators $\hat{M}'_i$ satisfy the anomalous commutation relations (10.2.4) and have the representation $(\hat{M}'_i)^{a'}{}_a{}^{b'}{}_b = (\hbar/2)\sigma_i^{\mathrm{T}b'}{}_b\delta^{a'}{}_a$. The Pauli matrices of course satisfy up to a constant the Clifford algebra (10.4.1) and the commutation relations (10.2.4), and likewise for the transposed Pauli matrices.

We now consider the passage to the Pauli equation by first of all taking the matrix representation of the $s = \frac{1}{2}$ Schrödinger equation (10.2.6) with respect to the basis functions (10.2.24). Expanding the operators, (10.2.6) becomes

$$\sum_{a,b} i\hbar\frac{\partial}{\partial t}(\psi^{ab}D^{1/2}_{ab}) = \sum_{a,b} D^{1/2}_{ab}\left\{\left(\frac{1}{2m}\right)\left[\hat{\mathbf{p}} - \left(\frac{e}{c}\right)\mathbf{A}\right]^2 + \frac{3\hbar^2}{8I} + \tfrac{1}{2}\mu'^2 I\mathbf{B}^2 + eA_0 + V\right\}$$
$$\times\,\psi^{ab} + \sum_{a,b}\mu' B_i(\hat{M}_i D^{1/2}_{ab})\psi^{ab}, \tag{10.4.4}$$

where we have substituted (10.2.13) with $s = \frac{1}{2}$. The constant $3\hbar^2/8I$ may be absorbed into an overall phase factor and will be ignored. Multiplying both sides of (10.4.4) by $D^{1/2*}_{a'b'}$ and using (10.2.19) and (10.2.4), we obtain

$$i\hbar\frac{\partial\psi^{ab}}{\partial t} = \left\{\left(\frac{1}{2m}\right)\left[\hat{\mathbf{p}} - \left(\frac{e}{c}\right)\mathbf{A}\right]^2 + \tfrac{1}{2}\mu'^2 I\mathbf{B}^2 + eA_0 + V\right\}\psi^{ab}$$
$$+ \sum_{a'}\left(\frac{\hbar}{2}\right)\mu' B_i\sigma_i{}^a{}_{a'}\psi^{a'b}. \tag{10.4.5}$$

To obtain the Pauli equation from (10.4.5) it simply remains to take the limit of a point particle ($I \to 0$) in such a way that the mass, charge and magnetic moment remain finite. Writing $\mu' = 2\mu/\hbar$, we then find

$$i\hbar\frac{\partial\psi^{ab}}{\partial t} = \sum_{a'}\left(\left\{\left(\frac{1}{2m}\right)\left[\hat{\mathbf{p}} - \left(\frac{e}{c}\right)\mathbf{A}\right]^2 + eA_0 + V\right\}\delta^a{}_{a'} + \mu(\mathbf{B}\cdot\boldsymbol{\sigma})^a{}_{a'}\right)\psi^{a'b}, \tag{10.4.6}$$

which yields two copies of the Pauli equation (9.3.5) corresponding to the two values of $b = \pm\frac{1}{2}$. Here

$$\psi^{ab}(\mathbf{x}, t) = \int D^{1/2*}_{ab}(\alpha, \beta, \gamma)\psi(\mathbf{x}, \boldsymbol{\alpha}, t)\,\mathrm{d}\Omega, \tag{10.4.7}$$

which for fixed b gives the components of a Pauli contravariant spinor field.

If we fix b and substitute the expression (10.2.23) for ψ into the normalization condition (10.2.8), we find using (10.2.19) the usual normalization for a Pauli spinor:

$$\sum_{a=1}^{2} \int |\psi^a|^2 \, d^3x = 1. \tag{10.4.8}$$

This establishes the formal relations between the Pauli and rigid rotator theories. The difference between the two is that in the latter we start by assuming the radius of the body is finite and this extra degree of freedom implies that a more detailed treatment of a spinning body motion is possible.

In fact, what we have shown in (10.4.5) is that whenever the term $\frac{1}{2}\mu'^2 I \mathbf{B}^2$ can be neglected (for whatever reason), the space and time dependent coefficients $\psi^{ab}(\mathbf{x}, t)$ necessarily satisfy the Pauli equation. This result is of considerable help in solving rigid rotator problems since it means that in many problems of interest one may take over well-established results in the Pauli theory.

10.4.2 Particle equations

Let us now look at the relation between the two theories as viewed in the causal interpretation. To begin with, we define the average of a quantity $A(\mathbf{x}, \alpha, \beta, \gamma, t)$ over the internal coordinates to be

$$\langle A \rangle = \int R^2 A \, d\Omega \Big/ \int R^2 \, d\Omega, \qquad d\Omega = \sin\alpha \, d\alpha \, d\beta \, d\gamma, \tag{10.4.9}$$

where $R^2 = |\psi|^2$. When ψ is given by (10.2.23) we have

$$\int R^2 \, d\Omega = \sum_{a,b} |\psi^{ab}|^2. \tag{10.4.10}$$

In the case that b is fixed, this is equal to

$$R'^2 \equiv \sum_a |\psi^a|^2, \tag{10.4.11}$$

i.e., the Pauli probability density. Considering only this case, (10.4.9) becomes

$$\langle A \rangle = R'^{-2} \int R^2 A \, d\Omega, \qquad \text{eigenstate of } \hat{M}'_3. \tag{10.4.12}$$

Applying formula (10.4.12) to the spin vector (10.3.5) yields

$$\langle I\omega_i \rangle = \sum_{a,a'} (\hbar/2R'^2)\psi^{a'*}\sigma_{ia'a}\psi^a + \mu' I B_i, \tag{10.4.13}$$

which, when $I = 0$, is the Pauli spin vector s_i, (9.3.3). A similar operation performed on the velocity field (10.3.4) gives

$$\langle \mathbf{v} \rangle = (\hbar/2miR'^2) \sum_a \psi^{a*} \overleftrightarrow{\nabla} \psi^a - (e/mc)\mathbf{A}, \tag{10.4.14}$$

which will be recognized as the Pauli velocity field (9.3.8).

To compare these functions in detail, we substitute into the expressions (10.3.22)–(10.3.26) the explicit representation (9.2.25) of the Pauli spinor components in terms of the Euler angles:

$$\left.\begin{aligned} R_+ = R' \cos(\theta/2), \qquad R_- = R' \sin(\theta/2) \\ S_+/\hbar = (\chi + \phi)/2, \qquad S_-/\hbar = (\chi - \phi)/2 + \pi/2 \end{aligned}\right\} \tag{10.4.15}$$

with R', θ, ϕ and χ functions of $\mathbf{x}$ and t. This gives

$$\psi(\mathbf{x}, \boldsymbol{\alpha}, t) = (2\sqrt{2}\,\pi)^{-1} R' \,\mathrm{e}^{\mathrm{i}(\chi - \gamma)/2} \left[\cos\frac{\theta}{2} \cos\frac{\alpha}{2} \,\mathrm{e}^{\mathrm{i}(\phi - \beta)/2} + \sin\frac{\theta}{2} \sin\frac{\alpha}{2} \,\mathrm{e}^{-\mathrm{i}(\phi - \beta)/2} \right] \tag{10.4.16}$$

$$\tan\left(\frac{S}{\hbar}\right) = \frac{\sin\left(\dfrac{\chi + \phi - \gamma - \beta}{2}\right) - \tan\dfrac{\alpha}{2}\tan\dfrac{\theta}{2}\sin\left(\dfrac{\phi - \chi + \gamma - \beta}{2}\right)}{\cos\left(\dfrac{\chi + \phi - \gamma - \beta}{2}\right) + \tan\dfrac{\alpha}{2}\tan\dfrac{\theta}{2}\cos\left(\dfrac{\phi - \chi + \gamma - \beta}{2}\right)} \tag{10.4.17}$$

$$\left.\begin{aligned} M_1 &= (\hbar/4q)(\sin\theta \sin\phi + \sin\alpha \sin\beta) \\ M_2 &= (\hbar/4q)(\sin\theta \cos\phi + \sin\alpha \cos\beta) \\ M_3 &= (\hbar/4q)(\cos\theta + \cos\alpha) \end{aligned}\right\} \tag{10.4.18}$$

and

$$\mathbf{v} = \left(\frac{\hbar}{2m}\right)\left[\nabla\chi + \frac{(\cos\theta + \cos\alpha)\nabla\phi}{2q} - \frac{\sin\alpha \sin(\phi - \beta)\nabla\theta}{2q}\right] - \left(\frac{e}{mc}\right)\mathbf{A}, \tag{10.4.19}$$

where

$$q = 2\left(\frac{2\pi R}{R'}\right)^2 = \cos^2\left(\frac{\alpha - \theta}{2}\right) - \sin\alpha \sin\theta \sin^2\left(\frac{\phi - \beta}{2}\right). \tag{10.4.20}$$

Eqs. (10.4.18) and (10.4.19) should be compared with the corresponding Pauli expressions (9.3.3) and (9.3.8) for the spin vector and velocity fields. In particular it is clear that the magnitude of the spin vector is variable and not restricted to the Pauli value of $\hbar/2$. In the case $\mu' I \mathbf{B} = 0$ we have from (10.4.18):

$$\mathbf{M}^2 = \left(\frac{\hbar}{4q}\right)^2 \left[\sin^2\theta + \sin^2\alpha + 4\cos^2\left(\frac{\theta+\alpha}{2}\right)\cos^2\left(\frac{\theta-\alpha}{2}\right) + 2\sin\theta\sin\alpha\cos(\phi-\beta)\right]. \quad (10.4.21)$$

Notice also that the velocity (10.4.19) of the extended theory contains a contribution from $\nabla\theta$.

Some remarks are in order at this stage on the meaning of the formulae just derived, so that it is clear how the physical models of the Pauli and the extended theories are related. In the latter we suppose that the coordinates $\mathbf{x}$ label the centre of mass of a rigid sphere, whose orientation is specified by the (Euler) angles α, β, γ made by a set of axes $\mathbf{e}'_i$ fixed in the body with respect to a set of space axes $\mathbf{e}_i$. The axis through the centre of mass about which the body is rotating is defined by the angular momentum vector with components $I\omega_i$ with respect to $\mathbf{e}_i$, or in the free case by (10.4.18). It will be observed that this latter vector is proportional to the sum of the Pauli spin vector and ($\hbar/2$ times) the third moving axis $\mathbf{e}'_3$:

$$M_i = (1/2q)[s_i + (\hbar/2)(\mathbf{e}'_3)_i], \quad (10.4.22)$$

where $\mathbf{s} = \mathbf{s}(\mathbf{x}, t)$ and $\mathbf{e}'_3 = \mathbf{e}'_3(\alpha, \beta, \gamma)$. In other words, the direction in space of the spin vector is given by the 'external' orientation vector $\mathbf{s}$ fixed to each space point together with the 'internal' orientation vector $\mathbf{e}'_3$. In the last chapter we assumed that $\mathbf{s}$ may be associated with one of the principal axes of the body. Here we see that that model is an idealization which is only valid if we neglect the internal motions of the body, i.e., ignore its extension. If we consider an ensemble of spinning bodies with density R^2 then at each space point $\mathbf{x}$ the Pauli spin vector is just the marginal spin vector which results from an average over all possible internal orientations. As we have said, the frame $\mathbf{e}'_i$ specifies the orientation of the body. The vector $\mathbf{s}$ rotates with respect to the body axes and, as a consequence, so does $\mathbf{M}$.

The connection between the two theories for many-body systems is established in an analogous and obvious way. For example, the mean of the spin correlation tensor (10.3.29) is just the corresponding Pauli tensor (cf. §9.6):

$$\langle M_{ij}\rangle = (\hbar/2)s_{ij}. \quad (10.4.23)$$

10.5 The meaning of 'spin up'

10.5.1 Free rotator

To examine what in the extended theory of spin corresponds to a free system having 'spin up', we assume that the centre of mass is at rest so that the wavefunction factorizes as in (10.3.10) and the **x**-dependent part may be ignored. There is therefore no spin–orbit coupling and the state will be the analogue of an eigenstate of the spin operator $(\hbar/2)\boldsymbol{\sigma}\cdot\mathbf{n}$ in some direction **n** in the Pauli theory. Consider the following stationary state solution to (10.2.6):

$$\psi = u_+ \mathrm{e}^{-\mathrm{i}Et/\hbar} = (2\sqrt{2}\,\pi)^{-1}\cos(\alpha/2)\,\mathrm{e}^{-\mathrm{i}[(\beta+\gamma)/2 + Et/\hbar]}, \tag{10.5.1}$$

which is a simultaneous eigenstate of $\hat{M}_3$ and $\hat{M}'_3$ corresponding to eigenvalues $\hbar a = \hbar b = \hbar/2$, with $E = 3\hbar^2/8I$. Writing $\psi = R\,\mathrm{e}^{\mathrm{i}S/\hbar}$ we have

$$\left.\begin{aligned} R &= (2\sqrt{2}\,\pi)^{-1}\cos(\alpha/2) \geqslant 0, \qquad 0 \leqslant \alpha \leqslant \pi \\ S/\hbar &= -\tfrac{1}{2}(\beta+\gamma) - Et/\hbar. \end{aligned}\right\} \tag{10.5.2}$$

The wavefunction has a node at $\alpha = \pi$ where the phase is undefined and hence the body cannot adopt such an orientation.

In the present instance $\psi_+ = \mathrm{e}^{-\mathrm{i}Et/\hbar}$, $\psi_- = 0$ in (10.2.25) so the corresponding Pauli spin vector is

$$\mathbf{s} = (\hbar/2)(0, 0, 1) \tag{10.5.3}$$

with respect to the space axes, i.e., $\mathbf{s} = (\hbar/2)\mathbf{e}_3$.

Substituting S into the guidance law (10.3.14) we find

$$\alpha = \text{constant}(\neq \pi), \qquad \beta = -ft + \beta_0, \qquad \gamma = -ft + \gamma_0, \tag{10.5.4}$$

where

$$f = \hbar/4I\cos^2(\alpha/2) = \mathbf{M}^2/\hbar I \tag{10.5.5}$$

is an angular frequency and β_0, γ_0 are constants which we choose to be zero. Substituting (10.5.4) into (10.3.8) gives for the angular momentum components

$$\left.\begin{aligned} M_1 &= (\hbar/2)\tan(\alpha/2)\sin\beta = -(\hbar/2)\tan(\alpha/2)\sin ft \\ M_2 &= (\hbar/2)\tan(\alpha/2)\cos\beta = (\hbar/2)\tan(\alpha/2)\cos ft \\ M_3 &= \hbar/2 \end{aligned}\right\} \tag{10.5.6}$$

$$\left.\begin{aligned} M'_1 &= (\hbar/2)\tan(\alpha/2)\sin\gamma = -(\hbar/2)\tan(\alpha/2)\sin ft \\ M'_2 &= -(\hbar/2)\tan(\alpha/2)\cos\gamma = -(\hbar/2)\tan(\alpha/2)\cos ft \\ M'_3 &= \hbar/2 \end{aligned}\right\} \tag{10.5.7}$$

and for the total angular momentum

$$|\mathbf{M}| = \hbar/2\cos(\alpha/2). \tag{10.5.8}$$

The body thus rotates at a uniform frequency $\hbar/2I\cos(\alpha/2)$ about **M**.

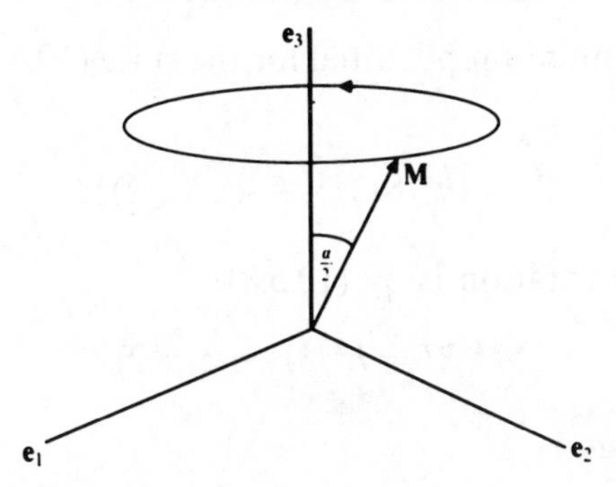

Fig. 10.1 'Spin up' means that **M** precesses in the domain $z > 0$.

Clearly, the magnitude of the spin angular momentum is not restricted to $\hbar/2$. Rather, this fixes a lower bound (when $\alpha = 0$).

The motion of the spin vector with respect to the space axes implied by (10.5.6) is depicted in Fig. 10.1. The origin of $\mathbf{e}_i$ is placed at the centre of mass. The spin vector makes an angle $\alpha/2$ with the $\mathbf{e}_3$-axis and precesses uniformly anticlockwise about this axis with frequency f on the surface of a cone of height $\hbar/2$ and radius $(\hbar/2)\tan(\alpha/2)$. If thus precesses about the Pauli spin vector (10.5.3). The motion as viewed in the body frame $\mathbf{e}'_i$ is by (10.5.7) the same except that now the precession is clockwise. 'Spin up' thus means that **M** is confined to the spatial domain $z > 0$.

The motion of the body may also be described by the time development of the orthonormal body axes $\mathbf{e}'_i(t)$ with respect to $\mathbf{e}_i$. Comparing with (9.2.22), (10.5.4) implies

$$\left.\begin{aligned} \mathbf{e}'_1 &= (\cos^2 ft - \cos\alpha\sin^2 ft, \tfrac{1}{2}\sin 2ft(1+\cos\alpha), -\sin ft\sin\alpha) \\ \mathbf{e}'_2 &= (-\tfrac{1}{2}\sin 2ft\ (1+\cos\alpha), \cos^2 ft\cos\alpha - \sin^2 ft, -\cos ft\sin\alpha) \\ \mathbf{e}'_3 &= (-\sin\alpha\sin ft, \sin\alpha\cos ft, \cos\alpha). \end{aligned}\right\} \quad (10.5.9)$$

We see that $\mathbf{e}'_3(t)$ precesses uniformly anticlockwise about $\mathbf{e}_3$ with frequency f on the surface of a cone of angle α, height $|\cos\alpha|$ and radius $|\sin\alpha|$. With respect to the space axes, the vector **M** and $\mathbf{e}'_3$ both precess at the same rate about the Pauli spin vector, but at angles $\alpha/2$ and α respectively. The component M'_3 of **M** with respect to $\mathbf{e}'_3$ is, of course, constant $(=\hbar/2)$.

Averaging over the internal orientation angles yields as expected the corresponding Pauli values (10.5.3):

$$\left.\begin{aligned} \langle M_i \rangle &= (\hbar/2)(0, 0, 1) \\ \langle M'_i \rangle &= (\hbar/2)(0, 0, 1). \end{aligned}\right\} \quad (10.5.10)$$

From (10.3.6), the quantum potential for the state (10.5.1) is constant along the orbit:

$$Q_s = (\hbar^2/8I)[3 - \sec^2(\alpha/2)]. \tag{10.5.11}$$

The kinetic energy of rotation is by (10.5.8)

$$\mathbf{M}^2/2I = \hbar^2/8I \cos^2(\alpha/2) \tag{10.5.12}$$

and so the total energy is

$$\mathbf{M}^2/2I + Q_s = 3\hbar^2/8I, \tag{10.5.13}$$

as expected from (10.5.2) ($E = -\partial S/\partial t$). We may also easily calculate the quantum torque (10.3.21):

$$\left.\begin{aligned} \mathbf{T} &= (\hbar^2/8I)\tan(\alpha/2)\sec^2(\alpha/2)(-\cos\beta, \sin\beta, 0)|_{\boldsymbol{\alpha}=\boldsymbol{\alpha}(t)} \\ &= -(\hbar/2)f\tan(\alpha/2)(\cos ft, \sin ft, 0) \end{aligned}\right\} \tag{10.5.14}$$

with respect to $\mathbf{e}_i$, where we have substituted (10.5.4) and (10.5.5). This result may also be recovered by differentiating M_i in (10.5.6) with respect to t. The torque $\mathbf{T}$ is thus a vector lying in the $\mathbf{e}_1$–$\mathbf{e}_2$ plane rotating uniformly clockwise with frequency f, about a circle of radius $(\hbar/2)f\tan(\alpha/2)$.

The classical analogue of this system is the torque-free motion of a rigid body with one point fixed. In the case of a sphere for which all three principal moments of inertia coincide the body is completely at rest (Goldstein, 1980). In the quantum case the body rotates due to the applied quantum torque. However, in both the classical and quantum cases the total angular momentum and energy are constants of the motion.

In an analogous way, the function

$$\psi = u_- e^{-iEt/\hbar} = -i(2\sqrt{2}\pi)^{-1}\sin(\alpha/2)\, e^{i[(\beta-\gamma)/2 - Et/\hbar]} \tag{10.5.15}$$

with $a = -\frac{1}{2}$, $b = \frac{1}{2}$ describes a state of 'spin down'. In this case we have

$$\left.\begin{aligned} R &= (2\sqrt{2}\pi)^{-1}\sin(\alpha/2) \geqslant 0, \qquad 0 \leqslant \alpha \leqslant \pi \\ S/\hbar &= \tfrac{1}{2}(\beta - \gamma) - 3\hbar t/8I - \pi/2 \end{aligned}\right\} \tag{10.5.16}$$

which implies

$$\alpha = \text{constant}(\neq 0), \qquad \beta = -\gamma = f't, \tag{10.5.17}$$

where the angular frequency is now $f' = \hbar/4I\sin^2(\alpha/2)$ and we choose β and γ to be initially zero. The angular momentum components are

$$\left.\begin{aligned}M_1 &= (\hbar/2)\cot(\alpha/2)\sin\beta = (\hbar/2)\cot(\alpha/2)\sin f't\\ M_2 &= (\hbar/2)\cot(\alpha/2)\cos\beta = (\hbar/2)\cot(\alpha/2)\cos f't\\ M_3 &= -\hbar/2\\ M'_1 &= -(\hbar/2)\cot(\alpha/2)\sin\gamma = (\hbar/2)\cot(\alpha/2)\sin f't\\ M'_2 &= (\hbar/2)\cot(\alpha/2)\cos\gamma = (\hbar/2)\cot(\alpha/2)\cos f't\\ M'_3 &= \hbar/2.\end{aligned}\right\} \quad (10.5.18)$$

The spin vector is inclined at an angle $(\alpha + \pi)/2$ to the $\mathbf{e}_3$-axis and precesses uniformly clockwise with angular frequency f' on the surface of a cone in both systems of axes. 'Spin down' thus means that the locus of $\mathbf{M}$ lies in a spatial domain $z < 0$.

The Pauli average values are now

$$\left.\begin{aligned}\langle M_i\rangle &= (\hbar/2)(0, 0, -1)\\ \langle M'_i\rangle &= (\hbar/2)(0, 0, 1),\end{aligned}\right\} \quad (10.5.19)$$

so that u_- describes the state 'spin down (up)' with respect to $\mathbf{e}_i$ ($\mathbf{e}'_i$).

Similar results to those above are obtained for the spin basis functions v_+ ($a = \frac{1}{2}$, $b = -\frac{1}{2}$) and v_- ($a = b = -\frac{1}{2}$). The former describes a state of 'spin up (down)' relative to $\mathbf{e}_i$ ($\mathbf{e}'_i$) and the latter a state of 'spin down' relative to both systems of axes.

Finally, we confirm that although half-angles appear in the formulae for the time development of the Euler angles and the angular momentum, the double-valued nature of functions such as u_+ under 2π rotations does not show up in the spatial representation of the motion since the half-angles appear in terms such as $\tan(\alpha/2)$ or $\cot(\alpha/2)$ (periodic in π) and in $\sin^2(\alpha/2)$ or $\cos^2(\alpha/2)$ (always $\geqslant 0$). In other words, whilst the configuration space of the wavefunction is $\mathbb{R}^3 \times SU(2)$, the configuration space of the physical quantities associated with the particle is $\mathbb{R}^3 \times SO(3)$.

10.5.2 *Effect of a uniform magnetic field*

Let us apply a uniform magnetic field in the z-direction, $\mathbf{B} = (0, 0, B)$, $B \geqslant 0$, and neglect the term in B^2 in the Hamiltonian (10.2.6). Separating out and ignoring the centre of mass motion, the Schrödinger equation becomes

$$i\hbar\,(\partial\psi/\partial t) = (\hat{\mathbf{M}}^2/2I + \mu' B\hat{M}_3)\psi, \qquad \hat{\mathbf{M}}^2\psi = (3\hbar^2/4)\psi. \quad (10.5.20)$$

To see how the free spin up state is modified by the external field, we consider the following stationary solution to (10.5.20):

$$\psi = u_+ \mathrm{e}^{-\mathrm{i}E_+ t/\hbar}, \qquad E_+ = 3\hbar^2/8I + (\hbar/2)\mu' B. \quad (10.5.21)$$

The amplitude and phase of this function are given by (10.5.2) (with E replaced by E_+) but now we have to take into account the magnetic field in the guidance formula (10.3.14). With initial conditions $\beta = \gamma = 0$, we find

$$\alpha = \text{constant}(\neq \pi), \qquad \beta = -(f + \mu' B)t, \qquad \gamma = -ft, \quad (10.5.22)$$

where f is given by (10.5.5). The angular momentum components are now

$$\left.\begin{aligned} I\omega_1 &= -(\hbar/2)\tan(\alpha/2)\sin(f + \mu' B)t \\ I\omega_2 &= (\hbar/2)\tan(\alpha/2)\cos(f + \mu' B)t \\ I\omega_3 &= \hbar/2 + IB\mu' \end{aligned}\right\} \quad (10.5.23)$$

with respect to the space axes. The spin vector makes an angle

$$\lambda = \cos^{-1}[(\tfrac{1}{2}\hbar + \mu' IB)\cos(\alpha/2)(\tfrac{1}{4}\hbar^2 + \hbar\mu' IB\cos^2(\alpha/2))^{-1/2}] \quad (10.5.24)$$

with the $\mathbf{e}_3$-axis and precesses uniformly anticlockwise about this axis with frequency $(f + B\mu')$ on the surface of a cone of height $(\frac{1}{2}\hbar + IB\mu')$ and radius $(\hbar/2)\tan(\alpha/2)$.

The rotation is generated in this case by the Larmor-type torque together with the quantum torque:

$$I\dot{\omega}_1 = T_1 - I\mu' B\omega_2, \qquad I\dot{\omega}_2 = T_2 + I\mu' B\omega_1, \qquad I\dot{\omega}_3 = T_3 \quad (10.5.25)$$

where

$$\mathbf{T} = -(\hbar/2)f\tan(\alpha/2)(\cos(f + \mu' B)t, \sin(f + \mu' B)t, 0) \quad (10.5.26)$$

along the orbit. Both torques depend on B. When $B = 0$ we recover the free system of §10.5.1. The body has rotational kinetic energy

$$\tfrac{1}{2}I\omega^2 = \hbar^2/8I\cos^2(\alpha/2) + (\hbar/2)\mu' B \quad (10.5.27)$$

and Q_s is independent of B.

The spin down state u_- is modified in a similar way by a uniform magnetic field. In that case the total energy is $E_- = 3\hbar^2/8I - (\hbar/2)\mu' B$.

A treatment of the Stern–Gerlach experiment may be developed in a straightforward way using the extended theory of spin. As in §9.5, the impulsive field produces two separating packets and the corpuscle joins one or the other. Eventually it attains a state of spin up or spin down as described in §10.5.1. Note that this interaction results in an eigenstate of angular momentum and does not constitute a measurement of the Euler angle operators. As we have seen, the Euler angles are not constant in this case. It seems we can invent interactions to measure functions of the angle operators by a straightforward application of the theory of Chap. 8. For example, this could be achieved for α through the interaction Hamiltonian $gx\cos\alpha$, where x is the meter coordinate (see also §11.4).

10.6 Quantum fields

10.6.1 Bosons

The model of spin $\frac{1}{2}$ systems developed in this chapter provides the key to applying the quantum theory of motion to fields quantized according to Fermi–Dirac statistics. To put the method in context, we first address the formal problem of how one might go about developing a causal interpretation of quantum field theory in the Bose–Einstein case. We shall treat here the simplest system that exhibits the essential features to be interpreted, the free Schrödinger field, along the lines of Bohm's (1952b) original treatment of the quantized electromagnetic field. There the field oscillators appearing in the normal mode expansion of the quantized field are assumed to have simultaneously well-defined values of their coordinates and momenta, and their motions are correlated by the quantum potential implied by the total quantum state (a different approach to the causal theory of fields is pursued by Bell (1987, Chap. 19), Sudbery (1987) and Roy & Singh (1990)). A fuller account of the following theory with applications is given for the relativistic case in Chap. 12.

We start with the Hamiltonian in the Schrödinger picture

$$H = \frac{-\hbar^2}{2m} \int \psi^*(\mathbf{x})\nabla^2\psi(\mathbf{x})\,\mathrm{d}^3x \tag{10.6.1}$$

and quantize the field $\psi(\mathbf{x})$ by imposing the commutation relations

$$\left.\begin{aligned} &[\psi(\mathbf{x}), \psi^*(\mathbf{x}')]_- = \delta(\mathbf{x} - \mathbf{x}'),\\ &[\psi(\mathbf{x}), \psi(\mathbf{x}')]_- = [\psi^*(\mathbf{x}), \psi^*(\mathbf{x}')]_- = 0. \end{aligned}\right\} \tag{10.6.2}$$

We make the usual normal mode expansion

$$\psi(\mathbf{x}) = \frac{1}{V^{1/2}} \sum_{\mathbf{k}} a_{\mathbf{k}}\,\mathrm{e}^{\mathrm{i}\mathbf{k}\cdot\mathbf{x}}, \tag{10.6.3}$$

where $E_{\mathbf{k}} = \hbar^2\mathbf{k}^2/2m$, V is the volume of a box of side L, and $k_i = 2\pi n_i/L$, $n_i \in \mathbb{Z}$, $i = 1, 2, 3$. The Hamiltonian becomes

$$H = \sum_{\mathbf{k}} E_{\mathbf{k}} a_{\mathbf{k}}^* a_{\mathbf{k}}, \tag{10.6.4}$$

with (10.6.2) replaced by

$$\left.\begin{aligned} &[a_{\mathbf{k}}, a_{\mathbf{k}'}^*]_- = \delta_{\mathbf{k}\mathbf{k}'},\\ &[a_{\mathbf{k}}, a_{\mathbf{k}'}]_- = [a_{\mathbf{k}}^*, a_{\mathbf{k}'}^*]_- = 0. \end{aligned}\right\} \tag{10.6.5}$$

We now introduce the Hermitian operators $q_{\mathbf{k}}$ and $p_{\mathbf{k}}$ defined by

$$\left.\begin{aligned} q_{\mathbf{k}} &= |\mathbf{k}|^{-1}(a_{\mathbf{k}} + a_{\mathbf{k}}^*), \\ p_{\mathbf{k}} &= \tfrac{1}{2}\mathrm{i}\hbar|\mathbf{k}|(a_{\mathbf{k}}^* - a_{\mathbf{k}}) \end{aligned}\right\} \tag{10.6.6}$$

and the relations (10.6.5) become

$$\left.\begin{aligned} [q_{\mathbf{k}}, p_{\mathbf{k}'}]_- &= \mathrm{i}\hbar\delta_{\mathbf{k}\mathbf{k}'}, \\ [q_{\mathbf{k}}, q_{\mathbf{k}'}]_- &= [p_{\mathbf{k}}, p_{\mathbf{k}'}]_- = 0. \end{aligned}\right\} \tag{10.6.7}$$

Thus we find for (10.6.4)

$$H = \sum_{\mathbf{k}} \tfrac{1}{2}E_{\mathbf{k}}\left(\frac{2p_{\mathbf{k}}^2}{\hbar^2\mathbf{k}^2} + \tfrac{1}{2}\mathbf{k}^2 q_{\mathbf{k}}^2 - 1\right). \tag{10.6.8}$$

In the position representation $q_{\mathbf{k}}$ is just a real number for each $\mathbf{k}$ and $p_{\mathbf{k}} = -\mathrm{i}\hbar\,\partial/\partial q_{\mathbf{k}}$. Eq. (10.6.8) becomes

$$H = \sum_{\mathbf{k}}\left(\frac{-\hbar^2}{2m}\frac{\partial^2}{\partial q_{\mathbf{k}}^2} + \tfrac{1}{2}m\omega_{\mathbf{k}}^2 q_{\mathbf{k}}^2 - \tfrac{1}{2}E_{\mathbf{k}}\right), \tag{10.6.9}$$

which is the Hamiltonian of a set of classically noninteracting harmonic oscillators of mass m and frequency $\omega_{\mathbf{k}} = E_{\mathbf{k}}/\hbar$. In this representation the Schrödinger equation is

$$\mathrm{i}\hbar\frac{\partial\Psi}{\partial t}[q_{\mathbf{k}}, t] = H\Psi[q_{\mathbf{k}}, t], \tag{10.6.10}$$

where $\Psi[q_{\mathbf{k}}, t] = \langle[q_{\mathbf{k}}]|\Psi(t)\rangle$ with $\langle[q_{\mathbf{k}}]| = \prod_{\mathbf{k}}\langle q_{\mathbf{k}}|$ and square brackets denote a functional,

$$|\Psi\rangle \in \bigoplus_{n=0}^{\infty} \mathscr{H}^{(n)},$$

$\mathscr{H}^{(n)}$ being the n-particle Hilbert space, and H is given by (10.6.9). Substituting $\Psi = R\,\mathrm{e}^{\mathrm{i}S/\hbar}$ into (10.6.10), where R and S are real amplitude and phase functionals respectively, we obtain a Hamilton–Jacobi-type equation and a conservation law:

$$\frac{\partial S}{\partial t} + \frac{1}{2m}\sum_{\mathbf{k}}\left(\frac{\partial S}{\partial q_{\mathbf{k}}}\right)^2 + \sum_{\mathbf{k}}\tfrac{1}{2}m\omega_{\mathbf{k}}^2 q_{\mathbf{k}}^2 - \sum_{\mathbf{k}}\tfrac{1}{2}E_{\mathbf{k}} + Q = 0, \tag{10.6.11}$$

$$\frac{\partial\rho}{\partial t} + \frac{1}{m}\sum_{\mathbf{k}}\frac{\partial}{\partial q_{\mathbf{k}}}\left(\rho\frac{\partial S}{\partial q_{\mathbf{k}}}\right) = 0, \tag{10.6.12}$$

where $\rho = R^2 = |\Psi|^2$ is the probability density and

$$Q = \frac{-\hbar^2}{2mR} \sum_{\mathbf{k}} \frac{\partial^2 R}{\partial q_{\mathbf{k}}^2} \tag{10.6.13}$$

is the quantum potential.

The postulate of the causal interpretation is that the normal coordinates are at all times well defined, and that their values $q_{\mathbf{k}}(t)$ may be found by solving the equation

$$\dot{q}_{\mathbf{k}} = \frac{1}{m} \frac{\partial S}{\partial q_{\mathbf{k}}}\bigg|_{q_{\mathbf{k}} = q_{\mathbf{k}}(t)}, \tag{10.6.14}$$

with $\dot{=} \mathrm{d}/\mathrm{d}t = \partial/\partial t + \sum_{\mathbf{k}} \dot{q}_{\mathbf{k}}\, \partial/\partial q_{\mathbf{k}}$, once the initial values $q_{\mathbf{k}}(0)$ are specified.

Applying the operator $\partial/\partial q_{\mathbf{k}}$ to (10.6.11) and using (10.6.14), we can write the guidance formula in the form

$$\ddot{q}_{\mathbf{k}} + \omega_{\mathbf{k}}^2 q_{\mathbf{k}} = -\frac{1}{m} \frac{\partial Q}{\partial q_{\mathbf{k}}}. \tag{10.6.15}$$

We thus find the equations of motion of a set of oscillators which are coupled by quantum force terms. The classical limit follows when the latter may be neglected and each oscillator vibrates independently. Eq. (10.6.15) is then equivalent to the free Schrödinger equation treated as a classical field theory:

$$i\hbar \frac{\partial \psi(\mathbf{x}, t)}{\partial t} = \frac{-\hbar^2}{2m} \nabla^2 \psi(\mathbf{x}, t). \tag{10.6.16}$$

This is easily shown by defining $\psi(\mathbf{x}, t)$ as in (10.6.3) where $a_{\mathbf{k}}$ is now time-dependent and using the relations (10.6.6) with $p_{\mathbf{k}} = m\dot{q}_{\mathbf{k}}$. In the general case of finite quantum forces, (10.6.15) translates into

$$i\hbar \frac{\partial \psi(\mathbf{x}, t)}{\partial t} = \frac{-\hbar^2}{2m} \nabla^2 \psi(\mathbf{x}, t) + \frac{1}{2}\left(\frac{\delta}{\delta\psi(-\mathbf{x})} + \frac{\delta}{\delta\psi^*(\mathbf{x})}\right) Q[\psi, \psi^*]|_{\psi(\mathbf{x}) = \psi(\mathbf{x}, t)}. \tag{10.6.17}$$

This shows how the 'classical' Schrödinger equation is modified by quantum effects arising in the procedure of second quantization.

10.6.2 Fermionic analogue of the oscillator picture of boson fields

Two key requirements of the causal version of the boson field theory just described are that the quantum theory should be presented in the Schrödinger picture and that the canonical commutation relations should be realized in a 'position representation' in which the conjugate momentum operator is a differential operator. It might be felt that this programme cannot be extended to include fermion fields because the anticommutation relations have no classical limit and do not permit a picture of continuous field variables of the kind used for boson systems (see, e.g., Bohm, Hiley & Kaloyerou (1987)).

In the model described here, the fermion field is represented as a collection of spherical rotators (Holland, 1988a). The anticommutative algebra makes contact with a description in terms of continuous variables by mapping it into an algebra of angular momentum operators realized in the angular coordinate representation. The Hamiltonian then describes a set of independent spin $\frac{1}{2}$ rigid rotators in an effective potential which brings about a Larmor-type precession of the spin vectors. The wavefunction depends on all the orientation parameters of the rotators, here the Euler angles, and brings about the correlated evolution of the angles via a spin-dependent quantum potential. As we shall see in §10.6.3, this approach allows one to give a consistent account of the classical limit of a quantum fermion field, and to see how it relates to the oscillator model of bosons.

A hint that the quantized fermion field is a set of spin $\frac{1}{2}$ rotators appears in the classic paper of Jordan & Wigner (1928), who observed that the generators of the anticommutative algebra could be expressed, in a matrix representation, in terms of the Pauli matrices (see also Brauer & Weyl (1935)). We shall use essentially the algebraic constructions of these references, but realize them in a different representation.

Eqs. (10.6.1), (10.6.3) and (10.6.4) remain valid, but (10.6.5) is replaced by the anticommutation relations

$$\left.\begin{aligned} [a_{\mathbf{k}}, a^*_{\mathbf{k}'}]_+ &= \delta_{\mathbf{k}\mathbf{k}'}, \\ [a_{\mathbf{k}}, a_{\mathbf{k}'}]_+ &= [a^*_{\mathbf{k}}, a^*_{\mathbf{k}'}]_+ = 0. \end{aligned}\right\} \tag{10.6.18}$$

To interpret the meaning of these relations physically, we pass to three sets of Hermitian operators $e_{i\mathbf{k}}$, $i = 1, 2, 3$. Writing $e^{\pm}_{\mathbf{k}} = e_{1\mathbf{k}} \pm ie_{2\mathbf{k}}$ we have

$$e^+_{\mathbf{k}} = 2\lambda_{\mathbf{k}} a^*_{\mathbf{k}}, \qquad e^-_{\mathbf{k}} = 2\lambda_{\mathbf{k}} a_{\mathbf{k}}, \qquad e_{3\mathbf{k}} = (2a^*_{\mathbf{k}} a_{\mathbf{k}} - 1), \tag{10.6.19}$$

where $\lambda_{\mathbf{k}} = \prod_{\mathbf{k}'<\mathbf{k}} (1 - 2a^*_{\mathbf{k}'} a_{\mathbf{k}'})$, with $\lambda^2_{\mathbf{k}} = 1$ and

$$\left.\begin{aligned} \lambda_{\mathbf{k}} a_{\mathbf{k}'} &= a_{\mathbf{k}'} \lambda_{\mathbf{k}}, &\quad \lambda_{\mathbf{k}} a^*_{\mathbf{k}'} &= a^*_{\mathbf{k}'} \lambda_{\mathbf{k}}, &\quad \mathbf{k} &\leqslant \mathbf{k}', \\ \lambda_{\mathbf{k}} a_{\mathbf{k}'} &= -a_{\mathbf{k}'} \lambda_{\mathbf{k}}, &\quad \lambda_{\mathbf{k}} a^*_{\mathbf{k}'} &= -a^*_{\mathbf{k}'} \lambda_{\mathbf{k}}, &\quad \mathbf{k} &> \mathbf{k}', \end{aligned}\right\} \tag{10.6.20}$$

and $\mathbf{k}' < \mathbf{k}$ means $k'_i < k_i$ for one $i = 1, 2, 3$, and $k'_j \leqslant k_j$ for $j \neq i$. The two independent sets of generators $e^{\pm}_{\mathbf{k}}$ are in one-to-one correspondence with $a^*_{\mathbf{k}}$ and $a_{\mathbf{k}}$. The inverse relations are

$$a^*_{\mathbf{k}} = \tfrac{1}{2}\lambda_{\mathbf{k}} e^+_{\mathbf{k}}, \qquad a_{\mathbf{k}} = \tfrac{1}{2}\lambda_{\mathbf{k}} e^-_{\mathbf{k}}, \tag{10.6.21}$$

with $\lambda_{\mathbf{k}} = \prod_{\mathbf{k}'<\mathbf{k}}(-e_{3\mathbf{k}'})$. The third set of generators $e_{3\mathbf{k}}$ is not independent of the first two sets: $e_{3\mathbf{k}} = \frac{1}{2}e_{\mathbf{k}}^{+}e_{\mathbf{k}}^{-} - 1$.

It is readily proved that the following commutation and anticommutation relations are equivalent to (10.6.18):

$$[e_{i\mathbf{k}}, e_{j\mathbf{k}}]_{-} = 2\mathrm{i}\varepsilon_{ijl}e_{l\mathbf{k}}, \tag{10.6.22a}$$

$$[e_{i\mathbf{k}}, e_{j\mathbf{k}}]_{+} = 2\delta_{ij}, \tag{10.6.22b}$$

$$[e_{i\mathbf{k}}, e_{j\mathbf{k}'}]_{-} = 0, \tag{10.6.22c}$$

for all $i, j, l = 1, 2, 3$ and $\mathbf{k} \neq \mathbf{k}'$. Relation (10.6.22a) shows that the $\mathbf{k}$th set of generators $e_{i\mathbf{k}}$, $i = 1, 2, 3$, form (apart from a constant) a set of angular momentum operators, and (10.6.22b) that their domain of application is restricted to spin $\frac{1}{2}$ states (cf. (10.4.1)). Eq. (10.6.22c) implies that the angular momentum generators for different $\mathbf{k}$ are independent.

Using (10.6.19), the Hamiltonian (10.6.4) becomes

$$H = \sum_{\mathbf{k}} \tfrac{1}{2}E_{\mathbf{k}}[1 + e_{3\mathbf{k}}]. \tag{10.6.23}$$

In the angular coordinate representation, this may be written as

$$H = \sum_{\mathbf{k}} [(1/2I_{\mathbf{k}})\hat{\mathbf{M}}_{\mathbf{k}}^{2} + B_{\mathbf{k}}\hat{M}_{3\mathbf{k}}], \tag{10.6.24}$$

which describes a set of noninteracting spin $\frac{1}{2}$ spherical rotators, each of mass m, effective moment of inertia $I_{\mathbf{k}} = 3m/2\mathbf{k}^2$, and acted upon by an effective 'magnetic field' $B_{\mathbf{k}} = E_{\mathbf{k}}/\hbar$ in the '$\mathbf{e}_3$-direction'. The Schrödinger equation is (cf. (10.5.20) with $\mu' = 1$)

$$\mathrm{i}\hbar\frac{\partial\Psi}{\partial t}[\boldsymbol{\alpha}_{\mathbf{k}}, t] = H\Psi[\boldsymbol{\alpha}_{\mathbf{k}}, t], \tag{10.6.25}$$

where Ψ is a function of all the spin $\frac{1}{2}$ rotator coordinates (Euler angles). The general solution of (10.6.25) is

$$\Psi[\boldsymbol{\alpha}_{\mathbf{k}}, t] = \sum_{n_{\mathbf{k}}=\pm} c[n_{\mathbf{k}}, t]\prod_{\mathbf{k}} u_{\mathbf{k}n_{\mathbf{k}}}(\boldsymbol{\alpha}_{\mathbf{k}}), \tag{10.6.26}$$

where each $u_{\mathbf{k}n_{\mathbf{k}}}(\boldsymbol{\alpha}_{\mathbf{k}})$, $n_{\mathbf{k}} = \pm$, describes a state of spin up or spin down for the $\mathbf{k}$th rotator.

To give the causal interpretation of this system we simply postulate that each rotator has a definite orientation $\boldsymbol{\alpha}_{\mathbf{k}}(t)$ at each time. The Hamilton–Jacobi and continuity equations are

$$\partial S/\partial t + \sum_{\mathbf{k}} \tfrac{1}{2} I_{\mathbf{k}} \boldsymbol{\omega}_{\mathbf{k}}^2 + Q = 0, \tag{10.6.27}$$

$$\partial \rho/\partial t + \sum_{\mathbf{k}} \mathrm{i}\hbar \hat{\mathbf{M}}_{\mathbf{k}} \cdot (\rho \boldsymbol{\omega}_{\mathbf{k}}) = 0, \tag{10.6.28}$$

where

$$Q = \sum_{\mathbf{k}} (1/2I_{\mathbf{k}} R) \hat{\mathbf{M}}_{\mathbf{k}}^2 R. \tag{10.6.29}$$

The formulae (10.3.5) and (10.3.14) remain valid for each rotator in the many-body system, but now the phase and amplitude functionals $S[\boldsymbol{\alpha}_{\mathbf{k}}, t]$, $R[\boldsymbol{\alpha}_{\mathbf{k}}, t]$ entering into the guidance equations depend on the coordinates of all the rotators. That is, the evolution of the $\mathbf{k}$th set of Euler angles depends on the total quantum state and on $\boldsymbol{\alpha}_{\mathbf{k}}(0)$ for all $\mathbf{k}$. The analogue of (10.6.15) for the bosonic field is here

$$I_{\mathbf{k}} \frac{\mathrm{d}\boldsymbol{\omega}_{\mathbf{k}}}{\mathrm{d}t} = \mathbf{T}_{\mathbf{k}} + I_{\mathbf{k}} \mathbf{B}_{\mathbf{k}} \times \boldsymbol{\omega}_{\mathbf{k}}, \qquad \mathbf{B}_{\mathbf{k}} = (0, 0, B_{\mathbf{k}}), \tag{10.6.30}$$

where

$$\mathbf{T}_{\mathbf{k}} = -(\mathrm{i}/\hbar) \hat{\mathbf{M}}_{\mathbf{k}} Q. \tag{10.6.31}$$

Only when the wavefunction factorizes in the form

$$\Psi[\boldsymbol{\alpha}_{\mathbf{k}}, t] = \prod_{\mathbf{k}} \Psi(\boldsymbol{\alpha}_{\mathbf{k}}, t) \tag{10.6.32}$$

do the rotators perform independent one-body motions. In this case the $\mathbf{k}$th rotator may perform a spin up or spin down motion of the kind described above for a single rotator, but where $B_{\mathbf{k}}$ is here restricted to be $B_{\mathbf{k}} = 3\hbar/4I_{\mathbf{k}}$. This implies that $E_{\mathbf{k}+} = E_{\mathbf{k}}$ and $E_{\mathbf{k}-} = 0$ (cf. (10.5.21) with $\mu' = 1$).

An example of factorizability is the vacuum state which is the normalized stationary state characterized by $\hat{M}_{\mathbf{k}}^- \Psi_0 = 0$ for all $\mathbf{k}$ (cf. (10.2.26)):

$$\Psi_0[\boldsymbol{\alpha}_{\mathbf{k}}] = \prod_{\mathbf{k}} u_{\mathbf{k}-}(\boldsymbol{\alpha}_{\mathbf{k}}). \tag{10.6.33}$$

All of the spin vectors independently precess on the surfaces of 'downward pointing' cones with frequency $(\hbar/4I_{\mathbf{k}})[3 + (\cos \frac{1}{2}\alpha_{\mathbf{k}})^{-2}]$. The total energy is zero, and the kinetic energy of rotation is balanced by the quantum potential energy. Another example is where n rotators are excited to a state of spin up:

$$\Psi^{(n)}[\boldsymbol{\alpha}_{\mathbf{k}}, t] = \mathrm{e}^{-\mathrm{i}(E_{\mathbf{k}1} + \cdots + E_{\mathbf{k}n})t/\hbar} \Psi_{\mathbf{k}_1 \ldots \mathbf{k}_n}[\boldsymbol{\alpha}_{\mathbf{k}}], \tag{10.6.34}$$

where

$$\Psi_{\mathbf{k}_1\ldots\mathbf{k}_n}[\alpha_\mathbf{k}] = u_{\mathbf{k}_1+}(\alpha_{\mathbf{k}_1})\ldots u_{\mathbf{k}_n+}(\alpha_{\mathbf{k}_n}) \prod_{\mathbf{k}'\neq\mathbf{k}_1,\ldots,\mathbf{k}_n} u_{\mathbf{k}'-}(\alpha_{\mathbf{k}'}).$$

A nonstationary state having the form of a wave packet for a single-quantum state may be constructed as follows:

$$\Psi_p[\alpha_\mathbf{k}, t] = \sum_\mathbf{k} f(\mathbf{k} - \mathbf{k}_0)\Psi^{(1)}[\alpha_\mathbf{k}, t], \tag{10.6.35}$$

where $f(\mathbf{k} - \mathbf{k}_0)$ is peaked around $\mathbf{k}_0$. To show that this indeed corresponds to a packet we can evaluate the expectation value of the density matrix in such a state:

$$\begin{aligned}\rho(\mathbf{x}, \mathbf{x}', t) &= \langle\Psi_p|\psi^*(\mathbf{x}')\psi(\mathbf{x})|\Psi_p\rangle\\ &= \int \Psi_p^*[\alpha_\mathbf{k}, t]\psi^*(\mathbf{x}')\psi(\mathbf{x})\Psi_p[\alpha_\mathbf{k}, t]\prod_\mathbf{k} \mathrm{d}\Omega_\mathbf{k}\\ &= g^*(\mathbf{x}', t)g(\mathbf{x}, t),\end{aligned} \tag{10.6.36}$$

where

$$g(\mathbf{x}, t) = \frac{1}{V^{1/2}}\sum_\mathbf{k} f(\mathbf{k} - \mathbf{k}_0)\,\mathrm{e}^{\mathrm{i}(\mathbf{k}\cdot\mathbf{x} - E_\mathbf{k}t/\hbar)}$$

is a packet function whose centre moves with velocity $\hbar\mathbf{k}_0/m$. The diagonal components give the mean particle density:

$$n(\mathbf{x}, t) = \rho(\mathbf{x}, \mathbf{x}, t) = |g(\mathbf{x}, t)|^2,$$

which is localized in space about the point $\mathbf{x} = \hbar\mathbf{k}_0 t/m$.

The number of spin up rotators corresponds in the particle representation to the number of particles present in the system. The general solution (10.6.26) is a superposition of states describing different numbers of particles. In this model, the creation and annihilation of particles is described by the continuous transformation of spin up rotators to spin down, and vice versa, a process that takes a finite time.

The picture we have developed of a set of spin $\frac{1}{2}$ rotators comes about from the Fermi–Dirac quantization of a spin 0 field. If we quantize a Pauli spinor field this would double the number of angular momentum generators, but the model would remain the same. A similar remark applies to the Dirac field.

10.6.3 The classical limit of a quantized fermion field

Since the system of rotators introduced above arises by quantizing the (first quantized) free Schrödinger equation treated as a classical field theory, we expect to return to the latter in the 'classical limit' of the second quantized fermion theory. Since we know the classical Schrödinger field in its normal mode representation may be pictured as a collection of oscillators, this implies that there is a second mathematically equivalent classical model, that of rotators.

To arrive at the classical limit we assume that there are circumstances in which one may neglect the quantum torque $\mathbf{T}_\mathbf{k}$ in the equation of motion of the $\mathbf{k}$th rotator. Written in component form, (10.6.30) reduces to

$$\left.\begin{aligned}\dot{\omega}_{1\mathbf{k}} &= -\omega_\mathbf{k}\omega_{2\mathbf{k}},\\ \dot{\omega}_{2\mathbf{k}} &= \omega_\mathbf{k}\omega_{1\mathbf{k}},\\ \dot{\omega}_{3\mathbf{k}} &= 0,\end{aligned}\right\} \tag{10.6.37}$$

where $\omega_\mathbf{k} = E_\mathbf{k}/\hbar = \hbar\mathbf{k}^2/2m$. This is the equation of motion for an angular momentum vector of constant magnitude precessing about the $\mathbf{e}_3$-direction with angular velocity $\omega_\mathbf{k}$. It is obvious that $\omega_{1\mathbf{k}}$ and $\omega_{2\mathbf{k}}$ each perform simple harmonic motion of frequency $\omega_\mathbf{k}$. Define for each $\mathbf{k}$ the new variables

$$q_\mathbf{k} = 3m\lambda_\mathbf{k}\omega_{1\mathbf{k}}/\hbar|\mathbf{k}|^3, \qquad p_\mathbf{k} = -3m\lambda_\mathbf{k}\omega_{2\mathbf{k}}/2|\mathbf{k}|, \tag{10.6.38}$$

where $\lambda_\mathbf{k}$ is a constant. The first two equations (10.6.37) are equivalent to

$$p_\mathbf{k} = m\dot{q}_\mathbf{k}, \qquad \dot{p}_\mathbf{k} = -m\omega_\mathbf{k}q_\mathbf{k} \tag{10.6.39}$$

and hence

$$\ddot{q}_\mathbf{k} + \omega_\mathbf{k}^2 q_\mathbf{k} = 0, \tag{10.6.40}$$

which is the equation of motion of a one-dimensional oscillator. Considering these relations for all $\mathbf{k}$, we see that in the classical limit of the quantized fermion system (a set of noninteracting rotators undergoing a Larmor-type precession) we fall upon the oscillator model of the quantized boson system in *its* classical limit (a collection of noninteracting oscillators).

This establishes the connection between the rotator and harmonic oscillator models of the classical Schrödinger field. The track traced out by the tip of the $\mathbf{k}$th spin vector with angular frequency $\omega_\mathbf{k}$ in a plane orthogonal to $\mathbf{e}_3$ is the phase space orbit of the corresponding harmonic oscillator of the same frequency. The conserved total energy of the oscillator, $(p_\mathbf{k}^2/2m) + \frac{1}{2}m\omega_\mathbf{k}^2 q_\mathbf{k}^2$, corresponds to the constant total kinetic energy of the rotator, $\frac{1}{2}I\omega_\mathbf{k}^2$. The Larmor-type torque corresponds to the oscillator force.

That both models are indeed equivalent to the Schrödinger equation (10.6.16) is easily shown, as mentioned in §10.6.1. We define the wavefunction as (10.6.3) where

$$a_\mathbf{k} = \tfrac{1}{2}|\mathbf{k}|q_\mathbf{k} + \mathrm{i}p_\mathbf{k}/\hbar|\mathbf{k}| = (3m\lambda_\mathbf{k}/2\hbar\mathbf{k}^2)(\omega_{1\mathbf{k}} - \mathrm{i}\omega_{2\mathbf{k}}) \tag{10.6.41}$$

in the two sets of variables, which are time-dependent. It is then a simple matter to check that the sets of equations (10.6.37) or (10.6.39) for all $\mathbf{k}$ imply that ψ so defined obeys the free Schrödinger equation.

In the limit, the quantum features embodied in the commutative or anticommutative algebras are not reflected in the field dynamics because the means by which they are manifested (quantum force or torque) are ineffective. Imposing these algebraic structures has no bearing on the 'reality' of the underlying continuous field. The oscillator and rotator models follow from different choices of the normal coordinates into which ψ may be expanded. Both are equally valid pictures of the classical Schrödinger field and correspond to the classical limit of fields quantized according to Bose–Einstein or Fermi–Dirac statistics respectively.

11
The Einstein–Podolsky–Rosen experiment and nonlocality

11.1 Assessment of Einstein's argument

11.1.1 Is quantum mechanics complete and local . . .

It was observed in §1.4 that Einstein's method of criticizing the orthodox view of quantum theory was to focus on examples whose special characteristics could be used to show that adherence to the completeness assumption compels one to adopt 'unnatural theoretical interpretations'. It was a way not of disproving the conventional interpretation by attempting to locate in it some logical flaw, but of pushing it into a corner that in Einstein's view no one would seriously want to inhabit. The paper of Einstein, Podolsky & Rosen (EPR) (1935) is the most famous component of this programme and the purpose of this chapter is to analyse the extent to which the arguments presented in it are compatible with the quantum theory of motion.

From contemporaneous correspondence it appears that the EPR paper was written by Podolsky and that Einstein did not consider that the point which was really troubling him had been made explicit (Fine, 1986, Chap. 3). According to Einstein's accounts, EPR had the following purpose: to show that if one believes the wavefunction exhausts all the statements that can be meaningfully asserted about a physical system, then one must also accept that the real physical state of the system depends on what befalls another system with which it has previously interacted, no matter how far apart the two systems may become. Einstein (1949, p. 682) expressed the view that one must relinquish one of the following two assertions:

(*a*) the description by means of the ψ-function is *complete*;
(*b*) the real states of spatially separated objects are independent of each other.

Which alternative he himself opted for is clear from the following passage (Einstein, 1949, p. 85): 'But on one supposition we should, in my opinion, absolutely hold fast: the real factual situation of the system S_2 is independent of what is done with the system S_1 which is spatially separated from the former.'

We now present a version of the EPR argument published by Einstein (1936, 1948). He considers a case where two systems 1 and 2 in known states mechanically (i.e., classically) interact for a definite period of time so that the wavefunction $\psi(\mathbf{x}_1, \mathbf{x}_2)$ becomes entangled (nonfactorizable). If subsequently the systems cease to interact classically, the free Schrödinger evolution preserves the entanglement. Suppose the systems fly apart† in space and are eventually separate in the sense that ψ is finite only where $\mathbf{x}_1$ belongs to a limited portion of space and $\mathbf{x}_2$ belongs to a portion separated from the first one. To be explicit, we shall work with the example given by Bohm (1951, Chap. 22), where the total system is a molecule of total spin 0 composed of two spin $\frac{1}{2}$ atoms in a singlet state:

$$\psi(\mathbf{x}_1, \mathbf{x}_2) = f_1(\mathbf{x}_1)f_2(\mathbf{x}_2)(1/\sqrt{2})(u_+v_- - u_-v_+), \tag{11.1.1}$$

where $f_1(\mathbf{x})$, $f_2(\mathbf{x})$ are nonoverlapping packet functions. Here $u_\pm$ ($v_\pm$) are the eigenfunctions of the spin operators $\hat{s}_{z_1}$ ($\hat{s}_{z_2}$) in the z-direction pertaining to particle 1 (2) (cf. §9.6):

$$\left.\begin{aligned} \hat{s}_{z_1}u_\pm &= \pm(\hbar/2)u_\pm \\ \hat{s}_{z_2}v_\pm &= \pm(\hbar/2)v_\pm . \end{aligned}\right\} \tag{11.1.2}$$

Given such a state, suppose we perform a spin measurement on system 1 in the z-direction and that we obtain the result spin up. Then, according to the usual quantum theory, the function (11.1.1) reduces to the first of its summands:

$$\psi \to f_1 f_2 u_+ v_- . \tag{11.1.3}$$

As a result, the final wavefunction is factorizable and we know the state of the unmeasured system 2, *viz.* v_-. But this outcome for system 2 depends on the kind of measurement carried out on 1. Thus, suppose instead that we measure the spin of atom 1 in the z'-direction (where z' is defined in §11.2). By the rotational symmetry of the singlet state, (11.1.1) can be written

$$\psi(\mathbf{x}_1, \mathbf{x}_2) = f_1(\mathbf{x}_1)f_2(\mathbf{x}_2)(1/\sqrt{2})(u'_+v'_- - u'_-v'_+) \tag{11.1.4}$$

where $u'_\pm$ and $v'_\pm$ are the eigenfunctions of the spin operators in the z'-direction. Then, if the result spin up is obtained in the new direction, the total wavefunction reduces to

$$\psi \to f_1 f_2 u'_+ v'_- \tag{11.1.5}$$

† Notice how in the language we use to set up the problem, which is that commonly employed, we invoke some informal image of objects in motion, a concept that is meaningful only in the causal interpretation.

and, in this case, we must attribute the state v'_- to the unmeasured atom 2. By performing different types of operation on atom 1 we have brought about distinct states of the atom 2, and this will, of course, be evident in the varying statistical predictions relating to the outcomes of subsequent spin measurements of 2 which will display *correlations* between the two atoms (the statistics of measurements pertaining to 2 alone are not altered by the operations performed on 1 – see §11.3).

Einstein considers that seen from the perspective of quantum mechanics alone this does not present any difficulty, for the different states of atom 2 refer to experimental procedures performed on different ensemble elements and so there is no question of attributing the different functions v_-, v'_- to one and the same system 2.

But Einstein now looks at the situation in the light of his 'locality' or 'separability' criterion (*b*) above. This may be stated more precisely as follows (Fine, 1986, p. 60):

Einstein-locality: The real, physical state of one system is not immediately influenced by the kinds of measurements directly made on a second system, which is sufficiently spatially separated from the first.

(to be even more precise, we might interpret 'sufficiently' to mean 'space-like'). In the case studied here, Einstein considers that for the ψ-function (11.1.1) a measurement on 1 represents a physical operation which only affects the region of space where f_1 is finite and can have no direct influence on the physical reality in the remote region of space inhabited by atom 2. Thus, the real state of affairs pertaining to atom 2 must be the same whatever action we carry out on 1 (including no measurement at all). Hence, the functions v_-, v'_- must be simultaneously attributable to atom 2. But this is impossible, for these states differ by more than a trivial phase factor and represent different *real* states of affairs for 2. Einstein concludes that the coordination of several ψ-functions with what should be a unique physical condition of 2 shows that ψ cannot be interpreted as a complete description of the physical condition of a system.

This conclusion would not follow, of course, if the localized disturbance of atom 1 did, in fact, bring about a correlated disturbance of atom 2. Einstein (1948) expressed the belief, however, that physics itself would become an impossible enterprise if such distant interconnectedness was admitted as a general property of nature, for it would deny the possibility of studying segments of matter in isolation, and physics would lose its empirical basis.

The analysis of the original EPR paper revolves around the notion of an 'element of reality', a concept that does not seem to appear in Einstein's own published accounts of the argument but which features in practically all subsequent treatments. In order to decide whether the quantum mechanical description of reality using wavefunctions provides a complete set of concepts, EPR propose the following 'criterion of completeness' which they regard as necessary: 'Every element of the physical reality must have a counterpart in the physical theory.' The EPR 'criterion of reality', which they regard as sufficient, is as follows: 'If, without in any way disturbing a system, we can predict with certainty (i.e., with probability equal to unity) the value of a physical quantity, then there exists an element of physical reality corresponding to this physical quantity.'

In the above example, the elements of reality correspond to the eigenvalues ($\pm\hbar/2$) of each of the two spin operators. EPR's argument is that when we perform the first measurement (of the spin of 1 in the z-direction) we can predict with certainty, and without disturbance, the value of the corresponding spin component of 2, and likewise when we carry out the second measurement (of the spin of 1 in the z'-direction). Hence, we should be able to simultaneously attribute these spin values to system 2. But quantum mechanics denies this possibility because the operators in question do not commute. Therefore, quantum mechanics is not a complete description of the physically real.

It seems that Einstein's own version of the EPR argument is more general than this in that it does not demand that the final states of system 2 are eigenfunctions of some physically relevant noncommuting operators (although this is the case in the example we have discussed), but merely requires them to be nontrivially distinct.

Bohr's (1935) response to EPR appears to be a reaffirmation of his view that the choice of different observables to be measured on 1 implies mutually exclusive experimental arrangements whose details cannot be compared with one another. In this view the debate regarding the locality or nonlocality of quantum mechanics is curtailed at the outset, as the notion of the separation of two 'subsystems' of a single quantum system is meaningless.

11.1.2 . . . or incomplete and nonlocal?

For Einstein, nonlocality is an expression of the incompleteness of the theoretical description afforded by the ψ-function. An implication of Einstein's reasoning is that if one foregoes the completeness assumption it should be possible to maintain locality, i.e., propagation via point-by-point contact in spacetime, as a basic principle governing all physical processes, in particular the many-body ones considered by EPR. But he does not seem to have entertained the possibility that by disturbing system 1, system 2 may indeed be influenced and that this is, in fact, a process to which he might try to apply his principle of locality to account for the influence by a subluminal exchange of signals.

With the advent of Bell's (1964) theorem it came to be believed that the locality programme (construed either in the sense that distant objects are completely unconnected or in the sense that they may be linked by a subluminal exchange of information) is unrealizable; no local theory of the type desired by Einstein, it is claimed, can reproduce exactly all of the (presumably correct) empirical predictions of quantum mechanics. Although experimental tests of Bell's inequality favour quantum mechanics over the class of theories embraced by Bell's theorem (§11.5), it is perhaps premature to claim on this basis alone that Einstein-local theories are now definitively ruled out. However, we want to explore in this chapter another possibility not envisaged in the dichotomous choice set up by Einstein ((*a*) and (*b*) in §11.1.1), for it does not follow that the only alternative to rejection of the completeness hypothesis is that one asserts the validity of some local theory. Instead, it may be that *quantum mechanics is both incomplete and nonlocal.*

In the causal interpretation no attempt is made to maintain the universal validity of the locality principle in all spacetime processes, and hence it falls outside the scope of Bell's theorem. Rather, the 'spooky action-at-a-distance' is embraced as a fact of life from the outset. In this way any possibility of conflict with the empirical content of quantum mechanics is avoided.

An advantage of the causal treatment is that it shows how Einstein's anxiety, that physics itself would become impossible if nonlocal connections were admitted, is unfounded. The world can in most ordinary circumstances be consistently regarded as composed of effectively disjoint systems but the conditions under which this is possible are more subtle than in classical physics; separation is not generally sufficient. EPR-type wavefunctions are an example of this fact. The criterion for physical independence is expressed through the wavefunction; if this is factorizable, no actions performed at the site of system 1 can possibly influence the physical state of 2, wherever this is located.

The causal interpretation questions the validity of both Einstein's criterion of separability and EPR's criterion of reality. We first consider the notion of separateness.

We saw in Chap. 7 how in a many-body system the properties of the parts, the particles, depend on properties of the whole, the wavefunction. The whole determines the organization of the parts. This feature is expressed through the definition of particle properties as the local expectation values of various operators in the given quantum state, especially when the latter is entangled.

In the light of this, we see that the assumption that 'measuring the spin of system 1' is, in fact, simply an operation relevant to 'system 1' alone because the spin operator depends only on variables pertaining to 1, is an unwarranted importation of classical conceptions into the quantum domain. For while our physical actions may be restricted to a limited portion of space where the subsystem 1 is supposed to 'be', there is nothing in quantum mechanics or in its causal interpretation that allows us to assert that, in fact, we are acting just on 1. The latter cannot be isolated in this way either conceptually in the conventional view or physically in the causal interpretation. Although ostensibly pertaining to one subsystem and localized in three-dimensional space, the interaction Hamiltonian probes the configuration space of the entire system. In this sense the nonlocality may be viewed as an artefact of mapping the local theory in configuration space into the lower-dimensional space.

Such distant action is to be expected for any nonproduct state. In the case of Bohm's wavefunction (11.1.1) the addends overlap and so it is not surprising that measurement outcomes on 1 and 2 are linked given that the particles are already correlated in the initial state (see §11.2). This initial correlation is most easily seen when the singlet state is written in the angular coordinate representation ((11.4.1) below). In that case the phase (11.4.3) is a nonadditive function of the two sets of particle coordinates.

But we expect similar nonlocal correlations to arise even in those cases where the initial wavefunction is effectively factorizable (§7.2) and the two particles are performing independent motions (the trajectory of 1 (2) is determined solely by $\mathbf{x}_1(0)$ ($\mathbf{x}_2(0)$)). As an example, consider the following wavefunction associated with two particles in one dimension:

$$\psi(x_1, x_2) = \psi_1(x_1)\phi_1(x_2) + \psi_2(x_1)\phi_2(x_2). \tag{11.1.6}$$

Here ϕ_1 and ϕ_2 overlap but ψ_1 and ψ_2 are sufficiently disjoint that the two summands are widely separated and nonoverlapping in configuration space. Moreover, it is assumed that whichever addend the system point lies in, the actual locations of the particles along the x-axis (according to the causal interpretation) are distant. If now we interact with this system in order to effect an ideal measurement of the position operator $\hat{x}_1$, we will find according to the treatment of Chap. 8 that the wavefunction is transformed in the following way, if particle 1 is found to lie in ψ_1:

$$\psi(x_1, x_2) \to \delta(x_1 - a)\phi_1(x_2). \tag{11.1.7}$$

According to the causal interpretation, this reveals that the premeasurement position of 1 was $x_1 = a$ and its state was ψ_1. Notice that any device designed to carry out this measurement inevitably interacts with both addends in

(11.1.6), no matter how far apart they are. From (11.1.7) we infer from the result for $\hat{x}_1$ that system 2 must be attributed the state $\phi_1(x_2)$ (and hence that this was the premeasurement state of 2).

Consider instead a measurement of the momentum operator $\hat{p}_1$ performed on the same initial state (11.1.6). In this case we find for the final wavefunction:

$$\psi(x_1, x_2) \to e^{ip_1x_1/\hbar}[\varphi_1(p_1)\phi_1(x_2) + \varphi_2(p_1)\phi_2(x_2)], \qquad (11.1.8)$$

where p_1 is the result found and φ_1, φ_2 are the Fourier components of ψ_1, ψ_2. As a consequence, we should now attribute to 2 the state defined in the square brackets. Because ϕ_1 and ϕ_2 overlap, this final state is essentially different from the state ϕ_1 found in the first measurement. Applying Einstein's argument of §11.1.1, we see that by performing distinct operations on 1 we have brought about distinct states of the remote particle 2, even though the particles started out as completely uncorrelated.

We conclude that in a two-body system nonlocality is to be expected for all nonproduct states, either because the motion of each particle already depends on both initial positions, or because the disturbance of one body brings about such a dependence in the case of initial effectively factorized states (this distinction depends on the notion that systems actually have definite locations in the unmeasured states and hence is not meaningful in the conventional view).

Finally, we consider the relevance of EPR's 'criterion of reality' to the causal interpretation. This definition is carefully honed to the problem at hand; demonstrating the incompleteness of quantum mechanics by analysing a process using concepts basic to the orthodox view *viz. prediction* (what we would find if we performed a measurement), *probability* (the likely outcome of the measurement) and *disturbance* (the inevitable intrusion of the observer). Yet outside that context there is something odd in defining something as fundamental as an 'element of reality' in terms of a high-level concept such as 'measurement' – surely the latter should be a function of the former. If pushed to do so, would a classical physicist have formulated a definition in quite these terms? For her, regularities in the phenomena might have played a role, but equally important in identifying an 'element of reality' would have been its contribution to a physical theory that explains how the stability, conductance etc. of matter come about. We shall not attempt an alternative definition, but to see the limitations of EPR's version we note that the idea of reality assumed in the causal interpretation does not fall within its scope. The 'elements of reality' are defined in this view by the properties of the wavefunction and the set of particles. These elements are not defined with reference to observation or probability because the latter notions are to be derived from them. In particular, the causal elements of reality *are* disturbed in the context of measurement processes.

11.2 Nonlocal correlations in a double Stern–Gerlach experiment performed on a singlet state

An advantage of Bohm's example is that the partial systems are clearly separated in space (in the conventional sense that the outcomes of position measurements on the two systems are widely separated, as well as in the causal sense that they *are* distant), yet they are correlated in their spin properties. *Entanglement in spin space implies nonlocality in Euclidean space.* This comes about because the spin measurements couple the spin and space variables.

We recall from Chap. 9 how the particles in a many-spin $\frac{1}{2}$ system governed by the Pauli equation may be ascribed continuously variable individual properties such as position, momentum and spin angular momentum for arbitrary quantum states. In a measurement of the spin of a single body these quantities are transformed in such a way that the particle joins one of the emerging beams while the spin vector becomes aligned with the analysing magnetic field. Here we shall extend the spin measurement theory to the case of two spin $\frac{1}{2}$ particles formed in the singlet state (11.1.1) (Dewdney *et al.*, 1987b; Holland, 1988b). To do this, we derive spacetime solutions to the Pauli equation describing the evolution of the two-body system when each particle passes through one of two distantly separated Stern–Gerlach devices, and evaluate the corresponding trajectories and spin vectors of the two corpuscles (the case considered by Einstein, disturbance of one particle only, is treated in §11.3). The correlations between these quantities are due to the nonlocal actions of the quantum potential and quantum torque. A key point is that the particles carry with them well-defined properties from the source, just as classical particles do; these do not come into being only through discontinuous projections of 'potentialities' when measurements occur. On the other hand, the magnitudes of one particle's properties may be steered into certain values solely by distant actions on the other coupled particle even though the systems do not classically interact. The subtlety of these correlations cannot be explained by any classical force.

Although the experiment described here is still of the *gedanken* variety, a conceptually similar experiment on the spin correlations of protons in the singlet state has been performed (Lamehi-Rachti & Mittig, 1976). Moreover, this example is conceptually closely connected with the experiments testing correlations in photon polarization (§11.5). It is therefore valuable to show

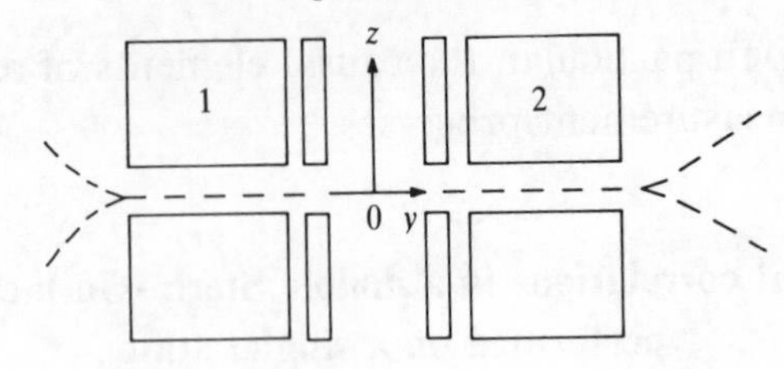

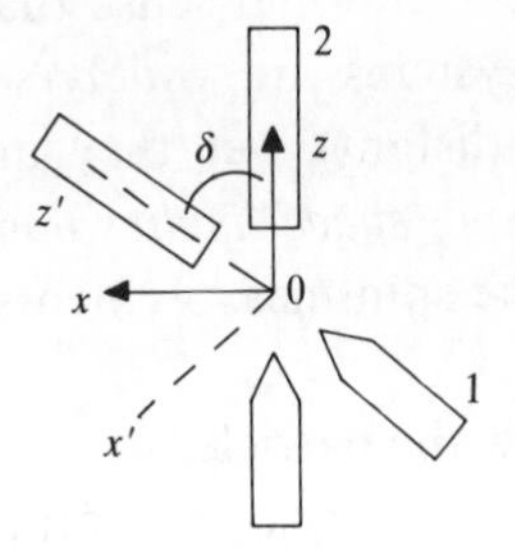

Fig. 11.1 Arrangement of two Stern–Gerlach magnets appropriate to the EPR experiment (from Dewdney *et al.* (1987b)).

how the quantum theory of motion can account for the double Stern–Gerlach experiment applied to spin $\frac{1}{2}$ particles.

The basic set-up is shown in Fig. 11.1. A pair of spin $\frac{1}{2}$ particles of mass m and magnetic moment μ is formed at 0 in a simultaneous eigenstate of the spin operator in the z-direction $(\hbar/2)(\sigma_{z_1} + \sigma_{z_2})$ and the total spin operator $(\hbar^2/4)(\boldsymbol{\sigma}_1 + \boldsymbol{\sigma}_2)^2$, with eigenvalue zero. The particles separate in the y-direction and pass through Gaussian slits oriented so as to produce packets in the directions of the analysing fields of two identical Stern–Gerlach devices. Magnet 2 is set to measure the spin in the z-direction, and magnet 1 has been rotated anticlockwise through an angle δ about the y-axis so that it has a gradient in the z'-direction.

At the entrance to the magnets the wavefunction is

$$\psi_0(z_1', z_2) = f_1(z_1')f_2(z_2)(1/\sqrt{2})(u_+v_- - u_-v_+), \tag{11.2.1}$$

where $f_1(z_1')$, $f_2(z_2)$ are normalized packets, z_1' and z_2 are the coordinates of particles 1 and 2 in the z'- and z-directions, respectively, and $\sigma_{z_1}u_\pm = \pm u_\pm$, $\sigma_{z_2}v_\pm = \pm v_\pm$. The outcome of the measurement is that the spin part of the wavefunction is left in one of the four states u'_+v_+, u'_+v_-, u'_-v_+, u'_-v_-. The state (11.2.1) predicts the following probabilities for the possible outcomes:

$$P(++) = P(--) = \tfrac{1}{2}\sin^2(\delta/2), \qquad P(+-) = P(-+) = \tfrac{1}{2}\cos^2(\delta/2). \tag{11.2.2}$$

In (11.2.1) we have suppressed the motion in the y-direction since this is not relevant to the measurement process. We only assume that the particles are sufficiently far apart on the y-axis so that they do not interact in the usual sense, and so that the measuring devices cannot influence one another.

Substituting (11.2.1) into (9.6.3) and (9.6.10) we have for the initial velocities and spin vectors:

$$\mathbf{v}_1 = \mathbf{v}_2 = \mathbf{s}_1 = \mathbf{s}_2 = 0. \tag{11.2.3}$$

These initial conditions provide a good example of how analogous quantities in the one- and two-body theories have quite different properties. The spin vectors are strictly zero, something that is not possible in the one-body case (cf. (9.3.4)). Notice in particular that the spins are not determined by either of the addends in (11.2.1), i.e., the particles are not in an initial state in which the spin of one is up (down) while the other is down (up), as one might expect in the analogous classical case. The usual informal way of speaking about the singlet state in terms of 'antiparallel spins' is, according to this model, misleading (but see §11.4). Nevertheless, the result that the particles are at rest is perfectly consistent, and evidently compatible with the zero total angular momentum eigenvalue.

The information in the singlet state implying that the corpuscles are initially correlated is contained not in the spin vectors (which are only trivially correlated in that both are zero) but is picked up by the spin correlation tensor (9.6.6). In the state (11.2.1) this becomes

$$s_{ij} = -(\hbar/2)\delta_{ij}. \tag{11.2.4}$$

This matrix indeed expresses maximum correlation; in an initial product state we would have $s_{ij} = (2/\hbar)s_{1i}s_{2j}$ (a matrix of rank one).

Eq. (11.2.1) describes a state in which the spin is independent of position and the purpose of the Stern–Gerlach devices is to introduce couplings between the spins (the measured variables) and the positions (the apparatus coordinates). The effect of the impulsive action of the inhomogeneous fields (assumed to act simultaneously) may be determined by solving (9.6.1) in the approximation where the kinetic energy operators are neglected. At the exit to each magnet two superposed packets form which separate in time according to motion described by the free Pauli equation along the directions of the analysing fields. This device then effects a simultaneous measurement of the commuting operators $\hat{s}_{z_1}$ and $\hat{s}_{z_2}$ in an obvious generalization of our treatment of §9.5. Details of the calculation are sketched in the Appendix.

Expanding in terms of products of eigenfunctions of the spin operators being measured, the normalized wavefunction at time t after the interaction with the fields is

$$\psi(z_1', z_2, t) = [\sigma(2\pi)^{1/2}s_t]^{-1} \exp\left\{-\left[\frac{z_1'^2 + z_2^2}{2} + \left(\frac{\hbar t\Delta'}{m}\right)^2\right](\varepsilon\sigma^2)^{-1}\right\}$$

$$\times \exp\left\{\frac{i\hbar t}{m\varepsilon}\left(\frac{z_1'^2 + z_2^2}{2} - \frac{\Delta'^2}{\sigma^4}\right)\right\}\left(\sin(\delta/2)\exp\left[-\frac{\hbar t\Delta'(z_1' + z_2)}{m\varepsilon\sigma^2}\right]\right.$$

$$\times \exp\left\{-i\left[2\Delta + \frac{(z_1' + z_2)\Delta'}{\varepsilon\sigma^4}\right]\right\}u'_+v_+$$

$$+ \cos(\delta/2)\exp\left[-\frac{\hbar t\Delta'(z_1' - z_2)}{m\varepsilon\sigma^2}\right]\exp\left[-\frac{i(z_1' - z_2)\Delta'}{\varepsilon\sigma^4}\right]u'_+v_-$$

$$- \cos(\delta/2)\exp\left[\frac{\hbar t\Delta'(z_1' - z_2)}{m\varepsilon\sigma^2}\right]\exp\left[\frac{i(z_1' - z_2)\Delta'}{\varepsilon\sigma^4}\right]u'_-v_+$$

$$\left. + \sin(\delta/2)\exp\left[\frac{\hbar t\Delta'(z_1' + z_2)}{m\varepsilon\sigma^2}\right]\exp\left\{i\left[2\Delta + \frac{(z_1' + z_2)\Delta'}{\varepsilon\sigma^4}\right]\right\}u'_-v_-\right), \tag{11.2.5}$$

where $s_t = (1/\sigma^2 + i\hbar t/m)$, $\sigma =$ constant, $\varepsilon = 1/\sigma^4 + (\hbar t/m)^2$, $\Delta = \mu B_0 T/\hbar$, $\Delta' = \mu B_0' T/\hbar$, B_0 is the homogeneous part of the field, B_0' is the field gradient and T is the period of interaction (s_t and σ here differ from those used in §4.7).

Immediately after the interaction with the magnets, however, the particles acquire nonzero z'- and z-components of their velocities and the quantum torques (9.6.14) act to start the particles spinning. At time t we have from (11.2.5) the density

$$\rho = (2\pi\sigma^2\varepsilon)^{-1}\, e^{-[z_1'^2 + z_2^2 + 2(\hbar t\Delta'/m)^2]/\varepsilon\sigma^2}\,\Omega, \tag{11.2.6}$$

where

$$\Omega = 2\sin^2(\delta/2)\cosh[2\hbar t\Delta'(z_1' + z_2)/m\varepsilon\sigma^2] + 2\cos^2(\delta/2)\cosh[2\hbar t\Delta'(z_1' - z_2)/m\varepsilon\sigma^2] \tag{11.2.7}$$

and for particle 1

$$\left.\begin{aligned} s_{x_1'} &= \frac{\hbar}{\Omega}\sin\delta\cos\left[2\left(\Delta + \frac{z_1'\Delta'}{\varepsilon\sigma^4}\right)\right]\sinh\left(\frac{2\hbar t\Delta' z_2}{m\varepsilon\sigma^2}\right) \\ s_{y_1'} &= \frac{\hbar}{\Omega}\sin\delta\sin\left[2\left(\Delta + \frac{z_1'\Delta'}{\varepsilon\sigma^4}\right)\right]\sinh\left(\frac{2\hbar t\Delta' z_2}{m\varepsilon\sigma^2}\right) \end{aligned}\right\} \tag{11.2.8}$$

(continued)

$$s_{z'_1} = \frac{\hbar}{\Omega}\left(\sin^2(\delta/2)\sinh\left[-\frac{2\hbar t\Delta'(z'_1 + z_2)}{m\varepsilon\sigma^2}\right] + \cos^2(\delta/2)\sinh\left[-\frac{2\hbar t\Delta'(z'_1 - z_2)}{m\varepsilon\sigma^2}\right]\right) \tag{11.2.8}$$

$$v_{z'_1} = \frac{z'_1\hbar^2 t}{m^2\varepsilon} - \frac{2\Delta'}{m\varepsilon\sigma^4}s_{z'_1} \tag{11.2.9}$$

while for particle 2

$$\left.\begin{aligned} s_{x_2} &= \frac{\hbar}{\Omega}\sin\delta\cos\left[2\left(\Delta + \frac{z_2\Delta'}{\varepsilon\sigma^4}\right)\right]\sinh\left(-\frac{2\hbar t\Delta' z'_1}{m\varepsilon\sigma^2}\right) \\ s_{y_2} &= \frac{\hbar}{\Omega}\sin\delta\sin\left[2\left(\Delta + \frac{z_2\Delta'}{\varepsilon\sigma^4}\right)\right]\sinh\left(-\frac{2\hbar t\Delta' z'_1}{m\varepsilon\sigma^2}\right) \\ s_{z_2} &= \frac{\hbar}{\Omega}\left(\sin^2(\delta/2)\sinh\left[-\frac{2\hbar t\Delta'(z'_1 + z_2)}{m\varepsilon\sigma^2}\right] + \cos^2(\delta/2)\sinh\left[\frac{2\hbar t\Delta'(z'_1 - z_2)}{m\varepsilon\sigma^2}\right]\right) \end{aligned}\right\} \tag{11.2.10}$$

$$v_{z_2} = \frac{z_2\hbar^2 t}{m^2\varepsilon} - \frac{2\Delta'}{m\varepsilon\sigma^4}s_{z_2} \tag{11.2.11}$$

with

$$\mathbf{s}_1^2 = \mathbf{s}_2^2 = \frac{\hbar^2}{4\Omega^2}(\Omega^2 - 4). \tag{11.2.12}$$

The components of $\mathbf{s}_1$ in the xyz system are easily found from

$$s_{x_1} = s_{x'_1}\cos\delta - s_{z'_1}\sin\delta, \qquad s_{y_1} = s_{y'_1}, \qquad s_{z_1} = s_{x'_1}\sin\delta + s_{z'_1}\cos\delta. \tag{11.2.13}$$

There are, of course, no additions to the initial components of the velocities in the y-direction (which we have ignored) from the impulsive action of the fields.

The general implications of these results are as follows. After leaving the magnets, each particle joins one of the separating packets. At first the spin vectors do not (in general) lie along the directions of the analysing fields, but eventually ($t \to \infty$) we obtain

$$\left.\begin{aligned} &s_{x'_1} = s_{y'_1} = 0, \quad s_{z'_1} = \pm\hbar/2, \quad s_{x_2} = s_{y_2} = 0, \quad s_{z_2} = \pm\hbar/2 \\ &s_{z'z} = \pm\hbar/2, \quad \text{all other } s_{i'j} = 0. \end{aligned}\right\} \tag{11.2.14}$$

Here we have referred the spin correlation tensor to two sets of space axes (whose z-axes coincide with the directions of the magnetic fields). The velocities (11.2.9) and (11.2.11) depend on the spins and the trajectories are correlated to these outcomes (e.g., motion of particle 1 in the $+z'$-direction corresponds to $s_{z'_1} = +\hbar/2$). The precise result obtained in each trial is fixed by the initial (uncontrollable) positions of the particles in the packets. Notice that *both* initial conditions contribute to the determination of *each* outcome, and that the correlations persist, in principle, to arbitrary interparticle distances. The probabilities (11.2.2) are reflected in the relative proportions of trajectories which move up or down on leaving the magnets. Changing the relative orientation of the magnets changes the total quantum state and hence implies different correlated motions. It is this feature that makes contact with Bohr's emphasis on the necessity for taking into account the complete experimental context in describing the outcomes of measurements, but we go on to explain how this context influences the results.

To gain further insight into this nonlocal action we consider two important special cases for which the trajectories and spin vectors are explicitly plotted – the case where the magnets are aligned ($\delta = 0$, this section) and the case of measurement on particle 1 only (next section).

In the case where $\delta = 0$ only the u_+v_-, u_-v_+ terms survive in (11.2.5) so that $s_{x_1} = s_{y_1} = s_{x_2} = s_{y_2} = 0$,

$$s_{z_1} = -s_{z_2} = \frac{\hbar}{2}\tanh\left[-\frac{2\hbar t\Delta'(z_1 - z_2)}{m\varepsilon\sigma^2}\right] \tag{11.2.15}$$

and

$$v_{z_1} = \frac{z_1\hbar^2 t}{m^2\varepsilon} - \frac{2\Delta'}{m\varepsilon\sigma^4}s_{z_1}, \qquad v_{z_2} = z_2\frac{\hbar^2 t}{m^2\varepsilon} - \frac{2\Delta'}{m\varepsilon\sigma^4}s_{z_2}. \tag{11.2.16}$$

The motion of each particle for any pair of trajectories depends sensitively on the choice of both initial positions at the entrance slits to the Stern–Gerlach devices. Eq. (11.2.15) shows, however, that the spins always come out to be opposite regardless of the initial positions. The results plotted in Fig. 11.2 were calculated by taking the initial position of particle 1 to be fixed in each case and then finding the correlated trajectories which develop for a representative range of initial positions of particle 2. When the initial position of particle 1, $z_1(0)$, is chosen to be equal to that of particle 2, $z_2(0)$, we obtain a bifurcation point. If $z_2(0) < z_1(0)$, then particle 2 has a negative velocity and s_{z_2} decreases from 0 to $-\hbar/2$ whilst the corresponding particle 1 has a positive velocity and s_{z_1} increases from 0 to $\hbar/2$. Analogous correlations are found if $z_2(0) > z_1(0)$. In Fig. 11.3 we illustrate the same phenomenon with a different choice of the constant $z_1(0)$.

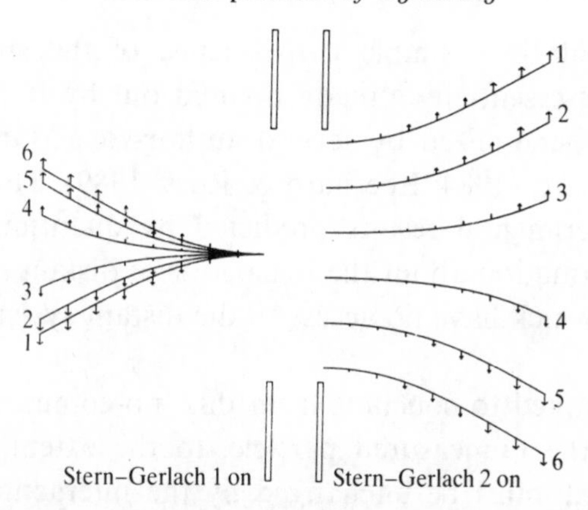

Fig. 11.2 Correlated pairs of trajectories and spin vector magnitudes (indicated by the length of the arrows) after impulsive spin measurements in the z-direction on both particles, for fixed $z_1(0)$ and variable $z_2(0)$ (from Dewdney *et al.* (1987b)).

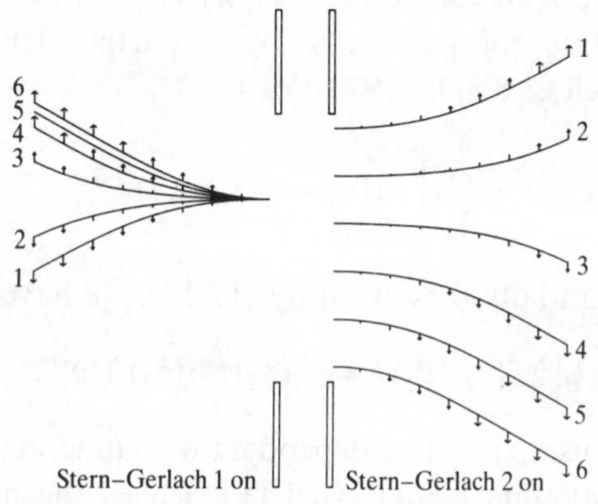

Fig. 11.3 The arrangement of Fig. 11.2 with a different $z_1(0)$ (from Dewdney *et al.* (1987b)).

11.3 The problem of signalling

To discuss the question of whether the correlations described above constitute in some sense a 'signal' linking two remote spatial regions, we now treat the case where a spin measurement is made on one particle only, say 1 (Holland, 1988b; Holland & Vigier, 1988). We first consider this question from the conventional point of view.

It is generally accepted that one cannot use EPR-type correlated systems to transfer information from one region of space to another. This may be demonstrated by showing that if one has a pair of nonclassically interacting particles 1, 2 in an entangled state, then the expectation value of any operator pertaining to just 2 is independent of any interaction that particle 1 may undergo. This result is a simple consequence of the standard quantum formalism and was essentially already pointed out by Bohm (1951). It has been restated and generalized by several authors (e.g., Ghirardi, Rimini & Weber, 1980; Shimony, 1984; Eberhard & Ross, 1989). Thus, at the level of the statistical experimental results predicted by quantum mechanics, one cannot derive information about the behaviour of distant systems purely by local experiments which have no access to the distant systems by some other means.

One may be tempted to conclude from this 'no-communication' theorem that the state of the unmeasured particle, to the extent that this can be individually defined, must be unchanged by the interaction carried out on the distant particle. Within the conventional interpretation (which only defines states in terms of wavefunctions or density matrices) that is indeed so, as we shall now see.

Given a pure-state wavefunction $\psi^{ab}(\mathbf{x}_1, \mathbf{x}_2, t)$ for a two-body system, one may define the 'state' of particle 2 by a partial density matrix, with components (Cantrell & Scully, 1978) (cf. §7.7)

$$\rho^{(2)ab}(\mathbf{x}_2, \mathbf{x}_2', t) = \sum_c \int \mathrm{d}^3x_1 \psi^{ca}(\mathbf{x}_1, \mathbf{x}_2, t)\psi^{*cb}(\mathbf{x}_1, \mathbf{x}_2', t). \tag{11.3.1}$$

In our case, where initially ψ is given by (11.2.1), we have at $t = 0$

$$\rho_0^{(2)ab}(z_2, (z_2)') = f_2(z_2)f_2^*((z_2)')\tfrac{1}{2}\delta^{ab}. \tag{11.3.2}$$

Explicit calculation using the time-dependent wavefunction derived in (11.3.5) below shows that at time t after particle 1 leaves the magnet, the partial density matrix (11.3.1) is given by

$$\rho^{(2)ab}(z_2, (z_2)', t) = f_2(z_2, t)f_2^*((z_2)', t)\tfrac{1}{2}\delta^{ab}. \tag{11.3.3}$$

This is independent of the interaction undergone by 1 and, comparing with (11.3.2), we see that its evolution is due purely to the spreading of the particle 2 wave packet. It is then indeed true that the state of the unmeasured particle, insofar as this is completely defined by (11.3.1), is unaltered by a measurement performed on a distant particle with which it enters into a total entangled state.

Because all the statistical properties of system 2 are embodied in the partial density matrix, this indeed constitutes a demonstration of the 'no-signalling' theorem of quantum mechanics. Note that the argument may be generalized to include any kind of disturbance of system 1, and is independent of the wave packet collapse hypothesis as applied to either system.

However, (11.3.3) can be written in the form

$$\rho^{(2)ab}(z_2,(z_2)',t)=\sum_{i=1,2} p_i\psi_i^a(z_2,t)\psi_i^{*b}((z_2)',t) \tag{11.3.4}$$

where $p_i=\frac{1}{2}$, $i=1,2$, and $\psi_i^a(z_2,t)=f_2(z_2,t)\delta_i^a$. The partial density matrix thus describes a mixture (of spin up and down states) and so contains less information than the total pure-state wavefunction out of whose components it is constructed. It is therefore not ruled out that a more precise specification of the state of a particle could pick up the further information in ψ not present in $\rho^{(2)}$, and that, in fact, this state *can* change as a result of distant interactions. This is, of course, precisely what happens according to the causal interpretation where the 'state' of a particle in a many-body system is specified by well-defined values of properties determined by the wavefunction of the total system. It is then possible that this more detailed characterization of a state will reflect the finer structure of ψ. The possibility of changes in this state brought about by local or nonlocal actions of the quantum potential and torque does not contradict the statistical noncommunication result stated above, of course. We can, however, say something more about the individual processes that build up the statistical ensemble.

When we measure the operator $\hat{s}_{z_1}=(\hbar/2)\sigma_{z_1}$, by passing particle 1 through a Stern–Gerlach device, but leave particle 2 alone, the wavefunction (11.2.1) develops into

$$\psi^{ab}(z_1,z_2,t)=f_2(z_2,t)c^{ab}(z_1,t), \tag{11.3.5}$$

where $c^{11}=c^{22}=0$, $c^{12}=U(z_1,t)$, $c^{21}=U'(z_1,t)$, and U and U' are packet functions depending on the details of the interaction. The initial conditions are given by (11.2.3). Explicitly, at time t after the magnetic interaction ceases, the wavefunction is given by

$$\begin{aligned}\psi(z_1',z_2,t)&=[(2\pi)^{1/2}\sigma s_t]^{-1}\exp\left\{-\left[z_1'^2+z_2^2+\left(\frac{\hbar t\Delta'}{m}\right)^2\right](2\varepsilon\sigma^2)^{-1}\right\}\\&\quad\times\exp\left[\frac{i\hbar t}{2m\varepsilon}\left(z_1'^2+z_2^2-\frac{\Delta'^2}{\sigma^4}\right)\right]\\&\quad\times\left\{\exp\left(-\frac{\hbar t\Delta' z_1'}{m\varepsilon\sigma^2}\right)\exp\left[-i\left(\Delta+\frac{z_1'\Delta'}{\varepsilon\sigma^4}\right)\right]u'_+v'_-\right.\\&\quad\left.-\exp\left(\frac{\hbar t\Delta' z_1'}{m\varepsilon\sigma^2}\right)\exp\left[i\left(\Delta+\frac{z_1'\Delta'}{\varepsilon\sigma^4}\right)\right]u'_-v'_+\right\}\end{aligned} \tag{11.3.6}$$

where $v'_+=\cos(\delta/2)v_+-\sin(\delta/2)v_-$, $v'_-=\sin(\delta/2)v_++\cos(\delta/2)v_-$, which implies that

$$s_{x'_1} = s_{y'_1} = 0, \qquad s_{z'_1} = \frac{\hbar}{2}\tanh\left(-\frac{2\hbar t \Delta' z'_1}{m\varepsilon\sigma^2}\right) \tag{11.3.7}$$

$$v_{z'_1} = \frac{z'_1 \hbar^2 t}{m^2 \varepsilon} - \frac{2\Delta'}{m\varepsilon\sigma^4} s_{z'_1} \tag{11.3.8}$$

$$s_{x_2} = \sin\delta\, s_{z'_1}, \quad s_{x'_2} = 0, \quad s_{y_2} = s_{y'_2} = 0, \quad s_{z_2} = -\cos\delta\, s_{z'_1}, \quad s_{z'_2} = -s_{z'_1} \tag{11.3.9}$$

$$v_{z_2} = z_2 \frac{\hbar^2 t}{m^2 \varepsilon}. \tag{11.3.10}$$

We see from (11.3.7) and (11.3.8) that the behaviour of particle 1 is independent of particle 2; the velocity and spin depend only on z'_1. The spin vector points along the direction of the field and changes continuously from 0 to $\hbar/2$ $(-\hbar/2)$ if the initial position on the z'-axis is above (below) $z' = 0$.

The fate of particle 2 on the other hand is dependent on the motion of 1. From (11.3.10) it is seen that the *trajectory* of 2 is unaffected by 1, and we may solve to find

$$z_2(t) = z_2(0)\sigma^2\left[\frac{1}{\sigma^4} + \left(\frac{\hbar t}{m}\right)^2\right]^{1/2} \tag{11.3.11}$$

i.e., the hyperbola which follows from the natural spread of the wave packet (§4.7). The component of the *spin vector* of 2 which lies along the z'-direction, however, changes continuously from zero to a finite value $(\mp\hbar/2)$ simultaneously with the commencement of rotation of 1, and is at all times equal and opposite to the spin of particle 1. The spin of 2 thus depends sensitively on the position of particle 1 through the spin vector of 1, as can be seen from (11.3.9), whereas the respective trajectories depend solely on the initial positions of the particles and the local environment.

When the beam containing particle 1 splits at the exit to the magnet and the particle enters one or other of the separating packets, the beam containing particle 2 does not split – particle 2 remains in the same packet, but now it is rotating about the z'-axis due to the nonlocal action of the quantum torque (9.6.14) (Fig. 11.4).

The sense of rotation is determined by the initial position of particle 1. If particle 1 is in the upper (lower) half of the bifurcating beam, it will rotate clockwise (anticlockwise) about the z'-axis and will be in a spin up (down) eigenstate of $\sigma_{z'}$. Particle 2 will have the opposite sense of rotation regardless of its position. If we subsequently pass particle 2 through a Stern–Gerlach

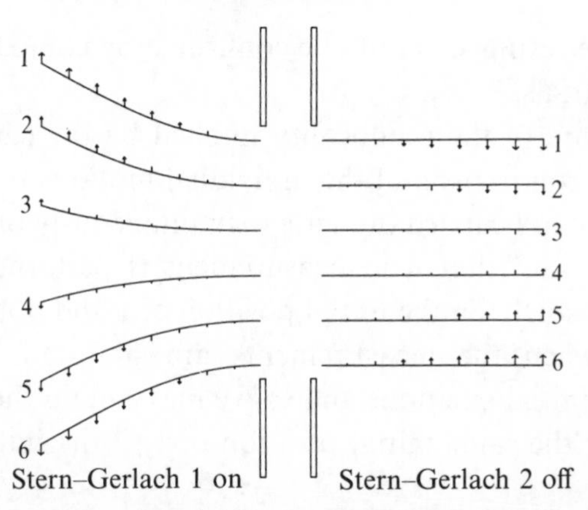

Fig. 11.4 Trajectories and correlated spin vectors for two particles after the impulsive measurement of the z'-component of the spin of particle 1 only (from Dewdney *et al.* (1987b)).

device oriented in the z'-direction, then we will, of course, obtain the opposite result to that found for particle 1, since all trajectories in a spin up (down) eigenstate join the upper (lower) beam (§9.5).

The state of particle 2, as we have defined it, has therefore changed as a result of a distant interaction undergone by another particle, both particles being guided by an entangled state. A demon sitting on 2 receives an instruction to start the ball rotating. It would thus appear that, at the level of the individual processes that it accounts for, the causal interpretation implies a form of 'signalling', via the quantum torque.

There is no way of extracting this transmission of information (spin angular momentum) between the sites of 1 and 2 experimentally, however. Because in quantum mechanics (and hence in the causal interpretation) we can do no more than predict the likely outcome of a measurement, we have no way of knowing what result would have been obtained for 2 had we made no measurement on 1. There is no standard against which we could detect a change. The kind of distant interconnections implied by the quantum potential approach (whether local or nonlocal) cannot be detected by any experiment that is also subject to the laws of quantum mechanics.

While it is hard to see in the conventional point of view how the instantaneous correlations in the results of spin measurements on 1 and 2 come about, given that the state (11.3.1) of 2 is unchanged, the causal interpretation provides an explanation in each individual case.

The act of measurement on 1 polarizes 2 (in the direction of the analysing field acting on 1) and in any subsequent measurement on 2, the results will come out in the way predicted by quantum mechanics. The existence of correlations in the results can only be confirmed by comparing the outcomes in a sequence of trials.

The subtle nature of the nonlocality implied by the causal interpretation may be seen by a comparison of the individual motions that result from spin measurements that are carried out either simultaneously or staggered in time. We have already said that if a measurement is performed on 1 only, the outcome depends solely on the initial position of 1 and not on 2. But we have also seen that when the measurements are simultaneous, any outcome depends on both initial positions and so we may obtain the *opposite* outcome for particle 1 (for the same initial position of 1) if the initial position of 2 is suitably chosen.

Consider, for example, the correlated trajectories 3 in the figures. In Fig. 11.4 particle 1 will always move upward regardless of the initial position of particle 2. In Figs. 11.2 and 11.3 on the other hand (§11.2) particle 2 moves up or down depending on the value of $z_1(0)$. The identical statistics obtained in simultaneous or staggered measurements result in part therefore from quite different evolutions at the level of the individual particles which make up an ensemble. The nonlocal mechanism at work in the two cases is not the same, but they cannot be experimentally distinguished.

It is notable that the failure of quantum correlations to provide a signalling mechanism at the empirical level is consistent with the requirement of special relativity that no signal be transmitted faster than the speed of light. Yet the statistical compatibility of quantum mechanics with relativity in this case seems to be something of an accident, for what is forbidden is signalling of *any* kind (sub- or super-luminal) and the exclusion of superluminal connections is just a special case of this general result. It seems indeed that the instantaneous connections between distant regions of space, which according to the quantum potential approach underlie the correlations, at least contradict the spirit of relativity which emphasizes local propagation by contiguous field elements. These features (nonlocality in the individual process, statistical locality) persist in the relativistic quantum theory of motion (Chap. 12).

To summarize, the quantum potential implies that a certain kind of 'signalling' does, in fact, take place between the sites of distantly separated spin $\frac{1}{2}$ particles in an entangled state, if one of the particles undergoes a local interaction. This transfer of information cannot, however, be extracted by any experiment which obeys the laws of quantum mechanics. The causal interpretation thus provides an explanation of how the correlations come about in each individual process, in a way that is consistent with the statistical noncommunication of information.

11.4 More general treatment: the rigid rotator

Referring to §10.3.2, we consider a system of two identical spheres in a singlet state. Neglecting the space dependence, the appropriate solution to (10.3.28) is

$$\psi = \mathrm{e}^{-\mathrm{i}Et/\hbar}\frac{1}{\sqrt{2}}(u_{1+}u_{2-} - u_{1-}u_{2+}), \tag{11.4.1}$$

where $E = 3\hbar^2/4I$. This is a simultaneous eigenstate of the operators $(\hat{M}_{z_1} + \hat{M}_{z_2})$ and $(\hat{\mathbf{M}}_1 + \hat{\mathbf{M}}_2)^2$, with eigenvalue zero. The amplitude and phase of the function (11.4.1) are given by

$$R = \frac{1}{8\sqrt{2}\pi^2}\left[\sin^2\frac{\alpha_1-\alpha_2}{2}\cos^2\frac{\beta_1-\beta_2}{2} + \sin^2\frac{\alpha_1+\alpha_2}{2}\sin^2\frac{\beta_1-\beta_2}{2}\right]^{1/2}, \tag{11.4.2}$$

$$\frac{S}{\hbar} = -\frac{Et}{\hbar} - \tfrac{1}{2}(\pi+\gamma_1+\gamma_2) + \tan^{-1}\left[\tan\frac{\beta_2-\beta_1}{2}\sin\frac{\alpha_1+\alpha_2}{2}\Big/\sin\frac{\alpha_2-\alpha_1}{2}\right]. \tag{11.4.3}$$

From the relations (10.3.8) applied to each body, we find that the spin vector fields $\mathbf{M}_1(\alpha_{1i}, \alpha_{2i})$ and $\mathbf{M}_2(\alpha_{1i}, \alpha_{2i})$ must satisfy the relation

$$\mathbf{M}_1 + \mathbf{M}_2 = 0 \tag{11.4.4}$$

with respect to the space axes. The spin vectors of the particles guided by the singlet state are therefore 'antiparallel', in accordance with the usual imagery. The quantum state does not, of course, fix which direction the vectors lie in. The latter is determined by the choice of initial angular coordinates.

Substituting S into the two sets of guidance equations (10.3.14) for $\dot{\alpha}_{1i}$ and $\dot{\alpha}_{2i}$, $i = 1, 2, 3$, we have to solve six simultaneous equations. The solutions to these depend sensitively on the initial conditions. Assuming at $t = 0$ that

$$\alpha_1 = \beta_1 = \beta_2 = \gamma_1 = \gamma_2 = 0, \qquad \alpha_2 = \pi, \tag{11.4.5}$$

we derive the following solution at time t:

$$\alpha_1 = \beta_1 = \beta_2 = 0, \qquad \alpha_2 = \pi, \qquad \gamma_1 = \gamma_2 = -\hbar t/2I. \tag{11.4.6}$$

Since only γ_1 and γ_2 are time dependent, the components of the two spin vectors $\mathbf{M}_1$ and $\mathbf{M}_2$ derived from (11.4.6) are constant. With respect to the space axes we have

$$\mathbf{M}_1 = (\hbar/2)(0, 0, 1), \qquad \mathbf{M}_2 = (\hbar/2)(0, 0, -1). \tag{11.4.7}$$

and with respect to the body axes $\mathbf{M}_1 = \mathbf{M}_2 = (\hbar/2)(0, 0, 1)$.

The centres of mass of the particles are at rest, and they are spinning about their respective third body axes in opposite senses with a constant angular velocity of $\hbar/2I$. In fact, if we choose $\alpha = 0$ (so that β and γ add linearly) in (10.5.4) and (10.5.6) we see that these respective motions correspond to one particle (1) being in a state of 'spin up' and the other (2) in a state of 'spin down'. The energy of the system comprises rotational kinetic energy and spin-dependent quantum potential energy, but the quantum torques are zero (since the angular velocities are constant). It is emphasized that this is only true for the specific choice of initial conditions (11.4.5). If initially $\alpha_1 \neq 0$, say, then the subsequent motion of each particle is quite different from that of spin up or spin down, as defined in §10.5 for a single body. Once again we see how the fate of an individual is influenced by the total context. A nonclassically interacting body which is part of a wider system is not 'free', since it is subjected to forces and torques derived from the total quantum state.

Suppose now that we measure the spin of particle 1 by passing it through a Stern–Gerlach device oriented in the z-direction (making the same assumptions as in §11.3). This may be treated by adding the term $(2\mu_1/\hbar)B_z\hat{M}_{z_1}$ to the Hamiltonian in (10.3.28). At time t after 1 leaves the magnet, the state (11.4.1) evolves into

$$\psi(z_1, z_2, \alpha_{1i}, \alpha_{2i}, t)$$
$$= e^{-iEt/\hbar}(\psi^{+-}(z_1, z_2, t)u_{1+}u_{2-} - \psi^{-+}(z_1, z_2, t)u_{1-}u_{2+}), \quad (11.4.8)$$

where the coefficients $\psi^{\pm\mp}$ satisfy the two-body Pauli equation and are given by (11.3.6). The general characteristics of the motion are the same as in the Pauli case, although the details differ. It is easy to see that the phase of (11.4.8) is of the form

$$S(z_1, z_2, \alpha_{1i}, \alpha_{2i}, t) = g(z_2, t) + h(z_1, \alpha_{1i}, \alpha_{2i}, t) \quad (11.4.9)$$

and therefore that the trajectory of 2 (the integral curves of $v_2 = (1/m)\,\partial S/\partial z_2$) is unaffected by the interaction undergone by 1. Independent of the initial conditions, the Euler angles α_{2i} describing the orientation of 2 on the other hand evolve so that they eventually define a state of rotation similar to that of a single body in a state of spin up or spin down (§10.5.1). The axis of the cone on which the spin vector lies points along the z-axis, and is opposite to that of particle 1. The particular outcome that is found for 1 is determined by *both* sets of initial orientation parameters (11.4.5) and by the initial position of 1 – but not the initial position of 2. The remarks made in §11.3 on the transfer of information in such a process apply here.

11.5 Is there a local explanation?

11.5.1 Bell's inequality

The preceding analysis enables us to see the manner in which the assumptions made by Bell (1964), in his derivation of an inequality that any local hidden-variables theory must apparently satisfy, are violated in the causal interpretation. In discussing the EPR spin experiment Bell supposed that the results of the two spin measurements are determined completely by a set of hidden variables λ and made two assumptions which he claimed should be satisfied by a local hidden-variables theory:

(i) The result A of measuring $\boldsymbol{\sigma}_1 \cdot \mathbf{a}$ on particle 1 is determined solely by $\mathbf{a}$ and λ, and the result B of measuring $\boldsymbol{\sigma}_2 \cdot \mathbf{b}$ on particle 2 is determined solely by $\mathbf{b}$ and λ, where $\mathbf{a}$ and $\mathbf{b}$ are unit vectors with $\mathbf{a} \cdot \mathbf{b} = \cos \delta$. Thus

$$A = A(\mathbf{a}, \lambda) = \pm 1, \qquad B = B(\mathbf{b}, \lambda) = \pm 1.$$

Possibilities such as $A = A(\mathbf{a}, \mathbf{b}, \lambda)$, $B = B(\mathbf{a}, \mathbf{b}, \lambda)$ are excluded.

(ii) The normalized probability distribution of the hidden variables depends only on λ: $\rho = \rho(\lambda)$. Possibilities such as $\rho = \rho(\lambda, \mathbf{a}, \mathbf{b})$ are excluded.

From assumptions (i) and (ii) and the definition of the correlation function of the product of the results A and B,

$$P(\mathbf{a}, \mathbf{b}) = \int \mathrm{d}\lambda\, \rho(\lambda) A(\mathbf{a}, \lambda) B(\mathbf{b}, \lambda), \tag{11.5.1}$$

Bell deduced the following inequality which is to be satisfied by the functions (11.5.1) ($\mathbf{c}$ is a further unit vector):

$$|P(\mathbf{a}, \mathbf{b}) - P(\mathbf{a}, \mathbf{c})| \leqslant 1 + P(\mathbf{b}, \mathbf{c}). \tag{11.5.2}$$

We now consider to what extent assumptions (i) and (ii) are valid in the causal interpretation, restricting attention to the Pauli case. The hidden variables λ are then the particle positions $\mathbf{x}_1, \mathbf{x}_2$ (the internal orientation spin vectors $\mathbf{s}_1, \mathbf{s}_2$ along the trajectories are determined by the positions and the wavefunction; in the rigid rotator case, the hidden variables include the Euler angles). In the case of staggered measurements, it follows from §11.3 that $A = A(\mathbf{x}_1, \mathbf{a})$ and $B = B(\mathbf{x}_1, \mathbf{x}_2, \mathbf{a} \cdot \mathbf{b})$. When the measurements are performed simultaneously, it follows from §11.2 that which of the eventual results $\pm\hbar/2$ is obtained for each of $s_{z_1'}$ and s_{z_2} is determined by the initial positions of both particles and by δ, i.e., $A = A(\mathbf{x}_1, \mathbf{x}_2, \mathbf{a} \cdot \mathbf{b})$, $B = B(\mathbf{x}_1, \mathbf{x}_2, \mathbf{a} \cdot \mathbf{b})$. Thus assumption (i) is not valid in either case. Neither is assumption (ii) satisfied. Initially, the singlet state (11.2.1) implies that $\rho(= \psi^\dagger \psi)$ is factorizable and independent of δ. But as the measuring process develops, the probability distribution of positions comes to depend on δ, that is $\rho(t) = \rho(\mathbf{x}_1, \mathbf{x}_2, \mathbf{a}\cdot\mathbf{b})$.

Using the probabilities (11.2.2), the quantum mechanical correlation function corresponding to the joint measurement of the two spin operators is given by

$$\begin{aligned} P_\psi(\mathbf{a}, \mathbf{b}) &= \tfrac{1}{2}\sin^2(\delta/2) + \tfrac{1}{2}\sin^2(\delta/2) - \tfrac{1}{2}\cos^2(\delta/2) - \tfrac{1}{2}\cos^2(\delta/2) \\ &= -\cos\delta. \end{aligned} \tag{11.5.3}$$

This function does not satisfy the inequality (11.5.2). In reproducing the formula (11.5.3), we see that the causal interpretation disobeys both of Bell's basic assumptions. Although we have considered only the singlet state, it may be shown that any nonproduct state violates a Bell-type inequality (Gisin, 1991), a result that holds true for high quantum numbers (Gisin & Peres, 1992).

The correlation function (11.5.3) is the expectation value of the product operator $(\boldsymbol{\sigma}_1\cdot\mathbf{a})(\boldsymbol{\sigma}_2\cdot\mathbf{b})$ in the singlet state (11.2.1), which is how (11.5.3) is usually quoted in the literature. Note that in this chapter we have described the joint measurement of the two commuting spin operators $(\hbar/2)\boldsymbol{\sigma}_1\cdot\mathbf{a}$ and $(\hbar/2)\boldsymbol{\sigma}_2\cdot\mathbf{b}$, and not the direct measurement of the product operator. In fact, for a given element of the ensemble of systems associated with the same initial state, the result of a direct measurement of the product operator (e.g., $+(\hbar/2)^2$) may not coincide with the product of the measurement outcomes when the individual operators are measured jointly on the same system (e.g., $-(\hbar/2)\times(\hbar/2)$). We can deduce the statistical properties of the results of measuring the product operator from measurements of the individual operators, but not conversely. If a physical interaction could be found to effect a measurement of the product operator, the process could be naturally incorporated as a further instance of the causal theory of measurement developed in Chap. 8, §9.5 and this chapter in which the relevant system property being 'measured' is the spin correlation tensor. The evolution of this quantity in the circumstances of such a measurement will be the same as given in (11.2.14).

The experimental tests of Bell's theorem have centred on investigations of correlations in the polarization properties of entangled two-photon states of the electromagnetic field (for reviews see Clauser & Shimony (1978), Duncan & Kleinpoppen (1988) and Redhead (1989)). A particularly significant test is that of Aspect, Dalibard & Roger (1982) in which the orientations of the polarizers were switched during the flight of the photons from the source to the detectors. To date, the experiments have largely confirmed the predictions of quantum mechanics in this case and hence demonstrated the empirical violation of Bell's inequality. One may reasonably conclude that there is presently good evidence that if an objective theory of individual quantum processes is to be at all possible, it must involve some kind of nonlocality or at least imply some radical revision of customary assumptions in physics (see below).

With regard to a causal treatment of the optical experiments testing Bell's inequality, we expect to find nonlocal actions in the electromagnetic field analogous to those described in this chapter. A detailed treatment of this problem starting from the causal interpretation of quantum field theory has not yet been given, and would require developing the methods described in Chap. 12 to include polarization phenomena.

11.5.2 Further remarks

It perhaps cannot be stated with absolute certainty that Bell's theorem rules out forever the theoretical possibility of a local objective model that is compatible with all the quantum mechanical predictions in these types of experiments ('. . . what is proved by impossibility proofs is lack of imagination.' (Bell, 1982)). For instance, one of the tacit assumptions in the proof is that systems which appear to be widely separated in space actually *are* separate. Given the conventional assumptions regarding the simply-connected topology of Euclidean space (or more accurately in this context, Minkowskian spacetime) this is a reasonable assumption. However, if we dropped the requirement of simple-connectedness and admitted multiply-connected spacetime topologies, the situation with regard to the implications of Bell's theorem might change. We can, for example, envisage the existence of handles or 'wormholes' embedded in the spacetime manifold so that regions which appear to be distantly (space-like) separated in the external space can be only a short distance apart along a route traversing the wormhole. We shall briefly pursue this point not as a serious suggestion for a candidate theory to explain quantum nonlocality, but as an illustration of the problems faced by a programme determined to hang on to the principle of locality in individual processes.

Suppose a source emits a pair of separating spin $\frac{1}{2}$ particles in what quantum mechanics would describe as a singlet state. To offer an alternative representation of this state, we first assume that each particle may be attributed its own wavefunction (and that single-body quantum mechanics is valid and 'local'). The wavefunctions are eigenfunctions of the two spin operators in some direction, say z, such that when one is spin up the other is spin down. To account for the statistics of spin measurements pertaining to each particle alone, we suppose that for an ensemble of pairs emitted by

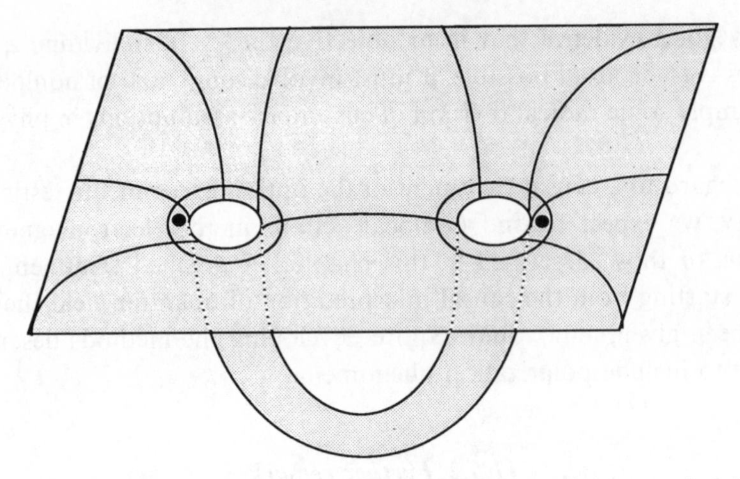

Fig. 11.5 Might entangled particles communicate via a wormhole?

the source the spins are randomly distributed over all directions with equal probability. Now we have to account for the correlations between the two sets of measurements. To do this, we propose that what in quantum mechanics would be expressed as entanglement of the total wavefunction may be represented here through a modification of the spacetime structure in which spacetime possesses a wormhole through which physical signals can propagate (Fig. 11.5). We suppose that the particles lie in the vicinity of the wormhole mouths (in a more sophisticated model we might identify the mouths with the 'particles'). If now particle 1 interacts with a Stern–Gerlach device, the impulse is required to set up a disturbance which propagates subluminally to the site of 2 via the wormhole. In order to reproduce the correlations predicted by quantum mechanics between the outcomes of spin measurements on 1 and 2, this disturbance would have to have the ability to impart to 2 a spin vector equal and opposite to that imparted to 1 by the analysing field (cf. §11.3). Since, in practice, two measurements will never be truly simultaneous, we suppose that we can always make the wormhole short enough so that by the time the measurement on 2 is carried out, it has already received the information from the site of 1 concerning which direction it should be polarized in.

Such a model obviously poses severe technical and conceptual difficulties. The nature of the propagating disturbance and the mechanism of its production would have to be established. To orient the spin vector of 2 in the opposite direction to that of 1 regardless of the initial polarization, the signal would presumably require the properties of some kind of retarded quantum potential. One would then need to know its equation of motion (and, of course, that of the underlying spacetime geometry). Next, topologies of this sort potentially imply violations of causality, the very property the model was designed to save, although this can perhaps be circumvented (Morris, Thorne & Yurtsever, 1988; Friedman *et al.*, 1990). And finally, if after the measurements are complete the systems are no longer correlated, some mechanism would have to be specified whereby the topology of spacetime may change (from multiply- to simply-connected).

Although this suggests an interesting possible connection between general relativity and quantum mechanics, this is all getting pretty daunting and one may reasonably ask what would be gained even if the theoretical consistency of such an approach could be demonstrated. For, outside the specific context of the spin correlation problem, any putative local theory would have to reproduce the entire range of phenomena successfully treated at present by many-body quantum mechanics (energy levels, molecular structure, superconductivity and so on). And the theoretical options on constructing a local model of many-body theory have been severely closed down by the observation that the nonlocal correlations grow considerably in strength as the number of particles in the system increases (Greenberger *et al.*, 1990; Mermin, 1990). One may also question the point of such a programme from the perspective that the 'locality' that is assumed to be intrinsic to the one-body theory is not itself definitively established and understood (see §11.6).

Another option is that quantum mechanics itself will break down in the circumstances of more refined EPR experiments (for example, if the inter-detector distance is large enough) and that if this were the case some kind of local theory may prevail. However, when it is recalled that this is but one application of the quantum mechanical many-body theory which has been verified in a great many other ways, and that there is no particular reason why quantum mechanics should fail if more sophisticated EPR-type experiments were performed, this does not appear to be a likely option at this stage. The current theoretical and experimental situation favours a nonlocal explanation of an unmodified quantum mechanics, if one is sought at all of course.

11.6 Nonlocality and the one-body problem

11.6.1 Meaning of nonlocality

The term 'nonlocal' appears in various guises in the quantal literature, beyond the specific meaning it has been given above as denoting instantaneous action-at-a-distance within nonclassically interacting many-body systems. Here we shall briefly assess the validity of some of these other usages, in connection with the one-body problem.

To begin with, we consider applications of this concept that are suggested by the purely probabilistic interpretation of the ψ-function. For example, the existence of regularities in empirical results appears to imply some kind of dependence of the individual on an ensemble of particles, in particular distant ones, which may seem like a kind of 'nonlocality'. In the same vein, the sudden change in our knowledge accompanying a measurement apparently entails a nonlocal collapse of the wavefunction from an extended to a localized region of space (cf. Chap. 8). Similar considerations arise in the conventional analysis of the two-slit experiment where an electron somehow 'knows' that the distant slit through which it does not pass is open.

'Nonlocality' is also applied in the sense that the properties of systems may be altered though apparently no forces act on them. For instance, the wavefunction of a particle that is initially and finally free may pick up between a phase factor by alteration of boundary conditions, such as temporary confinement in a box, which alters the observable properties of the system even though the 'particle' has not come 'near' the confining walls (Lévy-Leblond, 1987). A conceptually similar example is provided by moving one of the walls of a box containing a particle. Although the 'particle' may be confined in a region distant from the moving barrier, 'it' picks up momentum and hence this may be regarded as an instance of nonlocality (Greenberger, 1988). A final example of this type is the Aharonov–Bohm effect where an electron interference pattern is shifted although the interfering beams move in a region of zero Lorentz force (cf. §5.2).

Especially in the latter examples, there is an underlying assumption in these cases of alleged 'nonlocality' that there is something 'there' which can be acted upon and respond 'nonlocally'. We have another example of how physicists desire to discuss an individual process in terms of objective elements. But in assessing these claims it is difficult to see what meaning can be given to 'nonlocality' in this context in the absence of a definite model of matter. What *is* it that is influenced nonlocally? If it is simply probabilities, why go to the trouble of inventing particularly striking scenarios – we know in classical physics that a change in knowledge of distant events may occur instantaneously and without the intervention of forces.

In fact, most authors do discuss this issue in terms of some implicit model of what an electron is, but not openly and consistently. Indeed, it seems to be a prerequisite of any discussion on nonlocality that we can clearly define using the concepts of the theory the notions of separation (distance) and the time at which an event occurs. Yet neither of these can be precisely defined using the wavefunction alone. Suppose we attempt to characterize 'non-locality' in terms of the properties of wave packets. Aside from the ontological problem of what it is that is described by the packet, one is faced by the fact of dispersion and ill-defined boundaries. Hence one cannot give a precise meaning to the phrases 'the position of the electron in an interferometer' and 'the electron passes a given point at a certain time' in this way. It appears that in order to discuss nonlocality we must first assume some notion of locality, and to do that we must give meaning to the concept of distance.

From the vantage point of the causal interpretation, which does make a definite statement about the objective constitution of matter, the use of the term 'nonlocal' is not warranted in any of the examples cited so far. In the two-slit experiment the particle 'knows' about the presence of both slits because the local field in its vicinity carries their imprint. The wave negotiating the wave guide or interacting with a moving wall locally acts via the modified quantum potential on the particle, even though the latter may not itself come into direct contact with a physical boundary.

Let us focus for a moment on the particle in a box with a moving wall. Suppose we start with a box of length a_0 and a stationary state $\psi_0(x) = (2/a_0)^{1/2}\sin(n\pi x/a_0)$. If one wall moves longitudinally the whole spectrum of eigenfunctions is excited (ter Haar, 1964, p. 189; Doescher & Rice, 1969; see also: Schlitt & Stutz, 1970; Pinder, 1990) and the wavefunction takes on a complicated form. Suppose the box expands linearly outwards at a rate α so that the length at time t is $a(t) = a_0 + \alpha t$. A 'steady state' solution to the time-dependent Schrödinger equation satisfying the boundary conditions $\psi(0, t) = \psi(a, t) = 0$ is given by (cf. §6.5):

$$\psi_n(x, t) = (2/a)^{1/2}\sin(n\pi x/a)\, \mathrm{e}^{im\alpha x^2/2\hbar a}\, \mathrm{e}^{-\mathrm{i}\int E_n t\, \mathrm{d}t/\hbar}, \tag{11.6.1}$$

where $E_n = n^2\pi^2\hbar^2/2ma^2$ (this is not an instantaneous eigenfunction – it does not satisfy $\hat{H}\psi_n = E_n\psi_n$). The space-dependent part of the phase ensures that the particle moves in concert with the packet and stays between the nodes. Applying the guidance formula we find for the trajectory

$$x(t) = x_0(1 + \alpha t a_0^{-1}), \tag{11.6.2}$$

so that $\dot{x} = x_0\alpha/a_0 =$ constant. The speed varies linearly with the initial distance x_0 from $x = 0$. In this case the quantum force vanishes and the particle moves uniformly. But there is no 'nonlocal' action of the moving wall on the particle – the latter simply moves according to the initial momentum implied by the phase of the wavefunction. This conclusion is not modified if the initial state is a narrow packet whose centre lies far from the walls. In this case the particle will move under the influence of the quantum force determined by the interaction of the wave with the walls, which eventually occurs after the packet spreads.

As regards the AB effect we saw in §5.2 that the phase shift may be understood as the outcome of the local action of the quantum potential on each particle in the beam, Q being modified in a gauge invariant way by the vector potential. While the effect displays global features (statistical dependence of observations on a loop integral) we were able to give an entirely local explanation for it.

It is not therefore true that in these typical quantum processes the particle is free of forces and the appellation 'nonlocal' to describe such phenomena is unjustified, at least on these grounds. It is important to distinguish the context – or state-dependent character of quantum mechanics from the issue of locality and nonlocality. Naturally, if there is no appreciable overlap of a wave with a potential then no forces due to the latter can influence the particle – this is an idea at the root of physics which allows us to separate off the 'rest of the universe' and study the properties of systems in isolation.

According to the conventional treatment, part of the EPR problem involves the feature that there is apparently no causal mechanism to explain how an unmeasured particle acquires definite properties in the absence of interaction, but the term 'nonlocality' is usually reserved in this context to describe the other aspect, that this acquisition occurs instantaneously. Interpreted in the latter sense, it is, in fact, a relevant question to enquire whether nonlocal effects are implicit in time-dependent one-body processes, i.e., to address the problem of how information on changing boundary conditions or potentials is propagated to distant points in the field. The above examples are pertinent to such an investigation. It is, in fact, tacitly assumed in the EPR debate that the problem of action-at-a-distance arises basically in the context of many-body processes and that the one-body problem is in this sense 'local'. In examining this question one should bear in mind that a single particle in a prescribed external potential is an approximation to a many-body system in which the 'source' particles of the potential are neglected. It might therefore be expected that when the wave interacts with the potential, changes in the latter may in some circumstances propagate instantaneously, as they do in the case of many-body entangled states. The question at issue is not whether particles are subject to forces which are functions of distant boundary conditions, but how the forces in the vicinity of the particle respond to alterations in the boundary conditions.

The analogue of our problem in the AB effect would be to investigate what happens if during the course of an interference experiment the flux in the solenoid is suddenly altered – does the electron in the interferometer, wherever it is, instantaneously respond? There has been some discussion of this (van Kampen, 1984; Troudet, 1985; Peshkin & Tonomura, 1989, p. 19) but the essential point at issue relates to general properties of classical potentials in quantum mechanics and has nothing directly to do with the AB effect. It is possible to analyse the problem using conceptually simpler examples which do not involve subtle questions of topology, single-valuedness and gauge invariance which may tend to obscure the main point.

It is easy to find examples where information on changing boundary conditions is propagated at a finite speed to a distant point so there is no nonlocal effect. For a packet scattering from a barrier (Galilean-equivalent to a packet at rest and a barrier moving towards it) the potential has no effect on the wave and hence a particle located in it until the wave hits the barrier and an appreciable reflected wave has formed (Dewdney, 1988). As is evident from Fig. 5.11, particles in the rear of the incident packet are not deviated until the reflected wave reaches them. Neglecting spreading, the information on the presence of the potential is propagated with approximately the group velocity of the reflected packet. A similar effect may be seen for a packet bouncing back and forth in a box (Brandt & Dahmen, 1985, p. 98). The packet as a whole does not instantaneously rearrange itself so as to accommodate the time-varying boundary conditions and a particle near its centre will therefore not be nonlocally affected either in the sense of being free of forces or in the sense of action-at-a-distance.

On the other hand, it is possible to find an example of instantaneous transmission in the form of the free Green function, where an initially perfectly localized system spreads instantaneously over all space and the particle may move arbitrarily fast.

This property of the Green function may be an artefact of a highly idealized system and to see if it is likely to be a generic feature we consider in §11.6.3 a slightly less idealized example. Reasonable requirements for the possible existence of nonlocal effects are that the wave be initially in contact with a confining boundary and that the alteration in the latter be sudden. Given these conditions, we wish to investigate what happens to distant points in the field which are not in direct contact with the changing external agency.

11.6.2 *Quantitative description of propagation*

Can we quantify the rate at which $|\psi|^2$ propagates? We mentioned above that in some circumstances the group velocity gives a measure of the speed at which information on a boundary propagates to distant points if dispersion may be neglected. On the other hand, there are cases where the group velocity is negligible and yet a packet propagates by spreading. Here we might characterize the rate of propagation by the time dependence of the root mean square width, but this does not give a detailed picture of the spacetime dependence of the packet. What we require is a general characterization of how fast $|\psi|^2$ is changing at each point, which reduces to group velocity and root mean square width in appropriate cases.

A suitable choice is the speed of a surface of constant probability density. The equation $R^2(\mathbf{x}, t) = c$ where c is a constant defines at each instant a surface in space which traces out a volume as time elapses. The constant c may be characterized by any point $\mathbf{x}_0$ in space at which the initial function R_0^2 has the prescribed value:

$$R^2(\mathbf{x}, t) = R_0^2(\mathbf{x}_0) = c. \tag{11.6.3}$$

By identical arguments which led to the formula (2.4.5) for the speed of a surface of constant phase, the component of the velocity $\mathbf{u} = \mathrm{d}\mathbf{x}/\mathrm{d}t$ of an equiprobability surface along the normal $\mathbf{n}$ at each spacetime point is given by

$$\mathbf{n}\cdot\mathbf{u} = -\frac{\partial R^2/\partial t}{|\nabla R^2|}. \tag{11.6.4}$$

In one dimension we may write

$$u = \frac{\mathrm{d}x}{\mathrm{d}t} = -\frac{\partial R^2/\partial t}{\partial R^2/\partial x}. \tag{11.6.5}$$

In this case eq. (11.6.3) defines the trajectory pursued by a given initial value $R_0^2(x_0)$ (this should not be confused with the trajectory of a particle guided by the wave which starts from x_0).

To check that this is a useful definition we apply it to a few special cases.

(*a*) For stationary states the formula (11.6.4) gives a speed zero, as is intuitively reasonable.

(*b*) The most extreme example of confinement and sudden release is the point source described by the free Green function (6.3.6). Here R^2 is independent of $\mathbf{x}$ and so $\mathbf{n}\cdot\mathbf{u} = \infty$ for all $\mathbf{x}$, t, which is again obvious intuitively.

(*c*) For the one-dimensional harmonically oscillating packet (4.9.5) formula (11.6.5) gives $x(t) = x_0 + a(\cos \omega t - 1)$ which is appropriate to a coherent state.

(*d*) More interesting is the free Gaussian packet where each value of $|\psi|^2$ eventually disappears as the packet spreads, in order to conserve the norm. Formula (11.6.4) then shows finer detail of the spreading process than is contained in the root mean square width. Recall that the amplitude squared of a one-dimensional Gaussian packet of initial half-width σ_0 and group velocity v is

$$R^2 = (2\pi\sigma^2)^{-1/2}\, \mathrm{e}^{-(x-vt)^2/2\sigma^2}, \tag{11.6.6}$$

where $\sigma = \sigma_0(1 + \alpha t^2)^{1/2}$ with $\alpha = (\hbar/2m\sigma_0^2)^2$ (cf. §4.7). Substitution of (11.6.6) into (11.6.5) yields

$$u(x, t) = v + \dot{\sigma}[\sigma^{-1}(x - vt) - \sigma(x - vt)^{-1}], \tag{11.6.7}$$

where $\dot{\sigma} = \mathrm{d}\sigma/\mathrm{d}t$. The first term on the right hand side of (11.6.7) corresponds to the group velocity and the second to dispersion. For ease of visualization, let us consider a packet at rest ($v = 0$). Then

$$u = (\hbar/2m\sigma_0)^2(t/x\sigma^2)(x^2 - \sigma^2) \tag{11.6.8}$$

and (11.6.5) yields

$$x^2(t) = (1 + \alpha t^2)x_0^2 - \sigma_0^2(1 + \alpha t^2)\log(1 + \alpha t^2). \tag{11.6.9}$$

It will be observed from (11.6.8) that as the peak decreases points of constant probability lying instantaneously in the region $|x| < \sigma$ are moving towards the origin, and points outside are moving away, while the points $x = \pm\sigma$ are at rest (Fig. 11.6). The track of the root mean square width, $x = \pm\sigma$, does not follow the path of a point of constant probability (put $x_0 = \sigma_0$ in (11.6.9)). From (11.6.9) we see that the disappearance of each value of R^2 is indicated by the fact that $x(t)$ becomes imaginary. The peak of the packet, $x_0 = 0$, disappears instantly for $t > 0$.

The extension of formula (11.6.4) to the many-body case is obvious:

$$u(\mathbf{x}_1, \ldots, \mathbf{x}_n, t) = \frac{-\partial R^2/\partial t}{\left[\sum_{i=1}^{n} (\nabla_i R^2)^2\right]^{1/2}}. \tag{11.6.10}$$

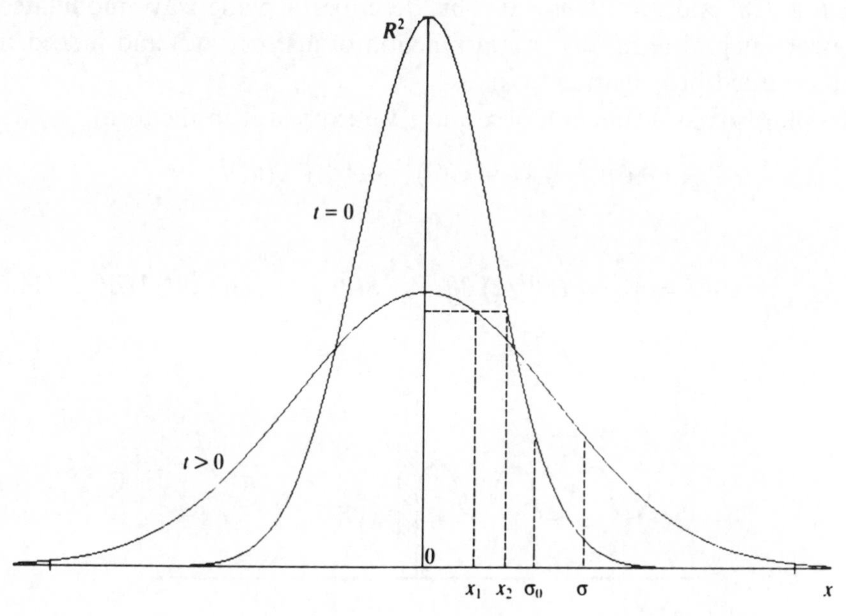

Fig. 11.6 Motion of points of constant probability density for a Gaussian packet. The point $x_2 < \sigma_0$ moves towards the origin to point x_1.

11.6.3 The shutter problem

We shall discuss the example of a one-dimensional plane wave $\psi_0 = \mathrm{e}^{ikx}$ confined to the region $x < 0$ and incident on an infinite barrier at $x = 0$ which is suddenly removed at $t = 0$ (Fig. 11.7). This 'shutter problem' is of interest for reasons other than our immediate concern for it provides the temporal analogue of spatial diffraction at a sharp edge (Moshinsky, 1952; Rosenfeld, 1965; Frahn, 1977).

Employing the free particle Green function, the wavefunction for all x at time t after removal of the shutter is given by

$$\psi(x, t) = (m/2\pi\mathrm{i}\hbar t)^{1/2} \int_{-\infty}^{0} \mathrm{e}^{iky}\, \mathrm{e}^{im(x-y)^2/2\hbar t}\, \mathrm{d}y. \tag{11.6.11}$$

We ignore spatial diffraction in the transverse directions. Eq. (11.6.11) is easily evaluated to give

$$\psi(x, t) = \tfrac{1}{2}\, \mathrm{e}^{\mathrm{i}(kx - Et/\hbar)}[1 + (1 - \mathrm{i})F(\theta)], \tag{11.6.12}$$

where

$$F(\theta) = \int_0^\theta \mathrm{e}^{\mathrm{i}\pi\theta^2/2}\, \mathrm{d}\theta, \tag{11.6.13}$$

$$\theta = (m/\pi\hbar t)^{1/2}(vt - x), \tag{11.6.14}$$

$E = \hbar^2 k^2/2m$ and $v = \hbar k/m > 0$. This describes a plane wave modulated by an error integral term. It is not a function of just $(x - vt)$ and indeed there is no retarded propagation at all.

To plot $|\psi|^2$ as a function of x and t we express it in the form

$$|\psi|^2 = \tfrac{1}{2}\{[\tfrac{1}{2} + C(\theta)]^2 + [\tfrac{1}{2} + S(\theta)]^2\}, \tag{11.6.15}$$

where

$$C(\theta) = \int_0^\theta \cos(\pi\theta^2/2)\, \mathrm{d}\theta, \qquad S(\theta) = \int_0^\theta \sin(\pi\theta^2/2)\, \mathrm{d}\theta \tag{11.6.16}$$

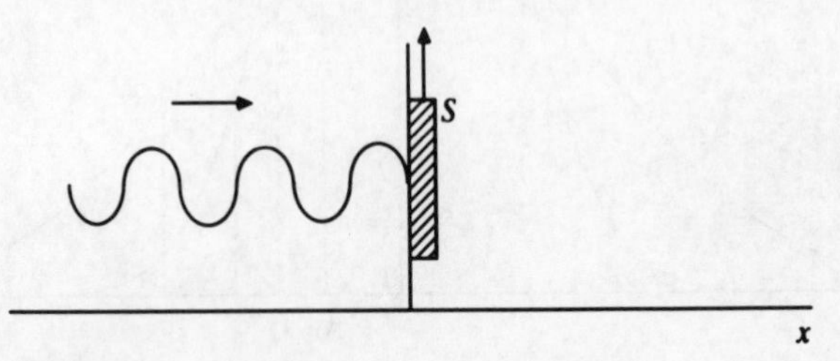

Fig. 11.7 The shutter problem. A plane wave in the region $x < 0$ is suddenly released at $t = 0$ by removal of shutter S.

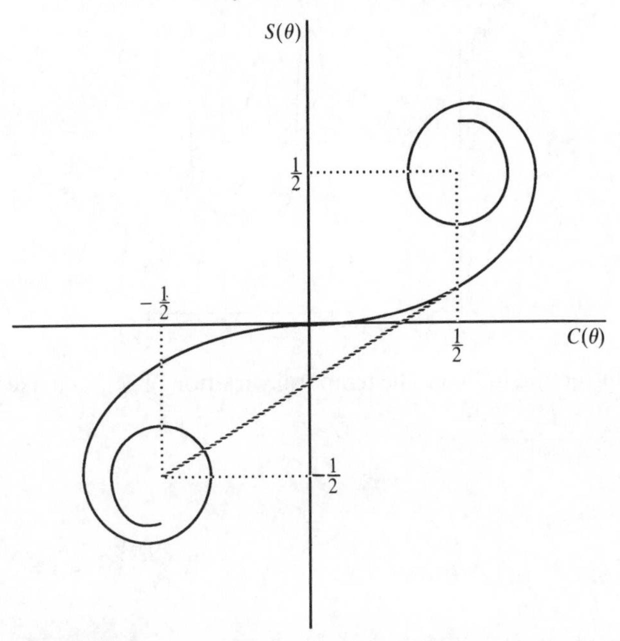

Fig. 11.8 The Cornu spiral. The points $\pm(\frac{1}{2}, \frac{1}{2})$ correspond to $\theta = \pm\infty$. The length of the dashed line is $2|\psi(\theta)|^2$.

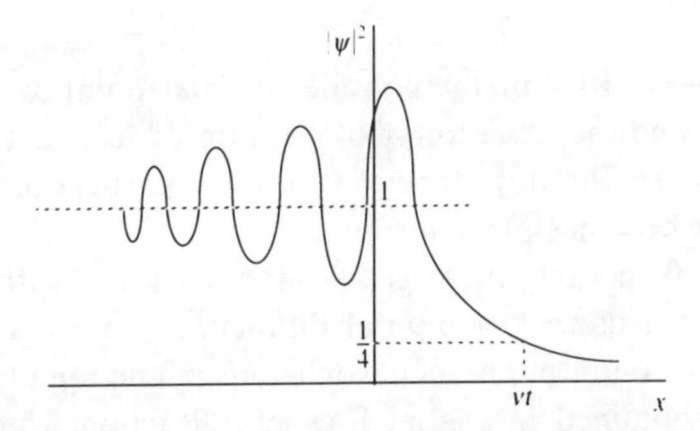

Fig. 11.9 Longitudinal diffraction pattern at a fixed time $t > 0$.

are the Fresnel integrals. The amplitude squared (11.6.15) may be evaluated using the properties of the Cornu spiral (Fig. 11.8), for it is one-half of the square of the radius vector through the point $(-\frac{1}{2}, -\frac{1}{2})$ to the point on the spiral whose distance from the origin, along the curve, is θ. Let us consider a fixed time $t > 0$. Then from (11.6.14) we see that when $-\infty < x \leqslant vt$, $0 \leqslant \theta < \infty$ and when $\infty > x \geqslant vt$, $-\infty < \theta \leqslant 0$. Using Fig. 11.8 we obtain for $|\psi|^2$ the spatial diffraction pattern shown in Fig. 11.9. On the other hand, consider a fixed space point $x > 0$ and the variation of $|\psi|^2$ with time. When $0 < t \leqslant x/v$, $-\infty < \theta \leqslant 0$ while when $\infty > t \geqslant x/v$, $0 \leqslant \theta < \infty$. The amplitude

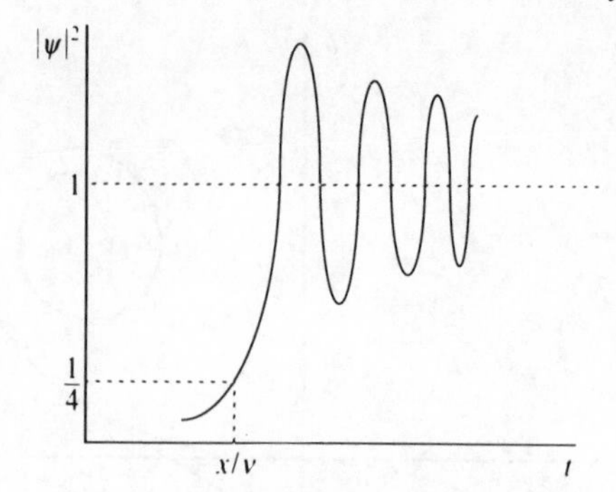

Fig. 11.10 Diffraction in time. The temporal variation of $|\psi|^2$ at a fixed point $x > 0$.

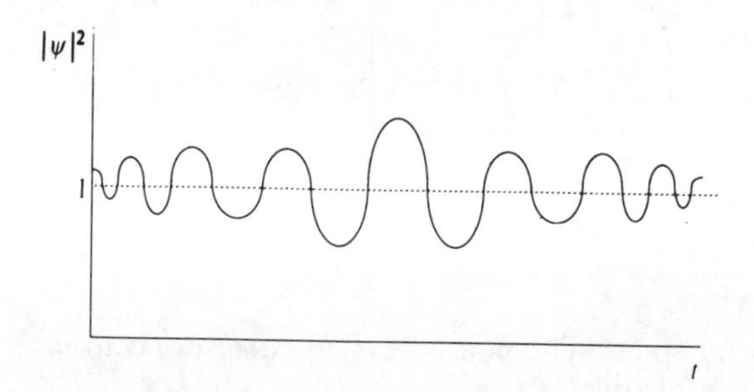

Fig. 11.11 The time dependence of $|\psi|^2$ at a fixed point $x < 0$.

is plotted in Fig. 11.10. Finally, consider a fixed point $x < 0$. Then when $t = 0, \infty, \theta = \infty$ and the variation of $|\psi|^2$ is plotted in Fig. 11.11. For points connected by $x = vt$ (the trajectory of a classical particle of speed v) $|\psi| = \frac{1}{2}$ and as $t \to \infty$, ψ becomes plane for all x.

At points $x > 0$ the amplitude grows monotonically with time and then oscillates to form a damped temporal diffraction pattern which eventually dies out and tends to unity. The shape of $|\psi|^2$ as a function of time is identical to the intensity obtained in spatial Fresnel diffraction from the edge of a semi-infinite plane (Pauli, 1973, p. 66). For $x < 0$ the amplitude again forms a temporal diffraction pattern.

It will be observed from the figures that for arbitrarily small $t > 0$ the amplitude is altered at all finite points x, however distant. The quantum potential varies likewise in a complicated way and a particle lying initially anywhere in the region $x < 0$ (and moving with velocity v) will respond effectively instantaneously to the removal of the shutter. Moreover, while initially there is zero probability of the particle being at any point $x > 0$, for $t > 0$ there is a finite probability of the particle being anywhere in this domain, so that a particle starting from a point $x < 0$ can move arbitrarily fast. Where it actually ends up depends on its initial position. In the analogous classical problem the particle has, of course, the unique velocity v and the relative probability is zero for $x > vt$ and unity for $x < vt$.

To see in more detail how $|\psi|^2$ is changing we consider the motion of the points of constant probability. From (11.6.14) and (11.6.15) these satisfy the equation

$$x(t) = vt + At^{1/2}, \tag{11.6.17}$$

where A is a constant. Differentiating (11.6.17) or directly using (11.6.5) yields the speed

$$u = \tfrac{1}{2}(v + xt^{-1}). \tag{11.6.18}$$

The first term on the right hand side of (11.6.17) corresponds to the classical contribution and the second to quantal spreading. The smaller $t > 0$ is, and the larger $|x|$ is, the larger u is. When $x < 0$, $u < 0$ for $|x/t| > v$.

In passing, we note that if one pursues the notion that the opening and closing of a shutter is the time analogue of an aperture in a screen in spatial diffraction, we expect to obtain a temporal version of a beam splitter by opening and closing a shutter twice in succession. This leads to the phenomenon of 'interference in time' (Gähler & Golub, 1984; Felber, Gähler & Golub, 1988; Felber *et al.*, 1990). Our explanation for this effect rests on the same principles as described in Chap. 5. Part of the incident wave passes through the shutter when it is first open and another part when it is subsequently open. The particle lies within one or other subwave. The two waves overlap at a point distant from the shutter and form a temporal interference pattern, the particle being channelled as usual into regions where $|\psi|^2$ is large by the quantum potential. For neutrons of de Broglie wavelength 30 Å, slit openings $\approx 2 \times 10^{-8}$ s, distance between slits $\approx 4 \times 10^{-8}$ s and an observation point ≈ 1.5 m from the shutter, the 'width' of the central peak is ≈ 6 μs.

To return to our main problem, it is possible that the instantaneous response of the wave and particle in regions distant from the shutter when the latter is removed may stem from the use of an initial infinitely extended plane wave. To remove this restriction we first take into account the reflected plane wave e^{-ikx} in the region $x < 0$. For an initial state $\sin kx$ in the region $x < 0$, sudden removal of the shutter results in the propagation of both component plane waves in both directions, as in (11.6.12), but we have to take into account their interference in forming $|\psi|^2$. We now confine the initial wave in a finite region by introducing a second shutter at the point $x = -a$. We thus obtain the stationary state of the particle in a box, where $k = n\pi/a$ (cf. §6.5). Simultaneous removal of both walls leads at time t to the wavefunction

$$\psi(x, t) = \left[\frac{(1 + \mathrm{i})}{2(2a)^{1/2}}\right] \mathrm{e}^{-\mathrm{i}Et/\hbar} \{\mathrm{e}^{\mathrm{i}kx}[F(\theta'_-) - F(\theta_-)] + \mathrm{e}^{-\mathrm{i}kx}[F(\theta'_+) - F(\theta_+)]\}, \tag{11.6.19}$$

where

$$\theta_\pm = (m/\pi\hbar t)^{1/2}(vt \pm x), \qquad \theta'_\pm = (m/\pi\hbar t)^{1/2}(vt \pm x \pm a). \tag{11.6.20}$$

Using again the properties of the Cornu spiral we find that the temporal packet depicted in Fig. 11.12 forms at each space point beyond the box (Gerasimov & Kazarnovskii, 1976). At each instant there are two spatial packets symmetrically placed about the box and moving apart. Eq. (11.6.19) is the wavefunction obtained in the time-of-flight method of measuring momentum described in §§6.5 and 8.6.

A further idealization we have made, that of an infinite shutter speed, can be relaxed by assuming a time-dependent shutter opening. However, this leads to complications in that the spatial diffraction at the moving edge depends on time and so can no longer be ignored as we have done above, but we shall not pursue this further (see Gähler & Golub (1984)).

The existence of action-at-a-distance effects in the one-body Schrödinger theory does not, of course, pose any conceptual or formal difficulties since our treatment is nonrelativistic. However, it is important to appreciate the distinction with EPR action-at-a-distance where there is no conflict with

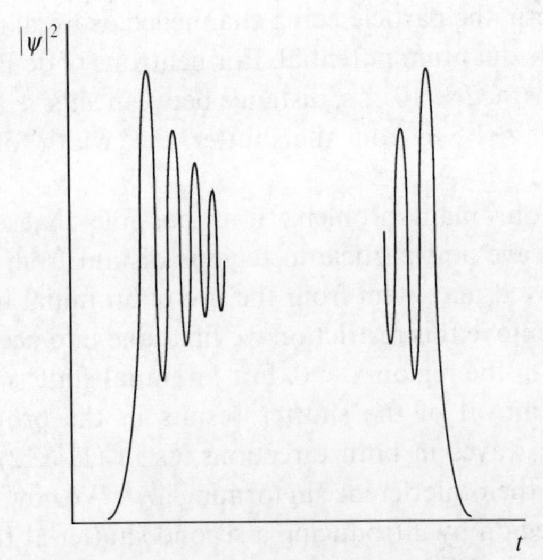

Fig. 11.12 Temporal packet at a point x outside the box after removal of the walls (adapted from Gerasimov & Kazarnovskii (1976)).

relativity at a statistical level due to the 'no-signalling' theorem (§11.3). In contrast, here we do have instantaneous statistical transmission of information in that local probabilities are altered by distantly changing boundary conditions. If these effects were found to persist in the relativistic domain this would actually pose a greater challenge to the peaceful coexistence of relativity and quantum mechanics than the EPR correlations. Some results have been obtained in this direction. Application of the Klein–Gordon equation to the shutter problem (Moshinsky, 1952) shows that the amplitude at a point $x > 0$ is nonzero only after a time $t > x/c$ (c = speed of light) so no contradiction with relativity arises there. However, the general treatment of this problem is not straightforward as it involves consideration of the appropriate definition of position operators and probability densities in relativistic quantum mechanics, and raises as yet unsolved difficulties with the relativistic theory of measurement. Apparent violations of Einstein causality have been theoretically demonstrated under fairly general conditions (Hegerfeldt, 1974, 1985; Hegerfeldt & Ruijsenaars, 1980; Rosenstein & Usher, 1987). On the other hand, there are theorems in relativistic quantum field theory that appear to forbid any violation of local causality (Eberhard & Ross, 1989; Bell, 1990b). The study of these relativistic effects is beyond the scope of this book (but see Chap. 12).

To conclude, rapidly varying boundary conditions can lead to statistical and individual action-at-a-distance effects in the quantum mechanical one-body problem and hence more generally in the configuration space of a many-body system. The nonlocality of EPR arises in a different way, as the mapping into ordinary space of the effect of locally disturbing the system point in configuration space. The one-body problem is local in the sense that the statistical patterns are an aggregate of individual motions subject to forces acting at the location of corpuscles rather than occurring without apparent cause, and local in that in some cases a localized disturbance will propagate to a distant point in a finite time. But because it does not generally imply a retarded propagation, implicit in the local differential Schrödinger evolution is the possibility of effectively instantaneous actions.

Appendix

In order to obtain (11.2.5) we first derive the effect of the impulsive measurements by solving

$$i\hbar\frac{\partial\psi}{\partial t} = [\mu(B_0 + z_1' B_0')\sigma_{z_1'} + \mu(B_0 + z_2 B_0')\sigma_{z_2}]\psi, \tag{A1}$$

where $\sigma_{z'} = \sigma_z \cos\delta - \sigma_x \sin\delta$, $z' = -x\sin\delta + z\cos\delta$, B_0 is the uniform field and B_0' is the field gradient in each Stern–Gerlach magnet. We ignore the effect of the fields in the x- and x'-directions, as explained in §9.5.

The initial wavefunction is (11.2.1). Since magnet 1 is not oriented in the z-direction we must expand the solution to (A1) in terms of all of the four basis functions:

$$\psi(z_1', z_2, t) = \sum_{a,b=\pm} \psi^{ab}(z_1', z_2, t) u_a v_b,$$

where $\sigma_{z_1} u_\pm = \pm u_\pm$, $\sigma_{z_2} v_\pm = \pm v_\pm$. Substituting this expression in (A1) it is easy to find second-order differential equations satisfied by each ψ^{ab}. Expressing the solution in terms of the basis functions which correspond to the possible outcomes of the experiment, we find

$$\begin{aligned}\psi(z_1', z_2, T) = \frac{1}{\sqrt{2}} f_1(z_1') f_2(z_2) (&\sin(\delta/2)\, \mathrm{e}^{-\mathrm{i}[2\Delta + (z_1' + z_2)\Delta']} u_+' v_+ \\ &+ \cos(\delta/2)\, \mathrm{e}^{-\mathrm{i}(z_1' - z_2)\Delta'} u_+' v_- - \cos(\delta/2)\, \mathrm{e}^{\mathrm{i}(z_1' - z_2)\Delta'} u_-' v_+ \\ &+ \sin(\delta/2)\, \mathrm{e}^{\mathrm{i}[2\Delta + (z_1' + z_2)\Delta']} u_-' v_-),\end{aligned} \tag{A2}$$

where $\sigma_{z_1'} u_\pm' = \pm u'$, $\Delta = \mu B_0 T/\hbar$, $\Delta' = \mu B_0' T/\hbar$ and T is the time that each particle spends in the field.

The subsequent motion proceeds according to the free equation

$$\mathrm{i}\hbar \frac{\partial \psi}{\partial t} = -\frac{\hbar^2}{2m}\left(\frac{\partial^2}{\partial z_1'^2} + \frac{\partial^2}{\partial z_2^2}\right)\psi \tag{A3}$$

with (A2) as initial wavefunction. To find the solution to (A3) we first Fourier analyse the initial packets f_1, f_2. The coefficient of $u_+' v_+$ in (A2), say, is then

$$\frac{1}{\sqrt{2}} \sin(\delta/2)\, \mathrm{e}^{-2\mathrm{i}\Delta} \frac{1}{2\pi} \int g_1(k_1') g_2(k_2)\, \mathrm{e}^{\mathrm{i}[(k_1' - \Delta')z_1' + (k_2 - \Delta')z_2]}\, \mathrm{d}k_1'\, \mathrm{d}k_2, \tag{A4}$$

where $g_1(k_1')$, $g_2(k_2)$ are normalized packets centred around $k_1' = 0$, $k_2 = 0$ respectively. As time passes, the (k_1', k_2)th Fourier component picks up a factor $\mathrm{e}^{-\mathrm{i}(\omega_{k_1'} + \omega_{k_2})t}$, where

$$\omega_{k_1'} = \frac{\hbar}{2m}(k_1' - \Delta')^2, \qquad \omega_{k_2} = \frac{\hbar}{2m}(k_2 - \Delta')^2.$$

(A4) thus becomes

$$\frac{1}{2\pi\sqrt{2}} \sin(\delta/2)\, \mathrm{e}^{-2\mathrm{i}\Delta} \int g_1(k_1') g_2(k_2) \times \mathrm{e}^{\mathrm{i}[(k_1' - \Delta')z_1' - (\hbar t/2m)(k_1' - \Delta')^2 + (k_2 - \Delta')z_2 - (\hbar t/2m)(k_2 - \Delta')^2]}\, \mathrm{d}k_1'\, \mathrm{d}k_2.$$

The centre of the configuration space wave packet occurs where the phase has an extremum, i.e., where

$$z_1' = -\hbar t\Delta'/m, \qquad z_2 = -\hbar t\Delta'/m$$

so that the centres of the wave packets emerging from the magnets move in the same direction (relative to the local analysing fields), as is to be expected from the $u'_+ v_+$ part of the solution. The other three terms in (A3) develop similarly and describe the other possible combinations for the outcome of the experiment. Each of the particles enters one or the other packets at the exits to the magnets depending on their initial positions in the packets at the entrances to the magnets. It is assumed that the momenta imparted to the packets by the inhomogeneous fields are sufficiently great that the spreading of the packets will not mask the spin-dependent deflection and so a classical separation of the two beams emerging from each magnet is ensured.

The final states of each of the apparatuses (packet coordinates) are thus nonoverlapping and those packets which the particles do not enter can be dropped from further attention. The measurement is complete when devices are placed beyond the magnets to detect which beams the particles actually enter. This irreversible stage of the process merely tells us what has already happened and there is no need to invoke the 'wavefunction collapse' hypothesis.

The final step in deriving (11.2.5) is to substitute explicit expressions for the packets g_1, g_2 in the four terms of the form (A4). Writing

$$g_1(k_1') = (\sigma\sqrt{\pi})^{-1/2}\,\mathrm{e}^{-k_1'^2/2\sigma^2}, \qquad g_2(k_2) = (\sigma\sqrt{\pi})^{-1/2}\,\mathrm{e}^{-k_2^2/2\sigma^2}$$

performing the integrations and rearranging, we deduce (11.2.5) ($\sigma = 1/\sqrt{2}\sigma_0$ where σ_0 is the width used in §4.7).

12

Relativistic quantum theory

12.1 Problems with the extension to relativity. The Klein–Gordon equation

The attempt to extend the causal interpretation to cover relativistic quantum mechanics raises several problems of technique and principle. These relate not only to difficulties peculiar to the quantum theory of motion but have to do with still unresolved issues in the underlying quantum mechanical formalism (such as the appearance of infinities in the computation of physical quantities and the lack of a relativistic theory of measurement). Some problems are common to both, such as the difficulty of defining positive densities. The subject has not been extensively investigated and in this chapter we shall discuss only some of the relevant issues. There are two particularly important, overlapping problems: Can the trajectory concept be retained in the relativistic quantum domain? And: how do we interpret relativistic quantum fields?

The unresolved issue underlying both questions is the extent to which we should require a quantum theory to be 'relativistic'. It is known that Lorentz covariant wave equations can have consequences in conflict with relativity. This, as we shall see, is indeed the case with the current relativistic formalism as interpreted, or completed, by the quantum theory of motion. Especially in field theory, the individual processes that build up and explain the covariant statistical predictions of the formalism exhibit features that are in conflict with our entrenched ideas regarding covariance and locality as necessary concomitants of relativity theory. That is, relativity is statistically valid but the individual events do not have an intrinsically relativistic character. This seems inevitable in a theory whose basic dynamical equations are defined in configuration space rather than ordinary spacetime, as already pointed out in connection with the EPR problem, but problems occur even when the configuration space coincides with spacetime. What these features portend for the future of physical theory is unknown.

We begin by reviewing why the notion that a physical system comprises a corpuscle pursuing a well-defined spacetime track under the influence of a de Broglie wave does not work when applied to the Klein–Gordon equation describing a spin 0 particle. The next two sections deal with the analogous problem in the Dirac theory which does not suffer from quite the same drawbacks. The remainder of the chapter addresses the field theory.

We adopt the following conventions. Greek indices $\mu, \nu, \ldots$ take the range 0, 1, 2, 3 and Latin indices $i, j, \ldots$ the range 1, 2, 3. The summation convention over repeated indices is understood and the Minkowski spacetime metric has signature $\eta_{\mu\nu} = \mathrm{diag}(1, -1, -1, -1)$. In a system of units in which $\hbar = c = 1$, the Klein–Gordon equation for a particle of mass m and charge e is:

$$(\mathrm{i}\partial_\mu - eA_\mu)(\mathrm{i}\partial^\mu - eA^\mu)\psi = m^2\psi, \tag{12.1.1}$$

where $\psi = \psi(\mathbf{x}, t)$ is a complex scalar function and A_μ is the external electromagnetic potential.

Making our usual polar decomposition of the wavefunction, $\psi = R\,\mathrm{e}^{\mathrm{i}S}$ where R, S are real Lorentz scalar functions, (12.1.1) splits up into two real equations:

$$(\partial_\mu S + eA_\mu)(\partial^\mu S + eA^\mu) = m^2(1 + Q) \tag{12.1.2}$$

$$\partial_\mu j^\mu = 0, \tag{12.1.3}$$

where

$$j^\mu = (1/2m)\psi^*(\mathrm{i}\overleftrightarrow{\partial^\mu} - 2eA^\mu)\psi = -(R^2/m)(\partial^\mu S + eA^\mu). \tag{12.1.4}$$

Eq. (12.1.2) has the form of the classical relativistic Hamilton–Jacobi equation for a charged particle modified by a quantum potential-like term $Q = \Box R/Rm^2$ and (12.1.3) is a conservation law for the real current j^μ. One is tempted to propose that the current defines a congruence of world lines of an ensemble of particles connected with the wave ψ, their density being given by j^0. The tangent to a world line would be given by the 4-velocity u^μ which is defined in terms of the 4-momentum p^μ via the relation

$$Mu^\mu = p^\mu = -(\partial^\mu S + eA^\mu), \tag{12.1.5}$$

where $M = m(1 + Q)^{1/2}$ is a 'variable mass' and (12.1.2) implies that $u^\mu u_\mu = 1$ (this definition of momentum implies, for example, that $\mathbf{p} = \nabla S - e\mathbf{A}$, the usual formula). Solving $u^\mu = \mathrm{d}x^\mu/\mathrm{d}\tau$ where τ is the proper time would then apparently yield a trajectory $x^\mu = x^\mu(\tau)$ once the initial position of a particle in the ensemble is specified. The particle acceleration implied by differentiating (12.1.2) is given by (Holland, 1987):

$$\mathrm{d}u^\mu/\mathrm{d}\tau = \tfrac{1}{2}(\eta^{\mu\nu} - u^\mu u^\nu)\partial_\nu \log(1 + Q) + (e/M)F^\mu{}_\nu u^\nu \tag{12.1.6}$$

where $d/d\tau = u^\mu \partial_\mu$. This contains a term due to the quantum force and a term due to the Lorentz force, as might be expected (cf. §3.11.2).

That the guidance formula (12.1.5) is reasonable in the case of plane waves is readily established. In the absence of external fields the 4-momentum is given by $p^\mu = -\partial^\mu S$ and the current by $j^\mu = -(R^2/m)\partial^\mu S$. For a positive energy plane wave

$$\psi = e^{-i(Et - \mathbf{p}\cdot\mathbf{x})}, \qquad E = (\mathbf{p}^2 + m^2)^{1/2}, \tag{12.1.7}$$

we have $p^\mu = (E, \mathbf{p})$ and the current is time-like ($j^\mu j_\mu > 0$) and future-pointing ($j^0 > 0$). One might hope that at least for solutions confined to the Hilbert space of positive energy states, and for sufficiently weak external fields, a single-particle trajectory interpretation is tenable.

There are two principal reasons why this proposal will not work, except in special cases. First, the fourth component of the current, j^0, is of indefinite sign and hence cannot be treated as a particle or probability density. This is generally so even for free solutions lying in the manifold of positive energy states and may be exhibited by the superposition of two positive energy solutions (see Blokhintsev (1973, p. 81) and below). Moreover, this is not a problem peculiar to superpositions which have a finite amplitude throughout space; it occurs, for example, for superpositions of exponentially damped atomic eigenfunctions (Gerlach, Gromes & Petzold, 1967). Indeed, if we view the classical limit of the Klein–Gordon equation as that domain where the quantum potential (or rather, force) may be neglected, the problem of j^0 being potentially negative may well persist in what should be the classical regime and prevent a probability interpretation there too. Second, j^μ is not generically time-like; the definition (12.1.5) of the '4-velocity' u^μ cannot be generally maintained since the 'variable mass' M may be imaginary. Again, one may see this for solutions of the free Klein–Gordon equation drawn from the positive frequency spectrum (de Broglie, 1956, p. 123 and below). A theory of material objects in which an initially time-like future-pointing trajectory may pass through the light cone to become space-like, and even move backwards in time, is clearly unacceptable.

A simple example displaying these pathological properties is provided by the following superposition of free positive energy states for a particle of unit mass in one space dimension (Kyprianidis, 1985):

$$\psi(x, t) = e^{-it} + (2/E)\, e^{-i(Et - px)}, \tag{12.1.8}$$

where $E^2 - p^2 = 1$ and $E > 2$. The energy and momentum of a particle guided by such a wave are, according to our definition,

$$\left.\begin{aligned} p^0 &= -\partial S/\partial t = (1/R^2)\{1 + (4/E) + (2/E)(1 + E)\cos[(1 - E)t + px]\} \\ p^1 &= \partial S/\partial x = (2p/ER^2)\{(2/E) + \cos[(1 - E)t + px]\}. \end{aligned}\right\} \tag{12.1.9}$$

In spacetime regions where the cos term $= +1$ we have $p^0 > 0$. But when $\cos = -1$, $p^0 < 0$, i.e., the energy is negative. Moreover, in the same domain ($\cos = -1$),

$$E^4R^2(p^{0^2} - p^{1^2}) = (2 - E)^2(4 - 3E^2) < 0 \tag{12.1.10}$$

showing that the '4-momentum' is space-like.

The oft-quoted statement that a satisfactory single-particle interpretation may be developed for free Klein–Gordon solutions built from the positive spectrum is not supportable. It has been proposed that (e times) j^0 be interpreted as a charge density (Feshbach & Villars, 1958; Bjorken & Drell, 1964, Chap. 9) but we see that the positive (negative) energy spectrum is not correlated with just positive (negative) values of j^0 (although one can establish such a correlation for the total 'charge' $\int j^0\, d^3x$). Thus one cannot achieve a clean partition of the positive and negative energy spectra into sets associated respectively with particles and antiparticles distinguished by the sign of the charge density.

We conclude that the Klein–Gordon equation does not have a consistent single-particle interpretation and the naive transcription of the trajectory interpretation of nonrelativistic Schrödinger quantum mechanics into this context does not work. Some insight has been gained in special cases of the one- and many-body theories (e.g. Vigier, 1982; Dewdney *et al.*, 1985; Cufaro-Petroni *et al.*, 1985; Holland, 1987) but the approach does not have a secure basis. There has always been a formal difficulty with the notions of position and localizability in relativistic quantum mechanics if we restrict the class of physically realizable states to the positive spectrum. For the argument '$\mathbf{x}$' of the Klein–Gordon (or Dirac) wavefunctions is not the eigenvalue of some operator that is Hermitian with respect to the scalar product in the Hilbert space of positive energy states (Schweber, 1961, p. 60). As a result one cannot interpret the wavefunction $\psi(\mathbf{x}, t)$ in accordance with the usual axioms as the probability amplitude for a particle to be found in an elementaty volume d^3x around the point $\mathbf{x}$ at time t (or in the causal interpretation as the probability of the particle *being* in the volume). To discover such a probability one has to define a Hermitian position operator. Various proposals have been made; the most natural, in that it follows on fairly reasonable grounds of covariance, is that of Newton & Wigner (1949). The eigenfunctions of the Newton–Wigner operator are not δ-functions (which cannot be expressed as a superposition of just positive energy solutions) but packets with a minimum width approximately equal to the Compton wavelength, $1/m$. This operator is employed in interpreting the configuration space approach to relativistic quantum field theory (Wightman & Schweber, 1955). How the Newton–Wigner operator connects with empirical determinations of position via some relativistic theory of measurement is, to our knowledge, unknown. Likewise the possibility of defining a particle trajectory from it, perhaps via the local expectation value, has not been investigated.

It should not perhaps surprise us that the conserved Lorentz vector j^μ, (12.1.4), can only be associated with a material current at the expense of violating basic requirements of relativity concerning the motion of physical objects. For the formulation of an intrinsically relativistic theory entails more than the mere imposition of the Lorentz covariance of local fields. Although causality implies the Lorentz group (Zeeman, 1964), it is known in other contexts that covariant wave equations can have noncausal consequences (Aharonov, Komar & Susskind, 1969; Velo & Zwanziger, 1969a,b; Rañada & Rodero, 1980; see also the remarks at the end of §11.6.3). To avoid superluminal transmission and the possibility of effects preceding causes it is necessary to supplement a wave equation with (relativistically covariant) constraints. In the context of the attempt to describe the motion of particles guided by a Klein–Gordon wave via the guidance formula (12.1.5) this would mean, in the free case, admitting as physically valid only those solutions for which $\partial_\mu S\, \partial^\mu S > 0$ and $\partial_0 S < 0$ for all $\mathbf{x}$ and t. But these are not constraints one would naturally impose on the free Klein–Gordon equation and indeed solutions that are otherwise acceptable disobey them.

It seems that the Klein–Gordon theory cannot be rescued by minor modifications or generalizations. For example, a five-dimensional generalization has been proposed in which the wavefunction $\psi(\mathbf{x}, t, \tau)$ depends on the usual spacetime coordinates x^μ and a new intrinsic evolution variable τ, and satisfies a Schrödinger-like wave equation in which the role of t is played by τ (Kyprianidis, 1987). This has the advantage of placing $\mathbf{x}$ and t on an equal footing and implies a nonnegative probability density of events, $|\psi|^2$. But it brings with it difficulties which make it untenable as regards the causal interpretation. Aside from the problem of interpreting τ, the relativistic Schrödinger equation suffers from the same problem as the primitive Klein–Gordon theory: if we define particle momentum through the gradient of the phase, there is no naturally occurring constraint in the theory that makes this future-pointing and time-like.

We mention also the theory of fusion of de Broglie (1954) in which he attempts to construct a spin 1 boson (a 'photon') from the fusion of two spin $\frac{1}{2}$ Dirac particles. This is done by forming a 4×4 matrix $\Psi_{ab} = \bar{\psi}_a \phi_b$ from the components of two Dirac spinors ψ, ϕ associated with each of the component particles (cf. §12.2) and expanding this in terms of a set of tensors C_A:

$$\Psi = \sum_A C_A \gamma_{Aab}, \tag{12.1.11}$$

where A denotes a set of tensor indices. The tensor components C_A may be associated with integer spin particles. Unfortunately, as noted by de Broglie (1954, p. 120), the fourth component of the conserved current J implied by the wave equation postulated for Ψ suffers from the classic drawback that plagues boson theories: it is not generically positive-definite (J is not simply the linear superposition of the two positive-definite, time-like Dirac currents associated with each of the spin $\frac{1}{2}$ particles). Nor is J time-like. In addition, we note that the tensors C_A must satisfy certain subsidiary conditions in order that the expansion (12.1.11) factorizes into a product of spinors (Holland, 1986) and this restricts the solutions of the theory to a subset of Maxwell's or Proca's equations (the latter are Maxwell's equations modified by a nonzero rest mass attributed to a 'photon'). The theory of fusion does not therefore lead to a satisfactory resolution of the problems outlined in this section.

Yet we should be careful not to draw a definitive conclusion regarding the illegitimacy of the trajectory concept in relativistic quantum mechanics from the above considerations alone. A successful interpretation of the Dirac equation along these lines turns out to be possible (see below) and meaningful spacetime pictures of quantized boson fields may be developed. In particular, one can define for the classical Klein–Gordon field an energy flow that is consistent with relativistic principles (§12.6.2) and this can be extended to the mean behaviour of the one-quantum sector of quantum field theory (§12.6.4). We shall see in the field theory that the tendency of the causal interpretation is away from the universal or intrinsic validity of relativity in individual processes towards the notion of statistical relativity in which the principles are obeyed only on the average.

For these reasons it seems advisable to approach the problem of a particle interpretation of bosonic theories from a more oblique angle, such as deriving it from some underlying field structure, rather than grafting the trajectory onto a theory whose foundations are already rather shaky.

12.2 Causal interpretation of the Dirac equation

The problems with the interpretation of the Klein–Gordon equation pointed out in §12.1 are to some extent ameliorated in the Dirac theory of spin $\frac{1}{2}$ particles, for the conserved current implied by the Dirac equation possesses a positive fourth component and is time-like. Then, if we treat the '$\mathbf{x}$' appearing in the argument of the Dirac wavefunction as the eigenvalue of a position operator so that ψ is a probability amplitude (this is valid if the spectrum of solutions contains negative as well as positive energies), we may define relativistically covariant particle world threads tangent to the lines of current flow whose density reproduces the probability distribution determined by ψ. At least for the applications we consider here, this leads to a consistent single-particle physical interpretation of Dirac's theory (Holland, 1992). Further developments, such as the hole theory and pair creation, are yet to be investigated.

The Dirac equation for a particle of mass m, charge e and spin $\frac{1}{2}$ is

$$\gamma^\mu(\mathrm{i}\partial_\mu - eA_\mu)\psi = m\psi. \tag{12.2.1}$$

Here $\psi = (\psi^a)$, $a = 1, 2, 3, 4$, is a 4-component spinor, γ^μ are Dirac matrices and we shall generally suppress spinor indices. One may proceed to the construction of a causal interpretation by extending the approach adopted in the Pauli theory (Chap. 9) and interpret the content of the wave equation (12.2.1) and the degrees of freedom in the spinor in terms of a set of objectively real spacetime tensors and a set of properties of a particle. Following §3.5, the latter will be determined by the 'local expectation values' of the associated operators. This task has been successfully carried through in detail by Takabayasi (1957; see also Holland, 1986).

A Dirac spinor field determines the following set of real relativistic tensors, the so-called 'bilinear covariants':

$$\left.\begin{aligned} a = \bar{\psi}\psi \text{ (scalar)}, \qquad & j^\mu = \bar{\psi}\gamma^\mu\psi \text{ (vector)}, \\ b^{\mu\nu} = \bar{\psi}\sigma^{\mu\nu}\psi \text{ (bivector)}, \qquad & s^\mu = \tfrac{1}{2}\bar{\psi}\gamma^\mu\gamma^5\psi \text{ (pseudo-vector)}, \\ b = \mathrm{i}\bar{\psi}\gamma^5\psi & \text{ (pseudo-scalar)}, \end{aligned}\right\} \tag{12.2.2}$$

where $\bar{\psi} = \psi^\dagger\gamma^0$, $\gamma^5 = \mathrm{i}\gamma^0\gamma^1\gamma^2\gamma^3$ and $\sigma^{\mu\nu} = (\mathrm{i}/2)(\gamma^\mu\gamma^\nu - \gamma^\nu\gamma^\mu)$ (we use the conventions of Bjorken & Drell (1964)). These quantities are connected by various identities induced by properties of the Dirac matrices and by the field equation (12.2.1). For example, the bivector may be expressed purely algebraically in terms of the other tensors via the relation:

$$(a^2 + b^2)b^{\mu\nu} = 2b(j^\mu s^\nu - j^\nu s^\mu) - a\varepsilon^{\sigma\rho\mu\nu}(j_\sigma s_\rho - j_\rho s_\sigma), \tag{12.2.3}$$

where $\varepsilon^{\sigma\rho\mu\nu}$ is the Levi–Civita symbol. The remaining entities satisfy the following relations among themselves:

$$j^\mu j_\mu = a^2 + b^2, \tag{12.2.4}$$

$$s^\mu s_\mu = -\tfrac{1}{4}(a^2 + b^2), \tag{12.2.5}$$

$$s^\mu j_\mu = 0. \tag{12.2.6}$$

The vector j^μ is the usual Dirac current density and obeys the conservation law

$$\partial_\mu j^\mu = 0. \tag{12.2.7}$$

It has a nonnegative fourth component ($j^0 = \psi^\dagger\psi \geqslant 0$ in each Lorentz frame) which allows the interpretation of j^0 as the probability density of a Gibbs ensemble of particles. The key property of the current is (12.2.4) which, because the right hand side is a real and positive scalar, states that j^μ is time-like. It may therefore be used to define the 4-velocity field of the ensemble:

$$u^\mu = (a^2 + b^2)^{-1/2} j^\mu, \qquad u^\mu u_\mu = 1, \qquad u^0 \geqslant 1 \tag{12.2.8}$$

(assuming a and b are nonzero).

It is the possibility of defining this future-time-like 4-velocity that allows a trajectory interpretation. We are led then to introduce a physical particle, with u^μ tangent to its world line, in addition to the objectively real spinor field as our model of the individual physical system described by the Dirac theory. The trajectories of the ensemble of particles associated with a given solution of Dirac's equation are the solutions $\mathbf{x}(t)$ to the differential equation

$$\mathrm{d}x^i/\mathrm{d}t = v^i(\mathbf{x}, t)|_{\mathbf{x}=\mathbf{x}(t)}, \tag{12.2.9}$$

where the 3-velocity field is given by

$$v^i(\mathbf{x}, t) = \frac{u^i}{u^0} = \frac{j^i}{j^0} = \frac{\psi^\dagger \alpha^i \psi}{\psi^\dagger \psi} \tag{12.2.10}$$

with $\alpha^i = \gamma^0\gamma^i$, $i = 1, 2, 3$. The probability that at time t the particle lies in the element d^3x around the point $\mathbf{x}$ is the scalar $j^0\,\mathrm{d}^3x$. The tracks are distinguished by specification of the initial point $\mathbf{x}_0 = \mathbf{x}(0)$ since, in virtue of the single-valuedness of the velocity field, they can never cross. Formula (12.2.9) is the guidance equation of the relativistic spin $\frac{1}{2}$ theory and remains valid even when $a = b = 0$. From (12.2.8) we have

$$|\mathbf{v}| \leqslant 1, \tag{12.2.11}$$

so that the velocity can never exceed the speed of light, a basic requirement of a relativistic theory of motion (equality obtains when $a = b = 0$).† States for which j^μ is null are presumably unattainable in practice although the subclass of such states comprising eigenfunctions of one component of α^i is supposed to result from velocity measurements, and we find instances of luminal speeds in the examples of §12.3.

† Bohm (1953b) defines the 4-velocity as j^μ/a which does not lead to the usual condition $u^\mu u_\mu = 1$. The formula (12.2.10) is the same, however.

We emphasize that these definitions and the result (12.2.11) remain valid for arbitrary solutions of the Dirac equation, in particular general superpositions of positive and negative energy basis functions in arbitrary admissible external potentials. In this respect the negative energy solutions do not pose any problems of interpretation. At least as regards defining particle orbits we do not need to invoke the Newton–Wigner or other such position operators – this only becomes necessary if we restrict the set of physically realizable states to the positive spectrum. We shall, in fact, take the view that a single particle may be guided by a wave drawn from the entire energy spectrum as this enables us to gain insight into some characteristic relativistic properties (cf. §12.3).

Notice from the last member of (12.2.10) that we have passed from the current density to the velocity of a particle in the ensemble via the local expectation value of the velocity operator α^i. The mean velocity in the ensemble coincides with the usual expression:

$$\langle v^i \rangle = \int j^0 v^i \, \mathrm{d}^3x = \int \psi^\dagger \alpha^i \psi \, \mathrm{d}^3x = \langle \alpha^i \rangle. \tag{12.2.12}$$

While the eigenvalues of α^i are ± 1, the velocity of the particle is defined for all states and is continuously variable (subject to (12.2.11)).

With the aid of the 4-velocity u^μ, we define the energy-momentum 4-vector of the particle guided by the wave ψ to be

$$p^\mu = m u^\mu, \qquad p_\mu p^\mu = m^2, \tag{12.2.13}$$

where m is the usual rest mass. Because $u^0 \geqslant 1$ we have that $p^0 \geqslant m$ so the actual energy for an arbitrary state, *including negative energy states*, is always positive and bounded from below by the rest mass (the positive and negative energy states are distinguished by the form of other quantities, such as the sign of the scalar; see §12.3). Moreover, the momentum always lies in the direction of the current j^μ. Defining the proper time derivative as $\mathrm{d}/\mathrm{d}\tau = u^\mu \partial_\mu$ we may deduce a variety of expressions for the force exerted on the particle, $\mathrm{d}p^\mu/\mathrm{d}\tau$, in terms of the tensor fields but we shall not do so (in this connection we note that (12.2.4) plays the role of the Hamilton–Jacobi equation in the Schrödinger and Pauli theories but it is a purely algebraic relation and not deduced from the wave equation).

Clearly, the definition of momentum (12.2.13) is quite different from a definition as the local expectation value of the momentum operator $\hat{p}_\mu = \mathrm{i}\partial_\mu$. The two definitions are connected by the Gordon decomposition of the Dirac current which in the free case ($A_\mu = 0$) is:

$$j^\mu = (1/2m)[\bar{\psi}(\hat{p}^\mu \psi) - (\hat{p}^\mu \bar{\psi})\psi] - (\mathrm{i}/2m)\hat{p}_\nu b^{\mu\nu}. \tag{12.2.14}$$

The first term in square brackets, the (conserved) convection current, is the familiar expression for current in the Schrödinger and Pauli theories but it is not, in general, a time-like vector with positive fourth component. It is the addition of the second spin contribution to (12.2.14) that gives the Dirac current its peculiar properties.

Other definitions of particle 4-momentum have been proposed. Takabayasi (1957) defines it as proportional to $T^{\mu\nu}u_\nu$ where $T^{\mu\nu}$ is the energy-momentum tensor of the Dirac field, and Vigier (1952) proposes the definition Mu^μ where $M = m(a^2 + b^2)^{1/2}/a$ although M has no particular sign so the momentum so defined could be opposite to the current (for example, for the negative energy solution (12.3.4) below).

The pseudo-vector s^μ, which from (12.2.5) is space-like, will be interpreted as the spin density vector of the ensemble of particles. The spin 3-vector per particle is then given by the expression:

$$w^i = \frac{s^i}{j^0} = \frac{\frac{1}{2}\psi^\dagger\alpha^i\gamma^5\psi}{\psi^\dagger\psi} \tag{12.2.15}$$

evaluated along the trajectory. This will be recognized as the local expectation value of the spin angular momentum operator of the Dirac theory, $\frac{1}{2}\alpha^i\gamma^5$. Thus, in addition to its spacetime track, we endow the particle with an intrinsic angular momentum. To avoid engaging the problem of defining an extended body in relativity we shall assume that the particle is a point with an attached body frame or tetrad of which one space-like vector is s^μ, similar to the model admitted by the Pauli theory (Chap. 9). Obviously, there is considerable scope for the further development of a theory of the detailed structure of the particle but this is not necessary for an interpretation of the basic one-body Dirac theory. In the course of the motion the spin vector precesses and we may deduce an equation for $\mathrm{d}s^\mu/\mathrm{d}\tau$ from the Dirac equation in analogy to the Pauli equation of precession but we shall not do so.

In the instantaneous rest frame ($v^i = j^i = 0$) we have from (12.2.6) that $s^0 = 0$ and from (12.2.4) and (12.2.5) that $(j^0)^2 = a^2 + b^2$ and $|\mathbf{s}|^2 = (1/4)(a^2 + b^2)$. Hence

$$|\mathbf{w}| = |\mathbf{s}|/j^0 = \tfrac{1}{2}\,(=\hbar/2), \tag{12.2.16}$$

which expresses the spin $\frac{1}{2}$ character of the intrinsic angular momentum.

Pursuing our aim of associating the tensor quantities determined by the Dirac spinor with particle properties, we now consider the bivector $b^{\mu\nu}$. Consideration of the effect of the external electromagnetic field (Schweber, 1961, p. 101) suggests that $b^{\mu\nu}$ is proportional to the tensor determining the magnetic moment density (components b^{ij}) and the electric moment density (components b^{0i}). The magnetic moment per particle would then be defined as proportional to b^{ij}/j^0. Finally, there is the problem of connecting the scalar a and the pseudo-scalar b with the properties of a particle. On their own these quantities do not, in fact, appear to describe such properties naturally although some combination of them may do so. They certainly have some physical significance, however, for a is proportional to the trace of the Dirac energy-momentum tensor and $b = (1/m)\partial_\mu s^\mu$ is the 'source' of the spin density.

The Dirac spinor possesses eight real degrees of freedom. It has been proposed above that its physical meaning may be understood through the tensors (12.2.2) but, because of the identities (12.2.3)–(12.2.6), these describe only seven of the degrees of freedom. The remaining information not picked up by the algebraic functions lies in the overall phase of the spinor (because if we change the phase of each spinor component by the same amount the tensors are invariant). Such an entity appears in tensor functions involving differential operators such as the convection current contribution to (12.2.14). There is, however, some ambiguity in what we mean by the 'overall phase'. Suppose we try to write the spinor in polar form:

$$\psi^a = R\,\mathrm{e}^{\mathrm{i}S}\phi^a, \qquad a = 1, 2, 3, 4, \tag{12.2.17}$$

where R is a real Lorentz scalar amplitude and S a real phase. ϕ will be a spinor containing six independent degrees of freedom. A suitable candidate for R^2 is the scalar quantity $(a^2 + b^2)^{1/2}$ since this is the probability density in the rest frame. Then we require that the corresponding quantity constructed from ϕ should be unity:

$$(\bar{\phi}\phi)^2 + (\bar{\phi}\mathrm{i}\gamma^5\phi)^2 = 1. \tag{12.2.18}$$

How do we define the 'overall phase' S in terms of ψ? We could take it as the average phase of the spinor components, $\frac{1}{4}\sum_a S^a$ (Takabayasi, 1957), but this is not a Lorentz scalar. A better definition would be the scalar that is common to the phases of all the ψ^as. For example, for the positive energy solution (12.3.2) below, $S = -k_\mu x^\mu$. The ϕ^as would be correspondingly restricted so that their common scalar phase is zero and the eight real degrees of freedom in ψ would be covariantly represented by R, S and the six degrees remaining in ϕ. Note that S so defined is not, in general, the average phase and that the gradient $\partial_\mu S$ does not give the actual momentum of the particle as we have defined it in (12.2.13).

The most characteristic property of a spinor, its sign change under a 2π-rotation, corresponds in our formalism to the transformation $S \to S + \pi$. The 'local observables' (tensors) constructed from ψ are insensitive to this transformation. However, if the rotated wave is coherently superposed with another wave, the total spinor and its associated tensors do depend on the 2π-rotation and hence this will be reflected in the particle properties.

In principle, the trajectory interpretation of Dirac's theory can be extended to many-fermion systems described for a fixed number of particles by a wavefunction $\psi^{a_1 \cdots a_n}(x_1, \ldots, x_n)$ (Bohm *et al.*, 1987). The 3-velocity of the rth particle in the system would be

$$v_r^i = \frac{\psi^\dagger \alpha_r^i \psi}{\psi^\dagger \psi}, \tag{12.2.19}$$

which satisfies the required condition $|\mathbf{v}_r| \leqslant 1$. Integration yields a set of spacetime paths $\mathbf{x}_r = \mathbf{x}_r(t)$, $r = 1, \ldots, n$. However, problems of covariance arise in the many-body theory that are not present in the one-body case and it is not known whether all the objections raised by Belinfante (1973, p. 113) can be satisfactorily answered. In any case, this version of many-fermion theory is subject to the same criticism we levelled against its nonrelativistic counterpart: one cannot account for the details of all the processes the system partakes in using just the spacetime trajectories of each particle alone as the 'hidden variables'. One must be able to define, for example, a spin vector attached to each particle. An alternative is to develop the relativistic theory of rotators, a subject that exists but would take us too far afield to discuss here.

Finally, we remark that other wave equations governing spin $\frac{1}{2}$ particles are conceivable but they lack the appealing properties of the Dirac equation. For example, the second order equation due to Feynman & Gell-Mann (1958) suffers from the same drawback as the Klein–Gordon equation *viz.* the lack of a time-like current with a positive fourth component.

12.3 Superposition effects in the Dirac theory

12.3.1 Elementary solutions

We apply here the definitions of particle properties given above to the elementary positive and negative energy solutions. We use the following representation for the Dirac matrices:

$$\left.\begin{aligned} &\gamma^0 = \begin{bmatrix} I & 0 \\ 0 & -I \end{bmatrix}, \quad \gamma^i = \begin{bmatrix} 0 & \sigma^i \\ -\sigma^i & 0 \end{bmatrix}, \quad I = \begin{bmatrix} 1 & 0 \\ 0 & 1 \end{bmatrix}, \\ &\sigma^1 = \begin{bmatrix} 0 & 1 \\ 1 & 0 \end{bmatrix}, \quad \sigma^2 = \begin{bmatrix} 0 & -\mathrm{i} \\ \mathrm{i} & 0 \end{bmatrix}, \quad \sigma^3 = \begin{bmatrix} 1 & 0 \\ 0 & -1 \end{bmatrix}. \end{aligned}\right\} \tag{12.3.1}$$

Consider first the positive energy state

$$\psi = [(E+m)/2E]^{1/2} \begin{pmatrix} 1 \\ 0 \\ \dfrac{k^3}{E+m} \\ \dfrac{k^1 + \mathrm{i}k^2}{E+m} \end{pmatrix} \mathrm{e}^{-\mathrm{i}k_\mu x^\mu} \tag{12.3.2}$$

where $E = k^0 = +(\mathbf{k}^2 + m^2)^{1/2}$. This function is normalized so that $j^0 = \psi^\dagger\psi = 1$. It is an eigenstate of the energy-momentum operator $\hat{p}^\mu$ corresponding to k^μ. Inserting (12.3.1) into (12.2.8) yields for the actual momentum (12.2.13)

$$p^\mu = (E, \mathbf{k}). \tag{12.3.3}$$

From (12.2.9) it is evident that the motion is uniform and rectilinear with 3-velocity $\mathbf{k}/E$ and energy E. In the case that $k^1 = k^2 = 0$, the spin density vector is $s^\mu = \frac{1}{2}(k^3/E, 0, 0, 1)$ so the 3-spin of the particle is $\mathbf{w} = (0, 0, \frac{1}{2})$. The particle is spinning about the z-axis, the direction of motion.

We pass now to the negative energy state

$$\psi = [(E+m)/2E]^{1/2} \begin{pmatrix} \dfrac{k^3}{E+m} \\ \dfrac{k^1 + \mathrm{i}k^2}{E+m} \\ 1 \\ 0 \end{pmatrix} \mathrm{e}^{\mathrm{i}k_\mu x^\mu}. \tag{12.3.4}$$

This function is an eigenstate of the operator $\hat{p}^\mu$ corresponding to eigenvalue $-k^\mu$. It is for this reason that it is called a negative energy state. Yet evaluation of (12.2.13) shows that the actual energy-momentum vector is identical to that derived from the positive energy state, i.e., (12.3.3), with positive energy and momentum $\mathbf{k}$. Moreover, when $k^1 = k^2 = 0$, the 3-spin is given by $\mathbf{w} = (0, 0, \frac{1}{2})$, as above.

In regard to its orbit and intrinsic rotation the particle guided by the wave (12.3.4) is therefore indistinguishable from that guided by (12.3.2). There is no problem of interpreting the 'antiparticle' solution because the *mathematical* property of having a negative eigenvalue of the energy operator is not translated into actual physical behaviour.

The two solutions are distinguished by the sign of the scalar, or equivalently the bivector, and by the overall phase. We have $a = \bar{\psi}\psi = m/E(-m/E)$ for the positive (negative) energy wavefunction and computing the bivector from (12.2.3) yields that the only nonzero component is $b^{12} = a = \pm m/E$. In both cases the pseudo-scalar is zero. Note that for each of these solutions the spin contribution to the total current in (12.2.14) vanishes and the convection current equals the scalar a times the momentum eigenvalue. Although both waves are 'plane', the gradient of the overall phase gives the actual momentum only in the first case.

12.3.2 The Klein paradox

As might be expected, the linear superposition of the elementary solutions described in §12.3.1 corresponding to different wave numbers implies fundamental new effects. As a first example we apply the guidance formula (12.2.9) to the so-called 'Klein paradox' which is intended to illustrate the problem of interpreting the single-body Dirac equation when one attempts to confine a particle (Klein, 1929; Bjorken & Drell, 1964, p. 40).

We consider a particle restricted to the z-axis moving in the region $z < 0$ and incident on a potential step $V > 0$ lying in the domain $z > 0$. V is treated as the fourth component of a vector field (e.g., electrostatic potential). In the domain $z < 0$ the total wavefunction is $\psi_{\mathrm{I+R}} = \psi_{\mathrm{I}} + \psi_{\mathrm{R}}$, where

$$\psi_{\mathrm{I}} = \begin{pmatrix} 1 \\ 0 \\ \dfrac{k}{E+m} \\ 0 \end{pmatrix} \mathrm{e}^{-\mathrm{i}(Et-kz)}, \qquad \psi_{\mathrm{R}} = A \begin{pmatrix} 1 \\ 0 \\ \dfrac{-k}{E+m} \\ 0 \end{pmatrix} \mathrm{e}^{-\mathrm{i}(Et+kz)} \tag{12.3.5}$$

are respectively the incident and reflected waves, with $k > 0$, $E = +(m^2 + k^2)^{1/2}$ and A a constant. In the domain $z > 0$ the transmission wavefunction is

$$\psi_{\mathrm{T}} = C \begin{pmatrix} 1 \\ 0 \\ \dfrac{q}{E-V+m} \\ 0 \end{pmatrix} \mathrm{e}^{-\mathrm{i}(Et-qz)}, \tag{12.3.6}$$

where the wave number $q = +[(E-V)^2 - m^2]^{1/2}$ may be real or imaginary depending on the magnitude of V, and the requirement of continuity at

$z = 0$ implies that

$$1 + A = C, \qquad 1 - A = q(E + m)C/k(E - V + m). \tag{12.3.7}$$

The currents associated with each of the elementary wavefunctions may be computed from $j = \psi^\dagger \alpha^3 \psi$ and are

$$j_\mathrm{I} = 2k/(E + m), \qquad j_\mathrm{R} = -2k|A|^2/(E + m), \tag{12.3.8}$$

$$j_\mathrm{T} = |C|^2(q + q^*)\,\mathrm{e}^{\mathrm{i}(q-q^*)z}/(E - V + m). \tag{12.3.9}$$

The probability density in the segment $z > 0$ is ($j^0 = \psi^\dagger \psi$)

$$j^0_\mathrm{T} = |C|^2\,\mathrm{e}^{\mathrm{i}(q-q^*)z}(1 + |q|^2(E - V + m)^{-2}). \tag{12.3.10}$$

The idea now is to attempt to confine the particle in the region $z < 0$ by raising the step height. When $|\mathrm{E} - V| < m$, q is imaginary and j^0_T is damped exponentially, as we would expect for a particle restricted to move only in $z < 0$. But as we increase V so that $V > E + m$, q becomes real and the probability density j^0_T is finite for all $z > 0$. In attempting to confine the particle the opposite has happened. Moreover, from (12.3.9) and (12.3.8) we see that the transmission current j_T is negative and the reflected current j_R exceeds the incident current j_I, since solving (12.3.7) yields that $|A|^2 > 1$. This complex of problems constitutes the Klein paradox.

There are two remarks to be made at this stage. The first is that in order that these results may be considered to violate our 'ordinary reasoning' in this case we must first believe that the Dirac current is describing in some way a state of affairs that actually exists. For if these currents merely refer to phenomena that would be observed should the system be subjected to a measuring process, the measurement interaction could very well disturb the system to a degree sufficient to bring about, for example, a negative transmission current. In that case, it is not clear that the results described above are paradoxical. If, however, we take the view that the Dirac current is giving insight into an objective process we are led to our second observation that as regards the motion in the region $z < 0$ the waves $\psi_\mathrm{I}, \psi_\mathrm{R}$ taken individually (from which we derive the currents $j_\mathrm{I}, j_\mathrm{R}$) have no particular significance since the actual wavefunction in this domain is $\psi_{\mathrm{I}+\mathrm{R}}$. Now this wavefunction does not, in fact, describe our private image of an initially free particle incident from $z = -\infty$ which interacts after only a certain period with the step, for the reflected wave is already defined at $t = 0$ over the entire domain $-\infty < z < 0$. Similarly, the transmitted wave is defined for all $z > 0$ at $t = 0$; there is, indeed, a finite probability of the particle being anywhere in space at $t = 0$. The waves are already arranged at $t = 0$ to accommodate the boundary condition of continuity at $z = 0$ – if ψ_I really described a free incident particle, how could it 'know' about the relative magnitude of ψ_T? A similar point was raised in the Schrödinger theory (§4.4). To describe a free incident particle we should use a packet. Thus, granted that we are not actually dealing with a free incident particle, it follows that there is not necessarily anything paradoxical in the appearance of a negative transmission current since the word 'transmission' may be misleading.

To gain insight into why this example does not present a paradox (although it may still imply difficulties of interpretation) we examine in detail the particle motion. This we do for a set of four steps of progressively increasing height. For an arbitrary value of V, and writing $A = |A|\,e^{i\varphi}$, the probability density and current in the region $z < 0$ are respectively given by

$$j^0_{I+R} = 2(E+m)^{-1}[(1+|A|^2)E + 2m|A|\cos(2kz-\varphi)], \quad (12.3.11)$$

which contains an interference term, and

$$j_{I+R} = 2k(E+m)^{-1}(1-|A|^2), \quad (12.3.12)$$

which is the sum of j_I and j_R and does not. The trajectory is obtained from the velocity fields in the two regions which are

$$v_{I+R} = k(1-|A|^2)[(1+|A|^2)E + 2m|A|\cos(2kz-\varphi)]^{-1}, \quad (12.3.13)$$

$$v_T = (q+q^*)(E-V+m)[(E-V+m)^2+|q|^2]^{-1} \quad (12.3.14)$$

(the x- and y-components of velocity are zero).

Suppose first that $V < E - m$. Then q and A are real, $|A|^2 < 1$ and the velocity is everywhere positive so the particle moves from left to right. In the domain $z < 0$ the orbit is

$$(2kz-\varphi) + \varepsilon\sin(2kz-\varphi) = \omega t + c, \quad (12.3.15)$$

where

$$\varepsilon = 2m|A|/E(1+|A|^2), \qquad \omega = 2k^2(1-|A|^2)/E(1+|A|^2) \quad (12.3.16)$$

with $0 < \varepsilon < 1$ and c is a constant determined by the initial position z_0. In the domain $z > 0$ the motion is uniform:

$$z = qt(E-V)^{-1} + c', \quad (12.3.17)$$

where c' is fixed by the requirement of continuity of the trajectory at $z = 0$. It will be observed that the expression (12.3.15) is identical to (4.3.9) in the case $k_1 = -k_2 = k$ and that z expressed as a function of t is an infinite sum of Bessel functions. For the choice of constants $c = c' = 0$, $m = k = \frac{1}{2}$ and

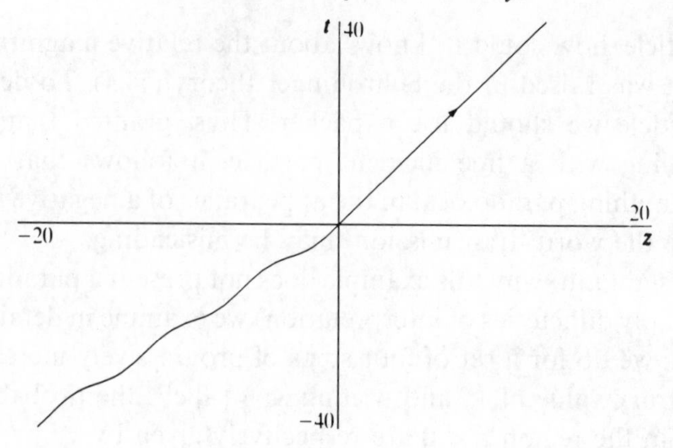

Fig. 12.1 The trajectory of a Dirac particle incident on a step of height $V < E - m$ located in the region $z > 0$.

$A = \sqrt{2} - 1$ (so $\varphi = 0$) which corresponds to $V = 1/4\sqrt{2}$, we have the orbit

$$\left.\begin{aligned} z + \tfrac{1}{2}\sin z &= \tfrac{1}{2}t \qquad (z < 0), \\ z &= \tfrac{1}{3}t \qquad (z > 0), \end{aligned}\right\} \tag{12.3.18}$$

which is depicted in Fig. 12.1. The particle moves in, oscillating about the line $z = t/2$, and the mean orbit is refracted at the origin; it is always transmitted. Notice that at points where $\cos z = -1$ the speed is luminal.

Increasing V to the range $E - m < V < E + m$, q becomes imaginary and $|A| = 1$; from (12.3.13) and (12.3.14) we see that the particle is at rest wherever it is placed along the z-axis.

Finally, we return to the case of the Klein paradox and let $V > E + m$. Then q and A are again real, with $|A| > 1$, and (12.3.13) and (12.3.14) imply that the velocity is negative in both regions. The trajectory is given by (12.3.15) and (12.3.17) but where now $\omega < 0$ and $E - V < 0$. Choosing the constants as above, except now $|A| = (2\sqrt{2} - 1)/(3 - \sqrt{2})$ and $\varphi = \pi$, which corresponds to $V = 7/4\sqrt{2}$, the motion is given by

$$\left.\begin{aligned} z - 0.7\sin z &= -0.1t \qquad (z < 0), \\ z &= -\tfrac{1}{3}t \qquad (z > 0), \end{aligned}\right\} \tag{12.3.19}$$

and is shown in Fig. 12.2. The particle moves out of the step along the negative z-axis. This is a perfectly consistent result and does not confound our intuition concerning this process if it is borne in mind that the use of elementary functions does not describe the process of scattering of an initially free particle from a localized barrier, and it is remembered that the direction of wavefront

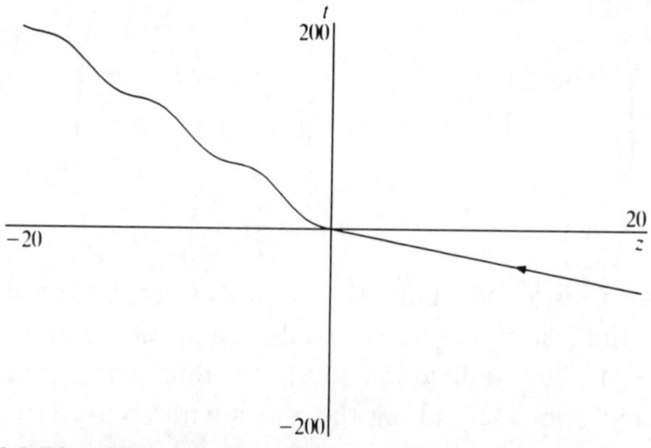

Fig. 12.2 The Klein paradox. The particle moves out of the step.

(overall phase) progression (always to the right in this example) is not directly related to the particle current. As remarked above, the particle is never 'free' since the influence of the barrier is already contained in the total wavefunction for all $\mathbf{x}$ at $t = 0$. In no case is a particle incoming and then turned around at the step. Our results should be compared with the usual explanation for the Klein phenomenon which invokes the antiparticle concept (Greiner, 1990, Chap. 13).

12.3.3 *Zitterbewegung*

In the Heisenberg picture the coordinate operator $\hat{\mathbf{x}}(t)$ satisfies the equation of motion $d\hat{x}^i/dt = \alpha^i$. Solving this relation shows that $\hat{\mathbf{x}}$ is a sum of two terms: an operator corresponding to the classical expression for the position coordinate in relativity, and a quantum term describing a high frequency ($> 2m$) oscillation about the classical motion of amplitude less than half the Compton wavelength (Schrödinger, 1930; Barut & Bracken, 1981). This oscillation may be seen in the time evolution of mean values arising from the inteference of positive and negative energy components making up a generic free Dirac wavefunction. The term *zitterbewegung* has been used to describe it, albeit in the context of a theory of motion that lacks any notion of an orbit that could be subjected to a microscopic 'trembling movement'. In fact, we shall see that this imagery is justified in the quantum theory of motion, at least in the simple examples we consider.

We start by linearly superposing the following positive and negative energy states:

$$\psi_+ = \begin{pmatrix} 1 \\ 0 \\ \dfrac{k}{E+m} \\ 0 \end{pmatrix} e^{-i(Et-kz)}, \qquad \psi_- = A \begin{pmatrix} \dfrac{-k}{E+m} \\ 0 \\ 1 \\ 0 \end{pmatrix} e^{i(Et+kz)}, \quad (12.3.20)$$

where $A = |A|\,e^{i\varphi}$ is a constant. ψ_+ describes a particle with 3-velocity $\mathbf{v} = (0, 0, k/E)$ and 3-spin $\mathbf{w} = (0, 0, \frac{1}{2})$ and ψ_- a particle with $\mathbf{v} = (0, 0, -k/E)$ and $\mathbf{w} = (0, 0, \frac{1}{2})$. The individual solutions thus correspond to uniform motions directed oppositely along the z-axis with spin up in that direction. Forming $\psi_+ + \psi_-$ we find that the motion is still in the z-direction but with a variable velocity:

$$dz/dt = [E(1 + |A|^2)]^{-1}[k(1 - |A|^2) + 2m|A|\cos(2Et + \varphi)]. \quad (12.3.21)$$

This may be integrated immediately to yield

$$z = [E(1 + |A|^2)]^{-1}[k^2(1 - |A|^2) + (m|A|/E)\sin(2Et + \varphi)] + c, \quad (12.3.22)$$

where c is fixed by the initial position. The first term on the right hand side represents the classical mean motion of the weighted oppositely directed elementary motions. The second term, the *zitterbewegung*, is due to quantum mechanical interference. The motion is displayed in Fig. 12.3 (for a choice of constants $k = m = 1/2\sqrt{2}$, $|A| = \sqrt{2} - 1$ and $c = \varphi = 0$ so that $2z = t + \sin t$). The orbit oscillates about the classical motion (the straight line $z = t/2$) with frequency $2E > 2m$ and amplitude $m|A|/E^2(1 + |A|^2)$ which, while depending on the relative amplitude of the positive and negative energy summands, is always less than $1/2m$ (half the Compton wavelength). Note that the orbit becomes light-like for times when $\cos t = 1$ and that for equal amplitudes ($|A| = 1$) or $k = 0$ the motion is pure *zitterbewegung*.

In the above example the particle spin in the superposed state is unaffected; it remains $\mathbf{w} = (0, 0, \frac{1}{2})$. To obtain a solution for which the spin vector undergoes a high frequency precession analogous to the trembling motion of the orbit we superpose the positive energy spin up state ψ_+ with a different negative energy spin down state given by

$$\psi'_- = A \begin{pmatrix} 0 \\ \dfrac{k}{E+m} \\ 0 \\ 1 \end{pmatrix} e^{i(Et+kz)}. \quad (12.3.23)$$

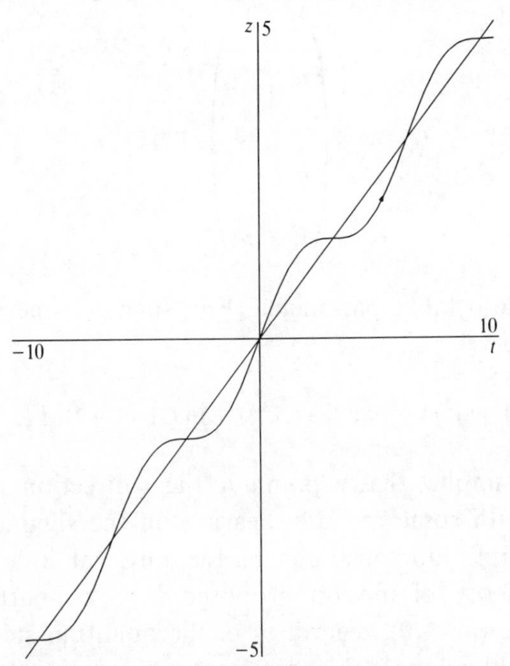

Fig. 12.3 *Zitterbewegung*. The orbit undergoes a microscopic oscillation about the classical motion (straight line).

ψ'_- corresponds to a motion $\mathbf{v} = (0, 0, -k/E)$ and spin $\mathbf{w} = (0, 0, -\frac{1}{2})$. The trajectory induced by the total wave $\psi_+ + \psi'_-$ moves in three dimensions:

$$\mathbf{x} = [E(1 + |A|^2)]^{-1}(|A| \sin(2Et + \varphi), -|A| \cos(2Et + \varphi), kt(1 - |A|^2)) + \mathbf{c}. \tag{12.3.24}$$

This is a helix directed along the z-axis and provides a good illustration of the decoupling of the velocity and momentum operators, for the state in question is an eigenfunction of the operator $\hat{\mathbf{p}}$ corresponding to eigenvalue $(0, 0, k)$. Accompanying this motion is a precession of the spin vector having a similar time dependence:

$$\mathbf{w} = [E(1 + |A|^2)]^{-1}(k|A| \cos(2Et + \varphi), k|A| \sin(2Et + \varphi), \tfrac{1}{2}E(1 - |A|^2)). \tag{12.3.25}$$

Independently of our interest in *zitterbewegung*, we mention for the sake of completeness that we may obtain a constant spin vector pointing in a range of directions in space, while the trajectory remains rectilinear, if we superpose the positive energy, spin up state ψ_+ with the positive energy spin down state

$$\psi'_+ = A \begin{pmatrix} 0 \\ 1 \\ 0 \\ \dfrac{-k}{E+m} \end{pmatrix} e^{-i(Et-kz)}, \qquad (12.3.26)$$

where A is an adjustable parameter. For such a superposition we find $\mathbf{v} = (0, 0, k/E)$ and

$$\mathbf{w} = [2E(1 + |A|^2)]^{-1}(m(A + A^*), -im(A - A^*), E(1 - |A|^2)). \qquad (12.3.27)$$

The choice $A = 1$ implies that $\mathbf{w}$ points in the x-direction, for example.

We conclude with some remarks bearing on the significance of positive and negative energy superpositions in the physical interpretation of the theory. In the theory of motion proposed here no particle ever travels backwards in time ($u^0 > 0$), regardless of the quantum state of the system. Future work should compare this treatment with the spacetime approach to electrodynamics (Feynman, 1949) in which particles (positive energy spectrum) travel forwards in time and antiparticles (negative energy spectrum) travel backwards. It is noteworthy that the Feynman diagrams, which correspond to the terms in a perturbation expansion of the propagator, depict world lines that are deemed meaningless in principle according to the conventional view of quantum mechanics. Nevertheless, physicists have found them useful as an *aide mémoire.* That, of course, is also the principal virtue of the particle track in the de Broglie–Bohm theory, yet a consistent introduction of the particle concept into quantum theory along de Broglie–Bohm lines shows that the paths of Feynman do not have any direct connection with the actual orbit (cf. §6.9). Feynman presents an aid to calculation but is what he describes merely 'virtual' rather than actually there? A superposed state may not have the character of either a 'particle' or an 'antiparticle' in the sense that the latter are usually defined in terms of energy eigenstates. We recall the situation in the Schrödinger case where we associate the entire state with a single particle and do not give the basis functions any particular physical significance. Likewise we propose as a hypothesis worth exploring that the Dirac wavefunction should not be viewed as a 'mixture' of particles and antiparticles connected with a special expansion of the wave selected from an infinite number of possible expansions into eigenfunctions of Hermitian operators. Rather, only the total spinor has physical reality in the individual case.

12.4 Quantum field theory in the Schrödinger picture and its interpretation

12.4.1 Space representation

In §10.6 we applied the causal interpretation to field quantization of nonrelativistic boson and fermion systems. In the normal mode representation of the field coordinate the method used there extends essentially unmodified to relativistic fields. Our aim here is to develop a causal interpretation of relativistic quantum field theory written in the Schrödinger picture starting from the space representation of the field coordinate. We consider boson fields and for simplicity restrict attention to a neutral, spin 0, massless field described classically by the real scalar function $\psi(\mathbf{x}, t)$. Our treatment follows that of Bohm *et al.* (1987) (for an early attempt see Freistadt (1955, 1956)). The extension to higher integer-spin fields is reasonably straightforward but involves unnecessary complications, at least as regards explaining the basic principles. The functional Schrödinger picture in position space for relativistic fermions can be treated by the methods of Grassmann calculus (DeWitt, 1984; Floreanini & Jackiw, 1988) but will not be discussed here.†

The Lagrangian density of the free classical field is

$$\mathscr{L} = \tfrac{1}{2}\partial_\mu\psi\,\partial^\mu\psi = \tfrac{1}{2}[\dot{\psi}^2 - (\nabla\psi)^2], \tag{12.4.1}$$

where $\dot{\psi} = \partial\psi/\partial t$ and its variation yields the wave equation

$$\Box\psi = 0. \tag{12.4.2}$$

Defining the momentum conjugate to the field coordinate ψ by the relation $\pi = \partial\mathscr{L}/\partial\dot{\psi} = \dot{\psi}$ the Hamiltonian density is

$$\mathscr{H} = \pi\dot{\psi} - \mathscr{L} = \tfrac{1}{2}[\pi^2 + (\nabla\psi)^2] \tag{12.4.3}$$

with the field Hamiltonian given by

$$H = \int \mathscr{H}\,\mathrm{d}^3x. \tag{12.4.4}$$

Replacing π by $\delta S/\delta\psi$ where $S[\psi]$ is a functional, the classical Hamilton–Jacobi equation of the field, $(\partial S/\partial t) + H = 0$, becomes

† The proposal of the causal interpretation is that the physical interpretation of quantum mechanics is best effected in the Schrödinger picture of the underlying theory. There is a problem in the way of this programme in that solutions may not exist to a quantum theory formulated in this picture because the state vector does not remain in Hilbert space (e.g., in QED; see Dirac (1965)). In that case it may be argued that the Schrödinger and Heisenberg pictures are not equivalent and the latter is preferred. This caveat does not impinge on our examples and the point will not be pursued.

$$\frac{\partial S}{\partial t} + \frac{1}{2}\int \mathrm{d}^3x\left[\left(\frac{\delta S}{\delta \psi}\right)^2 + (\nabla\psi)^2\right] = 0. \tag{12.4.5}$$

The term $\frac{1}{2}\int \mathrm{d}^3x(\nabla\psi)^2$ plays the role of an 'external potential'.

So far we have considered the classical canonical theory. To quantize this system we treat $\psi(\mathbf{x})$ and $\pi(\mathbf{x})$ as Schrödinger operators (time-independent) obeying the commutation rules

$$\left.\begin{aligned}[\psi(\mathbf{x}), \psi(\mathbf{x}')] &= [\pi(\mathbf{x}), \pi(\mathbf{x}')] = 0\\ [\psi(\mathbf{x}), \pi(\mathbf{x}')] &= \mathrm{i}\delta(\mathbf{x} - \mathbf{x}')\end{aligned}\right\} \tag{12.4.6}$$

and work in a representation $|\psi(\mathbf{x})\rangle$ in which the Hermitian operator $\psi(\mathbf{x})$ is diagonal. The operator π is then replaced by the functional derivative $-\mathrm{i}\delta/\delta\psi$. The Hamiltonian (12.4.4) becomes an operator $\hat{H}$ acting on a wavefunction $\Psi[\psi(\mathbf{x}), t] = \langle\psi(\mathbf{x})|\Psi(t)\rangle$ which is a functional of the real field coordinate ψ and a function of the time t. (This is *not* a point function of $\mathbf{x}$; Ψ depends on the variable ψ for all $\mathbf{x}$.) The Schrödinger equation for the field is

$$\mathrm{i}\,\partial\Psi/\partial t = \hat{H}\Psi \tag{12.4.7}$$

or, explicitly,

$$\mathrm{i}\frac{\partial\Psi}{\partial t} = \int \mathrm{d}^3x\,\frac{1}{2}\left[-\frac{\delta^2}{\delta\psi^2} + (\nabla\psi)^2\right]\Psi. \tag{12.4.8}$$

(The analogous wave equation for the electromagnetic field is obtained by replacing $E_i \to \mathrm{i}\delta/\delta A^i$ and $B^i \to (\nabla \times \mathbf{A})^i$ in the Hamiltonian density (12.6.12) below.) The usual conditions of finiteness, continuity and single-valuedness are imposed on Ψ.

Clearly, ψ here is playing the role of the space variable $\mathbf{x}$ in the particle Schrödinger equation, the continuous index $\mathbf{x}$ being analogous to the discrete index i on $\mathbf{x}_i$ in the many-particle theory. To arrive at a causal interpretation of the field system we write the wavefunction as usual in polar form, $\Psi = R\,\mathrm{e}^{\mathrm{i}S}$ where $R[\psi, t]$ and $S[\psi, t]$ are real functionals, and decompose (12.4.8) into two real equations:

$$\frac{\partial S}{\partial t} + \frac{1}{2}\int \mathrm{d}^3x\left[\left(\frac{\delta S}{\delta \psi}\right)^2 + (\nabla\psi)^2\right] + Q = 0, \tag{12.4.9}$$

$$\frac{\partial R^2}{\partial t} + \int \mathrm{d}^3x\,\frac{\delta}{\delta\psi}\left(R^2\frac{\delta S}{\delta\psi}\right) = 0, \tag{12.4.10}$$

where

$$Q[\psi, t] = -(1/2R)\int \mathrm{d}^3x\,\delta^2 R/\delta\psi^2. \tag{12.4.11}$$

Eq. (12.4.9) is the classical Hamilton–Jacobi equation (12.4.5) modified by the quantum potential (12.4.11) and (12.4.10) is a conservation law which justifies the assumption that, at time t, $R^2\,\mathrm{D}\psi$ is the probability for the field to lie in an element of 'volume' $\mathrm{D}\psi$ about the configuration $\psi(\mathbf{x})$ for all $\mathbf{x}$. The notation $\mathrm{D}\psi$ means the infinite product $\prod_{\mathbf{x}} \mathrm{d}\psi$ of field volume elements $\mathrm{d}\psi$ for each value of $\mathbf{x}$. The wavefunction may be assumed to be normalized:

$$\int |\Psi|^2\,\mathrm{D}\psi = 1. \tag{12.4.12}$$

Pursuing the analogy with the many-body theory, we see that the quantum potential involves a sum over all $\mathbf{x}$, analogous to a sum over all particles, which introduces the state Ψ as an active, nonclassical agent in the dynamics of the wave.

We now introduce the assumption that at each instant t the field ψ has a well-defined value for all $\mathbf{x}$, as in classical field theory, whatever the state Ψ of the field. The time evolution may be obtained from the solution of the guidance formula

$$\frac{\partial\psi(\mathbf{x},t)}{\partial t} = \left.\frac{\delta S[\psi(\mathbf{x}),t]}{\delta\psi(\mathbf{x})}\right|_{\psi(\mathbf{x})=\psi(\mathbf{x},t)} \tag{12.4.13}$$

(analogous to $m\dot{\mathbf{x}} = \nabla S$) once we have specified the initial function $\psi_0(\mathbf{x})$ (analogous to $\mathbf{x}_0$). This is, of course, the assumption of classical field theory in the Hamilton–Jacobi formalism.

To find the equation of motion of the field coordinate we apply the functional derivative $\delta/\delta\psi$ to the Hamilton–Jacobi equation (12.4.9). By simple techniques of functional calculus we obtain

$$\frac{\mathrm{d}}{\mathrm{d}t}\dot{\psi} = -\frac{\delta}{\delta\psi}\left[Q + \int \mathrm{d}^3x\,\tfrac{1}{2}(\nabla\psi)^2\right], \tag{12.4.14}$$

where we have identified $\delta S/\delta\psi$ with $\dot{\psi}$ and

$$\frac{\mathrm{d}}{\mathrm{d}t} = \frac{\partial}{\partial t} + \int \mathrm{d}^3x\,\frac{\partial\psi}{\partial t}\frac{\delta}{\delta\psi}. \tag{12.4.15}$$

Eq. (12.4.14) is analogous to the particle equation of motion $m\ddot{\mathbf{x}} = -\nabla(V+Q)$. Noting that $\mathrm{d}\dot{\psi}/\mathrm{d}t = \partial\dot{\psi}/\partial t$ and taking the classical 'external force' term over to the left hand side, we find†

† We might envisage including a 'genuine' external potential functional $V[\psi]$ on the right hand side of (12.4.8). Eq. (12.4.16) is then modified by the addition of $-\delta V/\delta\psi$ on the right hand side. To treat the massive spin 0 field one should add $\frac{1}{2}m^2\psi^2$ to the integrand in (12.4.8) and $m^2\psi^2$ to the left hand side of (12.4.16).

$$\Box\psi(\mathbf{x}, t) = -\frac{\delta Q[\psi(\mathbf{x}), t]}{\delta\psi(\mathbf{x})}\bigg|_{\psi(\mathbf{x})=\psi(\mathbf{x},t)}. \tag{12.4.16}$$

This wave equation, in which ψ is now time-dependent, should be compared with the classical wave equation (12.4.2). The 'quantum force' term on the right hand side is responsible for all the characteristic effects of quantum field theory.

We emphasize that $\psi(\mathbf{x}, t)$ is a c-number at each spacetime point. It is the eigenvalue of the Schrödinger field operator evaluated along a system 'trajectory', a notion that has no meaning in the conventional interpretation but is crucial to the recovery of the classical limit (see below). $\psi(\mathbf{x}, t)$ is not to be confused with the Heisenberg field operator $\hat{\psi}(\mathbf{x}, t)$ which satisfies the classical wave equation, $\Box\hat{\psi} = 0$.

As might be expected from the corresponding discussion in the particle theory (§3.9) the energy of the quantum field, $E = -\partial S/\partial t$, is continuously variable and not conserved in general. From (12.4.9) we deduce

$$\frac{\mathrm{d}E}{\mathrm{d}t} = \frac{\partial Q}{\partial t}\bigg|_{\psi(\mathbf{x})=\psi(\mathbf{x},t)} \tag{12.4.17}$$

(the analogue of the particle equation $\mathrm{d}E/\mathrm{d}t = \partial(V + Q)/\partial t$ although the 'classical potential' located above does not contribute on the right hand side in the field case).

The classical limit of the quantized field theory is readily established following the treatment of Chap. 6: When the magnitude and gradient of the quantum potential are negligible we obtain the classical Hamilton–Jacobi and wave equations. In addition, the energy conservation law of the classical field follows as a special case from (12.4.17). The conservation law (12.4.10) retains its significance as determining the statistical properties of the classical field although it is not an equation one customarily finds in field theory texts.

Returning to the quantum theory, it is apparent that the causal interpretation enables one to account for individual events in quantum field theory if the classical definition of an individual system, the field ψ, is supplemented by the state $\Psi[\psi]$. The two aspects of the individual deterministically evolve together in accordance with the laws of motion (12.4.8) and (12.4.16). This solves the problem of connecting classical and quantum field theory in a completely general and consistent way. We now wish to mention two key features of the causal theory's conceptual basis which flow from this state-dependent characterization of individuals and bear on its relation with relativity.

First, it is evident that although we started with a scalar classical field, the quantization process has resulted in a field $\psi(\mathbf{x}, t)$ that is *not* a Lorentz scalar function: $\psi'(\mathbf{x}', t') \neq \psi(\mathbf{x}, t)$ if $x'^{\mu} = x'^{\mu}(x^{\nu})$ is a Lorentz transformation. The field satisfies a noncovariant wave equation (12.4.16) whose right hand side cannot generally be expressed in terms of Lorentz invariants because t appears alone. In the classical limit where the right hand side disappears ψ is a scalar, and Lorentz covariance is restored. *The breaking of Lorentz covariance is a quantum effect of individual processes.* This property obviously stems from the initial 3 + 1 Schrödinger-picture formulation of the quantization procedure which picks out a preferred frame in which the field operator is time-independent and the state function depends on t. (We do not consider the possibility of a fully covariant formulation of the Schrödinger picture.)

Second, it also appears from (12.4.16) that the evolution of the field is governed by a highly nonlinear and nonlocal equation that, in principle, involves the state of the field over the entire universe. This is the field-theoretic version of the holistic aspect of quantum theory alluded to many times previously. It implies instantaneous connections between distant field elements which, in turn, implies the violation of relativity in individual processes (a remark presaged in Chap. 11). This feature is characteristic of the quantum domain and vanishes in the classical limit. *Nonlocality of fields is a quantum effect.* Note though that the nonlocal and noncovariant properties can disappear in situations where the quantum force is finite if it is a local and scalar function of the spacetime coordinates, so classical field theory is not uniquely characterized just by the absence of these typical quantum characteristics (for an example see §12.8).

We are in sympathy with the reader who feels uneasy when confronted with such blatant revisions of our customary views concerning covariance and locality. A defence is that it offers at least *some* kind of insight into the physics of quantum field theory in an area where comprehensive competitors do not exist. And the world really may be built like this; one cannot say with certainty that relativity has unlimited validity in all physical processes.

In any case, it is emphasized that these rather striking properties of the causal interpretation of field theory, namely noncovariance and nonlocality, do not lead to the kind of inconsistencies that arose in the Klein–Gordon particle theory (§12.1) and relate solely to the individual processes that go to make up the ensemble whose statistical properties are described by Ψ. They underpin rather than contradict the Lorentz covariance and locality of the experimental predictions of quantum field theory. The usual requirement of 'microscopic causality' that field operators at space-like separated points commute remains valid, as does the impossibility of sending signals between such regions (§11.3). The observable properties of the field system are basically statistical and embodied in the expectation values of operators,

$$\langle\Psi|\hat{A}|\Psi\rangle = \int \Psi^*[\psi](\hat{A}\Psi)[\psi]\,\mathrm{D}\psi, \tag{12.4.18}$$

which are tensorial objects. While the underlying individual events may, and in general do, disobey the principles of relativity, the relativity theory will be found to be valid in all experiments that test and confirm the probabilistic formalism. *Lorentz covariance and locality are statistical effects.*

The above account of what is novel in quantum as opposed to classical field theory differs markedly from the usual one which locates discreteness (appearance of field quanta) and irreducible probability laws as the fundamental nonclassical features. For us discreteness and probability are aspects of a basically continuous and deterministic description.

It is of interest to place the above formalism in the context of the 'method of the causal interpretation' (§3.12) in which one describes the physical properties of a system (here the field) in terms of the local expectation values of operators, extracted from the expression (12.4.18):

$$A[\psi] = \mathrm{Re}\,\frac{\Psi^*[\psi](\hat{A}\Psi)[\psi]}{|\Psi[\psi]|^2} \tag{12.4.19}$$

(this is 'local' relative to the configuration space of the field rather than physical space). Evaluation of this formula for a solution $\psi(\mathbf{x}, t)$ of the wave equation gives an 'actual property' of an individual field in the ensemble, corresponding to the operator $\hat{A}$. For example, if $\hat{A} = \hat{H}$, the Hamiltonian in (12.4.8), then using (12.4.9) we find $A = E = -\partial S/\partial t$, the actual total energy of the field. Similarly, if $\hat{A} = \hat{\mathbf{P}}$, where

$$\hat{\mathbf{P}} = \frac{\mathrm{i}}{2}\int \mathrm{d}^3x\left(\nabla\psi\frac{\delta}{\delta\psi} + \frac{\delta}{\delta\psi}\nabla\psi\right) \tag{12.4.20}$$

is the total momentum operator, then $A = \mathbf{P}$ with

$$\mathbf{P} = -\int \mathrm{d}^3x\,\dot{\psi}\nabla\psi, \tag{12.4.21}$$

which is the actual total momentum of the field, the same as the classical expression. Neither of these quantities has meaning in the conventional interpretation of quantum field theory which only ascribes physical significance to averages over all the field coordinates (i.e., (12.4.18)). In fact, because the operators one inserts may be local functions of $\mathbf{x}$ (e.g., the integrand in (12.4.20)), the expectation values (12.4.18) themselves can provide considerable insight into the structure of the field, for they can be expressed as a classical mean plus quantum corrections. For further discussion of these matters see §12.6.

Although some examples of the application of the theory have been given (Bohm, 1952b; Belinfante, 1973; Bohm *et al.*, 1987; and below) a systematic investigation of the solutions admitted by the integro-differential equation (12.4.16) has not been carried out. It is possible that it may play the role of the nonlinear equation sought for but never given by de Broglie (1956) in his theory of the 'double solution', where material objects are represented as localized concentrations of continuous fields. It is known that even the classical linear wave equation $\Box\psi = 0$ admits solutions describing the directed motion of localized finite energy pulses (Ziolkowski, 1989). We expect therefore that its quantum generalization will possess a rich variety of solutions, for example that a single field may exhibit a set of soliton-like entities moving and interacting without essential change of form as if they possessed 'mass', and so on. This would be similar to Einstein's concept of a unified field, but in a form rather different to that envisaged by him. A proposal that may have some relevance in this context is made in §12.8.

12.4.2 Normal mode representation

It will prove useful in the concrete examples to be discussed in §12.5. to discretize the field by imposing periodic boundary conditions. We seek an expansion of the field coordinate into normal modes in a box of side L and volume V:

$$\psi(\mathbf{x}) = V^{-1/2} \sum_{\mathbf{k}} q_{\mathbf{k}}\, e^{i\mathbf{k}\cdot\mathbf{x}}. \tag{12.4.22}$$

Here $q_{\mathbf{k}}$ are complex numbers subject to the condition $q_{\mathbf{k}}{}^* = q_{-\mathbf{k}}$ ($q_{-\mathbf{k}}$ does *not* mean $q_{\mathbf{k}}$ with $\mathbf{k}$ replaced by $-\mathbf{k}$) in virtue of the reality of $\psi(\mathbf{x})$, and $k^i = 2\pi n_i/L$, $i = 1, 2, 3$, n_i being a positive or negative integer including zero. Eq. (12.4.22) may be viewed as a coordinate transformation in the 'superspace' whose points are labelled by ψ. The inverse transformation is

$$q_{\mathbf{k}} = V^{-1/2} \int_V \psi(\mathbf{x})\, e^{-i\mathbf{k}\cdot\mathbf{x}}\, d^3x \tag{12.4.23}$$

and hence

$$\frac{\delta}{\delta\psi(\mathbf{x})} = V^{-1/2} \sum_{\mathbf{k}} e^{-i\mathbf{k}\cdot\mathbf{x}} \frac{\partial}{\partial q_{\mathbf{k}}}, \qquad \frac{\partial}{\partial q_{\mathbf{k}}} = V^{-1/2} \int d^3x\, e^{i\mathbf{k}\cdot\mathbf{x}} \frac{\delta}{\delta\psi(\mathbf{x})}. \tag{12.4.24}$$

In these coordinates the Schrödinger equation (12.4.8) becomes

$$\mathrm{i}\frac{\partial\Psi}{\partial t} = \sum_{\mathbf{k}/2}\left(-\frac{\partial^2}{\partial q_{\mathbf{k}}\,\partial q_{\mathbf{k}}^*} + \mathbf{k}^2 q_{\mathbf{k}}^* q_{\mathbf{k}}\right)\Psi, \tag{12.4.25}$$

where $\Psi = \Psi[q_{\mathbf{k}}, q_{\mathbf{k}}^*, t]$ and the notation $\mathbf{k}/2$ denotes that we have restricted the summation to half the **k**-lattice since the summands corresponding to **k** and $-\mathbf{k}$ are identical. Thus, if the wavefunction is a product state in the coordinate system $q_{\mathbf{k}}$,

$$\Psi = \prod_{\mathbf{k}/2} \psi_{\mathbf{k}}(q_{\mathbf{k}}, q_{\mathbf{k}}^*), \tag{12.4.26}$$

each $\psi_{\mathbf{k}}$ satisfies the wave equation

$$\mathrm{i}\frac{\partial\psi_{\mathbf{k}}}{\partial t} = \left(-\frac{\partial^2}{\partial q_{\mathbf{k}}\,\partial q_{\mathbf{k}}^*} + \mathbf{k}^2 q_{\mathbf{k}}^* q_{\mathbf{k}}\right)\psi_{\mathbf{k}} \tag{12.4.27}$$

with no factor $\frac{1}{2}$ on the right hand side.

The picture of the field as a collection of harmonic oscillators becomes evident in these coordinates. Writing $q_{\mathbf{k}}$ in terms of its real and imaginary parts, $q_{\mathbf{k}} = (1/\sqrt{2})(a_{\mathbf{k}} - \mathrm{i}b_{\mathbf{k}})$, we obtain

$$\mathrm{i}\frac{\partial\Psi}{\partial t} = \sum_{\mathbf{k}/2}\left[-\frac{1}{2}\left(\frac{\partial^2}{\partial a_{\mathbf{k}}^2} + \frac{\partial^2}{\partial b_{\mathbf{k}}^2}\right) + \frac{\mathbf{k}^2}{2}(a_{\mathbf{k}}^2 + b_{\mathbf{k}}^2)\right]\Psi \tag{12.4.28}$$

with $\Psi = \Psi[a_{\mathbf{k}}, b_{\mathbf{k}}, t]$ so that two oscillators (or one oscillator in two dimensions) correspond to each mode **k**. The normalization condition (12.4.12) becomes

$$\int_{-\infty}^{\infty} \cdots \int_{-\infty}^{\infty} |\Psi[a_{\mathbf{k}}, b_{\mathbf{k}}]|^2 \prod_{\mathbf{k}/2} \mathrm{d}a_{\mathbf{k}}\,\mathrm{d}b_{\mathbf{k}} = 1. \tag{12.4.29}$$

We may now translate the guidance formula (12.4.13) into the complex coordinates:

$$\frac{\mathrm{d}q_{\mathbf{k}}}{\mathrm{d}t} = \left.\frac{\partial S}{\partial q_{\mathbf{k}}^*}\right|_{q_{\mathbf{k}} = q_{\mathbf{k}}(t)} \tag{12.4.30}$$

for each **k**. Likewise, the equation of motion (12.4.16) becomes

$$\ddot{q}_{\mathbf{k}} + \mathbf{k}^2 q_{\mathbf{k}} = -\partial Q/\partial q_{\mathbf{k}}^*, \qquad Q = -(1/R)\sum_{\mathbf{k}/2} \partial^2 Q/\partial q_{\mathbf{k}}\,\partial q_{\mathbf{k}}^* \tag{12.4.31}$$

(for the generalization of this formula to include the matter–radiation interaction see Belinfante (1973, p. 209)). The initial Fourier coefficients $q_{\mathbf{k}}(0)$ are found from $\psi_0(\mathbf{x})$ via (12.4.23). Under circumstances where we may neglect the quantum force, (12.4.31) reduces to the representation of a classical field as a set of independent oscillators. In general, the quantum potential brings about interactions between the oscillators.

In this formalism the mode creation and annihilation operators are respectively given by

$$\left.\begin{aligned} a_{\mathbf{k}}^{\dagger} &= (2k)^{-1/2}(kq_{\mathbf{k}}^{*} - \partial/\partial q_{\mathbf{k}}), \\ a_{\mathbf{k}} &= (2k)^{-1/2}(kq_{\mathbf{k}} + \partial/\partial q_{\mathbf{k}}^{*}), \end{aligned}\right\} \tag{12.4.32}$$

where $k = |\mathbf{k}|$ and they satisfy the commutation relations:†

$$[a_{\mathbf{k}}, a_{\mathbf{k}'}] = [a_{\mathbf{k}}^{\dagger}, a_{\mathbf{k}'}^{\dagger}] = 0, \qquad [a_{\mathbf{k}}, a_{\mathbf{k}'}^{\dagger}] = \delta_{\mathbf{k}\mathbf{k}'}. \tag{12.4.33}$$

In terms of them ψ and the momentum operator π obeying the exchange rules (12.4.6) are given by

$$\left.\begin{aligned} \psi &= \psi^{+} + \psi^{-} = V^{-1/2}\sum_{\mathbf{k}}(2k)^{-1/2}(a_{\mathbf{k}}\,\mathrm{e}^{\mathrm{i}\mathbf{k}\cdot\mathbf{x}} + a_{\mathbf{k}}^{\dagger}\,\mathrm{e}^{-\mathrm{i}\mathbf{k}\cdot\mathbf{x}}), \\ \pi &= \pi^{+} + \pi^{-} = -\mathrm{i}V^{-1/2}\sum_{\mathbf{k}}(k/2)^{1/2}(a_{\mathbf{k}}\,\mathrm{e}^{\mathrm{i}\mathbf{k}\cdot\mathbf{x}} - a_{\mathbf{k}}^{\dagger}\,\mathrm{e}^{-\mathrm{i}\mathbf{k}\cdot\mathbf{x}}). \end{aligned}\right\} \tag{12.4.34}$$

The Hamiltonian in (12.4.8) may, following the substitution (12.4.34), be brought to either of the forms

$$\begin{aligned} \hat{H} &= \int \mathrm{d}^3x(\pi^{-}\pi^{+} + \nabla\psi^{-}\cdot\nabla\psi^{+}) + \tfrac{1}{2}\sum_{\mathbf{k}} k \\ &= \sum_{\mathbf{k}}(ka_{\mathbf{k}}^{\dagger}a_{\mathbf{k}} + \tfrac{1}{2}k). \end{aligned} \tag{12.4.35}$$

Similarly, the momentum and number of quanta operators have the following two expressions in the position and momentum spaces:

$$\hat{\mathbf{P}} = -\int \mathrm{d}^3x(\pi^{-}\nabla\psi^{+} + (\nabla\psi^{-})\pi^{+}) = \sum_{\mathbf{k}} \mathbf{k}a_{\mathbf{k}}^{\dagger}a_{\mathbf{k}}, \tag{12.4.36}$$

$$\hat{N} = \mathrm{i}\int \mathrm{d}^3x(\psi^{-}\pi^{+} - \pi^{-}\psi^{+}) = \sum_{\mathbf{k}} a_{\mathbf{k}}^{\dagger}a_{\mathbf{k}}. \tag{12.4.37}$$

Let us denote by $|0\rangle$ the vacuum ket and define the basis kets

$$|\mathbf{k}_1 \ldots \mathbf{k}_n\rangle = (n!)^{-1/2}a_{\mathbf{k}_1}^{\dagger}\ldots a_{\mathbf{k}_n}^{\dagger}|0\rangle. \tag{12.4.38}$$

† There are various conventions for the definitions of these operators in the literature. Schweber (1961, p. 160) has a factor k on the right hand side of (12.4.33) whereas Schiff (1968, p. 517) has a factor k^{-1}.

These form a complete set and are orthonormal:

$$\langle \mathbf{k}_1 \ldots \mathbf{k}_n | \mathbf{k}'_1 \ldots \mathbf{k}'_m \rangle = \delta_{mn}(n!)^{-1} \sum_P \delta(\mathbf{k}_1 - \mathbf{k}'_1) \ldots \delta(\mathbf{k}_n - \mathbf{k}'_n), \quad (12.4.39)$$

where P denotes a permutation over all $\mathbf{k}$, $\mathbf{k}'$. Then, at any instant, say $t = 0$, an arbitrary state of the field can be expanded as follows:

$$|\Psi(0)\rangle = \sum_{n=0}^{\infty} \sum_{\mathbf{k}_1 \ldots \mathbf{k}_n} f_{\mathbf{k}_1 \ldots \mathbf{k}_n} |\mathbf{k}_1 \ldots \mathbf{k}_n\rangle, \quad (12.4.40)$$

where $f_{\mathbf{k}_1 \ldots \mathbf{k}_n}$ denotes a set of complex coefficients. At time t the solution of the free Schrödinger equation is

$$|\Psi(t)\rangle = \sum_{n=0}^{\infty} \sum_{\mathbf{k}_1 \ldots \mathbf{k}_n} f_{\mathbf{k}_1 \ldots \mathbf{k}_n} \, \mathrm{e}^{-\mathrm{i}(k_1 + \cdots + k_n + E_0)t} |\mathbf{k}_1 \ldots \mathbf{k}_n\rangle, \quad (12.4.41)$$

where $E_0 = \frac{1}{2} \sum_{\mathbf{k}} k$ is the 'zero-point' or ground state energy. The normalization condition implies that

$$\langle \Psi | \Psi \rangle = \sum_{n=0}^{\infty} \sum_{\mathbf{k}_1 \ldots \mathbf{k}_n} |f_{\mathbf{k}_1 \ldots \mathbf{k}_n}|^2 = 1. \quad (12.4.42)$$

12.5 Preferred frame and nonlocal effects in quantized fields

12.5.1 Vacuum state. The Casimir effect

The vacuum state of the field, Ψ_0, is the stationary state defined by the requirement $a_{\mathbf{k}} \Psi_0 = 0$ for each $\mathbf{k}$ where $a_{\mathbf{k}}$ is the annihilation operator. Using (12.4.32) we have

$$(kq_{\mathbf{k}} + \partial/\partial q^*_{\mathbf{k}})\Psi_0 = 0 \qquad \text{for each } \mathbf{k} \quad (12.5.1)$$

or equivalently (Schweber, 1961, p. 194)

$$[(-\nabla^2)^{1/2}\psi(\mathbf{x}) + \delta/\delta\psi(\mathbf{x})]\Psi_0 = 0 \qquad \text{for each } \mathbf{x}, \quad (12.5.2)$$

where

$$(-\nabla^2)^{1/2}\psi = V^{-1/2} \sum_{\mathbf{k}} kq_{\mathbf{k}} \, \mathrm{e}^{\mathrm{i}\mathbf{k}\cdot\mathbf{x}} \quad (12.5.3)$$

in the limit of continuous $\mathbf{k}$. Solution of these differential equations yields

$$\Psi_0[q, q^*, t] = N \prod_{\mathbf{k}/2} \mathrm{e}^{-kq^*_{\mathbf{k}}q_{\mathbf{k}} - \mathrm{i}E_0 t} \quad (12.5.4)$$

or

$$\Psi_0[\psi, t] = N\, \mathrm{e}^{-\frac{1}{2}\int \psi(\mathbf{x})(-\nabla^2)^{1/2}\psi(\mathbf{x})\, \mathrm{d}^3x - \mathrm{i}E_0 t}, \tag{12.5.5}$$

where the normalization constant $N = \prod_{\mathbf{k}/2} (k/\pi)^{1/2}$ and the energy is found by direct substitution into (12.4.25) to be $E_0 = \sum_{\mathbf{k}/2} k$. Clearly, the latter is infinite.

The state (12.5.4), which is a product over the modes, is the ground state of an infinite set of independent oscillators. As we know from §4.9 each oscillator is at rest. To check this we note that up to a constant the phase of the wavefunction is given by $S = -E_0 t$. Thus, (12.4.13) becomes $\partial\psi/\partial t = \delta S/\delta\psi = 0$ or

$$\psi(\mathbf{x}, t) = \psi_0(\mathbf{x}) \tag{12.5.6}$$

and the field is static. The total 'zero-point' energy, $-\partial S/\partial t = E_0$, resides in just quantum plus 'classical' potential energy. The quantum potential is given by

$$Q = E_0 - \tfrac{1}{2}\int (\nabla\psi)^2\, \mathrm{d}^3x = E_0 - \sum_{\mathbf{k}/2} k|q_{\mathbf{k}}|^2 \tag{12.5.7}$$

and its effect is to cancel the classical contribution to leave just the zero-point energy. The word 'vacuum' is obviously something of a misnomer in this context since space is filled with a field which carries energy ('ground state' is not entirely appropriate either since by a change of boundary conditions we shall see below that we can attain a state of lower energy, an instance of the result proved in §4.1.2). We have become used to thinking of the vacuum as a sort of fluctuating 'sea', the more finely we probe the sea the greater the fluctuation (e.g., Wheeler, 1968). But this notion is not based on any definite model of what a quantized field system actually is and the present approach, which is so based, leads to a very different nondynamical picture.

It will be observed that the field is static, i.e., the relation (12.5.6) is valid, only in a preferred Lorentz frame. In this theory the vacuum is a noncovariant concept. But no experiment compatible with quantum theory can reveal in which frame it is that the vacuum field appears static. Still, although in principle unobservable according to the theory, this concept of vacuum is a valuable aid in the struggle to understand the physics of quantum field theory. It is akin to a reintroduction of the ether concept into physics, an idea that is neither terribly radical nor conservative. Einstein (1922) pointed out that the ether is compatible with special relativity if we deprive it of any mechanical qualities such as a state of motion. Dirac (1951) observed that quantum mechanics could reconcile the physical equivalence of all Lorentz frames with the preferred spacetime direction picked out by a mechanical ether at each point if the wavefunction ascribed equal probabilities to all ether velocities. Developing this theme, Dirac (1953) pointed out that the notions of absolute simultaneity and absolute time are also compatible with Lorentz covariance in a quantum context. While we have not uncovered a mechanical ether of the type envisaged by Dirac, it is clear that absolute time is playing a fundamental role in the causal theory of fields.

Although the (infinite) zero-point energy is not directly observable, the difference between it and the ground state energy corresponding to a redistribution of the field to accommodate new boundary conditions is, according to the theory, an observable quantity. A simple example of this phenomenon is due to Casimir (1948), who showed that a force of attraction exists between two parallel conducting plates in the electromagnetic vacuum. Forces of this type were subsequently observed (Tabor & Winterton, 1969).

To see how the quantum potential energy of the field contributes to this effect we consider a spin 0 field in a box defined as above by $0 < x < L$, $0 < y < L$ but where now $0 < z < a$, $a < L$. The normal expansion of the field, still with periodic boundary conditions, will be

$$\psi(\mathbf{x}) = (La^{1/2})^{-1} \sum_{\mathbf{k}}{}' q_{\mathbf{k}} \, \mathrm{e}^{\mathrm{i}\mathbf{k}\cdot\mathbf{x}} \tag{12.5.8}$$

where the prime indicates that the sum is over the range $k^1 = 2\pi n_1/L$, $k^2 = 2\pi n_2/L$, $k^3 = 2\pi n_3/a$, with n_1, n_2, n_3 any integers. Substitution of (12.5.8) into the Schrödinger equation (12.4.8) yields (12.4.25) but with $\sum$ replaced by $\sum'$. The new vacuum state is therefore given by

$$\Psi_0' = \prod_{\mathbf{k}/2}{}' (k/\pi)^{1/2} \, \mathrm{e}^{-kq_{\mathbf{k}}^* q_{\mathbf{k}} - ikt} \tag{12.5.9}$$

and the infinite zero-point energy is

$$E_0' = \sum_{\mathbf{k}/2}{}' k = Q' + \sum_{\mathbf{k}/2}{}' \mathbf{k}^2 q_{\mathbf{k}}^* q_{\mathbf{k}}, \tag{12.5.10}$$

where Q' is the new quantum potential. Obviously, the field in the smaller box is static.

Let us denote by δ the difference between the various energies in the old and new vacua. Then

$$\delta E_0 \equiv E_0' - E_0 = \delta Q + \delta V. \tag{12.5.11}$$

To evaluate this expression one goes to the limit of large L and continuous k^1 and k^2 but discrete k^3 in the sum for E_0' and continuous $\mathbf{k}$ in E_0. Calculation shows that $\lim \delta E = K/3a^3$, a finite quantity, where K is a negative constant.

The energy in the new vacuum is less than in the old one. The force acting between the walls at $z = 0$ and $z = a$ is therefore

$$-\mathrm{d}(\lim \delta E)/\mathrm{d}a = K/a^4, \tag{12.5.12}$$

which is attractive. It is evident from (12.5.11) that the origin of the force may be attributed to the nonclassical part of the differential quantum potential δQ, which remains after cancellation of the differential classical harmonic oscillator energy δV.

The modified vacuum Ψ'_0 is a state in the Fock space built on the old vacuum, as in (12.4.41). What we are effectively doing in the Casimir effect is renormalizing the mean stress tensor in the new state by subtracting the zero-point energy density of the 'true' vacuum and arguing (or hoping) that the residue is a finite, physical quantity (see §12.6). The quantum potential approach does not appear to throw any light on the conceptual issues raised by this procedure, such as why it is precisely the zero-point energy of the true Minkowski vacuum that must be subtracted to achieve a meaningful result.

A similar result is obtained if we start with the physically more reasonable Dirichlet boundary condition that the field vanishes at the plates $z = 0, a$. In that case (12.5.8) is replaced by the expansion

$$\psi(\mathbf{x}) = \mathrm{i}(La^{1/2})^{-1} \sum_{\mathbf{k}} q_{\mathbf{k}}\, \mathrm{e}^{\mathrm{i}(k^1 x + k^2 y)} \sin k^3 z, \tag{12.5.13}$$

where k^1, k^2 take the same range but $k^3 = \pi n_3/a$. Clearly, $\psi(x, y, 0) = \psi(x, y, a) = 0$. This is equivalent to summing (12.4.22) over this range subject to the condition $q_{k^1,k^2,k^3} + q_{k^1,k^2,-k^3} = 0$ and hence expressions for the vacuum state etc. are formally the same apart from summation over a different lattice. For a discussion of the differences between the Casimir effect calculated with scalar or electromagnetic fields and with different boundary conditions see DeWitt (1975) and Fulling (1989, Chap. 5).

12.5.2 Excited states and nonlocality

The previous example illustrates the noncovariance of the underlying process but not the nonlocality. To exhibit the latter it is sufficient to consider an excited state corresponding to one quantum (i.e., a Fock state, an eigenstate of the number operator (12.4.37)). This may or may not be stationary.

Applying the creation operator (12.4.32) to the vacuum (12.5.4) we obtain the normalized state

$$\Psi_{\mathbf{k}}[q, q^*, t] = (2k)^{1/2} q^*_{\mathbf{k}}\, \mathrm{e}^{-\mathrm{i}kt} \Psi_0[q, q^*, t]. \tag{12.5.14}$$

The time dependence may be readily checked by direct substitution into (12.4.25). This is still a product, and stationary, state and describes the case where one quantum has been excited in the **k**th mode while all the other oscillators remain in the ground state (the energy of the **k**th mode is $2k$ while that of the $\mathbf{k}'$th mode, $\mathbf{k}' \neq \mathbf{k}$, is k'). Actually, because each mode corresponds to two oscillators, (12.5.14) is not the most general one-quantum excitation of a single mode. From the one-quantum sector of (12.4.41) we see that the general solution for the **k**th mode which takes account of this degeneracy is

$$\Psi_{\mathbf{k}}[q, q^*, t] = (2k)^{1/2}(f_{\mathbf{k}} q_{\mathbf{k}} + f_{-\mathbf{k}} q^*_{\mathbf{k}})\, \mathrm{e}^{-ikt}\Psi_0[q, q^*, t], \qquad (12.5.15)$$

where the coefficients are arbitrary, except that $|f_{\mathbf{k}}|^2 + |f_{-\mathbf{k}}|^2 = 1$.

More generally still, we may excite a range of modes and write an arbitrary one-quantum state as

$$\Psi_1[q, q^*, t] = \sum_{\mathbf{k}} f_{\mathbf{k}}(2k)^{1/2} q_{\mathbf{k}}\, \mathrm{e}^{-ikt}\Psi_0[q, q^*, t], \qquad (12.5.16)$$

where $f_{\mathbf{k}}$ denotes a set of constant complex numbers. This is no longer a product or stationary state but involves interference between the oscillators and, for suitable $f_{\mathbf{k}}$, may describe a wave packet. The possibility of non-stationary single-quantum states suggests that one should not think of a 'quantum' as a parcel of energy with a definite value since the actual energy of the field, $-\partial S/\partial t$, fluctuates in time (such states have been prepared in the laboratory (Hong & Mandel, 1986)).

To see the behaviour of the field in spacetime we rewrite (12.5.16) in terms of ψ:

$$\Psi_1[\psi, t] = \int f(\mathbf{x}, t)\psi(\mathbf{x})\, \mathrm{d}^3x\, \Psi_0[\psi, t], \qquad (12.5.17)$$

where

$$f(\mathbf{x}, t) = V^{-1/2} \sum_{\mathbf{k}} f_{\mathbf{k}}(2k)^{1/2}\, \mathrm{e}^{-\mathrm{i}(kt + \mathbf{k}\cdot\mathbf{x})} \qquad (12.5.18)$$

is a complex scalar function that may describe a packet. Extracting the phase from (12.5.17) we obtain the instantaneous rate of change of the field at a point **x** (using (12.4.13)):

$$\frac{\partial\psi(\mathbf{x}, t)}{\partial t} = \left[2\mathrm{i}\left|\int f(\mathbf{x}, t)\psi(\mathbf{x}, t)\, \mathrm{d}^3x\right|^2\right]^{-1} \times \left[f(\mathbf{x}, t)\int f^*(\mathbf{x}', t)\psi(\mathbf{x}', t)\, \mathrm{d}^3x' - f^*(\mathbf{x}, t)\int f(\mathbf{x}', t)\psi(\mathbf{x}', t)\, \mathrm{d}^3x'\right]. \qquad (12.5.19)$$

The dependence of $\dot{\psi}$ on the values of ψ over the entire region of space where $|f|$ is appreciable is apparent. At any instant, the field values at distinct spatial points, however distant, are nonlocally connected. We may therefore expect that disturbing the field in a localized region of space will bring about an instantaneous rearrangement of the field everywhere.

To see this sort of thing explicitly, we solve the guidance formula for the stationary single-mode excitation (12.5.14). This is the extreme case where the packet $f(\mathbf{x}, t)$ is a plane wave extending over all space:

$$f(\mathbf{x}, t) = (2k/V)^{1/2}\, \mathrm{e}^{-\mathrm{i}(kt - \mathbf{k}\cdot\mathbf{x})}. \tag{12.5.20}$$

The phase is given by

$$S = (1/2\mathrm{i})\log(\Psi_{\mathbf{k}}/\Psi_{\mathbf{k}}^*) = (1/2\mathrm{i})\log(q_{\mathbf{k}}^*/q_{\mathbf{k}}) - kt - \sum_{\mathbf{k}/2} kt. \tag{12.5.21}$$

Hence, using (12.4.30),

$$\dot{q}_{\mathbf{k}} = \partial S/\partial q_{\mathbf{k}}^* = 1/2\mathrm{i}q_{\mathbf{k}}^*, \tag{12.5.22}$$

which has the solution

$$q_{\mathbf{k}}(t) = q_{\mathbf{k}0}\, \mathrm{e}^{-\mathrm{i}\omega_{\mathbf{k}}t}, \qquad \omega_{\mathbf{k}} = 1/2|q_{\mathbf{k}0}|^2. \tag{12.5.23}$$

The $\mathbf{k}$th normal coordinate is thus oscillating but with a frequency $\omega_{\mathbf{k}}$ that is not equal to the classical value k but depends rather on the initial coordinate $q_{\mathbf{k}0}$. This is in accord with the general result that stationary states give rise to nonclassical motions and is a consequence of the finite quantum force in (12.4.31). Indeed, we might have expected to obtain no motion at all in a stationary state given that this is the unique solution for a one-dimensional oscillator (cf. §4.9). A state having this property is $\Psi = k^{1/2}(q_{\mathbf{k}} + q_{\mathbf{k}}^*)\,\mathrm{e}^{-\mathrm{i}kt}\Psi_0$ (a particular case of (12.5.15)) for which $\dot{q}_{\mathbf{k}} = \partial\psi/\partial t = 0$. In general, though, the single-mode one-quantum field is dynamic.

The quickest way to find $\psi(\mathbf{x}, t)$ is to deduce it from (12.4.22) rather than substituting (12.5.20) into (12.5.19) and integrating. We have

$$\psi(\mathbf{x}) = V^{-1/2}(q_{\mathbf{k}}\,\mathrm{e}^{\mathrm{i}\mathbf{k}\cdot\mathbf{x}} + q_{\mathbf{k}}^*\,\mathrm{e}^{-\mathrm{i}\mathbf{k}\cdot\mathbf{x}} + \sum_{\mathbf{k}' \neq \pm\mathbf{k}} q_{\mathbf{k}'}\,\mathrm{e}^{\mathrm{i}\mathbf{k}'\cdot\mathbf{x}}). \tag{12.5.24}$$

Now the normal coordinates $q_{\mathbf{k}'}$ for $\mathbf{k}' \neq \pm\mathbf{k}$ are constants: $q_{\mathbf{k}'} = q_{\mathbf{k}'0}$. Hence, substituting from (12.5.23) and writing $q_{\mathbf{k}0} = |q_{\mathbf{k}0}|\,\mathrm{e}^{-\mathrm{i}\delta_{\mathbf{k}}}$ we find

$$\psi(\mathbf{x}, t) = 2|q_{\mathbf{k}0}|V^{-1/2}\cos(\omega_{\mathbf{k}}t - \mathbf{k}\cdot\mathbf{x} + \delta_{\mathbf{k}}) + g(\mathbf{x}), \tag{12.5.25}$$

where $g(\mathbf{x})$ is an arbitrary function. The latter is fixed by the initial configuration $\psi_0(\mathbf{x})$ and so we have finally

$$\psi(\mathbf{x}, t) = \psi_0(\mathbf{x}) + 2|q_{\mathbf{k}0}|V^{-1/2}[\cos(\omega_{\mathbf{k}}t - \mathbf{k}\cdot\mathbf{x} + \delta_{\mathbf{k}}) - \cos(-\mathbf{k}\cdot\mathbf{x} + \delta_{\mathbf{k}})], \tag{12.5.26}$$

where $|q_{\mathbf{k}0}|$ and $\delta_{\mathbf{k}}$ are given in terms of $\psi_0(\mathbf{x})$ by

$$q_{\mathbf{k}0} = V^{-1/2} \int \psi_0(\mathbf{x})\, \mathrm{e}^{-\mathrm{i}\mathbf{k}\cdot\mathbf{x}}\, \mathrm{d}^3x. \tag{12.5.27}$$

Eq. (12.5.26) describes a monochromatic wave running in the direction **k** with phase velocity $\omega_{\mathbf{k}}/k$, superimposed on the static field defined by $\psi_0(\mathbf{x})$. The solution (12.5.14) with $q_{\mathbf{k}}^*$ replaced by $q_{\mathbf{k}}$ implies a running wave in the $(-\mathbf{k})$-direction. Clearly, these waves are not Lorentz scalars.

The dependence of the field dynamics on the entire field distribution at $t = 0$ is evident. This is particularly striking for the initial field 'velocity':

$$\partial\psi(\mathbf{x}, 0)/\partial t = (V^{1/2}|q_{\mathbf{k}0}|)^{-1} \sin(-\mathbf{k}\cdot\mathbf{x} + \delta_{\mathbf{k}}). \tag{12.5.28}$$

Recall that ψ_0 is arbitrary but $\dot{\psi}_0$ is fixed by the initial state of the field, via $\dot{\psi}_0 = \delta S_0/\delta\psi$.

As expected, then, the one-quantum state is nonclassical, but in its nonlocal aspect rather than in the sense of discreteness. Actually, there is one case in which the field behaves classically, when the initial function $\psi_0 = (2/kV)^{1/2} \cos(-\mathbf{k}\cdot\mathbf{x} + \delta_{\mathbf{k}})$ which gives $|q_{\mathbf{k}0}| = 1/(2k)^{1/2}$ and $q_{\mathbf{k}'0} = 0$ for $\mathbf{k}' \neq \pm\mathbf{k}$ from (12.5.27). Then

$$\psi(\mathbf{x}, t) = (2/kV)^{1/2} \cos(kt - \mathbf{k}\cdot\mathbf{x} + \delta_{\mathbf{k}}), \tag{12.5.29}$$

which satisfies the classical wave equation $\Box\psi = 0$ (this is analogous in particle quantum mechanics to showing that a particle placed at the centre of a harmonic oscillator packet, for example, pursues a classical orbit).

There is one other feature of this field to which we wish to call attention, but it deserves its own section, §12.8.

12.5.3 Coherent states and the classical limit

As our final example we consider a class of states traditionally connected with the classical limit of quantum field theory, the coherent states. The properties of such states closely parallel those of the simple harmonic oscillator.

We begin with the single-mode coherent state corresponding to an excitation of the **k**th mode which may be defined as the eigenstate of the annihilation operator $a_{\mathbf{k}}$ with eigenvalue $\alpha_{\mathbf{k}}\,\mathrm{e}^{-ikt}$ where $\alpha_{\mathbf{k}}$ is an arbitrary complex constant (Glauber, 1963b). Using (12.4.32) we have

$$(kq_{\mathbf{k}} + \partial/\partial q_{\mathbf{k}}^*)\Psi_{\alpha_{\mathbf{k}}} = (2k)^{1/2}\alpha_{\mathbf{k}}\,\mathrm{e}^{-ikt}\Psi_{\alpha_{\mathbf{k}}}, \tag{12.5.30}$$

so that

$$\Psi_{\alpha_{\mathbf{k}}}[q, q^*, t] = N_{\mathbf{k}}\,\mathrm{e}^{(2k)^{1/2}\alpha_{\mathbf{k}}q_{\mathbf{k}}^*\mathrm{e}^{-ikt}}\Psi_0[q, q^*, t], \tag{12.5.31}$$

where $N_{\mathbf{k}}$ is a normalization constant. This is a product state over **k** but it is neither a stationary nor a number state. It is readily checked by differentiation that (12.5.31) satisfies the Schrödinger equation. The modes $\mathbf{k}' \neq \mathbf{k}$ remain unexcited so we may write:

$$\begin{aligned}\Psi_{\alpha_{\mathbf{k}}} = {} & N_{\mathbf{k}}\exp[-kq_{\mathbf{k}}^*q_{\mathbf{k}} + (2k)^{1/2}\alpha_{\mathbf{k}}q_{\mathbf{k}}^*\,\mathrm{e}^{-ikt} - ikt] \\ & \times\exp\left[-\sum_{(\mathbf{k}'\neq\mathbf{k})/2}(k'q_{\mathbf{k}'}^*q_{\mathbf{k}'} + ik't)\right].\end{aligned} \tag{12.5.32}$$

That this state is indeed coherent is obvious from the form of its amplitude. Writing $q_{\mathbf{k}} = (1/\sqrt{2})(a_{\mathbf{k}} - ib_{\mathbf{k}})$ and $\alpha_{\mathbf{k}} = |\alpha_{\mathbf{k}}|\,\mathrm{e}^{-i\theta_{\mathbf{k}}}$ we have:

$$\begin{aligned}|\Psi_{\alpha_{\mathbf{k}}}| = N_{\mathbf{k}}\exp\Bigg\{ & -(k/2)[a_{\mathbf{k}} - k^{-1/2}|\alpha_{\mathbf{k}}|\cos(kt + \theta_{\mathbf{k}})]^2 \\ & - (k/2)[b_{\mathbf{k}} - k^{-1/2}|\alpha_{\mathbf{k}}|\sin(kt + \theta_{\mathbf{k}})]^2 \\ & + \tfrac{1}{2}|\alpha_{\mathbf{k}}|^2 - \sum_{(\mathbf{k}'\neq\mathbf{k})/2} k'|q_{\mathbf{k}'}|^2\Bigg\}.\end{aligned} \tag{12.5.33}$$

This describes a two-dimensional Gaussian packet in $(a_{\mathbf{k}}, b_{\mathbf{k}})$-space oscillating in each direction without change of shape with frequency k, amplitude $k^{-1/2}|\alpha_{\mathbf{k}}|$ and phase $\theta_{\mathbf{k}}$.

As might be expected, the normal coordinate oscillates with the same frequency, amplitude and phase as the packet. To obtain this result we evaluate the phase of (12.5.32) which is

$$S = [(2k)^{1/2}/2\mathrm{i}](\alpha_{\mathbf{k}}q_{\mathbf{k}}^*\,\mathrm{e}^{-ikt} - \alpha_{\mathbf{k}}^*q_{\mathbf{k}}\,\mathrm{e}^{ikt}) - \sum_{\mathbf{k}/2} kt. \tag{12.5.34}$$

Hence, using (12.4.30),

$$\frac{\mathrm{d}q_{\mathbf{k}}}{\mathrm{d}t} = [(2k)^{1/2}/2\mathrm{i}]\alpha_{\mathbf{k}}\,\mathrm{e}^{-ikt}, \tag{12.5.35}$$

whence

$$q_{\mathbf{k}}(t) = [|\alpha_{\mathbf{k}}|/(2k)^{1/2}]\,\mathrm{e}^{-\mathrm{i}(kt+\theta_{\mathbf{k}})} + c_{\mathbf{k}}, \tag{12.5.36}$$

where $c_{\mathbf{k}}$ is fixed by $q_{\mathbf{k}0}$. The energy of the field, $-\partial S/\partial t$ evaluated with (12.5.36), is time-varying unless $c_{\mathbf{k}} = 0$.

The time dependence of $\psi(\mathbf{x})$ follows from (12.4.22):

$$\psi(\mathbf{x}, t) = (2/kV)^{1/2}|\alpha_{\mathbf{k}}| \cos(kt - \mathbf{k}\cdot\mathbf{x} + \theta_{\mathbf{k}}) + f(\mathbf{x}), \qquad (12.5.37)$$

where the arbitrary function $f(\mathbf{x})$ is fixed by the initial field $\psi_0(\mathbf{x})$. The same result may, of course, be derived by writing the phase (12.5.34) as

$$S = -(2k/V)^{1/2}|\alpha_{\mathbf{k}}| \int \sin(k_\mu x^\mu + \theta_{\mathbf{k}})\psi(\mathbf{x})\,\mathrm{d}^3x - \sum_{\mathbf{k}/2} kt \qquad (12.5.38)$$

and solving (12.4.13).

The first part of (12.5.37) describes a classical monochromatic running wave in the $\mathbf{k}$-direction, similar to the one-quantum state (12.5.25) but without the feature of nonlocality. The total wave still displays, however, the nonscalar character typical of the quantum domain, through $f(\mathbf{x})$. We can remove this feature by going to the large amplitude limit in which the quantum potential vanishes, as explained in §6.6.1. Since $|\alpha_{\mathbf{k}}|^2$ is the mean number of quanta in the coherent state this is tantamount to the high quantum number limit. We obtain

$$\psi(\mathbf{x}, t) = (2/kV)^{1/2}|\alpha_{\mathbf{k}}| \cos(k_\mu x^\mu + \theta_{\mathbf{k}}). \qquad (12.5.39)$$

Alternatively, we can recover the same result for any value of the amplitude by a judicious choice of ψ_0, namely one for which $f(\mathbf{x}) = 0$ (this is analogous to choosing the initial position to lie at the centre of the packet in the particle case). Either way, we recover a classical plane wave with wave vector k^μ and fixed amplitude and phase.

The usual discussion of the classical limit of the quantized field proceeds from a consideration of expectation values (Loudon, 1986, p. 150). It is easily proved that the mean field operator in the coherent state is just a classical plane wave of amplitude $A_{\mathbf{k}} = (2/kV)^{1/2}|\alpha_{\mathbf{k}}|$ and phase $\theta_{\mathbf{k}}$:

$$\langle\hat{\psi}\rangle = \langle\Psi_{\alpha_{\mathbf{k}}}(t)|\hat{\psi}(\mathbf{x})|\Psi_{\alpha_{\mathbf{k}}}(t)\rangle = A_{\mathbf{k}} \cos(k_\mu x^\mu + \theta_{\mathbf{k}}). \qquad (12.5.40)$$

The root mean square field operator is

$$\Delta\hat{\psi} = (\langle\hat{\psi}^2\rangle - \langle\hat{\psi}\rangle^2)^{1/2} = (2kV)^{-1/2}, \qquad (12.5.41)$$

where we have retained only the $\mathbf{k}$th contribution to the zero-point field and discarded the rest. One argues that in the high mean quantum number limit, the amplitude of $\langle\hat{\psi}\rangle$ becomes better defined relative to the 'uncertainty' $\Delta\hat{\psi}$ and hence in this regime the classical stable wave has been recovered.

Eq. (12.5.40) is an example of Ehrenfest's theorem (cf. §3.8.3) for the field which asserts that for sufficiently 'localized' states, the mean field satisfies the classical wave equation:

$$\Box\langle\hat{\psi}\rangle = 0. \tag{12.5.42}$$

But for reasons explained in §6.7.2 such a result does not constitute a solution to the conceptual problem of connecting quantum with classical field theory, for the latter starts from the objective existence of the wave $\psi(\mathbf{x}, t)$ whereas there is no counterpart to this notion in conventional quantum field theory. The mere fact that the mean quantum field propagates like a classical field does not allow one to claim that the objective classical wave has somehow been 'deduced'. In our approach we may pass continuously from one domain to the other by adopting states and parameter values for which the quantum potential vanishes, the field ψ being a real entity regardless of which regime we deem to be 'quantum' and which 'classical'.

Eq. (12.5.39) is a plane wave solution of the classical wave equation. To obtain an arbitrary classical wave we must independently excite all modes to a coherent state. The result is a product wavefunction in which each factor is a coherent state corresponding to the first factor in (12.5.32). This implies that the phase is additive over $\mathbf{k}$, and hence so is ψ:

$$\begin{aligned}\psi(\mathbf{x}, t) &= \sum_{\mathbf{k}} \psi_{\mathbf{k}}(\mathbf{x}, t) \\ &= \sum_{\mathbf{k}} (2/kV)^{1/2}|\alpha_{\mathbf{k}}| \cos(k_\mu x^\mu + \theta_{\mathbf{k}}) + f(\mathbf{x}).\end{aligned} \tag{12.5.43}$$

Once again, a special choice of large amplitudes $|\alpha_k|$ or of ψ_0 yields a classical solution, an arbitrary positive frequency superposition obeying the scalar wave equation.

This result may be compared with the high quantum number limit of a number state. Suppose, for example, that n modes are each excited to the first quantum state. The total wavefunction is a product over $\mathbf{k}$ and hence the field will be a sum of waves (12.5.25) corresponding to the different $\mathbf{k}$s. This is a nonclassical function in that it does not satisfy the classical wave equation, and this property is not modified in the limit of large n.

We conclude that the nonlocal and noncovariant properties of the causal model disappear for a suitable choice of states and initial conditions or parameters and this provides an intuitively simple transition to the classical limit. None of the conceptual problems that beset the usual approach, which is incapable of deducing classical physical waves from quantum mathematical operators, arise. Further properties of the number and coherent states are discussed in the following sections.

12.6 Light paths

12.6.1 Are there photon trajectories?

In the last two sections a causal interpretation of the quantized field theory was developed in fairly strict analogy to that of particle quantum mechanics, the field coordinates $\psi(\mathbf{x})$ replacing the particle variables $\mathbf{x}$. One may think that the task is thereby completed and there is no need, or perhaps even meaning, to seek further insight into the individual processes underlying the statistical predictions of quantum field theory. We are thinking particularly here of how one is to understand the photon concept. In quantum field theory the phrase 'a photon of energy k and momentum $\mathbf{k}$' formally means that the wavefunction of the electromagnetic field is a simultaneous eigenstate of the number operator corresponding to eigenvalue unity and the total (normal ordered) energy–momentum operator $\int d^3x{:}\ \hat{T}^{0\mu}{:}$ corresponding to eigenvalue k^μ. The quantum theory of motion applied to such a state implies that the field coordinate is a continuous harmonic wave in space and time (apart from an additive function of $\mathbf{x}$; cf. (12.5.25)). There is nothing in the free dynamics of the wave to indicate the presence of a discrete entity or 'photon'. In the causal interpretation as developed so far the 'photon' is an entity associated with the entire field (over all space); this is the thing with the 'photonic' property.

But what of Einstein's original conception of the photon as a quantum of energy localized at points in space and moving without dividing (Einstein, 1905)? Einstein thought that the particle (photon) was guided by a wave, obeying Maxwell's equations, into regions where the field was most intense (he apparently never published the idea because he thought it conflicted with energy–momentum conservation in individual processes: see Wigner (1980)). This was a view also held by de Broglie† (e.g., de Broglie & Andrade e Silva, 1968; de Broglie & Vigier, 1972) and, of course, is the essence of the causal interpretation as developed for nonrelativistic matter systems in this book.

Curiously, the 'photon as particle' concept has survived in modern conventional descriptions of electrodynamical processes which often conjure up such a quasi-classical image. One finds talk of 'a detector firing when a photon is at its position' when all that is justified from the formalism, and indeed the account of the causal interpretation given so far, is that the field exchanges energy with an atomic wave field, the exchange involving a change of quantum number because of the evolution implied by the Schrödinger

† An experiment of Wang, Zou & Mandel (1991) has apparently refuted a version of de Broglie's ideas on light. It should be noted that this has no bearing on the validity of the causal interpretation discussed in this book (Holland & Vigier, 1991).

equation. Or one speaks of operators 'creating' and 'annihilating' photons at a point **x** or with a momentum **k** when what is involved is a formal relation between wavefunctions rather than an accurate account of a physical transformation in which quanta are added to or subtracted from a field by interaction. And it is commonplace to talk of photons as travelling along rays in geometrical optics (Misner *et al.*, 1973, Chap. 22).

There is an imprecisely formulated assumption permeating the historical discussion on quantum optics that hints at the possibility of treating the photon as some kind of localized entity moving in spacetime, and even that quantized waves are somehow 'made up' of such things. It is ironic that the causal theory of fields, which seeks to make precise and rigorous our informal images of quantum processes, does not lead to such a conception. Yet there are elements in the formalism of quantum field theory that we have not explored so far and which may allow us to introduce further objective structures in addition to the field coordinates. It is premature to close the subject on the basis of what we have found out hitherto.

The basic question to be addressed is: using the current mathematical theory, can any means be found to give substance to the notion that a photon is a localized object pursuing a definite spacetime path? And if not, are there nevertheless other structures, such as flow of energy, that are meaningful?

The following six arguments may be advanced against the 'photon as particle' concept:

(1) The classical field that is quantized has zero rest mass. What material object is it that travels along the putative paths?
(2) The photon is a product of the quantization procedure – there is no evidence of discreteness in classical field theory. This is in contrast to the quantization of matter systems where the particle concept already applies in the unquantized theory.
(3) The discreteness only appears when the field interacts and exchanges energy with matter. In this sense the 'photon' is an artefact of the detection process.
(4) The photon disappears on detection – if it is a physical object what becomes of it during this process?
(5) It seems reasonable to require that we should be able to define the density of particles in a region of space. But it is not possible to find a nonnegative number density consistent with the current formalism. To see this, consider the expression (12.4.37) for the number of quanta operator as the space integral of the local operator

$$\hat{N}(\mathbf{x}) = \mathrm{i}(\psi^{-}\pi^{+} - \pi^{-}\psi^{+}). \tag{12.6.1}$$

A natural candidate for the mean number density of quanta, averaged over the ensemble of field coordinates, would be the expectation value of this operator:

$$N(\mathbf{x}, t) = \langle \Psi(t)|\hat{N}(\mathbf{x})|\Psi(t)\rangle. \tag{12.6.2}$$

That this expression is not generally positive may be seen by inserting for Ψ the single-quantum state (12.5.16). Expressing $\hat{N}(\mathbf{x})$ in terms of the creation and annihilation operators (12.4.32) we find

$$\langle \Psi_1|\hat{N}(\mathbf{x})|\Psi_1\rangle = (\mathrm{i}/2)(\phi^*\partial_0\phi - \phi\partial_0\phi^*), \tag{12.6.3}$$

where

$$\phi(\mathbf{x}, t) = V^{-1/2}\sum_{\mathbf{k}} k^{-1/2} f_{\mathbf{k}}\, \mathrm{e}^{-\mathrm{i}(kt - \mathbf{k}\cdot\mathbf{x})}. \tag{12.6.4}$$

The complex scalar field ϕ satisfies the wave equation $\Box\phi = 0$ and lies in the manifold of positive frequency solutions (we shall meet this function again in §12.6.4). The right hand side of (12.6.3) is just the fourth component of the conserved current associated with this wave equation, but as pointed out in the similar Klein–Gordon case in §12.1, it is not generally positive. Hence, one cannot give meaning to the notion of a particle density. The integral of (12.6.2) over the entire volume V has a meaning as the total number of quanta in the field at time t (in this case, unity) but one cannot attribute these to elementary spatial volumes. (Some meaning may be attached to the integral of (12.6.2) over a volume smaller than V if its linear dimensions are large compared with the wavelength of any occupied mode of the field (Mandel, 1966). In the case of (12.6.3) where ϕ is a localized packet this will effectively coincide with the integral over V.) Moreover, there is no position operator of the Newton–Wigner type for the massless spin 1 field so one cannot write down a probability amplitude for the outcome of position measurements.

(6) The free radiation field can exist in states for which the total number of quanta is, in the conventional view, undefined (i.e., a linear superposition of Fock states such as the coherent state). How could a set of individual trajectories, of which there are by definition a whole number, be assigned to such a system? If we draw an analogy between number states and their superpositions, and stationary states and their superpositions, and recall that in the causal interpretation the energy concept is well defined for all states and not just the stationary ones, we may try to apply the 'method of the causal interpretation' (cf. §3.12) and define an 'actual' number of particles in the field at any instant for any state. That is, in the state Ψ, the actual number of quanta would be the local expectation value

$$N(\psi(\mathbf{x}, t), t) = \mathrm{Re}\left.\frac{\Psi^*[\psi, t]\hat{N}\Psi[\psi, t]}{|\Psi[\psi, t]|^2}\right|_{\psi(\mathbf{x}) = \psi(\mathbf{x},t)}, \tag{12.6.5}$$

where $\hat{N}$ is the total number operator. But this is not a meaningful definition as it is easy to see that the number N is not only fractional but, in general, not even positive (exceptions are the number eigenstates where $N = n$ and the $\mathbf{k}$th coherent state where $N = |\alpha_{\mathbf{k}}|^2$). According to the quantum theory of motion, it is then not generally correct to suppose that there is a definite but perhaps unknown number of quanta in the system for all states (unlike the energy concept which applies equally in all states). The concept is simply not defined.

Objections (1)–(4) are of a conceptual character and can perhaps be overcome by developments of the theory. For example, giving the 'photon' (i.e., the field) a very small but finite rest mass is a respectable idea (in particular in cosmology; see Vigier (1990)). This would circumvent objection (1) although it is not clear how far it improves the feasibility of a particle interpretation. And, of course, as remarked before, one cannot rule out the possibility of soliton solutions to the wave equation which could model a kind of 'particle'. But, for the present, the technical objections (5) and (6) seem to imply that, although intuitively attractive, the notion that quanta of the field pursue individual spacetime tracks does not appear tenable because it is in conflict with the field-theoretic formalism. The language traditionally employed to inform the mathematics of optics is misleading.

So what then is a 'photon'? If it is an attribute of the entire field what is its nature? It cannot refer to the energy of the field since this is not unique; one can superpose a set of one-quantum stationary states covering a range of energies to obtain again a one-quantum state but with a time-dependent energy. In those number states where the energy is continually varying it is surely not appropriate to speak of the photon as a 'quantum of energy'. Usually in the causal interpretation we aim to associate with a Hermitian operator some physical attribute of the system relevant to the terminology normally used to describe the operator (e.g., 'actual energy' is connected with 'energy operator'). In the case of the number operator we tried this in (12.6.2) and failed. We confess we do not see how to carry through our usual programme in this case. It may be that in the further development of the causal interpretation the particle interpretation used in the bulk of this book will be supplanted by fields as basic entities. One may hope to construct the 'particle' as a nonlinear persistent aspect of the field, at least in quasi-free situations, with the particle perhaps losing its integrity during interactions. The causal interpretation is not after all a monolith and there is scope for the development of competing theories of light.

This conclusion does not rule out the possibility of augmenting our spacetime picture of quantum fields in some other way, in particular by tracing the energy flow in the field. This would establish a continuity of interpretation with the classical theory where the photon concept does not arise but energy most certainly flows across space. It is to the problem of defining the lines of energy transport that we now turn.

12.6.2 Energy flow in classical optics

The basic structure in classical Maxwell theory, at least in the absence of charges, is the continuous field. The notion of a ray, or the spacetime track along which light energy is transported, has a meaning in the geometrical optics limit but is not considered to have general applicability in the exact wave theory. Bearing in mind that in the causal interpretation of material systems the trajectory, or ray, concept is meaningful in the exact wave theory as well as in the 'geometrical optics' (classical) limit, it is of interest to examine whether the ray can be retained in the full electromagnetic theory. For the classical Maxwell theory is, of course, the 'first quantized' formulation of electromagnetism in which Maxwell's equations,

$$\partial_\mu F^{\nu\mu} = J^\nu, \qquad \partial_\mu {}^*F^{\mu\nu} = 0, \tag{12.6.6}$$

correspond to Schrödinger's equation. Here $F_{\mu\nu}$ is the antisymmetric Maxwell field tensor, ${}^*F^{\mu\nu} = \frac{1}{2}\varepsilon^{\mu\nu\sigma\rho}F_{\sigma\rho}$ is the dual field and J^μ is the material electric current. In fact, to bring out the similarity of Maxwell's equations with quantum wave equations, we note that they can be written in the form of Dirac's equation, albeit without any hint of Planck's constant intruding (Riesz, 1958).

Let us digress for a moment to see this. Defining the 'wavefunction' (complex matrix) $F = \frac{1}{2}F_{\mu\nu}\gamma^{\mu\nu}$ and writing $J = J_\mu\gamma^\mu$, the field equations (12.6.6) are equivalent to the single wave equation

$$\gamma^\mu\partial_\mu F + J = 0, \tag{12.6.7}$$

which is just the Dirac equation (with a source term) for a bivector F rather than a spinor. This may be proved by simple manipulations with the Dirac matrices. For a free field ($J = 0$) the Maxwell equations can be written in the Schrödinger form

$$\mathrm{i}\frac{\partial F}{\partial t} = \hat{H}F, \tag{12.6.8}$$

where the Hamiltonian operator $\hat{H} = -\alpha^i\hat{p}_i$ with $\hat{p}_i = \mathrm{i}\partial_i$. The conserved 'Dirac current' implied by (12.6.8) is the energy–momentum tensor of the electromagnetic field, (12.6.11) (Riesz, 1946; Hestenes, 1966).

The essential fact that accounts for the agreement of classical and quantum optics in many cases, such as the formation of interference patterns, is that Maxwell's theory admits a linear superposition principle.

The ray concept emerges in the usual treatment when the wavelength is short compared with characteristic lengths over which the fields appreciably vary. For the free field in vacuum considered here, the geometrical optics approximation coincides with a plane wave solution to the exact field equations. A solution of this type is described by an electric field of constant amplitude $\mathbf{E}_0$ and phase φ:

$$\mathbf{E}(\mathbf{x}, t) = \mathbf{E}_0 \cos(k_\mu x^\mu + \varphi), \tag{12.6.9}$$

where $k_\mu k^\mu = 0$ and $\mathbf{E}\cdot\mathbf{k} = 0$. The congruence of rays connected with this wave is defined as the set of curves orthogonal to the surfaces of constant phase, $k^\mu x_\mu + \varphi = \text{constant}$. Parametrizing the rays by coordinate time t, they are given by

$$x^i = (k^i/k^0)t + x_0^i, \qquad i = 1, 2, 3 \tag{12.6.10}$$

and are distinguished by the initial point $\mathbf{x}_0$. Energy flows in the direction $\mathbf{k}$ at the speed of light.

The general theory of rays in geometrical optics based on the complexified vector potential A_μ satisfying the wave equation $\Box A_\mu = 0$ (Misner *et al.*, 1973, Chap. 22) cannot be extended in a straightforward way to the full wave theory for the reasons already explained in connection with the Klein–Gordon current. One cannot thereby assert, however, that there is no conserved causal (time-like or null) vector at all associated with the general Maxwell field; there does not appear to be any basic physical principle that outlaws 'light paths'.

To see how one might define a conserved causal current for the electromagnetic field we first recall that when in relativity theory a system has energy-momentum 4-vector p^μ, and an observer has 4-velocity a^μ, then the energy of the system according to the observer is the *scalar* quantity $p^\mu a_\mu$, $a_\mu = \eta_{\mu\nu} a^\nu$, for in the rest frame this is numerically equal to p^0. We now generalize this method of generating covariant quantities to the energy-momentum tensor of the electromagnetic field.

The energy-momentum tensor, which is symmetric, is given by

$$T^{\mu\nu} = -F^\mu{}_\alpha F^{\nu\alpha} + \tfrac{1}{4}\eta^{\mu\nu} F^{\alpha\beta} F_{\alpha\beta}. \tag{12.6.11}$$

Identifying the components of the field F_{0i} with the electric vector E^i, $i = 1, 2, 3$, and (F_{32}, F_{13}, F_{21}) with the magnetic vector B^i, the stress tensor has the components

$$\left.\begin{aligned} T^{00} &= \tfrac{1}{2}(\mathbf{E}^2 + \mathbf{B}^2) = \text{energy density}, \\ T^{0i} &= (\mathbf{E} \times \mathbf{B})^i = \text{Poynting flux}, \\ T^{ij} &= -E^i E^j - B^i B^j + \tfrac{1}{2}\delta^{ij}(\mathbf{E}^2 + \mathbf{B}^2) = \text{Maxwell stress}, \end{aligned}\right\} \tag{12.6.12}$$

where $i, j = 1, 2, 3$. The energy-momentum tensor is conserved (in free space):

$$\partial_\mu T^{\mu\nu} = 0 \tag{12.6.13}$$

and has the property that its square as a matrix is a nonnegative scalar multiple of the unit matrix (Misner *et al.*, 1973, p. 481):

$$T^\mu{}_\nu T^\nu{}_\sigma = \lambda \delta^\mu{}_\sigma, \tag{12.6.14}$$

where

$$\lambda = \tfrac{1}{4} T^{\mu\nu} T_{\mu\nu} \geqslant 0. \tag{12.6.15}$$

The latter invariant can be expressed in terms of the two Lorentz scalars of the electromagnetic field,

$$\left.\begin{aligned} \tfrac{1}{2} F^{\mu\nu} F_{\mu\nu} &= \mathbf{B}^2 - \mathbf{E}^2, \\ \tfrac{1}{2} F^{\mu\nu} {}^*F_{\mu\nu} &= 2\mathbf{E}\cdot\mathbf{B}, \end{aligned}\right\} \tag{12.6.16}$$

as

$$\lambda = \tfrac{1}{4}(\mathbf{E}^2 - \mathbf{B}^2)^2 + (\mathbf{E}\cdot\mathbf{B})^2. \tag{12.6.17}$$

A field for which the two scalars, or equivalently λ, vanish is said to be 'null'.

From these properties we may deduce some further useful results. Let a^μ be any future-causal vector (i.e., $a^\mu a_\mu \geqslant 0$, $a^0 > 0$) and form the entity $j^\mu = T^{\mu\nu} a_\nu$ which is evidently a vector. Then a property of fundamental significance is that this vector is future-causal (Penrose & Rindler, 1984, p. 327):

$$j^\mu j_\mu \geqslant 0, \tag{12.6.18}$$

$$j^\mu a_\mu \geqslant 0. \tag{12.6.19}$$

Eq, (12.6.18) states that j^μ is time-like or null and is easily proved using the result (12.6.14), and (12.6.19) states that it is future-pointing. Finally, it is obvious from (12.6.13) that if a^μ is a constant vector in spacetime, the vector j^μ is conserved:

$$\partial_\mu j^\mu = 0. \tag{12.6.20}$$

Suppose now that a^μ is the (constant) 4-velocity of some observer. Then we may interpret j^μ as the electromagnetic energy-momentum density 4-vector measured by her and (12.6.20) as its law of transport. The fourth component j^0 is a nonnegative energy density and the space part $\mathbf{j}$ is a momentum density. The tracks of energy flow are defined by the solution $\mathbf{x} = \mathbf{x}(t, \mathbf{x}_0)$ to the 'guidance formula'

$$\frac{\mathrm{d}x^i}{\mathrm{d}t} = \left.\frac{j^i}{j^0}\right|_{\mathbf{x}=\mathbf{x}(t)} = \left.\frac{T^{i\mu} a_\mu}{T^{0\mu} a_\mu}\right|_{\mathbf{x}=\mathbf{x}(t)}. \tag{12.6.21}$$

Here the paths are parametrized by the time t measured by a Lorentz observer with respect to whom space points are assigned the coordinates x^i. This expression is a Lorentz 3-vector and it is clear from (12.6.18) that the speed of energy transport cannot exceed that of light:

$$\dot{x}^i\dot{x}^i \leqslant 1. \tag{12.6.22}$$

It reaches the speed of light when the field is null. We propose that these flow lines represent the generalization to physical optics of the rays of geometrical optics. Their undulations explain the formation of diffraction and interference patterns, for the trajectories bunch where the energy density is greatest and do not pass through 'nodes' (where $j^0 = 0$), if $\mathbf{x}_0$ does not lie in such regions, as a consequence of the conservation law (12.6.20) (for the proof see §3.3.8). When $a^\mu = \delta^\mu{}_0$ the formula (12.6.21) reduces to

$$\dot{x}^i = T^{i0}/T^{00}|_{\mathbf{x}=\mathbf{x}(t)}. \tag{12.6.23}$$

The law of motion in this form has been illustrated by application to typical diffraction problems (Prosser, 1976).

It may not have escaped the attention of the reader that the vector j^μ, being future-causal and conserved, has all the properties that we may require of the current 4-vector of a beam of particles. Indeed, we might propose that j^0 is proportional to the probability density and accordingly normalize it. Although this interpretation is formally tenable, we are reluctant to adopt it, at least if the 'particles' are supposed to be photons, for the reasons already discussed. It would mean, for example, that when $a^\mu = \delta^\mu{}_0$, T^{00} should be proportional to the probability density of photons which it manifestly is not in the analogous formulae of the second quantized theory. In addition, the classical wave is the limit of a coherent quantum state in which the number of photons is indefinite. Thus, for reasons of compatibility with quantum mechanics, it does not appear reasonable to suppose that the tracks $\mathbf{x} = \mathbf{x}(t)$ deduced from (12.6.21) are the orbits of 'photons'.

The formula (12.6.23) has been criticized on the grounds that it is not a 3-vector with respect to Lorentz transformations (Bohm *et al.*, 1987). We circumvent this objection by interpreting the formula according to (12.6.21), that is, under a Lorentz transformation only the first indices on the top and bottom of (12.6.23) are affected while the second indices are ignored.

For a plane wave (12.6.9) the definition (12.6.23) coincides with (12.6.10) since $T^{0i} = \mathbf{E}^2 k^i/k$ and $T^{00} = \mathbf{E}^2$. The rays are rectilinear. But in the general vacuum case the rays are curved by a 'force' due to $F_{\mu\nu}$ acting via $T_{\mu\nu}$. Note that all fields connected by a duality transformation $F'_{\mu\nu} = \cos b\, F_{\mu\nu} + \sin b\, {}^*F_{\mu\nu}$

where b is an arbitrary constant will determine the same congruence since $T_{\mu\nu}$ is unaffected by such substitutions (Misner *et al.*, 1973, p. 482).

To summarize so far, we have made a relativistically covariant proposal for the concept of a light ray in the electromagnetic field which is valid in the full wave theory as well as in the geometrical optics limit. Interpreting j^μ as an energy-momentum vector accords with the usual definition of the intensity of a light beam as the absolute value of the Poynting vector (Born & Wolf, 1970, p. 10). Note that we do not face the conceptual difficulty posed by the classical limit of quantum mechanics (Chap. 6) in passing to the geometrical optics limit of electromagnetism, for no one doubts the physical reality of electromagnetic fields in the exact classical theory (even though they are manifested only indirectly through their effect on charges) and the ray is simply a structure that may be discerned in the field.

The above results remain valid for the scalar field. In showing this we suppose for generality that the field has a nonzero rest mass which as we shall see does not destroy the causal property of the energy-momentum current. The energy-momentum tensor of a real massive scalar field $\phi(\mathbf{x}, t)$ obeying the Klein–Gordon equation is

$$T_{\mu\nu} = \partial_\mu\phi\,\partial_\nu\phi - \tfrac{1}{2}\eta_{\mu\nu}(\partial^\sigma\phi\,\partial_\sigma\phi - m^2\phi^2). \tag{12.6.24}$$

We form as above the vector $j^\mu = T^{\mu\nu}a_\nu$ where a^μ is an observer's 4-velocity. Then the scalar $j^\mu a_\mu \geqslant 0$ since in the rest frame ($a^\mu = \delta^\mu{}_0$) it is given by

$$T^{00} = \tfrac{1}{2}[(\partial_0\phi)^2 + (\nabla\phi)^2 + m^2\phi^2] \geqslant 0. \tag{12.6.25}$$

Similarly, the scalar $j^\mu j_\mu \geqslant 0$ since in the rest frame it reduces to

$$T^{0\mu}T_{0\mu} = m^2\phi^2(T^{00} - \tfrac{1}{4}m^2\phi^2) + \tfrac{1}{4}(\partial^\mu\phi\,\partial_\mu\phi)^2, \tag{12.6.26}$$

which from (12.6.25) is nonnegative.

Finally, of course, j^μ is a conserved current: $\partial_\mu j^\mu = 0$. We conclude that the association of j^μ with causal lines of energy flow in the massive scalar field is acceptable. Note that although there is a significant difference between the electromagnetic and scalar cases in that the former has a traceless energy-momentum tensor ($T^\mu{}_\mu = 0$) while the latter does not, this has not affected our central result.

We have assumed above that the field ϕ (or $F_{\mu\nu}$) is perfectly known. If it is not then we may introduce a probability functional of the field, $\rho[\phi] \geqslant 0$, to describe a classical distribution. We then define the mean energy-momentum tensor over the ensemble of fields as

$$\langle T^{\mu\nu}(\mathbf{x}, t)\rangle = \int \rho[\phi]T^{\mu\nu}(\mathbf{x}, t)\,\mathrm{D}\phi, \tag{12.6.27}$$

where the measure $\mathrm{D}\phi$ is that defined in §12.4.1. This is a function of the spacetime coordinates and the conservation law (12.6.13) induces the equation

$$\partial_\mu \langle T^{\mu\nu} \rangle = 0. \tag{12.6.28}$$

Using the quantity (12.6.27) we can define the spacetime tracks of mean energy flow as the congruence defined by the mean current $\langle j^\mu \rangle = \langle T^{\mu\nu} \rangle a_\nu$:

$$\langle j^\mu \rangle = \int \rho[\phi] j^\mu \, \mathrm{D}\phi. \tag{12.6.29}$$

Clearly, $\langle j^0 \rangle \geqslant 0$ and, extending the theorem that the sum of two future-causal vectors is again future-causal to the infinite sum of future-causal vectors appearing in the integral (12.6.29), we find that the mean current is future-causal. A similar result may be proved for the electromagnetic field.

We emphasize that the above considerations refer to the free-field case and we would expect modifications in the presence of interactions with matter. Also, we have not made any attempt to employ the spin angular momentum tensor in the electromagnetic case to give a spacetime picture of the 'spin' of the field, analogous to that given for energy transport.

12.6.3 Energy flow in quantum optics

In attempting to extend the results of the previous subsection to the quantum domain we shall restrict attention to the quantized massless spin 0 field in order not to obscure the technique with the computational complexities of the spin 1 theory. We recall that quantization of the field introduces at a stroke two concepts which go beyond the basic classical field theory: (1) nonlocal, nonlinear and noncovariant quantum behaviour of the field implied by the quantum potential derived from the state vector; and (2) a statistical mechanics of the field whose probability distribution is also determined by the state function. We expect that both these factors will play a role in defining energy flow in the quantum field.

Following the definition (12.4.19), we start by making the following tentative definition of the 'actual' energy-momentum functional of an ensemble of fields, having a range of initial forms in the state Ψ, as the local expectation value of the energy-momentum tensor operator:

$$T_q^{\mu\nu}[\psi] = \mathrm{Re}(\Psi^* \hat{T}^{\mu\nu} \Psi)/|\Psi|^2. \tag{12.6.30}$$

Evaluated along a field 'trajectory' $\psi(\mathbf{x}) = \psi(\mathbf{x}, t)$, i.e., the solution of $\partial_0 \psi = \delta S/\delta\psi$ for prescribed $\psi_0(\mathbf{x})$, we obtain the putative energy-momentum of a single field in the ensemble. In the representation in which $\hat{\psi}$ is diagonal the components of the energy-momentum density operator are given by

$$\left.\begin{aligned}\hat{T}^{00} &= \frac{1}{2}\left[-\frac{\delta^2}{\delta\psi^2} + (\nabla\psi)^2\right],\\ \hat{T}^{0i} &= (\mathrm{i}/2)\left[\partial_i\psi\,\frac{\delta}{\delta\psi} + \frac{\delta}{\delta\psi}\,\partial_i\psi\right],\\ \hat{T}^{ij} &= \partial_i\psi\,\partial_j\psi - \tfrac{1}{2}\delta_{ij}\left[\frac{\delta^2}{\delta\psi^2} + (\nabla\psi)^2\right],\end{aligned}\right\} \tag{12.6.31}$$

with $\eta_{\mu\nu} = (1, -\delta_{ij})$. Inserting these expressions into (12.6.30) and writing $\Psi = R\,\mathrm{e}^{\mathrm{i}S}$ yields

$$\left.\begin{aligned}T_q^{00} &= \frac{1}{2}\left[\left(\frac{\delta S}{\delta\psi}\right)^2 + (\nabla\psi)^2\right] + Q_\psi,\\ T_q^{0i} &= -\partial_i\psi\,\frac{\delta S}{\delta\psi},\\ T_{qij} &= \partial_i\psi\,\partial_j\psi + \tfrac{1}{2}\delta_{ij}\left[\left(\frac{\delta S}{\delta\psi}\right)^2 - (\nabla\psi)^2\right] + \delta_{ij}Q_\psi,\end{aligned}\right\} \tag{12.6.32}$$

where

$$Q_\psi = -(1/2R)\,\delta^2 R/\delta\psi^2. \tag{12.6.33}$$

This has the form of the classical energy-momentum tensor of an ensemble of fields ψ in the Hamilton–Jacobi formulation, in which $\partial_0\psi$ in (12.6.24) is replaced by $\delta S/\delta\psi$, modified by a quantum mechanical addition Q_ψ to the diagonal components. Comparison with (12.4.11) shows that Q_ψ is a 'quantum potential density' in that the usual quantum potential is obtained from it by integration over all space: $Q = \int \mathrm{d}^3x\, Q_\psi$. This is in accord with the meaning of $T_q^{\mu\nu}$ as an energy-momentum density.

As noted above, for $T_q^{\mu\nu}$ to become an 'actual' property of an individual wave in the ensemble we must evaluate it using a solution of the guidance equation. Substituting $\partial_0\psi$ for $\delta S/\delta\psi$ we obtain the classical expression (12.6.24) modified by the quantum potential density term. But we immediately run into difficulties which mitigate against using $T_q^{\mu\nu}$ as a covariant definition of energy flow. The most fundamental problem is that while $T_q^{\mu\nu}$ has the *form* of a modified classical energy-momentum tensor for the field ψ, it is *not* actually a tensor because ψ, being a solution to the quantum guidance equation (12.4.16), is not a scalar. It is only in the classical limit when the quantum correction vanishes and ψ becomes a scalar that we obtain a (classical) tensor. In any case, $T_q^{\mu\nu}$ lacks the required formal properties: the presence of Q_ψ means that it is not conserved, the 00-component is not nonnegative because Q_ψ is of indefinite sign, and it does not define a causal vector for the same reason. That is, the quantum potential density breaks the desirable properties of the classical energy-momentum tensor, that it is conserved, implies a positive energy density and defines a causal vector. The definition of lines of energy flow in an individual field that is valid in classical optics cannot be generalized to the case where the quantum state of the field is relevant.

However, what does turn out to be meaningful in many cases is the notion of the *mean* energy flow obtained by averaging the quantum expression (12.6.30) over the ensemble of fields:†

$$\langle T_q^{\mu\nu} \rangle = \int |\Psi|^2 T_q^{\mu\nu}\, \mathrm{D}\psi. \tag{12.6.34}$$

As we shall see below it is this quantity that is relevant to experiments designed to detect the intensity of the field. From the definition (12.6.30), (12.6.34) is obviously equivalent to the mean energy-momentum tensor operator in the state Ψ:

$$\langle T_q^{\mu\nu} \rangle = \langle \Psi | \hat{T}^{\mu\nu} | \Psi \rangle. \tag{12.6.35}$$

That this leads to a meaningful definition of energy flow may be seen as follows. First of all, (12.6.35) is a Lorentz tensor. Next, it is clear that it is conserved since the expectation value is picture-independent and working in the Heisenberg picture $|\Psi\rangle$ is time-independent and $\Box\hat{\psi} = 0$ implies $\partial_\mu \hat{T}^{\mu\nu} = 0$. Finally, we substitute (12.6.32) into (12.6.34) and perform a partial integration on the quantum correction:

$$\int |\Psi|^2 Q_\psi\, \mathrm{D}\psi = \frac{1}{2}\int (\delta R/\delta\psi)^2\, \mathrm{D}\psi. \tag{12.6.36}$$

Then, to the extent that neglect of the surface terms in the integration by parts is valid, the mean energy-momentum density can be written in the form

$$\langle \Psi | \hat{T}^{\mu\nu} | \Psi \rangle = \int (\pi^\mu \pi^\nu - \tfrac{1}{2}\eta^{\mu\nu}\pi^\sigma\pi_\sigma)|\Psi|^2\, \mathrm{D}\psi, \tag{12.6.37}$$

where

$$\pi^0 = [(\delta S/\delta\psi)^2 + (\delta R/\delta\psi)^2]^{\frac{1}{2}}, \qquad \pi^i = \partial^i\psi. \tag{12.6.38}$$

As regards computing expectation values, the effective local energy-momentum

† The actual *total* energy-momentum $\int \mathrm{d}^3x T_q^{\mu\nu}$ is also meaningful, the 00-component, for example, being the actual field energy $-\partial S/\partial t$.

density of the ensemble of fields may therefore be taken to be $\pi^\mu\pi^\nu - \frac{1}{2}\eta^{\mu\nu}\pi^\sigma\pi_\sigma$ instead of (12.6.32). This object clearly possesses as algebraic properties a nonnegative 00-component and a nonnegative square (as in (12.6.14)). The mean energy density $\langle \hat{T}^{00} \rangle$ is thus nonnegative.

Let us define a vector $j^\mu = a_\nu \langle \Psi | \hat{T}^{\mu\nu} | \Psi \rangle$ where a^μ is as usual an observer's 4-velocity. Then, contracting (12.6.37) with a_ν, the integration consists in attaching a positive weight $|\Psi|^2$ to a quantity having the properties of a future-causal vector and summing for all fields ψ. The vector j^μ is therefore future-causal and, moreover, conserved:

$$\partial_\mu j^\mu = 0. \tag{12.6.39}$$

This is the law of mean energy propagation in the quantum field. The causal trajectory of mean energy flow is as usual defined by the solution to the equation

$$\dot{x}^i = \left.\frac{j^i}{j^0}\right|_{\mathbf{x}=\mathbf{x}(t)} = \left.\frac{\langle \hat{T}^{i\mu} \rangle a_\mu}{\langle \hat{T}^{0\mu} \rangle a_\mu}\right|_{\mathbf{x}=\mathbf{x}(t)}. \tag{12.6.40}$$

On the face of it we have successfully generalized the covariant definition of energy transport applicable in the classical theory to the quantum theory, albeit to the case of a mean over the ensemble. Although much information on the field is lost in taking this mean, the result is not simply the classical expectation of an ensemble of scalar fields distributed with density $|\Psi|^2$ but includes an explicit quantum contribution in (12.6.38). In the classical limit we recover the mean classical flow (12.6.29).

The quantum term in (12.6.38) is, in fact, precisely that which arises in the analogue for the field of the Heisenberg relations (cf. §6.7.3). Assuming $\langle \hat{\psi} \rangle = \langle \hat{\pi} \rangle = 0$ we have

$$(\Delta\hat{\psi})^2 = \int \psi^2 |\Psi|^2 \, \mathrm{D}\psi$$

$$\begin{aligned}(\Delta\hat{\pi})^2 &= -\int \Psi^* (\delta^2\Psi/\delta\psi^2) \, \mathrm{D}\psi \\ &= \int |\Psi|^2 [(\delta S/\delta\psi)^2 + 2Q_\psi] \, \mathrm{D}\psi \\ &= \int |\Psi|^2 [(\delta S/\delta\psi)^2 + (\delta R/\delta\psi)^2] \, \mathrm{D}\psi\end{aligned}$$

by parts. Then

$$\Delta\hat{\psi} \, \Delta\hat{\pi} \geqslant \tfrac{1}{2} \tag{12.6.41}$$

and we see that the origin of this inequality is the quantum potential density of the field, as we expect by analogy with particle mechanics.

It is interesting that (12.6.35), being the expectation value of an operator, has a meaning in the usual interpretation of quantum field theory and that therefore a notion of energy that is continuously variable and not restricted to the eigenvalues of an energy operator is admitted into the conventional as well as the causal theory.

At this point we must sound a note of caution and admit that the theory is not as satisfactory as has been suggested above, for we have swept under the carpet the basic problem that plagues quantum field theory: the quantities we deal with are generally infinite. The usual prescription for dealing with this is to renormalize the mean stress tensor by subtracting the vacuum contribution (or 'normal ordering' the operators appearing in $\hat{T}^{\mu\nu}$) in the hope that the residue will be finite. Because the difference between two future-causal vectors is not generally future-causal this procedure will tend to destroy the properties which are useful to us. For example, we may get a negative net mean energy density (a celebrated example is the Casimir effect).

Let us look at the details of this. It is easy to verify, using the formulae (12.6.47) and (12.6.48) below, that for any normalized state $|\Psi\rangle$ we may write

$$\langle\Psi|\hat{T}^{00}|\Psi\rangle = \langle\Psi|:\hat{T}^{00}:|\Psi\rangle + V^{-1}\sum_{\mathbf{k}/2} k, \tag{12.6.42}$$

where the colons signify normal ordering, i.e., the movement of all creation operators to the left of all destruction operators. The second term on the right hand side of (12.6.42) is the infinite vacuum energy density $\langle\Psi_0|\hat{T}^{00}|\Psi_0\rangle$. We have then that the 'physical' mean energy density is given by

$$\langle\Psi|:\hat{T}^{00}:|\Psi\rangle = \langle\Psi|\hat{T}^{00}|\Psi\rangle - \langle\Psi_0|\hat{T}^{00}|\Psi_0\rangle. \tag{12.6.43}$$

Note that the normal ordering does not affect the mean momentum density:

$$\langle\Psi|\hat{T}^{0i}|\Psi\rangle = \langle\Psi|:\hat{T}^{0i}:|\Psi\rangle \tag{12.6.44}$$

(no 'zero-point momentum') since $\sum_{\mathbf{k}} \mathbf{k} = 0$. To see the type of states for which the expression (12.6.43) may be negative we expand $|\Psi\rangle$ in terms of an orthonormal basis of number states:

$$|\Psi\rangle = \sum_{n=0}^{\infty} c_n|\Psi_n\rangle$$

with $\langle\Psi_n|\Psi_{n'}\rangle = \delta_{nn'}$, $\sum_n |c_n|^2 = 1$. Then

$$\langle\Psi|:\hat{T}^{00}:|\Psi\rangle = \sum_n |c_n|^2\langle\Psi_n|:\hat{T}^{00}:|\Psi_n\rangle + \sum_{n'=n\pm 2} c_n^* c_{n'}\langle\Psi_n|:\hat{T}^{00}:|\Psi_{n'}\rangle. \tag{12.6.45}$$

The physical mean energy density splits into two parts. The first, involving the diagonal terms, is nonnegative (see below). The second, involving interference between number states differing by two quanta, may be negative, and we show by example in §12.6.4 that (12.6.45) is indeed not generically positive.

Moreover, even if the normally ordered mean energy density is positive, it is not guaranteed that the modified current vector defined by

$$j^\mu = a_\nu \langle \Psi | : \hat{T}^{\mu\nu} : | \Psi \rangle \tag{12.6.46}$$

will be causal (for an example see §12.6.4). However, for the states of most interest to us, the Fock and coherent states, this vector does turn out to be future-causal. In these cases at least the definition (12.6.40) of the tracks of mean energy flow is meaningful and we shall adopt (12.6.46) as our definition of the energy-momentum current according to the observer having 4-velocity a^μ. The entity $\langle \hat{T}^{0i} \rangle$ determines the mean energy crossing unit area per unit time and generalizing the classical definition this will be termed the *intensity* of a beam.

We have investigated the possibility of supplementing our study of the spacetime dynamics of the field $\psi(\mathbf{x}, t)$, which applies equally in the three levels of the spin 0 theory and its electromagnetic analogue (geometrical optics, physical optics and quantum optics), with the notion of energy transport in the field. In the classical case this was shown to be possible in general for an individual field but in the quantum theory the energy flow appears to have a covariant meaning only in the mean of a statistical aggregate of fields, and only then for a restricted class of states. We emphasize that in none of the cases do we suggest that the energy density may be interpreted as a number or probability density of quanta, or that the tracks are traversed by 'photons'.

12.6.4 Examples of mean energy flow

Evaluation of expectation values is most efficiently carried out by expressing the components of the energy-momentum operator in terms of the creation and annihilation operators (12.4.32) and using the commutation relations (12.4.33). There is no need to perform integrations over field coordinates. We have

$$\begin{aligned} :\hat{T}^{00}: &= :\tfrac{1}{2}(\pi^2 + (\nabla\psi)^2): \\ &= (4V)^{-1} \sum_{\mathbf{k},\mathbf{k}'} (kk')^{-1/2}(kk' + \mathbf{k}\cdot\mathbf{k}') \\ &\quad \times \{2a_{\mathbf{k}}^\dagger a_{\mathbf{k}'}\,\mathrm{e}^{\mathrm{i}(\mathbf{k}'-\mathbf{k})\cdot\mathbf{x}} - a_{\mathbf{k}} a_{\mathbf{k}'}\,\mathrm{e}^{\mathrm{i}(\mathbf{k}+\mathbf{k}')\cdot\mathbf{x}} - a_{\mathbf{k}}^\dagger a_{\mathbf{k}'}^\dagger\,\mathrm{e}^{-\mathrm{i}(\mathbf{k}+\mathbf{k}')\cdot\mathbf{x}}\} \end{aligned} \tag{12.6.47}$$

$$\hat{T}^{0i} = :\hat{T}^{0i}: = \tfrac{1}{2}[\pi\partial^i\psi + (\partial^i\psi)\pi]$$

$$= (4V)^{-1}\sum_{\mathbf{k},\mathbf{k}'} [k^i(k'/k)^{1/2} + k'^i(k/k')^{1/2}]\{\ \} \quad (12.6.48)$$

where the curly brackets in (12.6.48) are the same as in (12.6.47).

In the vacuum the mean energy and momentum densities vanish and hence from (12.6.46) there is no energy flow.

Consider next the general state (12.5.16) containing one quantum:

$$|\Psi_1\rangle = \sum_{\mathbf{k}} f_{\mathbf{k}}\, \mathrm{e}^{-ikt} a_{\mathbf{k}}^{\dagger}|\Psi_0\rangle. \quad (12.6.49)$$

The salient properties of the wave induced by this function, the noncovariance and nonlocality are not apparent at the level of the mean values. However, that the function (12.6.49) describes a packet state is readily demonstrated. After some manipulation we discover that the mean energy and momentum densities may be written in the forms

$$\langle\Psi_1|:\hat{T}^{00}:|\Psi_1\rangle = \tfrac{1}{2}(\partial_0\phi\,\partial_0\phi^* + \nabla\phi\cdot\nabla\phi^*), \quad (12.6.50)$$

$$\langle\Psi_1|:\hat{T}^{0i}:|\Psi_1\rangle = \tfrac{1}{2}(\partial_0\phi\,\partial^i\phi^* + \partial_0\phi^*\,\partial^i\phi), \quad (12.6.51)$$

where

$$\phi(\mathbf{x}, t) = V^{-1/2}\sum_{\mathbf{k}} k^{-1/2} f_{\mathbf{k}}\, \mathrm{e}^{-\mathrm{i}(kt - \mathbf{k}\cdot\mathbf{x})}. \quad (12.6.52)$$

The functions (12.6.50) and (12.6.51) will be recognized as the 0μ-components of the energy-momentum tensor of a complex massless scalar field:

$$t^{\mu\nu} = \tfrac{1}{2}[\partial^\mu\phi\,\partial^\nu\phi^* + \partial^\mu\phi^*\,\partial^\nu\phi - \eta^{\mu\nu}\,\partial^\sigma\phi^*\,\partial_\sigma\phi]. \quad (12.6.53)$$

We see from (12.6.52) that the function ϕ is a superposition of positive frequency solutions of the classical wave equation $\Box\phi = 0$. It is normalized with respect to the usual Klein–Gordon scalar product:

$$(\mathrm{i}/2)\int (\phi^*\,\partial_0\phi - \phi\,\partial_0\phi^*)\,\mathrm{d}^3x = 1, \quad (12.6.54)$$

a condition preserved by the wave equation $\Box\phi = 0$.

It seems reasonable to call ϕ the 'one-photon wavefunction' with the qualifying remark that, although normalized as in (12.6.54), ϕ cannot be interpreted as a probability amplitude for the 'position of a quantum' for the reasons explained in §12.6.1 (in particular, the fourth component of the current, the integrand in (12.6.54), is not positive). We might also call ϕ the 'mean wave packet' but note that it is not the mean *field*; indeed, the latter is zero: $\langle\Psi_1|\hat{\psi}|\Psi_1\rangle = 0$.

The interpretation of (12.6.54) is that the mean total number operator is unity. This may be contrasted with the mean total energy-momentum operator:

$$\left\langle \Psi_1 \middle| \int d^3x :\hat{T}^{0\mu}: \middle| \Psi_1 \right\rangle = \sum_{\mathbf{k}} |f_{\mathbf{k}}|^2 k^\mu. \tag{12.6.55}$$

This formula suggests that $|f_{\mathbf{k}}|^2$ may be interpreted as the probability of obtaining the result k^μ for the total energy-momentum of the field on measurement of such a quantity (the states $|\Psi_{\mathbf{k}}\rangle$, (12.5.15), are eigenstates of this operator corresponding to eigenvalues k^μ).

While a wave ψ in the ensemble connected with the state Ψ_1 propagates nonlocally, the mean energy in the field propagates locally and subluminally according to the dynamics of the spacetime packet ϕ. We readily confirm that the mean energy-momentum (12.6.50) and (12.6.51) generates a future-causal vector by a simple generalization of the proof given for the real scalar field at the end of §12.6.2.

For the special case of a single-mode excitation (12.5.14) we find

$$\langle :\hat{T}^{0\mu}: \rangle = k^\mu/V, \tag{12.6.56}$$

so that the rays are rectilinear

$$\mathbf{x} = (\mathbf{k}/k)t + \mathbf{x}_0 \tag{12.6.57}$$

and the mean energy propagates at the speed of light. Although the mean rays coincide with those associated with a classical plane wave of wave vector k^μ, (12.6.10), we do not, in fact, recover the classical result in its entirety because the distribution of rays in space differs in the two cases. To obtain the classical results in detail we must use a coherent state.

We obtain for the $\mathbf{k}$th mode coherent state (12.5.31)

$$\langle \Psi_{\alpha_{\mathbf{k}}} | :\hat{T}^{0\mu}: | \Psi_{\alpha_{\mathbf{k}}} \rangle = 2k^\mu |\alpha_{\mathbf{k}}|^2 V^{-1} \sin^2(kt - \mathbf{k}\cdot\mathbf{x} + \theta_{\mathbf{k}}). \tag{12.6.58}$$

Clearly, the rays are again given by (12.6.57) and in this respect the coherent and one-quantum states agree. But the mean coherent density exhibits a sinusoidal variation in spacetime which means that the congruence is of variable density and in particular will avoid zeroes of the energy density if $\mathbf{x}_0$ does not lie at such points. In keeping with the 'classical' nature of the coherent field, the expression (12.6.58) coincides with the 0μ-components of the energy-momentum tensor of a real classical plane wave

$$\psi(\mathbf{x}, t) = A_{\mathbf{k}} \cos(k_\mu x^\mu + \theta_{\mathbf{k}}) \tag{12.6.59}$$

of amplitude $A_{\mathbf{k}} = (2/kV)^{1/2}|\alpha_{\mathbf{k}}|$. This may be confirmed by substitution into (12.6.24) (with $m = 0$).

Results analogous to those found for the one-quantum state are obtained for an n-quantum state. The right hand sides of (12.6.50) and (12.6.51) are the same except that ϕ is replaced by a set of fields

$$\phi_{\mathbf{k}_1 \ldots \mathbf{k}_{n-1}}(\mathbf{x}, t) = V^{-1/2} \sum_{\mathbf{k}} k^{-1/2} f_{\mathbf{k}\mathbf{k}_1 \ldots \mathbf{k}_{n-1}} \, \mathrm{e}^{-\mathrm{i}k_\mu x^\mu}, \tag{12.6.60}$$

which are symmetric with respect to interchange of the indices, and one sums over $\mathbf{k}_1, \ldots, \mathbf{k}_{n-1}$. This may be called the 'n-photon wavefunction'. The future-causal property of the vector j^μ remains intact for this state; in particular, the mean energy density is positive. The mean total energy for a state in which the $\mathbf{k}$th mode is excited to the $n_{\mathbf{k}}$th level is the classic formula $E = n_{\mathbf{k}} k$. This formula has undoubtedly contributed to the feeling that a quantized wave is 'composed' of photons. But, of course, an n-quantum state may cover a range of modes as in (12.6.60) and the energy is a function of the state (cf. (12.6.55)) which contains information beyond the fact of being an eigenfunction of the number operator. Hence one cannot legitimately conclude that an n-quantum field is in some way a discrete aggregation of quanta.

Finally, we furnish an example of a state for which the normal ordered mean energy density is negative and the current space-like. The simplest such example is a superposition of the vacuum state and a 2-quantum state:

$$|\Psi\rangle = f|\Psi_0\rangle + f' a^\dagger_{\mathbf{k}_1} a^\dagger_{\mathbf{k}_2} \, \mathrm{e}^{-\mathrm{i}(k_1 + k_2)t} |\Psi_0\rangle, \tag{12.6.61}$$

where f, f' are real and positive and $f^2 + f'^2 = 1$. Suppose that $k_1 = k_2 = k$ and let δ be the angle between $\mathbf{k}_1$ and $\mathbf{k}_2$ with $\delta \neq 0$ or π. Then,

$$\langle\Psi|{:}\hat{T}^{00}{:}|\Psi\rangle = (2kf'/V)[f' - f \cos\xi \cos^2(\delta/2)], \tag{12.6.62}$$

where $\xi = 2kt - (\mathbf{k}_1 + \mathbf{k}_2)\cdot\mathbf{x}$. Clearly, choosing f to be nearly unity and f' almost negligible this expression will become negative at certain spacetime points (e.g., where $\cos\xi = 1$) due to the interference term. A perturbation of the vacuum of this kind, however slight, will tend to *decrease* the energy. It is this sort of thing that occurs in the Casimir effect (§12.5.1), where the modified vacuum state can be expanded in terms of the Fock basis appropriate to the old vacuum and the basis states interfere.

We have also for the state (12.6.61)

$$\langle\Psi|\hat{T}^{0i}|\Psi\rangle = V^{-1}(k^i_1 + k^i_2)(f'^2 - ff' \cos\xi). \tag{12.6.63}$$

Then from the definition (12.6.46) of the current we have in the observer's rest frame

$$\begin{aligned} j^\mu j_\mu &= \langle{:}\hat{T}^{00}{:}\rangle^2 - \langle\hat{T}^{0i}\rangle\langle\hat{T}^{0i}\rangle \\ &= 4k^2 V^{-2} f'^2 \sin^2(\delta/2)(f'^2 - f^2 \cos^2\xi \cos^2(\delta/2)). \end{aligned} \tag{12.6.64}$$

Let us choose again f' very small and f large but suppose now that $\cos\xi = -1$ so that the energy density (12.6.62) is positive. Then the expression (12.6.64) is negative. In such states, we cannot define a subluminal flow of mean energy, even if this is positive.

12.6.5 Remarks on the detection process

We shall briefly examine now the connection between our definition above of the intensity of a quantum optical light beam, $\langle \hat{T}^{0i} \rangle$, and the intensity determined by measurements.† A typical detector employs the photoelectric effect in which ionization of an atom caused by the absorption of a single quantum from the field brings about a cascade and a measurable electric current. Whatever the initial state of the field, measurements of field intensity involve discrete events (counting of 'photons') and patterns are built up over a period of time. But it would be a mistake to conclude that the process underlying the detection is therefore intrinsically discrete. On the contrary, continuous fields whose interactions are governed by the Schrödinger equation can give rise to single rapid transfers of energy. The principles of the atom–radiation interaction are those already spelled out in our previous treatments of many-body processes. One simply replaces the coordinates of a 'particle' in the interacting system by those of the field ($q_{\mathbf{k}}$ or $\psi(\mathbf{x})$).

A point to notice is that if the initial state of the field is, for example, a one-quantum packet (12.5.16) of time-varying energy sufficient to ionize an atom, *all* the energy above the zero-point will be transferred into the electron during detection and the field left in its ground state. The process is therefore discrete but not in the sense of the *quantity* of energy exchanged (this is not generally a multiple of some basic unit). Rather, it is discrete because all of the energy is transferred, as a concomitant of the change in quantum number. Given the state of the field and the atomic wavefunction, the outcome of the interaction is causally determined by the initial configuration of the field, ψ_0, and the initial position of the electron, $\mathbf{x}_0$. Details are given by Bohm (1952b), Belinfante (1973) and Bohm *et al.* (1987).

Let us return for a moment to the electromagnetic field. We let the state of the field be $|\Psi\rangle$ and denote by $\mathbf{E}^+(\mathbf{x}, t)$ ($\mathbf{E}^-(\mathbf{x}, t)$) the positive (negative) frequency part of the electric field operator in the Heisenberg picture. In the case where the mode sum is restricted to wave vectors having a common direction $\mathbf{k}$, analysis of the operation of an ideal detector reveals that the

† We are not concerned here with the problem of the measurability of field intensities over finite spacetime regions as treated by Bohr & Rosenfeld (1933, 1950).

probability per unit time of absorbing a photon at the point $\mathbf{x}$ at time t is proportional to the quantity (Glauber, 1963a)

$$\mathbf{I} = \langle\Psi|\mathbf{E}^-(\mathbf{x}, t)\cdot\mathbf{E}^+(\mathbf{x}, t)\hat{\mathbf{k}}|\Psi\rangle. \tag{12.6.65}$$

This prompts the following general definition of the intensity operator as the quantum counterpart of the classical definition of intensity (the Poynting vector) (Loudon, 1986, p. 185):

$$\hat{\mathbf{I}} = \mathbf{E}^- \times \mathbf{B}^+ + \text{hc}. \tag{12.6.66}$$

To compare with our definition of intensity above we write down the analogue of (12.6.66) for the scalar field in the Schrödinger picture. This appears to be

$$\hat{\mathbf{I}} = -\tfrac{1}{2}(\pi^-\nabla\psi^+ + \nabla\psi^-\pi^+), \tag{12.6.67}$$

where $\pi^\pm, \psi^\pm$ are defined by (12.4.34) (the minus sign enters because $\hat{\mathbf{I}}$ represents contravariant components and $\nabla \equiv \partial_i = -\partial^i$). When the mode sum is restricted to wave vectors of common direction $\hat{\mathbf{k}} = \mathbf{k}/k$ we have $\nabla\psi^\pm = -\pi^\pm\hat{\mathbf{k}}$ and (12.6.67) reduces to

$$\hat{\mathbf{I}} = \pi^-\pi^+\hat{\mathbf{k}} = \nabla\psi^-\cdot\nabla\psi^+\hat{\mathbf{k}}, \tag{12.6.68}$$

which is the spin 0 analogue of (12.6.65). The operator coefficients here will be observed to be parts of the energy density operator in (12.4.35) so that (12.6.68) is an operator analogue of the classical relation between energy density and 'Poynting vector' for a plane wave:

$$T^{0i} = T^{00}k^i/k. \tag{12.6.69}$$

The *operational* intensity for the scalar field is then, in general,

$$\mathbf{I} = \langle\Psi| -\tfrac{1}{2}(\pi^-\nabla\psi^+ + \nabla\psi^-\pi^+)|\Psi\rangle. \tag{12.6.70}$$

We observe that the creation and annihilation operators in this expression are normally ordered which expresses the fact that detectors do not respond to the zero-point field. But more than this, comparison with the 'actual' intensity as we have defined it,

$$\langle\Psi|\hat{T}^{0i}|\Psi\rangle = -\langle\Psi|\tfrac{1}{2}[\pi\nabla\psi + (\nabla\psi)\pi]|\Psi\rangle \tag{12.6.71}$$

shows that the two definitions of intensity differ by the expectation value of the operator $-\frac{1}{2}(\pi^+\nabla\psi^+ + \nabla\psi^-\pi^-)$. For some states (e.g., the number states) the definitions coincide but, in general, the spin 0 photoelectric detector is sensitive to only part of the total intensity of a light beam (at least, as we have defined it). When the mode sum is restricted to a unique direction $\hat{\mathbf{k}}$, our definition of intensity operator gives

$$\hat{T}^{0i} = \pi^2 \hat{k}^i = (\nabla\psi)^2 \hat{k}^i = \hat{T}^{00} k^i / k, \tag{12.6.72}$$

which is our version of the relation (12.6.68) and is an alternative operator analogue of the classical relation (12.6.69) for a plane wave.

The lack of coincidence of the actual and measurable intensities may indicate that we have adopted an incorrect quantum generalization of the (scalar equivalent of the) Poynting flux. Yet our definition has the advantage of making clear the connection with the classical limit, and the mere fact that a physical detector is only responsive to a certain quantity does not imply that further contributions to that quantity may not be physically relevant (it is after all already insensitive to the zero-point energy). In any case, both types of intensity display similar properties of interference and their integrated values over all space agree:

$$\left\langle \Psi \middle| \int \mathrm{d}^3x\, \hat{T}^{0i} \middle| \Psi \right\rangle = \left\langle \Psi \middle| \int \mathrm{d}^3x\, \hat{I}^i \middle| \Psi \right\rangle \tag{12.6.73}$$

with $\hat{I}^i$ given by (12.6.67).

The definition of intensity as a mean density over an ensemble of fields does not probe the detailed structure of the state $|\Psi\rangle$. For example, the transition probability measured by detectors should not be confounded with the field probability density $|\Psi[\psi]|^2$. Finer detail is revealed by taking into account correlations between photon detectors placed at distinct spacetime points, as determined by the expectation value of products of field operators at the various points (Glauber, 1963a). The equal-time n-point correlation function corresponding to our definition of intensity would take the form

$$\langle \Psi | : \hat{T}^{\mu_1 \nu_1}(\mathbf{x}_1) \ldots \hat{T}^{\mu_n \nu_n}(\mathbf{x}_n) : | \Psi \rangle, \tag{12.6.74}$$

but we shall not pursue this here.

When we speak of 'intensity' in our approach to quantum optics we mean a property of the light beam that is independent of whether or not an atomic detector is placed in its path. In the usual approach the quantum intensity refers to a transition rate for photons to be absorbed from the field and in that sense is a function of the atom–radiation interaction rather than the radiation field alone.

12.7 Two-slit interference of quantized fields

12.7.1 Single source

Experiments demonstrating the temporal build-up of optical interference patterns using low intensity sources have been performed since the earliest days of quantum theory (e.g., Taylor, 1909). But the first experiment to demonstrate 'single-photon self-interference' with a genuine quantum mechanical one-quantum state was apparently not published until 1986 (Grangier, Roger & Aspect, 1986). This is remarkable for much of the historical debate surrounding wave-particle duality is based on the empirical validity of the assertion that, as Dirac (1974, p. 9) put it, 'Each photon interferes only with itself. Interference between two different photons never occurs.' It seems, however, that previous experiments claiming to provide support for this prediction generally employed low intensity chaotic light sources which presumably gave the appearance of single-photon interference only because the detection process is discrete. As to whether in confirming the quantum formalism the genuine one-quantum interference experiments also confirm Dirac's statement or the implications that have been drawn from it regarding the nature of light is a matter to which we shall return below.

The field-theoretic analysis of Young's double-slit experiment that follows is conceptually analogous to that given for electron interference in Chap. 5. In the matter case the electron is defined by the position $\mathbf{x}(t)$ of a corpuscle (which passes through one slit) and the wavefunction $\psi(\mathbf{x})$. In the case under consideration here the physical system is defined by the coordinates of the field $\psi(\mathbf{x}, t)$ (which passes through both slits) and the wavefunction $\Psi[\psi]$. But there is a further component here that, with respect to this analogy, is missing in the matter case, namely the 'photon' as a supposed structure in the field. Of course, one usually thinks in terms of another analogy in which the photon plays the role of the corpuscle in the matter system but for the reasons set out in §12.6.1 it is hard to see how this could be reconciled with the quantum formalism.

In any case, in spite of its prominence in the debate over the true nature of quantum interference effects, we shall see that the 'photonic' aspect of the field is not actually relevant at all to the basic interference phenomenon and only enters in the detection process which reveals through a sequence of apparently discrete events a flow of energy that is in essence entirely continuous. The interference is a property of the objective continuous field distributed in spacetime and not of the detector as is sometimes suggested. Moreover, interference properties are displayed by many types of states including those where the number of quanta is undefined and it is therefore not clear why the single-photon state should be ascribed such singular theoretical importance.

The new feature not present in the classical optical treatment of two-slit interference, which likewise explains the formation of fringes through the

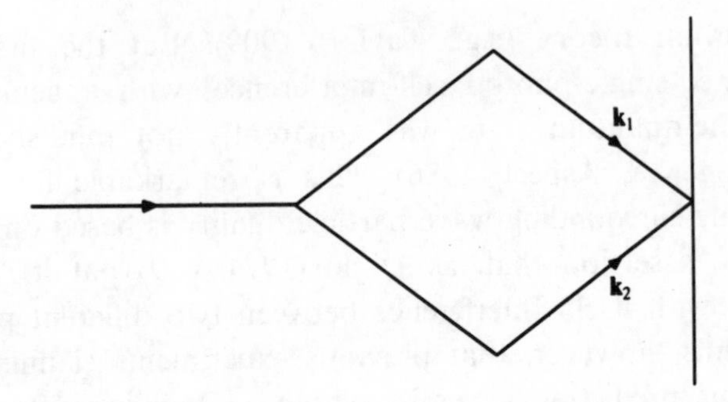

Fig. 12.4 Schematic depiction of single-quantum interference.

linear superposition principle, is that the quantum field is accompanied and guided by the wavefunction Ψ.

We consider the interference of two beams originating in a common source which generates a single-mode one-quantum state (although similar results will, in principle, be obtained with many other types of states). In the usual way we associate probability amplitudes with each of the paths taken by the field through the interferometer (see Fig. 12.4) and superpose them at the detecting screen (for a corresponding discussion in the Heisenberg picture see Walls (1977)). The two probability amplitudes are single-mode one-quantum excitations (12.5.14) corresponding to wavevectors $\mathbf{k}_1$ and $\mathbf{k}_2$ with $\mathbf{k}_1 \neq \mathbf{k}_2$ so that the states are orthogonal. The resultant wavefunction in the normal coordinate space of the field representation is therefore

$$\Psi[q, q^*, t] = [f_{\mathbf{k}_1}(2k_1)^{1/2} q^*_{\mathbf{k}_1} \,\mathrm{e}^{-ik_1 t} + f_{\mathbf{k}_2}(2k_2)^{1/2} q^*_{\mathbf{k}_2} \,\mathrm{e}^{-ik_2 t}]\Psi_0, \quad (12.7.1)$$

where $|f_{\mathbf{k}_1}|^2 + |f_{\mathbf{k}_2}|^2 = 1$. We shall write $f_{\mathbf{k}_i} = |f_{\mathbf{k}_i}| \,\mathrm{e}^{-i\varphi_{\mathbf{k}i}}$, $i = 1, 2$. Since the beams emanate from a common source they have the same frequency $k_1 = k_2 = k$ and it is convenient to suppose they are of equal amplitude, $|f_{\mathbf{k}_1}| = |f_{\mathbf{k}_2}| = 1/\sqrt{2}$, and phase. We have then

$$\Psi = k^{1/2} \,\mathrm{e}^{-i\varphi_{\mathbf{k}1}}(q^*_{\mathbf{k}_1} + q^*_{\mathbf{k}_2}) \,\mathrm{e}^{-ikt}\Psi_0. \quad (12.7.2)$$

The condition $\mathbf{k}_1 \neq \pm\mathbf{k}_2$ means that the beams meet at an angle $\delta \neq 0$.

Recall from (12.5.25) that the waves associated with each summand are, apart from additive functions of $\mathbf{x}$, simple harmonic of frequency $\omega_{\mathbf{k}_i} = 1/2|q_{\mathbf{k}_i 0}|^2$ and wave vector $\mathbf{k}_i$, $i = 1, 2$. We saw in eq. (12.6.57) that the mean energy in each wave flows along rectilinear rays $\mathbf{x} = (\mathbf{k}_i/k)t + \mathbf{x}_0$ at the speed of light.

For the superposition (12.7.2), solution of the guidance formula $\dot{q}^*_{\mathbf{k}} = \partial S/\partial q_{\mathbf{k}}$ for $\mathbf{k} = \mathbf{k}_1$ or $\mathbf{k}_2$ yields for the time dependence of the normal coordinates

$$\left.\begin{aligned} q_{\mathbf{k}_1}(t) &= \tfrac{1}{2}(q_{\mathbf{k}_1 0} + q_{\mathbf{k}_2 0}) \,\mathrm{e}^{-i\omega t} + \tfrac{1}{2}(q_{\mathbf{k}_1 0} - q_{\mathbf{k}_2 0}), \\ q_{\mathbf{k}_2}(t) &= \tfrac{1}{2}(q_{\mathbf{k}_1 0} + q_{\mathbf{k}_2 0}) \,\mathrm{e}^{-i\omega t} + \tfrac{1}{2}(q_{\mathbf{k}_2 0} - q_{\mathbf{k}_1 0}), \end{aligned}\right\} \quad (12.7.3)$$

where $\omega = |q_{\mathbf{k}_1 0} + q_{\mathbf{k}_2 0}|^{-2}$ and $q_{\mathbf{k}_1 0}, q_{\mathbf{k}_2 0}$ are the initial coordinates. Hence, writing $q = |q|\,\mathrm{e}^{-\mathrm{i}\beta} = \frac{1}{2}(q_{\mathbf{k}_1 0} + q_{\mathbf{k}_2 0})$ and substituting into (12.4.22), the field is given by

$$\psi(\mathbf{x}, t) = 2|q|V^{-1/2}[\cos(\omega t - \mathbf{k}_1\cdot\mathbf{x} + \beta) + \cos(\omega t - \mathbf{k}_2\cdot\mathbf{x} + \beta)] + f(\mathbf{x}), \tag{12.7.4}$$

where $f(\mathbf{x})$ is an arbitrary function and $\omega = 1/4|q|^2$. This displays the expected properties of noncovariance, through ω and $f(\mathbf{x})$, and nonlocality, through

$$q = (1/2V^{1/2}) \int \psi_0(\mathbf{x})(\mathrm{e}^{-\mathrm{i}\mathbf{k}_1\cdot\mathbf{x}} + \mathrm{e}^{-\mathrm{i}\mathbf{k}_2\cdot\mathbf{x}})\,\mathrm{d}^3x. \tag{12.7.5}$$

But, neglecting f, ψ is essentially the linear superposition of two equal frequency plane waves such as we might contemplate in classical optics (although they do not satisfy the classical wave equation because $\omega \neq k$ in general). Writing it in the form

$$\psi(\mathbf{x}, t) = 4|q|V^{-1/2}\cos[\tfrac{1}{2}(\mathbf{k}_1 - \mathbf{k}_2)\cdot\mathbf{x}]\cos[\omega t - \tfrac{1}{2}(\mathbf{k}_1 + \mathbf{k}_2)\cdot\mathbf{x} + \beta] + f(\mathbf{x}) \tag{12.7.6}$$

we see that the field is a harmonic plane wave travelling in the direction $\frac{1}{2}(\mathbf{k}_1 + \mathbf{k}_2)$ and spatially modulated (apart from the effect of f).

Still neglecting $f(\mathbf{x})$, we pass to a complex representation of the wave, as is usual in optics:

$$\psi = \mathrm{Re}\,\psi_c = \mathrm{Re}\,2|q|V^{-1/2}[\mathrm{e}^{\mathrm{i}(\omega t - \mathbf{k}_1\cdot\mathbf{x} + \beta)} + \mathrm{e}^{\mathrm{i}(\omega t - \mathbf{k}_2\cdot\mathbf{x} + \beta)}]. \tag{12.7.7}$$

Then the amplitude squared of the wave is, as usual, given by

$$|\psi_c|^2 = 8|q|^2V^{-1}[1 + \cos(\mathbf{k}_1 - \mathbf{k}_2)\cdot\mathbf{x}]. \tag{12.7.8}$$

This displays the expected interference pattern, for a nonclassical field guided by a one-quantum state function.

We now consider how this compares with the spatial pattern exhibited in the mean energy flow which determines the observable light intensity. For a field in the general state (12.7.1) the 'one-quantum wavefunction' (12.6.52) becomes

$$\phi(\mathbf{x}, t) = V^{-1/2}[k_1^{-1/2}f_{\mathbf{k}_1}\,\mathrm{e}^{-\mathrm{i}(k_1 t - \mathbf{k}_1\cdot\mathbf{x})} + k_2^{-1/2}f_{\mathbf{k}_2}\,\mathrm{e}^{-\mathrm{i}(k_2 t - \mathbf{k}_2\cdot\mathbf{x})}] \tag{12.7.9}$$

and we obtain the following expressions for the mean energy and momentum densities:

$$\langle:\hat{T}^{00}:\rangle = V^{-1}[k_1|f_{\mathbf{k}_1}|^2 + k_2|f_{\mathbf{k}_2}|^2 + (k_1k_2)^{-1/2}(k_1k_2 + \mathbf{k}_1\cdot\mathbf{k}_2)|f_{\mathbf{k}_1}||f_{\mathbf{k}_2}|\cos\xi], \tag{12.7.10}$$

$$\langle \hat{T}^{0i} \rangle = V^{-1}[k_1^i|f_{\mathbf{k}_1}|^2 + k_2^i|f_{\mathbf{k}_2}|^2 + [(k_1/k_2)^{1/2}k_2^i + (k_2/k_1)^{1/2}k_1^i]|f_{\mathbf{k}_1}||f_{\mathbf{k}_2}|\cos\xi], \quad (12.7.11)$$

with

$$\xi = (k_1 - k_2)t - (\mathbf{k}_1 - \mathbf{k}_2)\cdot\mathbf{x} + \varphi_{\mathbf{k}_1} - \varphi_{\mathbf{k}_2}.$$

Both quantities display characteristic interference terms in addition to the first two 'classical' terms. Let us now specialize to the case of equal frequency, amplitude and phase as in (12.7.2). Then

$$\langle :\hat{T}^{00}: \rangle = (k/V)\{1 + \cos^2(\delta/2)\cos[(\mathbf{k}_1 - \mathbf{k}_2)\cdot\mathbf{x}]\}, \quad (12.7.12)$$

$$\langle \hat{T}^{0i} \rangle = [(k_1^i + k_2^i)/2V][1 + \cos(\mathbf{k}_1 - \mathbf{k}_2)\cdot\mathbf{x}], \quad (12.7.13)$$

where δ is the angle between the beams. Comparing with (12.7.8) we see that the intensity (12.7.13) displays the same pattern of spatial interference as each wave in the underlying ensemble connected with the state (12.7.2) (insofar as we neglect the arbitrary function $f(\mathbf{x})$). Since in the case of number states $\langle \hat{T}^{0i} \rangle$ coincides with the intensity measured by a (spin 0) detector, we conclude that in this case experiment would reveal the detail of the actual state of affairs and no essential information is lost because we are dealing with an average over the ensemble.

The energy flows uniformly along straight lines at a speed less than light. To see this we note that the relations

$$\frac{dx}{dy} = \frac{\langle \hat{T}^{01} \rangle}{\langle \hat{T}^{02} \rangle} = \frac{k_{1x} + k_{2x}}{k_{1y} + k_{2y}}, \quad \frac{dx}{dz} = \frac{\langle \hat{T}^{01} \rangle}{\langle \hat{T}^{03} \rangle} = \frac{k_{1x} + k_{2x}}{k_{1z} + k_{2z}} \quad (12.7.14)$$

imply that, along a trajectory,

$$(\mathbf{k}_1 - \mathbf{k}_2)\cdot\mathbf{x} = (\mathbf{k}_1 - \mathbf{k}_2)\cdot\mathbf{x}_0 = \text{constant} \quad (12.7.15)$$

where we have used $\mathbf{k}_1^2 - \mathbf{k}_2^2 = 0$. Hence, the equation of motion is

$$\frac{d\mathbf{x}}{dt} = \frac{\langle \hat{T}^{0i} \rangle}{\langle :\hat{T}^{00}: \rangle} = \left(\frac{\mathbf{k}_1 + \mathbf{k}_2}{2k}\right)\{1 + \cos[(\mathbf{k}_1 - \mathbf{k}_2)\cdot\mathbf{x}_0]\} \times \left\{1 + \cos^2\left(\frac{\delta}{2}\right)\cos[(\mathbf{k}_1 - \mathbf{k}_2)\cdot\mathbf{x}_0]\right\}^{-1} \quad (12.7.16)$$

from which we readily deduce that the flow is uniform. It is easy to show that $|\dot{\mathbf{x}}| < 1$; indeed, $\mathbf{x}_0$ can be chosen so that $\dot{\mathbf{x}} = 0$. The congruence is distributed in space with density $\langle :\hat{T}^{00}: \rangle$ and avoids the zeroes of this quantity.

Field energy courses through both arms of the interferometer in a continuous way, but one does not detect the entire field in one go. Rather, our knowledge of its structure is built up through localized, discrete events. In each trial the source generates a single field ψ and its guiding wavefunction Ψ and these contribute to a single detection event. This is a basic difference with classical field theory and reflects the peculiar characteristics of the Schrödinger equation governing the detector–radiation interaction. Nevertheless, the nonclassical method of gathering information may tend to lead one away from an appreciation of the underlying process as involving linear superposition of continuous fields, much as in classical optics.

What, then, is the status of 'wave-particle duality' in this picture, and of Dirac's statement quoted above that 'Each photon interferes only with itself...'? The latter assertion is formally true in the sense that the linear superposition of two one-quantum states yields again a one-quantum state and not, say, the ground state or a two-quantum state; 'photons' do not combine or annihilate one another by superposition. But it seems that emphasizing the 'photon' as the main point of interest when examining quantal interference focusses on the wrong aspect of the field. For what is happening is the coherent superposition, or 'self-interference', of quantum states which, in turn, implies the linear superposition of fields and the appearance of interference terms in the objective intensity of the light. This is a property of a variety of states. Problems relating to which 'path' the 'photon' takes through the interferometer are in this view meaningless since such a structure cannot be discerned independently of the detection. A key feature in our treatment of electron interference in Chap. 5 was that the effect of the wave on the particle and the resultant channelling of the latter into certain regions of space where the wave amplitude is large was not causally influenced by the detecting screen – placing the latter in the path of the interfering beams merely told us what had already happened and played no role in the basic 'self-interference' phenomenon that was to be explained. Likewise here, the interference effect is not contingent on the functioning of the detector. The 'wave-particle duality' of light is rather a 'wave–wave synthesis' (of ψ and Ψ). Our explanation for the quantum mechanical version of Young's experiment therefore operates at the same conceptual level as the one we gave for material particles in Chap. 5.

Dirac's statement could, of course, be construed as an assertion that no single conceptual model of light can be formulated that covers all its multitudinous interactions with matter. This view is explicitly refuted by the model described here.

The above formulae can be adapted unmodified to describe an experiment in which a one-quantum beam emanating from a source is incident on a beam splitter and the two emerging orthogonal beams arrive at detectors rather than being recombined. The state of the field after the beam splitter is given by the state entangled with the vacuum (12.7.1) and its spacetime structure by (12.7.4) ($\mathbf{k}_1$ lies along one arm, $\mathbf{k}_2$ along the other). In this case one finds perfect anticoincidence in the firing of the detectors (Grangier *et al.*, 1986). An experiment of this kind using a particular type of beam splitter composed of two prisms separated by a gap has been proposed as an example where wave and particle concepts must both be invoked in the description of a single experimental phenomenon (Ghose, Home & Agarwal, 1991).

12.7.2 Independent sources

Suppose now we have two physically independent sources each generating a coherent state of the field (number states from independent sources do not form visible interference patterns). The independence implies that the total state is a product:

$$\Psi = \Psi_{\alpha_{\mathbf{k}_1}} \Psi_{\alpha_{\mathbf{k}_2}} \tag{12.7.17}$$

where the individual wavefunctions are given by (12.5.31) (with only one overall factor Ψ_0). In many-body quantum mechanics we are used to the idea that factorization of the wavefunction implies the total system is composed of a set of noninterfering subsystems. The 'subsystems' need not be 'physical' (particles); a two-body wavefunction may factorize into centre of mass and relative coordinates, for example. Similarly here, the phase of the complex function (12.7.17) is an additive function of the normal coordinates and hence for each $\mathbf{k}$ these will evolve independently. But a product in this space can translate into a linear sum at the level of the physical spacetime field coordinates, which will therefore exhibit interference.

The wave corresponding to the state (12.7.17) is a special case of (12.5.43):

$$\begin{aligned}\psi(\mathbf{x}, t) = {} & (2/k_1 V)^{1/2}|\alpha_{\mathbf{k}_1}| \cos(k_1 t - \mathbf{k}_1 \cdot \mathbf{x} + \theta_{\mathbf{k}_1}) \\ & + (2/k_2 V)^{1/2}|\alpha_{\mathbf{k}_2}| \cos(k_2 t - \mathbf{k}_2 \cdot \mathbf{x} + \theta_{\mathbf{k}_2}) + f(\mathbf{x}).\end{aligned} \tag{12.7.18}$$

Ignoring the arbitrary function $f(\mathbf{x})$, the amplitude squared of this wave is, following (12.7.7),

$$\begin{aligned}|\psi_c|^2 = {} & (2/V)\{k_1^{-1}|\alpha_{\mathbf{k}_1}|^2 + k_2^{-1}|\alpha_{\mathbf{k}_2}|^2 \\ & + 2(k_1 k_2)^{-1/2}|\alpha_{\mathbf{k}_1}||\alpha_{\mathbf{k}_2}| \cos[(k_1 - k_2)t - (\mathbf{k}_1 - \mathbf{k}_2)\cdot\mathbf{x} + \theta_{\mathbf{k}_1} - \theta_{\mathbf{k}_2}]\}.\end{aligned} \tag{12.7.19}$$

The amplitudes of the component waves in (12.7.18) are proportional to the mean number of quanta in each wave. The expectation value of the number operator for the total state (12.7.17) is simply the sum:

$$\langle \hat{N} \rangle \equiv \left\langle \sum_{\mathbf{k}} a^\dagger_{\mathbf{k}} a_{\mathbf{k}} \right\rangle = |\alpha_{\mathbf{k}_1}|^2 + |\alpha_{\mathbf{k}_2}|^2. \tag{12.7.20}$$

Now the fringe visibility, or contrast, implied by the function (12.7.19) is defined by

$$C = 2|\alpha_{\mathbf{k}_1}||\alpha_{\mathbf{k}_2}|/(k_1 k_2)^{1/2}[k_1^{-1}|\alpha_{\mathbf{k}_1}|^2 + k_2^{-1}|\alpha_{\mathbf{k}_2}|^2] \tag{12.7.21}$$

with $0 \leqslant C \leqslant 1$. Clearly C varies according to the relative magnitude of the amplitudes $|\alpha_{\mathbf{k}_1}|$, $|\alpha_{\mathbf{k}_2}|$ and is not responsive to their absolute values. Thus, however small the mean number of quanta (12.7.20) becomes, e.g., $\langle \hat{N} \rangle \ll 1$, the interference effect will persist; the contrast will be maximum, for example, if $|\alpha_{\mathbf{k}_1}|/k_1^{1/2} = |\alpha_{\mathbf{k}_2}|/k_2^{1/2}$, independently of their actual value.

We now compare these results with the patterns implied by the mean energy-momentum. It is straightforward to evaluate these:

$$\begin{aligned}\langle :\hat{T}^{00}: \rangle = (2/V)\{&k_1|\alpha_{\mathbf{k}_1}|^2 \sin^2(k_1 t - \mathbf{k}_1 \cdot \mathbf{x} + \theta_{\mathbf{k}_1}) \\ &+ k_2|\alpha_{\mathbf{k}_2}|^2 \sin^2(k_2 t - \mathbf{k}_2 \cdot \mathbf{x} + \theta_{\mathbf{k}_2}) \\ &+ (k_1 k_2)^{-1/2}|\alpha_{\mathbf{k}_1}||\alpha_{\mathbf{k}_2}|(k_1 k_2 + \mathbf{k}_1 \cdot \mathbf{k}_2) \\ &\times \sin(k_1 t - \mathbf{k}_1 \cdot \mathbf{x} + \theta_{\mathbf{k}_1}) \sin(k_2 t - \mathbf{k}_2 \cdot \mathbf{x} + \theta_{\mathbf{k}_2})\},\end{aligned} \tag{12.7.22}$$

$$\begin{aligned}\langle \hat{T}^{0i} \rangle = (2/V)\{&k_1^i|\alpha_{\mathbf{k}_1}|^2 \sin^2(k_1 t - \mathbf{k}_1 \cdot \mathbf{x} + \theta_{\mathbf{k}_1}) \\ &+ k_2^i|\alpha_{\mathbf{k}_2}|^2 \sin^2(k_2 t - \mathbf{k}_2 \cdot \mathbf{x} + \theta_{\mathbf{k}_2}) \\ &+ |\alpha_{\mathbf{k}_1}||\alpha_{\mathbf{k}_2}|[(k_2/k_1)^{1/2}k_1^i + (k_1/k_2)^{1/2}k_2^i] \\ &\times \sin(k_1 t - \mathbf{k}_1 \cdot \mathbf{x} + \theta_{\mathbf{k}_1}) \sin(k_2 t - \sin \mathbf{k}_2 \cdot \mathbf{x} + \theta_{\mathbf{k}_2})\}.\end{aligned} \tag{12.7.23}$$

Each of these expressions is a sum of the corresponding quantities for the individual coherent states as given in (12.6.58) together with an interference term. The latter involves not only the phase difference between the component waves but also the phase sum. This is an indication that, unlike the case of the number states, the mean momentum density (12.7.23) differs somewhat from the intensity measured by a physical detector. The latter, defined by (12.6.70), is determined only by the $a^\dagger_{\mathbf{k}} a_{\mathbf{k}'}$ terms in the energy-momentum operator and results in an expression similar to (12.7.19), whereas the former includes contributions from the terms $a^\dagger_{\mathbf{k}} a^\dagger_{\mathbf{k}'}$ and $a_{\mathbf{k}} a_{\mathbf{k}'}$. Nevertheless, the intensity (12.7.23) possesses the important property that interference persists even in the case of low mean quantum number. It will be noted that the expressions (12.7.19), (12.7.22) and (12.7.23) are identical to those obtained from the superposition of two classical waves of definite amplitude and phase.

Thus, for certain states of the radiation field we expect that light emanating from independent sources will form interference fringes, in particular when the mean quantum number is low, and this effect has indeed been observed (Pfleegor & Mandel, 1967; for a review see Paul, 1986). In a sense the possibility of achieving this with coherent states further undermines the invocation of the 'photon' as a relevant explanatory concept when discussing quantal interference phenomena. For in such states there is no meaning at all to statements such as 'each photon interfers only with itself' since the overlapping fields are associated with an indefinite number of quanta (even though the mean number may be low). Once again we should say that the photon concept is only relevant to the detection process and does not bear at all on the underlying linear superposition of two continuous coherent fields and the formation of a spatio-temporal interference pattern.

12.8 Quantum potential as the origin of mass?

The equation of motion (12.4.16) of a massless quantum field,

$$\Box\psi = -\delta Q[\psi, t]/\delta\psi|_{\psi(\mathbf{x})=\psi(\mathbf{x},t)}, \tag{12.8.1}$$

generally implies noncovariant and nonlocal properties of the field. In fact, these features characterize the extremes of quantum behaviour and, in principle, there exist states for which the right hand side of (12.8.1) is a scalar and local function of the spacetime coördinates. The fact that the quantum force is still finite means that the wave will be essentially nonclassical but will obey the type of equation we might postulate for a classical field, in which the scalar wave equation is equated to some (linear or nonlinear) function of the field. A general question then arises whether the quantum force can simulate these types of classical equation for special choices of states and initial conditions. The specific problem addressed here is whether through quantization of a massless field the resulting quantum force can give 'mass' to the field in the sense that the quantum wave obeys the classical massive Klein–Gordon equation:

$$(\Box + m^2)\psi = 0 \tag{12.8.2}$$

as a special case of (12.8.1), where m is a real constant.

To see how this possibility may be realized we recall the equation of motion (12.5.23) of the $\mathbf{k}$th normal mode for a one-quantum state (12.5.14) which we write in the form

$$\ddot{q}_{\mathbf{k}} + \mathbf{k}^2 q_{\mathbf{k}} = (\mathbf{k}^2 - \omega_{\mathbf{k}}^2) q_{\mathbf{k}}, \tag{12.8.3}$$

where $\omega_{\mathbf{k}} = 1/2|q_{\mathbf{k}0}|^2$ and $q_{\mathbf{k}0} = |q_{\mathbf{k}0}|\, \mathrm{e}^{-\mathrm{i}\delta_{\mathbf{k}}}$ is arbitrary. Suppose we choose $|q_{\mathbf{k}0}| = [4(m^2 + \mathbf{k}^2)]^{-1/4}$ where m is a real constant and leave $\delta_{\mathbf{k}}$ unspecified. Then (12.8.3) becomes

$$\ddot{q}_{\mathbf{k}} + (\mathbf{k}^2 + m^2) q_{\mathbf{k}} = 0. \tag{12.8.4}$$

Translating (12.8.4) into spacetime via (12.5.24) (with $q_{\mathbf{k}'0} = 0$ for $\mathbf{k}' \neq \pm\mathbf{k}$) we find that the field $\psi(\mathbf{x}, t)$ satisfies the massive Klein–Gordon equation (12.8.2). Notice that this is true only for the class of waves having the above initial coordinates.

Thus, the quantum force associated with a one-quantum state and evaluated for this choice of initial conditions acts so that the massless quantum field behaves as if it were a classical field with mass. The same result is obtained if we use a product n-quantum state in which n modes $\mathbf{k}$ are independently excited and the initial coordinates of each are chosen as above.

The attempt to construct particle attributes from pure fields has a long pedigree, including Einstein's unified field theory and the programme of Wheeler (1962) who attempted to treat mass and charge as aspects of fields integrated over topological holes in spacetime. Our demonstration here is comparatively modest. We have shown that the attribute of mass may be viewed as a quantum effect, but not for arbitrary solutions of the Klein–Gordon equation. The reason for this restriction is that the parameter m is associated with $\psi_0(\mathbf{x})$ rather than with the quantum state. This is analogous to showing in particle quantum mechanics that a special choice of initial position yields the motion of some known classical system. Note also that m is not associated with a localized object but is a property of the entire field. Still, perhaps the demonstration is of some interest and can be extended to cover more general cases.

12.9 Beyond space-time-matter. Wavefunction of the universe

The classical analytical approach to the theory of matter assumes that a complex physical system may be broken down into a collection of subsystems obeying relatively simple laws which govern their interactions, and that the state of the whole is defined by no more than a summation of the states of the parts. In the quantum theory of motion this procedure is turned on its head. Here the basic notion is that of an objectively real state of an individual system that lies beyond its material components (particles and fields in their classical conception), and even beyond the spacetime manifold. In the sense that its law (the Schrödinger equation) governs the law of the elements (the guidance formula), we may say that *the state of the whole is prior to that of the parts* (in the model studied in this book the parts are not *physically* determined as aspects of the whole, as they would be in a unified field theory, for instance). Including the laws of the parts in the quantum mechanical theory allows one to overcome what we have located as the central interpretative dilemma posed by quantum mechanics – the failure to provide sufficient concepts to furnish a complete description of the processes with which quantum mechanics deals in circumstances where we are sure something more could be said that is not contained in the formalism (e.g., the positions of meters).

The holistic concept represents a first step towards realizing Einstein's programme of freeing microphysics from its reliance on the classical paradigm, although not in a way that Einstein approved of. It is, perhaps paradoxically, anticipated most forcibly in Bohr's analysis. Like Bohr, the mode of being of the parts is a function of the whole, but unlike him they can be conceptually analysed, including when they are examined empirically. Such a view tends to contradict the classical notion of a mechanism and is more suggestive of a self-regulating organism. Its ability to form and maintain subtle, stable patterns of matter, and to bring about transitions to qualitatively new stable structures, implies a role for the wavefunction akin to a kind of multi-dimensional field of *self-organization.* And because an infinite variety of organized forms can be generated through the linear superposition principle, there is here the hint of a primitive description of creativity or novelty in nature.

The emphasis it lays on the primacy of the whole implies two special properties of the quantum theory of motion. First, it forces into the open the issue of how far the individual events whose mean behaviour is grasped by quantum theory should obey the principles of relativity, and whether the latter have merely a statistical validity. If the description of the quantum theory of motion applied to fields is correct, the Lorentz symmetry exhibited by the formalism represents not so much a *unification* of the quantum and relativity theories as an expression merely of their empirical *compatibility*. In the realm of the theory of matter, i.e., the structure and motion of individual systems, true unification of the two theories cannot be said to have yet been achieved. Strict observance of the relativity principles is restored only in the classical limit and for some other nonclassical special cases (§12.8).

The second special property of the theory relates to attempts to apply quantum mechanics to the universe as a whole, a currently fashionable subject known as 'quantum cosmology'. This has been widely interpreted according to the many-worlds picture of quantum mechanics but there is no need for this. Because it is first and foremost a theory of individuals and does not rely on the ensemble or probability concepts for its formulation, the quantum theory of motion is eminently suited to a description of systems that are essentially unique, such as the universe. We now briefly indicate how one may consistently maintain the notion of a uniquely determined and objective quantum universe, in the context of current studies in quantum cosmology (for a related approach see Kowalski-Glikman & Vink (1990) and Squires (1992)).

The dynamical field in classical general relativity is the spacetime metric $g_{\mu\nu}$ ($\mu, \nu = 1, 2, 3, 4$) which obeys Einstein's equations

$$R_{\mu\nu} - \tfrac{1}{2} g_{\mu\nu} R = 0, \tag{12.9.1}$$

where $R_{\mu\nu}$ is the Ricci tensor and R is the curvature scalar. For simplicity we consider only the vacuum case. In the canonical Hamiltonian theory, one makes the following (3 + 1) decomposition of the metric (Misner *et al.*, 1973, Chap. 21) (in this section we assume a signature $-+++$):

$$\begin{aligned} ds^2 &= g_{\mu\nu}\, dx^\mu\, dx^\nu \\ &= (N_i N^i - N^2)\, dt^2 + 2N_i\, dx^i\, dt + g_{ij}\, dx^i\, dx^j, \end{aligned} \tag{12.9.2}$$

$i, j = 1, 2, 3$; $g_{ij}(\mathbf{x})$ is the 3-metric of a 3-surface embedded in spacetime. The dynamics of spacetime is now described in terms of the evolution of $g_{ij}(\mathbf{x}, t)$ in superspace, the space of all 3-geometries.

Quantizing the Hamiltonian constraint of general relativity in the standard way one obtains the Wheeler–DeWitt equation, the Schrödinger equation of the gravitational field (DeWitt, 1967; Wheeler, 1968):

$$\left[G_{ijkl} \frac{\delta^2}{\delta g_{ij}\, \delta g_{kl}} + g^{1/2}\, {}^3R \right] \Psi = 0 \tag{12.9.3}$$

(we ignore factor-ordering problems). Here $g = \det g_{ij}$, ${}^3R$ is the intrinsic curvature, $G_{ijkl} = \tfrac{1}{2} g^{-1/2}(g_{ik}g_{jl} + g_{il}g_{jk} - g_{ij}g_{kl})$ is the supermetric, and $\Psi[g_{ij}(\mathbf{x})]$ is a wavefunctional in the superspace, the 'wavefunction of the universe'.

There are a variety of technical and conceptual problems connected with the wave equation (12.9.3), in particular the role of time which it will be noted does not appear explicitly, and the definition of probability (for reviews see Padmanabhan (1989), Zeh (1989) and Halliwell (1991)). However, it seems a formulation of the theory along the lines of the causal interpretation is unproblematic and may be constructed in an entirely obvious way.

Substituting $\Psi = A\,e^{iS}$ where A and S are real functionals into (12.9.3) we obtain as usual a conservation law and a modification of the Einstein–Hamilton–Jacobi equation which includes a quantum potential contribution in addition to the usual external matter sources (neglected here):

$$G_{ijkl}\frac{\delta}{\delta g_{ij}}\left(A^2\frac{\delta S}{\delta g_{kl}}\right) = 0, \tag{12.9.4}$$

$$G_{ijkl}\frac{\delta S}{\delta g_{ij}}\frac{\delta S}{\delta g_{kl}} - g^{1/2}\,{}^3R + Q = 0, \tag{12.9.5}$$

where

$$Q = -A^{-1}G_{ijkl}\delta^2 A/\delta g_{ij}\delta g_{kl}. \tag{12.9.6}$$

Q is invariant under 3-space diffeomorphisms. We make the assumption that the universe whose quantum state is governed by (12.9.3) has a definite 3-geometry at each instant, represented by the 3-metric $g_{ij}(\mathbf{x}, t)$. The law of evolution of the quantum 3-geometry is that of the classical Hamilton–Jacobi theory,

$$\frac{\partial g_{ij}(\mathbf{x}, t)}{\partial t} = \partial_i N_j + \partial_j N_i + 2NG_{ijkl}\left.\frac{\delta S}{\delta g_{kl}}\right|_{g_{ij}(\mathbf{x}) = g_{ij}(\mathbf{x},t)}, \tag{12.9.7}$$

but where S is now the phase of the quantum wavefunctional. Solving (12.9.7) requires specifying the initial data $g_{ij0}(\mathbf{x})$.

From this relation, (12.9.5) and the other constraint equations of the theory, we recover Einstein's equations (12.9.1) but with an additional term on the right hand side of quantum mechanical origin (analogue of the 'quantum force' term in (12.4.16)). This term is responsible for all the geometrical effects of quantum gravity.

Thus, in this view the primary significance of Ψ is that it organizes the dynamics of the gravitational field (and other fields and matter) via the law (12.9.7). Assuming the breakdown of covariance that appeared in the flat space theory (§12.4) does not imply special complications here, it seems that we are able in this way to give a consistent treatment of the dynamics of the quantum universe conceived as an objective, unique and autonomous physical system. Time enters the theory through (12.9.7) as in classical Einstein–Hamilton–Jacobi theory and the expression (12.9.2) for the line element of the universe remains intact. In particular, the theory is independent of any subsidiary probability interpretation one may like to attach to Ψ. This is in the spirit of the declaration of Einstein (1953, quoted by Freistadt, 1957) that 'Nature as a whole can only be viewed as an individual system, existing only once, and not as a collection of systems.'

The classical limit, i.e., the limit of classical cosmology as expressed in the usual Einstein equations, is obtained in circumstances where one may neglect the additional quantum potential contribution (12.9.6). The remarks made in Chap. 6 regarding the problems of deriving this limit via techniques such as the variation of parameters and the WKB approximation apply in this new context. As in ordinary quantum mechanics, a real Ψ generally implies highly nonclassical behaviour and the approach to the classical limit requires consideration of complex wavefunctionals.

Contemplation of examples such as this suggests two observations regarding the current situation in quantum physics. The first is that the precise manner in which quantum mechanics constitutes a break with the past is not yet understood, having been obscured behind a probabilistic smokescreen. The second is that we are in a period of transition between two great world views – the universal machine of the classicists and a new holistic universe whose details we are only beginning to glimpse. The end is not in sight for theoretical physics.

References

Aharonov, Y. (1984). In *Proc. Int. Symp. Foundations of Quantum Mechanics, Tokyo, 1983*, ed. S. Kamefuchi *et al.*, pp. 10–19. Tokyo: Japanese Physical Society.
Aharonov, Y. & Albert, D. Z. (1987). In *Quantum Implications*, ed. B. J. Hiley & F. D. Peat, pp. 224–6. London: Routledge & Kegan Paul.
Aharonov, Y. & Bohm, D. (1959). *Phys. Rev.*, **115**, 485–91.
Aharonov, Y. & Bohm, D. (1961). *Phys. Rev.*, **123**, 1511–24.
Aharonov, Y. & Bohm, D. (1962). *Phys. Rev.*, **125**, 2192–3.
Aharonov, Y. & Bohm, D. (1963). *Phys. Rev.*, **130**, 1625–32.
Aharonov, Y., Komar, A. & Susskind, L. (1969). *Phys. Rev.*, **182**, 1400–3.
Aharonov, Y. & Susskind, L. (1967). *Phys. Rev.*, **158**, 1237–8.
Aharonov, Y. & Vardi, M. (1980). *Phys. Rev. D*, **21**, 2235–40.
Allcock, G. R. (1969). *Ann. Phys. (US)*, **53**, 253–85, 286–310, 311–48.
Arnold, V. I. (1978). *Mathematical Methods of Classical Mechanics.* New York: Springer-Verlag.
Arthurs, E. & Kelly, J. L. (1965). *Bell System Tech. J.*, **44**, 725–9.
Asanov, G. S. (1985). *Finsler Geometry, Relativity and Gauge Theories.* Dordrecht: D. Reidel.
Aspect, A., Dalibard, J. & Roger, G. (1982). *Phys. Rev. Lett.*, **49**, 1804–7.
Badurek, G., Rauch, H. & Summhammer, J. (1983). *Phys. Rev. Lett.*, **51**, 1015–18.
Badurek, G., Rauch, H., Summhammer, J., Kischko, U. & Zeilinger, A. (1983). *J. Phys. A: Math. Gen.*, **16**, 1133–9.
Badurek, G., Rauch, H. & Tuppinger, D. (1986). *Phys. Rev. A*, **34**, 2600–8.
Ballentine, L. E. (1970). *Rev. Mod. Phys.*, **42**, 358–81.
Ballentine, L. E. (1972). *Am. J. Phys.*, **40**, 1763–71.
Barut, A. O. & Bracken, A. J. (1981). *Phys. Rev. D*, **23**, 2454–63.
Beck, G. & Nussenzveig, H. M. (1958). *Nuovo Cimento*, **9**, 1068–76.
Belinfante, F. J. (1973). *A Survey of Hidden-Variables Theories.* Oxford: Pergamon Press.
Bell, J. S. (1964). *Physics*, **1**, 195–200.
Bell, J. S. (1966). *Rev. Mod. Phys.*, **38**, 447–52.
Bell, J. S. (1980). *Int. J. Quant. Chem.*, **14**, 155–9.
Bell, J. S. (1981). In *Quantum Gravity 2*, ed. C. J. Isham, R. Penrose & D. Sciama, pp. 611–37. Oxford: Clarendon Press.
Bell, J. S. (1982). *Found. Phys.* **12**, 989–99.

Bell, J. S. (1987). *Speakable and Unspeakable in Quantum Mechanics.* Cambridge: University Press (contains reprints of preceding papers).
Bell, J. S. (1990a). *Physics World*, **3**, August, 33–40.
Bell, J. S. (1990b). In *Between Science and Technology*, ed. A. Sarlemijn & P. Kroes, pp. 97–115. Amsterdam: North-Holland.
Berkowitz, D. B. & Skiff, P. D. (1972). *Am. J. Phys.*, **40**, 1625–8.
Bernstein, H. J. (1967). *Phys. Rev. Lett.*, **18**, 1102–3.
Berry, M. V. (1980). *Ann. NY Acad. Sci.*, **357**, 183–202.
Berry, M. V. (1981). In *Physics of Defects, Les Houches, 1980*, ed. R. Balian *et al.*, pp. 453–543. Amsterdam: North-Holland.
Berry, M. V. (1982). *J. Phys. A: Math. Gen.*, **15**, L385–8.
Berry, M. V. (1987). *Proc. Roy. Soc. Lond. A*, **413**, 183–98.
Berry, M. V. (1991). In *Chaos and Quantum Physics, Les Houches, 1989*, ed. M.-J. Giannoni, A. Voros & J. Zinn-Justin, pp. 251–303. Amsterdam: North-Holland.
Berry, M. V. & Balazs, N. L. (1979). *Am. J. Phys.*, **47**, 264–7.
Berry, M. V. & Mount, K. E. (1972). *Rep. Prog. Phys.*, **35**, 315–97.
Bjorken, J. D. & Drell, S. D. (1964). *Relativistic Quantum Mechanics.* New York: McGraw-Hill.
Blokhintsev, D. I. (1973). *Space and Time in the Microworld.* Dordrecht: D. Reidel.
Blümel, R. & Smilansky, U. (1990). *Physics World*, **3**, February, 30–4.
Bohm, D. (1951). *Quantum Theory.* New York: Prentice-Hall.
Bohm, D. (1952a). *Phys. Rev.*, **85**, 166–79 (reprinted in Wheeler & Zurek (1983)).
Bohm, D. (1952b). *Phys. Rev.*, **85**, 180–93 (reprinted in Wheeler & Zurek (1983)).
Bohm, D. (1952c). *Phys. Rev.*, **87**, 389–90.
Bohm, D. (1552d). *Phys. Rev.*, **89**, 319–20.
Bohm, D. (1953a). *Phys. Rev.*, **89**, 458–66.
Bohm, D. (1953b). *Prog. Theor. Phys.*, **9**, 273–87.
Bohm, D. (1953c). In *Scientific Papers Presented to Max Born*, pp. 13–19. Edinburgh: Oliver & Boyd.
Bohm, D. (1957). *Causality and Chance in Modern Physics.* London: Routledge & Kegan Paul.
Bohm, D. (1962). *Brit. J. Phil. Soc.*, **12**, 265–80.
Bohm, D. (1987). In *Quantum Implications*, ed. B. J. Hiley & F. D. Peat, pp. 33–45. London: Routledge & Kegan Paul.
Bohm, D., Dewdney, C. & Hiley, B. J. (1985). *Nature*, **315**, 294–7.
Bohm, D. & Hiley, B. J. (1984). *Found. Phys.*, **14**, 255–74.
Bohm, D. & Hiley, B. J. (1985). *Phys. Rev. Lett.*, **55**, 2511–14.
Bohm, D. & Hiley, B. J. (1987). *Phys. Rep.*, **144**, 323–48.
Bohm, D. & Hiley, B. J. (1988). In *Quantum Mechanics Versus Local Realism*, ed. F. Selleri, pp. 235–56. New York: Plenum Press.
Bohm, D. & Hiley, B. J. (1989). *Phys. Rep.*, **172**, 93–122.
Bohm, D., Hiley, B. J. & Kaloyerou, P. N. (1987). *Phys. Rep.*, **144**, 349–75.
Bohm, D. & Peat, F. D. (1989). *Science, Order and Creativity.* London: Routledge.
Bohm, D. & Schiller, R. (1955). *Suppl. Nuovo Cimento*, **1**, 67–91.
Bohm, D., Schiller, R. & Tiomno, J. (1955). *Suppl. Nuovo Cimento*, **1**, 48–66.
Bohm, D. & Vigier, J. P. (1954). *Phys. Rev.*, **96**, 208–16.
Böhm, A. (1986). *Quantum Mechanics*, 2nd edn. Berlin: Springer-Verlag.
Bohr, N. (1932). *J. Chem. Soc.*, 349–83.
Bohr, N. (1934). *Atomic Theory and the Description of Nature.* Cambridge: University Press.
Bohr, N. (1935). *Phys. Rev.*, **48**, 696–702 (reprinted in Wheeler & Zurek (1983)).

Bohr, N. (1948). *Dialectica*, **2**, 312–19.
Bohr, N. (1949). In *Albert Einstein: Philosopher-Scientist*, ed. P. A. Schilpp, pp. 199–241. La Salle, Ill: Open Court (reprinted in Wheeler & Zurek (1983)).
Bohr, N. (1958). *Atomic Physics and Human Knowledge*. New York: Wiley.
Bohr, N. & Rosenfeld, L. (1933). *Mat. Fys. Medd. Dan. Vid. Selsk.*, **12**, no. 8 (English translation in Wheeler & Zurek (1983)).
Bohr, N. & Rosenfeld, L. (1950). *Phys. Rev.*, **78**, 794–8 (reprinted in Wheeler & Zurek (1983)).
Bolker, E. D. (1973). *Am. Math. Mthly.*, November, 977–84.
Bonse, U. & Wroblewski, T. (1983). *Phys. Rev. Lett.*, **51**, 1401–4.
Bopp, F. & Haag, R. (1950). *Z. Naturforschg.*, **5a**, 644–53.
Born, M. (1926). *Z. Phys.*, **37**, 863–7 (English translation in Wheeler & Zurek (1983)).
Born, M. (1955). *Mat. Fys. Medd. Dan. Vid. Selsk.*, **30**, no. 2, 1–26.
Born, M. (1971). ed. *The Born–Einstein Letters*. London: MacMillan.
Born, M. & Hooton, D. J. (1956). *Proc. Camb. Phil. Soc.*, **52**, 287–300.
Born, M. & Ludwig, W. (1958). *Z. Phys.*, **150**, 106–17.
Born, M. & Wolf, E. (1970). *Principles of Optics*, 4th edn. Oxford: Pergamon Press.
Brandt, S. & Dahmen, H. D. (1985). *The Picture Book of Quantum Mechanics*. New York: Wiley.
Brandt, S. & Dahmen, H. D. (1989). *Quantum Mechanics on the Personal Computer*. Berlin: Springer-Verlag.
Brauer, R. & Weyl, H. (1935). *Am. J. Math.*, **57**, 425–49.
Brillouin, L. (1926). *C.R. Acad. Sci. Paris*, **183**, 24–6.
de Broglie, L. (1926a). *C.R. Acad. Sci. Paris*, **183**, 447–8.
de Broglie, L. (1926b). *Nature*, **118**, 441–2.
de Broglie, L. (1927a). *C.R. Acad. Sci. Paris*, **184**, 273–4.
de Broglie, L. (1927b). *C.R. Acad. Sci. Paris*, **185**, 380–2.
de Broglie, L. (1927c). *J. de Phys.*, **8**, 225–41.
de Broglie, L. (1930). *Introduction à l'Etude de la Mécanique Ondulatoire*. Paris: Hermann (English translation: *An Introduction to the Study of Wave Mechanics*, 1930, New York: Dutton).
de Broglie, L. (1953a). *La Physique Quantique Restera-t-elle Indéterministe?* Paris: Gauthier-Villars (contains reprints of de Broglie (1926a, 1927a,c)).
de Broglie, L. (1953b). In *Scientific Papers Presented to Max Born*, pp. 21–8. Edinburgh: Oliver & Boyd.
de Broglie, L. (1954). *Théorie Générale des Particules à Spin* (*Méthode de Fusion*), 2nd edn. Paris: Gauthier-Villars.
de Broglie, L. (1956). *Une Tentative d'Interprétation Causale et Non-Linéaire de la Mécanique Ondulatoire*. Paris: Gauthier-Villars (English translation: *Nonlinear Wave Mechanics*, 1960, Amsterdam: Elsevier).
de Broglie, L. (1963). *Etude Critique des Bases de l'Interprétation Actuelle de la Mécanique Ondulatoire*. Paris: Gauthier-Villars (English translation: *The Current Interpretation of Wave Mechanics*, 1964, Amsterdam: Elsevier).
de Broglie, L. & Andrade e Silva, J. (1968). *Phys. Rev.*, **172**, 1284–5.
de Broglie, L. & Vigier, J. P. (1972). *Phys. Rev. Lett.*, **28**, 1001–4.
Busch, P. & Lahti, P. J. (1984). *Phys. Rev. D*, **29**, 1634–46.
Busch, P. & Schroeck, F. E. (1989). *Found. Phys.*, **19**, 807–72.
Cabrera, G. G. & Kiwi, M. (1987). *Phys. Rev. A*, **36**, 2995–8.
Cantrell, C. D. & Scully, M. O. (1978). *Phys. Rep.*, **43**, 499–508.
Carnal, O. & Mlynek, J. (1991). *Phys. Rev. Lett.*, **66**, 2689–92.
Carruthers, P. & Nieto, M. M. (1968). *Rev. Mod. Phys.*, **40**, 411–40.

Carruthers, P. & Zachariasen, F. (1983). *Rev. Mod. Phys.*, **55**, 245–85.
Cartan, E. (1966). *The Theory of Spinors.* Cambridge MA: MIT (reprinted by Dover, New York, 1981).
Casimir, H. B. G. (1948). *Proc. Kon. Ned. Akad. Wet.*, **51**, 793–5.
Clauser, J. F. & Shimony, A. (1978). *Rep. Prog. Phys.*, **41**, 1881–927.
Cohen, L. (1990). *Found. Phys.*, **20**, 1455–73.
Colella, R., Overhauser, A. W. & Werner, S. A. (1975). *Phys. Rev. Lett.*, **34**, 1472–4.
Croca, J. R. (1987). *Found. Phys.*, **17**, 971–80.
Cufaro-Petroni, N., Dewdney, C., Holland, P., Kyprianidis, A. & Vigier, J. P. (1985). *Phys. Rev. D*, **32**, 1375–83.
Dačić Gaeta, Z. & Stroud, C. R. (1990). *Phys. Rev. A*, **42**, 6308–13.
Dahl, J. P. (1983). In *Energy Storage and Redistribution in Molecules*, ed. J. Hinze, pp. 557–71. New York: Plenum Press.
Dankel, T. G. (1970). *Arch. Ration. Mech. Anal.*, **37**, 192–221.
Davies, E. B. (1976). *Quantum Theory of Open Systems.* London: Academic Press.
Dewdney, C. (1985). *Phys. Lett. A*, **109**, 377–84.
Dewdney, C. (1988). In *Problems in Quantum Physics: Gdańsk '87*, ed. L. Kostro, A. Posiewnik, J. Pykacz & M. Zukowski, pp. 62–85. Singapore: World Scientific.
Dewdney, C. & Hiley, B. J. (1982). *Found. Phys.*, **12**, 27–48.
Dewdney, C. & Holland, P. R. (1988). In *Quantum Mechanics Versus Local Realism*, ed. F. Selleri, pp. 301–25. New York: Plenum Press.
Dewdney, C., Holland, P. R. & Kyprianidis, A. (1986). *Phys. Lett. A*, **119**, 259–67.
Dewdney, C., Holland, P. R. & Kyprianidis, A. (1987a). *Phys. Lett. A*, 121, 105–10.
Dewdney, C., Holland, P. R. & Kyprianidis, A. (1987b). *J. Phys. A: Math. Gen.*, **20**, 4717–32.
Dewdney, C., Holland, P. R., Kyprianidis, A. & Vigier, J. P. (1985). *Phys. Rev. D*, **31**, 2533–8.
Dewdney, C., Holland, P. R., Kyprianidis, A. & Vigier, J. P. (1986). *Phys. Lett. A*, **114**, 365–70.
Dewdney, C., Holland, P. R., Kyprianidis, A. & Vigier, J. P. (1988). *Nature*, **336**, 536–44.
Dewdney, C., Kyprianidis, A. & Vigier, J. P. (1984). *J. Phys. A: Math. Gen.*, **17**, L741–4.
DeWitt, B. S. (1967). *Phys. Rev.*, **160**, 1113–48.
DeWitt, B. S. (1975). *Phys. Rep.*, **19**, 295–357.
DeWitt, B. S. (1984). *Supermanifolds.* Cambridge: University Press.
Dirac, P. A. M. (1951). *Nature*, **168**, 906–7.
Dirac, P. A. M. (1953). *Physica*, **19**, 888–96.
Dirac, P. A. M. (1965). *Phys. Rev.*, **139B**, 684–90.
Dirac, P. A. M. (1974). *The Principles of Quantum Mechanics*, 4th edn (revised). Oxford: Clarendon Press.
Doescher, S. W. & Rice, M. H. (1969). *Am. J. Phys.*, **37**, 1246–9.
Donoghue, J. F. & Holstein, B. R. (1987). *Eur. J. Phys.*, **8**, 105–13.
Dubovik, V. M. & Shabanov, S. V. (1989). *Phys. Lett. A*, **142**, 211–14.
Dubovik, V. M. & Shabanov, S. V. (1990). *J. Phys. A: Math. Gen.*, **23**, 3245–55.
Duncan, A. J. & Kleinpoppen, H. (1988). In *Quantum Mechanics Versus Local Realism*, ed. F. Selleri, pp. 175–218. New York: Plenum Press.
Dürr, D., Goldstein, S. & Zanghi, N. (1992). *J. Stat. Phys.*, **67**, 843–907.
Eberhard, Ph. H. & Ross, R. R. (1989). *Found. Phys. Lett.*, **2**, 127–49.
Einstein, A. (1905). *Ann. Phys.*, **17**, 132–48 (English translation: A. B. Arons & M. B. Peppard (1965), *Am. J. Phys.*, **33**, 367–74).

Einstein, A. (1922). *Sidelights on Relativity*. New York: Dutton (reprinted by Dover, New York, 1983).
Einstein, A. (1925). *Sitzungsber. Preuss. Akad. Wiss. Phys.–Math. Kl.*, 3–14.
Einstein, A. (1936). *J. Franklin Inst.*, **221**, 349–82 (reprinted in Einstein (1950)).
Einstein, A. (1940). *Science*, **91**, 487–92 (reprinted in Einstein (1950)).
Einstein, A. (1948). *Dialectica*, **2**, 320–3 (English translation in Born (1971, pp. 168–73)).
Einstein, A. (1949). In *Albert Einstein: Philosopher-Scientist*, ed. P. A. Schilpp, pp. 1–95 & 663–88. La Salle, Ill: Open Court.
Einstein, A. (1950). *Out of My Later Years*. New York: Philosophical Library.
Einstein, A. (1953). In *Scientific Papers Presented to Max Born*, pp. 33–40. Edinburgh: Oliver & Boyd.
Einstein, A. (1989). *Oeuvres Choisies 4, Correspondances Françaises*. Editions du Seuil.
Einstein, A., Podolsky, B. & Rosen, N. (1935). *Phys. Rev.*, **47**, 777–80 (reprinted in Wheeler & Zurek (1983)).
Electrons et Photons (1928). *Rapports et Discussions du Cinquième Conseil de Physique, 1927*. Paris: Gauthier-Villars.
Englert, B. G., Schwinger, J. & Scully, M. O. (1988). *Found. Phys.*, **18**, 1045–56.
Epstein, S. T. (1952). *Phys. Rev.*, **89**, 319.
Epstein, S. T. (1953). *Phys. Rev.*, **91**, 985.
d'Espagnat, B. (1976). *Conceptual Foundations of Quantum Mechanics*, 2nd edn. Reading, MA: W.A. Benjamin.
d'Espagnat, B. (1983). *In Search of Reality*. Berlin: Springer-Verlag.
Felber, J., Gähler, R. & Golub, R. (1988). *Physica B*, **151**, 135–9.
Felber, J., Müller, G., Gähler, R. & Golub, R. (1990). *Physica B*, **162**, 191–6.
Fermi, E. (1926). *Z. Phys.*, **40**, 399–402.
Feshbach, H. & Villars, F. (1958). *Rev. Mod. Phys.*, **30**, 24–45.
Feynman, R. P. (1948). *Rev. Mod. Phys.*, **20**, 367–87.
Feynman, R. P. (1949). *Phys. Rev.*, **76**, 749–59.
Feynman, R. P. & Gell-Mann, M. (1958). *Phys. Rev.*, **109**, 193–8.
Feynman, R. P. & Hibbs, A. R. (1965). *Quantum Mechanics and Path Integrals*. New York: McGraw-Hill.
Feynman, R. P., Leighton, R. B. & Sands, M. (1965). *The Feynman Lectures on Physics*, vol. 3. Reading, MA: Addison-Wesley.
Fine, A. (1986). *The Shaky Game. Einstein, Realism and the Quantum Theory*. Chicago: University of Chicago Press.
Floreanini, R. & Jackiw, R. (1988). *Phys. Rev. D*, **37**, 2206–24.
Folse, H. J. (1985). *The Philosophy of Niels Bohr*. Amsterdam: North-Holland.
Frahn, W. E. (1977). *Riv. Nuovo Cimento*, **7**, 499–542.
Freistadt, H. (1955). *Phys. Rev.*, **97**, 1158–61.
Freistadt, H. (1956). *Phys. Rev.*, **102**, 274–8.
Freistadt, H. (1957). *Suppl. Nuovo Cimento*, **5**, 1–70.
Friedman, J., Morris, M. S., Novikov, I. D., Echeverria, F., Klinkhammer, G., Thorne, K. S. & Yurtsever, U. (1990). *Phys. Rev. D*, **42**, 1915–30.
Fröbrich, P., Lipperheide, R. & Thoma, G. (1977). *Ann. Phys. (US)*, **104**, 478–502.
Fulling, S. A. (1989). *Aspects of Quantum Field Theory in Curved Space-Time*. Cambridge: University Press.
Furry, W. H. (1963). In *Lectures in Theoretical Physics*, vol. 5, ed. W. E. Brittin, B. W. Downs & J. Downs, pp. 1–112. New York: Interscience.
Gähler, R. & Golub, R. (1984). *Z. Phys. B.*, **56**, 5–12.
Gale, W., Guth, E. & Trammell, G. T. (1968). *Phys. Rev.*, **165**, 1434–6.

Garbaczewski, P. (1988). In *Problems in Quantum Physics: Gdańsk '87*, ed. L. Kostro, A. Posiewnik, J. Pykacz & M. Zukowski, pp. 494–509. Singapore: World Scientific.
George, C. & Prigogine, I. (1979). *Physica A*, **99**, 369–82.
George, C., Prigogine, I. & Rosenfeld, L. (1972). *Mat. Fys. Medd. Dan. Vid. Selsk.*, **38**, no. 12, 1–44.
Gerasimov, A. S. & Kazarnovskii, M. V. (1976). *Sov. Phys. JETP*, **44**, 892–7.
Gerlach, B., Gromes, D. & Petzold, J. (1967). *Z. Phys.*, **202**, 401–11.
Gerry, C. C. & Kiefer, J. (1988). *Phys. Rev. A*, **37**, 665–71.
Ghirardi, G. C., Rimini, A. & Weber, T. (1980). *Lett. Nuovo Cimento*, **27**, 293–8.
Ghose, P., Home, D. & Agarwal, G. S. (1991). *Phys. Lett. A*, **153**, 403–6.
Girardeau, M. D. (1965). *Phys. Rev.*, **139**, B500–8.
Gisin, N. (1991). *Phys. Lett. A*, **154**, 201–2.
Gisin, N. & Peres, A. (1992). *Phys. Lett. A*, **162**, 15–17.
Glauber, R. J. (1963a). *Phys. Rev.*, **130**, 2529–39.
Glauber, R. J. (1963b). *Phys. Rev.*, **131**, 2766–88.
Goldberg, A., Schey, H. M. & Schwartz, J. L. (1967). *Am. J. Phys.*, **35**, 177–86.
Goldstein, H. (1980). *Classical Mechanics*, 2nd edn. Reading, MA: Addison-Wesley.
Grangier, P., Roger, G. & Aspect, A. (1986). *Europhys. Lett.*, **1**, 173–9.
Greenberger, D. (1968). *Ann. Phys. (US)*, **47**, 116–26.
Greenberger, D. M. (1983). *Rev. Mod. Phys.*, **55**, 875–905.
Greenberger, D. M. (1988). *Physica B*, **151**, 374–7.
Greenberger, D. M., Horne, M. A., Shimony, A. & Zeilinger, A. (1990). *Am. J. Phys.*, **58**, 1131–43.
Greenberger, D. M. & Overhauser, A. W. (1979). *Rev. Mod. Phys.*, **51**, 43–78.
Greenberger, D. M. & Yasin, A. (1988). *Phys. Lett. A*, **128**, 391–4.
Greiner, W. (1990). *Relativistic Quantum Mechanics, Wave Equations*. Berlin: Springer-Verlag.
Gudder, S. (1970). *J. Math. Phys.*, **11**, 431–6.
Gurtler, R. & Hestenes, D. (1975). *J. Math. Phys.*, **16**, 573–84.
Gutschick, V. P. & Nieto, M. M. (1980). *Phys. Rev. D*, **22**, 403–18.
Gutzwiller, M. C. (1990). *Chaos in Classical and Quantum Mechanics*. Berlin: Springer-Verlag.
ter Haar, D. (1964). ed. *Selected Problems in Quantum Mechanics*. London: Infosearch Ltd.
Halbwachs, F. (1960). *Théorie Relativiste des Fluides à Spin*. Paris: Gauthier-Villars.
Halliwell, J. J. (1989). *Phys. Rev. D*, **39**, 2912–23.
Halliwell, J. J. (1991). In *Proceedings of the 1989 Jerusalem Winter School on Quantum Cosmology and Baby Universes*, ed. S. Coleman, J. B. Hartle, T. Piran & S. Weinberg. Singapore: World Scientific.
Halpern, O. (1952). *Phys. Rev.*, **87**, 389.
Hanson, N. R. (1969). *Patterns of Discovery*. Cambridge: University Press.
Hauge, E. H. & Støvneng, J. A. (1989). *Rev. Mod. Phys.*, **61**, 917–36.
Hegerfeldt, G. C. (1974). *Phys. Rev. D*, **10**, 3320–1.
Hegerfeldt, G. C. (1985). *Phys. Rev. Lett.*, **54**, 2395–8.
Hegerfeldt, G. C. & Ruijsenaars, S. N. M. (1980). *Phys. Rev. D*, **22**, 377–84.
Heisenberg, W. (1927). *Z. Phys.*, **43**, 172–98 (English translation in Wheeler & Zurek (1983)).
Heisenberg, W. (1930). *The Physical Principles of the Quantum Theory*. Chicago: University of Chicago Press (reprinted by Dover, New York, 1949).

Heisenberg, W. (1955). In *Niels Bohr and the Development of Physics*, ed. W. Pauli, pp. 12–29. Oxford: Pergamon Press.
Heisenberg, W. (1962). *Physics and Philosophy*. New York: Harper & Row (reprinted by Penguin, Harmondsworth, 1989).
Heitler, W. (1956). *Elementary Wave Mechanics*, 2nd edn. Oxford: Clarendon Press.
Heitler, W. & London, F. (1927). *Z. Phys.*, **44**, 455–72.
Heller, E. J. (1991). In *Chaos and Quantum Physics, Les Houches, 1989*, ed. M.-J. Giannoni, A. Voros & J. Zinn-Justin, pp. 547–663. Amsterdam: North-Holland.
Hepp, K. (1974). *Commun. Math. Phys.*, **35**, 265–77.
Hestenes, D. (1966). *Space-Time Algebra*. New York: Gordon & Breach.
Hestenes, D. (1971). *Am. J. Phys.*, **39**, 1013–27.
Hilgevoord, J. & Uffink, J. B. M. (1988). In *Microphysical Reality and Quantum Formalism*, ed. F. Selleri, A. van der Merwe & G. Tarozzi, pp. 91–114. Dordrecht: D. Reidel.
Hillery, M., O'Connell, R. F., Scully, M. O. & Wigner, E. P. (1984). *Phys. Rep.*, **106**, 121–67.
Hirschfelder, J. O., Christoph, A. C. & Palke, W. E. (1974a). *J. Chem. Phys.*, **61**, 5435–55.
Hirschfelder, J. O., Goebel, C. J. & Bruch, L. W. (1974b). *J. Chem. Phys.*, **61**, 5456–9.
Hirschfelder, J. O. & Tang, K. T. (1976a). *J. Chem. Phys.*, **64**, 760–85.
Hirschfelder, J. O. & Tang, K. T. (1976b). *J. Chem. Phys.*, **65**, 470–86.
Holland, P. R. (1982). *Phys. Lett. A*, **91**, 275–8.
Holland, P. R. (1986). *Found. Phys.*, **16**, 701–19.
Holland, P. R. (1987). *Found. Phys.*, **17**, 345–63.
Holland, P. R. (1988a). *Phys. Lett. A*, **128**, 9–18.
Holland, P. R. (1988b). *Phys. Rep.*, **169**, 293–327.
Holland, P. R. (1988c). In *Microphysical Reality and Quantum Formalism*, ed. A. van der Merwe *et al.*, pp. 89–101. Dordrecht: D. Reidel.
Holland, P. R. (1989). *Found. Phys. Lett.*, **2**, 471–85.
Holland, P. R. (1992). *Found. Phys.*, **22**, 1287–301.
Holland, P. R. & Kyprianidis, A. (1988). *Ann. Inst. Henri Poincaré*, **49**, 325–39.
Holland, P. R., Kyprianidis, A., Marić, Z. & Vigier, J. P. (1986). *Phys. Rev. A*, **33**, 4380–3.
Holland, P. R. & Vigier, J. P. (1988). *Found. Phys.*, **18**, 741–50.
Holland, P. R. & Vigier, J. P. (1991). *Phys. Rev. Lett.*, **67**, 402.
Home, D. & Kaloyerou, P. N. (1989). *J. Phys. A: Math. Gen.*, **22**, 3253–66.
Home, D. & Sengupta, S. (1984). *Nuovo Cimento*, **82B**, 214–23.
Home, D. & Whitaker, M. A. B. (1992). *Phys. Rep.*, **210**, 223–317.
Hong, C. K. & Mandel, L. (1986). *Phys. Rev. Lett.*, **56**, 58–60.
Horne, M. A., Shimony, A. & Zeilinger, A. (1989). *Phys. Rev. Lett.*, **62**, 2209–12.
Jammer, M. (1974). *The Philosophy of Quantum Mechanics*. New York: Wiley.
Jönsson, C. (1961). *Z. Phys.*, **161**, 454–74 (English translation: D. Brandt & S. Hirschi (1974), *Am. J. Phys.*, **42**, 4–11).
Jordan, P. & Wigner, E. (1928). *Z. Phys.*, **47**, 631–51.
Kaempffer, F. A. (1965). *Concepts in Quantum Mechanics*. New York: Academic Press.
van Kampen, N. G. (1984). *Phys. Lett. A*, **106**, 5–6.
Kasevich, M. & Chu, S. (1991). *Phys. Rev. Lett.*, **67**, 181–4.
Keith, D. W., Ekstrom, C. R., Turchette, Q. A. & Pritchard, D. E. (1991). *Phys. Rev. Lett.*, **66**, 2693–6.
Keller, J. B. (1953). *Phys. Rev.*, **89**, 1040–1.

Kemble, E. C. (1937). *The Fundamental Principles of Quantum Mechanics.* New York: McGraw-Hill.
Klein, O. (1929). *Z. Phys.*, **53**, 157–65.
Knoll, J. & Schaeffer, R. (1976). *Ann. Phys. (US)*, **97**, 307–66.
Kobe, D. (1979). *Ann. Phys. (US)*, **123**, 381–410.
Koester, L. (1976). *Phys. Rev. D*, **14**, 907–9.
Koopman, B. O. (1931). *Proc. Nat. Acad. Sci.*, **17**, 315–18.
Körner, S. (1957). ed. *Observation and Interpretation in the Philosophy of Physics.* London: Constable.
Korsch, H. J. & Möhlencamp, R. (1978a). *Phys. Lett. A*, **67**, 110–12.
Korsch, H. J. & Möhlencamp, R. (1978b). *J. Phys. B: Atom. Molec. Phys.*, **11**, 1941–52.
Kowalski-Glikman, J. & Vink, J. C. (1990). *Class. Quantum Grav.*, **7**, 901–18.
Kramers, H. A. (1926). *Z. Phys.*, **39**, 828–40.
Kramers, H. A. (1957). *Quantum Mechanics.* Amsterdam: North-Holland.
Kyprianidis, A. (1985). *Phys. Lett. A*, **111**, 111–16.
Kyprianidis, A. (1987). *Phys. Rep.*, **155**, 1–27.
Kyprianidis, A. (1988a). *Phys. Lett. A*, **131**, 411–18.
Kyprianidis, A. (1988b). *Found. Phys.*, **18**, 1077–91.
Kyprianidis, A. & Vigier, J. P. (1987). In *3rd Workshop on Hadronic Mechanics, Patras, 1986. Hadronic J.*
Lam, M. M. & Dewdney, C. (1990). *Phys. Lett. A*, **150**, 127–35.
Lamb, W. E. (1969). *Physics Today*, **22**, April, 23–8.
Lamehi-Rachi, M. & Mittig, W. (1976). *Phys. Rev. D*, **14**, 2543–55 (reprinted in Wheeler & Zurek (1983)).
Lanczos, C. (1970). *The Variational Principles of Mechanics*, 4th edn. Toronto: University of Toronto Press.
Landau, L. D. & Lifschitz, E. M. (1958). *Quantum Mechanics.* Oxford: Pergamon Press.
Landau, L. D. & Lifschitz, E. M. (1959). *Fluid Mechanics.* Oxford: Pergamon Press.
Landau, L. D. & Lifschitz, E. M. (1960). *Mechanics.* Oxford: Pergamon Press.
Langhoff, P. W. (1971). *Am. J. Phys.*, **39**, 954–7.
Leavens, C. R. (1990a). *Solid State Comm.*, **74**, 923–8.
Leavens, C. R. (1990b). *Solid State Comm.*, **76**, 253–61.
Leavens, C. R. & Aers, G. C. (1991). *Solid State Comm.*, **78**, 1015–23.
Lee, H. W. (1990). *Phys. Lett. A*, **146**, 287–92.
Lee, H. W. & Scully, M. O. (1983). *Found. Phys.*, **13**, 61–72.
Leggett, A. J. (1980). *Suppl. Prog. Theor. Phys.*, No. 69, 80–100.
Leggett, A. J. (1987). In *Proc. 2nd. Int. Symp. Foundations of Quantum Mechanics, Tokyo, 1986*, ed. M. Namiki *et al.*, pp. 287–97. Tokyo: Japanese Physical Society.
Leinaas, J. M. & Myrheim, J. (1977). *Nuovo Cimento*, **37B**, 1–23.
León, M. de & Rodrigues, P. R. (1985). *Generalized Classical Mechanics and Field Theory.* Amsterdam: North-Holland.
Leubner, C., Alber, M. & Schupfer, N. (1988). *Am. J. Phys.*, **56**, 1123–9.
Lévy-Leblond, J. M. (1987). *Phys. Lett. A*, **125**, 441–2.
Lévy-Leblond, J. M. & Balibar, F. (1990). *Quantics. Rudiments of Quantum Physics.* Amsterdam: North-Holland.
Liboff, R. L. (1975). *Found. Phys.*, **5**, 271–93.
Liboff, R. L. (1984). *Physics Today*, **37**, February, 50–5.
Lichte, H. (1986). *Ann. NY Acad. Sci.*, **480**, 175–89.
Lighthill, Sir J. (1986). *Proc. Roy. Soc. Lond. A*, **407**, 35–50.

London, F. & Bauer, E. (1939). *La Théorie de l'Observation en Mécanique Quantique.* Paris: Hermann (English translation in Wheeler & Zurek (1983)).
Loudon, R. (1986). *The Quantum Theory of Light*, 2nd edn. Oxford: Clarendon Press.
Madelung, E. (1926). *Z. Phys.*, **40**, 322–6.
Mandel, L. (1966). *Phys. Rev.*, **144**, 1071–7.
McCarthy, I. E. & Weigold, E. (1983). *Am. J. Phys.*, **51**, 152–5.
McCullough, E. A. & Wyatt, R. E. (1971). *J. Chem. Phys.*, **54**, 3578–91.
McDonald, S. W. & Kaufman, A. N. (1979). *Phys. Rev. Lett.*, **42**, 1189–91.
Mermin, N. D. (1990). *Phys. Rev. Lett.*, **65**, 1838–40.
Messiah, A. (1961). *Quantum Mechanics*, vol. 1. Amsterdam: North-Holland.
Misner, C. W., Thorne, K. S. & Wheeler, J. A. (1973). *Gravitation.* San Francisco: W.H. Freeman.
Misra, B. & Prigogine, I. (1983). In *Long-Time Prediction in Dynamics*, ed. C. W. Horton, L. E. Reichl & A. G. Szebehely, pp. 21–43. New York: Wiley.
Mittelstaedt, P., Prieur, A. & Schieder, R. (1987). *Found. Phys.*, **17**, 891–903.
Morris, M. S., Thorne, K. S. & Yurtsever, U. (1988). *Phys. Rev. Lett.*, **61**, 1446–9.
Morse, P. M. & Feshbach, H. (1953). *Methods of Theoretical Physics*, part 2. New York: McGraw-Hill.
Moshinsky, M. (1952). *Phys. Rev.*, **88**, 625–31.
Mott, N. F. (1929a). *Proc. Roy. Soc. Lond. A*, **124**, 425–42.
Mott, N. F. (1929b). *Proc. Roy. Soc. Lond. A*, **126**, 79–84 (reprinted in Wheeler & Zurek (1983)).
Mott, N. F. & Massey, H. S. W. (1965). *The Theory of Atomic Collisions*, 3rd edn. Oxford: Clarendon Press.
Mugur-Schächter, M. (1964). *Etude du Caractère Complet de la Théorie Quantique.* Paris: Gauthier-Villars.
Murdoch, D. (1989). *Niels Bohr's Philosophy of Physics.* Cambridge: University Press.
de Muynck, W. M. (1987). In *Symposium on the Foundations of Modern Physics*, ed. P. Lahti & P. Mittelstaedt, pp. 419–28. Singapore: World Scientific.
de Muynck, W. M., Janssen, P. A. E. M. & Santman, A. (1979). *Found. Phys.*, **9**, 71–122.
Nassar, A. B. (1990). *Phys. Lett. A*, **146**, 89–92.
Nelson, E. (1966). *Phys. Rev.*, **150**, 1079–85.
Nelson, E. (1985). *Quantum Fluctuations.* Princeton: University Press.
von Neumann, J. (1955). *Mathematical Foundations of Quantum Mechanics.* Princeton: University Press.
Newton, T. D. & Wigner, E. P. (1949). *Rev. Mod. Phys.*, **21**, 400–6.
Nieto, M. M. (1980). *Phys. Rev. D*, **22**, 391–402.
Nieto, M. M. & Simmons, L. M. (1979). *Phys. Rev. D*, **20**, 1321–31, 1332–41, 1342–50.
Normand, J. M. (1980). *A Lie Group: Rotations in Quantum Mechanics.* Amsterdam: North-Holland.
Nye, J. F. & Berry, M. V. (1974). *Proc. Roy. Soc. Lond. A*, **336**, 165–90.
Olariu, S. & Popescu, I. I. (1985). *Rev. Mod. Phys.*, **57**, 339–436.
Padmanabhan, T. (1989). *Int. J. Mod. Phys. A*, **4**, 4735–818.
Pais, A. (1982). '*Subtle is the Lord . . .' The Science and the Life of Albert Einstein.* Oxford: Clarendon Press.
Park, J. L. & Margenau, H. (1968). *Int. J. Theor. Phys.*, **1**, 211–83.
Paul, H. (1986). *Rev. Mod. Phys.*, **58**, 209–31.
Pauli, W. (1953). In *Louis de Broglie, Physicien et Penseur*, ed. A. George, pp. 33–42. Paris: Albin-Michel.

Pauli, W. (1958). In *Handbuch der Physik*, ed. S. Flügge, vol. 5, pp. 1–168. Berlin: Springer-Verlag.
Pauli, W. (1973). *Pauli Lectures on Physics: vol. 2. Optics and the Theory of Electrons.* Cambridge, MA: MIT Press.
Payne, W. T. (1952). *Am. J. Phys.*, **20**, 253–62.
Penrose, R. & Rindler, W. (1984). *Spinors and Space-Time*, vol. 1. Cambridge: University Press.
Peshkin, M. & Tonomura, A. (1989). *The Aharonov-Bohm Effect*, Lecture Notes in Physics 340. Berlin: Springer-Verlag.
Pfleegor, R. L. & Mandel, L. (1967). *Phys. Rev.*, **159**, 1084–8.
Philippidis, C. (1980). *Ph.D. Thesis.* University of London (unpublished).
Philippidis, C., Bohm, D. & Kaye, R. D. (1982). *Nuovo Cimento*, **71B**, 75–87.
Philippidis, C., Dewdney, C. & Hiley, B. J. (1979). *Nuovo Cimento*, **52B**, 15–28.
Pinch, T. J. (1977). In *The Social Production of Scientific Knowledge*, ed. E. Mendelsohn, P. Weingart & R. Whitley, pp. 171–215. Dordrecht: D. Reidel.
Pinch, T. J. (1979). *Phys. Educ.*, **14**, 48–52.
Pinder, D. N. (1990). *Am. J. Phys.*, **58**, 54–8.
Pitzer, K. S. (1954). *Quantum Chemistry.* New York: Prentice-Hall.
Popper, K. R. (1982). *Quantum Theory and the Schism in Physics.* London: Hutchinson.
Prigogine, I. (1962). *Nonequilibrium Statistical Mechanics.* New York: Wiley.
Prigogine, I. (1980). *From Being to Becoming.* San Francisco: W.H. Freeman.
Prizbram, K. (1967). ed. *Letters on Wave Mechanics.* New York: Philosphical Library.
Prokhorov, L. V. (1988). *Sov. Phys. Usp.*, **31**, 151–62.
Prosser, R. D. (1976). *Int. J. Theor. Phys.*, **15**, 169–80.
Rañada, A. F. & Rodero, G. S. (1980). *Phys. Rev. D*, **22**, 385–90.
Rauch, H. (1990). In *Proc. 3rd. Int. Symp. Foundations of Quantum Mechanics*, ed. S. Kobayashi *et al.* Tokyo: Physical Society of Japan.
Rauch, H. & Summhammer, J. (1984). *Phys. Lett. A*, **104**, 44–6.
Rauch, H., Treimer, W. & Bonse, U. (1974). *Phys. Lett. A*, **47**, 369–71.
Rauch, H. & Vigier, J. P. (1990). *Phys. Lett. A*, **151**, 269–75.
Rauch, H., Zeilinger, A., Badurek, G., Wilfing, A., Bauspiess, W. & Bonse, U. (1975). *Phys. Lett. A*, **54**, 425–7.
Redhead, M. (1989). *Incompleteness, Nonlocality, and Realism.* Oxford: Clarendon Press.
Reid, J. L. & Ray, J. R. (1984). *Z. Augew. Math. u. Mech.*, **64**, 365–6.
Riehle, F., Kisters, Th., Witte, A., Helmcke, J. & Bordé, Ch. J. (1991) *Phys. Rev. Lett.*, **67**, 177–80.
Riesz, M. (1946). In *C.R. 10ème Cong. Math. Scand.*, pp. 123–48. Copenhague.
Riesz, M. (1958). *Clifford Numbers and Spinors*, Lecture Series no. 38. The Institute for Fluid Dynamics and Applied Mathematics, University of Maryland.
Rosen, N. (1945). *J. Elisha Mitchell Sci. Soc.*, **61**, 67–73.
Rosen, N. (1951). *Phys. Rev.*, **82**, 621–4.
Rosen, N. (1964a). *Am. J. Phys.*, **32**, 377–9.
Rosen, N. (1964b). *Am. J. Phys.*, **32**, 597–600.
Rosen, N. (1965). *Am. J. Phys.*, **33**, 146–50.
Rosen, N. (1974). *Nuovo Cimento*, **19B**, 90–8.
Rosen, N. (1984). *Found. Phys.*, **14**, 579–605.
Rosen, N. (1986). *Found. Phys.*, **16**, 687–700.
Rosenfeld, L. (1958). *Nature*, **181**, 658.
Rosenfeld, L. (1965). *Nucl. Phys.*, **70**, 1–27.

Rosenstein, B. & Usher, M. (1987). *Phys. Rev. D*, **36**, 2381–4.
Rowe, E. G. P. (1987). *J. Phys. A: Math. Gen.*, **20**, 1419–31.
Roy, S. M. & Singh, V. (1990). *Phys. Lett. B*, **234**, 117–20.
Royer, A. (1989). *Found. Phys.*, **19**, 3–32.
Schiff, L. I. (1968). *Quantum Mechanics*, 3rd edn. Tokyo: McGraw-Hill.
Schiller, R. (1962a). *Phys. Rev.*, **125**, 1100–8.
Schiller, R. (1962b). *Phys. Rev.*, **125**, 1109–15.
Schiller, R. (1962c). *Phys. Rev.*, **125**, 1116–23.
Schlitt, D. W. & Stutz, C. (1970). *Am. J. Phys.*, **38**, 70–5.
Schönberg, M. (1954). *Nuovo Cimento*, **11**, 674–82.
Schrödinger, E. (1926a). *Ann. Phys.*, **79**, 489–527.
Schrödinger, E. (1926b). *Naturwissenschaften*, **14**, 664–6.
Schrödinger, E. (1926c). *Phys. Rev.*, **28**, 1049–70.
Schrödinger, E. (1928). *Collected Papers on Wave Mechanics*. London: Blackie (contains English translations of Schrödinger (1926a,b)).
Schrödinger, E. (1930). *Sitzungsber. Preuss. Akad. Wiss. Phys.-Math. Kl.*, **24**, 418–28.
Schrödinger, E. (1935a). *Proc. Camb. Phil. Soc.*, **31**, 555–62.
Schrödinger, E. (1935b). *Naturwissenschaften*, **23**, 807–12, 823–8, 844–9 (English translation in Wheeler & Zurek (1983)).
Schrödinger, E. (1936). *Proc. Camb. Phil. Soc.*, **32**, 446–52.
Schweber, S. S. (1961). *An Introduction to Relativistic Quantum Field Theory*. New York: Harper & Row.
Schwinger, J., Scully, M. O. & Englert, B. G. (1988). *Z. Phys. D*, **10**, 135–44.
Scully, M. O., Englert, B. G. & Schwinger, J. (1989). *Phys. Rev. A*, **40**, 1775–84.
Scully, M. O., Englert, B. G. & Walther, H. (1991). *Nature*, **351**, 111–16.
Scully, M. O., Lamb, W. E. & Barut, A. (1987). *Found. Phys.*, **17**, 575–83.
Scully, M. O., Shea, R. & McCullen, J. D. (1978). *Phys. Rep.*, **43**, 485–98.
Scully, M. O. & Walther, H. (1989). *Phys. Rev. A*, **39**, 5229–36.
Selleri, F. (1982). *Found. Phys.*, **12**, 1087–112.
She, C. Y. & Heffner, H. (1966). *Phys. Rev.*, **152**, 1103–10.
Shimony, A. (1984). In *Proc. Int. Symp. Foundations of Quantum Mechanics, Tokyo, 1983*, ed. S. Kamefuchi *et al.*, pp. 225–30. Tokyo: Japanese Physical Society.
Shimony, A. (1989). *Found. Phys.*, **19**, 1425–9.
Silverman, M. P. (1990). *Phys. Lett. A*, **148**, 154–7.
Spiller, T. P., Clark, T. D., Prance, R. J. & Prance, H. (1990). *Europhys. Lett.*, **12**, 1–4.
Squires, E. (1986). *The Mystery of the Quantum World*. Bristol: Adam Hilger.
Squires, E. J. (1992). *Phys. Lett. A*, **162**, 35–6.
Sudbery, A. (1987). *J. Phys. A: Math. Gen.*, **20**, 1743–50.
Summhammer, J., Badurek, G., Rauch, H. & Kischko, U. (1982). *Phys. Lett. A*, **90**, 110–12.
Summhammer, J., Rauch, H. & Tuppinger, D. (1987). *Phys. Rev. A*, **36**, 4447–55.
Summhammer, J., Rauch, H. & Tuppinger, D. (1988). *Physica B*, **151**, 103–7.
Synge, J. L. (1954). *Geometrical Mechanics and de Broglie Waves*. Cambridge: University Press.
Tabor, D. & Winterton, R. H. S. (1969). *Proc. Roy. Soc. Lond. A*, **312**, 435–50.
Takabayasi, T. (1952) *Prog. Theor. Phys.*, **8**, 143–82.
Takabayasi, T. (1953) *Prog. Theor. Phys.*, **9**, 187–222.
Takabayasi, T. (1954) *Prog. Theor. Phys.*, **11**, 341–73.
Takabayasi, T. (1955) *Prog. Theor. Phys.*, **14**, 283–302.
Takabayasi, T. (1957) *Suppl. Prog. Theor. Phys.*, No. 4, 1–80.
Takabayasi, T. (1983) *Prog. Theor. Phys.*, **69**, 1323–44.

Takabayasi, T. & Vigier, J. P. (1957). *Prog. Theor. Phys.*, **18**, 573–90.
Taylor, G. I. (1909). *Proc. Camb. Phil. Soc.*, **15**, 114–15.
Tipler, F. J. (1984). *Phys. Lett. A*, **103**, 188–92.
Tipler, F. J. (1987). *Class. Quantum Grav.*, **4**, L189–95.
Tolman, R. C. (1938). *The Principles of Statistical Mechanics.* Oxford: University Press (reprinted by Dover, New York, 1979).
Tonomura, A., Osakabe, N., Matsuda, T., Kawasaki, T., Endo, J., Yano, S. & Yamada, H. (1986). *Phys. Rev. Lett.*, **56**, 792–5.
Tonomura, A., Endo, J., Matsuda, T., Kawasaki, T. & Ezawa, H. (1989). *Am. J. Phys.*, **57**, 117–20.
Troudet, T. (1985). *Phys. Lett. A*, **111**, 274–6.
Uffink, J. B. M. & Hilgevoord, J. (1985). *Found. Phys.*, **15**, 925–44.
Valentini, A. (1991). *Phys. Lett. A*, **156**, 5–11; **158**, 1–8.
Velo, G. & Zwanziger, D. (1969a). *Phys. Rev.*, **186**, 1337–41.
Velo, G. & Zwanziger, D. (1969b). *Phys. Rev.*, **188**, 2218–22.
van Vleck, J. H. (1928). *Proc. Nat. Acad. Sci.*, **14**, 178–88.
Vigier, J. P. (1952). *C.R. Acad. Sci. Paris*, **235**, 1107–9.
Vigier, J. P. (1956). *Structure des Micro-Objets dans l'Interprétation Causale de la Théorie des Quanta.* Paris: Gauthier-Villars.
Vigier, J. P. (1982). *Astron. Nachr.*, **303**, 55–80.
Vigier, J. P. (1990). *IEEE Transactions on Plasma Science*, **18**, 64–72.
Vigier, J. P., Dewdney, C., Holland, P. R. & Kyprianidis, A. (1987). In *Quantum Implications*, ed. B. J. Hiley & F. D. Peat, pp. 169–204. London: Routledge & Kegan Paul.
Walls, D. F. (1977). *Am. J. Phys.*, **45**, 952–6.
Wan, K. K. & Sumner, P. J. (1988). *Phys. Lett. A*, **128**, 458–62.
Wang, L. J., Zou, X. Y. & Mandel, L. (1991). *Phys. Rev. Lett.*, **66**, 1111–14.
Watson, G. N. (1944). *A Treatise on the Theory of Bessel Functions*, 2nd edn. Cambridge: University Press.
Wentzel, G. (1926). *Z. Phys.*, **38**, 518–29.
Werner, S. A., Colella, R., Overhauser, A. W. & Eagen, C. F. (1975). *Phys. Rev. Lett.*, **35**, 1053–5.
Wheeler, J. A. (1962). *Geometrodynamics.* New York: Academic Press.
Wheeler, J. A. (1968). In *Battelle Rencontres*, ed. C. M. DeWitt & J. A. Wheeler, pp. 242–307. New York: Benjamin.
Wheeler, J. A. (1978). In *Mathematical Foundations of Quantum Theory*, ed. A. R. Marlow, pp. 9–48. New York: Academic Press.
Wheeler, J. A. & Zurek, W. H. (1983). ed. *Quantum Theory and Measurement.* Princeton: University Press.
Wightman, A. S. & Schweber, S. S. (1955). *Phys. Rev.*, **98**, 812–37.
Wigner, E. (1932). *Phys. Rev.*, **40**, 749–59.
Wigner, E. P. (1963). *Am. J. Phys.*, **31**, 6–15.
Wigner, E. P. (1980). In *Some Strangeness in the Proportion*, ed. H. Woolf, pp. 461–8. Reading, MA: Addison-Wesley.
Wigner, E. P. (1983a). In Wheeler & Zurek (1983), pp. 260–314.
Wigner, E. P. (1983b). In *Quantum Optics, Experimental Gravity, and Measurement Theory*, ed. P. Meystre & M. O. Scully, pp. 43–63. New York: Plenum Press.
van Winter, C. (1954). *Physica*, **20**, 274–92.
ten Wolde, A., Noordam, L. D., Lagendijk, A. & van Linden van den Heuvell, H. B. (1988). *Phys. Rev. Lett.*, **61**, 2099–101.

Wootters, W. K. & Zurek, W. H. (1979). *Phys. Rev. D*, **19**, 473–84 (reprinted in Wheeler & Zurek (1983)).
Wu, T. T. & Yang, C. N. (1975). *Phys. Rev. D*, **12**, 3845–57.
Yeazell, J. A. & Stroud, C. R. (1988). *Phys. Rev. Lett.*, **60**, 1494–7.
Zeeman, E. C. (1964). *J. Math. Phys.*, **5**, 490–3.
Zeh, H. D. (1988). *Found. Phys.*, **18**, 723–30.
Zeh, H. D. (1989). *The Physical Basis of the Direction of Time*. Berlin: Springer-Verlag.
Zeilinger, A., Gähler, R., Shull, C. G., Treimer, W. & Mampe, W. (1988). *Rev. Mod. Phys.*, **60**, 1067–73.
Ziolkowski, R. W. (1989). *Phys. Rev. A*, **39**, 2005–33.
Zurek, W. H. (1982). *Phys. Rev. D*, **26**, 1862–80.

Index